P9-CFO-447

Radio System Design for Telecommunications

WILEY SERIES IN TELECOMMUNICATIONS AND SIGNAL PROCESSING

John G. Proakis, Editor
Northeastern University

Worldwide Telecommunications Guide for the Business Manager
Walter H. Vignault

Expert System Applications for Telecommunications
Jay Liebowitz

Business Earth Stations for Telecommunications
Walter L. Morgan and Denis Rouffet

Introduction to Communications Engineering, 2nd Edition
Robert M. Gagliardi

Satellite Communications: The First Quarter Century of Service
David W. E. Rees

Synchronixation in Digital Communications, Volume I
Heinrich Meyr and Gerd Ascheid

Digital Telephony, 2nd Edition
John Bellamy

Elements of Information Theory
Thomas M. Cover and Joy A. Thomas

Telecommunications Transmission Handbook, 3rd Edition
Roger L. Freeman

Digital Signal Estimation
Robert J. Mammone, Editor

Telecommunication Circuit Design
Patrick D. van der Puije

Meteor Burst Communications: Theory and Practice
Donald L. Schilling, Editor

Mobile Communications Design Fundamentals, 2nd Edition
William C. Y. Lee

Fundamentals of Telecommunication Networks
Tarek N. Saadawi, Mostafa Ammar, with Ahmed El Hakeem

Optical Communications, 2nd Edition
Robert M. Gagliardi and Sherman Karp

Wireless Information Networks
Kaveh Pahlavan and Allen H. Levesque

Practical Data Communications
Roger L. Freeman

Active Noise Control Systems: Algorithms and DSP Implementations
Sen M. Kuo and Dennis R. Morgan

Telecommunication System Engineering: 3rd Edition
Roger L. Freeman

Radio System Design for Telecommunications, 2nd Edition
Roger L. Freeman

Radio System Design for Telecommunications

Second Edition

Roger L. Freeman

A Wiley-Interscience Publication
JOHN WILEY & SONS, INC.
New York • Chichester • Weinheim • Brisbane • Singapore • Toronto

Printed on acid-free paper.

Published by John Wiley & Sons, Inc.

Library of Congress Cataloging in Publication Data:

Freeman, Roger L., 1928–
Radio system design for telecommunications/
Roger L. Freeman.—2nd ed.
p. cm.—(Wiley series in telecommunications and signal
processing)
Rev. ed. of: Radio system design for telecommunications
(1-100GHz).
ISBN 0-471-16260-4 (cloth : alk. paper)
1. Radio relay systems—Design and construction. I. Freeman,
Roger L., 1928– Radio system design for telecommunications (1-100
GHz). II. Title. III. Series.
TK6553.F722254 1997
621.384'156—dc20 96-44722
CIP

Printed in the United States of America

10 9 8 7 6 5 4 3 2

For my daughter
Rosalind

CONTENTS

PREFACE TO THE SECOND EDITION

This book provides essential design techniques for radio systems that operate at frequencies of 3 MHz to 100 GHz and which will be employed in the telecommunication service. We may also call these *wireless systems*, wireless being synonymous with radio.

The first edition of *Radio System Design for Telecommunications* only treated line-of-sight microwave, troposcatter/diffraction and satellite communication links. In this second edition we have broadened the scope to include:

1. Cellular radio
2. Personal communication system (PCS)
3. Very small aperture (VSAT) satellite communication networks
4. High-frequency links
5. Meteor burst communications

Cellular radio has more than 40 million subscribers in the United States according to a 1997 estimate. It has become a vital part of our culture. PCS is a major extension of cellular radio. The International Telecommunications Union Radio Communications Sector (ITU-R) places the two technologies together under one futuristic name: Future Public Land Mobile Telecommunication System (FPLMTS). The ITU-R estimates that there will be more than 100 million PCS subscribers by the year 2000. There is also a satellite adjunct to cellular/PCS such as Motorola's IRIDIUM system. We devote a long chapter to these technologies. Although many books would break this topic into two chapters, we did not because the topics are so intertwined.

VSAT systems have also blossomed during this period. Many private digital networks worldwide are built around the VSAT concept. The rationale for their employment varies. They can of course be classified as an economic alternative to connectivity via the public switched telecommunications network (PSTN). An example of such use is a bank in an emerging nation where we offered a satellite course. This bank had 120 branches with no direct electrical communication among branches whatsoever. They solved this problem by introducing VSAT connections with the hub located at the bank's headquarters in the nation's capital. Another reason for introducing VSAT connections may be intentional bypass of the PSTN. In some countries, it is argued that such networks are more cost effective and efficient, and in others they are needed just to achieve an acceptable quality of service.

For completeness, we have added a chapter on HF (3 – 30 MHz) systems. HF has been around since the days of Marconi and is in wide use today, particularly by the world's Armed Forces.

Meteor burst communications had its start in a classified context back in the 1950s. Its primary applications today is a report-back means for sensors, such as the approximately 1000 terminals reporting snow accumulations in the U.S. Rocky Mountains.

We have also improved the discussion in various places, such as the E_b/N_0 vs. BER curves, where we present a new view of fading and dispersion effects on high level m-ary systems, jitter accumulation, coordination contours, the penalties for not meeting "line-of-sight" conditions, and bandwidth requirements for digital radiolinks. Scintillation effects are touched on briefly. INTELSAT standards have been upgraded and calculations of excess attenuation due to rainfall have been modified. Reference is also made to 900 MHz SCADA systems.

A new Chapter 4 has been added covering forward error correction and advanced digital modulation waveforms such as trellis-coded modulation. Code Division Multiplex Access (CDMA) is discussed in the cellular radio chapter because of its promise for improved spectral usage, its capability to operate in high interference environments, and its ability to mitigate multipath effects both with cellular and PCS systems.

The 14 chapters are arranged in a hierarchical manner. For example, Chapter 1 deals with general propagation problems. Propagation peculiar to a certain radio transmission medium is treated in the chapter discussing that topic such as troposcatter, HF, meteor burst and cellular radio. Chapters 2 and 3 deal with line-of-sight microwave transmission. Chapters 5, 6, 7, and 8 are outgrowths of this topic. Chapters 6, 7, and 8 depend on information discussed in Chapter 4 (forward error correction).

The reader is only expected to have a working knowledge of electrical communication, algebra, trigonometry, logarithms, and time distributions. Decibel and its many derivative forms are used widely throughout. The more difficult communication system concepts are referenced to other texts or standards for further reading and clarification. Nearly every key formula is followed by at least one worked example. At the end of every chapter there is a set of review questions and problems. These are meant as a review and can be worked easily without going to other texts.

On January 1, 1993 CCIR changed its name to the ITU Radiocommunication Sector, abbreviated ITU-R. In this text documents from this group published after January 1, 1993 are called ITU-R (recommendations). We use the older CCIR nomenclature for documents published before that date.

ACKNOWLEDGMENTS

I am indebted to many people for assisting me in the preparation of the second edition of this text. Among this large group let me cite: Marshall Cross, Chairman, Megawave Systems of Boylston, MA.; Dr. Enric Vilar, Department of Telecommunications, Portsmouth Polytechnic University (UK) regarding rainfall loss and scintillation; Dr. Len Wagner, Naval Research Laboratories for basic HF data; Gerry Einig, Director of Presale Systems, Scientific-Atlanta, Norcross, GA for link budgets; Bob Egri of Ma-Com, Lowell, MA for data on receiver noise figures; Tom Adcock, Engineering Director, Lucas, McGowan, Guitierrez and Nace, Washington, D.C. for cellular reuse patterns; Dr. Donald Schilling, Professor Emeritus of City College (NY) for CDMA on PCS systems; Susan Hoyler of TIA, Arlington, VA for help with cellular standards; Sam Compton (deceased), Vice President of Link Plus, Columbia, MD for new information on Lincompex; Ron Jost, System Engineering Manager/Chief Engineer for Motorola Satellite Systems Division, Chandler, AZ; John Ballard, a dear friend, president of TCI International, Inc., Sunnyvale, CA for ionograms; and Don Marsh, Principal, Marsh Consultants for VSAT information. Finally I am grateful to the students in my various telecommunications seminars that I have been giving at the University of Wisconsin, Madison since 1981 for helping me shape several of my books, and to Fran Drake and Dan Danbeck, Engineering Professional Development program directors at the university whose critique has been invaluable.

ROGER L. FREEMAN

Scottsdale, Arizona
March 1997

1

RADIO PROPAGATION

1.1 INTRODUCTION

The purpose of this book is to describe methods of the design of radio systems for telecommunications that operate approximately from 3 MHz to 300 GHz. These systems include line-of-sight (LOS) microwave, diffraction/scatter links, satellite links, and very small aperture terminal (VSAT) systems (Refs. 1–5). Such systems may be described as broadband. The book also covers narrowband systems such as high frequency (HF; i.e., in the 3–30-MHz band). Also a chapter is devoted to cellular radio systems and personal communication systems (PCSs). Such systems are commonly part of a larger system, the public switched telecommunication network (PSTN).

A common thread throughout such systems is propagation. Free-space loss and fading are examples. There are also some notable differences such as the propagation behavior of HF compared to LOS microwave. This chapter covers topics in propagation that have commonality between at least two of the subject areas. However, those propagation issues peculiar to each of the transmission types are dealt with in the appropriate chapter.

We have made an arbitrary division of the radio spectrum of interest at 10 GHz. In general, for those systems operating below 10 GHz, we can say that atmospheric absorption and precipitation play a secondary role, and with many systems these issues are neglected. For those systems operating above 10 GHz, excess attenuation due to rainfall and losses due to atmospheric gases may become principal issues.

In the following discussion we use the *isotropic* as the reference antenna.* The IEEE dictionary (Ref. 1) defines an isotropic radiator as "a hypothetical lossless antenna having equal radiation intensity in all directions." An isotropic

*The reader must take care in this instance. Almost without exception in cellular systems, the reference antenna is a dipole. Such a reference may also be encountered in PCS and HF systems. A reference dipole has a 2.15-dB gain over the isotropic antenna.

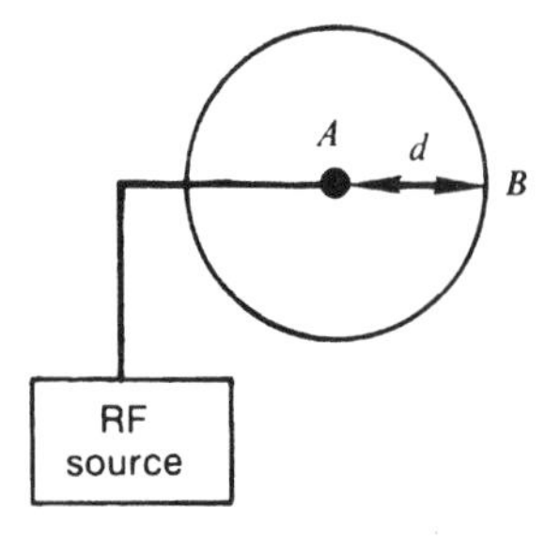

Figure 1.1. A point source isotropic radiator located at the center of a sphere.

antenna has a gain of 1 (in decimal units) or a gain of 0 dB. The decibel unit of antenna gain (or loss) is the dBi (decibels referenced to an isotropic).

1.2 LOSS IN FREE SPACE

Loss in free space is a function of frequency squared plus distance squared plus a constant. Let's see how this relationship is developed.

Consider power P_T radiated from an isotropic transmitting antenna. This transmitting antenna is a point source radiating power uniformly in all directions. Let's imagine a sphere with radius d, which is centered on that point source. This concept is shown in Figure 1.1.

If free-space transmission is assumed, meaning transmission with no absorption or reflection of energy by nearby objects, the radiated power density will be uniform at all points on the sphere's surface. The total radiated power P_T will pass outward through the sphere's surface. The radially directed power density at any point on the surface of the sphere is

$$\text{Power density} = P_T/4\pi d^2 \tag{1.1}$$

If a receiving antenna with an effective area A_R is located on the surface of the sphere, the total receive power P_R is equal to the power density times the area of the antenna. This is expressed as

$$P_R = P_T \times A_R/4\pi d^2 \tag{1.2}$$

A transmitting antenna with an effective area of A_T, which concentrates its radiation within a small solid angle or beam, has an on-axis transmitting antenna gain, with respect to an isotropic radiator, of

$$g_T = 4\pi A_T/\lambda^2 \tag{1.3}$$

where λ is the wavelength of the emitted signal. Let's use this antenna in place of the familiar isotropic radiator. The receive power relationship in equation (1.2) now becomes

$$P_R = P_T(4\pi A_T/\lambda^2)(A_R/4\pi d^2) \tag{1.4}$$

We rearrange equation (1.4) to describe the transmitting and receiving antennas in terms of their gains relative to an isotropic antenna. This then gives us

$$P_R = P_T(4\pi A_T/\lambda^2)(4\pi A_R/\lambda^2)(\lambda/4\pi d)^2 \tag{1.5a}$$

$$= P_T(g_T)(g_R)(\lambda/4\pi d)^2 \tag{1.5b}$$

where $g_R = 4\pi A_R/\lambda^2$ is the receiving antenna gain referenced to an isotropic radiator. Restating in decibels, the ratio of P_T to P_R is

$$10\log(P_T/P_R) = 20\log(4\pi d/\lambda) - 10\log g_T - 10\log g_R \quad \text{dB} \tag{1.6}$$

The distance- and frequency-dependent term in equations (1.5a) and (1.5b) is called the *free-space path loss* between isotropic radiators. Expressed in decibel form, the free-space loss is

$$\text{Free-space loss (FSL)}_{\text{dB}} = 20\log(4\pi d/\lambda) \tag{1.7}$$

or

$$\text{FSL}_{\text{dB}} = 21.98 + 20\log(d/\lambda) \tag{1.8}$$

Converting λ to the more familiar frequency term and using kilometers (km) for distance and megahertz (MHz) for frequency, we have

$$\text{FSL}_{\text{dB}} = 32.45 + 20\log D_{\text{km}} + 20\log F_{\text{MHz}} \tag{1.9a}$$

The same formula, but using miles (D in statute miles), can be written

$$\text{FSL}_{\text{dB}} = 36.58 + 20\log D_{\text{sm}} + 20\log F_{\text{MHz}} \tag{1.9b}$$

If D is measured in nautical miles (nm), we have

$$\text{FSL}_{\text{dB}} = 37.80 + 20\log D_{\text{nm}} + 20\log F_{\text{MHz}} \tag{1.9c}$$

If D is measured in feet, change the constant in equation (1.9) to -37.87; if D is in meters, change the constant to -27.55.

Return to equation (1.8). Substitute λ for D. This means that if we go out one wavelength (λ) from an antenna, the loss is just under 22 dB.

These free-space loss formulas are useful on point-to-point links assuming no nearby obstacles and clear weather. The ITU-R organization in CCIR Rec. 525.1 provides the following formula to calculate field strength on point-to-area links:

$$E = \sqrt{\frac{30p}{D}} \tag{1.10}$$

where E is the root mean square (rms) field strength, in volts per meter; p is the isotropically radiated power (EIRP) of the transmitter in the direction of the point in question, in watts; and D is the distance from the transmitter to the point in question, in meters.

Equation (1.10) is often replaced by equation (1.11), which uses practical units:

$$E_{\mathrm{mV/m}} = 173\sqrt{p_{\mathrm{kW}}/D_{\mathrm{km}}} \quad \text{(linear polarization only)} \tag{1.11}$$

It should be noted that there is no frequency term.

Example. If the EIRP were 100 watts and the distance (D) were 10 km, then the field strength, E, would be 5.47 mV/m.

If the transmitter isotropic power were 1 kW and the distance were 1 km, then the field strength would be 173 mV/m.

1.3 ATMOSPHERIC EFFECTS ON PROPAGATION

1.3.1 Introduction

If a radio beam is propagated in free space, where there is no atmosphere (by definition), the path followed by the beam is a straight line. The transmission loss in free space was derived in Section 1.2.

However, a radio ray propagated through the earth's atmosphere encounters variations in the atmospheric refractivity index along its trajectory that causes the ray path to become curved (Ref. 6). Atmospheric gases will absorb and scatter the radio path energy, the amount of absorption and scattering being a function of frequency and altitude above sea level. Absorption and scattering do become serious contributors to transmission loss above 10 GHz and are discussed in Chapter 8. The principal concern in this section is the effect of the atmospheric refractive index on propagation. Refractivity of the atmosphere will affect not only the curvature of the ray path (expressed by a factor K) but will also give some insight into the fading phenomenon.

1.3.2 Refractive Effects on Curvature of Ray Beam

1.3.2.1 K-Factor. The K-factor is a scaling factor (actually assumed as a constant for a particular path) that helps quantify curvature of an emitted ray path. Common radiolinks, which are described as line-of-sight (LOS), incor-

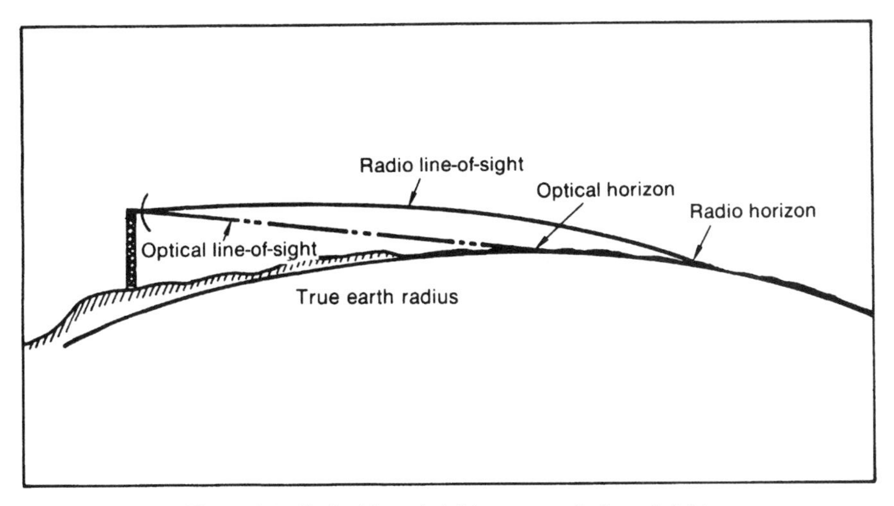

Figure 1.2. Optical line-of-sight versus radio line-of-sight.

rectly suggest that effective communications are limited by the optical horizon (i.e., $K = 1$). In most cases radiolinks are not restricted to LOS propagation. In fact, we often can achieve communications beyond the optical horizon by some 15% (i.e., $K = 1.33$). Figure 1.2 shows this concept in a simplified fashion, and Figure 1.3 shows the effects of various K-factors on the bending of the radio ray beam. This bending is due to angular refraction.

Angular refraction through the atmosphere occurs because radio waves travel with differing velocities in different parts of a medium of varying dielectric constant. In free space the group velocity is maximum, but in the nonionized atmosphere, where the dielectric constant is slightly greater due to the presence of gas and water molecules, the radio wave travels more slowly. In what radiometeorologists have defined as a standard atmosphere, the pressure, temperature, and water vapor content (humidity) all decrease with increasing altitude. The dielectric constant, being a single parameter combining the resultant effect of these three meteorological properties, also decreases with altitude (Refs. 7–9). Since electromagnetic waves travel faster in a medium of lower dielectric constant, the upper part of a wavefront tends to travel with a greater velocity than the lower part, causing a downward deflection of the beam. In a horizontally homogeneous atmosphere where the vertical change of dielectric constant is gradual, the bending or refraction is continuous, so that the ray is slowly bent away from the thinner density air toward the thicker, thus making the beam tend to follow the earth's curvature. This bending can be directly related to the radii of spheres. The first sphere, of course, is the earth itself (i.e., radius = 6370 km) and the second sphere is that formed by the curvature of the ray beam with its center coinciding with the center of the earth. The K-factor can now be defined as

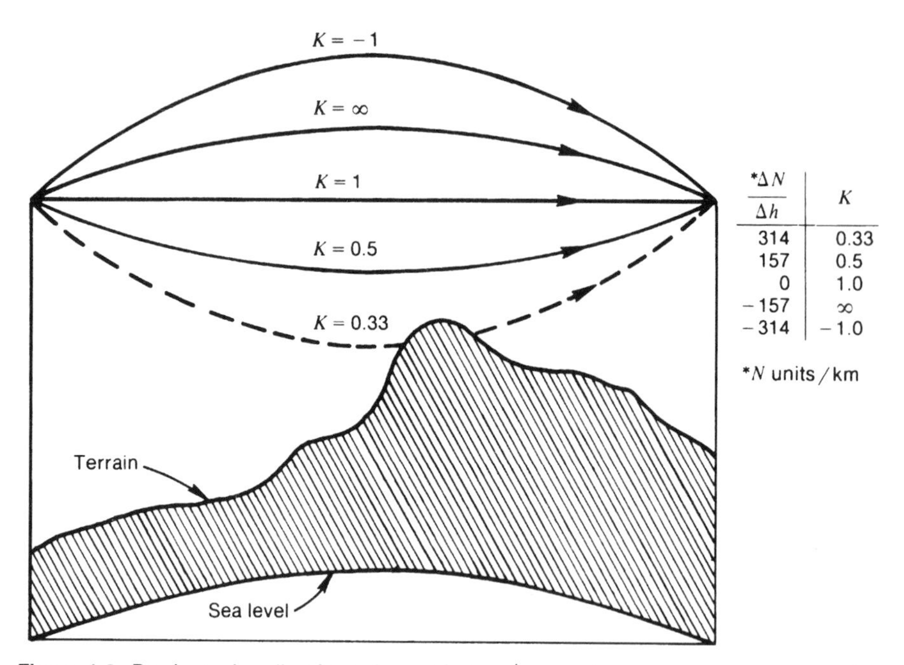

Figure 1.3. Ray beam bending for various *K*-factors (linear refractivity gradients assumed).

the ratio of the radius, r, of the ray beam curvature to the true radius of the earth, r_0, or

$$K \approx \frac{r}{r_0} \tag{1.12}$$

where K is often called the effective earth radius factor and r is the effective earth radius.

1.3.2.2 Refractivity*. The radio refractive index is defined as the ratio of the velocity of propagation of a radio wave in free space to the velocity in a specified medium. At standard atmosphere conditions near the earth's surface, the radio refractive index, n, has a value of approximately 1.0003.

The atmospheric radio refractive index, n, can be calculated by the following formula:

$$n = 1 + N \times 10^{-6} \tag{1.13}$$

* Section 1.3.2.2 is based on Ref. 10.

where N, the radio refractivity, is expressed by

$$N = \frac{77.6}{T}\left(P + 4810\frac{e}{T}\right) \tag{1.14}$$

where P = atmospheric pressure (hPa)*
e = water vapor pressure (hPa)*
T = absolute temperature (K)

This expression may be used for all radio frequencies; for frequencies up to 100 GHz, the error is less than 0.5%.

For ready reference, the relationship between water vapor pressure e and relative humidity is given by

$$e = \frac{He_s}{100} \tag{1.15}$$

with

$$e_s = a \exp\left(\frac{bt}{t + c}\right) \tag{1.16}$$

where H = relative humidity (%)
t = Celsius temperature (°C)
e_s = saturation vapor pressure (in hPa) at the temperature t (in *C)

The coefficients $a, b, c,$ are as follows:

For Water	For Ice
$a = 6.1121$	$a = 6.1115$
$b = 17.502$	$b = 22.452$
$c = 240.97$	$c = 272.55$
(valid between −20°C and +50°C, with an accuracy of ±0.20%)	(valid between −50°C and 0°C, with an accuracy of ±0.20%)

Vapor pressure e is obtained from the water vapor density ρ using the equation

$$e = \frac{\rho T}{216.7} \quad \text{hPa} \tag{1.17}$$

*It should be noted that the World Meteorological Organization has recommended the adoption of hPa, which is numerically identical to mb, as the unit of atmospheric pressure.

where ρ is given in g/m^3. Representative values of ρ are given in Recommendation 836.

1.3.3 Refractivity Gradients

Probably of more direct interest to the radiolink design engineer is refractivity gradients. If we assume that the refractive index, n, of air varies linearly with the height h for the first few tenths of a kilometer above the earth's surface and does not vary in the horizontal direction, then we can restate the K-factor in terms of the gradient $\Delta n/\Delta h$ by (Ref. 5)

$$\frac{r}{r_0} = K \approx \left(1 + \frac{r_0 \, \Delta n}{\Delta h}\right)^{-1} \tag{1.18}$$

again where $r_0 \approx 6370$ km and h is the height above the earth's surface.

As in equation (1.9), $N \approx (n - 1)10^6$, so that

$$\frac{\Delta n}{\Delta h} = \frac{\Delta N}{\Delta h}(10^{-6}) \quad N\text{-units/km} \qquad (1.19)^*$$

and

$$K \approx \left[1 + \left(\frac{\Delta N}{\Delta h}\right)\Big/157\right]^{-1} \tag{1.20}$$

Return to Figure 1.3 where several values of K and $\Delta N/\Delta h$ are listed to the right. The figure also illustrates subrefractive conditions, $0 < K < 1.0$, where the refractivity gradients are positive. The worst case in Figure 1.3 is where the ray beam is interrupted by the surface, where $K = 0.33$, placing the receiving terminal out of normal propagation range.

The more commonly encountered situation is where $\infty \geq K \geq 1.0$ or $-157 \leq \Delta N/\Delta h \leq 0$. In this case the ray is bent toward the earth. When $\Delta N/\Delta h = -157$, the ray has the same curvature as the earth and the ray path acts like straight-line propagation over a flat earth.

As we can now see, the bending of a radio ray beam passing through the atmosphere is controlled by the gradient refractive index. For most purposes the horizontal gradient is so small that it can be neglected. The vertical change under standard atmospheric conditions is approximately -40 N-units/km, which approximates the value at noon on a clear, summer day at sea level, in the temperate zone with a well-mixed atmosphere. However, in the very lowest levels of the atmosphere, the vertical gradient may vary between extreme values as large as $+500$ to -1000 N-units/km over height intervals of several hundred feet. The variance is a function of climate, season, time of day, and/or transient weather conditions. It is also affected by terrain, vegetation, radiational conditions, and atmospheric stratification. The more extreme stratification tends to occur in layers less than 100 m in

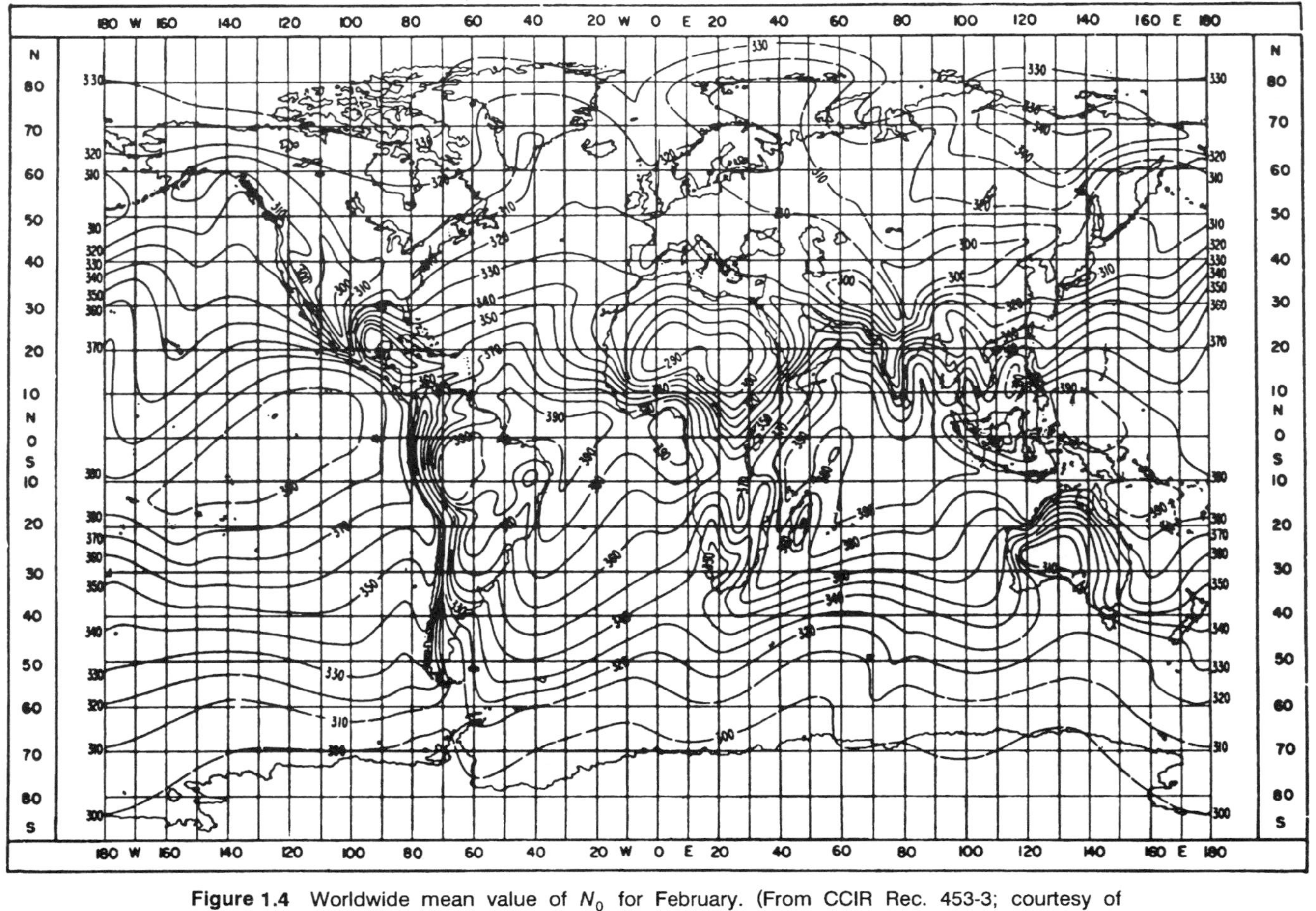

Figure 1.4 Worldwide mean value of N_0 for February. (From CCIR Rec. 453-3; courtesy of ITU-CCIR, Geneva.)

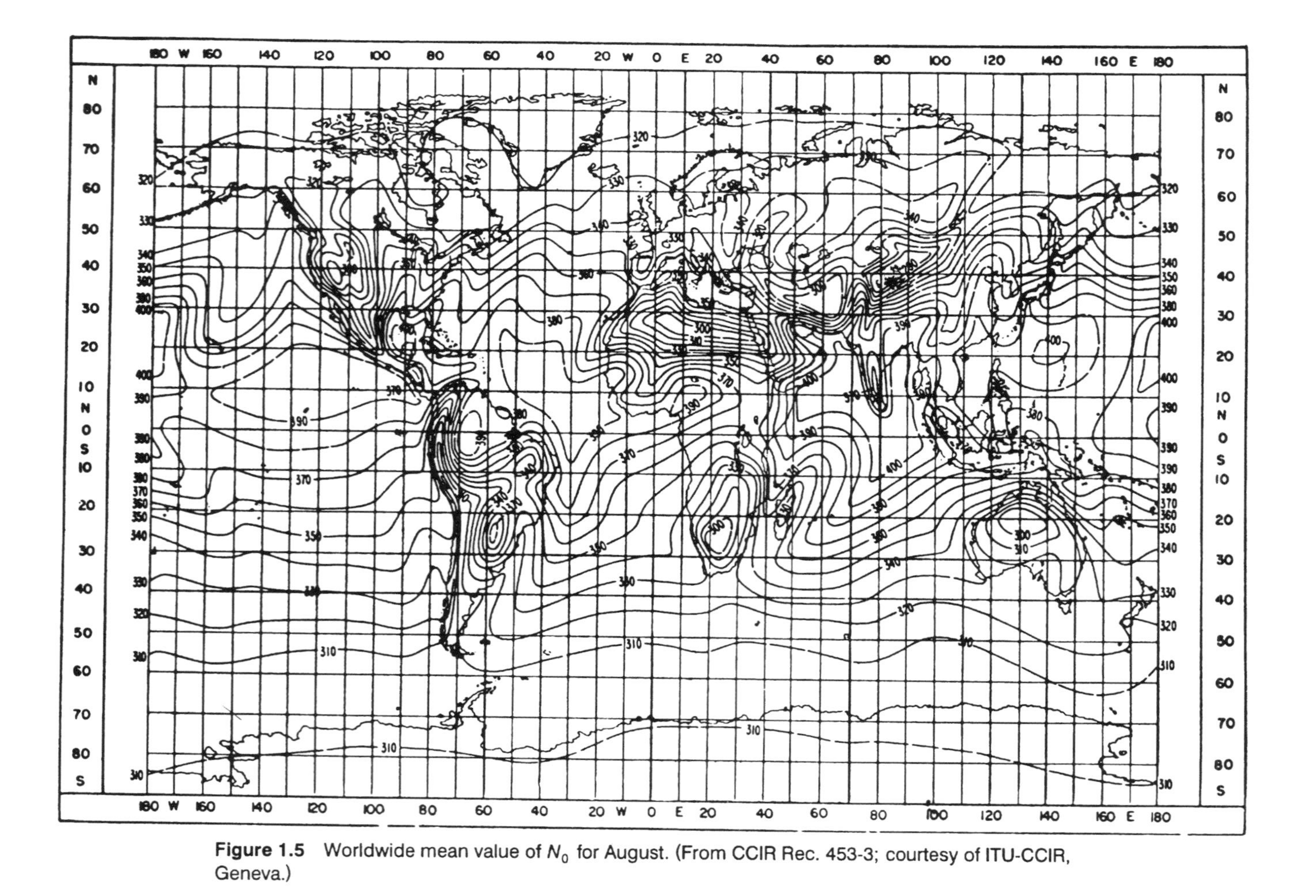

Figure 1.5 Worldwide mean value of N_0 for August. (From CCIR Rec. 453-3; courtesy of ITU-CCIR, Geneva.)

thickness and which at times extend over long distances. When averaged over 500–1000-m heights above ground, the radio refractivity gradient is likely to vary between 0 and -300 N-units/km. However, during a large percentage of the year, the standard value of -40 N-units/km is more likely.

Commonly we will encounter two refractivity parameters used in estimating radio propagation effects; these are surface refractivity N_s and surface refractivity reduced to sea level N_0. Figures 1.4–1.6 give data on mean N_0 values for the world. Figure 1.4 provides worldwide mean values of N_0 for February; Figure 1.5 gives similar information for August; and Figure 1.6 provides data on mean monthly values of N_0 in excess of 350 N-units. High values of N are usually due to the wet term in the refractivity equation.

For long-term median estimates, an empirical relation has been established between the average mean refractivity gradient $\Delta N/\Delta h$ for the first kilometer above the surface and the value of the average monthly mean refractivity N_s at the surface. For the continental United States the relationship is

$$\frac{\Delta N}{\Delta h} = -7.32 \exp\left(0.005577\overline{N}_s\right) \tag{1.21}$$

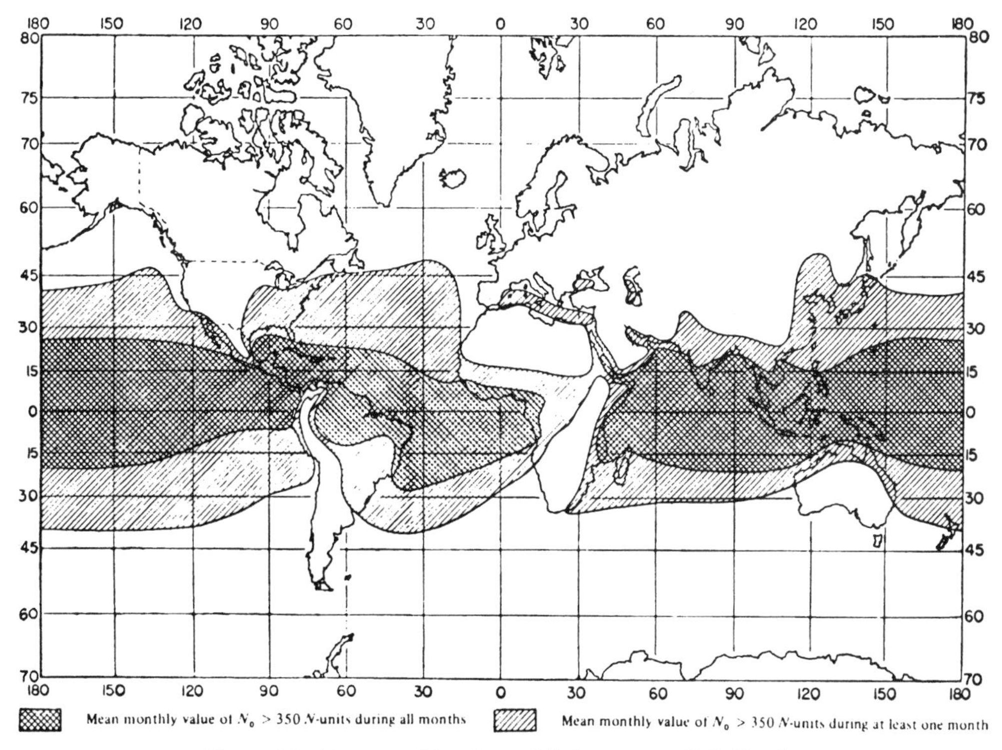

Figure 1.6. Mean monthly values of N_0 in excess of 350 N-units.

where $\Delta N/\Delta h$ is in N-units/km and $\overline{N}_s$ is in N-units. For Germany:

$$\frac{\Delta \overline{N}}{\Delta h} = -9.30 \exp(0.004565 \overline{N}_s) \tag{1.22}$$

and for the United Kingdom

$$\frac{\Delta \overline{N}}{\Delta h} = -3.95 \exp(0.0072 \overline{N}_s) \tag{1.23}$$

These relationships are valid for $250 \leq \overline{N}_s \leq 400$ N-units and are only applicable to average negative gradients close to the surface. Often, particularly in transhorizon communication, it is more convenient to use values of surface refractivity N_s because information for this parameter is more readily available. The value of N_s is a function of temperature, pressure, and humidity and therefore decreases on the average with elevation. For a particular link the applicable values are read for $\overline{N}_0$ from a map (such as Figures 1.4 and 1.5) and are converted to values of N_s by

$$N_s = N_0 \exp(-0.1057 h_s) \qquad \text{(Ref. 5)} \tag{1.24a}$$

where h_s is the height above mean sea level (in kilometers) of the radio horizon in the direction of the far end of the link. For h_s in kilofeet, the following expression may be used:

$$N_s = N_0 \exp(-0.03222 h_s) \qquad \text{(Ref. 7)} \tag{1.24b}$$

Figure 1.7 is a nomogram to convert N_0 to N_s for values of h_s in thousands of feet (kilofeet).

The effective earth radius (Section 1.3.2) can be calculated from N_s with the following formula:

$$r = r_0 \left[1 - 0.04665 \exp(0.005577 \overline{N}_s)\right]^{-1} \qquad \text{(Ref. 9)} \tag{1.25}$$

where $r_0 = 6370$ km (as in Section 1.3.2).

Figure 1.8 is a curve to derive effective earth radius from surface refractivity N_s, where the earth radius is given in kilometers.

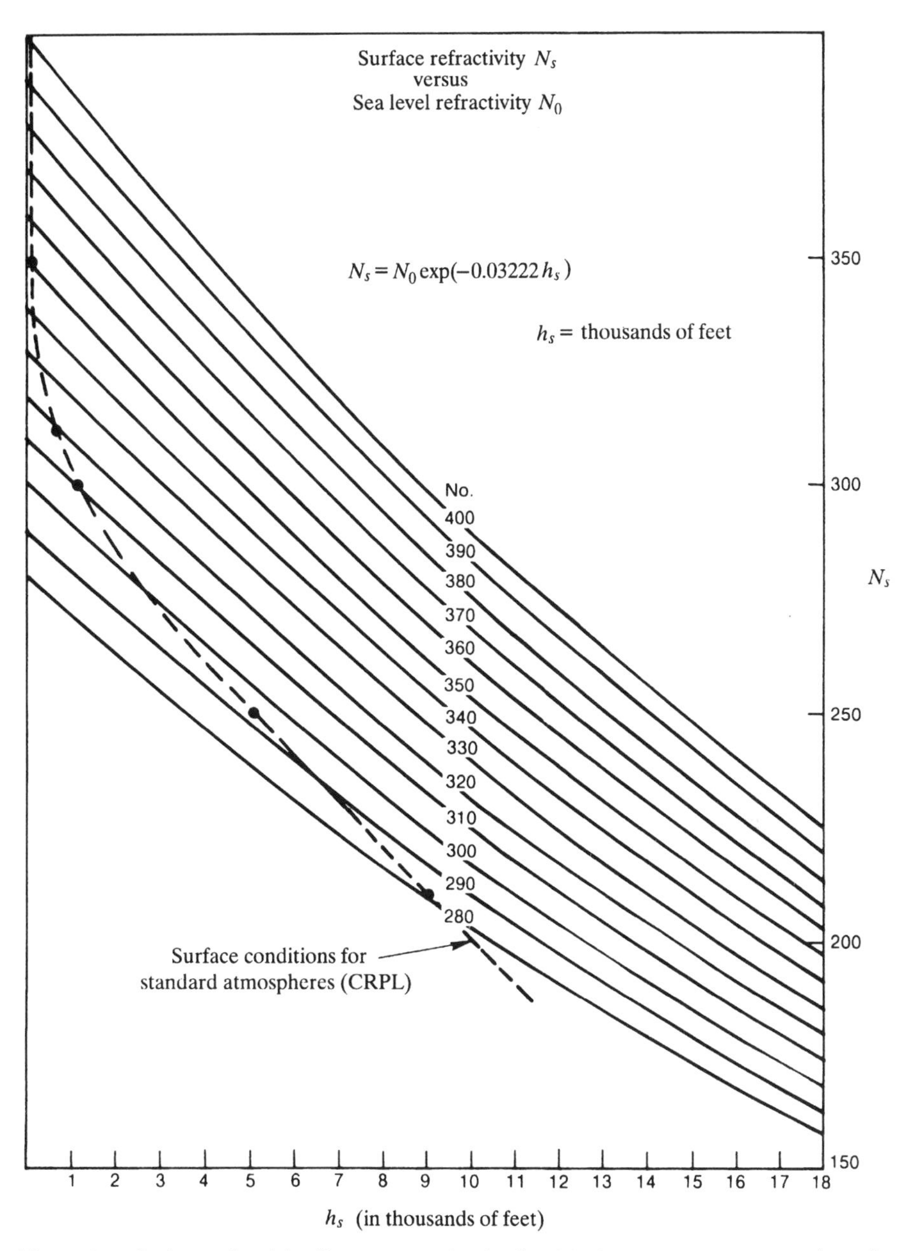

Figure 1.7. Surface refractivity N_s versus sea level refractivity N_0, based on equation (1.24b). (From Navelex 0101, 112; Ref. 7.)

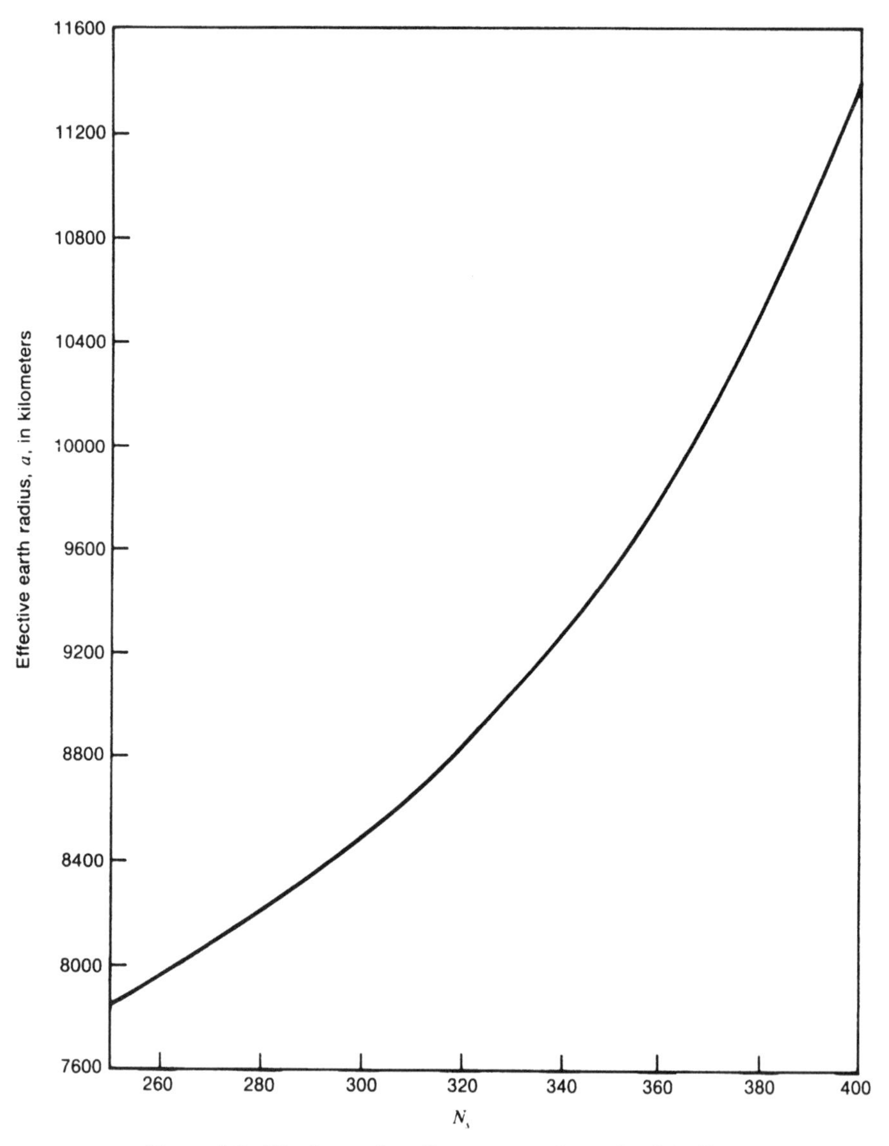

Figure 1.8. Effective earth radius versus surface refractivity N_s.

1.4 DIFFRACTION EFFECTS—THE FRESNEL ZONE PROBLEM

Diffraction of a radio wavefront occurs when the wavefront encounters an obstacle that is large compared to the wavelength of the ray. Below about 1000 MHz there is diffraction or bending from an obstacle with increasing attenuation as a function of obstacle obstruction. Above about 1000 MHz, with increasing obstruction of an obstacle, the attenuation increases even more rapidly such that the path may become unusable by normal transmission means (see Chapter 5). The actual amount of obstruction loss is dependent on the area of the beam obstructed in relation to the total frontal area of the energy propagated and to the diffraction properties of the obstruction.

Under normal transmission conditions (i.e., nondiffraction paths, Chapters 2 and 3), the objective for the system designer is to provide sufficient clearance of the obstacle without appreciable transmission loss due to the obstacle. To calculate the necessary clearance we must turn to wave physics, Huygen's principle, and the theory developed by Fresnel. When dealing with obstacle diffraction, we will assume that the space volume is small enough that gradient effects can be neglected so that the diffraction discussion can proceed as though in a homogeneous medium.

Consider Figure 1.9. The Huygens–Fresnel wave theory states that the electromagnetic field at a point S_2 is due to the summation of the fields caused by reradiation from small incremental areas over a closed surface about a point source S_1, provided that S_1 is the only source of radiation. The field at a constant distance r_1 from S_1, which is a spherical surface, has the same phase over the entire surface since the electromagnetic wave travels at a constant phase velocity in all directions in free space. The constant phase surface is called a *wavefront.* If the distances r_2 from the various points on the wavefront to S_2 are considered, the contributions to the field at S_2 are seen to be made up of components that will add vectorially in accordance with their relative phase differences. Where the various values of r_2 differ by a half-wavelength ($\lambda/2$), the strongest cancellation occurs. Fresnel zones distinguish between the areas on a closed surface about S_1 whose components add in phase.

Let us consider a moving point P_1 in the region about the terminal antenna locations S_1 and S_2 such that the sum of the distances r_1 and r_2

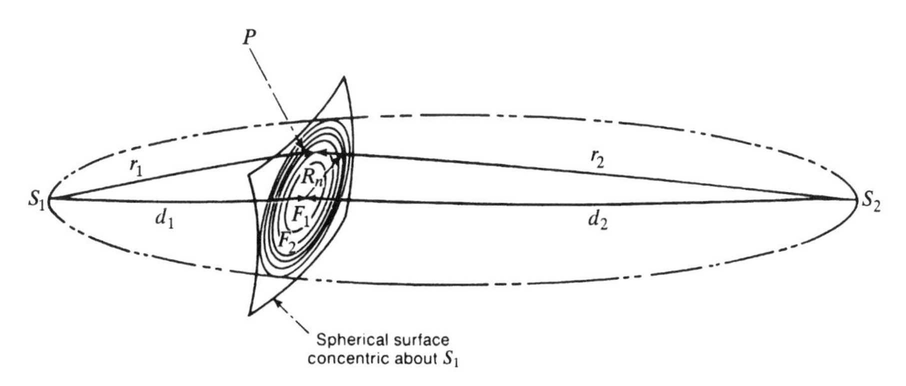

Figure 1.9. Fresnel zone geometry.

from the antennas to P is constant. Such a point will generate an ellipsoid with S_1 and S_2 as its foci. We now can define a set of concentric ellipsoidal shells so that the sum of the distances r_1 and r_2 differs by multiples of a half-wavelength ($\lambda/2$). The intersection of these ellipsoids defines Fresnel zones on the surface as shown in Figure 1.9. Thus, on the surface of the wavefront, a first Fresnel zone F_1 is defined as bounded by the intersection with the sum of the straight-line segments r_1 and r_2 equal to the distance d plus one-half wavelength ($\lambda/2$). Now the second Fresnel zone F_2 is defined as the region where $r_1 + r_2$ is greater than $d + \lambda/2$ and less than $d + 2(\lambda/2)$. Thus the general case may now be defined where F_n is the region where $r_1 + r_2$ is greater than $d + (n - 1)\lambda/2$ but less than $d + n\lambda/2$. Field components from even Fresnel zones tend to cancel those from odd zones since the second, third, fourth, and fifth zones (etc.) are approximately of equal area.

Fresnel zone application to path obstacles may only be used in the far field. The minimum distance d_F where the Fresnel zone is applicable may be roughly determined by $d_F > 2D^2/\lambda$, where D is the antenna aperture measured in the same units as λ.

To calculate the radius of the nth Fresnel zone R_n on a surface perpendicular to the propagation path, the following equation provides a good approximation:

$$R_n \simeq \sqrt{n\lambda\left(\frac{d_1 d_2}{d_1 + d_2}\right)} \tag{1.26a}$$

where R_n and d are in the same units, or

$$R_n \simeq 17.3\sqrt{\frac{n}{F_{\mathrm{GHz}}}\left(\frac{d_1 d_2}{d_1 + d_2}\right)} \tag{1.26b}$$

where d_1 is the distance to the near-end antenna and d_2 is the distance to the far-end antenna from the obstacle, and in equation (1.26b) all distances are in kilometers, the frequency of the emitted signal is in gigahertz, and R_n is in meters.

If R_1 is the first Fresnel zone, then

$$R_n = R_1\sqrt{n} \qquad \text{(Ref. 4)} \tag{1.27}$$

To calculate the radius of the first Fresnel zone in feet with d is measured in

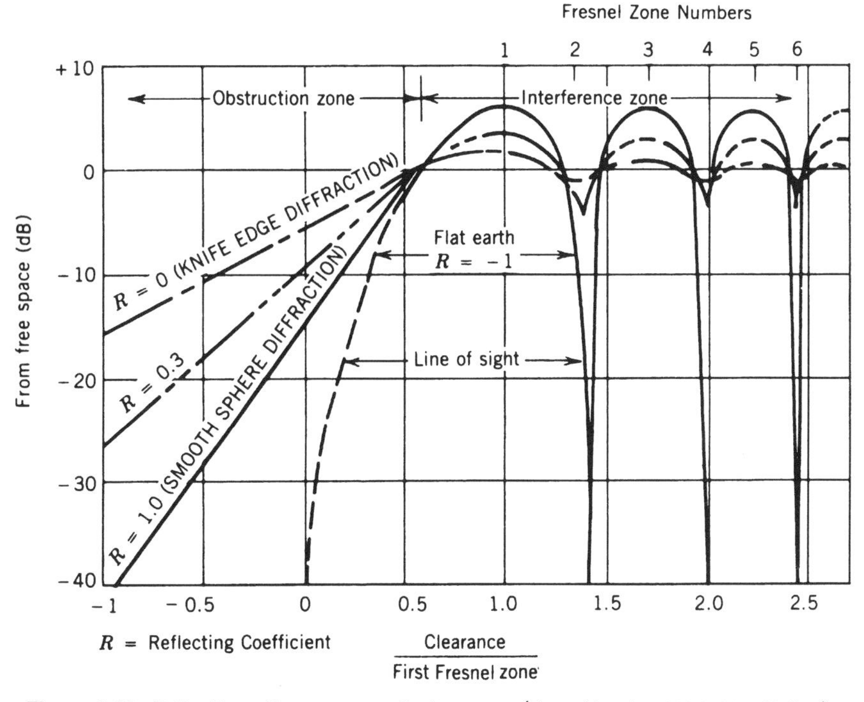

Figure 1.10. Path attenuation versus path clearance. (From Navelex 0101, 112; Ref. 7.)

statute miles,

$$R_1 = 72.1\sqrt{\frac{d_1 d_2}{F_{\text{GHz}}(d_1 + d_2)}} \qquad \text{(Ref. 4)} \qquad (1.28)$$

Conventionally, we require 0.6 Fresnel zone clearance of the beam edge (3-dB point) due to obstacles in the path. Figure 1.10 shows path attenuation versus path clearance. Providing 0.6 Fresnel zone clearance usually is sufficient to ensure that attenuation due to an obstacle in or near the ray beam path is negligible.

Reference 4 gives some practical Fresnel zone clearance guidelines related to K-factor:

> Although there are some variations, there are two basic sets of clearance criteria which are in common use in microwave communications systems. One is the "heavy route" set used for those systems with the most stringent reliability requirements, the other a "light route" set used for systems where some slight

relaxation of requirements can be made. The following are typical clearance criteria:

For "heavy route," highest-reliability systems: At least $0.3F_1$ at $K = \frac{2}{3}$, and $1.0F_1$ at $K = \frac{4}{3}$, whichever is greater. In areas of very difficult propagation, it may be necessary to ensure a clearance of at least grazing at $K = \frac{1}{2}$ (for 2 GHz paths above 36 miles, substitute $0.6F_1$ at $K = 1.0$).

Note that the evaluation should be carried out along the entire path and not just at the center. Earth bulge and Fresnel zone radii vary in a different way along the path, and it often happens that one criterion is controlling for obstacles near the center of the path and the other is controlling for obstacles near one end of the path.

For "light-route" systems with slightly less reliability requirements at least $0.6F_1 + 10$ feet at $K = 1.0$. At points quite near the ends of the paths, the Fresnel zones and earth bulge become vanishingly small, but it is still necessary to maintain some minimum of perhaps 15 to 20 feet above obstacles to avoid near field obstructions.

Experience has shown that, to avoid diffraction and scattering complications on a radio path, a minimum clearance of 0.6 of the first Fresnel zone *plus 10 feet* (or 3 m) is advisable. Of course, on LOS microwave paths, the more clearance provided, the higher antenna towers must be. The resulting economic impact on system cost must also be taken into consideration.

1.5 GROUND REFLECTION

When a radio wave is incident on the earth's surface, it is not actually reflected from a point on the surface, but from a sizable area. The area of reflection may be large enough to encompass several Fresnel zones or it may have a small cross-sectional area such as a ridge or peak encompassing only part of a Fresnel zone.

The significance of ground-reflected Fresnel zones is similar to free-space Fresnel zones. However, radio waves reflected from the earth's surface are generally changed in phase depending on the polarization of the signal and the angle of incidence. Horizontally polarized waves in our band of interest are reflected from the earth's surface and are shifted in phase very nearly 180°, effectively changing the electrical path length by approximately one-half wavelength ($\lambda/2$). For vertically polarized waves, on the other hand, the phase shift varies between 0° and 180° depending on the angle of incidence and the reflection coefficient, which depends largely on ground conditions. For the horizontally polarized case, if the reflecting surface is large enough to encompass the total area of any odd-numbered Fresnel zones, the resulting reflections will arrive at the receiving antenna out of phase with the direct wave, causing fading. In some cases a similar phenomenon has been observed for vertically polarized signals.

To mitigate ground reflections on LOS paths, tower heights can be adjusted (i.e., low–high technique) to effectively move the reflection point to a portion of the intervening path that is on rough terrain where the reflected signal will be broken up. Methods of adjusting the reflection point are discussed in Chapter 2.

1.6 FADING*

1.6.1 Introduction

Fading is defined as any time varying of phase, polarization, and/or level of a received signal. The most basic definitions of fading are in terms of the propagation mechanisms involved: refraction, reflection, diffraction, scattering, attenuation, and guiding (ducting) of radio waves. These are basic because they determine the statistical behavior with time of measurable field parameters including amplitude (level), phase, and polarization, as well as frequency and spatial selectivity of the fading. Once these mechanisms are understood, remedies can be developed to avoid or mitigate the effects.

Fading is caused by certain terrain geometry and meteorological conditions that are not necessarily mutually exclusive. Basic background information has been established on these conditions in previous sections of this chapter. All radio transmission systems in the 0.3–300-GHz frequency range can suffer fading including satellite earth terminals operating at low elevation angles and/or in heavy precipitation.

1.6.2 Multipath Fading

Multipath fading is the most common type of fading encountered, particularly on LOS radiolinks. It is the principal cause of dispersion, which is particularly troublesome on digital troposcatter and high-bit-rate LOS links.

For an explanation of atmospheric multipath fading, we must turn to the refractive index gradient (Section 1.3.3). As the gradient varies, multipath fading results owing to (1) the interference between direct rays (Figure 1.11) and the specular component of a ground-reflected wave; (2) the nonspecular component of the ground-reflected wave; (3) partial reflections from atmospheric sheets or elevated layers; or (4) additional direct wave paths (i.e., nonreflected paths).

Of interest to the radiolink design engineer is the fading rate, meaning the number of fades per unit of time, and the fading depth, meaning how much

*Section 1.6 has been adapted from Ref. 3.

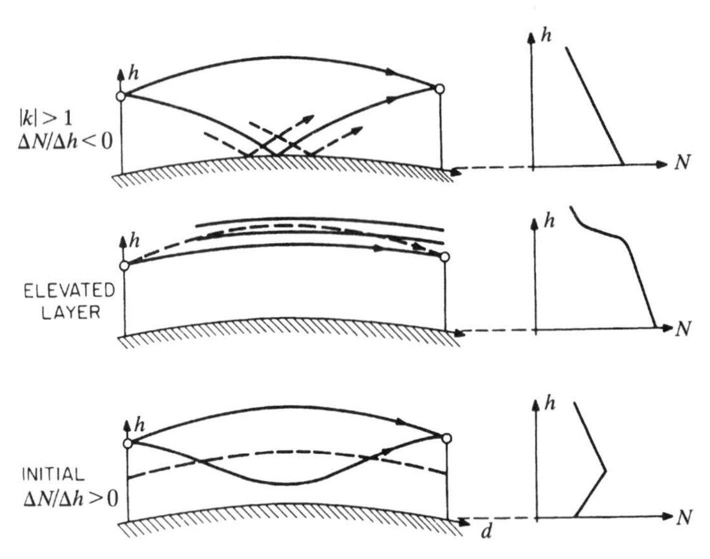

Figure 1.11. Mechanisms of multipath fading.

the signal intensity at the receiver varies from its free-space value, generally expressed in decibels.

The four multipath fading mechanisms previously listed can operate individually or concurrently. Fade depths can exceed 20 dB, particularly on longer LOS paths and more than 30 dB on longer troposcatter paths. Fade durations of up to several minutes or more can be expected.

Often multipath fading is frequency selective and the best technique for mitigation is frequency diversity. For effective operation of frequency diversity, sufficient frequency separation is required between the two transmit frequencies to provide sufficient decorrelation. On most systems a 5% frequency separation is desirable. However, on many installations such a wide separation may not be feasible owing to frequency congestion and local regulations. In such cases it has been found that a 2% separation is acceptable. Frequency diversity design is described further in Chapters 2, 3, 8, and 10–12.

1.6.3 Power Fading

Power fading results from a shift of the beam from the receiving antenna due to one or several of the following:

- Intrusion of the earth's surface or atmospheric layers into the propagation path
- Antenna decoupling due to variation of the refractive index gradient (variation of *K*-factor)

- Partial reflection from elevated layers that have been interpositioned in the ray beam path
- Where one of the terminal antennas is in a ducting formation
- Precipitation in the propagation path (discussed in Chapter 8)

Examples of the mechanisms of power fading are given in Figure 1.12.

1.6.3.1 Fading Due to Earth Bulge. When there is a positive gradient (subrefractive) of the refractive index, power fading may be expected owing to diffraction by the earth's surface, as shown in Figure 1.12a. Fade depths of 20–30 dB can be expected with fade durations lasting for several hours or more. This type of fading normally may not be mitigated by frequency diversity but may be reduced or completely avoided by the proper adjustment of antenna tower heights.

The guidelines for Fresnel zone clearance listed in Section 1.4 must be modified on those microwave LOS paths where subrefractive index gradients occur (i.e., where K is less than 1). Clearances greater than one Fresnel zone are recommended particularly where the intervening path approaches smooth earth. In mountainous regions where antennas are mounted on dominating ridges or peaks, a single Fresnel zone clearance would be sufficient. Similar guidelines may be used where a limited range of refractive index gradients is encountered.

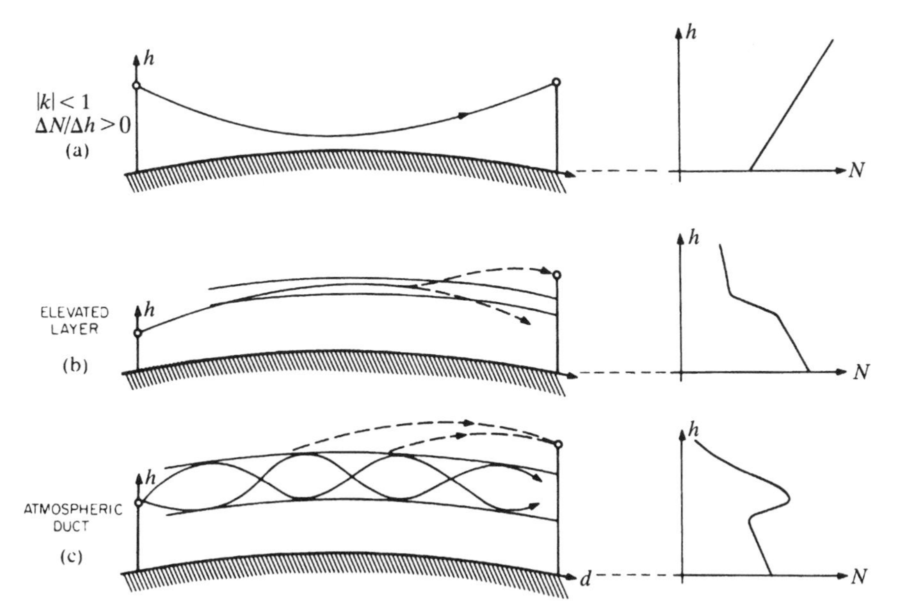

Figure 1.12. The mechanisms of power or attenuation fading. (From MIL-HDBK-416, Ref. 5.)

1.6.3.2 Duct and Layer Fading. Fades of about 20 dB or more can occur due to atmospheric ducts and elevated layers. These fades can persist for hours or days and are more prevalent during darkness. Neither space nor frequency diversity mitigates this type of fading.

An elevated duct is often characterized as a combination of a superrefractive layer above a subrefractive layer. Such a condition has the effect of guiding or focusing the signal along the duct. The reverse condition, namely, a subrefractive layer above a superrefractive layer, will tend to defocus signal energy introduced with the layer combination. The defocusing effect produces power fading.

One obvious cure for this type of fading is repositioning one or both antennas. Another would be to select other sites.

1.6.4 *K*-Factor Fading

This type of fading involves either multipath fading from direct ray and ground reflections or diffraction power fading that depends on the *K*-factor value. These two types of fading can supplement one another and cause fading throughout a wide range of refractive index gradients (values of *K*-factor). *K*-factor fading can be expected when the intervening terrain is comparatively smooth such as over-water paths, maritime terrain, or gently rolling terrain.

Figure 1.13, from Ref. 3, illustrates the resulting signal variations, the spherical earth transmission loss versus the refractive index gradient. The figure also shows the effect of terrain roughness, expressed by σ/λ and the divergence–convergence factor under the dynamic influence of the refractive index gradient. Reference 3 describes the ratio σ/λ, where σ is the standard deviation of the surface irregularities about a median spherical surface and λ is the transmission wavelength. Smooth earth is then defined as $\sigma/\lambda = 0$. For the smooth earth case, the fading, marked by nulls of interference between the direct wave and the specularly reflected wave, is serious only over a limited range of refractive index gradients. For the path parameters of Figure 1.13 as an example, the fades due to interference nulls can exceed 20 dB only within the range of -115 to -195 *N*-units/km and for gradients in excess of 300 *N*-units/km; as the surface roughness is increased from the smooth earth case or where $\sigma/\lambda = 0$, the critical region of gradients shifts to more negative values. In Figure 1.13, the critical region for negative gradients shifts to the range of -180 to -290 *N*-units/km for $\sigma/\lambda = 10$. Irregularities or roughness that would cause the median terrain surface to depart slightly from a sphere could also shift the range of critical negative gradients in either direction. These critical ranges, as well as those due to the diffraction fade (at values greater than 300 *N*-units/km), depend on specific link parameters such as transmission frequency, antenna heights, and path lengths. Figure 1.13 clearly shows that unless the terrain roughness is sufficient to shift the critical range of negative gradients outside the range of refractive

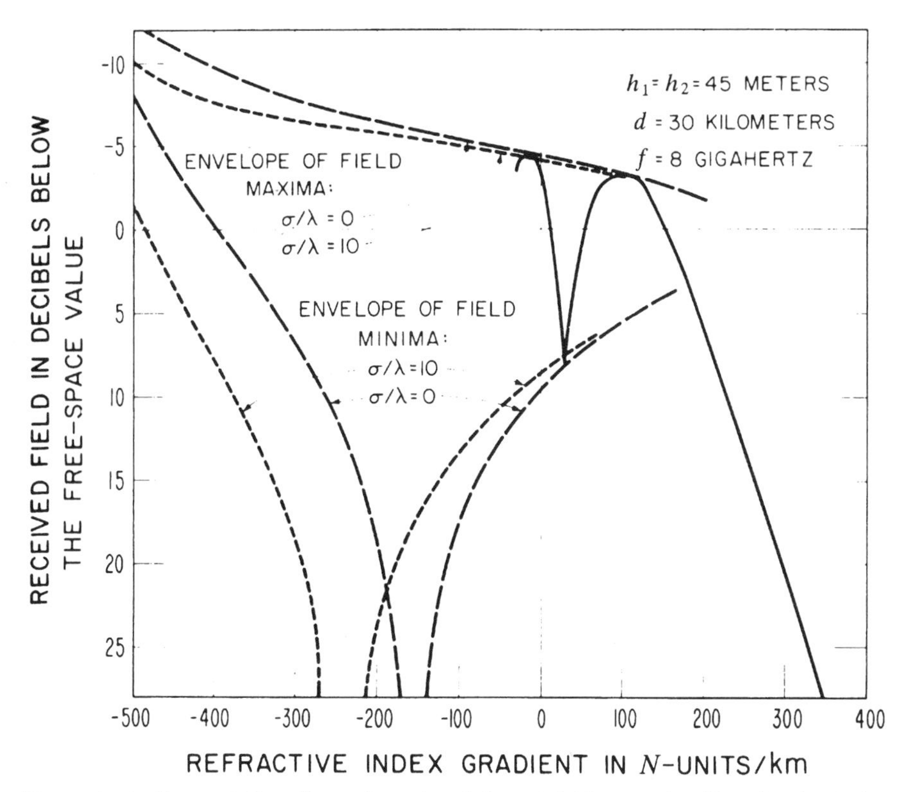

Figure 1.13. *K*-type fading, illustration of variation of field strength with refractive index gradient. (From Ref. 3.)

index gradients expected to occur at a particular location, reflections from the terrain surface cannot be neglected. Similarly, high points of the terrain cannot be considered to eliminate terrain reflection unless they also partially obstruct the reflected wave over the critical range of refractive index gradients.

The effects of *K*-type fading can be reduced by (1) increasing the terminal antenna heights to provide adequate protection against the diffraction fading for the expected extreme positive gradients of refractive index and (2) diversity reception that effectively reduces the attenuation due to multipath out of the expected extreme negative gradients of refractive index.

1.6.5 Surface Duct Fading on Over-Water Paths

Long LOS paths over water can encounter a special type of fading due to the presence of surface ducts. Such surface ducts can be a semipermanent condition, particularly in high-pressure regions such as the Bermuda High, which is in the Atlantic Ocean between 10° and 30° N latitude. In this case,

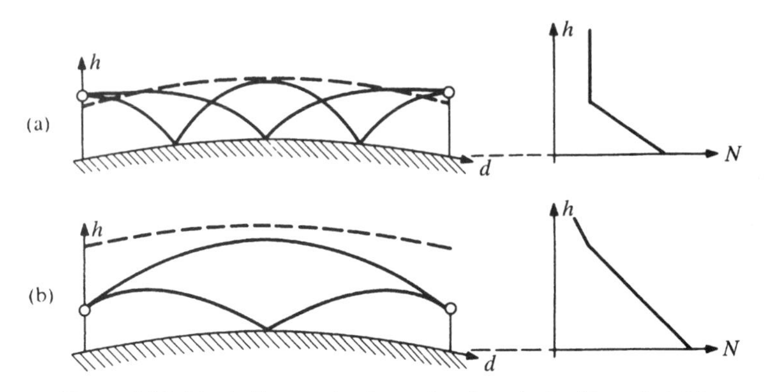

Figure 1.14. The fading mechanisms of surface ducts. (From Ref. 3.)

the ducts are formed less than 2 km from the shoreline and extend along the sea surface up to heights from 7 to 20 m for wind velocities from 15 to 55 km/h. They persist during fair weather and generally reform after rain showers and squalls. The resulting fading is due to a combination of multipath fading caused by sea reflections and power fading in the presence of the surface duct. Figure 1.14, from Ref. 3, illustrates two typical situations.

Because of the continual disturbance of the sea surface, a reflected wave consists of a diffuse or randomly distributed component superimposed on a specular component. This time distribution of the reflected wave is a Beckmann distribution (a constant plus a Hoyt distribution). This constitutes the received field of Figure 1.14a. In Figure 1.14a an addition of the direct wave produces an enhanced or reduced constant component due to phase interference. One effect of a surface duct on the multipath situation is to provide an increase in effective angle of incidence. This increases the ratio of the diffused to specular amplitudes, and increases the rapidly varying component of the reflected signal. The net result is a total signal whose distribution approaches the Nakagami–Rice distribution (a constant plus a Rayleigh-distributed variable) (Ref. 3).

These surface or ground-based ducts guide or trap the radio waves by the combination of a strong negative refractive index gradient (i.e., superrefraction, $\Delta N/\Delta h \leq -157$ N-units/km) and a reflecting sea or ground. As such, propagation within the duct can be described in terms of an equivalent linear gradient of refractive index. The corresponding equivalent earth representation is illustrated in Figure 1.15. The field results from phase interference between a direct wave, one to three singly reflected waves, and, for sufficiently strong superrefraction, doubly reflected waves.

Surface duct fading can be reduced by choosing terminal antenna heights to provide adequate Fresnel zone clearance above the ducting layer. This will tend to avoid the situation in Figure 1.14a. Likewise, lower antenna heights could achieve the situation in Figure 1.14b. In this latter case, diversity reception would also tend to mitigate the problem.

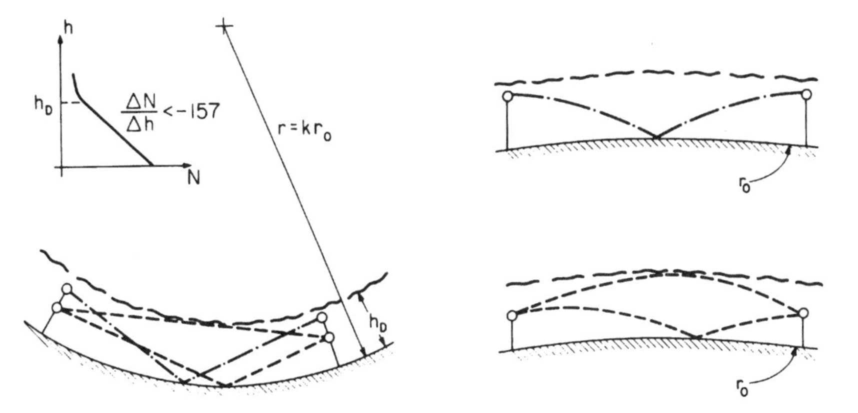

Figure 1.15. Examples of surface duct propagation for effective earth radius and true earth radius. (From Ref. 3.)

1.7 FROM ANOTHER PERSPECTIVE—A DISCUSSION OF FADING*

1.7.1 Comparison of Some Common Fading Types

Figures 1.16 and 1.17 illustrate selective and nonselective fading types that can affect the time availability of an LOS microwave path. Two or more of these types of fading can occur simultaneously. For example, multipath fading often accompanies other types of fading.

Consider Figure 1.16. Fading types (a) through (e) in the figure result from one or more interference rays arriving at the receive antenna in and out of phase with the main ray beam. We call these types of fades "selective" in that different frequencies within the transmitted band are affected differently. Any type of diversity (i.e., frequency or space) is almost totally effective in eliminating baseband hits or outages if adequate fade margins are provided.

The power or attenuation fading of Figure 1.17 is generally nonselective. In such cases normal frequency and space diversity techniques provide little or no improvement over nondiversity configurations regarding time availability.

Atmospheric multipath fading can appear in both stable and turbulent forms. Stable multipath fading is shown in Figure 1.16a. It occurs when only a small number (i.e., 2 or 3) of secondary reflected or refracted paths† are received simultaneously within the desired path. The resulting fading characteristic is relatively slow. But occasional fast, deep fades can occur. A turbulent multipath is shown in Figure 1.16b. Such a multipath causes fast but shallow fades with fewer outages. Both types of atmospheric multipath fading have a time-versus-depth fade distribution. For every 10-dB increase in fade margin outage time is reduced by a factor of 10. Multipath fading is

*Section 1.7 is based on material from Ref. 11.

†These can come from the atmosphere or from multiple low-amplitude ground reflection points.

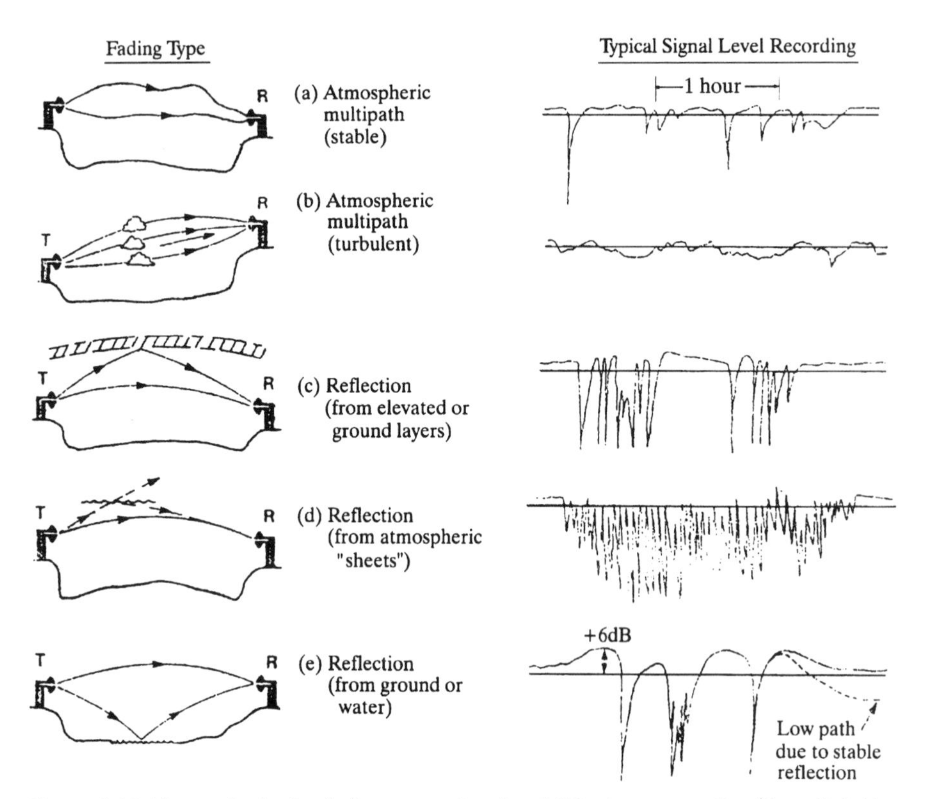

Figure 1.16. Types of selective fading encountered on LOS microwave paths. (From Ref. 11. Courtesy of Siemans, Inc., reprinted with permission.)

also quite sensitive to antenna orientation and size. Larger antennas typically have better selectivity against slightly off-path secondary multipath rays.

Figures 1.16c and 1.16d show the results of reflections from boundaries between elevated atmospheric layers and sheets. Elevated layer boundaries resulting from sharp changes in either temperature or humidity (or both) may form nearly perfect reflection planes at LOS microwave frequencies. Such layers are usually quite stable, at times moving slowly in a vertical direction. In-phase and out-of-phase reflections occur as the layer moves.

Sheets are high-altitude undulating layers perhaps some 10 feet thick and some 6 miles long. Such sheets are constantly changing, and the wild, fast, deep fades shown in Figure 1.16d occurring at night are often attributed to them. During severe reflection fading, the direct path is usually depressed by an obstruction or antenna decoupling, increasing its susceptibility to an interference ray. To mitigate such problems, larger antennas using vertical polarization may help. Either type of diversity with comparatively wide*

**Wide* in this context means greater than 2% frequency separation for frequency diversity and wider-than-normal antenna separation for space diversity.

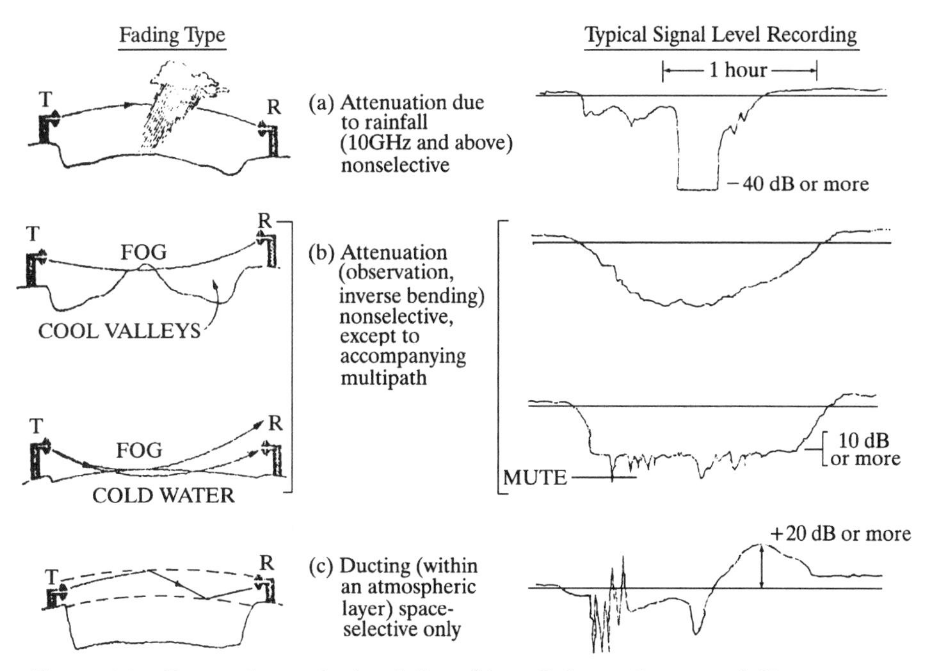

Figure 1.17. Types of nonselective fading. (From Ref. 11. Courtesy of Siemens, Inc., reprinted with permission.)

spacing can essentially eliminate baseband outages with fading from reflections from layers or sheets.

Figure 1.16e shows the results of specular reflections from water or ground. A stable signal reflected from such reflection points can combine with the desired main ray beam to increase the RSL (receive signal level) by 6 dB or completely suck out the desired signal with a fade perhaps > 50 dB. In a varying atmosphere, the RSL chart displays a rolling pattern. A path can get "stuck" on a stable reflection for days and weeks, representing a fade between 10 and 20 dB. Increasing the fade margin can help somewhat. A better solution is to move the reflection point as discussed in Chapter 2. Failing that alternative, the reflection could possibly be screened, planting trees in the sensitive area or moving the antennas so no severe reflection point occurs. Also, antennas can be tilted slightly upward, providing increased discrimination to the reflected ray. Another measure to be considered is to use larger antennas with smaller beamwidth. These can be especially attractive on high/low, where one antenna is considerably elevated over the other. Diversity can nearly eliminate the effects of this type of fading. Vertical polarization can also be effective if the reflection (grazing) angle is greater than 0.1 degree.

Figure 1.17a shows the effects of rainfall fading. Such fading usually occurs above 10 GHz. Even so, there are instances where rainfall fading has had deleterious effects on frequencies below 10 GHz. The mitigation of this type of fading is discussed in Chapter 9. Severe outages are usually caused by the blockage of the path by the passage of rain cells. Such cells typically have severe rainfall over short periods of time (i.e., 10–20 minutes) and are often accompanied by thunder. These rain cells are often 4–8 miles in diameter. The RSL chart shown in Figure 1.17a shows fairly slow, erratic level changes, with apparent rapid signal "suck-out" as the rain cell intercepts the signals path. Rainfall fades are nonselective, meaning that all paths in either direction are affected the same.

In rainfall fade situations, increased fade margin is useful. In some cases 45–60-dB fade margins may be required. In the case of satellite communications, excess attenuation due to rainfall versus time availability is a function of takeoff angle. One way of relieving the situation is to raise the minimum takeoff angle to 10, 15, or 20 degrees. In both the LOS microwave and satellite communication cases, we have the path averaging factor, which ameliorates or lessens some of the fade margin required. This is based on the concept that downpour rain cells are limited in radius, and except for very short paths, the entire path length should not be penalized for cells of limited radius (e.g., 3–7 km across). In the LOS microwave situation, route diversity can prove very effective if the parallel routes are separated by at least 5 miles (8 km).

Attenuation fading due to partial path obstruction in a substandard atmosphere* is nonselective. Using alternatives such as increased fade margin and greater antenna heights have had notable positive results. It is pointed out that the actual outages are caused by the accompanying severe multipath fade and not by the diffraction fading alone. The lower frequency band (e.g., 2 GHz) exhibits less obstruction or diffraction loss than the higher frequency bands for a given amount of inverse bending, so these alternatives are recommended where tower heights and fade margins cannot be increased.

Figure 1.17c shows the results of *ducting*. Here paths trapped in elevated or in ground ducts show enhanced signal levels. They exhibit wide, slow signal level changes, even up to 30 dB above the median in some instances. Ducts are often compared to direct waveguide coupling between transmitter and receiver. This can be seen in Figure 1.17c. Space diversity does mitigate the effects of ducting; frequency diversity does not.

1.7.2 Blackout Fading

According to Ref. 11, blackout fading is rare compared to other types of fading, but when it takes place, its effects are radical and catastrophic. Traditional measures to mitigate fading, such as increasing fade margin and the use of diversity, are usually ineffective. Blackout fading is not a new phenomenon, having been reported for more than 45 years.

*Substandard atmosphere refers to the strange condition in which atmospheric density increases with altitude, rather than decreasing as one might expect (Ref. 11).

The following is a list of the unique properties of a blackout fade:

1. The propagation failure is absolute, and no reasonable increase in fade margin can resolve it.
2. The LOS microwave links involved cross bodies of shallow warm water, swamps, shallow bays, and irrigated farmland and may be located parallel to coast lines.
3. Often the weather, just prior to a blackout fade, is unseasonably warm and humid. The fade coincides with a marked change in temperature and an increase in humidity. Often these blackout fades take place in the late evening, although some take place in daylight hours with the passage of a cold front.
4. Path lengths are usually comparatively long, in the 20–30-mile (32–48-km) range. Although these paths have sufficient diffraction clearance to prevent a complete diffraction outage during subrefractive periods, they are usually provided with less than 150 ft (46 m) of clearance over water or moist ground during normal propagation periods.
5. The catastrophic outages occur simultaneously in both directions of transmission and in both diversity paths except in certain very rare instances when the fade exhibits specific height selectivity on space diversity paths.

Figures 1.18 and 1.19a show RSL recording levels during a blackout fade. The figures show the unique pattern typical of blackout fading. Blackout fading depends only on the thickness, refractive index, and frequency of occurrence of the ground-based atmospheric layer, which, in turn, establishes whether or not the receive antenna is within the radio horizon of the transmit antenna. The more common types of fading, such as multipath fading,

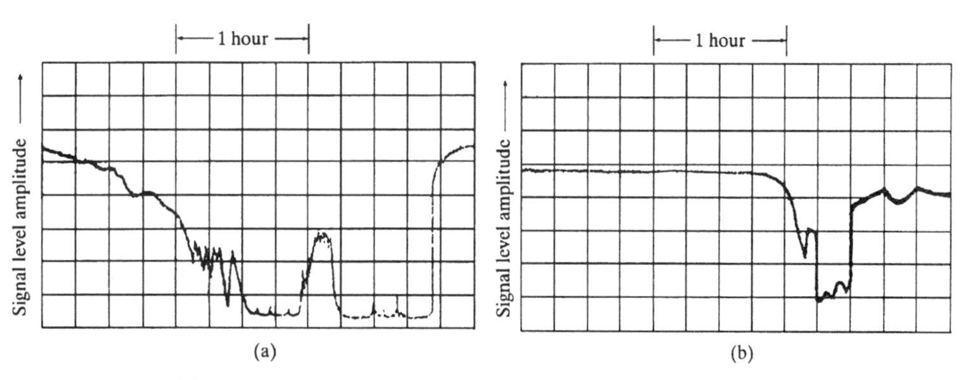

Figure 1.18. (a) Blackout fading on a low-clearance path over shallow, warm water (super-refractive atmospheric layer). (b) Diffraction fading on a low-clearance path over deep, cool water (subrefractive atmospheric layer). (From Ref. 11. Courtesy of Siemens, Inc., reprinted with permission.)

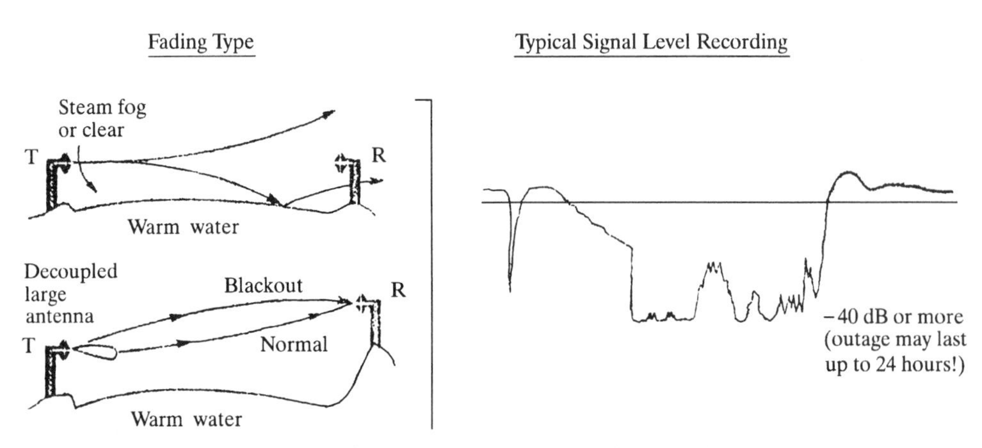

Figure 1.19. During blackout fading, the microwave path is trapped or arrives outside the main lobe of the receive antenna. This type of fading is usually nonselective but may occasionally be space-selective. (From Ref. 11. Courtesy of Siemens, Inc., reprinted with permission.)

depend on terrain reflectivity or water roughness, rainfall intensity, path clearance, amount of obstruction in a path, atmospheric turbulence, and/or transmitted frequency.

Blackout fading is generally a go/no-go phenomenon. Either the path is operational or it is totally lost. Normally, blackout fade depths are far greater than any other fading we have encountered.

Blackout fading is often confused with diffraction (obstruction) fading that takes place with inverse bending as shown in Figures 1.17b and 1.18b, although the atmospheric refraction characteristics of these two fade types are reversed. It should be emphasized that blackout fading results from the presence of an intervening superrefractive atmosphere, which is sometimes invisible except for boundary haze, and sometimes visible in the form of warm-water "steam" fog or mist that refracts the microwave wavefront downward or into the ground (or water) before it reaches the receive antenna. Usually no part of the transmit signal reaches the receive antenna for the duration of the fade event, as shown in Figure 1.19.

In contrast to this, inverse bending is the product of a subrefractive atmosphere, which often manifests itself as cold-water fog. This subrefractive atmosphere may refract the wavefront upward before it reaches the receive antenna. But in this type of fading mechanism, implying an obstructed LOS microwave path, the link continues to operate, although with degraded performance. On the contrary, with blackout fading, there is complete path outage. With obstruction fading, typically from earth bulge, there is rapid multipath fading, which can be mitigated by a space or frequency diversity scheme. Blackout fading with its path instability, on the other hand, shows little or no multipath fading although some of its characteristics appear as though they were multipath in nature.

If the radio horizon during a blackout fade occurs just in front of the receive antenna, small changes in the thickness or refractive index gradient of the layer may cause the RSL to whip in and out of the range of the receive antenna. The resulting fade pattern would be similar to interference fading with slightly varying path clearance with changes occurring in the K-factor value. However, with a blackout fade, the RSL does not exhibit typical 6-dB rolling upfade (signal improvement) characteristic of two-path reflection fading.

Blackout fading is brought about by an anomalous atmospheric condition. It results from the formation of unusally steep, negative atmospheric density gradients, a dramatic drop in humidity, or an increase in temperature with height. As a microwave ray beam is propagated through a blackout atmosphere, the lower part of the wavefront is slowed by the dense air with respect to the upper part, causing a bending of the ray beam toward the ground or water. The amount of bending is small, but over longer paths (i.e., more than 20 mi or 32 km), the accumulated bending could refract the ray beam into the ground before it reaches the receive antenna. As the ray beam approaches ground or grazes ground it can be absorbed by foliage or other verdure or can be scattered by rough terrain. It is often specularly reflected by smooth terrain or water.

1.8 FADE DEPTH AND FADE DURATION

The most common type of fading is multipath fading. System design requires knowledge of the annual amount of time during which multipath fading reduces the RSL to a value that is much lower than the nominal level. RSL, of course, is a power level measured in dBW or dBm. Let p, then, denote the RSL in the presence of fading, and p_0 the power level (RSL) without fading. The time of interest is that during which the ratio p/p_0 is less than a quantity L^2, where $20 \log L$ describes the fade level in decibels relative to the unfaded RSL. Reference 12 suggests that in a heavy fading month the probability (fraction of time) that $p/p_0 < L^2$ is

$$P = c(f/4)d^3L^2 \times 10^{-5} \tag{1.29}$$

where f is the frequency in gigahertz, and d is the path length in miles. A description of climate and terrain is contained in the factor c, which has a value of unity for average climate and terrain and c can exceed 6 on paths with smooth terrain in southern (U.S.) coastal areas. On the other hand, c can be less than 0.2 on paths with rough terrain and dry climate. The length of the fading season, which is related to the average annual temperature, determines the amount of annual fading. The annual fade time is three times that occurring in a heavy fading month when the climate is average.

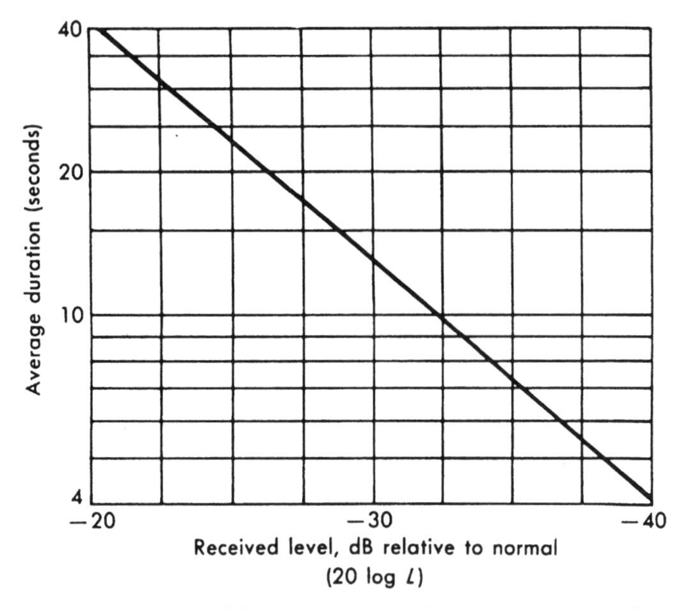

Figure 1.20. Average duration of fades on typical LOS microwave paths. (From Ref. 12; reprinted with permission.)

The variation of multipath fading with time is described by fade durations. Typically, a 20-dB fade lasts about 40 seconds, and the average duration of a 40-dB fade is about 4 seconds. This is shown in Figure 1.20. At any given fade depth, 1% of the fades may have a duration that is more than ten times the average.

1.9 PENALTY FOR NOT MEETING OBSTACLE CLEARANCE CRITERIA

If a radiolink path does not meet the obstacle clearance criteria established in Section 1.4, there is a penalty of excess attenuation due to diffraction loss.

Diffraction loss depends on the type of terrain and vegetation. For a given path ray clearance, the diffraction loss will vary from a minimum value for a single knife-edge obstruction to a maximum value for a spherical smooth earth. Methods of calculating diffraction loss are given in Chapter 4. The upper and lower limits of diffraction loss are given in Figure 1.21.

The diffraction loss over average terrain can be approximated for losses greater than about 15 dB by the formula

$$A_d = -20h/F_1 + 10 \tag{1.30}$$

where h is the height in meters of the most significant path blockage above

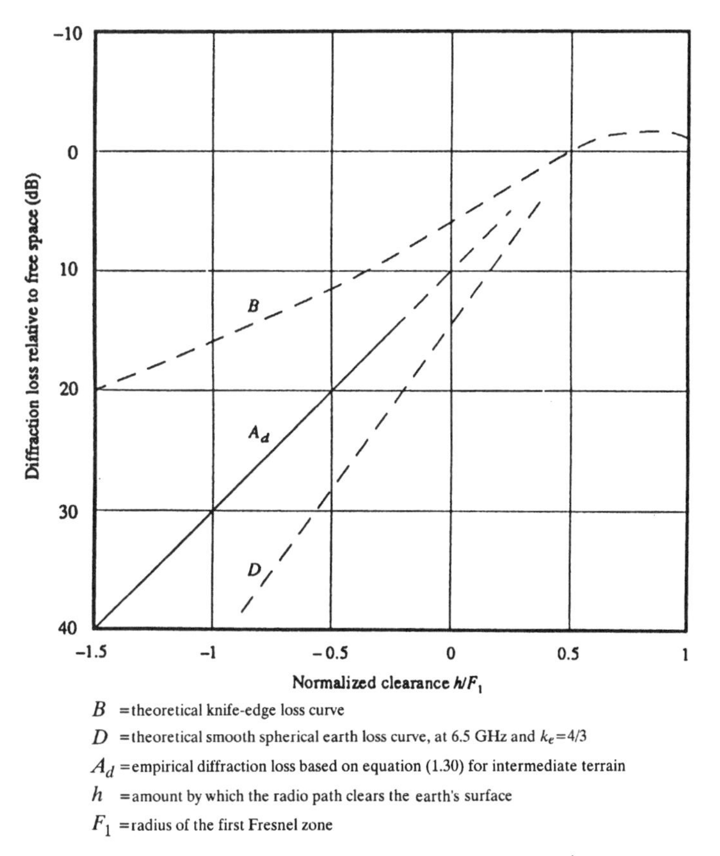

Figure 1.21. Diffraction loss for obstructed LOS microwave paths. (From Figure 1, p. 216, CCIR Rec. 530-4; Ref. 13.)

the path trajectory, h is negative if the top of the obstruction of interest is above the virtual line-of-sight, h is negative. F_1 is the radius of the first Fresnel zone, which can be calculated by formula (1.26b). The curve in Figure 1.21 is strictly valid for losses greater than 15 dB. It has been extrapolated up to 6-dB loss as an aid to link designers.

1.10 ATTENUATION THROUGH VEGETATION

Figure 1.22 gives attenuation data on microwave ray beams passing through woodland. The curves represent an approximate average for all types of woodland for frequencies up to about 3 GHz. When the attenuation inside such woodland becomes large (i.e., > 30 dB), the possibility of diffraction or surface modes has to be considered (Ref. 14).

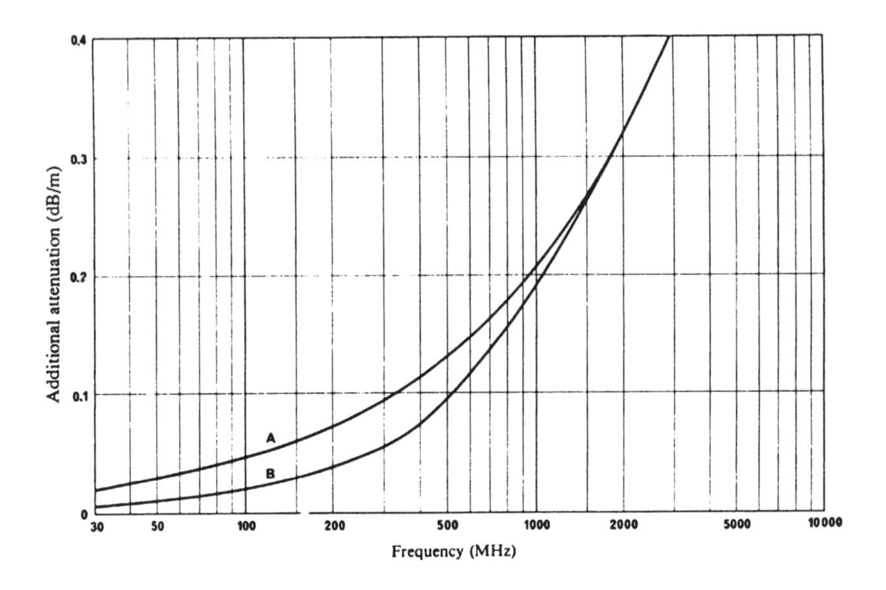

Figure 1.22. Specific attenuation of woodland. A = vertical polarization; B = horizontal polarization. (From Figure 1, p. 171, CCIR Rec. 833, RPN Series, CCIR 1992; Ref. 14.)

PROBLEMS AND EXERCISES

1 When dealing with radio propagation, why is there a demarcation line at about 10 GHz?

2 Define an isotropic antenna regarding gain and directivity. Can an isotropic antenna be fabricated and be truly isotropic?

3 Calculate free-space loss for a LOS radiolink 29 statute miles long, where the operating frequency is 6.135 GHz. Double the frequency and calculate the free-space loss. Halve the distance and calculate the free-space loss. What is the lesson learned from the last two steps?

4 Calculate the free-space loss to a satellite 3200 nautical miles distant and operating at a frequency of 14 GHz. What additional loss is there if the satellite is moved to a range of 3450 nautical miles?

5 A transmitter radiates (EIRP) 1 kW in a desired direction. What is the field strength in mV/m at a point 2 km distant in that direction?

6 What causes a radio beam to curve when passing through the atmosphere?

7 Relating to question 6, what mathematical tool is used to describe the curvature?

8 How does the atmospheric dielectric constant vary with altitude?

9 Differentiate between radio line-of-sight and optical line-of-sight.

10 Relate K-factor to true and effective earth radius.

11 If the effective earth radius were 8500 km, what would the equivalent K-factor be?

12 What determines K-factor?

13 Convert the N_0 value of 301 N-units to its equivalent N_s value at 1.2-km altitude.

14 A line-of-sight radiolink is 12 miles long and operates at 6.0 GHz. Calculate the 0.6 Fresnel zone clearance for an obstacle 3.6 mi from one end.

15 A SCADA link operates at 952 MHz and is 4 mi long. What 0.6 (+10 ft) Fresnel zone clearance is required at 1.9 mi from one end?

16 Why should we try to avoid having a LOS microwave ray beam hit reflective surfaces (e.g., bodies of water) along its path? If it does, what tends to result?

17 Describe at least three causes of multipath fading.

18 Describe at least three causes of power fading.

19 What are some of the mitigation techniques that can be used to counter multipath fading? Name at least three.

20 Define *selective* fading.

21 Relate blackout fading as to where we might expect to encounter it, types of path, and weather conditions. What can be done to avoid it?

22 Fading is most sensitive to: operating frequency, terrain, climate, or path length?

23 Suppose a so-called line-of-sight path does not meet required obstacle clearance criteria (e.g., $0.6F_1 + 10$ ft). In this case we have to pay a penalty in excess attenuation over the free-space value. The amount of penalty varies directly with _____ and inversely with _____.

24 How long should we expect a 30-dB fade to last (in seconds)?

REFERENCES

1. D. C. Livingston, *The Physics of Microwave Propagation*, GTE Technical Monograph, General Telephone & Electronics Laboratories, Bayside, NY, May 1967.
2. Military Handbook, *Facility Design for Tropospheric Scatter*, MIL-HDBK-417, U.S. Department of Defense, Washington, DC, Nov. 1977.

3. H. T. Dougherty, *A Survey of Microwave Fading Mechanisms, Remedies and Applications*, U.S. Department of Commerce, ESSA Tech. Rept. ERL-WPL 4, Boulder, CO, Mar. 1968.
4. *Engineering Considerations for Microwave Communications Systems*, GTE-Lenkurt, Inc., San Carlos, CA, 1975.
5. Military Handbook, *Design Handbook for Line of Sight Microwave Communication Systems*, MIL-HDBK-416, U.S. Department of Defense, Washington, DC, Nov. 1977.
6. B. R. Bean and E. J. Dutton, *Radio Meteorology*, Dover Publications, New York, 1968.
7. Naval Shore Electronics Criteria, *Line-of-Sight Microwave and Tropospheric Scatter Communication Systems*, Navelex 0101, 112, U.S. Department of the Navy, Washington, DC, May 1972.
8. *Radio Wave Propagation: A Handbook of Practical Techniques for Computing Basic Transmission Loss and Field Strength*, ECAC-HDBK-82-049 ADA 122-090, U.S. Department of Defense, Electromagnetic Compatibility Analysis Center, Annapolis, MD, Sept. 1982.
9. USAF Technical Order 31Z-10-13, *General Engineering Beyond-Horizon Radio Communications*, U.S. Air Force, Washington, DC, Oct. 1971.
10. *The Radio Refractive Index: Its Formula and Refractivity Data*, CCIR Rec. 453-3, RPN Series, CCIR, Geneva, 1992.
11. *Anomalous Propagation Parts 1 and 2*, GTE-Lenkurt Demodulator, July/Aug. 1975.
12. *Transmission Systems for Communications*, 5th ed., Bell Telephone Laboratories, Holmdel, NJ, 1982.
13. *Propagation Data and Prediction Methods Required for the Design of Terrestrial Line-of-Sight Systems*, CCIR Rec. 530-4, RPN Series, CCIR, Geneva, 1992.
14. *Attenuation in Vegetation*, CCIR Rec. 833, RPN Series, CCIR, Geneva, 1992.

2

LINE-OF-SIGHT MICROWAVE RADIOLINKS

2.1 OBJECTIVE AND SCOPE

Line-of-sight (LOS) microwave radiolinks, in the context of this book, provide broadband connectivity* for telecommunications using radio equipment with carrier frequencies above 900 MHz. For most applications these radiolinks are considered a subsystem of a telecommunications network. They carry one or a mix of the following:

- Telephone channels
- Data information
- Facsimile, now usually in a digital data format
- Video, especially conference TV
- Program channels
- Telemetry (which is a specialized subset of data)

The emitted waveform may be analog or digital. This chapter deals with analog LOS microwave using FM for its modulation; Chapter 3 deals with what we call digital microwave, which really is an analog carrier digitally modulated.

In highly industrialized nations, almost without exception, new LOS microwave links are digital. The exception is possibly the studio-to-transmitter link (STL) used to connect a TV studio to its transmitter. The position we take is as follows:

1. The reader should be made aware of FM analog LOS microwave, particularly its advantages and limitations. Such links are still being installed in many parts of the world.

*There are two exceptions to the broadband context: and one is SCADA systems, which are comparatively narrowband, providing data channels up to about 12 kHz of RF bandwidth, and another is narrowband tropo.

2. Once we feel comfortable with the design of FM microwave links and systems, the transition to digital microwave is fairly easy and straightforward.

LOS implies a terrestrial connectivity. Of course a satellite link, by definition, is also LOS from all of its associated earth terminals. On such a radiolink, the distinguishing feature is line-of-sight and, in particular, "radio line-of-sight" (see Section 1.3.2 and Figure 1.2). This requires sufficient clearance of intervening terrain on a LOS link such that the emitted ray beam from the transmitting antenna fully envelops its companion, far-end, receiving antenna. Typically, LOS microwave radiolinks, also called hops, are 10–100 km long. There are many exceptions. I designed a link in Pennsylvania less than 0.7 km long and another in the Andes more than 125 km long.

This chapter describes procedures to design LOS microwave links individually, as a series of links in tandem, and also as a subsystem that is part of an overall telecommunication network. The design of such a radiolink, whether analog or digital, is a four-step process with iteration among the steps:

1. Initial planning and site selection.
2. The drawing of a path profile.
3. Path analysis.
4. Site survey.

Commonly, resiting is necessary when a path is shown not to be feasible because of terrain, performance, and/or economic factors.

2.2 INITIAL PLANNING AND SITE SELECTION

A LOS microwave route consists of one, several, or many hops. It may carry analog or digital traffic. The design engineer will want to know if the LOS subsystem to be installed is an isolated system on its own, such as

- A private microwave (radiolink) system
- A studio-to-transmitter link
- An extension of a CATV headend

or is part of a larger telecommunication network where the link may be part of a backbone route or a "tail" from the backbone.

2.2.1 Requirements and Requirements Analyses

Let us suppose that we are to design a microwave LOS subsystem to provide telecommunication connectivity. The design criteria will be based on the current transmission plan available from the local telecommunication administration. For military systems, the appropriate version of MIL-STD-188 series would be imposed as a system standard. For systems carrying video and related program channel information, if no other standard is available, consult EIA RS-250 latest version and applicable CCIR recommendations.

A transmission plan will state, as a minimum, for analog systems:

- Noise accumulation in the voice channel for FDM telephony (see example from CCIR, Table 2.1)
- S/N for video and program channel information [CCIR Rec. 555 (Ref. 2) on 2500-km hypothetical reference circuit, on luminance signal to rms weighted noise ratio: 57 dB for more than 20% of a month and 45 dB for more than 0.1% of a month]

TABLE 2.1 Noise Accumulation in a FDM Voice Channel Due to Radio Portion on 2500-km Hypothetical Reference Circuit—CCIR Rec. 395-2

1. That, in circuits established over real links do not differ appreciably from the hypothetical reference circuit, the psophometrically weighted[a] noise power at a point of zero relative level in the telephone channels of frequency-division multiplex radio-relay systems of length L, where L is between 280 and 2500 km, should not exceed:

1.1 $3L$ pW 1-min mean power for more than 20% of any month;

1.2 47,500 pW 1-min mean power for more than $(L / 2500) \times 0.1\%$ of any month; it is recognized that the performance achieved for very short periods of time is very difficult to measure precisely and that in a circuit carried over a real link, it may, after installation, differ from the planning objective;

2. That circuits to be established over real links, the composition of which, for planning reasons, differs substantially from the hypothetical reference circuit, should be planned in such a way that the psophometrically weighted noise power at a point of zero relative level in a telephone channel of length L, where L is between 50 and 2500 km, carried in one or more baseband sections of frequency-division multiplex radio links, should not exceed:

2.1 for $50 \leq L \leq 840$ km:

2.1.1 $3L$ pW + 200 pW 1-min mean power for more than 20% of any month,

2.1.2 47,500 pW 1-min mean power for more than $(280 / 2500) \times 0.1\%$ of any month when L is less than 280 km, or more than $(L / 2500) \times 0.1\%$ of any month when L is greater than 280 km;

2.2 for $840 < L \leq 1670$ km:

2.2.1 $3L$ pW + 400 pW 1-min mean power for more than 20% of any month,

2.2.2 47,500 pW 1-min mean power for more than $(L / 2500) \times 0.1\%$ of any month;

2.3 for $1670 < L \leq 2500$ km:

2.3.1 $3L$ pW + 600 pW 1-min mean power for more than 20% of any month,

2.3.2 47,500 pW 1-min mean power for more than $(L / 2500) \times 0.1\%$ of any month;

3. That the following Note should be regarded as part of the Recommendation:

Note 1. Noise in the frequency-division multiplex equipment is excluded. On a 2500-km hypothetical reference circuit the CCITT allows 2500 pW mean value for this noise in any hour.

[a]The level of uniform-spectrum noise power in a 3.1-kHz band must be reduced by 2.5 dB to obtain the psophometrically weighted noise power.

Source: CCIR Rec. 395-2 (Ref. 1).

Traffic over the route must be quantified in number of voice channels, video channels, program channels, or gross bit rates for digital systems. Similarly, where applicable, location, routing, and quantification of traffic will be required for drops and inserts along the proposed route. For the case of video and program channels, information bandwidth will be required and, for video, limits of differential phase and gain as well as the manner of handling aural and cue channels.

Once the type of traffic has been stated, an orderwire and telemetry doctrine can be established.

Commonly, the life of a transmission system is 15 years, although many systems remain in operation, often with upgrades, for longer periods. System planning should include future growth out to 15 years with 5-year incremental milestones. Thoughtful provision for future growth during initial installation may well involve a greater first cost but can end up with major savings through the life of the system. These growth considerations will impact the following:

- Building size, space requirements, floor loading, prime power, air conditioning
- Frequency planning
- Installation (wired but not equipped)

Another important factor that may well drive design is compatibility with existing equipment.

2.2.2 Route Layout and Site Selection

We wish to interconnect points X and Y on the earth's surface by means of LOS microwave. The route between X and Y may be short, requiring only one microwave hop, or it may be long, requiring many hops in tandem. In any event a route must be laid out between those two points and terminal/repeater sites and/or repeater sites with drops and inserts (now called add-drop) are identified. For route layout and site selection, accurate topographic maps are used. It is advisable to carry out initial route layout on small-scale topographic (topo) maps such as 1 : 250,000 or 1 : 100,000. Terminal/repeater sites must be in apparent line-of-sight of each other. It is incumbent on the design engineer to minimize the number of repeaters along the route. Cost, of course, is a major driver of this requirement. We should not lose sight of some other, equally important, reasons. For analog systems, each additional relay inserts noise into the system. The distant receiver is very noise constrained (see Table 2.1). For digital systems, each relay site adds jitter to the signal and deteriorates error performance.

Drop and insert points, such as telephone exchanges (central offices), are first-choice relay locations for public switched telecommunication network

(PSTN) systems, although terrain may not permit this luxury. Local zoning may be another factor. In such cases, wire, fiberoptic, or radio spurs may be required. Other good candidate locations are existing towers (typically tall TV or FM broadcast towers) and tall buildings where space may be leased. Economy often limits tower heights to no more than 300 ft. On a smooth earth profile, this limits distance between relay sites to about 45 miles assuming $\frac{4}{3}$ earth ($K = \frac{4}{3}$ or 1.33). Figure 2.1 illustrates a simple method of making a rough calculation of distance (d) to the radio horizon, assuming smooth earth, $K = \frac{4}{3}$, where h is the height of the antenna. Equations (2.1a) and (2.1b) apply.

$$d = \sqrt{2h} \tag{2.1a}$$

where d is measured in statute miles and h is measured in feet. The following relation can be used for the metric system, where d is measured in kilometers and h in meters:

$$d = 2.9\sqrt{2h} \tag{2.1b}$$

The designer, of course, will take advantage of natural terrain features for relay sites, using prominent elevated terrain.

A final route layout is carried out using large-scale topographic maps with scales of 1 : 25,000 to 1 : 63,000 and contour intervals around 10 ft (3 m). For the United States these maps may be obtained from the following sources (Ref. 2):

Director
Defense Mapping Agency
Topographic Center
Washington, DC 20315

Sales Office
U.S. Geological Survey
Washington, DC 20305

Sales Office
U.S. Geological Survey
Denver, Colorado 80225

Map Information
U.S. Coast and Geodetic Survey*
Washington, DC 20350

U.S. Navy Hydrographic Office
Washington, DC 20390

*Computerized topographic data also available.

Figure 2.1. Illustration of factors for the calculation of distance to the horizon (grazing): smooth earth assumed.

In Canada topographic maps may be obtained from the following source (Ref. 2):

Department of Energy,
Mines and Resources
Surveys and Mapping Branch
615 Booth Street
Ottawa, Ontario, Canada

Other map sources include foreign government services, such as the British Ordnance Survey, and national agricultural, forest, and soil conservation departments.

In most situations large-scale maps, such as those with scale 1:25,000, must be joined together with careful alignment such that an entire hop (link) appears on one sheet. This is done on an open floor space by folding back or clipping map borders and joining them using blank paper backing and double-sided sticky tape.

Once joined, a straight line is drawn on the topo map connecting the two adjacent sites making up a single hop. This line forms the basis (database) for the path profile to determine required tower heights. The initial site selection was based on LOS clearance of the radio ray beam and drop and insert points. The path profile will confirm the terrain clearance. However, before final site selection is confirmed, the following factors must also be considered:

- Availability of land. Is the land available where we have selected the sites?
- Site access. Can we build a road to the site that is cost effective? This may be a major cost driver and may force us to turn to the use of another site where access is easier, but where a taller tower is needed.
- Construction restrictions, zoning regulations, nearby airport restrictions.
- Level ground for tower and shelter.

- Climatic conditions (incidence of snow and ice).
- Possibilities of anomalous propagation conditions such as in coastal regions, reflective desert, or over-water paths.

2.3 PATH PROFILES

A path profile is a graphical representation of a path between two adjacent radiolink sites in two dimensions. From the profile, tower heights are derived, and, subsequently, these heights can be adjusted (on paper) so that the ray beam reflection point will avoid reflective surfaces. The profile essentially ensures that the proper clearances of path obstructions are achieved.

There are three recognized methods to draw a path profile:

1. Fully Linear Method. Common linear graph paper is used where a straight line is drawn from the transmitter site to the receiver site, giving tangential clearance of equivalent obstacle heights. A straight line is also drawn from the receiver site to the transmitter site. The bending of the radio beam (see Section 1.3.4) is represented by adjustment of each obstacle height by equivalent earth bulge using the equation

$$h = \frac{d_1 d_2}{1.5K} \tag{2.2a}$$

where h is the change in vertical distance in feet from a horizontal reference line, d_1 is the distance in statute miles from one end of the path to the obstacle height in question, and d_2 is the distance from the other end of the path to the same obstacle. K is the selected K-factor (Section 1.3.2.1).

2. $\frac{4}{3}$ Earth Method. $\frac{4}{3}$ Earth graph paper is required. In this case true values of obstacle height may be used. An example of a profile using $\frac{4}{3}$ graph paper is shown in Figure 2.2. Of course, with this method, the value of K is fixed at $\frac{4}{3}$.

3. Curvature Method. Linear graph paper is used. True values of obstacle heights are employed from a reference line or mean sea level (MSL) and a curved line is drawn from the transmitter site (arbitrarily one end) to the receiver site and vice versa. The curved line has a curvature KR, where K is the applicable K-factor and R is the geometric radius of the earth or 3960 statute miles (6370 km) assuming the earth is a perfect sphere.

Method 1 is recommended because it (1) permits investigation and illustration of the conditions of several values of K to be made on one chart, (2) eliminates the need for special earth curvature graph paper, and (3) does not require plotting curved lines—only a straight edge is needed, thus

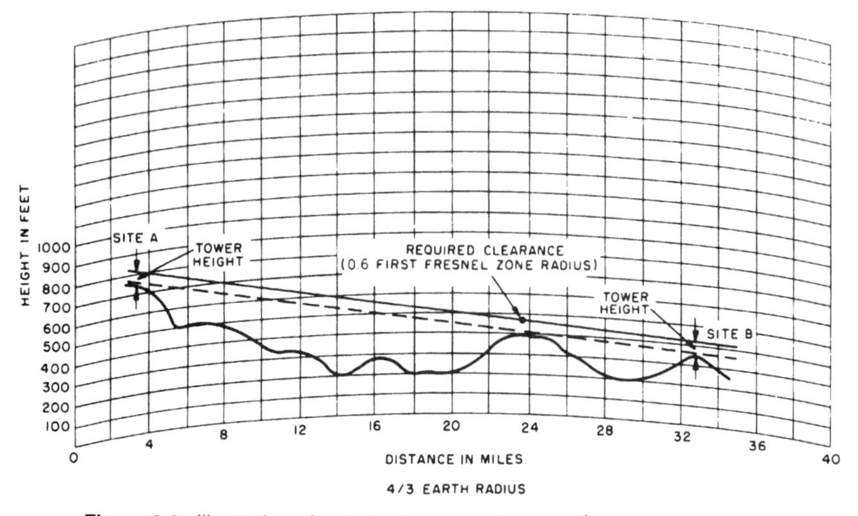

Figure 2.2. Illustration of path profile method 2 using $\frac{4}{3}$ earth graph paper.

facilitating the task of profile plotting. For this method it is convenient to plot on regular 10 × 10 divisions to the inch (or millimeter) graph paper, and *B* size (11 in. × 17 in.) paper is recommended.

The paper requires scaling. For the horizontal scale, usually 2 miles to the inch is satisfactory, permitting a 30-mile path to be plotted on one *B*-size sheet.

The scale to be used vertically will depend on the type of terrain for the path in question. Where changes in path elevation do not exceed 600–800 ft, a basic elevation scale of 100 ft to the inch will suffice. For a path profile involving hilly country, 200 ft to the inch may be required. Over mountainous country scales of 500 or 1000 ft to the inch should be used. It should be noted that if the distance scale is doubled, the height scale should be quadrupled to preserve the proper relationship.

Return now to the large-scale topo map. The design engineer follows the route line* drawn between the two adjacent sites, which may be called arbitrarily transmitter and receiver sites. Obstacles on the line that will or potentially would interfere with the ray beam are now identified. It is good practice to label each on the map with a simple code name such as letters of the alphabet. For each obstacle annotate on a table the distances d_1 and d_2

*It will be noted that the line drawn is a straight line, which navigators call a rhumb line. Purists in the radio propagation field will correct the author that in fact the beam follows a great circle path. However, for short paths up to 100 km, the rhumb line presents a sufficient approximation. For longer paths, such as in Chapter 4, great circle distances and bearings are used.

TABLE 2.2 Path Profile Database—Path Able Peak to Bakersville[a]

Obstacle	d_1 (mi)	d_2 (mi)	0.6 Fresnel (ft)	EC (ft)[b]	Vegetation	Total Height Extended (ft)
A	7.5	28.5	43	152	50	245
B	19.4	16.6	53	233	50	336
C	27.0	9.0	46	176	50	272
D	30.0	6.0	39	130	50	219

[a]Frequency band: 6 GHz; *K*-factor 0.92; $D = d_1 + d_2 = 36$ mi; vegetation: tree conditions 40 ft plus 10 ft growth.
[b]EC = earth curvature or earth bulge.

and the height determined from the contour line. d_1 is the distance from the transmitter site to the obstacle and d_2 is the distance from the obstacle to the receiver site. Optionally, the site location in latitude and longitude or grid coordinates may also be listed. A typical table is shown in Table 2.2, which applies to an example profile shown in Figure 2.3.

Earth curvature for a particular obstacle point is determined by

$$h = \frac{d_1 d_2}{1.5K} \tag{2.2a}$$

where h is in feet and d is in statute miles, or

$$h = \frac{d_1 d_2}{12.75K} \tag{2.2b}$$

where h is in meters and d is in kilometers.

The Fresnel zone clearance may be determined by the equation

$$F_1 = 72.1\sqrt{\frac{d_1 d_2}{DF_{\mathrm{GHz}}}} \tag{2.3}$$

where F_1 is the radius of the first Fresnel zone; d_1, d_2, and D are in statute miles; and F_1 is in feet. For the metric system,

$$F_1 = 17.3\sqrt{\frac{d_1 d_2}{DF_{\mathrm{GHz}}}} \tag{2.4}$$

where d_1, d_2, and D are in kilometers and F_1 is in meters.

In Table 2.2, the conventional value of $0.6F_1$ has been used. Section 1.4 shows the rationale for other values such as $0.3F_1$ or $0.6F_1$ with a 10-ft safety factor.

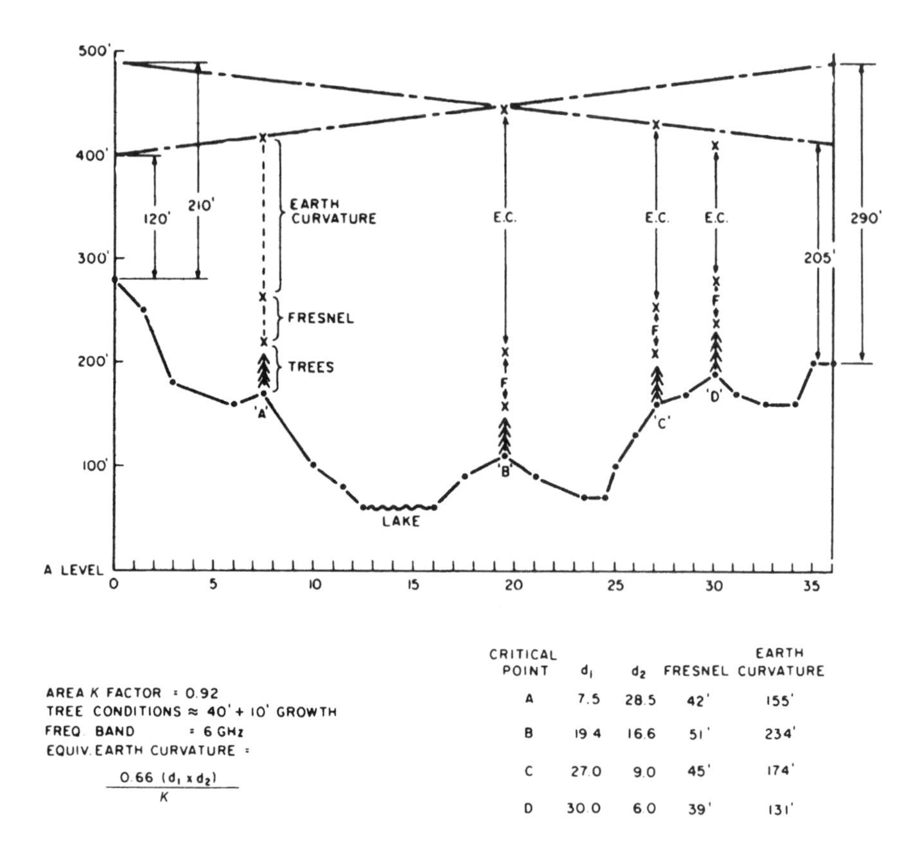

CRITICAL POINT	d_1	d_2	FRESNEL	EARTH CURVATURE
A	7.5	28.5	42'	155'
B	19.4	16.6	51'	234'
C	27.0	9.0	45'	174'
D	30.0	6.0	39'	131'

Figure 2.3. Example path profile using method 1. (See Table 2.3.)

2.3.1 Determination of Median Value for *K*-Factor

When refractive index gradients $\Delta N/\Delta h$ (defined below) are known, the *K*-factor or effective earth radius factor can be closely approximated from the relationship

$$K \approx \left(1 + \frac{1}{157}\frac{\Delta N}{\Delta h}\right)^{-1} \tag{2.5}$$

where ΔN may be found in *A World Atlas of Atmospheric Radio Refractivity* (Ref. 3).

N_s at altitude h_s above MSL can be derived from surface refractivity (N_0) gradients (Figures 1.3 and 1.4) by the following formula:

$$N_s = N_0 \exp(-0.1057 h_s) \tag{1.24a}$$

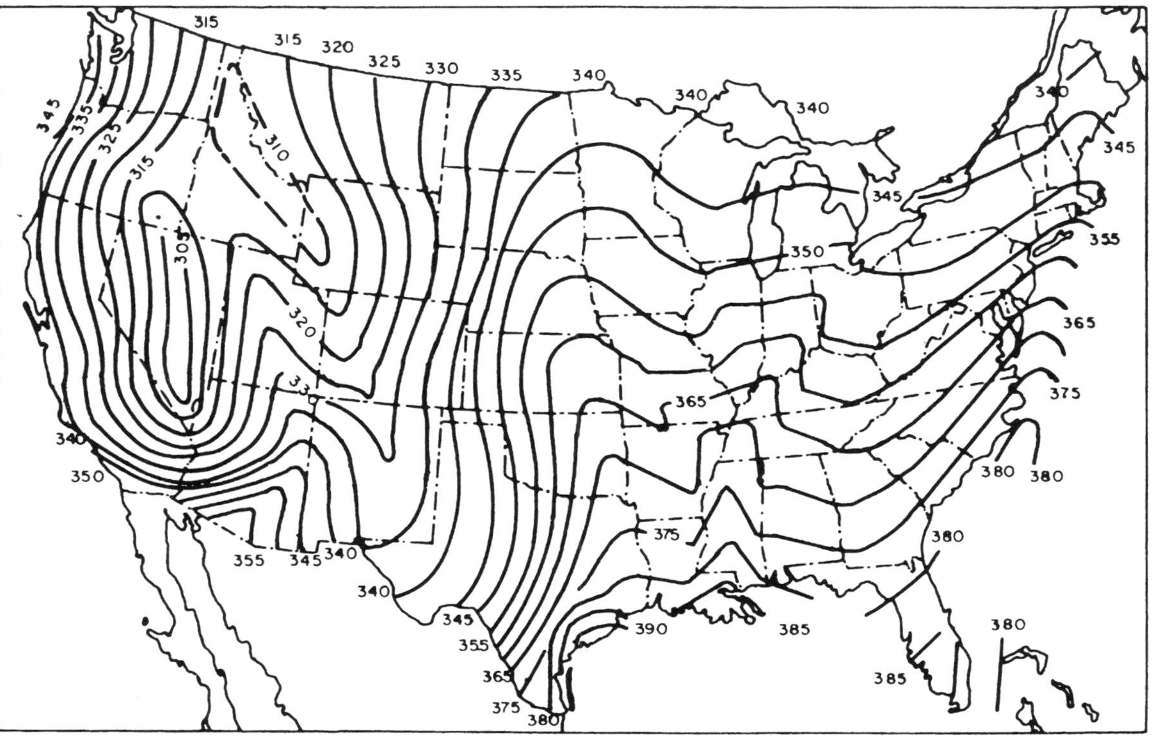

Figure 2.4. Sea level refractivity (N_0) maximum for August.

TABLE 2.3 Approximate Values of *R* for Various Terrain

Type of Terrain	R^a	Approximate Depth of Even Fresnel Zone Fade (dB)
Heavily wooded, forest land	0 to −0.1	0–2
Partially wooded (trees along roads perpendicular to path, etc.)	−0.1 to −0.4	2–5
Sagebrush, high grassy areas	−0.5 to −0.7	5–10
Cotton with foilage, rough seawater, low grassy areas	−0.7 to −0.8	10–20
Smooth seawater, salt flats, flat earth	−0.9 +	20–40 +

[a]The values of *R* given in this table are approximate, of course, but they do give an indication of signal degradation to be expected over various terrain should even-numbered Fresnel zone reflections occur.

Source: Reference 4.

where h_s, the altitude above MSL, is measured in kilometers. For conditions near the ground use the following empirical relationship between N_s and the difference in refractivity ΔN between N_s and N at 1 km above the earth's surface (i.e., $\Delta N/\Delta h$ where $\Delta h = 1$ km):

$$\Delta N(1 \text{ km}) = -7.32 \exp(0.005577 N_s) \tag{1.21}$$

Figure 2.4 gives values of N_0 maxima for August for the contiguous United States. The K-factor can also be calculated. See Chapter 1, equation (1.25).

2.4 REFLECTION POINT

As discussed in Chapter 1, ground reflections are a major cause of multipath fading. These reflections can be reduced or eliminated by adjustment of tower heights, effectively moving the reflection point from an area along the path of greater reflectivity (such as a body of water) to one of lesser reflectivity (such as an area of heavy forest). Of course, for those paths that are entirely over water or over desert, the designer will have to resort to other methods such as vertically spaced space diversity or frequency diversity to mitigate multipath fading. Table 2.3 provides a guide to the coefficient of reflectivity R for various types of terrain and the fading depths that may be expected for each value when the reflection point is located on a particular type of terrain.

There are several methods to determine the location of the reflection point. The simplest is a graphical method derived from Ref. 5. Here we use Figure 2.5, and an example is given in Figure 2.6.

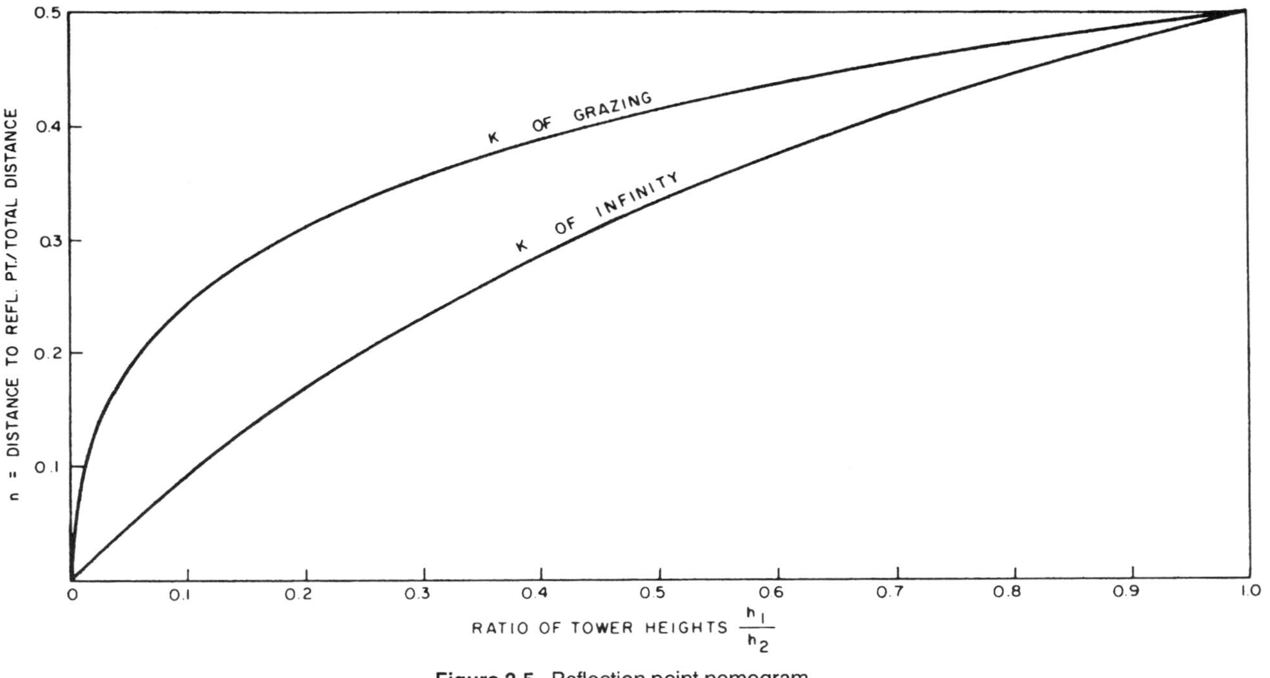

Figure 2.5. Reflection point nomogram.

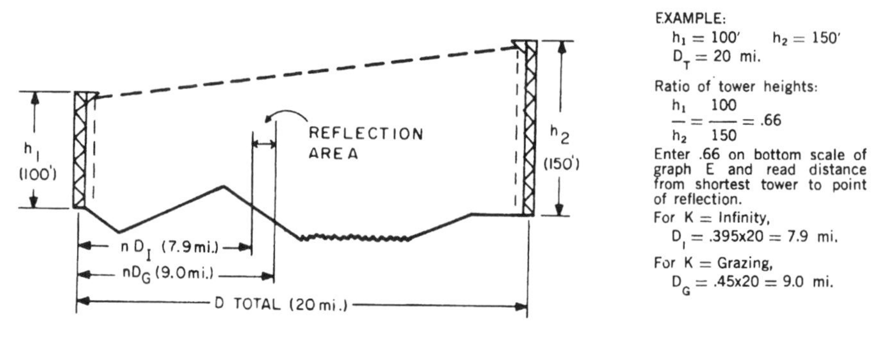

Figure 2.6. Example path.

The term reflection "point" is somewhat confusing. It would indeed be a singular point if we could assume a fixed K-factor. Such an assumption is only valid for a fully homogeneous atmosphere stable across the total path 365 days a year. This is not true because the atmosphere is dynamic, with constantly changing temperature, atmospheric pressure, and humidity across the path with time. For this reason, we assume that the reflection "point" covers a locus on the profile route line whose extremes are determined by extremes of K-factor (i.e., from $K = \infty$ to $K =$ grazing).*

From the path profile, we take the tower heights, h_1 for the transmitter site and h_2 for the receiver site, and determine the ratio h_1/h_2, which is entered on the x-axis of Figure 2.5. On the y-axis on the figure two values of n are taken, the first for $K = \infty$ and the second for $K =$ grazing. The distances from the nearest site defining the reflection locus are determined from the values nD for each value of n, where D is the total path length.

Example 1. Assume $h_1 = 100$ ft, $h_2 = 150$ ft, and the path length $D = 20$ mi.

$$\text{Ratio of tower heights:} \quad h_1/h_2 = 0.66$$

Enter the value 0.66 on the x-axis of Figure 2.6 and read the values for n for $K = \infty$ and for $K =$ grazing. Multiply each value by the path length D. Thus

$$D_i = 0.39 \times 20 = 7.9 \text{ mi} \quad (\text{value for } K = \infty)$$

$$D_g = 0.45 \times 20 = 9.0 \text{ mi} \quad (\text{value for } K = \text{grazing})$$

The reflection "area" is that "area" along the line between 7.9 and 9.0 mi from the shorter tower.

*K = grazing is its value where the boresight ray grazes the earth for the path of interest.

2.5 SITE SURVEY

2.5.1 Introduction

Once the path profile has been completed, the designer should verify the results by a field survey of the sites at each end of the path (hop) and the intervening terrain. Of primary importance is verification of site locations and conditions—that, indeed, the LOS criteria developed on the profile have been met. The designer should be particularly vigilant of structures that have been erected since the preparation of the topographic maps used to construct the profile. Also, maps, especially for emerging nations, may be in error.

2.5.2 Information Listing

The following is a general listing of information required from the field survey for a repeater site.

1. Precise Location of Site. At least two permanent survey monuments should be placed at each site and the azimuth between them recorded. Their locations should be indicated on a site survey sketch and site photograph. Geographical coordinates of the markers should be determined within $\pm 1''$ and the elevation of each to ± 1.5 m (± 5 ft). All elevation data should be referenced to mean sea level (MSL).

2. Site Layout Plan. A sketch of the site should include antenna locations with respect to monument markers as well as shelter location.

3. Site Description. Include soil type, vegetation, existing structures, access requirements, leveling or grading requirements, drainage, and so on. Use a sketch to show distances to property lines, benchmarks, roads, and the like. A detailed topographic map of the site should be made. Photographs should be taken to show close-up details as well as general location with respect to surrounding terrain features. For manned sites, an estimate should be made of suitability regarding water supply and sewage disposal. Helpful information in this regard may be obtained from local well drillers, agricultural agents, plumbing contractors, and nearby farmers or ranchers.

4. Description of Path. Ideally, the survey team should "walk" the path with a copy of the topo map being utilized. A general description of terrain and vegetation should be annotated while proceeding along the path. Critical points (obstacles) or new critical points (obstacles) are noted, their location (i.e., d_1 and d_2) should be determined within ± 0.2 km (± 0.1 mi), azimuths to $1'$ of arc, elevations to $\pm 1°$ of path centerline. The surveyor should show by sketch and photograph new construction or other features not correctly indicated on the topo map. From each site, sketches and photographs should be made on path centerline showing azimuth angles to prominent features (references on topo map) and noting elevation angles. A running diary of the survey should be kept showing data obtained and method of verification.

5. Power Availability. Give location of nearest commercial transmission line with reference to each site. List name and address of the utility firm. State voltages, phases, line frequency, and main feeder size.

6. Fuel Supply. List local sources of propane, diesel fuel, heating oils, and natural gas. Estimate cost of item(s) delivered to site.

7. Local Materials and Contractors. List local sources of lumber (if any) and ready-mixed concrete and list names and addresses of local general construction contractors and availability of cranes and earth-moving equipment.

8. Local Zoning Restrictions. Inquiries should be made as to national regulations or local zoning restrictions that might affect the use of the site or the height of the antenna tower. Give distance from site to nearest airport and determine if site is in a runway approach corridor.

9. Geologic and Seismic Data. Determine load-bearing qualities of soil at site, and depth to rock and groundwater. Obtain soil samples. Check with local authorities on the frequency and severity of seismic disturbances.

10. Weather Data. General climatological data should have been assembled during the initial design studies. This information should be verified/modified by local input information during the site survey stage. Check with local authorities to obtain the following data:

- Average monthly maximum and minimum temperature
- Average monthly precipitation, extreme short-period totals (day, hour, 5-min intervals, if possible), and days/month with rain
- Average wind direction and velocity, and direction and velocity of peak gusts
- Average and extreme snow accumulation, packing
- Flooding data
- Occurrence of hurricanes, typhoons, tornadoes
- Cloud and fog conditions
- Probability of extended periods of very light winds
- Icing conditions, freezing rain in particular
- Average dewpoint temperature and diurnal variation of relative humidity

11. Survey of Electromagnetic Interference (EMI). Show sources of "foreign" EMI such as similar system crossing or running parallel to the proposed route line; show locations of "foreign" repeater or terminal sites, EIRPs, antenna patterns, frequencies, bandwidths, and spurious emissions (specified). Care must also be exercised to assure that the proposed system does not interfere with existing radio facilities.

2.5.3 Notes on Site Visit

A sketch should be made of the site in a field notebook. The sketch should show the location of the tower and buildings, trees, large boulders, ditches, and so on. Measurements are then made with a steel tape and are recorded on the sketch. The sketch should be clearly identified by name, site number, geographical coordinates, and quadrangle map name or number. The sketch should be accompanied by a written description of soil type, vegetation, number, type and size of trees, number of trees to be removed, and approximate location and extent of leveling required. If soil samples are taken, the soil sample points with identifiers should be marked on the sketch.

Photographs should be made of the horizon to the north, east, south, and west from the tower location and along the centerline of the path. The use of a theodolite is recommended to determine azimuths of prominent features on the horizon of each photograph. Indicate the elevation angle to prominent features on the path centerline photograph(s).

A magnetic compass corrected for local declination is suitable for rough site layouts only. Final azimuth references should be determined or confirmed by celestial observation or by accurate radio navigation device.

The latitude and longitude of the tower location on the site should be determined to $\pm 1''$. This can be done by map scaling, traverse survey, triangulation survey, and celestial observations. In most cases sufficient accuracy can be obtained by careful scaling from a $7\frac{1}{2}'$ quadrangle with a device such as the Gerber Variable Scale.

The location should also be described with reference to nearby roads and cities so that contractors and suppliers can readily be directed to the site.

The elevation of the ground at the tower location should be determined to ± 1.5 m (± 5 ft). Differential leveling, or an extension of a known vertical control point (benchmark) by a series of instrument setups, is the most accurate method and is recommended for the final survey or when there is a benchmark close to the site. Trigonometric leveling is used to determine elevations over relatively long distances; it is not ordinarily recommended for site surveys but is useful to determine the elevation of obstacles along the path. Aerial photogrammetry and airborne profile recorders can also be used but tend to be expensive. Barometric leveling or surveying altimetry is the simplest method to determine relative ground elevations and, when carefully done, can provide sufficient accuracy for the final survey.

Access and commercial power entry to the site can be major cost drivers for the system first cost. The sketch map should show the route of the proposed road with reference to the site and existing roads. It is important to show the type of soil, number of trees that will probably have to be removed, degree of slope, rock formations, and approximate length of the road compared to crow line distance to existing road(s). The probability of all-year access must also be determined.

Site Name and Number
Latitude_______ Longitude_______ (Degrees, Min, Sec)_______
Map reference (most detailed topographic)_______
Nearest town (post office)_______
Access route: (all year?)

Property owner; local contact:
Site sketch_______ Site photograph_______ General description_______
Reference baseline_______ By Polaris_______ Other_______
Antenna No._______ True bearing_______
Ground elev. MSL_______ Takeoff angle (beam centerline)_______
Takeoff angles to 45° right and left of centerline_______
(Significant changes in horizon)
Critical Points: (include horizon)
Distance_______ Map elev._______ Survey elev._______
Tree height_______ Required clearance_______
Description:
Horizon sketch_______ Horizon photograph_______

Power availability:
a. Nearest transmission line_______ b. Voltage_______
c. Frequency_______ d. Phase_______ e. Operating utility_______
Drinking water source_______ Estimated depth to groundwater_______
Sewage disposal_______ Type and depth of soil on and near site_______
Nearest airport_______ railroad_______ highway_______
navigable river_______

Figure 2.7. Sample checklist for site survey (Ref. 5).

The map sketch should also include the location of the nearest source of commercial prime power, and the proposed route of the prime power extension should also be included. Prior to the survey an estimate should be made of the prime power requirements of the site including tower lighting and air conditioning (where required). This demand requirement will assist local power company officials in determining power extension cost.

Figure 2.7 is a sample checklist for a site survey.

2.6 PATH ANALYSIS

2.6.1 Objective and Scope

The path analysis or link power budget task provides the designer with the necessary equipment parameters to prepare a block diagram of the terminal or repeater configuration and to specify equipment requirements both quantitatively and qualitatively. We assume that frequency assignments have been made as we did in Section 2.3 or at least assume the frequency band in which assignments will be made by the appropriate regulatory authority.

The discussion that follows in this section is directed primarily toward analog radiolinks. Much of the same approach is also valid for digital radiolink design except that some units of measure will differ. In this section one might say that the bottom line for task output is noise (in dBrnC or pWp) and S/N in the standard voice channel or video channel, whereas in Chapter 3, where digital radiolinks are discussed, we will be dealing with signal energy per bit to noise spectral density ratio (E_b/N_0) and BER on the link. We also assume here that the modulation waveform is conventional FM.

This section will provide us with the tools to derive antenna aperture, receiver front end characteristics, FM deviation, IF/RF bandwidth, transmitter output power, diversity arrangements (if any), and link availability due to propagation. This latter will involve the necessary system overbuild to meet propagation availability requirements in a fading environment. The use of NPR (noise power ratio) as a tool to measure link noise performance will be described. Frequency/bandwidth assignments for analog systems will also be covered.

The analysis described in this chapter is valid for radiolinks in the 1–10-GHz band. Radiolinks operating on frequencies above 10 GHz begin to be impacted by excess attenuation due to rainfall and gaseous absorption. Below 10 GHz the effects of rainfall are marginal, and in many design situations they can be neglected. Chapter 8 describes the design of radiolinks above 10 GHz.

2.6.2 Unfaded Signal Level at the Receiver

It will be helpful to visualize a LOS radiolink by referring to the simplified model shown in Figure 2.8 where a radiolink transmitter and radiolink receiver are separated by a distance D.

Given the path from transmitter to receiver in Figure 2.8 and the transmitter's assigned frequency, we calculate the free-space loss (FSL) represented by the attenuator. Equations (1.9a), (1.9b), and (1.9c) from Section 1.2 apply and are restated here for convenience.

$$\mathrm{FSL}_{\mathrm{dB}} = 32.45 + 20 \log D_{\mathrm{km}} + 20 \log F_{\mathrm{MHz}} \qquad (1.9a)$$

where D is measured in kilometers;

$$\mathrm{FSL}_{\mathrm{dB}} = 36.58 + 20 \log D_{\mathrm{sm}} + 20 \log F_{\mathrm{MHz}} \qquad (1.9b)$$

where D is measured in statute miles; and

$$\mathrm{FSL}_{\mathrm{dB}} = 37.80 + 20 \log D_{\mathrm{nm}} + 20 \log F_{\mathrm{MHz}} \qquad (1.9c)$$

where D is measured in nautical miles. If F is stated in gigahertz, add 60 to the value of the constant term.

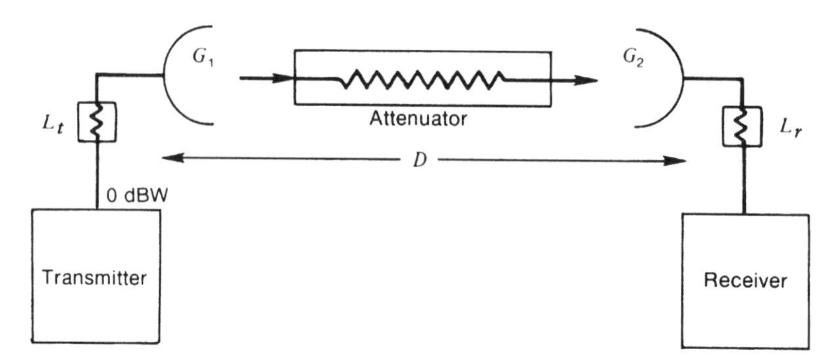

Figure 2.8. Simplified model, radiolink path analysis. L_t and L_r are the transmission line losses; G_1 and G_2 are the antenna gains.

Example 2. Compute the free-space loss for a path 31 statute miles long with a transmit frequency of 6135 MHz.

$$\begin{aligned}\text{FSL}_{\text{dB}} &= 36.58 + 20\log(31) + 20\log(6135) \\ &= 36.58 + 29.83 + 75.76 \\ &= 142.17 \text{ dB}\end{aligned}$$

Example 3. Compute the free-space loss for a path 43 km long with a transmit frequency of 4.041 GHz.

$$\begin{aligned}\text{FSL}_{\text{dB}} &= 92.45 + 20\log(43) + 20\log(4.041) \\ &= 92.45 + 32.67 + 12.13 \\ &= 137.25 \text{ dB}\end{aligned}$$

It should be noted that for a fixed-length path, if the frequency is doubled, approximately 6 dB is added to the free-space loss. Conversely, if the transmit frequency is halved, approximately 6 dB is then subtracted from the free-space loss. Using similar reasoning, for a fixed-frequency path, double the distance between transmitter and receiver and approximately 6 dB is added to the free-space loss; halve the distance and the free-space loss is 6 dB less.

Turning again to the model in Figure 2.8, the effective isotropic radiated power (EIRP) can be calculated as follows:

$$\text{EIRP}_{\text{dBW}} = P_0 + L_t + G_1 \tag{2.6}$$

where P_0 is the RF power output of the transmitter at the waveguide flange, L_t is the transmission line losses, and G_1 is the gain of the transmit antenna. It should be noted that gains are treated as positive numbers and losses, conventionally, are negative numbers in equation (2.6) and in the following discussion. This permits simple algebraic addition.

Example 4. A microwave radiolink transmitter has an output of 1 W at the waveguide flange; transmission line losses from the flange to the antenna feed are 3 dB, and the gain of the antenna is 31 dB. Compute the EIRP in dBW.

$$\begin{aligned} \text{EIRP} &= 0\ \text{dBW} + (-3\ \text{dB}) + 31\ \text{dB} \\ &= +28\ \text{dBW} \end{aligned}$$

Example 5. Compute the EIRP in dBm where the transmitter output power is 200 mW at the waveguide flange, the transmission line losses are 4.7 dB, and the antenna gain is 37.3 dB.

$$\begin{aligned} \text{EIRP} &= 10 \log 200 + (-4.7\ \text{dB}) + 37.3\ \text{dB} \\ &= +23\ \text{dBm} - 4.7\ \text{dB} + 37.3\ \text{dB} \\ &= +55.6\ \text{dBm} \end{aligned}$$

To calculate the signal power produced by an isotropic antenna at the receiving terminal location in Figure 2.8, the EIRP is algebraically added to the free-space loss (FSL) and the gaseous absorption loss L_g. This power level is called the isotropic receive level (IRL). Or, turning to the model in Figure 2.8, the EIRP is the input to the attenuator, and we simply calculate the output of the attenuator or

$$\text{IRL} = \text{EIRP} + \text{FSL}_{\text{dB}} + L_g \tag{2.7}$$

Example 6. Calculate the IRL of a radiolink where the EIRP is +28 dBW, the FSL is 137.25 dB, and the atmospheric gaseous absorption loss is 0.6 dB.

$$\begin{aligned} \text{IRL}_{\text{dBW}} &= +28\ \text{dBW} + (-137.25\ \text{dB}) + (-0.6\ \text{dB}) \\ &= -109.85\ \text{dBW} \end{aligned}$$

The unfaded receive signal level, RSL, at the receiver input terminal (Figure 2.8) is calculated by algebraically adding the isotropic receive level, the receive antenna gain, G_2, and the transmission line losses at the receiving terminal, L_r, or

$$\text{RSL} = \text{IRL} + G_r + L_r \tag{2.8}$$

where G_r and L_r are the generalized values of the receive antenna gain and the receive system transmission line losses.

To calculate RSL directly, the following formulas apply:

$$\text{RSL} = \text{EIRP} + \text{FSL} + L_g + G_r + L_r \tag{2.9}$$

or

$$\text{RSL} = P_0 + L_t + G_1 + \text{FSL} + L_g + G_2 + L_r \tag{2.10}$$

Example 7. Calculate the RSL in dBW where the transmitter output power to the waveguide flange is 750 mW, the transmission line losses at each end are 3.4 dB, the distance between the transmitter and receiver site is 17 statute miles, and the operating frequency is 7.1 GHz; the antenna gains are 30.5 dB at each end. Assume a gaseous absorption loss of 0.3 dB. First calculate FSL using equation (1.9b):

$$\begin{aligned}\text{FSL}_{\text{dB}} &= 36.58 + 20\log 17 + 20\log 7100\\ &= 36.58 + 24.61 + 77.02\\ &= 138.21\ \text{dB}\end{aligned}$$

Next calculate RSL using equation (2.10):

$$\begin{aligned}\text{RSL} &= 10\log 0.75 + (-3.4) + 30.5 + (-138.21)\\ &\quad +(-0.3) + 30.5 + (-3.4)\\ &= -1.25 - 3.4 + 30.5 - 138.21 - 0.3 + 30.5 - 3.4\\ &= -85.56\ \text{dBW}\end{aligned}$$

2.6.3 Receiver Thermal Noise Threshold

2.6.3.1 Objective and Basic Calculation. One waypoint objective in the path analysis is to calculate the unfaded carrier-to-noise ratio (C/N). With the RSL determined from Section 2.6.2 and with the receiver thermal noise threshold, we can simply calculate the C/N, where

$$\left(\frac{\text{C}}{\text{N}}\right)_{\text{dB}} = \text{RSL} - P_t \tag{2.11}$$

where P_t is the receiver thermal noise threshold. Note that RSL and P_t must be in the same units, conventionally in dBm or dBW.

The equipartition law of Boltzmann and Maxwell states that the available power per unit bandwidth of a thermal noise source is

$$P_n(f) = kT \quad \text{W/Hz} \tag{2.12}$$

where k is Boltzmann's constant (1.3805×10^{-23} J/K) and T is the absolute temperature of the source in kelvins. At absolute zero the available power in a 1-Hz bandwidth is −228.6 dBW. At room temperature, usually specified as 17°C or 290 K,* the available power in a 1-Hz bandwidth is −204 dBW or −228.6 + 10 log(290). Thus

$$P_a = kTBW \quad \text{W} \tag{2.13}$$

where BW is bandwidth expressed in Hz. Expressed in dBW at room temperature

$$P_a = -204 + 10\log(BW) \quad \text{dBW} \tag{2.14}$$

Expressed in dBm at room temperature

$$P_a = -174 + 10\log(BW) \quad \text{dBm} \tag{2.15}$$

The thermal noise level is frequently referred to as the thermal noise threshold. For a receiver operating at room temperature it is a function of the bandwidth of the receiver [in practical systems this is taken as the receiver's IF bandwidth (B_{IF})] measured in Hz and the noise figure NF (in dB) of the receiver. Thus the thermal noise threshold P_t of a receiver can be calculated as follows:

$$P_t = -204 \text{ dBW} + 10\log B_{\text{IF}} + \text{NF}_{\text{dB}} \tag{2.16}$$

Example 8. Compute the thermal noise threshold of a receiver with a 12-dB noise figure and an IF bandwidth of 4.2 MHz.

$$\begin{aligned} P_t &= -204 \text{ dBW} + 10\log(4.2 \times 10^6) + 12 \text{ dB} \\ &= -204 \text{ dBW} + 66.23 \text{ dB} + 12 \text{ dB} \\ &= -125.77 \text{ dBW} \quad \text{or} \quad -95.77 \text{ dBm} \end{aligned}$$

Example 9. Compute the thermal noise threshold of a receiving system with a 3.1-dB noise figure and an IF Bandwidth of 740 kHz.

$$\begin{aligned} P_t &= -204 \text{ dBW} + 10\log(740 \times 10^3) + 3.1 \text{ dB} \\ &= -204 \text{ dBW} + 58.69 \text{ dB} + 3.1 \text{ dB} \\ &= -142.2 \text{ dBW} \quad \text{or} \quad -112.2 \text{ dBm} \end{aligned}$$

*CCIR often uses the value 288 K for room temperature.

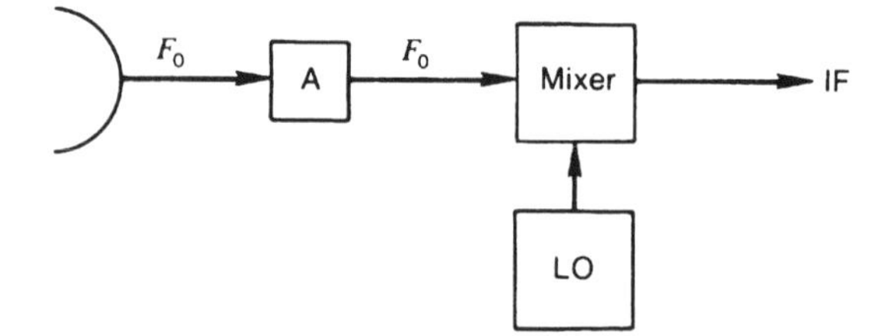

Figure 2.9. Simplified functional block diagram of a conventional LOS radiolink (microwave) receiver front end. F_0 = RF operating frequency; A = waveguide devices such as circulator and preselector; LO = local oscillator; IF = intermediate frequency.

(Note the change in terminology between Examples 8 and 9, "receiver" and "receiving system." This will be explained in the next section.)

2.6.3.2 *Practical Applications.* The majority of LOS radiolink receivers use a mixer for the receiver front end. Such an arrangement is shown (simplified) in Figure 2.9.

To calculate the receiver noise threshold of the mixer, use equation (2.16), where NF is the noise figure of the mixer. Under certain circumstances a path analysis may show a link to be marginal. One means of improvement is to add a low-noise amplifier (LNA), usually a FET-based device, in front of the mixer. This alternative configuration is shown in Figure 2.10. Mixers display noise figures on the order of 8–12 dB; GaAs FET and HEMT LNAs display noise figures from 0.4 to 2 dB for systems operating below 10 GHz.

Where front end noise figures are comparatively high, other noise sources, such as antenna noise and ohmic noise, can be neglected. However, if a configuration, like Figure 2.10, is used, where the receiver noise sources can be approximated by the noise figure of the LNA with NF on the order of 1–3 dB, other noise sources may not be neglected. Thus good practice dictates adding 2 dB to the noise figure of the LNA to approximate the noise contributed by other sources. A much more exact method of calculating total noise of a receiving system is given in Chapter 5.

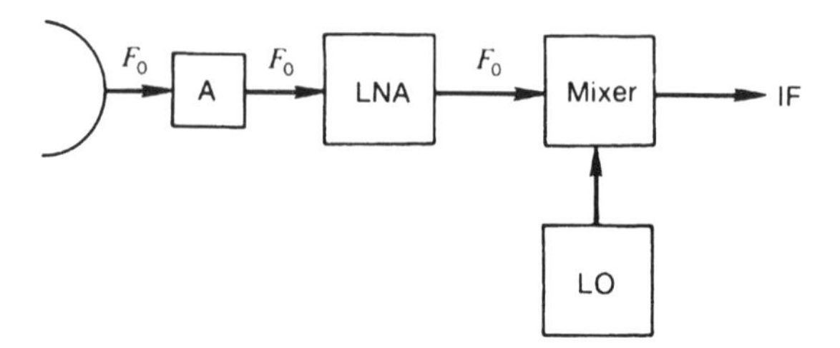

Figure 2.10. Alternative receiver front end configuration.

Example 10. A radiolink receiving system has a LNA with a noise figure of 1.3 dB and the IF has a bandwidth of 30 MHz. Compute the thermal noise threshold.

$$P_t = -204 \text{ dBW} + 10\log(30 \times 10^6) + 1.3 \text{ dB} + 2 \text{ dB}$$
$$= -204 + 74.77 + 3.3$$
$$= -125.93 \text{ dBW} \quad \text{or} \quad -95.93 \text{ dBm}$$

2.6.4 Calculation of IF Bandwidth and Peak Frequency Deviation

2.6.4.1 IF Bandwidth. The IF of a FM receiver must accommodate the RF bandwidth, which consists of the total peak deviation spread and a number of generated sidebands. The IF bandwidth can be estimated from Carson's rule,

$$B_{\text{IF}} = 2(\Delta F_p + F_m) \tag{2.17}$$

where ΔF_p is the peak frequency deviation and F_m is the highest modulating frequency given in the middle column of Table 2.5.

2.6.4.2 Frequency Deviation. The value of ΔF_p in equations (2.17) and (2.18) (below) is peak deviation. The peak deviation for a particular link should be taken from equipment manuals for the proposed load (i.e., number of voice channels or TV video). If this information is not available, peak deviation may be calculated using CCIR specified per channel deviation from CCIR Rec. 404-2. See Table 2.4. CCIR recommends a peak deviation of ± 4 MHz for video systems (CCIR Rec. 276-2) without pre-emphasis.

CCIR Rec. 404-2 further states that where pre-emphasis is used, the pre-emphasis characteristic should preferably be such that the effective rms

TABLE 2.4 Frequency Deviation Without Pre-emphasis

Maximum Number of Channels	rms Deviation per Channel[a] (kHz)
12	35
24	35
60	50, 100, 200
120	50, 100, 200
300	200
600	200
960	200
1260	140, 200
1800	140
2700	140

[a]For 1-mW, 800-Hz tone at a point of zero reference level.

Source: CCIR Rec. 404-2 (Ref. 6).

deviation due to the multichannel signal is the same with and without pre-emphasis. (For a discussion of pre-emphasis, see Section 2.6.5.)

To calculate peak deviation when rms per channel deviation Δf (as in Table 2.4) is given, the following formula applies:

$$\Delta F_p = \Delta f(\mathrm{pf})(\mathrm{NLR}_n) \tag{2.18}$$

where pf is the numerical ratio of peak-to-rms baseband voltage or

$$\mathrm{pf} = \mathrm{antilog}(\tfrac{1}{20}\mathrm{PF}) \tag{2.19}$$

where PF (peak factor) is the baseband peak-to-rms voltage in decibels. Some texts use 12 dB for PF, others use 13 dB. If 12 dB is used for the value of PF, then this represents a value where peaks are not exceeded more than 0.01% of the time, and if 13 dB, then 0.001% of the time. Normally, NLR_n (NLR = noise load ratio) will be specified at the outset for the number of channels (N) that the system will carry.

CCIR recommends, for FDM telephony baseband,

$$\mathrm{NLR}_{\mathrm{dB}} = -1 + 4 \log N \tag{2.20}$$

for FDM configurations from 12 to 240 voice channels and

$$\mathrm{NLR}_{\mathrm{dB}} = -15 + 10 \log N \qquad (2.21)^*$$

for FDM configurations where N is greater than 240. For U.S. military systems, the value

$$\mathrm{NLR}_{\mathrm{dB}} = -10 + 10 \log N \tag{2.22}$$

is used.

NLR_n will be given (or calculated) in decibels, but in equation (2.18) the numerical equivalent value is used or

$$\mathrm{NLR}_n = \mathrm{antilog}(\tfrac{1}{20}\mathrm{NLR}_{\mathrm{dB}}) \tag{2.23}$$

if we assume a PF of 13 dB; then with CCIR loading of FDM telephony channels

$$\Delta F_p = 4.47d\left[\log^{-1}\left(\frac{-1 + 4\log N}{20}\right)\right] \tag{2.24}$$

for values of N from 12 to 240 voice channels and

$$\Delta F_p = 4.47d\left[\log^{-1}\left(\frac{-15 + 10\log N}{20}\right)\right] \tag{2.25}$$

for values of N greater than 240. d is the rms test tone deviation from Table 2.4.

Through the use of Figure 2.11, $B_{\mathrm{IF}} = B_{\mathrm{RF}}$ can be derived graphically given the rms per channel deviation from Table 2.4. (The reader will note the variance with Carson's rule for $B_{\mathrm{IF}} = B_{\mathrm{RF}}$.)

*For North American practice: $\mathrm{NLR}_{\mathrm{dB}} = -16 + 10 \log N$ (2.21a).

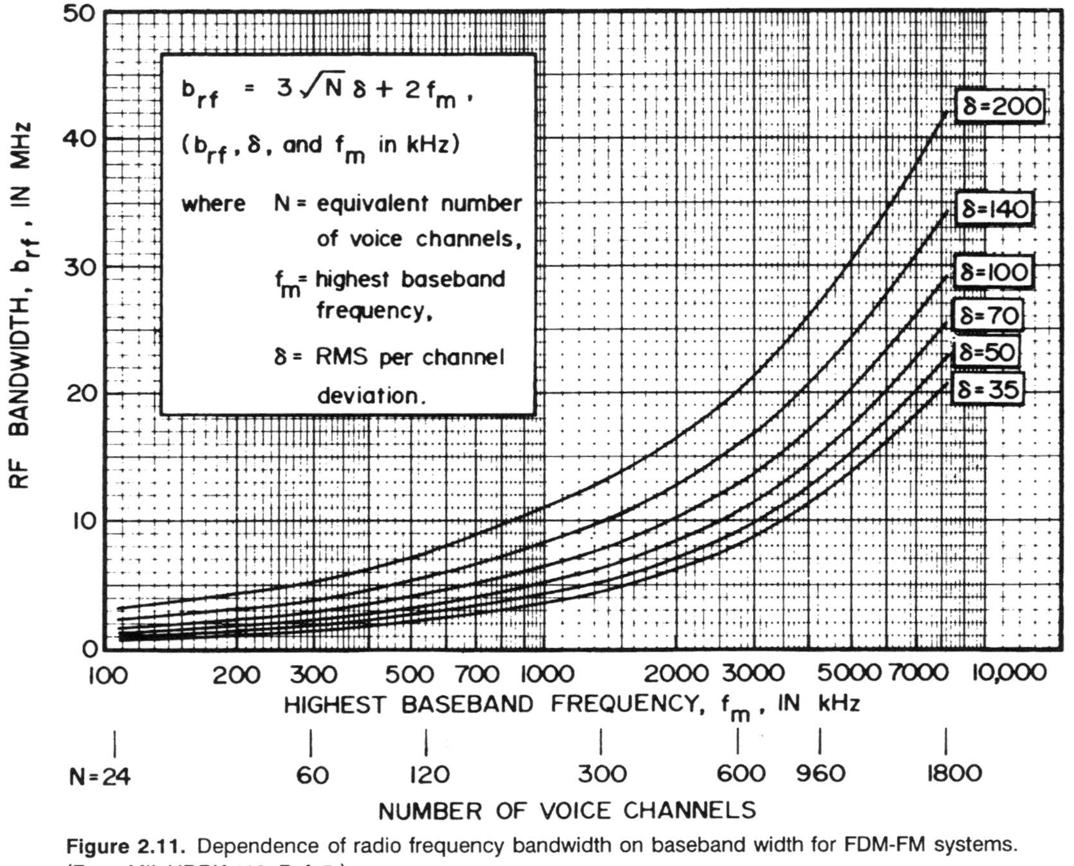

Figure 2.11. Dependence of radio frequency bandwidth on baseband width for FDM-FM systems. (From MIL-HDBK-416: Ref. 5.)

Example 11. Compute the peak deviation for a radiolink being designed to transmit 1200 voice channels of FDM telephony.

First calculate the value in dB of the noise load ratio (NLR_{dB}) assuming CCIR loading. Use equation (2.21):

$$\begin{aligned} NLR_{dB} &= -15 + 10\log(1200) \\ &= 15.79 \text{ dB} \end{aligned}$$

Calculate the numerical equivalent of this value using equation (2.23):

$$\begin{aligned} NLR_n &= \log^{-1}(15.79/20) \\ &= 6.16 \end{aligned}$$

Calculate the peak deviation using equation (2.25):

$$\Delta F_p = 4.47(200)(6.16)$$

where the value for d was derived from Table 2.4,

$$\Delta F_p = 5507 \text{ kHz} \quad \text{or} \quad 5.507 \text{ MHz}$$

2.6.5 Pre-emphasis / De-emphasis

After demodulation in a FM system, thermal noise power (in some texts called "idle noise") is minimum for a given signal at the lowest demodulated baseband frequency and increases at about 6 dB per octave as the baseband frequency increases. This effect is shown in Figure 2.12, which compares thermal noise in an AM system with that in a FM system.

To equalize the noise across the baseband, a pre-emphasis network is introduced ahead of the transmitter modulator to provide increasing attenuation at the lower baseband frequencies. The transmitting baseband gain is then increased so that baseband frequencies above a crossover frequency are increased in level, those below the crossover frequency are lowered in level, and the total baseband energy presented to the modulator is approximately unchanged relative to FM without pre-emphasis.

In the far end receiver of the FM radiolink, a de-emphasis network providing more attenuation with increasing frequency is applied after the demodulator. This network removes both the test-tone slope produced by pre-emphasis and the variable idle noise (thermal noise) slope to provide a more even distribution of signal-to-noise. The two networks must be complementary to ensure a nearly flat frequency versus level response across the baseband.

There are two types of networks that may be encountered in multichannel (FDM) FM systems: the CCIR network and the 6-dB octave network. The CCIR network has much greater present-day application, and henceforth in

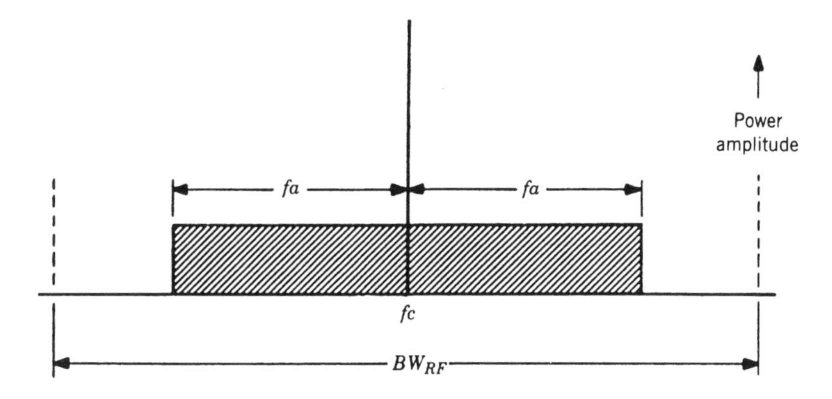

Noise power, AM System

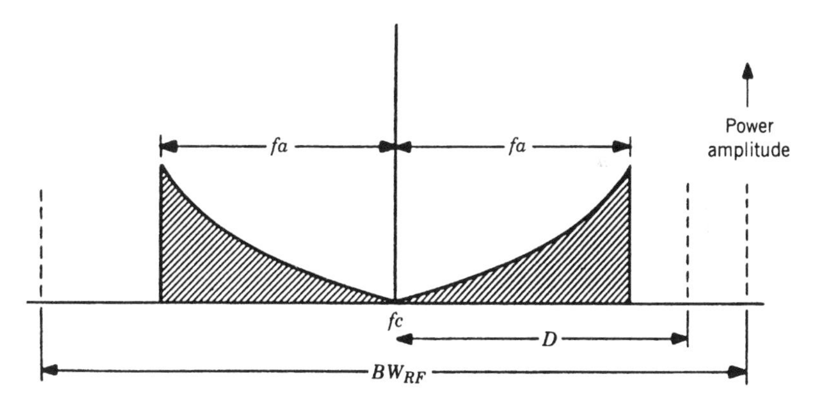

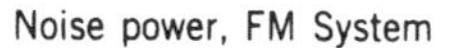

Noise power, FM System

Figure 2.12. Illustration of noise power distribution in AM and FM systems.

this text all reference to pre-emphasis will be that meeting CCIR Rec. 275-2 (Ref. 7) for multichannel FDM telephony links and CCIR Rec. 405-1 (Ref. 8) for television transmission. Figure 2.13 gives the pre-emphasis characteristic for telephony, and Table 2.5, which is associated with the figure, shows the characteristic frequencies of pre-emphasis/de-emphasis networks for standard CCITT FDM multiplex configurations.* Figure 2.14 shows the pre-emphasis characteristic for television (video). Figure 2.15 gives values in

*The reader should consult CCIR Rec. 938 (Ref. 32).

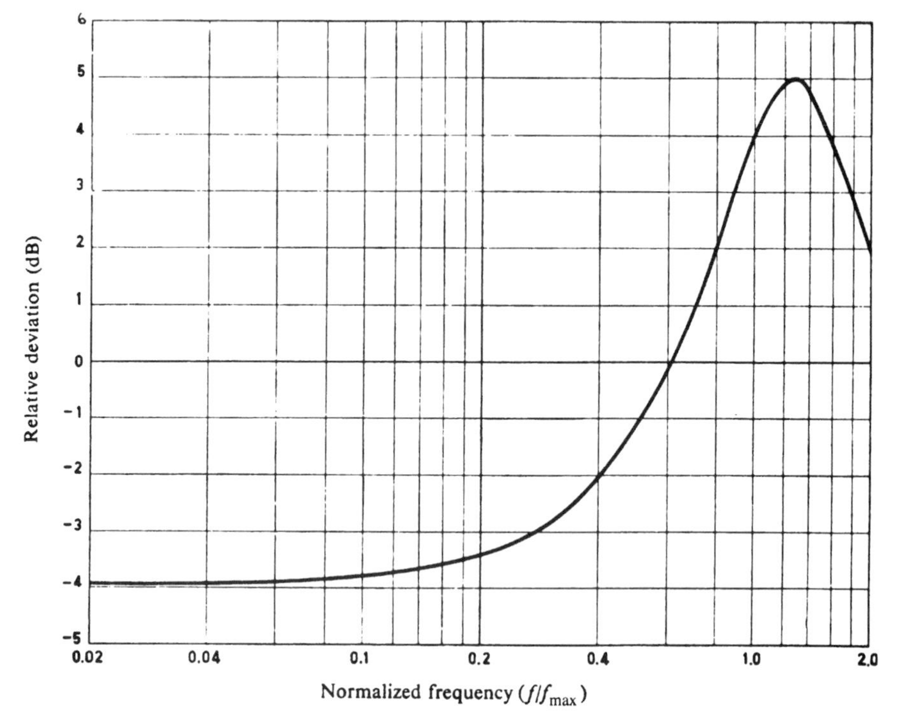

Figure 2.13. Pre-emphasis characteristic for telephony. (From CCIR Rec. 275-3; Ref. 7.)

TABLE 2.5 Characteristic Frequencies for Pre-emphasis and De-emphasis Networks for Frequency-Division Multiplex Systems

Maximum Number of Telephone Traffic Channels[a]	f^b_{max} (kHz)	f^c_r (kHz)
24	108	135
60	300	375
120	552	690
300	1,300	1,625
600	2,660	3,325
960	4,188	5,235
1,260	5,636	7,045
1,800	8,204	10,255
2,700	12,388	15,485

[a]This figure is the nominal maximum traffic capacity of the system and applies also when only a smaller number of telephone channels are in service.

[b]f_{max} is the nominal maximum frequency of the band occupied by telephone channels.

[c]f_r is the nominal resonant frequency of the pre-emphasis or de-emphasis network.

Source: CCIR Rec. 275-3 (Ref. 7).

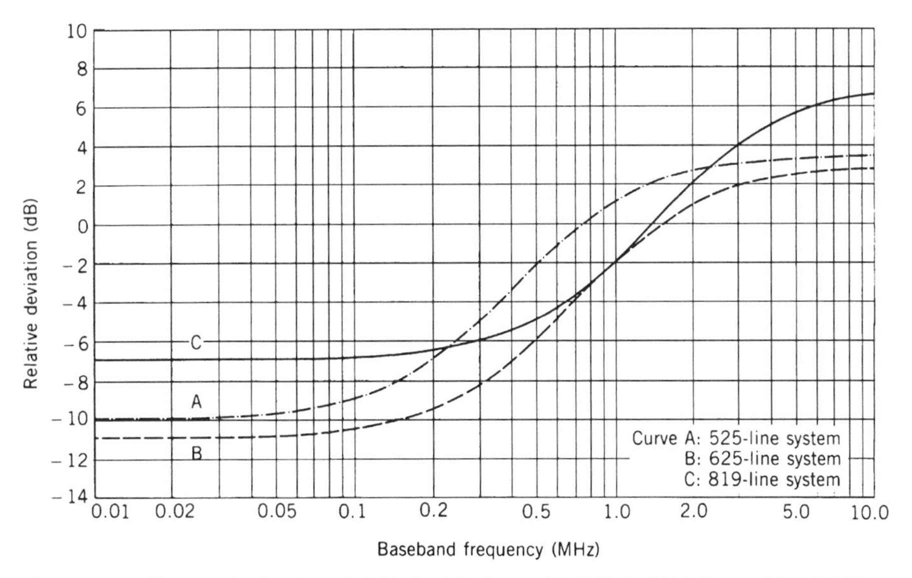

Figure 2.14. Pre-emphasis characteristic for television of (A) 525-, (B) 625-, and (C) 819-line systems (Ref. 8).

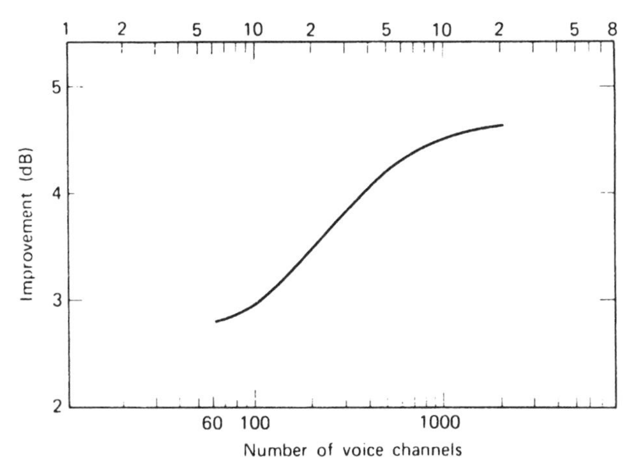

Figure 2.15. Pre-emphasis improvement in decibels as a function of the number of FDM telephone channels transmitted. (From Ref. 10.)

decibels for pre-emphasis improvement for FDM telephony links as a function of the number of FDM voice channels transmitted. These values will be used in link analysis to determine signal-to-noise ratios.

2.6.6 Calculation of Median Carrier-to-Noise Ratio (Unfaded)

The carrier-to-noise ratio (C/N) of an analog radiolink is a function of the receive signal level (RSL) (Section 2.6.2) and the thermal noise threshold (P_t)

of the receiver (Section 2.6.3) or

$$\left(\frac{C}{N}\right)_{dB} = RSL - P_t \tag{2.26}$$

Of course, RSL and P_t must be expressed in the same units, either dBm or dBW.

Example 12. Compute the unfaded C/N in decibels of a radiolink transmitting 960 voice channels of FDM telephony. The design of the link is based on CCIR recommendations and the link is 35 km long. The transmitter output to the waveguide flange is 750 mW on a frequency of 6.1 GHz, and the receiver noise figure is 9 dB. Assume antenna gains on each end of 30 dB and the transmission line losses at each end to be 2.1 dB. Let L_g (gaseous atmospheric loss) equal 0.3 dB.

Calculation of RSL: use equation (2.10),

$$\begin{aligned} RSL_{dBW} = {} & +10\log(0.750) + (-2.1\text{ dB}) + 30\text{ dB} + FSL + (-0.3\text{ dB}) \\ & + 30\text{ dB} + (-2.1\text{ dB}) \end{aligned}$$

Compute FSL: use equation (1.9a),

$$\begin{aligned} FSL_{dB} &= 32.45\text{ dB} + 20\log(35) + 20\log(6100) \\ &= 32.45 + 30.88 + 75.71 \\ &= 139.04\text{ dB} \\ RSL_{dBW} &= -1.25\text{ dBW} - 2.1\text{ dB} + 30\text{ dB} - 139.04\text{ dB} \\ &\quad - 0.3\text{ dB} + 30\text{ dB} - 2.1\text{ dB} \\ &= -84.79\text{ dBW} \end{aligned}$$

Note that in an equation, losses are conventionally expressed with a minus sign and gains are expressed with a plus sign.

Calculation of P_t in dBW: use equation (2.16),

$$P_t = -204\text{ dBW} + 9\text{ dB} + 10\log B_{IF}$$

Compute B_{IF} using Carson's rule, equation (2.17). The maximum modulating baseband frequency is 4188 kHz for 960 voice channels from Table 2.5,

$$B_{IF} = 2(\Delta F_p + 4188\text{ kHz})$$

Calculate ΔF_p: use equation (2.25). First calculate NLR.

$$NLR_{dB} = -15 + 10\log N$$

where $N = 960$,

$$\text{NLR}_{\text{dB}} = -15 + 29.82$$
$$= 14.82 \text{ dB}$$

Convert NLR to its numerical equivalent:

$$\text{NLR}_n = \log^{-1}(14.82/20)$$
$$= 5.51$$

$$\Delta F_p = 4.47(200)(5.51)$$ (*Note*: the value 200 kHz/channel is derived from Table 2.4)

$$= 4926 \text{ kHz}$$

$$B_{\text{IF}} = 2(4926 + 4188) \quad (\text{kHz})$$
$$= 18.228 \text{ MHz}$$

$$P_t = -204 + 9 + 10\log(18.228 \times 10^6)$$
$$= -204 + 9 + 72.61$$
$$= -122.39 \text{ dBW}$$

$$\left(\frac{\text{C}}{\text{N}}\right)_{\text{dB}} = \text{RSL} - P_t$$
$$= -84.79 \text{ dBW} - (-122.39 \text{ dBW})$$
$$= 37.6 \text{ dB}$$

2.6.7 Calculation of Antenna Gain

To achieve a required C/N for a radiolink, a primary tool at the designer's disposal is the sizing of the link antennas. For nearly all applications described in this text, the antenna will be based on the parabolic reflector.

The gain efficiencies of most commercially available parabolic antennas are on the order of 55–65%. It is generally good practice on the first cut link analysis to assume the 55% value. In this case the antenna gain can be calculated from the reflector diameter and the operating frequency by the following formula(s):

$$G_{\text{dB}} = 20\log B + 20\log F + 7.5 \tag{2.27a}$$

where B is the parabolic reflector diameter in feet and F is the operating frequency in gigahertz. In the metric system, the following formula applies:

$$G_{\text{dB}} = 20\log B + 20\log F + 17.8 \tag{2.27b}$$

For formulas (2.27a) and (2.27b), the antenna efficiency is assumed to be 55%.

Calculate antenna gain for a parabolic dish antenna with any efficiency η.

$$G_{\text{dB}} = 20 \log B_{\text{ft}} + 20 \log F_{\text{MHz}} + 10 \log \eta - 49.92 \text{ dB} \qquad (2.27\text{c})$$

B is the aperture diameter in feet; F is the operating frequency in megahertz.

For η value, use the decimal equivalent for the efficiency percentage. If it is 65%, insert 0.65, for example.

Example 13. A LOS microwave radiolink requires an antenna with a 32-dB gain. The link operates at 6 GHz. What diameter parabolic dish is required? Assume an aperture efficiency of 55%.

$$\begin{aligned} 32 \text{ dB} &= 20 \log B + 20 \log 6.0 + 7.5 \text{ dB} \\ 20 \log B &= 24.5 - 15.6 \\ \log B &= 8.9/20 \\ B &= 2.8 \text{ ft} \end{aligned}$$

Example 14. What is the gain of a high-efficiency parabolic dish antenna with an aperture of 8 ft, an operating frequency of 12 GHz, and an aperture efficiency of 70%?

$$\begin{aligned} G_{\text{dB}} &= 20 \log(8) + 20 \log(12{,}000) + 10 \log(0.7) - 49.92 \text{ dB} \\ &= 18.06 + 81.58 - 1.55 - 49.92 \\ &= 48.17 \text{ dB} \end{aligned}$$

2.7 FADING, ESTIMATION OF FADE MARGIN, AND MITIGATION OF FADING EFFECTS

2.7.1 Discussion of LOS Microwave Fading

Up to this point we have been dealing with the calculation of unfaded signal levels at the far end receiver. A fixed FSL was assumed. On most short links, on the order of 3 mi (5 km) or less, only FSL need be considered.* As link length (hop distance) increases, fading becomes a major consideration.

Fading is defined (Ref. 11) as "the variation with time of the intensity or relative phase, or both, of any of the frequency components of a received radio signal due to changes in the characteristics of the propagation path with time." During a fade the RSL decreases. This results in a degradation of C/N, thus a reduction of signal-to-noise ratio in the demodulated signal,

*In this case we recommend a minimum of 5-dB margin to be applied for equipment aging and miscellaneous other losses.

and, finally, an increase in noise in the derived voice channel. On digital systems, fading degrades the BER, causing burst errors.

The various causes of fading were described in Sections 1.6 and 1.7 of Chapter 1. This section gives several methods of quantifying fading so that the link design engineer can overbuild the link to maintain the C/N above a stated level for a specified percentage of time.

We define *time availability* (A_v) of a link/hop or several hops in tandem as the period of time during which a specified noise or BER performance is equaled or exceeded. Some texts use the term propagation reliability to describe time availability. For LOS microwave radiolinks, the time availability is commonly in the range from 0.99 to 0.99999 or 99% to 99.999% of the time.

Time unavailability or just *unavailability* (U_{nav}) is just contrary to the above definition. It is the time that a link or hop does *not* meet its performance requirements. We now have the relationship

$$U_{\text{nav}} = 1 - A_v \tag{2.28}$$

Example 15. A LOS microwave link has a time availability of 99.97%. What is its unavailability?

$$\begin{aligned} U_{\text{nav}} &= 1 - 0.9997 \\ &= 0.0003 \quad \text{or} \quad 0.03\% \end{aligned}$$

There are essentially four methods available to the link design engineer to mitigate the effects of fading, listed below in declining order of desirability:

1. Overbuild the link by use of (a) larger antennas, (b) improved receiver front ends, or (c) higher transmitter power output.
2. Use diversity.
3. Use FEC (see Chapter 4) on digital links.
4. Resite or shorten distance between sites.

2.7.2 Calculating Fade Margin

2.7.2.1 Using the Rayleigh Fading Assumption. If we assume fading is entirely due to multipath conditions, we then can describe the worst-case fading with a Rayleigh distribution. A Rayleigh fading distribution is plotted in Figure 5.2.

For Rayleigh fading, the link margins given in Table 2.6 are required versus time availability. Extrapolation from Table 2.6 can be made for any time availability. For example, a link requiring 99.95% time availability would require a 33-dB margin. This tells us that if the minimum unfaded C/N for the link were specified as 20 dB, the link would require 20 dB + 33 dB or a

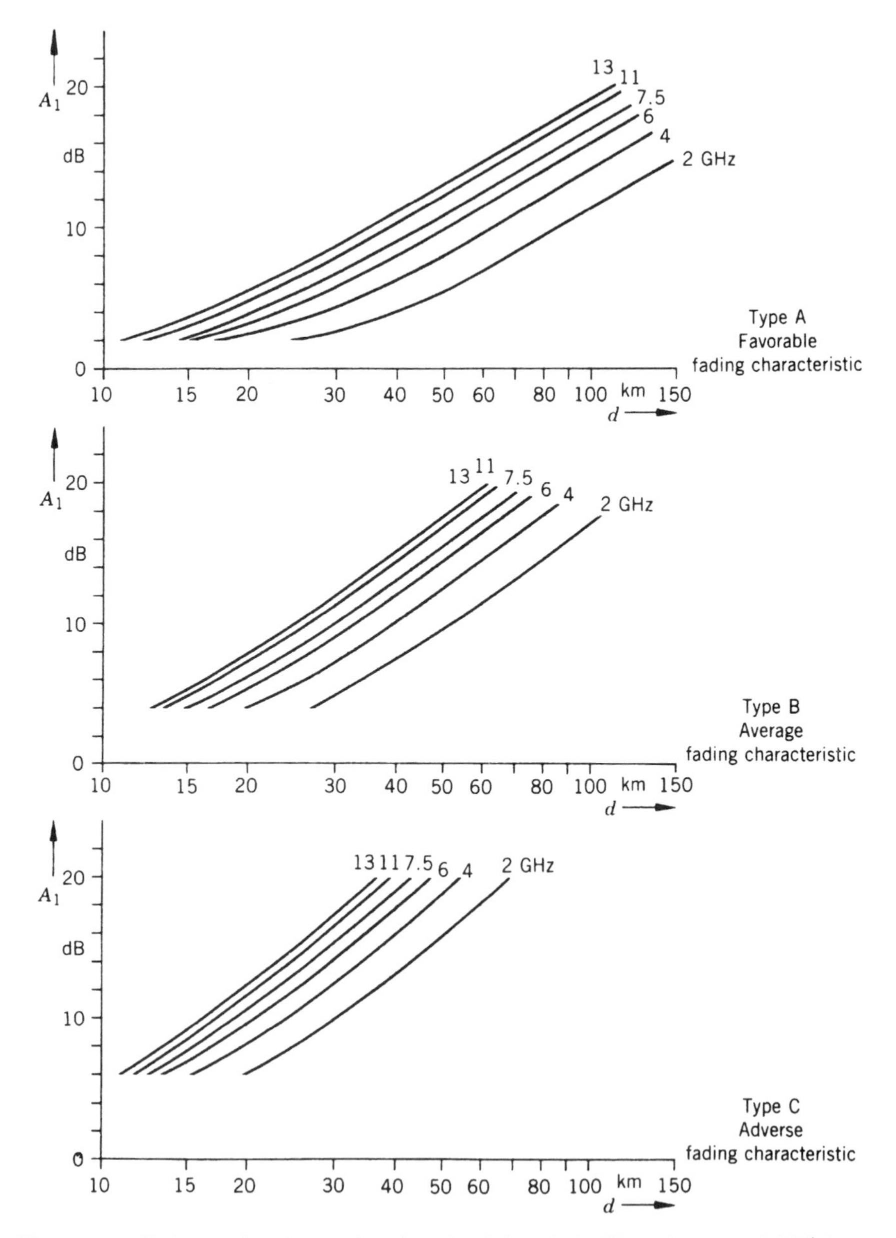

Figure 2.16. Fade margins A_1 as a function of path length d with unobstructed LOS (without precipitation attenuation). Typical values are for 1% of the time of a month with severe fading. (From *Planning and Engineering of Radio Relay Links*, 8th ed., Siemens-Heyden; copyright ©1978 Siemens Aktiengesellschaft, FRG; Ref. 12.)

TABLE 2.6 Fade Margins for Rayleigh Fading

Time Availability (%)	Fade Margin (dB)
90	8
99	18
99.9	28
99.99	38
99.999	48

C/N = 53 dB (unfaded) to meet the objective of 99.95% time availability. Allowing the validity of the pure Rayleigh assumption, the unavailability of the link is $1 - 0.9995$ or 0.0005. A year has 8760 h or 8760×60 min. Thus the total time in a year when the C/N would be less than 20 dB would be $0.0005 \times 8760 \times 60$ or 262.8 min.

2.7.2.2 Path Classification Method 1. This method of quantifying fade margin is based on Ref. 7 and applies only to those paths that are over land and with unobstructed LOS conditions. The method is empirical and was developed from CCIR reports as well as test data from the numerous LOS radiolinks installed by Siemens.

Siemens has classified LOS radiolink paths into three types—A, B, and C—depending on the characteristics of the paths. The curves presented in Figure 2.16 provide required fade margins for fading depths exceeded during 1% of the time in any month with severe fading as a function of frequency and path length. Conversion to lower probabilities of exceeding fading depth can be made using Figure 2.17.

Type A paths have comparatively favorable fading characteristics, where the formation of tropospheric layers is a rare occurrence and where calm weather is a relatively rare occurrence. Type A paths are over hilly country, but not over wide river valleys and inland waters; and in high mountainous country with paths high above valleys. Type A paths are also characterized as being between a plain or a valley and mountains, where the angle of elevation relative to a horizontal plane at the lower site exceeds about 0.5°.

Type B paths with average fading characteristics are typically over flat or slightly undulating country where tropospheric layers may occasionally occur. They are also over hilly country, but not over river valleys or inland waters. Type B paths are also characterized as being in coastal regions with moderate temperatures, but not over the sea, or also over those steeply rising paths in hot and tropical regions.

Type C paths have adverse fading and are characterized over humid areas where ground fog is apt to occur, particularly those paths that are low over flat country, such as wide river valleys and moors. They are also typically near the coast in hot regions and generally are those paths in tropical regions without an appreciable angle of elevation.

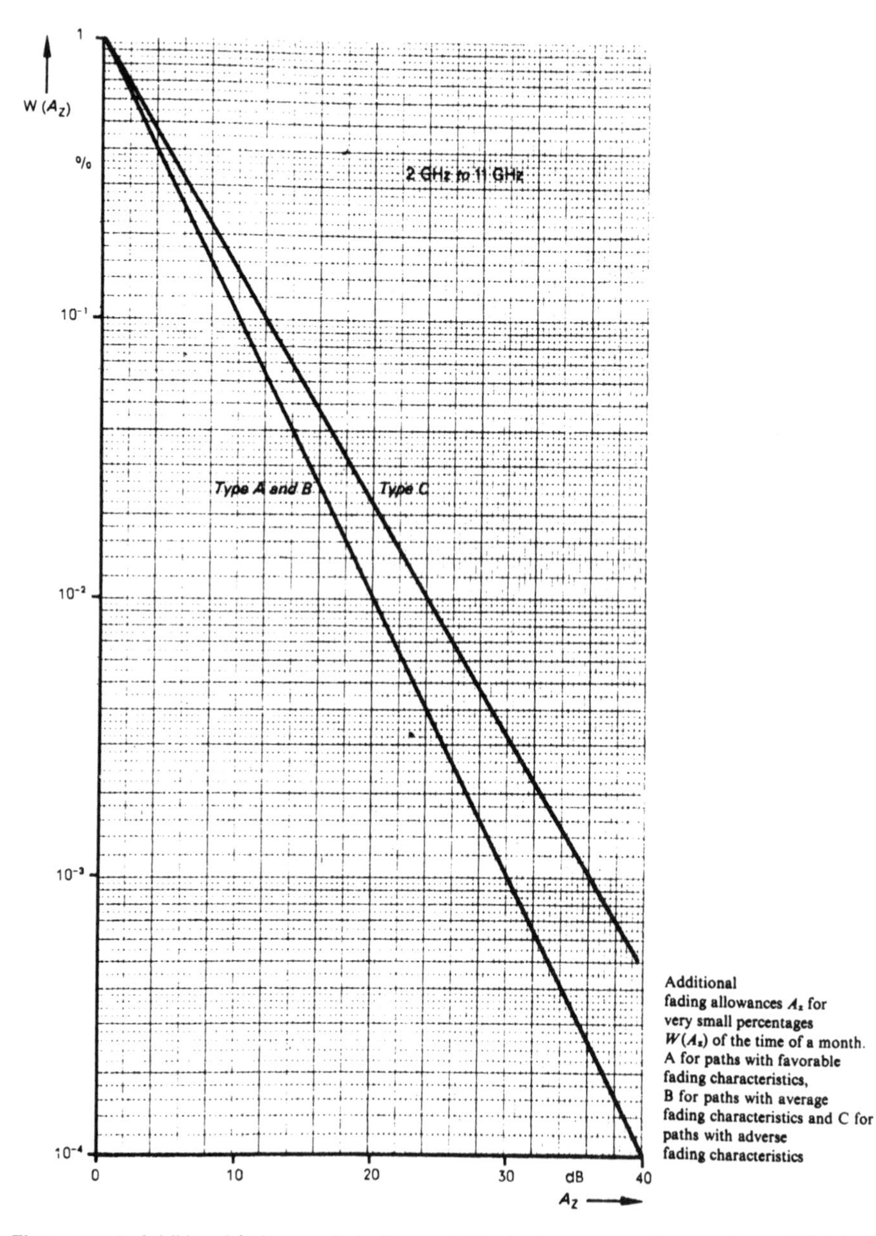

Figure 2.17. Additional fade margin to Figure 2.16, A_z for very small percentages $W(A_z)$ of the time of a month for paths type A, B, and C. (From *Planning and Engineering of Radio Relay Links*, 8th ed., Siemens-Heyden; copyright©1978 by Siemens Aktiengesellschaft, FRG; Ref. 12.)

Rather than use Figures 2.16 and 2.17, the following formulas apply for rough estimations of the probability of fading exceeding the fading depth A:

$$\text{Type A:} \quad W = 16 \times 10^{-7} f d^2 \times 10^{-A/10} \tag{2.29a}$$

$$\text{Type B:} \quad W = 8 \times 10^{-7} f d^{2.5} \times 10^{-A/10} \tag{2.29b}$$

$$\text{Type C:} \quad W = 2 \times 10^{-7} f d^3 \times 10^{-A/12} \tag{2.29c}$$

where W = probability that a fade depth A is exceeded during 1 year
f = radio carrier frequency in gigahertz
d = path length in kilometers
A = fading depth in decibels

The probability of exceeding a specified fading depth during an unfavorable month is

$$W_{\text{month}} = (12/M) \times W_{\text{year}} \tag{2.30}$$

where M is the number of months in any year with intense multipath fading (for Europe, $M \approx 3$).

The approximations for formulas (2.29a, b, and c) are valid with the following assumptions: $W \leq 10^{-2}$; $f = 2$–15 GHz; $d = 20$–80 km; and $A \geq 15$ dB. Over-sea paths may be characterized (with caution) as Type C paths.

Example 16. A particular path has been specified to have a time availability of 99.95%; the operating frequency is 4.0 GHz and the path is 40 km long. What fade margin is required?

Use equation (2.29b). It should be noted that "the probability of exceeding a specified fading depth" or W is really the path unavailability. This equals $1 - 0.9995$ or $W = 0.0005$. Then

$$0.0005 = 8 \times 10^{-7} \times 4 \times 40^{2.5} \times 10^{-A/10}$$

Solve for A and

$$\begin{aligned} 10^{-A/10} &= 0.0005/(8 \times 10^{-7} \times 4 \times 40^{2.5}) \\ 10^{-A/10} &= 0.01544 \\ -A/10 &= \log_{10}(0.01544) \\ -A/10 &= -1.8114 \\ A &= 18.11 \text{ dB} \end{aligned}$$

which fairly well conforms to Figure 2.17.

2.7.2.3 *Path Classification Method 2.* This method is based on an empirical formula developed by Barnett (Ref. 13). It is similar in many respects to Method 1 in that it classifies paths by terrain and climate, and it is used to estimate the percentage of time within a year P_{mf} that fades *exceed* a specified depth below free space (M_f) for a given path and frequency. The formula applies only to paths in the United States and does not specifically consider beam penetration angle through the atmosphere or beam clearance of terrain:

$$P_{mf}(\%) = 6.0 \times 10^{-5} abfd^3 \times 10^{-M_f/10} \tag{2.31}$$

where

$$a = \begin{cases} 4 & \text{for very smooth terrain including over water} \\ 1 & \text{for average terrain with some roughness} \\ \frac{1}{4} & \text{for mountainous, very rough, or very dry terrain} \end{cases}$$

$$b = \begin{cases} \frac{1}{2} & \text{Gulf Coast or similar hot, humid areas} \\ \frac{1}{4} & \text{normal interior temperate or northern climate} \\ \frac{1}{8} & \text{mountainous or very dry climate} \end{cases}$$

f = frequency in gigahertz

d = path length in kilometers

M_f = fading depth exceeded below free-space level, in decibels

Example 17. What will the path unavailability be for a path 50 km long over flat terrain in a relatively humid region, operating at 6 GHz with a 40-dB fade margin?

$$P_{mf}(\%) = 6.0 \times 10^{-5}(4)(\tfrac{1}{2})(6.0)(50^3) \times 10^{-40/10}$$
$$= 0.009\%$$

The path availability then is (1.00000 − 0.00009) or 99.991%.

2.7.2.4 *An ITU-R Method**. This method is for determining fade margin at small time percentages. A geoclimatic factor† K is introduced. There are four K categories, two of which are for overland links and two for over-water links. K can be estimated from the contour maps given in Figures 2.18–2.21 for the percentage of time p_L that the average refractivity gradient in the lowest 100 m of the atmosphere is less than −100 N-units/km, and the

*Section 2.7.2.4 is based on ITU-R Rec. PN.530-5 (Ref. 15).

†Not to be confused with the K for K-factor, which is used to determine ray beam bending.

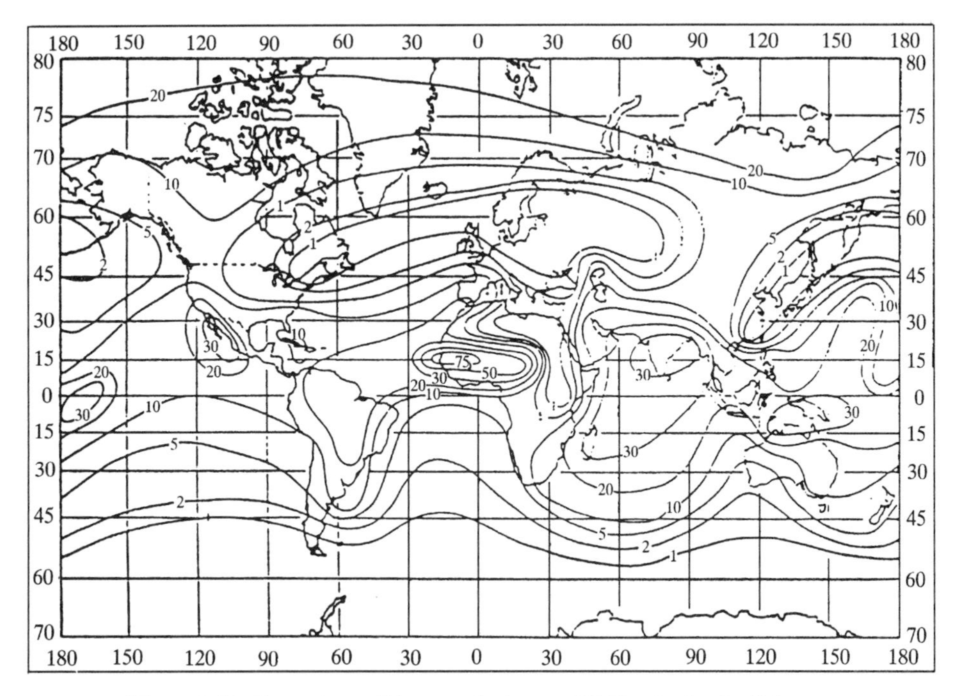

Figure 2.18. Percentage of time gradient ≤ -100 N-units/km for February.

following empirical relations:

$K = 10^{-(6.5 - C_{\text{Lat}} - C_{\text{Lon}})} p_L^{1.5}$ overland links for which the lower of the transmitting and receiving antennas is less than 700 m above mean sea level (see Note 2) (2.32)

$K = 10^{-(7.1 - C_{\text{Lat}} - C_{\text{Lon}})} p_L^{1.5}$ overland links for which the lower of the transmitting and receiving antennas is higher than 700 m above mean sea level (see Notes 1 and 2) (2.33)

$K = 10^{-(5.9 - C_{\text{Lat}} - C_{\text{Lon}})} p_L^{1.5}$ links over medium-sized bodies of water, coastal areas beside such bodies of water, or regions of many lakes (2.34)

$K = 10^{-(5.5 - C_{\text{Lat}} - C_{\text{Lon}})} p_L^{1.5}$ links over large bodies of water or coastal areas beside such bodies of water (2.35)

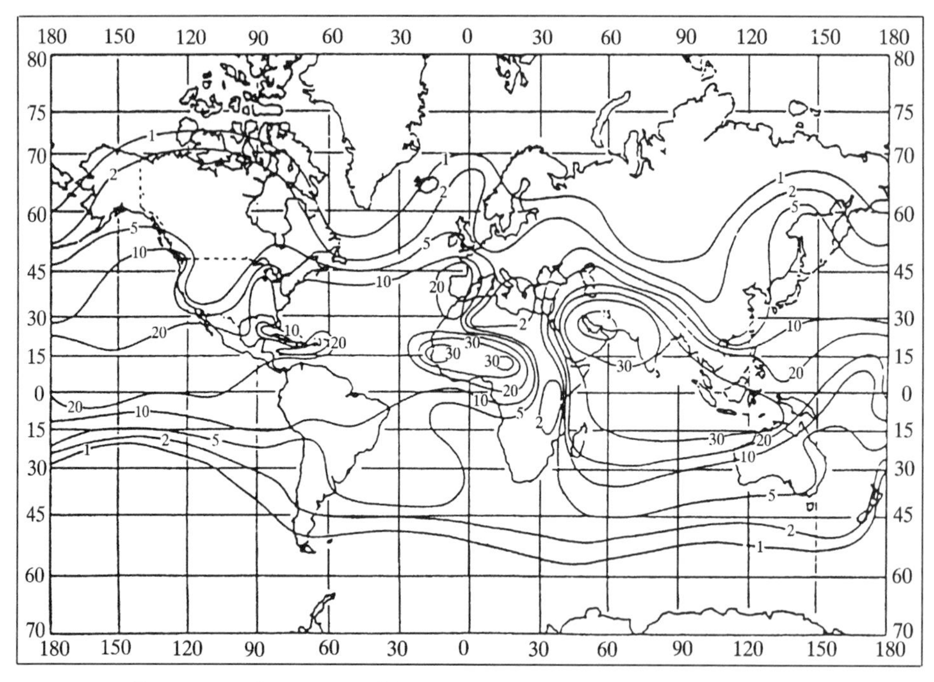

Figure 2.19. Percentage of time gradient ≤ -100 N-units/km for May.

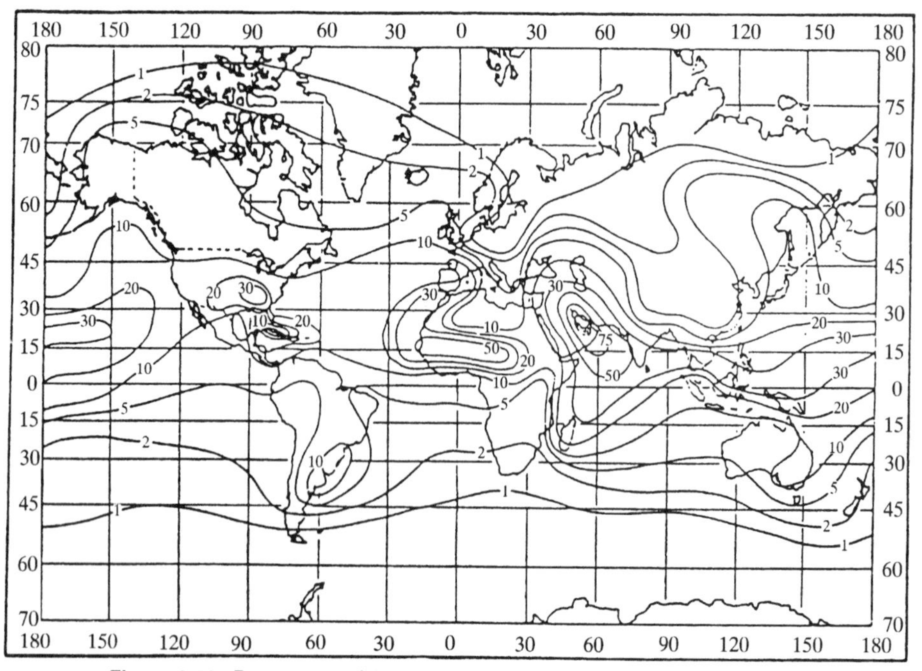

Figure 2.20. Percentage of time gradient ≤ -100 N-units/km for August.

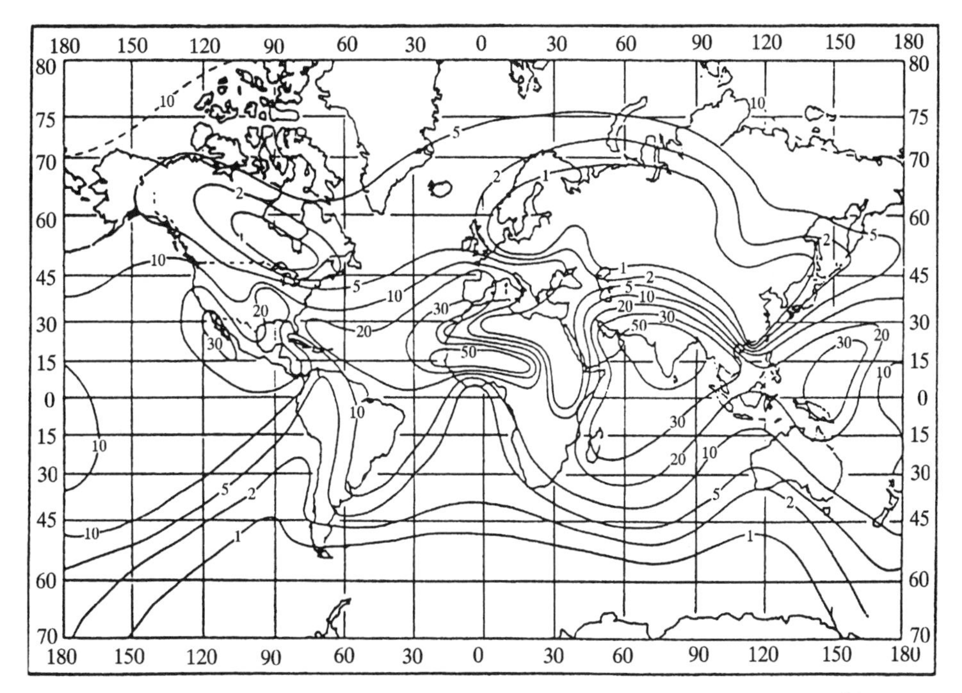

Figure 2.21. Percentage of time gradient ≤ -100 *N*-units/km for November. (Figures 2.18–2.21 are taken from ITU-R PN.453-4, Figures 7–10, pages 212 and 213, 1994 PN Series Volume; Ref. 14.)

where the coefficient C_{Lat} of latitude ξ is given by

$$C_{\text{Lat}} = 0 \qquad \text{for} \qquad 53^\circ\text{ S} \geq \xi \leq 53^\circ\text{N} \tag{2.36}$$

$$C_{\text{Lat}} = -5.3 + \xi/10 \qquad \text{for} \quad 53^\circ\text{ N or S} < \xi < 60^\circ\text{ N or S} \tag{2.37}$$

$$C_{\text{Lat}} = 0.7 \qquad \text{for} \qquad \xi \geq 60^\circ\text{ N or S} \tag{2.38}$$

and the longitude coefficient C_{Lon} is given by

$$C_{\text{Lon}} = 0.3 \qquad \text{for longitudes of Europe and Africa} \tag{2.39}$$

$$C_{\text{Lon}} = -0.3 \qquad \text{for longitudes of North and South America} \tag{2.40}$$

$$C_{\text{Lon}} = 0 \qquad \text{for all other longitudes} \tag{2.41}$$

The month that has the highest value of p_L is selected from the four seasonally representative months of February, May, August, and November for which maps are given in Figures 2.18–2.21. However, for latitudes greater than 60° N and S, only the maps of May and August should be used.

Note 1. In mountainous areas for which the data used to prepare the maps in Figures 2.18–2.21 were nonexistent or very sparse, these maps have insufficient detail and the value of K estimated from equation (2.32) will tend to be an upper bound. Such regions include the western mountains of Canada, the European Alps, and Japan. The adjustment contained in equation (2.33) can be used until a more detailed correction is available from the ITU-R Organization.

Note 2. Links passing over a small lake or river should normally be classified as overland links. In cases of uncertainty, the first coefficient in the exponent of equation (2.32) should be replaced by 6.2 (also see Note 4).

Note 3. Medium-sized bodies of water include the Bay of Fundy (on the east coast of Canada), the Strait of Georgia (west coast of Canada), the Gulf of Finland, and other bodies of water of similar size. Large bodies of water include the English Channel, the North Sea, and the larger reaches of the Baltic and Mediterranean Seas, Hudson Strait, and other bodies of water of similar size or larger. In cases of uncertainty as to whether the size of the body of water in question should be classed as medium or large, the first coefficient in the exponent of equations (2.34) and (2.35) should be replaced by 5.7.

Note 4. The link may be considered crossing a coastal area if a section of the path profile is less than 100 m (328 ft) above mean sea level and within 50 km (31 mi) of the coastline of a medium or large body of water, and if there is no height of land above 100 m (328 ft) altitude between the link and the coast. If the entire path profile is less than 100 m (328 ft) above mean sea level, then K should be obtained from equations (2.34) and (2.35) as appropriate. If only a fraction, r_c, of the path profile is below 100 m (328 ft) altitude and within 50 km (31 mi) of the coastline, then the coefficient 5.9 in the exponent of equation (2.34) can be replaced with $6.5 - 0.6r_c$, and the exponent 5.5 in equation (2.35) by $6.5 - r_c$.

Note 5. Regions (not otherwise in coastal areas) in which there are many lakes over a fairly large area are known to behave like coastal areas. The region of lakes in southern Finland provides the best known example. Until such regions can be better defined, K can be obtained from equation (2.34). In cases of uncertainty, the coefficient 5.9 in the exponent of equation (2.34) can be replaced by 6.2.

If the antenna heights h_e and h_r (meters above sea level or some other reference height) are known, calculate the magnitude of the path inclination $|\varepsilon_p|$ in milliradians (mrad) from

$$|\varepsilon_p| = |h_r - h_e|/d \tag{2.42}$$

where the path length d is in kilometers.

Calculate the percentage of time p_w that fade depth A (dB) is exceeded in the average worst month from

$$p_w = Kd^{3.6} f^{0.89} \left(1 + |\varepsilon_p|\right)^{-1.4} \times 10^{-A/10} \quad \% \tag{2.43}$$

where the frequency f is in gigahertz.

Example 18. A 40-mi (64-km) path operates over Lake Michigan (one of the U.S./Canada Great Lakes, where the region is comparatively low and flat). The latitude is 45° N and longitude is 87° W; $f = 7$ GHz. The fade margin is 40 dB, and $\varepsilon_p = 3$ mrad. What is the resulting time availability?

First, calculate the geoclimatic factor K using equation (2.35). Use Figure 2.20 and $p_L = 10$. The exponent in equation (2.35) is $-(5.5 - 0 + 0.3)$ and $K = 4.996 \times 10^{-5}$.

Now calculate the results from equation (2.43) solving for p_w:

$$\begin{aligned} p_w &= 4.996 \times 10^{-5} (64)^{3.6} (7)^{0.89} (4)^{-1.4} \times 10^{-4} \\ &= 0.01284 \quad \text{(unavailability)} \quad (\%) \\ \text{Time availability} &= 1 - 0.01284 \\ &= 0.98716\% \quad \text{or} \quad 0.0098716 \end{aligned}$$

Suppose in Example 18, we cut the link length in half and leave all other factors as they are. This procedure is meant to emphasize how sensitive fading is to link (hop) length.

Example 19. Use the same conditions as in Example 18 but the link length is now 32 km (20 mi).

$$\begin{aligned} p_w &= 4.996 \times 10^{-5} (32)^{3.6} (7)^{0.89} (4)^{-1.4} \times 10^{-4} \\ &= 0.00106249\% \quad \text{(unavailability)} \quad (\%) \\ \text{Time availability} &= 1 - 0.0000106249 \\ &= 0.999989376 \quad \text{or} \quad 99.9989376\% \end{aligned}$$

2.7.3 Notes on Path Fading Range Estimates

Three methods have been described in Section 2.7.2 for first-order rough estimations of fade margin. Other methods are available using Hoyt, Nakagami–Rice, and other distributions (Ref. 5). Even with these tools at our disposal, experience has shown that fade margins and other design improvements based on fading estimates are insufficient. The best tool is long-term testing, over a full year. For most projects, such testing is not economical. It is not unknown to install a link that subsequently does not meet specifications because fade depth and fading rate are greater than predicted. Such links require upgrading, often at the expense of the contractor.

Satisfactory methods of estimating fading probability on specific paths from climatological statistics are not available at present. Statistical studies of refractivity (Section 1.3) can, however, provide information on relative gradient probabilities for different areas and indicate the seasonal changes to be expected (Ref. 16). The low-level refractivity gradients that are of most importance to LOS radiolink propagation are very sensitive to variations in local weather conditions. Therefore the information obtained on site surveys can be very useful during this stage of link design. Operating experience on many microwave links (Ref. 5) shows the following:

- Fading is more likely on paths across flat ground than on paths over rough terrain.
- There is less fading on paths across dry ground than on paths across river valleys, wet or swampy terrain, or irrigated fields.
- Calm weather favors atmospheric stratifications that may result in deep fading. These conditions occur more often in broad, protected river valleys than over open country.
- Fading is likely near the center of large, slow-moving anticyclones (high-pressure areas). These are more likely to occur in summer and fall, in the northern hemisphere, than in winter and spring.
- Expect fading to occur more frequently and with greater severity in summer than in winter for midlatitudes (temperate zones).
- Paths with takeoff angles greater than about 0.5° are less susceptible to fading, and the effect of refractivity gradients is negligible where takeoff angles exceed 1.5°.

2.7.4 Diversity as a Means to Mitigate Fading

2.7.4.1 General. The first and most economic step to achieve required fade margins is to overbuild the link by using larger aperture antennas, improved receiver noise performance (i.e., use of a LNA in front of the mixer), and/or a higher transmitter output power. Under certain circumstances, the link design engineer will find that it is uneconomic to overbuild the link further. The next step, then, is to examine diversity as another method to mitigate fading and, at the same time, add a "diversity improvement factor," which can be translated directly to some 3 dB or more of system gain.

Diversity is based on providing separate paths to transmit redundant information. These paths may be in the domain of space, frequency, or time. In essence, the idea is that a fade occurring at time T_a on one of the redundant paths does not occur on the other. The ability to mitigate fading effects by diversity reception is a function of the correlation coefficient of the signals on the redundant paths. The higher the decorrelation, the more effective the diversity. Generally, when the correlation coefficient is ≤ 0.6, full diversity improvement can be entirely achieved (Ref. 16).

The most commonly used methods of diversity are frequency and space diversity to minimize the effects of multipath fading. (In Chapter 9 another form of "space" diversity is introduced called "spatial diversity" or path diversity.) Troposcatter and diffraction links (Chapter 5) often use a combination of frequency and space diversity, achieving still greater protection against the effects of fading.

Frequency diversity uses two different frequencies to transmit the same information. With space diversity, the same frequency is used, but two receive antennas separated vertically on the same tower receive the information over two different physical paths separated in space. These two types of diversity are shown conceptually in Figure 2.22.

Compared to space diversity, frequency diversity is somewhat less expensive to implement and has some operational and maintenance advantages. Its principal drawback is regulatory. As an example, the U.S. Federal Communications Commission (FCC) (Ref. 17) rules prohibit frequency diversity for common carriers unless sufficient evidence can be shown that frequency diversity is the only way to obtain the required system reliability. This ruling was established to preserve centimetric frequencies for working radio channels owing to frequency congestion in the centimetric bands and the demand for working frequencies. Therefore, in the United States, the first alternative should be space diversity.

2.7.4.2 Frequency Diversity. Frequency diversity offers two advantages. Not only does it provide a full order of diversity and resulting diversity gain, but it also provides a fully redundant path, improving equipment reliability. (See Appendix 1.)

To achieve maximum fading decorrelation, the separation in frequency of the two transmit frequencies must be on the order of 3–5%. However, because of congestion and lack of frequency assignments in highly developed countries, separations of 2% are more common, and some systems operate satisfactorily with separations under 1%. Figure 2.23 shows approximate worst-case multipath fading (Rayleigh) and diversity improvement for frequency diversity systems with different frequency separations. The figure shows that a 14–19-dB improvement may be expected for a link with a time

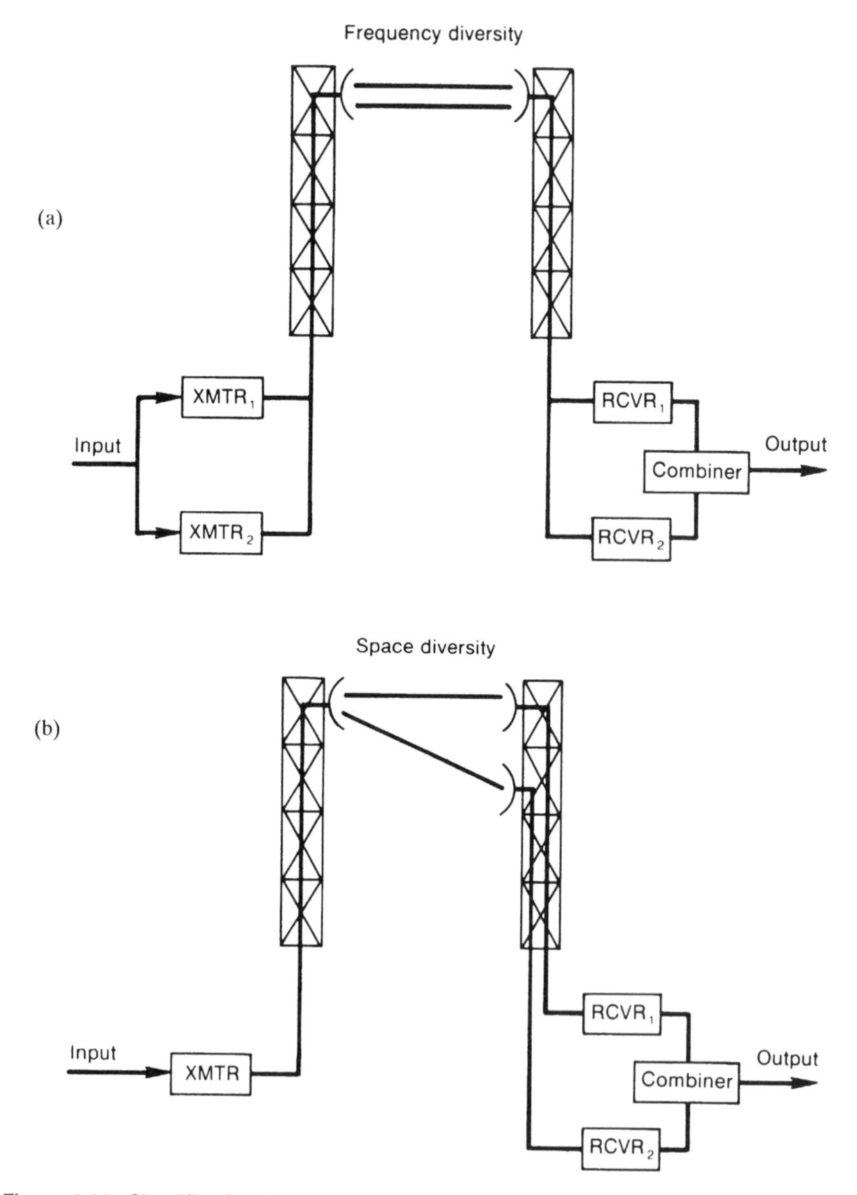

Figure 2.22. Simplified functional block diagrams distinguishing frequency and space diversity operation on LOS radiolink.

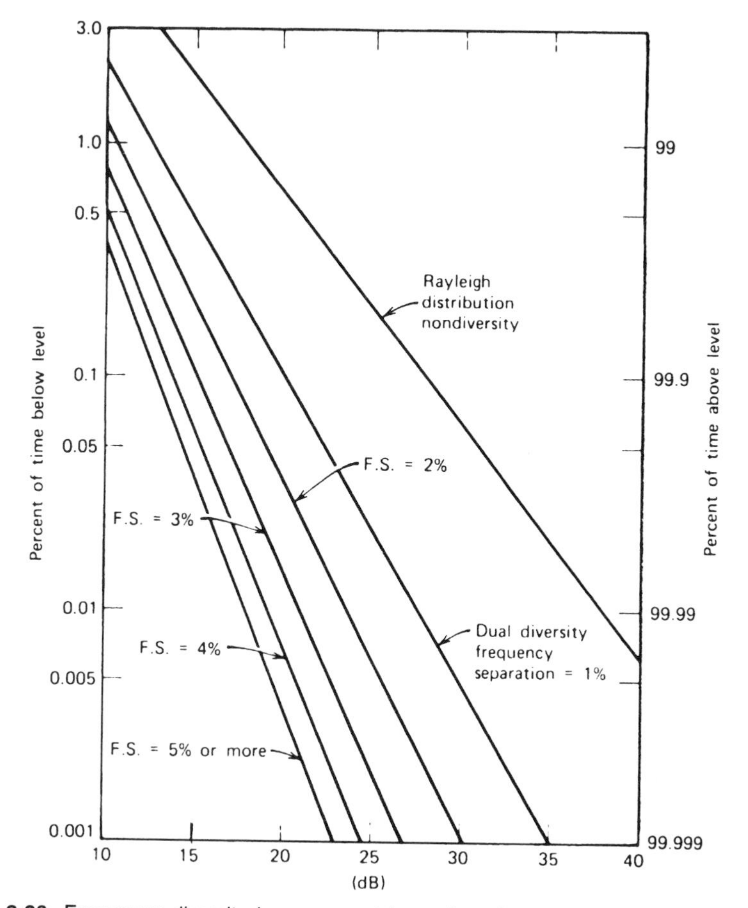

Figure 2.23. Frequency diversity improvement for various frequency spacings (in percent) compared to an equivalent nondiversity path. Rayleigh fading is assumed. (From Ref. 10.)

availability requirement of 99.99% over the same link with no diversity, assuming Rayleigh fading.

2.7.4.3 Space Diversity. Because of the difficulty of obtaining the second diversity frequency to transmit redundant information, vertical space diversity may be the easier of the two alternatives, and in some cases it may be the only diversity alternative open to the designer. In fact, experience is now showing that space diversity has considerably lower correlation coefficients, with consequently much larger diversity improvements than was earlier believed.

It should be noted that a space diversity arrangement can also provide full equipment redundancy when automatically switched hot standby transmitters are used (see Section 14.3.5). However, this arrangement does not provide a separate end-to-end operational path as does the frequency diversity arrangement.

Of principal concern in the design of space diversity on a particular path is the amount of vertical separation. Reference 5 suggests a rule-of-thumb separation distance of 200 wavelengths or more. One wavelength at 6 GHz is 5 cm. The required separation, then, would be 5 × 200 cm or 10 m (about 33 ft). Reference 13 suggests 60 ft at 2 GHz, 45 ft at 4 GHz, 30 ft at 6 GHz, and 15–20 ft at 12 GHz. It will be appreciated that both antennas must meet the path profile clearance criteria, and the result will be taller towers.

Reference 13 provides a formula (modified from Vigants) to calculate the space diversity improvement factor I_{sd}:

$$I_{sd} = \frac{7.0 \times 10^{-5} f_S^2 \times 10^{\bar{F}/10}}{D} \tag{2.44}$$

where f = frequency in gigahertz

s = vertical antenna spacing in feet between antenna centers

D = path length in statute miles

$\bar{F}$ = fade margin in decibels associated with the second antenna. The barred F factor is introduced to cover the situation where the fade margins are different on the upper and lower paths of the vertically spaced antennas. In such a case F will be taken as the larger of the two fade margins and will be used to calculate the unavailability (U_{nd}) in the computation for the nondiversity path. $\bar{F}$ in equation (2.44) will be taken as the smaller fade margin of the two, if different.

Reference 13 recommends that one first calculates the nondiversity path unavailability P_{mf} [from equation (2.31)] and then one calculates the space diversity improvement factor I_{sd}. The diversity outage (unavailability) or fade probability (U_{div}) is given by

$$U_{div} = \frac{P_{mf}}{I_{sd}} \tag{2.45}$$

Example 20. Consider a 30-mi (48.3-km) path with average terrain that includes some roughness and where the climate is inland temperate. The operating frequency of the radiolink is 6.7 GHz and the fade margin incorporated is 40 dB. Calculate the unavailability and path availability for the nondiversity case and for the space diversity case with 40-ft vertical spacing.

Calculate P_{mf} using equation (2.31):

$$P_{mf}\,(\%) = 6.0 \times 10^{-5}(1)(\tfrac{1}{4})(6.7)(48.3)^3(10)^{-40/10}$$
$$= 6.0 \times 10^{-5} \times 1.675 \times 112678.6 \times 10^{-4}$$
$$= 0.0011\% \quad \text{or} \quad 0.000011$$

This corresponds to a path availability of 1 − 0.00011 or 99.9989%. Calculate I_{sd} from equation (2.44):

$$I_{sd} = \frac{7.0 \times 10^{-5}(6.7)(40)^2 \times 10^4}{30}$$
$$\approx 250$$

Substitute this value into equation (2.45):

$$U_{div} = 0.000011/250$$
$$= 0.000000044$$

The path availability for space diversity is then 1 − 0.000000044 = 0.999999956 or 99.9999956%.

Figure 2.24 is a useful nomogram to determine the space diversity improvement factor denoted I_{sd}. The nomogram is taken from CCIR Rep. 376-3.

2.8 ANALYSIS OF NOISE ON A FM RADIOLINK

2.8.1 Introduction

In the design of a radiolink or system of radiolinks, a noise requirement will be specified. In the case of a system carrying FDM telephony, noise power in the derived voice channel will be specified. For a system carrying video, the requirement will probably be stated as a weighted signal-to-noise ratio.

Because C/N at the receiver input varies with time, noise power is specified statistically. Specifications are typically based on Table 2.1 and derive from the following excerpt from CCIR Rec. 393-4 (Ref. 19):

> The noise power at a point of zero relative level in any telephone channel on a 2500 km hypothetical reference circuit for frequency division multiplex radio relay systems should not exceed the values given below, which have been chosen

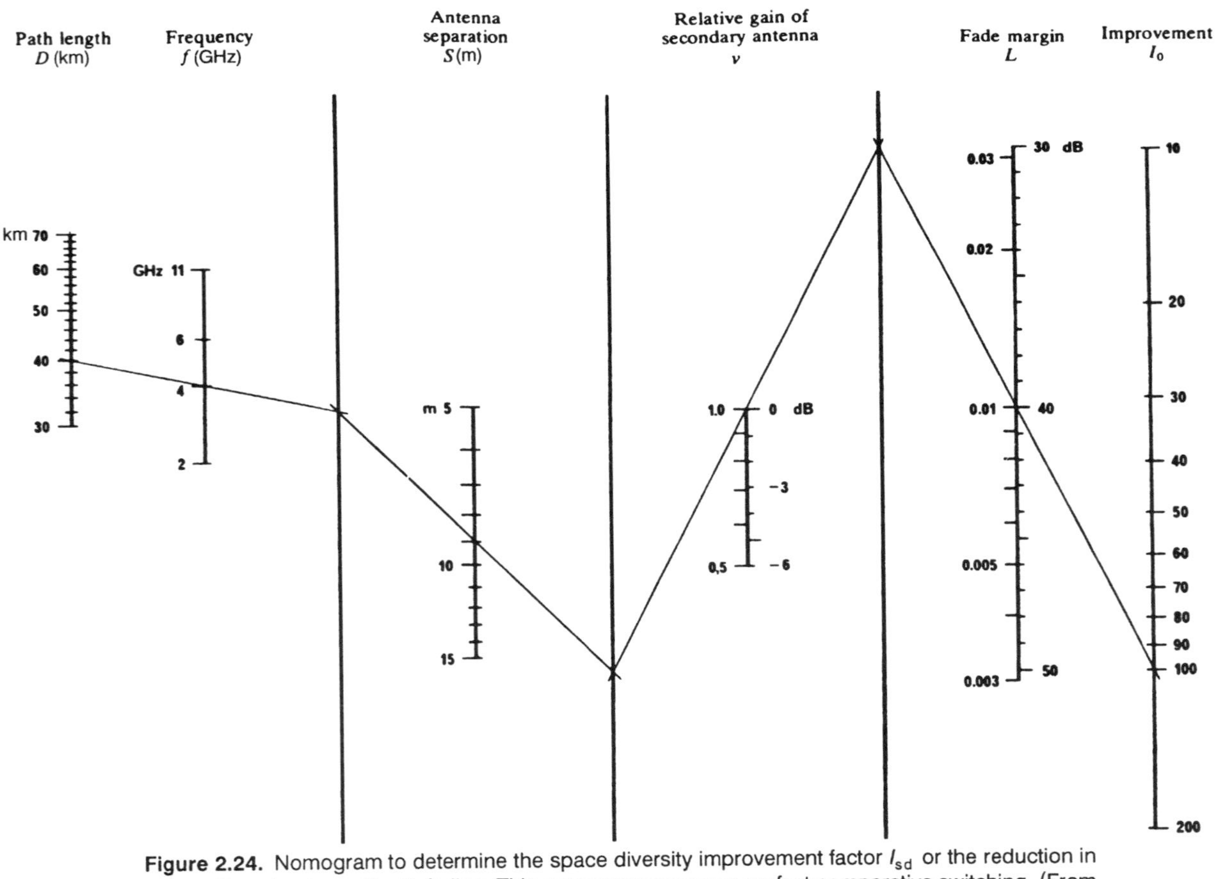

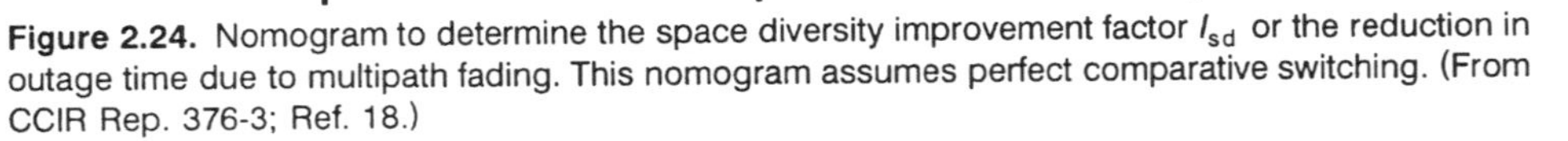

Figure 2.24. Nomogram to determine the space diversity improvement factor I_{sd} or the reduction in outage time due to multipath fading. This nomogram assumes perfect comparative switching. (From CCIR Rep. 376-3; Ref. 18.)

to take account of fading:

- 7500 pW0p, psophometrically weighted one minute mean power for more than 20% of any month;
- 47,500 pW0p, psophometrically weighted mean power for more than 0.1% of any month;
- 1,000,000 pW0, unweighted (with an integrating time of 5 ms) for more than 0.01% of any month.

If we were to connect a psophometer at the end of a 2500-km (reference) circuit made up of homogeneous radio relay sections, in a derived voice channel, we should read values no greater than these plus 2500 pW0p mean noise power due to the FDM equipment contribution, or, for example, not to exceed 10,000 pW0p total noise for more than 20% of any month.

The problem addressed in this section is to ensure that the radiolink design can meet this or other similar criteria.

2.8.2 Sources of Noise in a Radiolink

Figure 2.25 shows three basic noise sources assuming voice channel input (insert) on one end of the link and voice channel output (drop) at the other

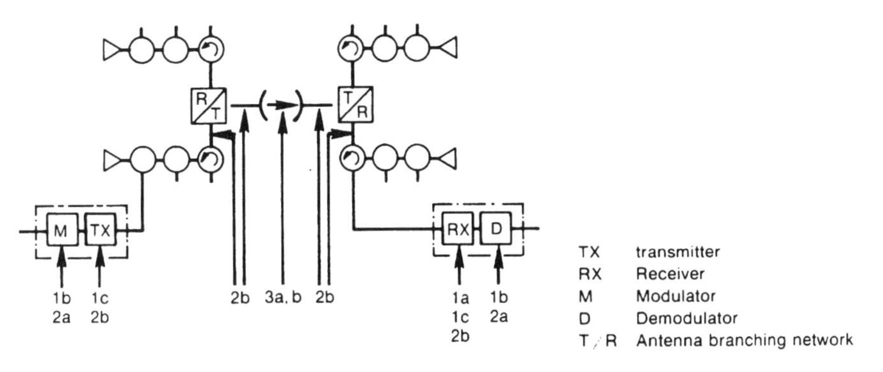

1 Load-invariant noise (basic noise):
1a Receiver thermal noise (cause of loss-dependent noise in the signal channel)
1b Basic noise of modem equipment
1c Basic noise of RF equipment
2 Load-dependent noise
2a Intermodulation noise of modem equipment
2b Intermodulation noise of RF equipment and antenna systems
3 Noise from other RF channels of the same system
3a Noise due to cochannel interference
3b Noise due to adjacent-channel interference

Figure 2.25. Noise contributors on a radiolink. (From *Planning and Engineering of Radio Relay Links*, 8th ed., Siemens-Heyden; copyright ©1978 Siemens Aktiengesellschaft, FRG; Ref. 12.)

end of the link. These are:

1. Load-invariant noise (thermal noise).
2. Load-dependent noise (intermodulation noise).
3. Interference noise.

The modem in the figure refers to the FDM modulator and demodulator.

2.8.3 FM Improvement Threshold

FM is wasteful of bandwidth when compared to AM-SSB, for instance. However, this "waste" of bandwidth is compensated for by an improvement in thermal noise power when the input signal level (RSL) reaches FM improvement threshold (i.e., when $C/N \simeq 10$ dB). In other words, we are giving up bandwidth for a thermal noise improvement.

If we were to draw a curve of signal-to-noise power ratio at the output of the FM demodulator versus the carrier-to-noise ratio at the input (Figure 2.26), we will note three important points of reference:

1. Thermal noise threshold (see Section 2.6.3).
2. FM improvement threshold (10 dB above thermal noise threshold P_{FM}).
3. Saturation (where compression starts to take place).

The first reference point, thermal noise threshold (P_t), is only a waypoint in link calculations. The second point, FM improvement threshold, where

$$P_{\mathrm{FM}} = P_t + 10\ \mathrm{dB} \tag{2.46}$$

will be used as a reference on which link calculations will be based. In equation (2.46) substitute the value of P_t taken from equation (2.16) or

$$P_{\mathrm{FM}} = -204\ \mathrm{dBW} + 10 \log B_{\mathrm{IF}} + NF_{\mathrm{dB}} + 10\ \mathrm{dB} \tag{2.47}$$

Example 21. A radiolink receiver has a noise figure of 8 dB and an IF bandwidth of 12 MHz. What is the FM improvement threshold in dBW? Use equation (2.47).

$$\begin{aligned} P_{\mathrm{FM}}(\mathrm{dBW}) &= -204\ \mathrm{dBW} + 10 \log(12 \times 10^6) + 8\ \mathrm{dB} + 10\ \mathrm{dB} \\ &= -115.21\ \mathrm{dBW} \end{aligned}$$

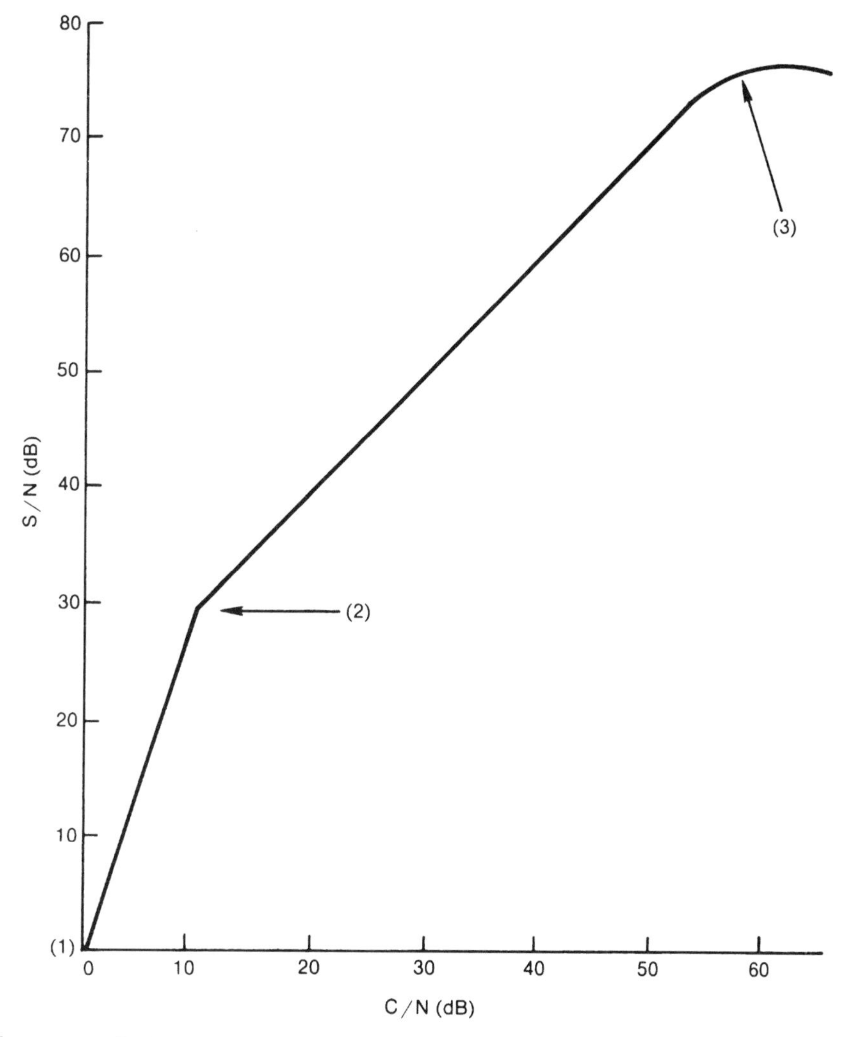

Figure 2.26. Typical plot of signal-to-noise power ratio at the output of a FM demodulator versus the carrier-to-noise ratio at the input of the demodulator. Point 1 is the thermal noise threshold, point 2 is the FM improvement threshold, and point 3 is saturation.

2.8.4 Noise in a Derived Voice Channel

2.8.4.1 Introduction. On a FM radiolink, noise is examined during conditions of fading and for the unfaded or median noise condition. For the low RSL or faded condition, Ref. 13 states that "the noise in a derived (FDM) voice channel at FM threshold, falls approximately at, or slightly higher than, the level considered to be the maximum tolerable noise for a telephone channel in the public network. By present standards, this maximum is considered to be 55 dBrnc0 (316,200 pWp0). In industrial systems, a value of 59 dBrnc0 (631,000 pWp0) is commonly used as the maximum acceptable noise level." FM threshold, therefore, is the usual point where fade margin is added (Section 2.7.2) to achieve the unfaded RSL. In many systems, however, the CCIR guideline of 3 pWp/km may not be achieved at the high-signal-level condition, and the unfaded RSL may have to be increased still further.

At low RSL, the primary contributor to noise is thermal noise. At high-signal-level conditions (unfaded RSL), there are three contributors:

1. Thermal noise (discussed below).
2. Radio equipment IM noise (Section 2.8.5.4.2).
3. IM noise due to antenna feeder distortion (Section 2.8.6).

Each are calculated for a particular link for unfaded RSL and each value is converted to noise power in mW, pW, or pWp and summed. The sum is then compared to the noise apportionment for the link from Table 2.1 and Section 2.1.

2.8.4.2 *Calculation of Thermal Noise.* The conventional formula (Ref. 20) to calculate signal-to-noise power ratio on FM radiolink (test tone to flat weighted noise ratio for thermal noise) is

$$\frac{\mathrm{S}}{\mathrm{N}} = \frac{\mathrm{RSL}}{(2ktbF)(\Delta f/f_c)^2} \tag{2.48}$$

where b = channel bandwidth (3.1 kHz)
Δf = channel test tone peak deviation (as adjusted by pre-emphasis)
f_c = channel (center) frequency in baseband in kilohertz
F = the noise factor of the receiver (i.e., the numeric equivalent of the noise figure in decibels)
$kt = 4 \times 10^{-18}$ mW/Hz
RSL = receive signal level in milliwatts
S/N = test tone-to-noise ratio (numeric equivalent of S/N in decibels)

In the more useful decibel form, equation (2.48) is

$$\left(\frac{\mathrm{S}}{\mathrm{N}}\right)_{\mathrm{dB}} = \mathrm{RSL}_{\mathrm{dBm}} + 136.1 - \mathrm{NF}_{\mathrm{dB}} + 20\log(\Delta f/f_c) \quad \text{(flat)} \tag{2.49}$$

In a similar fashion (Ref. 13) the noise power in the derived voice channel can be calculated as follows:

$$P_{\mathrm{dBrnc0}} = -\mathrm{RSL}_{\mathrm{dBm}} - 48.1 + \mathrm{NF}_{\mathrm{dB}} - 20\log(\Delta f/f_c) \tag{2.50}$$

and

$$P_{\mathrm{pWp0}} = \log_{10}^{-1}\left(\frac{-\mathrm{RSL}_{\mathrm{dBm}} - 48.6 + \mathrm{NF}_{\mathrm{dB}} - 20\log(\Delta f/f_c)}{10}\right) \quad (2.51)$$

Note: Values for Δf, the *channel* test tone *peak* deviation (Table 2.4): 200 kHz rms deviation for 60 through 960 VF channel loading and 140 kHz rms deviation for transmitters loaded with more than 1200 voice channels. Per channel peak deviation is 282.8 and 200 kHz, respectively.

Example 22. A 50-hop is to be designed to CCIR noise recommendations based on Table 2.1. What level RSL is required for 47,500 pWp0? (It should be noted that this level must be met 99.99% of the time.) The link is designed for 300 FDM channels and the highest channel center frequency is 1248 kHz. The noise figure of the receiver is 10 dB.

Use equation (2.50) and set 47,500 pWp0 equal to the value of the right-hand side of that equation:

$$47{,}500\ \mathrm{pWp0} = \log_{10}^{-1}\left(\frac{-\mathrm{RSL} - 48.6 + 10\ \mathrm{dB} - 20\log(\Delta f/f_c)}{10}\right)$$

Calculate the value of

$$\begin{aligned} 20\log(\Delta f/f_c) &= 20\log(282.8/1248) \\ &= -12.9\ \mathrm{dB} \end{aligned}$$

Then

$$\begin{aligned} 10\log(47{,}500) &= -\mathrm{RSL} - 48.6 + 10 + 12.9 \\ \mathrm{RSL} &= -72.47\ \mathrm{dBm} \end{aligned}$$

Table 2.9 presents equivalent values of dBrnc, pWp, and signal-to-noise power ratio for the standard voice channel.

2.8.4.3 Notes on Noise in the Voice Channel. Flat noise must be distinguished from weighted noise. Theoretically, flat noise power is evenly distributed across a band of interest such as the voice channel. As a theoretical example, suppose the noise power in a voice channel at 1000 Hz was measured at −31 dBm. We would then expect to measure this value at all points in the channel. In practice, this is impossible because of band-limiting characteristics of the medium such as FDM channel bank filters.

A weighted channel does not display a uniform response. A weighted channel displays certain special frequency and transient response characteristics. Two such "weightings" reflect response of the typical human ear. These are C-message and psophometric weightings.

In North America, C-message weighting is used, and its reference frequency is 1000 Hz. In Europe and many other parts of the world, psophometric weighting is used where the reference frequency is 800 Hz. The characteristics of each weighting type are slightly different as shown in the response curves in Figure 2.27. Weighting networks are used in test equipment to simulate these weighting characteristics, either C-message or psophometric.

As was previously pointed out, the practical voice channel response is not flat but has rolloff characteristics that are a function of the media's bandpass characteristics. To simulate this quasi-flat response for the voice channel, a 3-kHz flat network is used to measure power density (under "flat" conditions) of Gaussian noise. The network has a nominal low-pass response, which is down 3 dB at 3 kHz and rolls off at 12 dB per octave (Ref. 33).

The noise measurement unit used nearly universally in North America is the dBrnc0. The "c" implies C-message weighting and the "0" means that the measurement is referenced to the 0 test level point (0 TLP).

In Europe and in much of the CCITT/CCIR documentation, the weighted noise measurement unit is the pWp, picowatts psophometrically weighted (pWp0 or pW0p when referenced to the 0 TLP). Alternative measurement units are dBp and dBmp. The final "p" implies psophometric weighting.

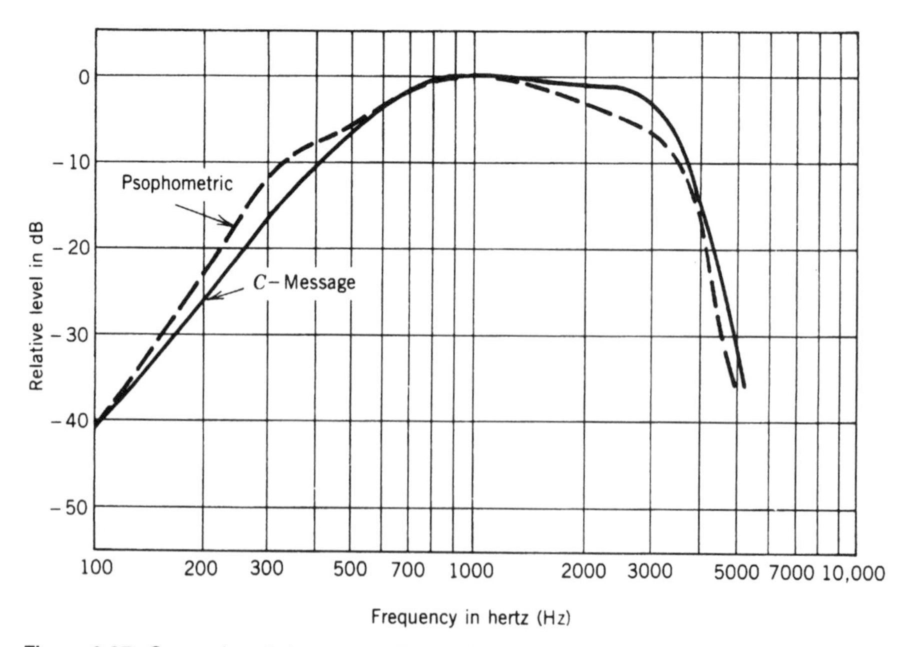

Figure 2.27. Comparison between psophometric and C-message weighting. (From Ref. 21.)

Some guidelines follow:

- A voice channel with psophometric weighting has a 2.5-dB noise improvement over a flat channel.
- dBrnC (dB above reference noise with C-message weighting): The reference frequency/level is a 1000-Hz tone at −90 dBm (Ref. 22).
- pWp (picowatts of noise power with psophometric weighting): The noise power reference frequency/level is an 800-Hz tone where 1 pWp = −90 dBm (pWp = pW × 0.56) (Ref. 22).
- dBmp (psophometrically weighted noise power measured in dBm), where, with an 800-Hz tone, 0 dBmp = 0 dBm. For flat noise in the band 300–3400 Hz, dBmp = dBm − 2.5 dB (Ref. 22).
- 0 dBrnC = −88.5 dBm. Commonly, the −88.5-dBm value is rounded off to −88 dBm; thus 0 dBrnc = −88.0 dBm (Ref. 22).
- dBrnC = dBmp + 90 dB: It should be noted that C-message weighting and psophometric weighting in fact vary by 1 dB (−1.5 dB and −2.5 dB) and this equivalency has an inherent error of 1 dB. However, the equivalency is commonly accepted in the industry (Ref. 22).
- $\text{dBrnC0} = 10 \log \text{pWp0} + 0.8 \text{ dB} = \text{dBmp} + 90.8 \text{ dB} = 88.3 \text{ (dB)} - (S/N)_{\text{dB (flat)}}$ (Ref. 13).

2.8.5 Noise Power Ratio (NPR)

2.8.5.1 Introduction. Up to this point we have dealt only with thermal noise in a radiolink. In an operational analog radiolink a second type of noise can be equally important. This is intermodulation (IM) noise.

IM noise is caused by nonlinearity when information signals in one or more channels give rise to harmonics or intermodulation products that appear as unintelligible noise in other channels. In a FDM/FM radiolink, nonlinear noise in a particular channel varies as the multiplex signal level and the position of the channel in the multiplex baseband spectrum. For a fixed multiplex signal level and for a specific FDM channel, nonlinear noise is constant.

For low receive signal levels at the far end FM receiver, such as during conditions of deep fades, thermal noise limits performance. During the converse condition, when there are high signal levels, IM noise may become the limiting performance factor.

In a FM radio system nonlinear (IM) noise may be attributed to three principal factors: (1) transmitter nonlinearity, (2) multipath effects of the medium, and (3) receiver nonlinearity. Amplitude and phase nonlinearity are equally important in contribution to total noise and each should carefully be considered.

2.8.5.2 *Methods of Measurement.* There are three ways to measure IM noise:

1. One-tone test where harmonic distortion is measured.
2. Two-tone test where specific IM products can be identified with a spectrum analyzer and levels accurately measured with a frequency-selective voltmeter.
3. White noise loading test where, among other parameters, NPR can be measured.

Method 3 will subsequently be described and its application discussed.

2.8.5.3 *White Noise Method of Measuring IM Noise in FDM/FM Radiolinks.* A common method of measuring total noise on a FM radiolink under maximum (traffic) loading conditions consists of applying a "white noise" signal at the baseband input port of the FM transmitter. A white noise generator produces a noise spectrum approximating that produced by the FDM equipment. The output noise level of the generator is adjusted to a desired multiplex composite baseband level (composite noise power). Then a notched filter is switched in to clear a narrow slot in the spectrum of the noise signal and a noise analyzer is connected to the output of the system. The analyzer can be used to measure the ratio of the composite noise power to the noise power in the cleared slot. The noise power is equivalent to the total noise (e.g., thermal plus IM noise) present in the slot bandwidth. Conventionally, the slot bandwidth is made equal to that of a single FDM voice channel. A typical white noise test setup is shown in Figure 2.28.

The most common unit of noise measurement in white noise testing is NPR, which is defined as follows (Ref. 22):

> NPR is the decibel ratio of the noise level in a measuring channel with the baseband fully loaded . . . to . . . the level in that channel with all of the baseband noise loaded except the measuring channel.

The notched (slot) filters used in white noise testing have been standardized for radiolinks by CCIR in Rec. 399-3 (Ref. 23) for ten common FDM baseband configurations. Available measuring channel frequencies and high and low baseband cutoff frequencies are shown in Table 2.7. Table 2.8 gives the stop-band (slot) filter rolloff characteristics. In a NPR test usually three different slots are tested separately: high frequency, midband, and low frequency.

When a NPR measurement is made at high RF signal levels, such as when the measurement is made in a back-to-back configuration, the dominant noise component is equipment IM noise. This parameter can be used as an approximation of the equipment IM noise contribution. This value together

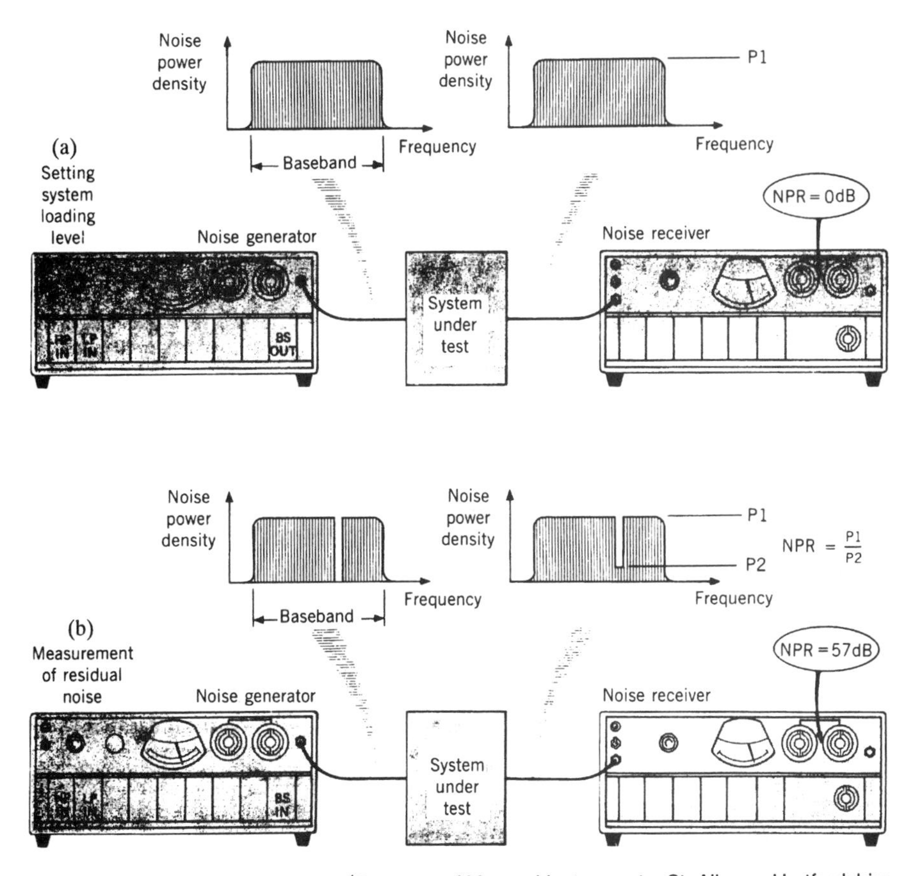

Figure 2.28. White noise testing. (Courtesy of Marconi Instruments, St. Albans, Hertfordshire, England; Ref. 22.)

with stated equivalent noise loading should also be available from manufacturer's published specifications on the equipment to be used. Modern, new radiolink terminal equipment (i.e., equipment that accepts an information baseband for modulation and demodulates a RF signal to baseband) should display a NPR of at least 55 dB when tested back-to-back. It should be noted, however, that diversity combining can improve NPR. This is because in equal gain and maximal ratio combiners the signal powers are added coherently, whereas the IM noise contribution, which is similar (for this discussion) to other noise, is added randomly. Reference 5 allows a 3-dB improvement in NPR when diversity combining is used on a link.

Up to this point NPR has been treated for terminal radio equipment or baseband repeaters. If the designer is concerned with heterodyne repeaters (Chapter 10), the white noise test procedure as previously described cannot be carried out per se, and the designer should rely on manufacturer's

TABLE 2.7 CCIR Measurement Frequencies for White Noise Testing

System Capacity (Channels)	Limits of Band Occupied by Telephone Channels (kHz)	Effective Cutoff Frequencies of Band-Limiting Filters (kHz)		Frequencies of Available Measuring Channels (kHz)
		High Pass	Low Pass	
60	60–300	60 ± 1	300 ± 2	70 270
120	60–552	60 ± 1	552 ± 4	70 270 534
300	{ 60–1,300 64–1,296 }	60 ± 1	1,296 ± 8	70 270 534 1,248
600	{ 60–2,540 54–2,660 }	60 ± 1	2,600 ± 20	70 270 534 1,248 2,438
960	{ 60–4,028 64–4,024 }	60 ± 1	4,100 ± 30	70 270 534 1,248 2,438 3,886
900	316–4,188	316 ± 5	4,100 ± 30	534 1,248 2,438 3,886
1,260	{ 60–5,636 60–5,564 }	60 ± 1	5,600 ± 50	70 270 534 1,248 2,438 3,886 5,340
1,200	316–5,564	316 ± 5	5,600 ± 50	534 1,248 2,438 3,886 5,340
1,800	{ 312–8,120 312–8,204 316–8,204 }	316 ± 5	8,160 ± 75	534 1,248 2,438 3,886 5,340 7,600
2,700	{ 312–12,336 316–12,388 312–12,388 }	316 ± 5	12,360 ± 100	534 2,438 1,248 3,886 5,340 7,600 11,700

Source: CCIR Rec. 399-3 (Ref. 23).

TABLE 2.8 CCIR Stop-Band Filter Rolloff Characteristics

Center Frequency f_c (kHz)	Bandwidth (kHz), in Relation to f_c, Over Which the Discrimination Should Be at Least				Bandwidth (kHz), in Relation to f_c, Outside Which the Discrimination Should Not Exceed	
	70 dB	55 dB	30 dB	3 dB	3 dB	0.5 dB
	{	±2.2	±3.5	—	±12	±18
70	±1.5	±1.7	±2.0	—	±5	±10
270	±1.5	±2.3	±2.9	—	±8	±24
534	±1.5	±3.5	±7.0	—	±15	±48
1,248	±1.5	±4.0	±11.0	—	±35	±110
2,438	±1.5	±4.5	±19	—	±60	±220
	{ ±1.5	±15.0	±30.0	—	±110	±350
3,886	±1.5	±1.8	±3.5	±8.0	±12	±100
5,340	±1.5	±2.2	±4.0	±8.5	±14	±150
7,600	±1.5	±2.4	±4.6	±9.5	±16	±200
11,700	±1.5	±3.0	±7.0	±11.0	±20	±300

Source: CCIR Rec. 399-3 (Ref. 23).

specifications. Alternatively, about 4-dB improvement (Ref. 5) in NPR over baseband radio equipment may be assumed, or for a new IF repeater of modern design, a NPR of at least 59 dB should be achieved.

2.8.5.4 *Application*

2.8.5.4.1 Thermal Noise Measurement. A distinction has been made between thermal (idle) noise, which is not a function of traffic loading level, and IM noise, which is caused by system nonlinearity. It can be seen that in Figure 2.29, with the removal of the white noise loading, the remaining noise being recorded in the noise receiver is thermal noise (assuming that there are no spurious signals being generated).

Thermal or idle noise in a test channel is defined by BINR (baseband intrinsic noise ratio), which is the decibel ratio of noise in a test channel with the baseband fully loaded and stop band or slot filters disconnected to the noise in the test channel with all noise loading removed.

The difference between NPR and BINR therefore indicates the amount of noise present in the system due to IM noise and crosstalk.

2.8.5.4.2 Deriving Signal-to-Noise Ratio and IM Noise Power. Specifying the noise in a test channel by NPR provides a relative indication of IM noise and crosstalk. An alternative is to express the noise in decibels relative to a specified absolute signal level in a test channel. In this case we can define the signal-to-noise power ratio (S/N) as the decibel ratio of the level of the standard test tone to the noise in a standard channel bandwidth (3100 Hz)

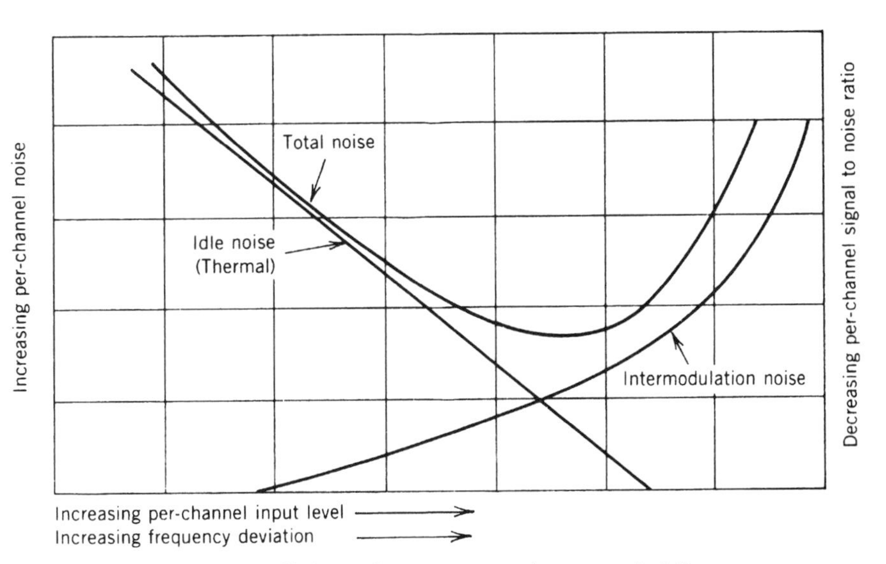

Figure 2.29. Noise performance versus frequency deviation.

within the test channel or

$$\left(\frac{\mathrm{S}}{\mathrm{N}}\right)_{\mathrm{dB}} = \mathrm{NPR} + \mathrm{BWR} - \mathrm{NLR} \tag{2.52}$$

where NLR (noise load ratio) is defined in formulas (2.20)–(2.22) or

$$\mathrm{NLR} = 10\log\left(\frac{\text{equivalent noise test load power}}{\text{voice channel test tone power}}\right) \tag{2.53}$$

$$\mathrm{BWR} = 10\log\left(\frac{\text{occupied baseband of white noise test signal}}{\text{voice channel bandwidth}}\right) \tag{2.54}$$

where BWR is bandwidth ratio.

Example 23. A particular FM radiolink transmitter and receiver back-to-back display a NPR of 55 dB. They have been designed and adjusted for 960 VF channel operation and will use CCIR loading. What is the S/N in a voice channel?

Consult Table 2.7. The baseband occupies 60–4028 kHz.

$$\begin{aligned}
\mathrm{BWR} &= 10\log[(4028 - 60)/3.1] \quad (\mathrm{kHz}) \\
&= 10\log(3968/3.1) \\
&= 31.07\ \mathrm{dB} \\
\mathrm{NLR} &= -15 + 10\log(960) \\
&= 14.82\ \mathrm{dB} \\
\mathrm{S/N} &= 55\ \mathrm{dB} + 31.07\ \mathrm{dB} - 14.82\ \mathrm{dB} \\
&= 71.25\ \mathrm{dB}
\end{aligned}$$

To calculate flat noise in the test channel, the following expression applies:

$$P_{\mathrm{tcf}} = \log^{-1}\left(\frac{90 - (\mathrm{S/N})_{\mathrm{dB}}}{10}\right) \quad (\mathrm{pW0}) \tag{2.55}$$

and to calculate psophometrically weighted noise:

$$P_{\mathrm{tcp}} = 0.56\log^{-1}\left(\frac{90 - (\mathrm{S/N})_{\mathrm{dB}}}{10}\right) \quad (\mathrm{pWp0}) \tag{2.56}$$

Example 24. If S/N in voice channel is 71.25 dB, what is the noise level in that channel in pWp0?

Use equation (2.56).

$$\begin{aligned} P_{\text{tcp}} &= 0.56 \log^{-1}\left(\frac{90 - 71.25}{10}\right) \\ &= 0.56 \times 74.99 \\ &= 41.99 \text{ pWp0} \end{aligned}$$

The results of the calculations in this subsection provide the input for the radio equipment IM noise contribution for total noise in the voice channel under unfaded conditions.

Table 2.9 gives approximate equivalents for S/N in the standard voice channel and its respective values of noise power in dBrnc0 and pWp0.

2.8.5.5 *Balance Between IM Noise and Thermal Noise.* After a path analysis is completed, the designer may find that the specified worst channel noise requirements have not been met. One alternative would be to move the reference operating point as described in Section 2.8.3. Another alternative would be to adjust the frequency deviation to ensure its optimum operating point.

We know that as the input level to a FM transmitter is increased, the deviation is increased. As the deviation is increased, the FM improvement threshold becomes more apparent. In effect, we are trading off bandwidth for thermal noise improvement. However, with increasing input levels, the IM noise of the system increases. There is some point of optimum input to a wideband FM transmitter where the thermal noise in the system and the IM noise level have been balanced so the total noise has been optimized. This concept is illustrated in Figure 2.29.

2.8.5.6 *Standardized FM Transmitter Loading for FDM Telephony.* Several standards have evolved, mostly based on the work done by Holbrook and Dixon of Bell Telephone Laboratories, for wideband FM transmitter loading of FDM baseband signals. These standard formulas assume that the FDM voice channels will carry telephony speech traffic with a 25% activity factor. From the formula(s) we derive the total system loading in dBm0, which would be used for white noise testing and for link analysis. The formulas are

$$P_{\text{tl}} = -15 + 10 \log N \quad \text{(CCIR)} \tag{2.57}$$

$$P_{\text{tl}} = -16 + 10 \log N \quad \text{(North American practice)} \tag{2.58}$$

where P_{tl} is the test load power level in dBm0 and N is the number of FDM channels. Equations (2.57) and (2.58) are valid only when N is equal to or greater than 240 channels.

TABLE 2.9 Approximate Equivalents for S / N and Common Noise Level Values[a]

dBrnc0	pWp0	dBm0p	S/N (flat)	NPR	dBrnc0	pWp0	dBm0p	S/N (flat)	NPR
0	1.0	−90	88	71.6	30	1,000	−60	58	41.6
1	1.3	−89	87	70.6	31	1,259	−59	57	40.6
2	1.6	−88	86	69.6	32	1,585	−58	56	39.6
3	2.0	−87	85	68.6	33	1,995	−57	55	38.6
4	2.5	−86	84	67.6	34	2,520	−56	54	37.6
5	3.2	−85	83	66.6	35	3,162	−55	53	36.6
6	4.0	−84	82	65.6	36	3,981	−54	52	35.6
7	5.0	−83	81	64.6	37	5,012	−53	51	34.6
8	6.3	−82	80	63.6	38	6,310	−52	50	33.6
9	7.9	−81	79	62.6	39	7,943	−51	49	32.6
10	10.0	−80	78	61.6	40	10,000	−50	48	31.6
11	12.6	−79	77	60.6	41	12,590	−49	47	30.6
12	15.8	−78	76	59.6	42	15,850	−48	46	29.6
13	20.0	−77	75	58.6	43	19,950	−47	45	28.6
14	25.2	−76	74	57.6	44	25,200	−46	44	27.6
15	31.6	−75	73	56.6	45	31,620	−45	43	26.6
16	39.8	−74	72	55.6	46	39,810	−44	42	25.6
17	50.1	−73	71	54.6	47	50,120	−43	41	24.6
18	63.1	−72	70	53.6	48	63,100	−42	40	23.6
19	79.4	−71	69	52.6	49	79,430	−41	39	22.6
20	100	−70	68	51.6	50	100,000	−40	38	21.6
21	126	−69	67	50.6	51	125,900	−39	37	20.6
22	158	−68	66	49.6	52	158,500	−38	36	19.6
23	200	−67	65	48.6	53	199,500	−37	35	18.6
24	252	−66	64	47.6	54	252,000	−36	34	17.6
25	316	−65	63	46.6	55	316,200	−35	33	16.6
26	398	−64	62	45.6	56	398,100	−34	32	15.6
27	501	−63	61	44.6	57	501,200	−33	31	14.6
28	631	−62	60	43.6	58	631,000	−32	30	13.6
29	794	−61	59	42.6	59	794,300	−31	29	12.6

[a] This table is based on the following commonly used relationships, which include some rounding off for convenience: Correlations between columns 3 and 4 are valid for all types of noise. All other correlations are valid for white noise only. dBrnc0 = $10 \log_{10}$ pWp0 = dBm0p + 90 = 88 − S/N (flat) = 7.16 − NPR.

Source: Extracted from EIA 252-A (Ref. 20).

For systems with 12–240 FDM channels, the following formula applies:

$$P_{\mathrm{tl}} = -1 + 4 \log N \quad \text{(CCIR)} \tag{2.59}$$

It should be stressed that the loading level values derived are for systems carrying speech telephony with the 25% activity factor. There is allowance for pilot tones and a small number of data channels. These data channels would have a test tone level of −13 dBm0, constant amplitude, and 100% activity factor.

The U.S. Department of Defense (MIL-STD-188-100, Ref. 24) (also see Ref. 22) uses a loading formula that permits unrestricted data loading:

$$P_{\mathrm{tl}} = -10 + 10 \log N \quad (\mathrm{dBm0}) \tag{2.60}$$

If a system is to be designed for CCIR or North American standard loading and will carry a significant number of data/telegraph channels, it is good practice to calculate each category separately and sum the power levels. Assume -13 dBm0 loading for each data and composite telegraph channel. Now suppose a system were to carry 900 FDM VF channels of which 800 are dedicated to conventional telephony and 100 for data/telegraph; then for the telephony portion

$$\begin{aligned} P_{\mathrm{tl}} &= -15 + 10 \log 800 \quad (\mathrm{dBm0}) \\ &= +14\ \mathrm{dBm0} \quad \text{or} \quad 25\ \mathrm{mW} \end{aligned}$$

and for the data/telegraph portion

$$\begin{aligned} P_{\mathrm{tl}} &= -13 + 10 \log 100 \quad (\mathrm{dBm0}) \\ &= +7\ \mathrm{dBm0} \quad \text{or} \quad 5\ \mathrm{mW} \end{aligned}$$

Add the milliwatt values (i.e., 25 + 5 = 30); then

$$\begin{aligned} P_{\mathrm{tl}} &= 10 \log 30 \\ &= +14.77\ \mathrm{dBm0} \quad \text{composite loading} \end{aligned}$$

2.8.5.7 *Standardized FM Transmitter Loading for TV.* The FM transmitter is loaded with an input power level necessary to achieve a peak deviation of ± 4 MHz (8 MHz total) referred to the nominal peak-to-peak amplitude of the video-frequency signal. (Refer to CCIR Rec. 276-2, Ref. 25.)

2.8.6 Antenna Feeder Distortion

Antenna feeder distortion or echo distortion is caused by mismatches in the transmission line connecting the radio equipment to the antenna. These mismatches cause echos or reflections of the incident wave. Similar distortion can be caused by long IF runs; however, in most cases, this can be neglected.

Echo distortion actually results from a second signal arriving at the receiver but delayed in time by some given amount. It should be noted that multipath propagation may also cause the same effect. In this case, though, the delay time is random and continuously varying, thus making analysis difficult, if not impossible.

The level of the echo signal is an inverse function of the return loss at each end of the transmission line and its terminating device (i.e., the antenna

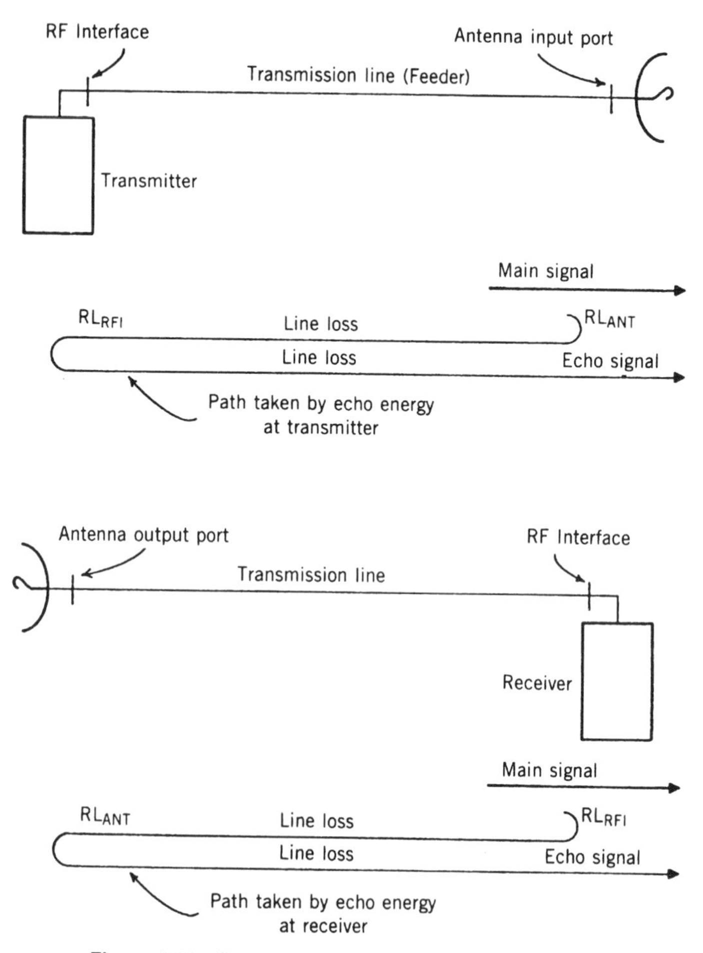

Figure 2.30. Illustration of principal signal echo paths.

at one end and the communication equipment at the other). An echo signal so generated will be constant, since the variables that established it are constant. Thus the distortion created by the echo will be constant but contingent on modulation. In other words, if the carrier were unmodulated, there would be no distortion due to echo. When the carrier is modulated, echo appears. Figure 2.30 illustrates echo paths on a transmission line.

Estimation of echo distortion, expressed as signal-to-distortion ratio, is at best an approximation. One of the most straightforward methods, as described in Ref. 13, is presented below. Inputs required for its calculation are the type and length of transmission line, usually waveguide for the applications covered in this text, the return losses (or VSWR) of the communications equipment, and the antenna return loss (or VSWR). These values should be provided in the equipment specifications, but typically are

Antenna VSWR:	1.05:1 to 1.2:1
Waveguide VSWR:	1.03:1 to 1.15:1
Equipment return loss:	26–32 dB

Calculations for echo distortion are performed separately for each end of the link. Procedures for its calculation are as follows.

From Figure 2.31, using the value for the corresponding length of waveguide run in feet and FDM channel configuration, determine the echo distortion gross contribution to noise. The value must now be reduced by two times the waveguide losses. (The echo signal travels up to the antenna and back down again as in Figure 2.31.) This noise value (in dBrnc0 or dBmp0) is further reduced by the equipment return loss and the composite antenna waveguide return loss.

To calculate return loss (dB) from VSWR, proceed as follows:

$$\mathrm{RL}_{\mathrm{dB}} = 20 \log(1/\rho) \tag{2.61}$$

where ρ is the reflection coefficient, and

$$\rho = \frac{\mathrm{VSWR} - 1}{\mathrm{VSWR} + 1} \tag{2.62}$$

Example 25. The signal-to-distortion ratio (S/D) expressed in dBrnc0 for a 1200-channel system operating at 6 GHz with 60 ft of rectangular waveguide is 66 dBrnc0; waveguide loss is 1.05 dB; equipment return loss is 30 dB; and the waveguide/antenna composite return loss is 26 dB (equivalent to a VSWR of 1.1:1). Calculate the noise contribution in dBrnc0 for the worst VF channel (top channel) for this end of the link.

P_e is the value for the echo distortion noise contribution:

$$\begin{aligned} P_e &= 66 - 2.1\ \mathrm{dB} - 30\ \mathrm{dB} - 26\ \mathrm{dB} \\ &= 7.9\ \mathrm{dBrnc0} \end{aligned}$$

Calculate the value in dBm0p. The value from Figure 2.31 is -24.5 dBm0p:

$$\begin{aligned} P_e &= -24.5 - 2.1\ \mathrm{dB} - 30\ \mathrm{dB} - 26\ \mathrm{dB} \\ &= -82.6\ \mathrm{dBm0p} \end{aligned}$$

Calculate the equivalent value in pWp:

$$\begin{aligned} P_e &= \log^{-1}\left(\frac{-82.6}{10}\right) \times 10^9 \\ &= 5.62\ \mathrm{pWp} \end{aligned}$$

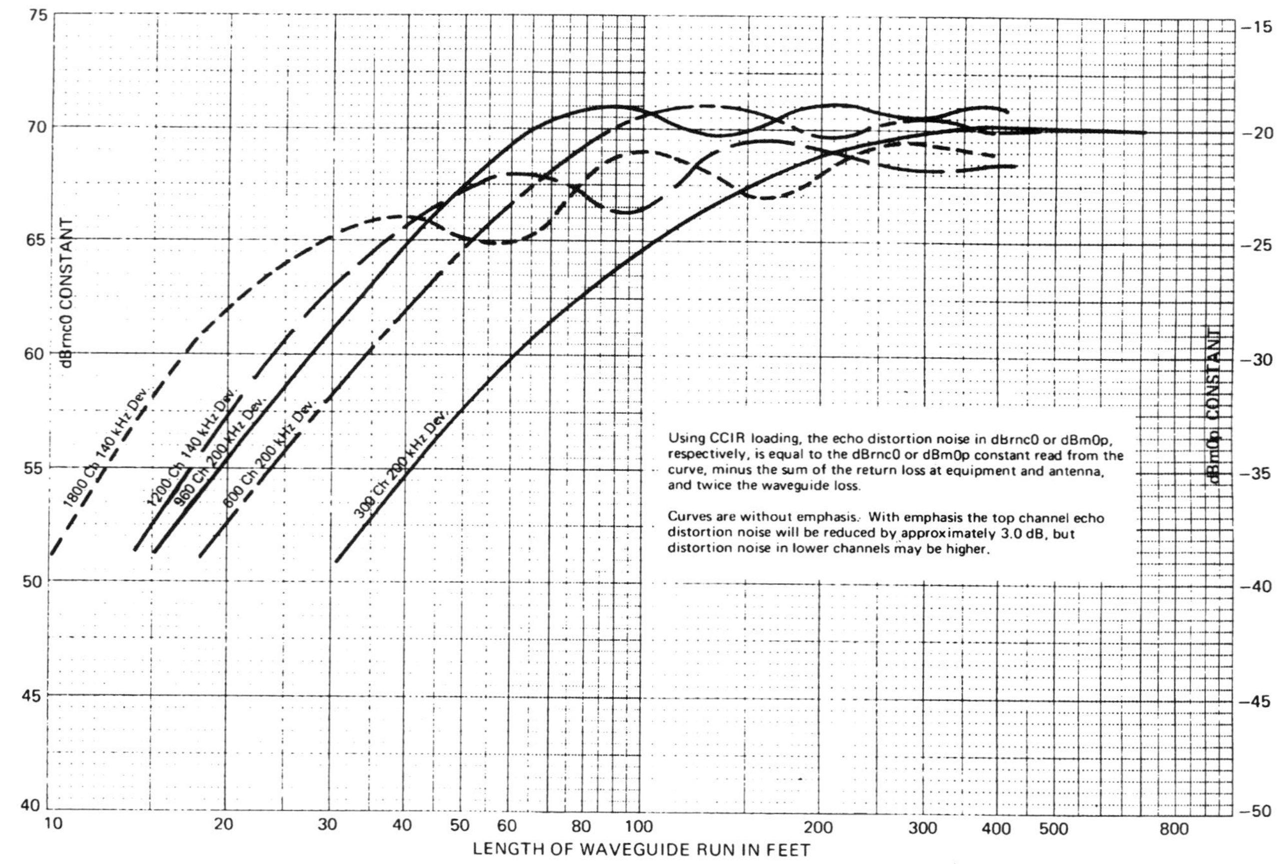

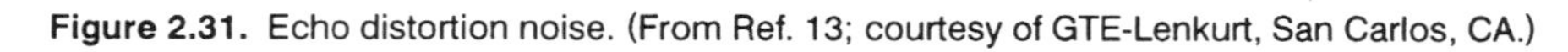

Figure 2.31. Echo distortion noise. (From Ref. 13; courtesy of GTE-Lenkurt, San Carlos, CA.)

If exactly the same parameters were valid for the other end of the link, we could double the value above (i.e., 5.62×2 pWp) and 11.24 pWp would then be added arithmetically to the value calculated for the worst VF channel noise in Section 2.8.4 or 2.8.5.

2.8.7 Total Noise in the Voice Channel

To calculate the total noise in the voice channel, convert the thermal noise at the unfaded RSL (Section 2.8.4) and intermodulation noise (Sections 2.8.5 and 2.8.6) to absolute values of noise power such as mW, pW, or pWp and add, reconverting the sum to decibels if desired.

2.8.8 Signal-to-Noise Ratio for TV Video

Radiolinks carrying exclusively video or video and a program channel are generally thermal noise limited. Here we are dealing with a broadband S/N, which is defined as the ratio of the peak-to-peak signal to the rms thermal noise in the video baseband. The value for S/N is a function of the RSL, the receiver noise figure, the video bandwidth and peak deviation, emphasis (when used), and the weighting function.

The following relations give a value of S/N in decibels. In all cases, assume a peak deviation of ± 4 MHz. For North American systems, the video baseband width is 4.3 MHz and for other systems the baseband bandwidth varies from 4 to 10 MHz.

For North American video systems, the following relations may be used (Ref. 13):

$$S/N = \text{RSL} - \text{NF}_{\text{dB}} + 118 \quad \text{(unweighted, unemphasized)} \tag{2.63}$$

$$S/N = \text{RSL} - \text{NF}_{\text{dB}} + 126.5 \quad \text{(EIA emphasis, EIA color weighting)} \tag{2.64}$$

The following was taken from Ref. 13, which states that these relations are valid for monochrome TV transmission only and cover television systems likely to be found in countries outside North America:

$$S/N = \text{RSL} - \text{NF}_{\text{dB}} + A \tag{2.65}$$

where RSL is the receive signal level in dBm and NF is the receiver noise figure in dB. A is made up of two terms (constants). The first constant represents the unemphasized, unweighted S/N value. The second constant term is the combined effect of emphasis and weighting:

$$A = 115.7 + 17.3 \quad \text{(525 lines, 4 MHz—Japan)} \tag{2.65a}$$

$$A = 112.8 + 16.2 \quad \text{(625 lines, 5 MHz)} \tag{2.65b}$$

$$A = 110.5 + 18.1 \quad \text{(625 lines, 6 MHz)} \tag{2.65c}$$

$$A = 112.8 + 13.5 \quad \text{(819 lines, 5 MHz)} \tag{2.65d}$$

$$A = 103.8 + 16.1 \quad \text{(819 lines, 10 MHz)} \tag{2.65e}$$

2.9 PATH ANALYSIS WORKSHEET AND EXAMPLE

2.9.1 Introduction

A path analysis worksheet is a useful tool for carrying out a path analysis or "link budget." The bottom line, so to speak, of the worksheet for an analog radiolink is the noise in the worst FDM channel, usually the highest or top voice channel in the baseband.

The worksheet sets out in tabular form the required calculations starting with transmitter power output, the various losses and gains from the transmitter outward through the medium, receive antenna system to the input port of the far end receiver, thence the receiver characteristics, pre/de-emphasis and diversity improvements, and then carries on down to the noise in the worst channel. An equivalent worksheet could be used for TV transmission to derive weighted S/N. These values are then compared to the specified values (Section 2.2.1) and the link parameters are then reviewed and, if necessary, adjusted accordingly.

2.9.2 Sample Worksheet

A sample worksheet is provided in Table 2.10. It is filled in for a hypothetical path where it is assumed that a path profile has been completed beforehand and site surveys have been carried out. The far right-hand column of the table gives reference section numbers in the text. These explanatory notes follow the table.

Specific equipment implementation for such a configuration is described in Chapter 14.

TABLE 2.10 Path Analysis Worksheet

No.	Item	Value	Unit	Reference
	System Identifier *Amber* Link *Charlie* to *Delta*			
	Signal type (TV, FDM, composite) *FDM* FDM chan. *1200*			
	Loading *+15.79* (dBm0) Baseband config. *60-5564*			(2.8.5.6)
	Frequency *6100* MHz			(2.10)
	Spec worst chan median noise − S/N *400* (dBrnc0/pWp0)			(2.2)
1.	SITE (A) *Charlie* TO SITE (B) *Delta*			
2.	Lat/Long (A) *43° 43′ N/90° 50′ W* Lat/Long (B) *44° 11′ N/90° 20′ W*			
3.	Path length *41.29* (~~km~~/sm)			(2.3)
4.	Site elevation (A) *185* (~~m~~/ft) (B) *250* (~~m~~/ft)			(2.3)
5.	Tower Height (A) *110* (~~m~~/ft) (B) *95* (~~m~~/ft)			(2.3)
6.	Azimuth from true north (A) *37° 21′* (B) *217° 42′*			(2.3)
7.	Transmitter power output at flange (A) *1.0* W; *0* dBW			(2.6.2)
8.	Transmission line losses (A):			
	8A. W/G type, *EW-64* W/G length *110* W/G loss	*1.67*	dB	(14.3.2.2)
	8B. Flex guide loss	*0.1*	dB	(14.3.2.2)
	8C. Transition/connector losses	*0.4*	dB	(14.3.2.3)
	8D. Directional cplr loss	*0.2*	dB	(14.3.2.3)
	8E. Circulator or hybrid losses	*0.5*	dB	(14.3.2.3)
	8F. Radome loss	*0.6*	dB	(14.3.2.3)
	8G. Other losses	*0.0*	dB	(14.3.2.3)
	8H. Total transmission line losses	*3.47*	dB	(2.6.2)
9.	Antenna (A) diameter *8* (~~m~~/ft) gain	*41.25*	dB	(2.6.7)
10.	EIRP	*+37.38*	dBW	(2.6.2)
11.	Free-space loss (FSL)	*144.6*	dB	(1.2; 2.6.2)
12.	Atmospheric absorption	*0.7*	dB	(2.6.2)
13.	Unfaded isotropic rec. level (B)	*−107.52*	dBW	(2.6.2)
14.	Antenna (B) diameter *8* (~~m~~/ft) gain	*41.25*	dB	(2.6.7)
15.	Transmission line losses (B):			
	15A. W/G type *EW-64* W/G length *95′* W/G loss	*1.45*	dB	(14.3.2.2
	15B. Flexguide loss	*0.13*	dB	(14.3.2.2)
	15C. Transition/connector losses	*0.4*	dB	(14.3.2.3)
	15D. Directional coupler loss	*0.0*	dB	(14.3.2.3)
	15E. Circulator or hybrid losses	*0.5*	dB	(14.3.2.3)
	15F. Radome loss	*0.6*	dB	(14.3.2.3)
	15G. Other losses	*0.0*	dB	(14.3.2.3)
	15H. Total transmission line losses	*3.08*	dB	(2.6.2)
16.	Unfaded RSL	*−69.35*	dBW	(2.6.2)
	Receiver noise threshold calculation (B):			(2.6.3)
	17A. Receiver noise figure	*8*	dB	(2.6.3)
	17B. RMS per channel deviation	*140*	kHz	(2.6.4)(Table 2.4)
	17C. Peak channel deviation	*200*	kHz	(2.9.4)
	17D. Carrier peak deviation	*3855*	MHz	(2.6.4.2)
	17E. Highest baseband frequency	*5.564*	MHz	(Table 2.7)
	17F. B_{IF}	*18.836*	MHz	(2.6.4.1)
	17G. Receiver thermal noise threshold	*−123.25*	dBW	(2.6.3)
18.	Receiver FM improvement threshold	*−113.25*	dBW	(2.8.3)

TABLE 2.10 (*continued*)

19.	Reference threshold	*−103.71* dBW	(2.8.4)
	19A. Diversity improvement/fading	*0* dB	(2.7.4)
20A.	Fade margin w/o diversity	*34.3* dB	(2.7.2)
20B.	Fade margin with diversity	*0* dB	(2.7.4)
21.	C/N unfaded	*53.9* dB	(2.6.6)
22.	Link unavail to ref level *0.01*%, link availability	*99.99* %	(2.7.2)
23.	Calculation of S/N in worst channel		
	24A. Pre-emphasis improvement	*4.5* dB	(2.6.5)
	24B. Diversity unfaded improvement	*0* dB	(2.8.3.4)
	24C. Link calculated S/N	*64.36* dB	(2.8.4)
	24D. NPR	*55* dB	(2.8.5)
	24E. IM noise contribution	*37.43* pWp0	(2.9.5)
	24F. Sum of echo noise contribution	*8.3* pWp0	(2.8.6)
	24G. Thermal noise contribution	*206.1* pWp0	(2.8.4)
	24H. Total noise worst channel	*251.83* pWp0	(2.8.4)
	24I. Specified noise worst channel	*400* pWp0	(2.2)
	24J. Margin	*2.0* dB	

EXPLANATORY NOTES TO TABLE 2.10 (Referenced to title and/or line number)

Loading. This is the level of the input test signal to the FM transmitter. See Section 2.8.5.6. For TV loading, see Section 2.8.5.7. In this case for FDM loading with 1200 VF channels, formula (2.56) was used.

$$P_{tl} = -15 + 10\log 1200$$
$$= +15.79 \text{ dBm0}$$

Spec worst channel median noise. This is the noise allotment for the link that is specified at the outset. The link is 41.29 mi long or 66.4 km. From Section 2.2, 200 pWp + (3 pWp)(66.4) = 400 pWp. The reader is cautioned regarding the initial 200 pWp. It can only be used once for short and medium haul sections up to 840 km long made up of one or more hops or links. Sections between 840 and 1670 km, use 400 pWp + 3 pWp/km and from 1670 to 2500 km, use 600 pWp + 3 pWp/km.

3, 6. Path length and azimuth from each site. For ordinary LOS radiolinks, rhumb line distance and azimuths suffice. Nevertheless, we used great circle distance and azimuths because, in the United States, among other locations, the FCC license application requires great circle azimuths and distances. It will be appreciated that final antenna alignment will be done by a rigger at the direction of the site installation engineer where the antenna on each end will be adjusted for peak RSL before being bolted down.

7. Transmitter output power in dBW (or dBm). This must be specified at a point in the transmitter subsystem where it can be measured conveniently.

8, 15. Transmission lines and related devices are described in Chapter 14. In this case, EW-64 elliptical waveguide was used for the principal waveguide run with about 0.015 dB/ft. Other losses are typical for the example. *8H* and *15H* are the sums of the transmission line losses in decibels.

10. EIRP [equation (2.6)]. The work must be done consistently with either dBm or dBW. In this case sum items 7, 8H, and 9:

$$\text{EIRP}_{\text{dBW}} = 0 \text{ dBW} - 3.47 \text{ dB} + 41.25 \text{ dB} = +37.78 \text{ dBW}$$

11. Free-space loss [equation (1.7)].

$$\text{FSL}_{\text{dB}} = 36.58 + 20\log(6100 \text{ MHz}) + 20\log(41.29)$$
$$= 144.6 \text{ dB}$$

13. Isotropic receive level [equation (2.7)]. Algebraically add items 10, 11, and 12:

$$\text{IRL}_{\text{dBW}} = +37.78 \text{ dBW} - 144.6 - 0.7 = -107.52 \text{ dBW}$$

TABLE 2.10 *(continued)*

EXPLANATORY NOTES TO TABLE 2.10 (Referenced to title and/or line number)

16. Unfaded RSL [equation (2.8)]. Algebraically add items 13, 14, and 15H:

$$\mathrm{RSL}_{\mathrm{dBW}} = -107.52 + 41.25\ \mathrm{dB} - 3.08\ \mathrm{dB} = -69.35\ \mathrm{dBW}$$

17. Receiver noise threshold calculation. First calculate in 17D the peak carrier deviation using rms deviation of 140 kHz/channel from Table 2.4. Use equation (2.25).

$$\begin{aligned}\Delta F_p &= 4.47(140)\left[\log^{-1}\left(\frac{-15 + 10\log 1200}{20}\right)\right]\\ &= 4.47(140)(6.160)\\ &= 3855\ \mathrm{kHz}\end{aligned}$$

Calculate B_{IF} using equation (2.17):

$$\begin{aligned}B_{\mathrm{IF}} &= 2(3855 + 5564)\\ &= 18.838\ \mathrm{MHz}\end{aligned}$$

where 5564 kHz is the highest modulating frequency. Calculate receiver noise threshold using equation (2.16), NF = 8 dB:

$$\begin{aligned}P_t &= -204\ \mathrm{dBW} + 10\log(18.838 \times 10^6) + 8\ \mathrm{dB}\\ &= -204 + 72.75 + 8\\ &= -123.25\ \mathrm{dBW}\end{aligned}$$

18. The FM improvement threshold is the value in 17G + 10 dB or −113.25 dBW.
19. Reference threshold. It is upon this threshold that we add fade margin (item 20). CCIR Rec. 393-1 states that "the additional objective on the 2500 km reference circuit should not exceed 1,000,000 pWO unweighted... for more than 0.01% of any month." 1,000,000 pWO corresponds to 562,000 pWpO (see Section 2.8.4.1). We can calculate the RSL for this value of noise in the voice channel by using equation (2.51) and add the value for emphasis for a 1200-channel system from Figure 2.15. This is 4.5 dB; thus

$$\begin{aligned}562{,}000 &= \log^{-1}\left(\frac{-\mathrm{RSL} - 48.6 + 8 - 4.5 + 28.89}{10}\right)\\ 57.5 &= -\mathrm{RSL} - 48.6 + 8 - 4.5 + 28.89\\ \mathrm{RSL} &= -73.71\ \mathrm{dBm} \quad \text{or} \quad -103.71\ \mathrm{dBW}\end{aligned}$$

20. Calculate fade margin (in this case without diversity). Turn to Section 2.7.2.3 and use equation (2.31). Path length 41.29 mi or 66.4 km = d. d^3 = 292,755. $a = 1$, $b = \frac{1}{4}$, $f = 6.1$, and $P_{mf} = 0.00995\%$. The fade margin is 34.3 dB and then the unfaded RSL (item 16) should be −69.41 dBW. However, it was calculated (using 8-ft aperture antennas) as −69.35 dBW. This latter value will be shown to be the appropriate value for the link in item 23. Availability = 1 − 0.0000995 or 99.99%.
21. C/N unfaded. Use equation (2.26):

$$\begin{aligned}\frac{\mathrm{C}}{\mathrm{N}} &= -69.35\ \mathrm{dBW} - (-123.25\ \mathrm{dBW})\\ &= 53.9\ \mathrm{dB}\end{aligned}$$

or item 16 minus item 17G.
22. The link unavailability from item 20 was established as 0.01%. Link availability is then 1 − 0.0001 or 99.99%. It should be remembered that link time availability/unavailability is system related and is discussed further in Appendix 2.
24A. Pre-emphasis improvement for 1200-channel FDM operation. From Figure 2.15. The value is 4.5 dB, and this value will be used below.

TABLE 2.10 (*continued*)

EXPLANATORY NOTES TO TABLE 2.10 (Referenced to title and/or line number)

24C. Link calculated S/N. This is S/N of this one link alone, not part of the system. Use equation (2.49). This equation does not reflect pre-emphasis improvement:

$$\frac{S}{N} = -39.35 \text{ dBm} + 136.1 - 8 + 20\log(\Delta f/f_c)$$

$$20\log(\Delta f/f_c) = 20\log(200/5564)$$

$$= -28.89$$

$$\frac{S}{N} = -39.35 \text{ dBm} + 136.1 - 8 - 28.89$$

$$= 59.86 \text{ dB}$$

and with emphasis of 4.5 dB,

$$\frac{S}{N} = 64.36 \text{ dB}$$

RSL from equation (2.49) is the unfaded RSL from item 16.
24D. NPR is given by the equipment manufacturer (or measured). Given as 55 dB.
24E. IM noise contribution. Use equation (2.52) where NLR is 15.79 dB.

$$\frac{S}{N} = 55 + 10\log(5564/3.1) - 15.79$$

$$= 55 + 32.54 - 15.79$$

$$= 71.75 \text{ dB}$$

Now use equation (2.56) and substitute 71.75 dB for S/N:

$$P_{tcp} = 0.56\log^{-1}\frac{90 - 71.75}{10}$$

$$= 37.43 \text{ pWp}$$

24F. Sum of echo noise contributions. Use Figure 2.31 to obtain values for the waveguide lengths given, 1200 VF channels with 140 kHz/channel rms deviation. For transmit, the value is −22 (dBm0p) and for receive the value is −23. Take transmission line losses from items 8H and 15H. Use the methodology of Section 2.8.6.

$$P_{et} = -22 - 30 - 26 - 5.74 \text{ dB}$$

$$= -83.74 \text{ dBm0p} \quad \text{or} \quad 4.23 \text{ pWp}$$

For the receive side the value is 4.07 pWp. Sum the two values and the total is 8.3 pWp.

Note: Values for equipment return loss are given by manufacturer as 30 dB and for composite waveguide/antenna return loss as 26 dB.

24G. Thermal noise contribution. Use equation (2.51). Value of RSL is unfaded value. Use pre-emphasis improvement value from 24A or 4.5 dB and modify equation (2.49):

$$P_{pWp0} = \log^{-1}\left(\frac{39.35 - 48.6 + 8 - 4.5 + 28.89}{10}\right)$$

$$= \log^{-1}(23.14/10)$$

$$= 206.1 \text{ pWp0}$$

24H. Total noise worst channel. Sum the values in pWp0 of 24E, 24F, and 24G or total noise = 37.43 + 8.3 + 206.1 = 251.83 pWp. Compare this against the value allotted for the link or 400 pWp0. The margin is derived by calculating the RSL when given the noise value of 400 pWp minus the value from 24H or 251.83 pWp0. The margin is 2.0 dB.

2.10 FREQUENCY ASSIGNMENT, COMPATIBILITY, AND FREQUENCY PLAN

2.10.1 Introduction

A major task in the planning and implementation of a radiolink involves radio frequency assignment, granting of a radio license, EMI (electromagnetic interference) studies for compatibility, and development of a frequency plan. Frequency assignments and licenses to operate radio transmitters are granted by a national regulatory authority such as the Federal Communications Commission in the United States.

Nearly all countries in the world are members of the International Telecommunications Union (ITU) and thus are signatories to the "Radio Regulations" issued by the ITU. The assignment of frequency bands is governed by the Radio Regulations. Because many radiolinks are transnational or are integral parts of extensions of an international network, national regulatory authorities take guidance from CCIR Recommendations.

Table 2.11 from ITU-R Rec. F.746-1 gives basic guidelines on available frequency bands up to 55 GHz and cross-references applicable ITU-R Recommendations.

2.10.2 Frequency Planning—Channel Arrangement*

A radiolink system may be a single-thread, low-capacity system or a multiple-thread, high-capacity system with spurs, or, initially, a single-thread system with future growth requirements. The system must also coexist with other nearby systems, which will be discussed further in Chapter 13.

One objective when drawing up the frequency plan is to minimize interchannel interference or what is also called cosystem interference. Also, the national regulatory authority will require that the system be spectrum conservative.

Figure 2.32 shows a generalized CCIR frequency arrangement for a multiple-thread (i.e., multiple RF channel) link. It will be appreciated that in most applications (some video links may prove the exception, such as a STL) we are working with frequency pairs, a "go" and a "return" channel. A frequency pair is designated, arbitrarily, with f_1 for a "go" channel, and its companion "return" channel is designated f'_1, f_2 and f'_2, and so on.

CCIR practice divides a band (Table 2.11 and Figure 2.33) into two halves separated by a center guard band, which is larger than or equal to the separation or spacing between the center frequencies of two adjacent channels. Thus, from a single site in one direction (i.e., site B to site C), one-half of the band is for transmit ("go") channels and one-half for receive ("return") channels.

*Portions of Section 2.10.2 have been extracted from ITU-R Rec. F.746-1 (Ref. 26).

TABLE 2.11 Radio Frequency Channel Arrangements for Radio-Relay Systems

Band (GHz)	Frequency Range (GHz)	Recommendation ITU-R F-Series	Channel Spacing (MHz)
1.5	1.427–1.53	746, Annex 1	0.5; 1; 2; 3.5
2	1.427–2.69	701	0.5 (pattern)
	1.7–2.1; 1.9–2.3	382	29
	1.7–2.3	283	14
	1.9–2.3	1098	3.5; 2.5 (patterns)
	1.9–2.3	1098, Annexes 1, 2	14
	1.9–2.3	1098, Annex 3	10
	2.3–2.5	746, Annex 2	1; 2; 4; 14; 28
	2.5–2.7	283	14
4	3.8–4.2	382	29
	3.6–4.2	635	10 (pattern)
	3.6–4.2	635, Annex 1	90; 80; 60; 40
5	4.4–5.0	746, Annex 3	28
	4.4–5.0	1099	10 (pattern)
	4.4–5.0	1099, Annex 1	40; 60
	4.54–4.9	1099, Annex 2	40; 20
L6	5.925–6.425	383	29.65
	5.85–6.425	383, Annex 1	90; 80; 60
U6	6.425–7.11	384	40; 20
7	7.425–7.725	385	7
	7.425–7.725	385, Annex 1	28
	7.435–7.75	385, Annex 2	5
	7.11–7.75	385, Annex 3	28
8	8.2–8.5	386	11.662
	7.725–8.275	386, Annex 1	29.65
	7.725–8.275	386, Annex 2	40.74
	8.275–8.5	386, Annex 3	14; 7
10	10.38–10.68	746, Annex 4	5; 2
	10.5–10.68	747, Annex 1	7; 3.5 (patterns)
	10.55–10.68	747, Annex 2	5; 2.5; 1.25 (patterns)
11	10.7–11.7	387, Annexes 1 and 2	40
	10.7–11.7	387, Annex 3	67
	10.7–11.7	387, Annex 4	60
12	11.7–12.5	746, Annex 5, §3	19.18
	12.2–12.7	746, Annex 5, §2	20 (pattern)
13	12.75–13.25	497	28; 7; 3.5
	12.75–13.25	497, Annex 1	35
	12.7–13.25	746, Annex 5, §1	25; 12.5
14	14.25–14.5	746, Annex 6	28; 14; 7; 3.5
	14.25–14.5	746, Annex 7	20
15	14.4–15.35	636	28; 14; 7; 3.5
	14.5–15.35	636, Annex 1	2.5 (pattern)
	14.5–15.35	636, Annex 2	2.5
18	17.7–19.7	595	220; 110; 55; 27.5
	17.7–21.2	595, Annex 1	160
	17.7–19.7	595, Annex 2	220; 80; 40; 20; 10; 6; 5
	17.7–19.7	595, Annex 3	13.75; 20; 110

TABLE 2.11 *(continued)*

Band (GHz)	Frequency Range (GHz)	Recommendation ITU-R F-Series	Channel Spacing (MHz)
23	21.2–23.6	637	3.5; 2.5 (patterns)
	21.2–23.6	637, Annex 1	112 to 3.5
	21.2–23.6	637, Annex 2	28; 3.5
	21.2–23.6	637, Annex 3	28; 14; 7; 3.5
	21.2–23.6	637, Annex 4	50
	21.2–23.6	637, Annex 5	112 to 3.5
	22.0–23.6	637, Annex 1	112 to 3.5
27	24.25–25.25	748	3.5; 2.5 (patterns)
	24.25–25.25	748, Annex 3	56; 28
	25.25–27.5	748	3.5; 2.5 (patterns)
	25.25–27.5	748, Annex 1	112 to 3.5
	27.5–29.5	748	3.5; 2.5 (patterns)
	27.5–29.5	748, Annex 2	112 to 3.5
	27.5–29.5	748, Annex 3	112; 56; 28
31	31.0–31.3	746, Annex 8	25; 50
38	36.0–40.5	749	3.5; 2.5 (patterns)
	36.0–37.0	749, Annex 3	112 to 3.5
	37.0–39.5	749, Annex 1	140; 56; 28; 14; 7; 3.5
	38.6–40.0	749, Annex 2	50
	39.5–40.5	749, Annex 3	112 to 3.5
55	54.25–58.2	1100	3.5; 2.5 (patterns)
	54.25–57.2	1100, Annex 1	140; 56; 28; 14
	57.2–58.2	1100, Annex 2	100

Source: Tables 1 and 2, pp. 92 and 93, ITU-R Rec. F.746-1, 1994 F Series Volume, Part 1 (Ref. 26).

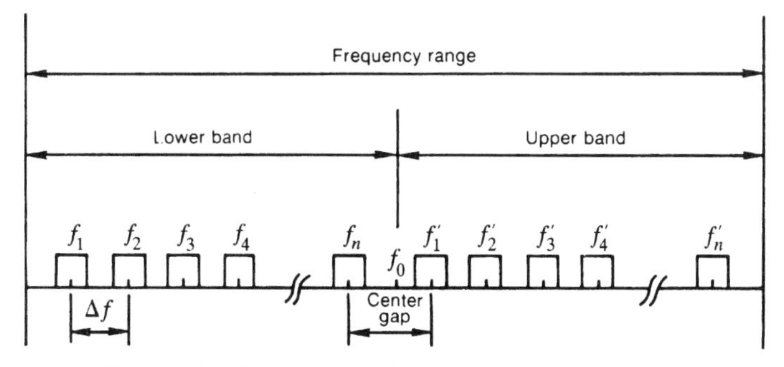

Figure 2.32. A generalized CCIR frequency arrangement.

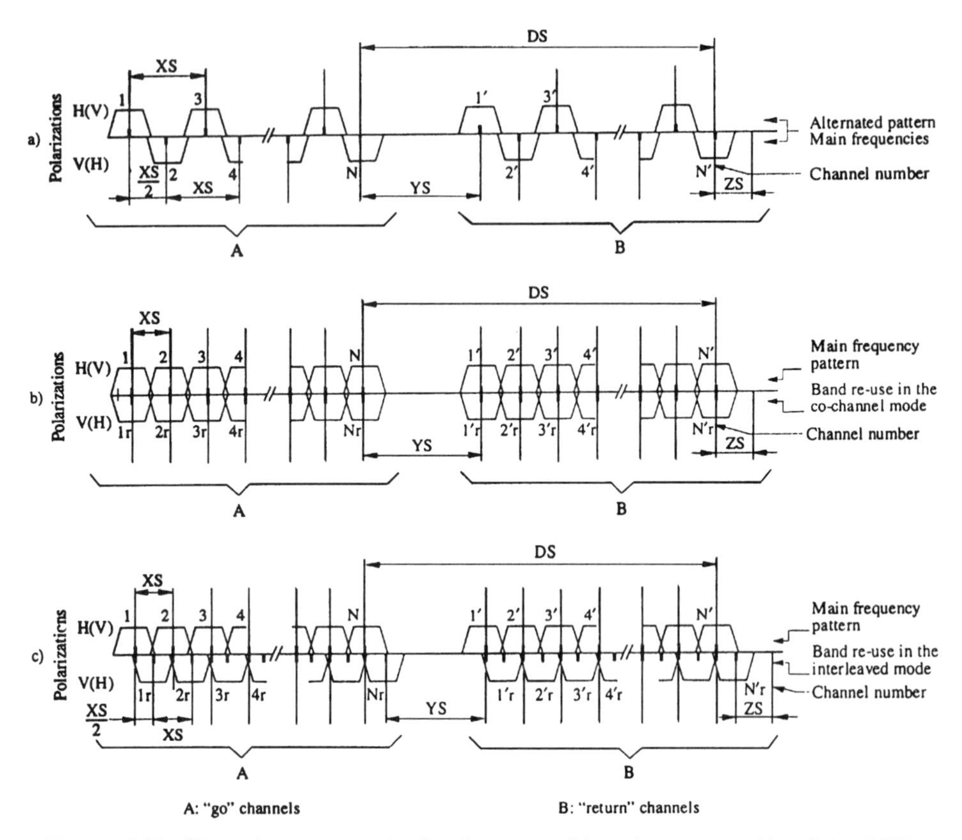

Figure 2.33. Channel arrangements for three possible schemes considered in CCIR radio-frequency channel arrangement. (From Figure 1, CCIR Rec. F.746-1, 1994 F Series Volume, Part 1; Ref. 26.)

Where two or more RF channels are to be provided over a route, frequencies should first be assigned from the odd-numbered group of channels or from the even-numbered group of channels, but not from both, since this would require the use of two antennas at each end of each section. As an example, when all the channels from the odd-numbered group have been assigned, further expansion from the even-numbered group would be provided by means of a second antenna with polarization orthogonal to that of the first antenna.

The main parameters affecting the choice of radio-frequency channel arrangements are as follows (see Figure 2.33):

XS Defined as the radio-frequency separation between the center frequencies of adjacent radio-frequency channels on the same polarization and in the same direction of transmission.

YS Defined as the radio-frequency separation between the center frequencies of the go and return radio-frequency channels that are nearest to each other. In the case where go and return frequency subbands are not contiguous, such that there is a (are) band(s) allocated for (an)other service(s) in the gap between, *YS* shall be considered to include the band separation (BS) equal to the total width of the allocated band(s) used by this (these) service(s).

ZS Defined as the radio-frequency separation between the center frequencies of the outermost radio-frequency channels and the edge of the frequency band. In the case where the lower and upper separations differ in value, Z_1S refers to the lower separation and Z_2S refers to the upper separation. In the case where go and return frequency subbands are not contiguous, such that there is a (are) band(s) allocated for (an)other service(s) in the gap between, ZS_i will be defined for the innermost edges of both subbands and will be included in *YS*.

DS Tx/Rx duplex spacing, defined as the radio-frequency separation between corresponding go and return channels, constant for each couple of ith and i'th frequencies, within a given channel arrangement.

The choice of radio-frequency channel arrangement depends on the values of cross-polar discrimination (XPD) and on the net filter discrimination (NFD) where these parameters are defined as follows:

$$\mathrm{XPD}_{H(V)} = \frac{\begin{array}{c}\text{power received on polarization } H(V)\\ \text{transmitted on polarization } H(V)\end{array}}{\begin{array}{c}\text{power received on opposite polarization } V(H)\\ \text{transmitted on polarization } H(V)\end{array}} \tag{2.66}$$

$$\mathrm{NFD} = \frac{\text{adjacent channel received power}}{\begin{array}{c}\text{adjacent channel power received by the main receiver}\\ \text{after RF, IF, and BB filters}\end{array}} \tag{2.67}$$

The XPD and NFD parameters (dB) contribute to the value of carrier-to-interference ratio.

If $\mathrm{XPD}_{\min}$ is the minimum value reached for the percentage of time required, from this value and from the adjacent channel NFD, the total amount of interfering power can be evaluated, and this result must be compared with the minimum value of carrier-to-interference $(C/I)_{\min}$ acceptable to the modulation adopted.

Alternated channel arrangements can be used (neglecting the co-polar adjacent channel interference contribution) if:

$$\mathrm{XPD}_{\min} + (\mathrm{NFD} - 3) \geq (\mathrm{C/I})_{\min} \quad \mathrm{dB} \tag{2.68}$$

Co-channel arrangements can be used if:

$$10 \log \frac{1}{\dfrac{1}{10^{\frac{XPD+XIF}{10}}} + \dfrac{1}{10^{\frac{NFD_a-3}{10}}}} \geq \left(\frac{\mathrm{C}}{\mathrm{I}}\right)_{\min} \quad \mathrm{dB} \tag{2.69}$$

Interleaved channel arrangements can be used if:

$$10 \log \frac{1}{\dfrac{1}{10^{\frac{XPD+(NFD_b-3)}{10}}} + \dfrac{1}{10^{\frac{NFD_a-3}{10}}}} \geq \left(\frac{\mathrm{C}}{\mathrm{I}}\right)_{\min} \quad \mathrm{dB} \tag{2.70}$$

where NFD_a = net filter discrimination evaluated at XS frequency spacing

NFD_b = net filter discrimination evaluated at $XS/2$ frequency spacing

XIF = XPD improvement factor of any cross-polar interference countermeasure, if implemented in the interfered receiver

The channel arrangements reported in Figure 2.33 may be used for digital radio-relay systems either with single-carrier or multi-carrier transmission.

2.10.3 Some Typical ITU-R Channel Arrangements

2.10.3.1 Channel Arrangements Based on a Homogeneous Pattern for Systems Operating in the 4-GHz Band. Table 2.12 provides an overview of channel arrangements for the 3600–4200-MHz band. Figure 2.34 shows channel arrangements for the band 3700–4200 MHz for six go and six return channels (Group 1) and an interleaved pattern of six go and six return channels (Group 2). Each channel accommodates up to 1260 telephone channels or equivalent, or up to 90 Mbps.

Definitions are as follows:

f_r Frequency of the lower edge of the band of frequencies occupied (MHz)

f_n Center frequency of one radio-frequency channel in the go (return) channel of the band (MHz)

f_n' Center frequency of one radio-frequency channel in the return (go) channel of the band (MHz)

TABLE 2.12 Radio-Frequency Channel Arrangements for the 4-GHz Band

Modulation (Capacity per Channel)	16-QAM[a] (200 Mbit/s)	16-QAM (155.52 Mbit/s) 256-QAM (311.04 Mbit/s)	256-QAM (311.04 Mbit/s)
Frequency band (MHz)	3600–4200	3600–4200	3600–4200
Center frequency of the band f_0 (MHz)	3900	3900	3900
Center frequency of the carriers f_n (MHz)	$f_0 \pm (30 + 40n)$ $n = 0, 1, \ldots, 6$	$f_0 \pm 20n$ $n = 1, 2, \ldots, 14$	$f_0 \pm (15 + 10n)$ $n = 0, 1, \ldots, 27$
Interleaved or co-channel	Interleaved	Co-channel	Co-channel
Transmission method	Single carrier transmission method Sbandwidth/carrier)	Three-carrier transmission method (20 MHz bandwidth/carrier)	Six-carrier transmission method (10 MHz
Number of channels	7	10[b]	10[b]
Channel bandwidth			
XS (MHz)	80	60	60
X	1.6	1.54	1.54
Center gap			
YS (MHz)	60	60	60
Y	1.2	1.54	1.54
Guard space			
ZS (MHz)	30	40	40
Z	0.6	1.03	1.03

[a] *M*-QAM systems are discussed in Chapter 3.
[b] The capacity of the innermost radio-frequency channels is limited to two-thirds of the full capacity.
Source: Table 1, p. 115, CCIR Rec. 635-2, (Ref. 27).

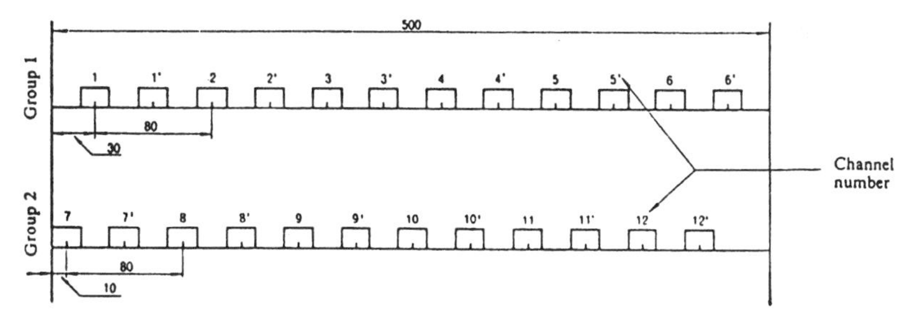

Figure 2.34. Radio-frequency channel arrangements for the 3.7–4.2-GHz band. All frequencies are in megahertz.

The frequencies (in MHz) of individual channels are expressed by the following relationships:

Group 1

go (return) channel, $f_n = f_r - 50 + 80n$

return (go) channel, $f'_n = f_r - 10 + 80n$

where $n = 1, 2, 3, 4, 5,$ and 6.

Group 2

go (return) channel, $f_n = f_r - 70 + 80(n - 6)$

return (go) channel, $f'_n = f_r - 30 + 80(n - 6)$

where $n = 7, 8, 9, 10, 11$, and 12.

2.10.3.2 Radio-Frequency Channel Arrangements for High-Capacity Radio-Relay Systems Operating in the Lower 6-GHz Band. Figure 2.35 shows the RF channel arrangement for radiolinks operating in the lower portion of the 6-GHz band as recommended by CCIR Rec. 383-5 (Ref. 28). It provides for up to eight go and eight return channels with each channel being either an analog channel accommodating up to 1800 telephone channels, or a digital channel with a capacity up to 140 Mbps, or SDH bit rates.

Definitions are as follows:

- f_0 Frequency of the center of the band of frequencies occupied (MHz)
- f_n Center frequency of one radio-frequency channel in the lower-half of the band
- f'_n Center frequency of one radio-frequency channel in the upper-half of the band.

Then the frequencies of the individual channels are expressed by the following relationships:

Lower-half of the band: $f_n = f_0 - 259.45 + 29.65n$ (MHz)

Upper-half of the band: $f'_n = f_0 - 7.41 + 29.65n$ (MHz)

where $n = 1, 2, 3, 4, 5, 6, 7$, or 8.

The go and return channels on a given section should preferably use the polarizations as shown below:

	Go								Return							
$H(V)$	1		3		5		7			2′		4′		6′		8′
$V(H)$		2		4		6		8	1′		3′		5′		7′	

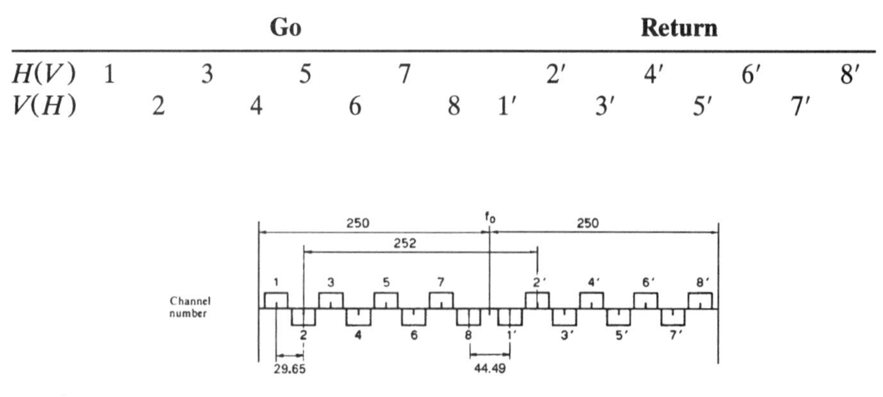

Figure 2.35. RF channel arrangement for systems operating in the 6-GHz band for use on international connections. (From Figure 1, p. 124, CCIR Rec. 383-5; Ref. 28.)

The following alternative arrangement of polarization may be used by agreement between the administrations concerned:

	Go								Return							
$H(V)$	1		3		5		7		1′		3′		5′		7′	
$V(H)$		2		4		6		8		2′		4′		6′		8′

When common transmit–receive antennas for double polarization are used and not more than four channels are accommodated on a single antenna, it is preferred that the channel frequencies be selected by either making $n = 1, 3, 5$, and 7 in both halves of the band or making $n = 2, 4, 6$, and 8 in both halves of the band.

When additional radio-frequency channels, interleaved between those of the main pattern, are required, the values of the center frequencies of these radio-frequency channels should be 14.825 MHz below those of the corresponding main channel frequencies; in systems for 1800 channels, or the equivalent, and digital high-capacity digital systems, it may not be practical, because of the bandwidth of the modulated carrier, to use interleaved frequencies.

The preferred center frequency is 6175.0 MHz.

Table 2.13 describes channel arrangements for the band 5925–6425 MHz, which are used for 16-QAM or 256-QAM systems. For a discussion on M-QAM systems, see Chapter 3.

2.10.3.3 Configuration for the 11-GHz Band for Analog Television and FDM Telephony Transmission for 600–1800 VF Channels and Digital Systems of Equivalent Bandwidth. CCIR Rec. 387-6 (Ref. 29) covers a configuration for the 11-GHz band where a 1-GHz transmission bandwidth is available permitting up to 12 go and 12 return analog channels, or up to 22 go and 22 return digital channels. Figures 2.36 and 2.37 show the CCIR recommended configurations for this band for analog transmission.

For radiolink systems with a maximum capacity of 1800 FDM telephony channels or the equivalent operating in the 11-GHz band, the preferred RF channel arrangement for analog transmission is derived as follows. Let f_0 be the frequency in megahertz of the center of the operating band; let f_n be the center frequency of one RF channel in the lower-half of the band and f_n' be the center frequency of one RF channel in the upper-half of the band. The frequencies (in MHz) of the individual channels are then expressed by the following relationship:

Lower-half of the band: $f_n = f_0 - 525 + 40n$ (MHz)
Upper-half of the band: $f_n' = f_0 + 5 + 40n$ (MHz)

where $n = 1, 2, 3, 4, 5, 6, 7, 8, 9, 10, 11$, or 12. Figure 2.36 applies.

TABLE 2.13 Radio-Frequency Channel Arrangements for the 6-GHz Band

Modulation (Capacity per Channel)	16-QAM[a] (200 Mbit/s)	16-QAM (155.52 Mbit/s) 256-QAM (311.04 Mbit/s)	256-QAM (311.04 Mbit/s)
Frequency band (MHz)	5925–6425	5925–6425	5925–6425
Center frequency of the band f_0 (MHz)	6175	6175	6175
Center frequency of the carriers f_n (MHz)	$f_0 \pm (50 + 80n)$ $n = 0, 1, 2$	$f_0 \pm 20n$ $n = 1, 2, \ldots, 12$	$f_0 \pm (15 + 10n)$ $n = 0, 1, \ldots, 23$
Interleaved or co-channel	Co-channel	Co-channel	Co-channel
Transmission method	Single-carrier transmission method	Three-carrier transmission method (20 MHz band-width/carrier)	Six-carrier transmissionmethod (10 MHz bandwidth /carrier)
Number of channels	6	8	8
Channel bandwidth			
XS (MHz)	80	60	60
X	1.6	1.54	1.54
Center gap			
YS (MHz)	100	80	80
Y	2.0	2.06	2.06
Guard space			
ZS (MHz)	40	30	30
Z	0.8	0.77	0.77

[a]*M*-QAM systems are described in Chapter 3.
Source: Table 1, CCIR Rec. 383-5 (Ref. 28).

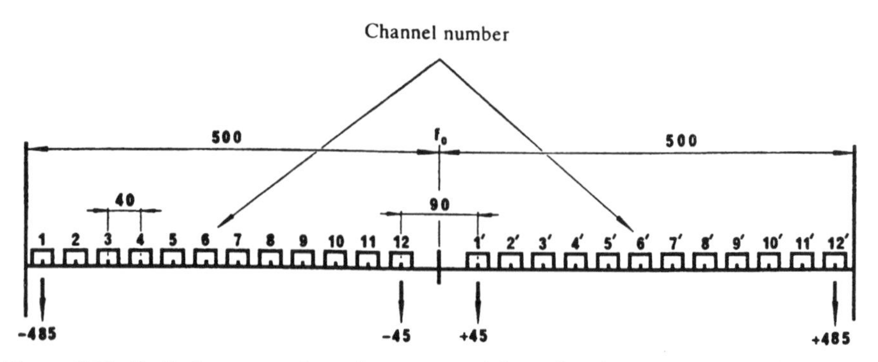

Figure 2.36. Radio-frequency channel arrangement for radio-relay systems operating in the 11-GHz band (main pattern). All frequencies are in megahertz.

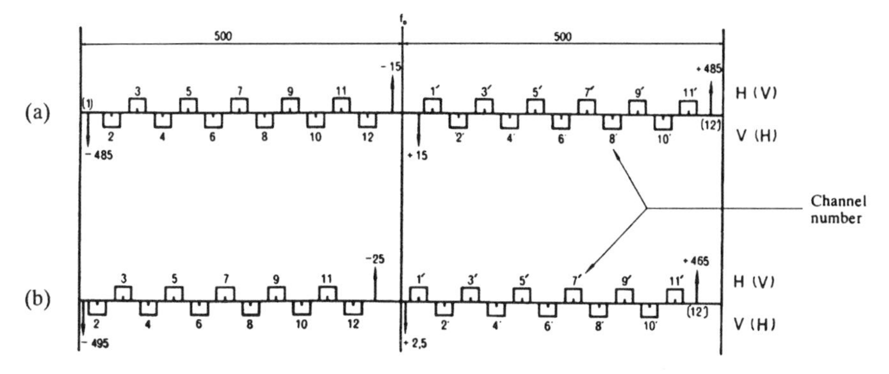

Figure 2.37. Radio-frequency channel arrangement for main and auxiliary radio-relay systems operating in the 11-GHz band. All frequencies are in megahertz. (a) Main pattern. (b) Interleaved pattern. (From CCIR Rec. 387-6; Ref. 29.)

When additional analog RF channels, interleaved between those of the main pattern, are required, the values of the center frequencies of these RF channels should be 20 MHz below those of the corresponding main channel frequencies (Figure 2.37). It should be noted that channel 1 of the interleaved pattern in the lower-half of the band is beyond the lower extremity of a 1000-MHz band and may therefore not be available for use.

When analog RF channels are also required for auxiliary radiolink systems, the preferred frequencies for 11 go and 11 return channels including two pairs of auxiliary channels in both the main and interleaved patterns should be derived by making $n = 2, 3, 4, \ldots, 12$ in the lower-half of the band and $n = 1, 2, 3, \ldots, 11$ in the upper-half of the band. The radio frequencies in megahertz for the auxiliary systems should be selected as shown below:

	Main Pattern	**Interleaved Pattern**
Lower-half of the band	$f_0 - 485$	$f_0 - 495$
	$f_0 - 15$	$f_0 - 25$
Upper-half of the band	$f_0 + 15$	$f_0 + 2.5$
	$f_0 + 485$	$f_0 + 465$

Figure 2.37 applies; it also shows a possible polarization arrangement.

If only three go and three return channels are accommodated on a common transmit–receive antenna, it is preferable that the channel frequencies (MHz) be selected by making

$n = 1, 5, 9$ or

$n = 2, 6, 10$ or

$n = 3, 7, 11$ or

$n = 4, 8, 12$

(all combinations in both halves of the band). Otherwise, for channel arrangements where more than three go and three return channels are required, all go channels should be in one half of the band and all return channels should be in the other half of the band.

It is further recommended that for adjacent analog RF channels in the same half of the band, different polarizations should be used alternately.

The preferred center frequency in the 11-GHz band is 11,200 MHz.

When high-capacity digital radio systems of up to 155 Mbps are used in the 11-GHz band, the radio-frequency channel arrangements should use the center frequencies shown above.

The preferred RF channel arrangement providing 12 go and 12 return channels based on the main pattern shown in Figure 2.36 is defined by:

$n = 1, 2, 3, \ldots, 12$ in the lower-half of the band
$n = 1, 2, 3, \ldots, 12$ in the upper-half of the band

The preferred RF channel arrangement providing 11 go and 11 return channels based on the main pattern shown in Figure 2.36 is defined by:

$n = 2, 3, 4, \ldots, 12$ in the lower-half of the band
$n = 1, 2, 3, \ldots, 11$ in the upper-half of the band

This corresponds to the main RF channels shown in Figure 2.37a.

The preferred RF channel arrangement providing 11 go and 11 return channels based on the interleaved pattern shown in Figure 2.37b is defined by:

$n = 2, 3, 4, \ldots, 12$ in the lower-half of the band
$n = 1, 2, 3, \ldots, 11$ in the upper-half of the band (see Figure 2.37b)

or:

$n = 2, 3, 4, \ldots, 12$ in the upper-half of the band

The preferred RF channel arrangement providing 12 go and 12 return channels is based on the second preferred RF channel arrangement above with two additional channels as shown in Figure 2.38 and defined by the following relationships:

Lower-half of the band: $f_n = f_0 - 505 + 40n$
Upper-half of the band: $f'_n = f_0 - 15 + 40n$

Note: Channels 1 and 12′ in the main pattern with a guard band of 15 MHz are generally considered unsuitable for high-capacity digital radio systems with a symbol rate of more than 25–30 megabaud.

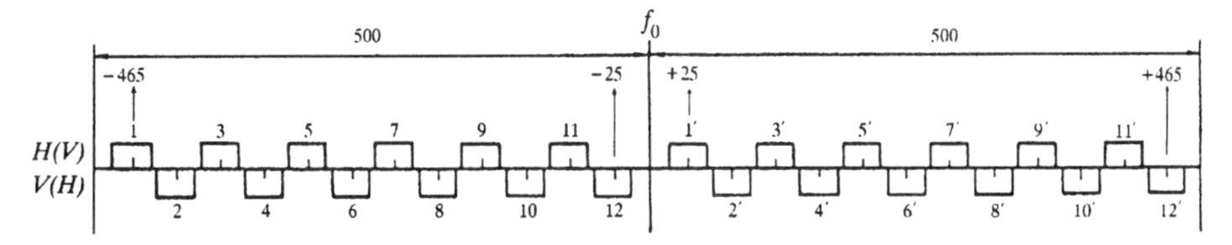

Figure 2.38. Radio-frequency channel arrangement for high-capacity digital radio-relay systems operating in the 11-GHz band. (From Figure 4, p. 150, CCIR Rec. 387-6, 1994 F Series Volume, Part 1; Ref. 29.)

2.10.3.4 Radio-Frequency Channel Arrangements for Radio-Relay Systems Operating in the 18-GHz Band. This RF band provides a 2-GHz bandwidth between 17.7 and 19.7 GHz. The preferred radio-frequency channel arrangement for digital radio systems with a capacity on the order of 280 Mbps, 140 Mbps, and 34 Mbps should be derived as follows:

- f_0 Center frequency of the band of frequencies occupied (MHz)
- f_n Center frequency of a radio-frequency channel in the lower-half of the band (MHz)
- f_n' Center frequency of a radio-frequency channel in the upper-half of the band (MHz)

Then the frequencies (MHz) of individual channels are expressed by the following relationships:

Co-channel Arrangement

For systems with a capacity of 280 Mbps:

$$\text{Lower-half of the band:} \quad f_n = f_0 - 1110 + 220n \quad \text{(MHz)}$$
$$\text{Upper-half of the band:} \quad f_n' = f_0 + 10 + 220n \quad \text{(MHz)}$$

where $n = 1, 2, 3$, or 4. The frequency arrangement is shown in Figure 2.39a.

For systems with a capacity on the order of 140 Mbps:

$$\text{Lower-half of the band:} \quad f_n = f_0 - 1000 + 110n \quad \text{(MHz)}$$
$$\text{Upper-half of the band:} \quad f_n' = f_0 + 10 + 110n \quad \text{(MHz)}$$

where $n = 1, 2, 3, 4, 5, 6, 7$, or 8. The frequency arrangement is shown in Figure 2.39b.

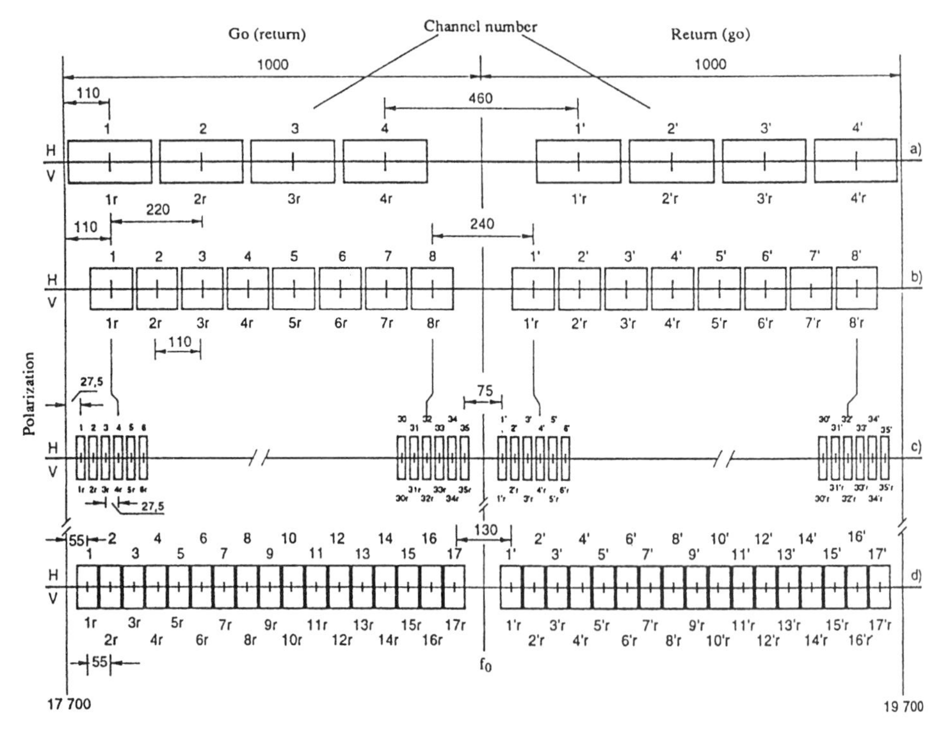

Figure 2.39. Radio-frequency channel arrangement for radio-relay systems operating in the 17.7–19.7 GHz band, co-channel arrangement, with all frequencies in megahertz. (From Figure 1, p. 162, CCIR Rec. 595-3, 1994 F Series Volume, Part 1; Ref. 30.)

For systems with a capacity of 34 Mbps:

Lower-half of the band: $f_n = f_0 - 1000 - 27.5n$ (MHz)
Upper-half of the band: $f'_n = f_0 + 10 + 27.5n$ (MHz)

where $n = 1, 2, 3, \ldots, 35$. The frequency arrangement is shown in Figure 2.39c.

2.10.4 Several FCC Frequency Plans

Figures 2.40–2.43 give current FCC (U.S. Federal Communications Commission) frequency plans for the frequency bands 1850–1990 MHz, 2130–2150 and 2180–2200 MHz, 6525–6875 MHz, and 12,200–12,700 MHz, respectively. Table 2.14 provides standard parameters applicable to the figures (from Ref. 31).

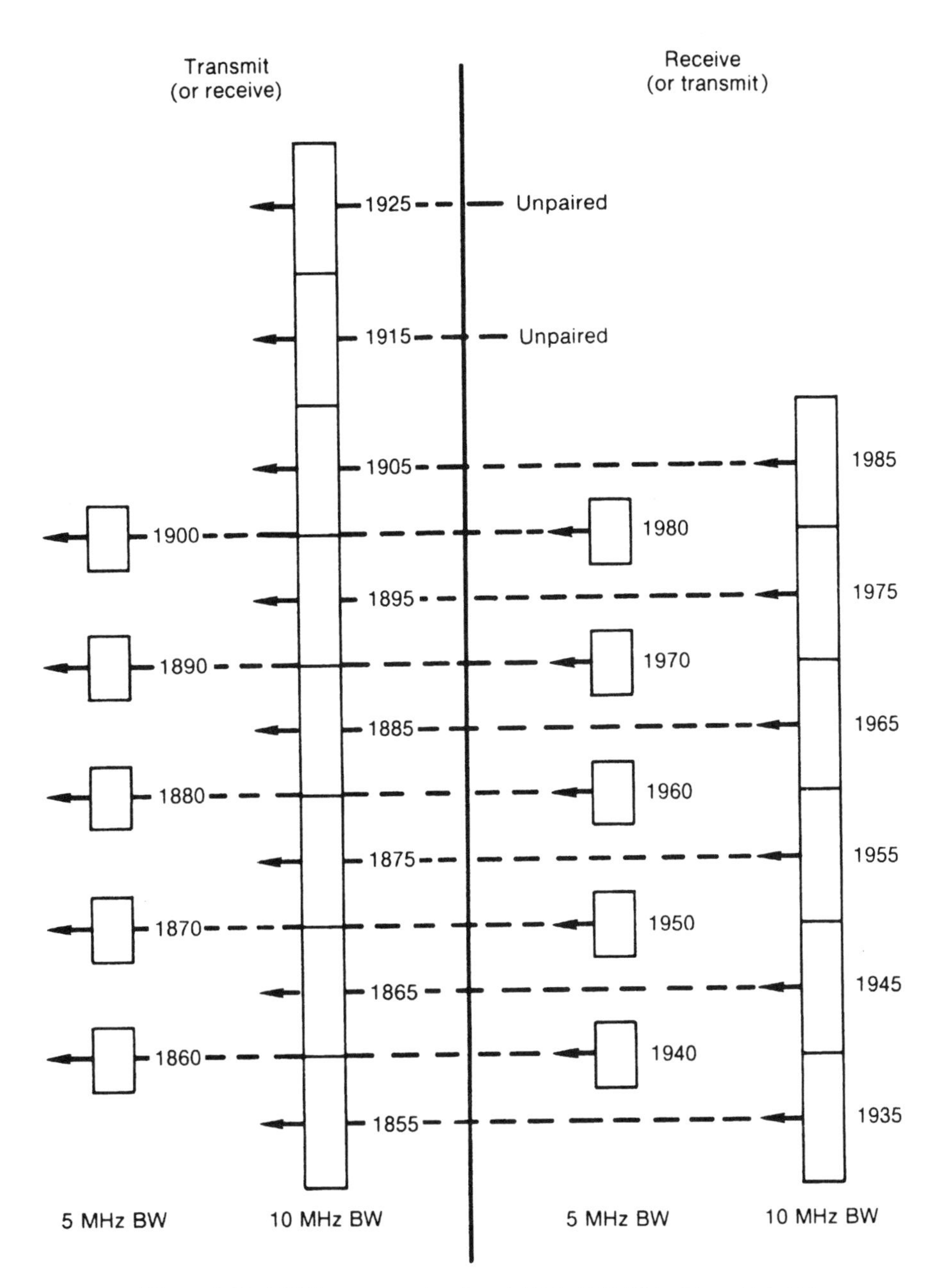

Figure 2.40. FCC frequency plan, band 1850–1990 MHz. (From Ref. 31.)

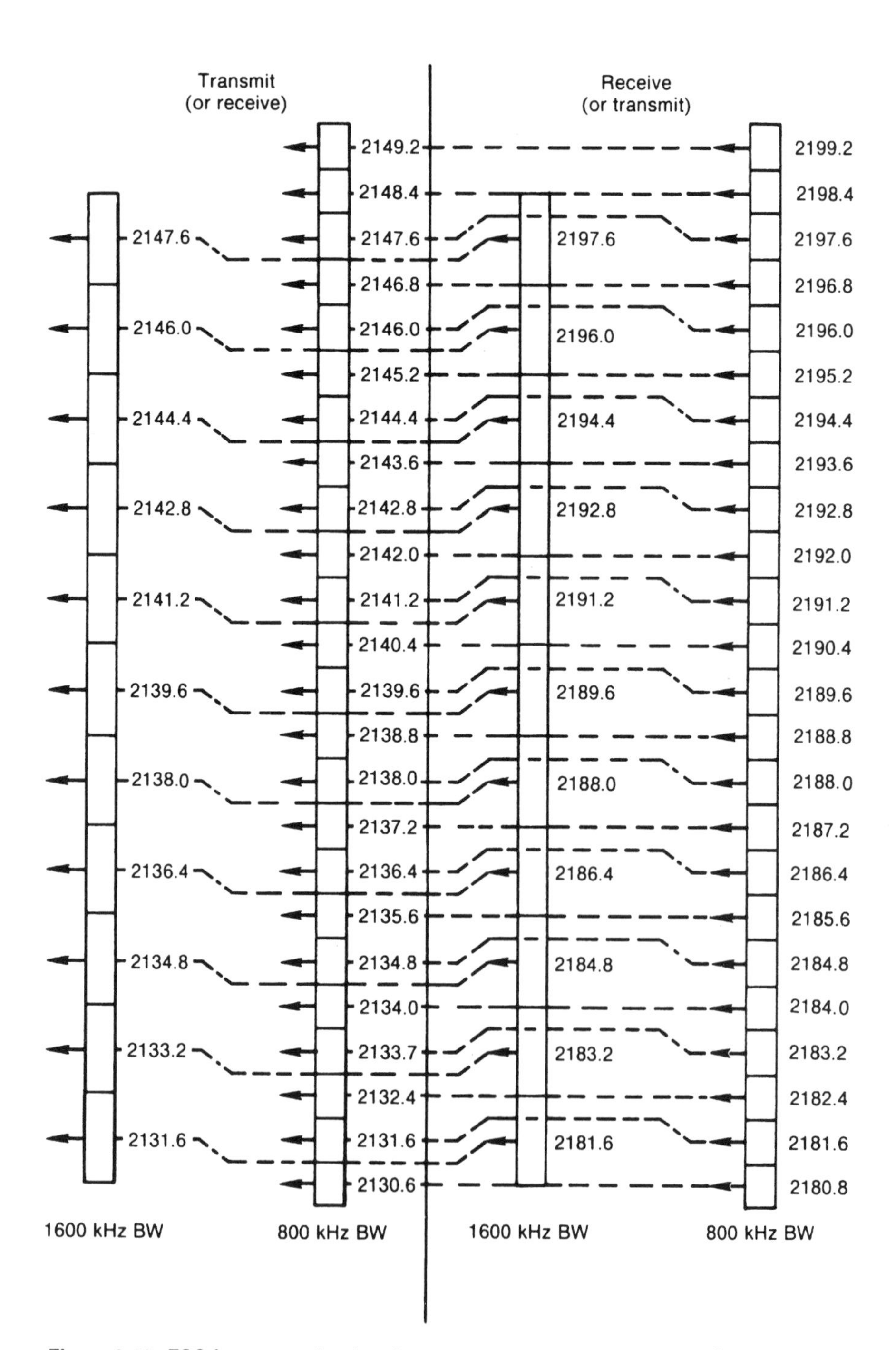

Figure 2.41. FCC frequency plan, bands 2130–2150 and 2180–2200 MHz. (From Ref. 31.)

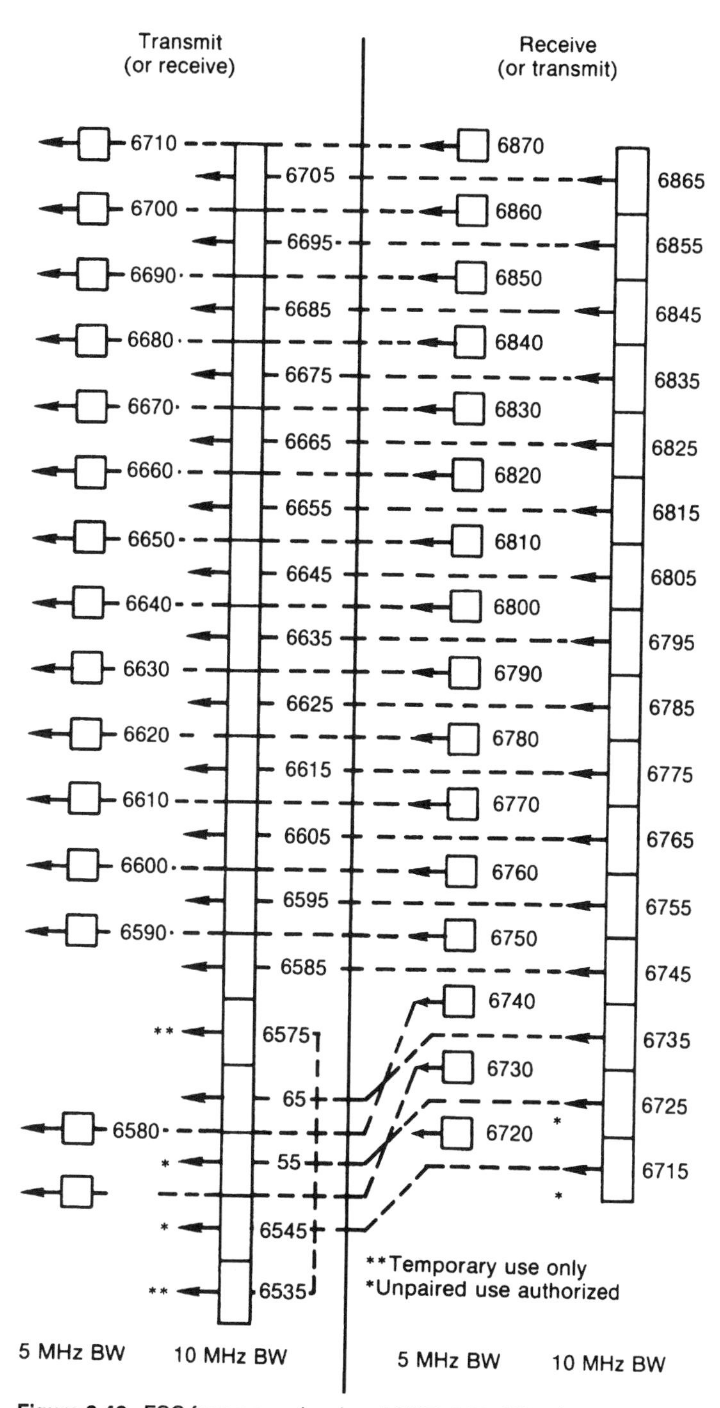

Figure 2.42. FCC frequency plan, band 6525–6875 MHz. (From Ref. 31.)

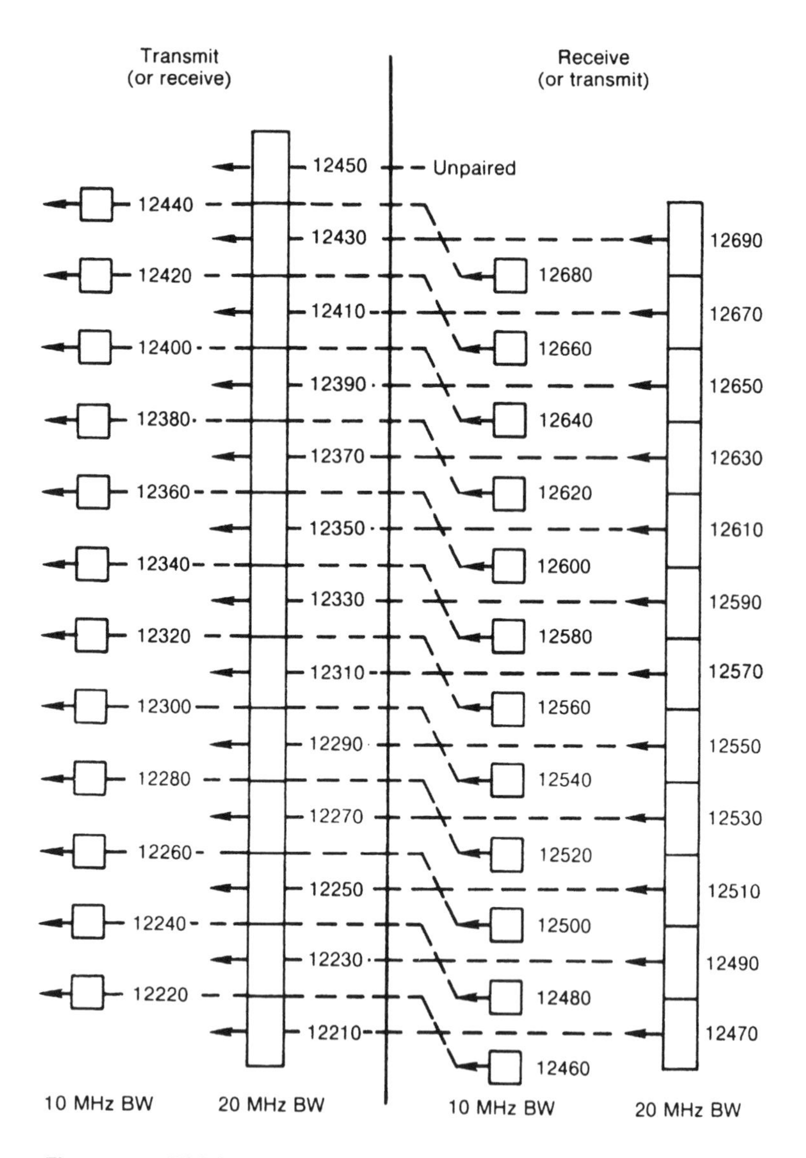

Figure 2.43. FCC frequency plan, band 12,200–12,700 MHz. (From Ref. 31.)

TABLE 2.14 FCC Standard Parameters (Applicable to Figures 2.40–2.43)

Frequency Band		1.85–1.990 GHz		2.13–2.15 and 2.18–2.2 GHz		6.525–6.875 GHz		12.2–12.7 GHz	
Carrier frequency and spacing		10 MHz with 10 MHz interstitial (Figure 2.40)		0.8 MHz with 1.6 MHz Co-channel (Figure 2.41)		10 MHz with 10 MHz interstitial (Figure 2.42)		20 MHz with 20 MHz interstitial (Figure 2.43)	
Bandwidths		5 MHz	10 MHz 300	0.8 MHz	1.6 MHz	5 MHz	10 MHz 300	10 MHz	20 MHz 300
Traffic capacity	FDM	120	480 600	24 48	72 96	120	480 600	300	600 1200
	Other	Digital	Digital	Digital	Digital	Digital	Digital	Digital	Video/digital
FDM per channel rms deviation (kHz)		200	200 200 140	35 25	60 47	200	200 200 140	200	200 200 140
Frequency stability[a]									
operational		±0.002%		±0.0003%		±0.002%		±0.002%	
lcw stability		±0.02%				±0.02%		±0.02%	
Frequency tolerance (FCC requirements)		0.002%		0.001%		0.005%		0.005%	

[a]Operational stability used is based on most probable performance of modern transmitters over normal environmental range. Low stability represents potentials from grandfathered systems.

Source: Reference 31.

PROBLEMS AND EXERCISES

1 For a 31-mi radiolink hop determine the additional values in feet to be added to obstacle heights at obstacle locations A, B, and C given the following properties:

$$A = 800 \text{ ft above MSL, 5 mi from east terminal site}$$

$$B = \text{midpath at 520 ft above MSL}$$

$$C = 720 \text{ ft, 11 mi from west site}$$

Assume $K = 1.2$, frequency = 7.1 GHz.

2 For an initial planning exercise, determine LOS distance assuming smooth earth, grazing, and $K = \frac{4}{3}$. A structure on one end of the path is 420 ft high, and there is space available on a TV tower at the other end at the 465-ft level.

3 The surface refractivity at a certain location is given as 301. What is the refractivity at this location at 8000-ft altitude?

4 What is the free-space loss (FSL) for a radiolink 9 mi long operating at 3.955 GHz?

5 What is the free-space loss (FSL) for a radiolink 31 km long operating at 37.05 GHz?

6 What is the free-space loss (FSL) to the moon at 6 GHz and at 12 GHz?

7 In reference to question 6, how important is it to state where on the earth's surface we measure from assuming LOS conditions?

8 What is the EIRP of a transmitting subsystem operating at 15 GHz with 4.5 dB of line losses? The output of the transmitter at its flange is 20 mW and the antenna gain is 41 dB.

9 A 21-mi radiolink operating at 6.1 GHz has the following characteristics: at the transmitter 120 ft of EW-64 waveguide (1.7 dB/100 ft), transition loss of 0.2 dB, a 4-ft antenna at the transmit end, 0 dBW output at the transmitter waveguide flange. What is the isotropic receive level at the distant end?

10 From question 9, on the receive end of the link there is 143 ft of EW-64 waveguide, a similar 4-ft antenna, and same transition. What is the unfaded RSL?

11 A low-noise amplifier (LNA) has a 3-dB noise figure, operates at room temperature, and incorporates a bandpass filter in its front end with 120-MHz passband. What is the thermal noise threshold of the receiver in dBm?

12 The first active stage of a radiolink receiver is a mixer with an 8.5-dB noise figure, and B_{RF} of the link is 30 MHz. What is the FM improvement threshold?

13 Calculate the peak deviation of a FM radio transmitter designed to CCIR recommendations carrying 600 VF channels in a FDM/FM configuration? Use a peaking factor of 13 dB.

14 Calculate the unfaded carrier-to-noise ratio (C/N) for a FDM/FM radiolink 34 mi long based on CCIR recommendations, where the receiver noise figure is 7.5 dB. The transmitter output power is 2.5 W and the operating frequency is 2.1 GHz. The link carries 1200 VF channels in a standard CCITT FDM configuration. Bandwidth is to be calculated according to Carson's rule. Use conventional 4-ft antennas at each end. Line losses are 2 dB at each end.

15 What is the gain of a 2-ft radiolink parabolic reflector antenna assuming the conventional efficiency?

16 What fade margin would be assigned to a radiolink with a path having 0.003% unavailability due to propagation where Rayleigh type fading is assumed?

17 Using the Barnett method, calculate path unavailability for a 45-km path operating at 4 GHz over very smooth terrain on the U.S. Gulf of Mexico coast with a 36-dB fade margin.

18 A certain radiolink path requires a 38-dB fade margin without diversity. What fade margin will be required (worst case) using frequency diversity with a frequency separation of 2%?

19 When calculating total noise at a radiolink receiver, what are the three types of noise to be considered? We may add these directly if the noise units are measured in _____. When operating near threshold, what type of noise predominates?

20 One hop of a 2500-km radiolink system is 45 mi long. What total noise is permitted at the receiver following CCIR recommendations (pWp0)?

21 Assume a NPR of 55 dB; calculate the noise components to reach the value in question 20. Let pre-emphasis improvement be 3.5 dB and diversity improvement be 3 dB. The link carries 600 VF channels in a standard FDM/FM configuration. Feeder distortion is 6 pWp. The link operates at 6 GHz. Set the necessary parameters with the design following CCIR recommendations.

REFERENCES

1. *Noise Accumulation in an FDM Voice Channel Due to Radio Portion on a 2500-km Hypothetical Reference Circuit*, CCIR Rec. 395-2, Vol. IX, XVIIth Plenary Assembly, Dusseldorf, 1990.
2. *Permissible Noise in the Hypothetical Reference Circuit of Radio-Relay Systems for Television*, CCIR Rec. 555, Geneva, 1982.
3. B. R. Bean et al., *A World Atlas of Atmospheric Radio Refractivity*, U.S. Department of Commerce, ESSA, Boulder, CO, 1966.
4. *Naval Shore Electronics Criteria, Line-of-Sight Microwave and Tropospheric Scatter Communication Systems*, Navelex 0101, 112, U.S. Department of the Navy, Washington, DC, May 1972.
5. *Design Handbook for Line-of-Sight Microwave Communication Systems*, MIL-HDBK-416, U.S. Department of Defense, Washington, DC, Nov. 1977.
6. *Frequency Deviation of Analog Radio-Relay Systems for Telephony Using Frequency-Division Multiplex*, CCIR Rec. 404-2, Vol. IX, XVIIth Plenary Assembly, Dusseldorf, 1990.
7. *Pre-emphasis Characteristic for Frequency Modulation Radio-Relay Systems for Telephony Using Frequency-Division Multiplex*, CCIR Rec. 275-3, Vol. IX, XVIIth Plenary Assembly, Dusseldorf, 1990.

8. *Pre-emphasis Characteristics for Frequency Modulation Radio-Relay Systems for Television*, CCIR Rec. 405-1, Vol. IX, XVIIth Plenary Assembly, Dusseldorf, 1990.
9. *Interconnection at Baseband Frequencies of Radio-Relay Systems for Telephony Using Frequency-Division Multiplex*, CCIR Rec. 380-4, Vol. IX, XVIIth Plenary Assembly, Dusseldorf, 1990.
10. Roger L. Freeman, *Telecommunication Transmission Handbook*, 3rd ed., Wiley, New York, 1991.
11. *The New IEEE Standard Dictionary of Electrical and Electronic Terms*, 5th ed., IEEE Std 100-1992, IEEE Press, New York, 1992.
12. H. Brodhage and W. Hormuth, *Planning and Engineering of Radio Relay Links*, 8th ed., Siemens-Heyden & Son Ltd., London, 1978.
13. *Engineering Considerations for Microwave Communication Systems*, GTE-Lenkurt, San Carlos, CA, 1975.
14. *The Radio Refractive Index: Its Formula and Refractivity Data*, ITU-R Rec. PN.453-4, 1994 PN Series Volume, ITU, Geneva, 1994.
15. *Propagation Data and Prediction Methods Required for the Design of Terrestrial Line-of-Sight Systems*, ITU-R Rec. PN.530-5, 1994 PN Series Volume, ITU, Geneva, 1994.
16. H. T. Dougherty, *A Summary of Microwave Fading Mechanisms, Remedies and Applications*, U.S. Department of Commerce, ESSA Technical Report ERL69-WPL 4, Boulder, CO, Mar. 1968.
17. FCC Rules and Regulations Part 21, Vol. VII, FCC, Washington, DC, Sept. 1982.
18. Recommendations and Reports of the CCIR, 1982, XVth Plenary Assembly, Geneva, 1982.
19. *Allowable Noise Power in the Hypothetical Reference Circuit for Radio-Relay Systems for Telephony Using Frequency-Division Multiplex*, CCIR Rec. 393.4, ITU, Geneva, 1982.
20. *Standard Microwave Transmission Systems*, EIA-252A, Electronics Industries Association, Washington, DC, Sept. 1972.
21. Roger L. Freeman, *Reference Manual for Telecommunication Engineering*, 2nd ed., Wiley, New York, 1993.
22. M. J. Tant, *The White Noise Handbook*, Marconi Instruments Ltd., Albans, Herts, UK, 1974.
23. *Measurement of Noise Using a Continuous Uniform Spectrum Signal on Frequency Division Multiplex Telephony Radio-Relay Systems*, CCIR Rec. 399-3, Vol. IX, XVIIth Plenary Assembly, Dusseldorf, 1990.
24. U.S. Military Standard, *Common Long-Haul and Tactical Communication System Technical Standards*, MIL-STD-188-100, U.S. Department of Defense, Washington, DC, Nov. 1972.
25. *Frequency Deviation and the Sense of Modulation for Analog Radio-Relay Systems for Television*, CCIR Rec. 276-2, Vol. IX, XVIIth Plenary Assembly, Dusseldorf, 1990.
26. *Radio-Frequency Channel Arrangements for Radio-Relay Systems*, ITU-R Rec. F.746-1, 1994 F Series Volume, Part 1, ITU, Geneva, 1994.

27. *Radio-Frequency Channel Arrangements Based on a Homogeneous Pattern for Radio-Relay Systems Operating in the 4 GHz Band*, CCIR Rec. 635-2, 1994 F Series Volume, Part 1, ITU, Geneva, 1994.
28. *Radio-Frequency Channel Arrangements for High-Capacity Radio-Relay Systems Operating in the Lower 6 GHz Band*, CCIR Rec. 383-5, 1994 F Series Volume, Part 1, ITU, Geneva, 1994.
29. *Radio-Frequency Channel Arrangements for Radio-Relay Systems Operating in the 11 GHz Band*, CCIR Rec. 387-6, 1994 F Series Volume, Part 1, ITU, Geneva, 1994.
30. *Radio-Frequency Channel Arrangements for Radio-Relay Systems Operating in the 18 GHz Frequency Band*, CCIR Rec. 595-3, 1994 F Series Volume, Part 1, ITU, Geneva, 1994.
31. *Telecommunications Systems Bulletin No. 10D—Interference Criteria for Microwave Systems in the Private Radio Services*, Electronics Industries Association, Washington, DC, Aug. 1983.
32. *Baseband Interconnection of Digital Radio-Relay Systems*, CCIR Rep. 938, Annex to Vol. IX, XVIIth Plenary Assembly, Dusseldorf, 1990.
33. *Transmission Systems for Communications*, 5th ed., Bell Telephone Laboratories, Holmdel, NJ, 1982.

3

DIGITAL LINE-OF-SIGHT MICROWAVE RADIOLINKS

3.1 INTRODUCTION

The implementation of digital LOS radiolinks is accelerating primarily due to the transition of the telephone network to an all-digital network. Many available texts (Refs. 1 and 2) demonstrate the rationale for the transition in that thermal and IM noise accumulation can be disregarded over the network, and noise becomes an isolated problem between points of regeneration. This is an overwhelming advantage over analog transmission, where the primary concern of the transmission engineer is noise accumulation.

Other arguments presented for going all-digital are based on its compatibility with digital information transmission requirements such as telephone signaling, data transmission, digitized voice, programming information, and facsimile.

Several very important factors representing the other side of the coin must also be highlighted. The digital radiolink engineer must not only be cognizant of these factors but also must understand them well. The digital network is based on a PCM waveform, which, when compared to its analog FDM counterpart, is wasteful of bandwidth. A nominal 4-kHz voice channel on a FDM baseband system occupies about 4 kHz of bandwidth (Ref. 1, Chapter 1). On a FDM/FM radiolink, by rough estimation, we can say it occupies about 16 kHz.

In conventional PCM baseband system, allowing 1 bit per hertz of bandwidth, a 4-kHz voice channel roughly requires 64 kHz (64 kbps) of bandwidth. This is derived using the Nyquist sampling rate of 8000/s (4000 Hz $\times$ 2) and each sample is assigned an 8-bit code word, thus 8000 $\times$ 8 bits per second or 64 kbps.

RF bandwidth is at a premium and, as a result, it is incumbent on the digital radiolink engineer to select a waveform that conserves bandwidth, achieving, essentially, more bits per hertz. He/she will also find that many national regulatory agencies require a minimum number of digital voice channels per unit of bandwidth.

This section first introduces some sample regulatory requirements and then discusses modulation techniques that are bandwidth conservative so that these national requirements can be met. It then describes methods of link analysis to achieve specified digital network performance. The discussion will rely heavily on previous sections, demonstrating that much of the approach used on analog radiolink design is also applicable to digital radiolink design. The unit of digital radiolink performance is BER rather than S/N and noise accumulation, which were the measures for analog radiolink design.

3.1.1 Energy per Bit per Noise Density Ratio E_b/N_0

The efficiency of a digital communication system in the presence of wideband noise with a single-sided noise spectral density N_0 is commonly measured by the received information bit energy to noise density ratio (E_b/N_0) required to achieve a specified error rate.

We can express E_b by relating it to the total received power (RSL or C*). E_b, or energy per bit, can be expressed as the RSL divided by the bit rate. In the domain of dBs, we write

$$E_{b(\text{dBW})} = \text{RSL}_{\text{dBW}} - 10\log(\text{bit rate}) \tag{3.1}$$

Example 1. Suppose the unfaded RSL of a certain microwave radiolink was -81 dBW and the *bit rate* was 1.544 Mbps. What is E_b?

$$\begin{aligned} E_b &= -81\text{ dBW} - 10\log(1.544 \times 10^6) \\ &= -81\text{ dBW} - 61.886\text{ (dB)} \\ &= -142.886\text{ dBW} \end{aligned}$$

N_0 is simple to calculate. It is the thermal noise level of a perfect receiver in 1 Hz of bandwidth (in dBW or dBm) plus the noise figure (NF) in decibels.

Now we write

$$N_0 = -204\text{ dBW} + \text{NF}_{\text{dB}} \tag{3.2}$$

Example 2. The noise figure of a particular receiver front end is 8 dB. What is N_0?

$$\begin{aligned} N_0 &= -204\text{ dBW} + 8\text{ dB} \\ &= -196\text{ dBW/Hz} \end{aligned}$$

*RSL and C (carrier level) are synonymous for this discussion.

Now we can write an expression for E_b/N_0 by bringing together equations (3.1) and (3.2).

$$E_b/N_0 = E_{b(\mathrm{dBW})} - N_{0(\mathrm{dBW})} \tag{3.3}$$

$$E_b/N_0 = \mathrm{RSL}_{\mathrm{dBW}} - 10\log(\text{bit rate}) - (-204\ \mathrm{dBW} + \mathrm{NF}_{\mathrm{dB}})$$

Simplifying,

$$E_b/N_0 = \mathrm{RSL}_{\mathrm{dBW}} - 10\log(\text{bit rate}) + 204\ \mathrm{dBW} - \mathrm{NF}_{\mathrm{dB}} \tag{3.4}$$

Example 3. A digital microwave radiolink is transmitting at 155.520 Mbps (e.g., SONET STS-3), and the unfaded RSL at the far end receiver front end is -76.3 dBW; the noise figure of the receiver front end is 3 dB. What is the value of E_b/N_0 under these conditions?

Use equation (3.4):

$$\begin{aligned} E_b/N_0 &= -76.3\ \mathrm{dBW} - 10\log(155.520 \times 10^6\ \mathrm{bps}) + 204\ \mathrm{dBW} - 3\ \mathrm{dB} \\ &= -76.3 - 81.92 + 204 - 3 \\ &= 42.78\ \mathrm{dB} \end{aligned}$$

3.2 REGULATORY ISSUES

The U.S. regulatory agency, the FCC, has long recognized that conventional digital modulation schemes (such as FSK and BPSK/QPSK) were not bandwidth conservative and ruled in Part 21.122 of "Rules and Regulations" (Ref. 3) that:

> Microwave transmitters employing digital modulation techniques and operating below 15 GHz shall, with appropriate multiplex equipment, comply with the following additional requirements:
>
> 1. The bit rate in bits per second shall be equal to or greater than the bandwidth specified by the emission designator in hertz (e.g., to be acceptable, equipment transmitting at a 20-Mb/s rate must not require a bandwidth greater than 20 MHz). Except that the bandwidth used to calculate the minimum rate shall not include any authorized guardband.
> 2. Equipment to be used for voice transmission shall be capable of satisfactory operation within the authorized bandwidth to encode at least the following

number of voice channels:

Frequency Band (MHz)	Allowable Bandwidth (MHz)	Minimum Capacity of Encoded Voice Channels
2,110–2,130	3.5	96
2,160–2,180	3.5	96
3,700–4,200	20	1,152
5,925–6,425	20	1,152
10,700–11,700	40	1,152

The FCC has the following rule on emission limitation (Part 21.106) (Ref. 3):

When using transmissions employing digital modulation techniques

(2) (i) For operating frequencies below 15 GHz, in any 4-kHz band, the center frequency of which is removed from the assigned frequency by more than 50% up to and including 250% of the authorized bandwidth: As specified by the following equation but in no event less than 50 dB:

$$A = 35 + 0.8(P - 50) + 10 \log_{10} B \qquad (3.5)$$

where A = attenuation (dB) below the mean output power level
P = percent removed from carrier frequency
B = authorized bandwidth (MHz)

(attenuation greater than 80 dB is not required).

(ii) For operating frequencies above 15 GHz, in any 1-MHz band, the center frequency of which is removed from the assigned frequency by more than 50% up to and including 250% of the authorized bandwidth: As specified by the following equation but in no event less than 11 dB:

$$A = 11 + 0.4(P - 50) + 10 \log_{10} B \qquad (3.6)$$

(attenuation greater than 56 dB is not required).

(iii) In any 4-kHz band, the center frequency of which is removed from the assigned frequency by more than 250% of the authorized bandwidth: At least $43 + 10 \log_{10}$ (mean power output in watts) dB or 80 dB, whichever is the least attenuation.

The reference FCC Part 21.106 describes the "FCC mask" often referenced in the literature as the FCC Docket 19311 mask. This spectral mask is shown in Figure 3.1.

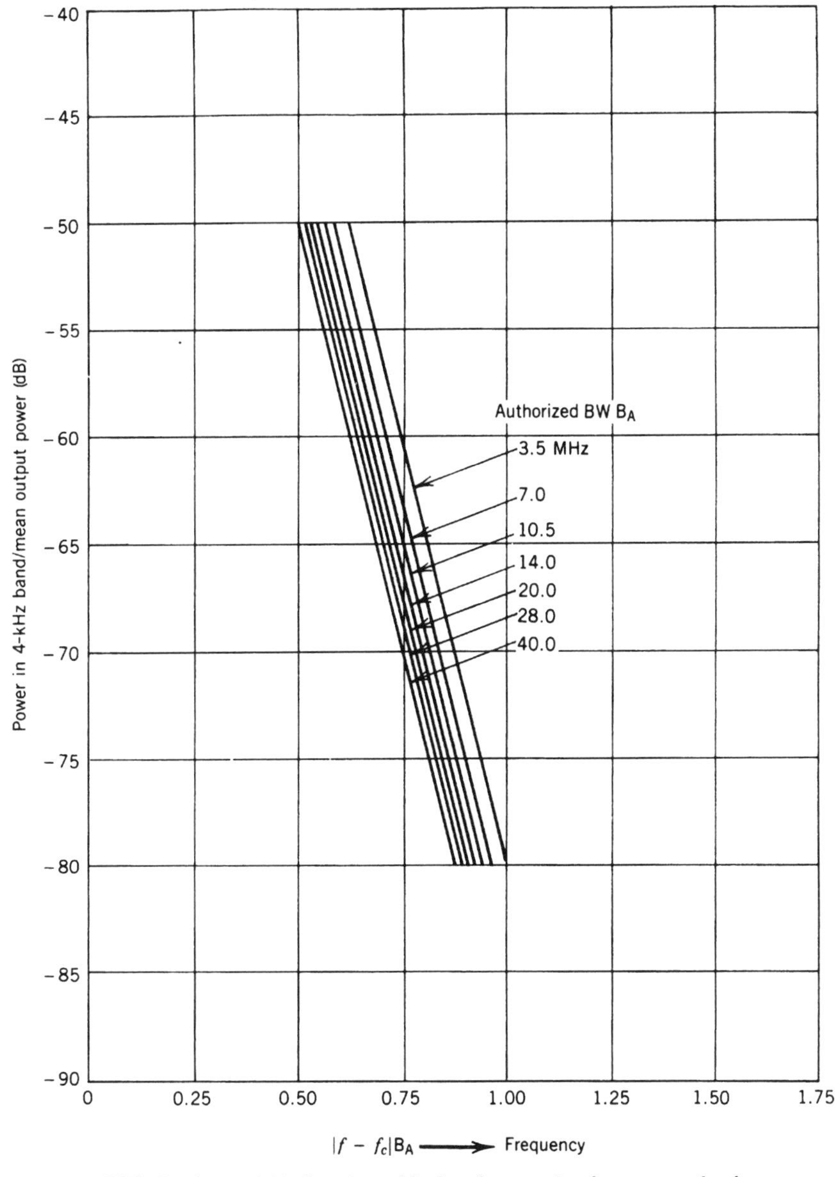

Figure 3.1. FCC Docket 19311 Spectrum Masks: f_c = center frequency; f = frequency of interest. (From Ref. 3.)

Whereas, conventionally, bandwidths most often describe 3-dB points, the 19311 mask describes 50-dB points at about plus or minus half the authorized bandwidth. As can clearly be seen, a packing ratio considerably greater than 1 bit/Hz must be achieved to comply with the FCC spectral mask. In fact, the bit packing ratio is ≥ 4.5 bits per Hz.

3.3 MODULATION TECHNIQUES, SPECTRAL EFFICIENCY, AND BANDWIDTH*

3.3.1 Introduction

The goal of the FCC was to achieve similar information transmission properties between conventional FM transmitting a FDM waveform (Chapter 2) and a digital radio transmitting conventional 8-bit PCM. Thus such a digital radio must provide a minimum of 4.5 bits per hertz of radio bandwidth for 99% spectral power containment. This would be a *practical* value.

We must explain the difference between theoretical bit-packing values and the practical. The theoretical and practical would essentially be equal if we could design efficient filters where their effective bandwidth equated to the baud rate.† We call that the Nyquist bandwidth. As we shall point out, this is impractical and the required bandwidth is almost always greater than the so-called Nyquist bandwidth except for partial response systems. These latter are more often called QPRS, standing for quadrature partial response systems. What we gain in smaller bandwidth requirements, we pay for in increased output power to maintain a certain bit error rate with QPRS techniques.

3.3.2 Bit Packing

We sometimes estimate the required bandwidth for a binary transmission system by assuming 1 bit/Hz. In most practical cases this is somewhat optimistic. Our goal, remember, is to achieve better than 4.5 bits/Hz.

For digital transmission, phase modulation has a number of excellent attributes. Among these are robustness and improved noise immunity. When we describe "phase," we usually employ a circle to represent 360° of phase. In the case of BPSK (binary phase-shift keying), we describe two phase states. The first phase state we can call a binary 1 and the second, a binary 0. The decision "distance" should be as large as possible. In the binary phase case, this distance is 180°. For instance, 0° can be assigned the value of binary 1 and 180° the value of binary 0. However, there is no reason why we cannot assign 45° as the binary 1 and 225° as binary 0, as long as there is 180°

*Section 3.3 is based on Refs. 4, 5, and 6.

†By baud rate we mean transitions per second or changes of state per second. Some call this the "symbol rate," but I don't care for that definition. I like to use "symbol rate" as the pulse rate at the output of a FEC coder.

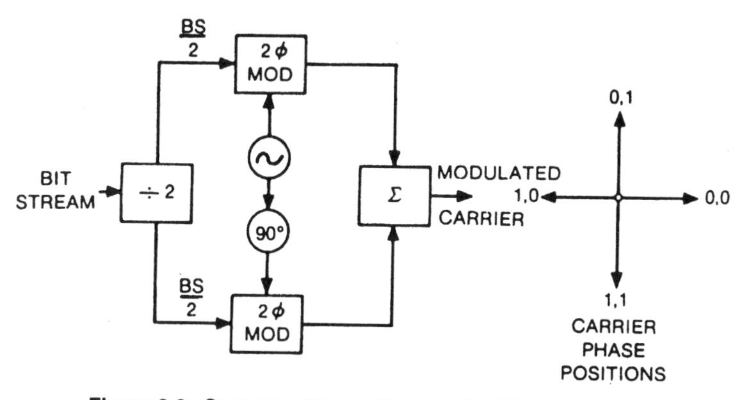

Figure 3.2. Conceptual block diagram of a QPSK modulator.

distance between the two states. Of course, what we describe here is the familiar BPSK.

Carry this thinking one step further. Instead of two phase states, we'll use four, each separated by 90°. Now assign the binary values to each of the four phase states: $0° = 0, 1$; $90° = 0, 0$; $180° = 1, 1$; and $270° = 1, 0$. We call this quadrature phase-shift keying or QPSK. A very important point here is that bandwidth is a function of transitions per second or changes of state per second. Of course, in the binary regime, bits per second and transitions per second are the same. Figure 3.2 is a conceptual block diagram of a QPSK (4-PSK) modulator.

With QPSK, we can *theoretically* achieve bit packing of 2 bits/Hz. Of course, such thinking can be carried yet further as shown in Figure 3.3, which illustrates some signal state diagrams for PSK. In the case of 8-PSK, sometimes called 8-ary PSK, the signal states are separated by 45°, and with 16-ary PSK, by 22.5°. The 8-ary PSK theoretically achieves 3 bits/Hz, and 16-ary PSK achieves 4 bits/Hz.

The generalized name for this type of modulation is M-ary PSK, where the M points are distributed uniformly on a circle. Note that as M increases, the distance between adjacent states decreases. As that distance decreases, with noise and group delay corrupting the received signal, it gets more and more difficult for the receiver demodulator to make a correct decision. System designers, for the PSK case, do not generally let M be greater than 8.

One should also note how the bit packing improves: with QPSK we get 2 bits/Hz (theoretical), 8-ary PSK derives 3 bits/Hz, and 16-ary PSK achieves 4 bits/Hz. That bit-packing value (in bits/Hz) is the square root of the number of states (i.e., $\sqrt{16} = 4$).

3.3.2.1 *M-QAM, a Hybrid Scheme.* QAM stands for quadrature amplitude modulation. It is a hybrid modulation scheme where both amplitude-shift

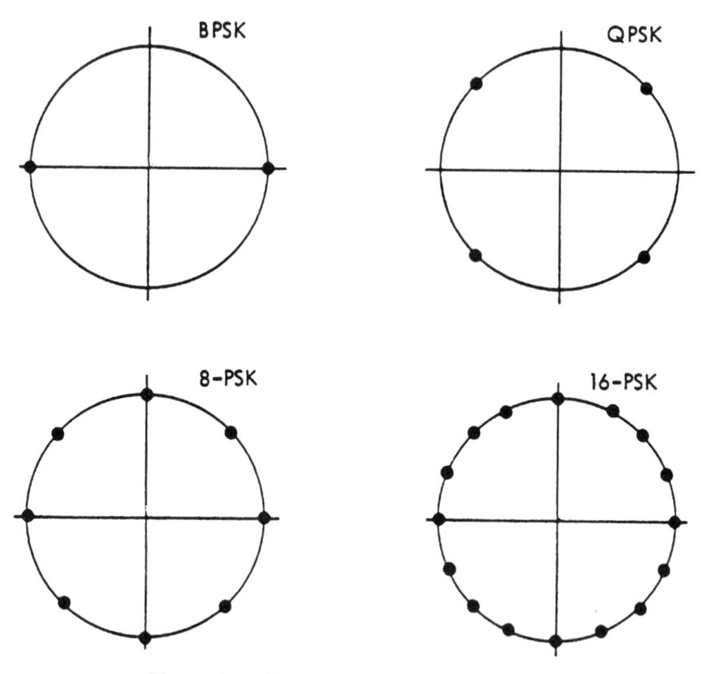

Figure 3.3. Signal state diagrams for PSK.

keying (ASK) and phase-shift keying are used in conjunction with one another.

One example might be QPSK using four amplitude levels. This is 16-QAM and its state diagram is illustrated in Figure 3.4. The amplitude levels are +3, +1, −1, and −3. 16-QAM provides the same bit packing as 16-ary PSK but is more robust because signal states are derived from two distinct modulation techniques: PSK and ASK. Also, it requires less E_b/N_0 for a given error performance than 16-ary PSK.

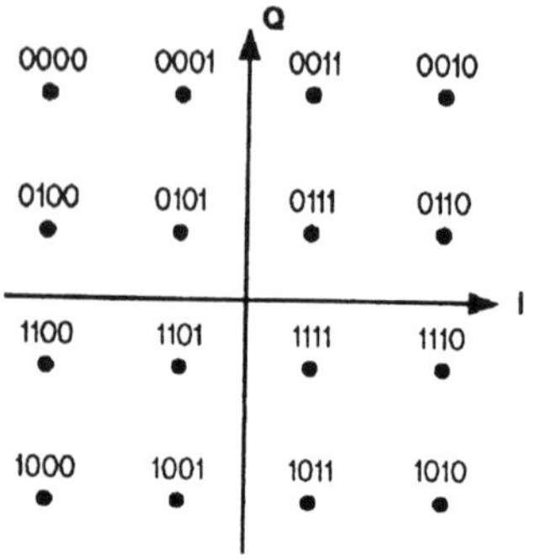

Q
0000 0001 0011 0010
0100 0101 0111 0110
I
1100 1101 1111 1110
1000 1001 1011 1010

Figure 3.4. 16-QAM state diagram. I = in-phase, Q = quadrature.

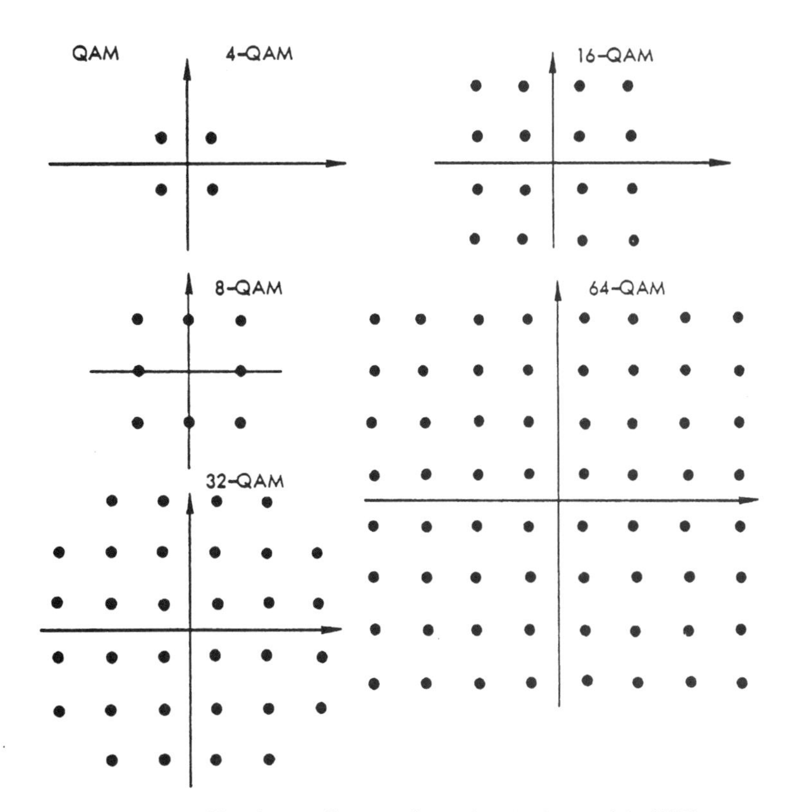

Figure 3.5. Signal state diagrams for 4-, 8-, 16-, 32-, and 64-QAM.

This can be carried yet further to 32-QAM and 64-QAM, shown in Figure 3.5. 32-QAM can be derived from six levels of amplitude modulation by dropping the corner state, achieving bit packing of 5 bits per Hz theoretical ($2^5 = 32$).

In a similar fashion there are 128-QAM and 256-QAM; and some systems have been fielded using 512-QAM.

The following is a relationship between the transmitted bit rate (R_b), baud rate (symbol rate) $(1/T)$, and the value of M:

$$R_b = (1/T) 10 \log_2 M \quad \text{(bps)} \tag{3.7}$$

This formula shows that the bit rate grows linearly with the baud rate (symbol rate) and logarithmically with M.

3.3.3 Spectral Efficiency

Let's assume we are assigned a RF bandwidth W. Common bandwidth in the 4-, 6-, 7-, and 11-GHz bands are 20, 30, or 40 MHz. One of these bandwidths may be the value of W.

Let η be the spectral efficiency; then

$$\eta = R_b/W \tag{3.8}$$

Substituting from equation (3.7) we have

$$\eta = (1/WT)\log_2 M \tag{3.9}$$

Reference 4 states that, in theory, WT can be as low as 1 without adjacent channel interference. Consider now that we will use a raised cosine filter. Using cosine rolloff shaping, this could be achieved by using a rolloff factor of $\alpha = 0$ or the Nyquist bandwidth. α defines the excess bandwidth, meaning in excess of the Nyquist bandwidth, usually expressed as $1/T$. The Nyquist bandwidth plus the excess bandwidth is commonly expressed as $(1 + \alpha)/T$. Figure 3.6 shows some typical transmitted spectra for cosine rolloff shaping.

Selecting the value for α is very important. If $\alpha = 1$, we compromise spectral efficiency because we require double the Nyquist bandwidth. Manufacturing is made more difficult if $\alpha = 0$, as well as having a system more vulnerable to impairments. In modern digital radios, α is commonly selected as 0.5. It can be shown, with $\frac{1}{T} = \frac{3}{4}$W and α near 0.5, that FCC mask requirements are complied with. The resulting η from (3.9) is $\frac{3}{4}\log_2 M$ so

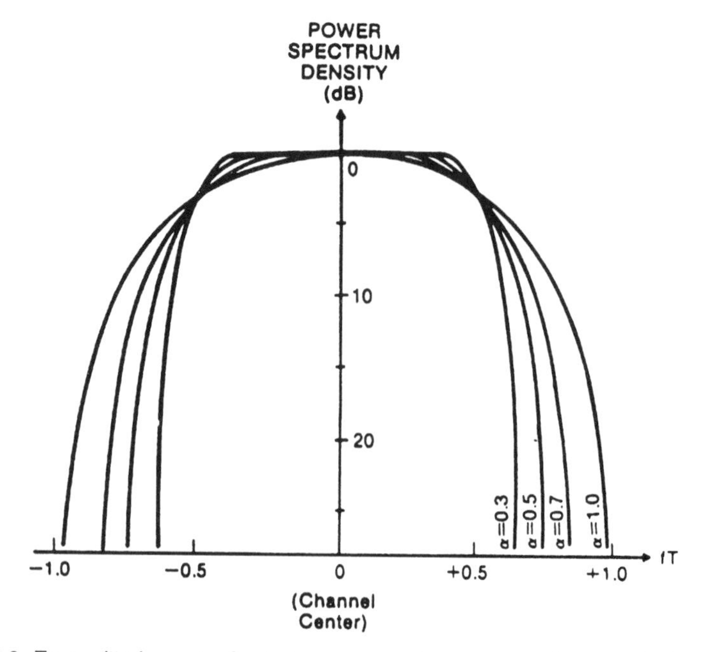

Figure 3.6. Transmitted spectra for typical cosine rolloff shaping. (From *IEEE Communications Magazine*, Oct. 1986, Figure 5, reprinted with permission; Ref. 4.)

that systems using $M = 4$, 16, 64, and 256 have spectral efficiencies of 1.5, 3.0, 4.5, and 6.0 bits/Hz respectively.

3.3.4 Power Amplifier Distortion

There is a serious distortion consideration for M-QAM waveforms due to the fact that digital microwave transmitter power amplifiers are peak-power-limited devices. These devices become increasingly nonlinear as they approach saturation.

The problem here is the peak power. However, at the receiver we are interested in the average power to achieve a specified bit error rate. Figure 3.7 shows peak instantaneous power relative to QPSK with square pulses as a function of the cosine rolloff factor, α. The figure is for a BER of 1×10^{-6}.

For $M > 8$, as M increases so does the ratio of peak to average power. To accommodate a given M-ary waveform and rolloff factor, the saturation

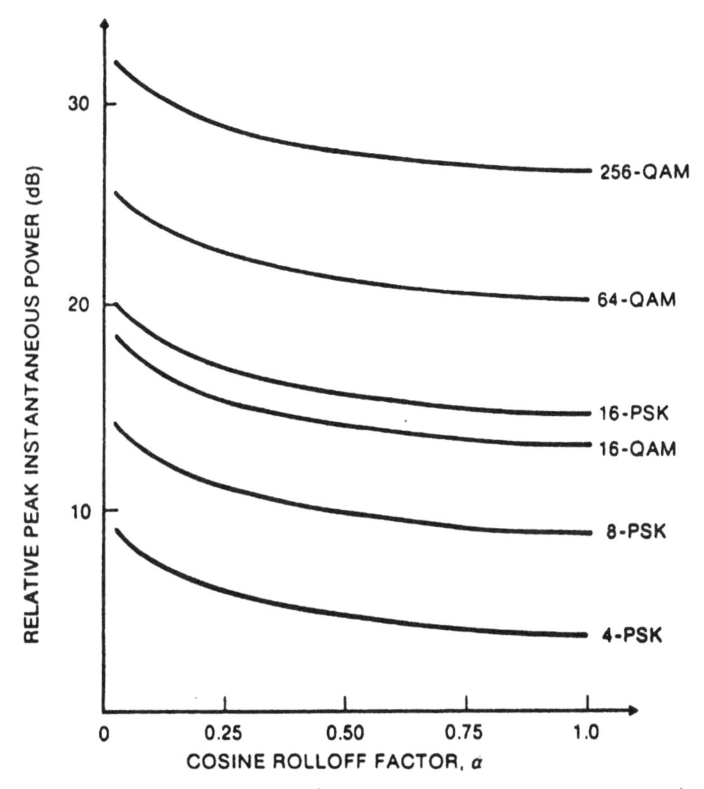

Figure 3.7. Peak instantaneous power (relative to QPSK with square pulses) for M-ary waveforms for several values of M. (From *IEEE Communications Magazine*, Oct. 1986, Figure 8, reprinted with permission; Ref. 4.)

power of the transmitter power amplifier must be sufficiently large that the peak power input lies in its linear range.

If power peaks push the transmitter into its nonlinear range, then nonlinear distortion results. It causes spectral spreading of the transmitter output. Predistortion and postdistortion of the transmitter signal help.

3.4 COMPARISON OF SEVERAL TYPES OF MODULATION

3.4.1 Objective

Our interest here, of course, is spectral efficiency as well as BER versus carrier-to-noise ratio performance. The section is based on ITU-R F.1101, 1994 F-Series, Vol. 1 (Ref. 7).

3.4.2 Definitions and Notation

There are three concepts of power level:

- W_{in} Received maximum steady-state signal power (i.e., the value of the carrier mean level related to the highest state of the modulation format)
- W_{av} Received average signal power
- W_p Received absolute peak of the mean power (i.e., peak of the signal envelope)

Figure 3.8 relates ideal W_p to W_{in} signal power ratio for multilevel QAM-type formats where half of the rolloff shaping is done at the transmitter and half on the receiver side.

Two normalized carrier-to-noise ratio concepts are introduced:

$$W_{(\text{dB})} = 10 \log\left(\frac{W_{\text{in}}}{W_n \cdot f_n}\right) \tag{3.10}$$

$$\text{S/N}_{(\text{dB})} = 10 \log\left(\frac{W_{\text{av}}}{W_n \cdot b_n}\right) \tag{3.11}$$

where f_n = bandwidth numerically equal to the bit rate (B) of a binary signal before modulation

$b_n = f_n/n$, a bandwidth numerically equal to the symbol rate bandwidth of a binary signal before the modulation process of 2^n states

W_n = noise power (spectral) density at receiver ($= N_0$ in our previous discussions)

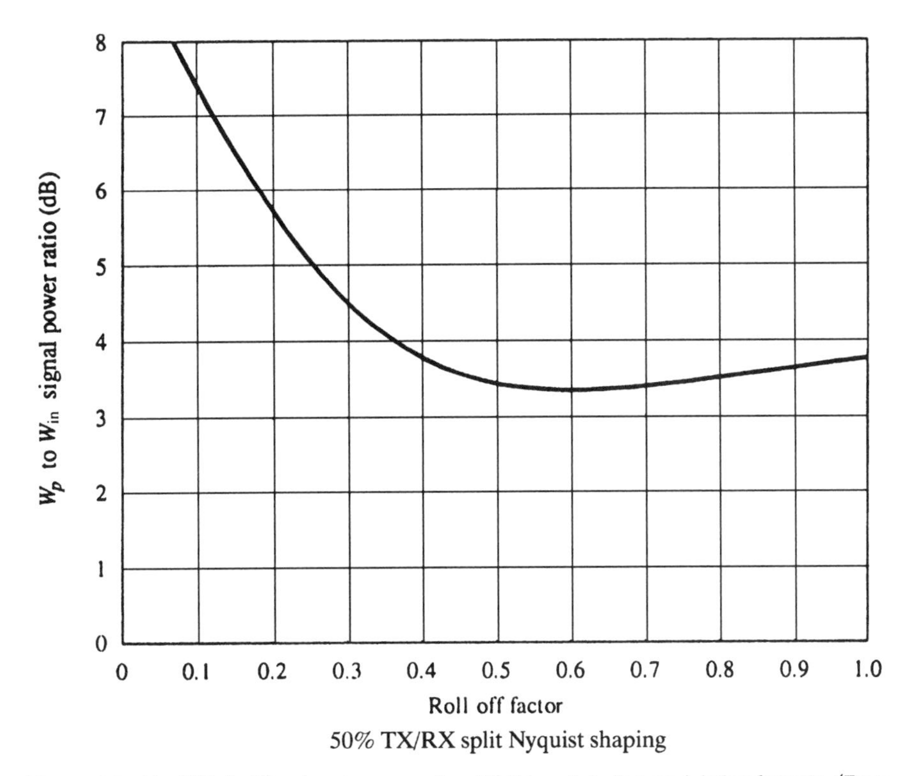

Figure 3.8. Ideal W_p to W_{in} signal power ratio of QAM and similar modulation formats. (From ITU-R Rec. F.1101, Figure 3, p. 239, ITU-R F Series, 1995; Ref. 7.)

The bit rate B is the gross bit rate and takes into account the redundancy possibly introduced for service or supervisory channels, error control, and so on.

It is noted, of course, that formula (3.10) actually expresses E_b/N_0.

The above carrier-to-noise ratio definitions are related through simple scaling factors as follows:

$$W = \mathrm{S/N} - 10 \log n + 10 \log(W_{\text{in}}/W_{\text{av}}) \tag{3.12}$$

3.4.3 Modulation Format Comparison

Table 3.1 shows carrier-to-noise ratios based on a BER of 1×10^{-6} for some common modulation formats and their bandwidth efficiencies based on Nyquist bandwidth (i.e., a bandwidth numerically equal to the bit rate).

The parameters listed in Table 3.1 are, in practical terms, idealistic because the only source of errors is due to thermal noise in the receiver. No modulation implementation loss is considered. See Section 3.3.4.

TABLE 3.1 Comparison of Different Modulation Schemes[a]

System	Variants	$W(=E_b/N_0)$ (dB)	S/N (dB)	Nyquist Bandwidth (b_n)
	Basic Modulation Schemes			
FSK	2-state FSK with discriminator detection	13.4	13.4	B
	3-state FSK (duo-binary)	15.9	15.9	B
	4-state FSK	20.1	23.1	$B/2$
PSK	2-state PSK with coherent detection	10.5	10.5	B
	4-state PSK with coherent detection	10.5	13.5	$B/2$
	8-state PSK with coherent detection	14.0	18.8	$B/3$
	16-state PSK with coherent detection	18.4	24.4	$B/4$
QAM	16-QAM with coherent detection	17.0	20.5	$B/4$
	32-QAM with coherent detection	18.9	23.5	$B/5$
	64-QAM with coherent detection	22.5	26.5	$B/6$
	128-QAM with coherent detection	24.3	29.5	$B/7$
	256-QAM with coherent detection	27.8	32.6	$B/8$
	512-QAM with coherent detection	28.9	35.5	$B/9$
QPR[b]	9-QPR with coherent detection	13.5	16.5	$B/2$
	25-QPR with coherent detection	16.0	20.8	$B/3$
	49-QPR with coherent detection	17.5	23.5	$B/4$
	Basic Modulation Schemes with Forward Error Correction			
QAM with block codes[c]	16-QAM with coherent detection	13.9	17.6	$B/4 \times (1+r)$
	32-QAM with coherent detection	15.6	20.6	$B/5 \times (1+r)$
	64-QAM with coherent detection	19.4	23.8	$B/6 \times (1+r)$
	128-QAM with coherent detection	21.1	26.7	$B/7 \times (1+r)$
	256-QAM with coherent detection	24.7	29.8	$B/8 \times (1+r)$
	512-QAM with coherent detection	25.8	32.4	$B/9 \times (1+r)$

[a] Theoretical W and S/N values at 10^{-6} BER; calculated values may differ slightly due to different assumptions.
[b] QPR = quadrature partial response.
[c] As an example, BCH error correction with a redundancy of 6.7% ($r = 6.7\%$) is used for calculations in this table.
Source: Table 1a, p. 241, ITU-R Rec. F.1101, ITU-R F-Series, 1995 (Ref.1).

The actual signal-to-noise parameter is related to the average RSL (receive signal level), for the relevant BER, through the noise figure of the receiver and the bit rate relationship:

$$\mathrm{RSL}_{\mathrm{BER}} = 10 \log kT^* + 10 \log b_n + \mathrm{NF}_{(\mathrm{dB})} + (\mathrm{S/N})_{\mathrm{BER}} \qquad (3.13)$$

3.4.4 Notes on Implementation and BER Performance

An ideal digital radio system using quadrature modulation is shown in Figure 3.9. The system is ideal in the sense that the filters used are bandlimited to

*In many texts this is taken as −204 dBW or −174 dBm, the noise temperature of an uncooled (room temperature) "perfect" receiver in 1 Hz of bandwidth.

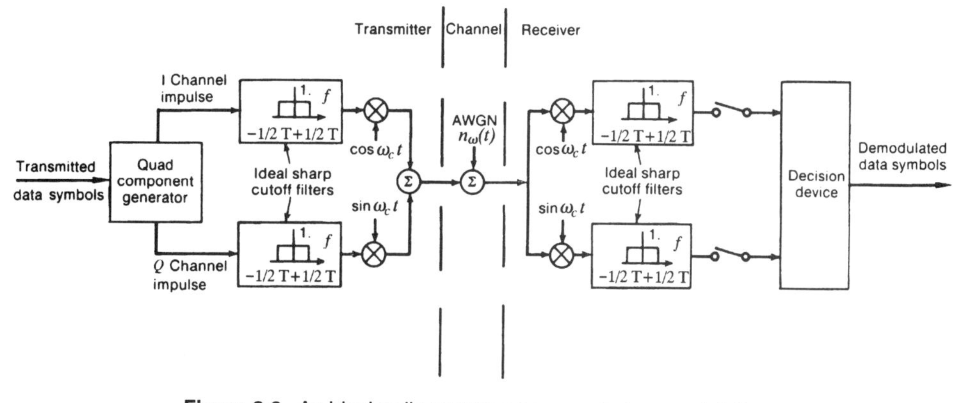

Figure 3.9. An ideal radio system using quadrature modulation.

the Nyquist bandwidth (i.e., $\alpha = 0$). The transmitted RF spectrum resulting from the use of such a filter is thus ideally band-limited to a bandwidth of $1/T$, where T is the transmitted symbol (baud) rate.

As shown in Figure 3.9 (see also Figure 3.2), the two baseband signals resulting from a pair of data bits are used (in its simplest configuration, QPSK) to modulate quadrature carriers. At the far end receiver, the RF signal is down converted back to baseband by the use of coherent quadrature mixer references. Because of the ideal Nyquist character of the channel, samples of the baseband signals will be transmitted with impulse weights corrupted by the channel noise, but with zero intersymbol interference (ISI). The sampled quadrature components at the receiver are fed to a decision device that optimally decides which quadrature pair was transmitted given the receiver pair of samples.

Figure 3.10 is a conceptual block diagram of a 16-QAM modulator. It is similar to the QPSK modulator shown in Figure 3.9 except that the I

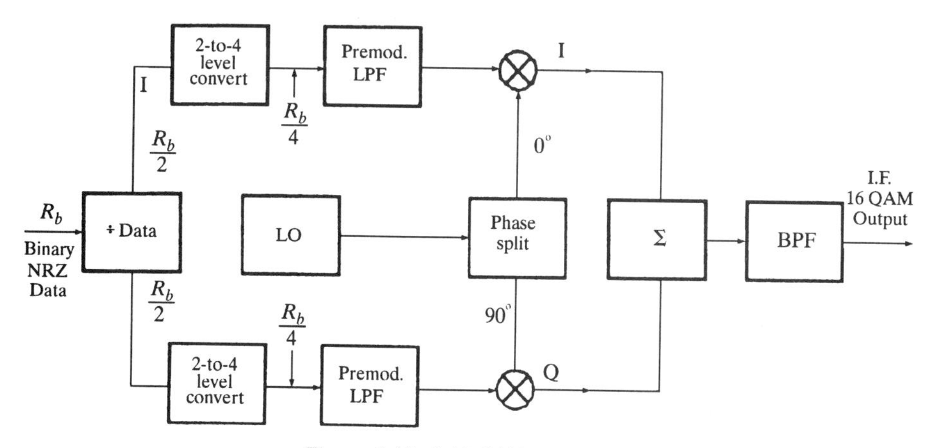

Figure 3.10. A 16-QAM modulator.

(in-phase) and Q (quadrature) carrier are each modulated by 4-level signals. Two serial bits are converted to one of four voltage levels, which control the output of the balanced modulators. This shows how the symbol rate (baud rate) becomes $R_b/4$. In 64-QAM, 3 bits are taken to generate one of eight voltage levels on both the I and Q streams. For such systems it is necessary to use linear modulators; otherwise, the phase state diagram will be distorted since the four voltage levels will not translate directly to the correct carrier level.

Bit error rate performance for various QAM and PSK waveforms for a BER of 1×10^{-6} is shown in Table 3.1. The column W in the table is defined the same way as E_b/N_0, equation (3.10). Some channel capacity limits are shown in Figure 3.11. Typical error performance curves are shown in Figure 3.12.

Figure 3.11 shows plots comparing ideal M-QAM and M-PSK systems. The figure gives values of C_I (bit packing) in bits/s/Hz versus E_b/N_0 for error rates of 1×10^{-5}, 1×10^{-7}, and 1×10^{-9}. At the left in the figure is a plot of Shannon's channel-capacity curve. Shannon's curve represents a theoretical bound on the absolute maximum capacity at zero error rate for

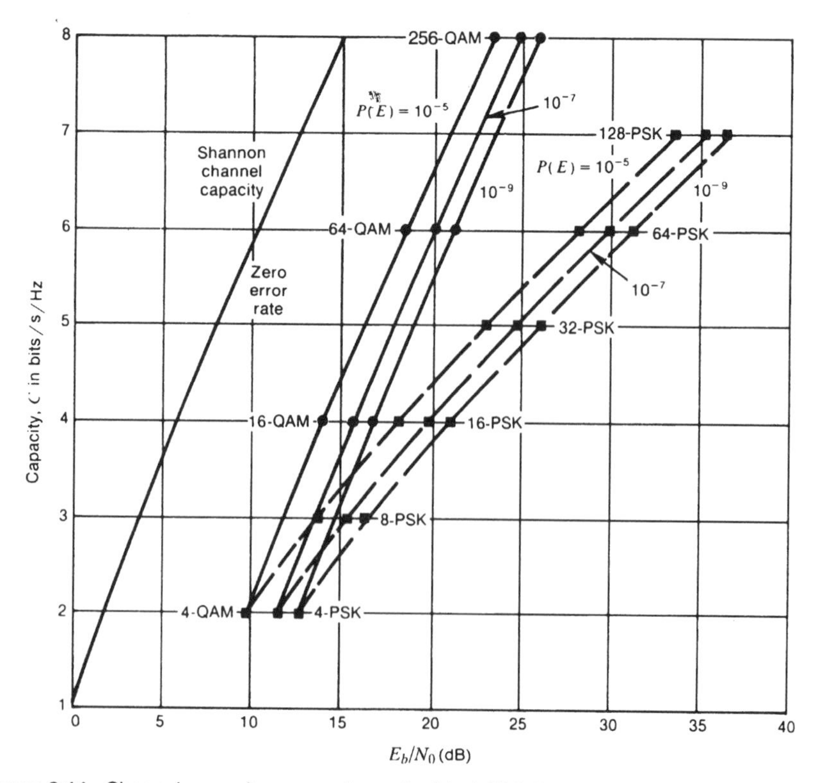

Figure 3.11. Channel capacity comparisons for ideal M-QAM and M-PSK systems (Ref. 8).

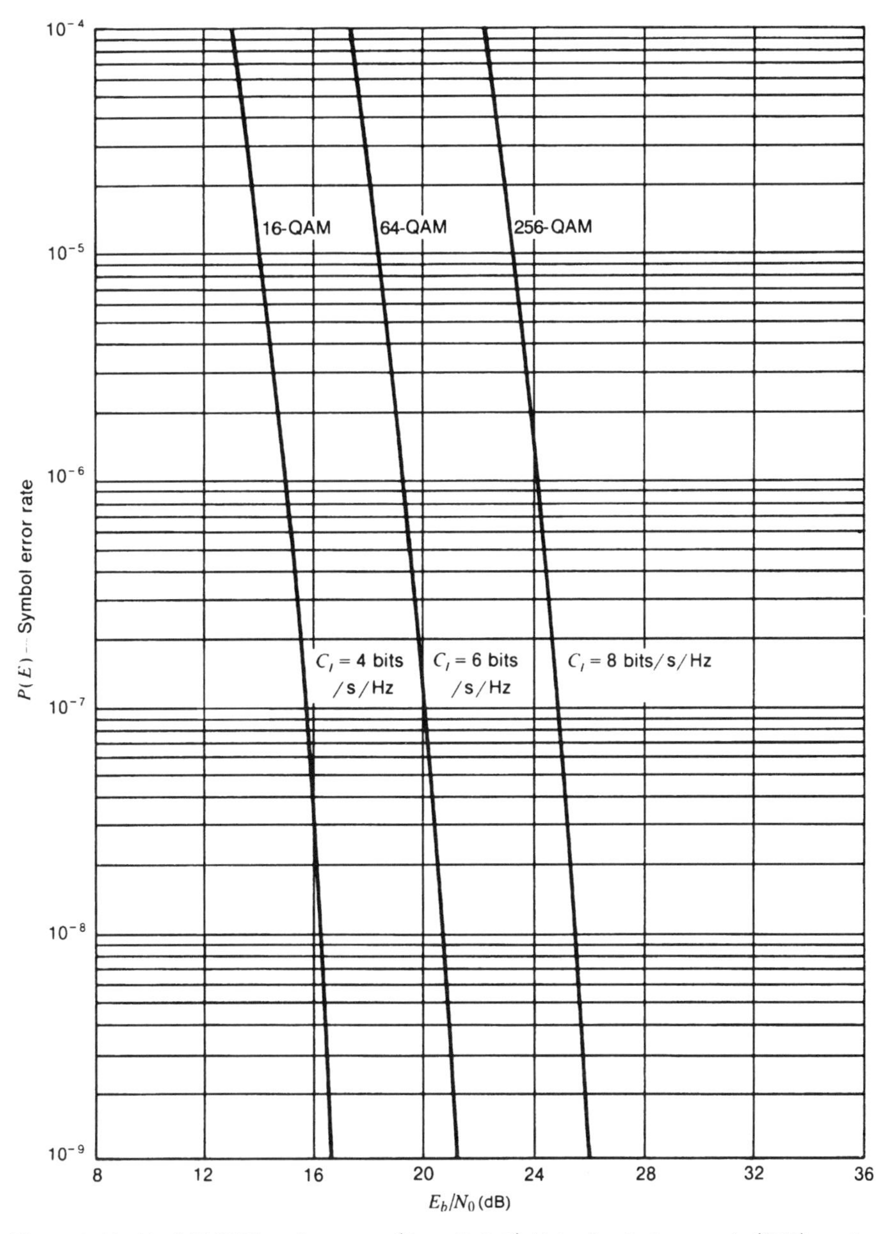

Figure 3.12. Ideal M-QAM performance. (From Ref. 8.) *Note:* Symbol error rate (SER) can be converted to bit error rate (BER) assuming Gray coded state assignment by:

$$\text{BER} = (1/\log M) \times \text{SER}$$

Thus for 16-QAM, BER = SER/4; for 64-QAM, BER = SER/6; and for 256-QAM, BER = SER/8. For example, for 64-QAM where the SER is 1×10^{-9}, the BER = 1.667×10^{-10} (Ref. 9).

a given E_b/N_0 for an infinitely complex digital modulation/demodulation transmission system, but it represents unattainable performance in a real-world practical system. It should be noted in the figure that the ideal M-QAM schemes track parallel to Shannon's capacity bound, being about 8 dB away at 1×10^{-5} BER.

We must be sure to distinguish between theoretical and practical values for E_b/N_0. The practical values are always higher than the equivalent theoretical values. The amount that they are higher (in dB) is called *modulation implementation loss*. This loss takes into account degradation parameters not included in an ideal system.

Items that may contribute to the modulation implementation loss budget are the following:

1. The effect of nonideal sharp cutoff spectral shaping filters.
2. The effect of additional linear distortion created by realistic filters.
3. The effect of signaling with a peak-power-limited amplifier.
4. The relative efficiency between preamplifier filtering and postamplifier filtering.
5. The effect of practical nonlinearities encountered with real amplifiers on system performance.
6. Technique(s) required for adapting realistic amplifier nonlinearities in order to render the amplifier linear for the purpose of supporting highly bandwidth-efficient digital data radio transmission.
7. The effect on performance of practical imperfections in an implementation of a bandwidth-efficient modem including a baseband equalizer for counteracting linear distortion caused by realistic filter characteristics.

Values for modulation implementation loss can be as high as over 5 dB (Ref. 10); my own experience suggests values of 2 dB and less.

3.5 SOME SYSTEM IMPAIRMENTS PECULIAR TO DIGITAL OPERATION

One particular concern we must be aware of is signal space distance, briefly described in Section 3.3.2. There we showed that as the value of M increases, the distance between points on the signal constellation (state diagram, see Figure 3.5) decreases. For instance, 16-QAM has 16 points in the constellation; 64-QAM has 64 points; 256-QAM has 256 points; and so forth. A digital transmitter and its companion far end receiver must be capable of maintaining good resolution among 256 signal points on a 16×16 grid; a 512-QAM system has twice as many signal points. Of the two, certainly the receiver is more vulnerable because it must resolve which signal point is transmitted. Outside disturbances tend to deteriorate the system by masking or otherwise confusing the correct signal point, degrading error performance.

One such outside disturbance is additive white Gaussian noise (AWGN, see Figure 3.9). The bandwidth-efficient modem consequently requires a higher E_b/N_0 for a given symbol rate as the number of bits per second per hertz is increased.

Another type of outside disturbance is created by co-channel and/or adjacent-channel interference. Here the bandwidth-efficient modem is more vulnerable to interference since less interference power is required to push the transmitted signal point selection to an adjacent point, thus resulting in a *hit* and causing error(s) at the receiver.

For the system engineer, one of the most perplexing impairments is caused by the transmission medium itself. This is multipath distortion causing signal dispersion. The problem is dealt with, in part, in the next subsection.

3.5.1 Mitigation Techniques for Multipath Fading*

In analog radiolink systems, multipath fading results in an increase in thermal noise as the RSL drops. In digital radio systems, however, there is a degradation in BER during periods of fading that is usually caused by intersymbol interference due to multipath. Even rather shallow fades can cause relatively destructive amounts of intersymbol interference. This interference results from frequency-dependent amplitude and group delay changes. The degradation depends on the magnitude of in-band amplitude and delay distortion. This, in turn, is a function of fade depth and time delay between the direct and reflected signals.

Five of the most common methods to mitigate the effects of multipath (Ref. 20) on digital radiolinks are:

1. System configuration (i.e., adjusting antenna height to avoid ground reflection; implementation of space and/or frequency diversity).
2. Use of IF combiners in diversity configurations.
3. Use of baseband switching combiners in diversity configuration.
4. Adaptive IF equalizers.
5. Adaptive transversal equalizers.

System configuration techniques have been described previously in Chapter 2, such as sufficient clearance to avoid obstacle diffraction, high–low antennas to place a reflection point on "rough" ground, and particularly the use of diversity. We will narrow our thinking here to space diversity. We now expand on several of the items listed above for mitigation of the effects of multipath.

*Section 3.5.1 is based on Section 3.2/3.3 of ITU-R Rec. F.1093 (Ref. 11).

An optimal IF combiner for digital radio receiving subsystems can be designed to adjust adaptively to path conditions. One such combining technique, the maximum power IF combiner, vectorially adds the two diversity paths to give maximum power output from the two input signals. This is done by conditioning the signal on one path with an endless phase shifter, which rotates the phase on this path to within a few degrees of the signal on the other diversity path prior to combining. The output of this type of combiner can display in-band distortion that is worse than the distortion on either diversity path alone, but functions well to keep the signal at an acceptable level during deep fades on one of the diversity paths.

A minimum distortion IF combiner operation is similar in most respects to the maximum power IF combiner but uses a different algorithm to control the endless phase shifter. The output spectrum of the combiner is monitored for flatness such that the phase of one diversity path is rotated and, when combined with the second diversity path, produces a comparatively flat spectral output. The algorithm also suppresses the polarity inversion on the group delay, which is present during nonminimum phase conditions. One disadvantage of this combiner is that it can cancel two like signals such that the signal level can be degraded below threshold.

Reference 12 suggests a dual algorithm combiner that functions primarily as a maximum power combiner and automatically converts to a minimum distortion combiner when signal conditions warrant. Using space diversity followed by a dual-algorithm combiner can give improvement factors better than 150.

Adaptive IF equalizers attempt to compensate directly at IF for multipath passband distortion. Digital radio transmitters emit a transmit spectra of relatively fixed shape. Thus various points on the spectrum can be monitored, and when distortion is present, corrective action can be taken to restore spectral fidelity. The three most common types of IF adaptive equalizers are shape-only equalizers, slope and fixed notch equalizers, and tracking notch equalizers.

Another equalizer is the adaptive transversal equalizer, which is efficient at canceling intersymbol interference due to signal dispersion caused by multipath. The signal energy dispersion can be such that energy from a digital transition or pulse arrives both before and after the main bang of the pulse. The equalizer uses a cascade of baud delay sections that are analog elements to which the symbol or baud sequence is inputted. The "present" baud or symbol is defined as the output of the *N*th section. Sufficient sections are required to encompass those symbols or bauds that are producing the distortion. These transversal equalizers provide both feedforward and feedback information. There are both linear and nonlinear versions. The nonlinear version is sometimes called a decision feedback equalizer. Reference 12 reports that both the IF and transversal equalizers show better than three times improvement in error rate performance over systems without such equalizers.

3.5.2 ITU-R Guidelines on Combating Propagation Effects

3.5.2.1 Space Diversity. Space diversity is one of the most effective methods of combating multipath fading. For digital radio systems, where performance objectives are difficult to meet owing to waveform distortion caused by multipath effects, system designs must often be based on the use of space diversity.

By reducing the effective incidence of deep fading, space diversity can reduce the effects of various types of interference. In particular, it can reduce the short-term interference effects from cross-polar channels on the same or adjacent channel frequencies, and the interference from other systems and from within the same system.

Linear amplitude dispersion (LAD) is an important component of waveform distortion and quadrature crosstalk effects and can be reduced by the use of space diversity. Diversity combining used specifically to minimize LAD is among the methods that are particularly effective in combating this type of distortion.

The improvement derived from space diversity depends on how the two signals are processed at the receiver (combiner). Two examples of techniques are "hitless" switching and variable phase combining. The "hitless" switch selects the receiver with the greatest eye opening or the lowest error rate, and the combiners use either co-phase or various types of dispersion-minimizing control algorithms. "Hitless" switching and co-phase combining provide very similar improvements.

3.5.2.2 Adaptive Channel Equalizing. Some form of receiver equalizing is usually necessary in the radio channel(s). As propagation conditions vary, an equalizer must be adaptively controlled to follow the variations in transmission characteristics. Such equalizers work in either the frequency domain or the time domain.

FREQUENCY DOMAIN EQUALIZATION. Equalizers operating in the frequency domain are comprised of one or more linear networks that are designed to produce amplitude and group delay responses. They compensate for transmission impairments, which are most likely to produce system performance degradation during periods of multipath fading. Table 3.2 shows several alternative equalizer structures that may be considered by the system engineer.

TIME DOMAIN EQUALIZATION. Time domain equalizers combat intersymbol interference directly. With these equalizers control information is derived by correlating the interference that appears at the instant of decision with the various adjacent symbols producing it, and this result is used to adjust tapped delay line networks to provide appropriate cancellation signals. Such an equalizer is able to handle simultaneous and independent types of distortion

TABLE 3.2 Comparison of Adaptive Equalizers

Generic Type		Description of Equalizer	Complexity of Implementation	Fade Characteristic and Position of Maximum Attenuation[a]: Minimum Phase, Out-of-Band	Minimum Phase, In-Band	Nonminimum Phase, Out-of-Band	Nonminimum Phase, In-Band
Frequency domain equalizers	F1	Amplitude tilts	Simple	2	1	2	1
	F2	F1 + parabolic amplitude	Simple	2	2	2	2
	F3	F2 + group delay tilt	Complex (moderately complex)	3	2	3(1)	2(1)
	F4	F3 + parabolic group delay (for F3 and F4 ratings in brackets apply "minimum phase" control assumptions)	Complex (moderately complex)	3	3	3(1)	3(0)
	F5	Single tuned circuit ("agile notch")	Simple	3	3	1	0
Time domain equalizers	T1	Two-dimensional linear transversal equalizer	Moderately complex/complex	3	2	3	2
	T2	Cross-coupled decision feedback equalizer	Moderately complex	3	3	2	1
	T3	T1 + T2 full-time domain equalization	Complex	3	3	3	2

[a]Effectiveness of equalization: 3, produces a well-equalized response; 2, produces a moderately equalized response; 1, produces a partially equalized response; and 0, not effective.

Source: Table 1, p. 53, ITU-R Rec. F.1093 (Ref. 11).

that arise from amplitude and group delay deviations in the fading channel. It therefore can provide either minimum phase or nonminimum phase characteristic compensation.

Quadrature distortion compensation is very important in QAM-type systems. Significant destructive effects are associated with crosstalk generated by channel asymmetries, for which a time domain equalizer must correct.

ADAPTIVE EQUALIZERS USED IN COMBINATION WITH SPACE DIVERSITY COMBINING. By far the best method of combating performance deterioration and outage due to multipath effects is to use adaptive equalization in conjunction with space diversity. There are synergistic effects between the two. It has been shown that the total improvement usually exceeds the product of the corresponding individual improvements of space diversity and adaptive equalization considered separately. The product of the space diversity improvement and the square of the equalizer improvement equals the total improvement of the combination of space diversity and adaptive equalization.

SUMMARY. There are three principal degradations that can cause deteriorated performance or system outages on digital radio systems: interference, thermal noise, and waveform distortion. Equalization is really only effective against waveform distortion. Adaptive equalizer performance must be judged in this light, especially where link degradation or outage is traced principally to waveform distortion, especially dispersion.

3.6 PERFORMANCE REQUIREMENTS AND OBJECTIVES FOR DIGITAL RADIOLINKS

3.6.1 Introduction

The principal objective of this section is to define error performance for digital radio systems and links. The section will also cover jitter and wander for the digital radio portion of a network.

3.6.2 Five Definitions*

Bit Error Ratio (Rate) (BER). For a binary digital signal, the ratio of the number of errored bits received to the number of bits received over a given time interval.

Residual Bit Error Ratio (Rate) (RBER). The bit error ratio in the absence of fading, including allowance for system inherent errors, environment, aging effects, and long-term interference.

Errored Second (ES). The time interval of 1 second during which a given digital signal is received with one or more errors. (*Note:* According to ITU-T recommendations, an errored second is defined for each direction of a 64-kbps circuit-switched connection.)

Severely Errored Second (SES). The time interval of 1 second during which a digital signal is received with an error ratio greater than a specified value. (*Note*: According to ITU-T recommendations, a severely errored second is defined for each direction of a 64-kbps circuit-switched connection and the specified BER value is 1×10^{-3}.)

Degraded Minute (DM). The time intervals of m seconds, 60 of them being not severely errored seconds but for which the error ratio is greater than a specified value. (*Note*: According to ITU-T recommendations, a degraded minute is defined for each direction of a circuit-switched 64-kbps connection and the specified BER value is 1×10^{-6}.)

3.6.3 Hypothetical Reference Digital Path (HRDP) for Radio-Relay Systems with a Capacity Above the Second Hierarchical Level†

To define error performance, a hypothetical reference digital path is often used to apportion error accumulation. The path is 2500 km long and

*From CCIR Rec. 592-2, 1994 F Series, Part 1 (Ref. 13).
†Section 3.6.3 is based on ITU-R Rec. F.634-3 (Ref. 14).

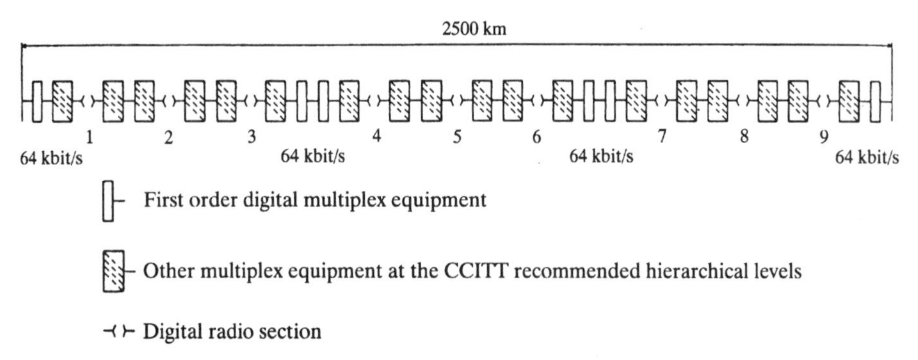

Figure 3.13. Hypothetical reference digital path for radio-relay systems with capacity above the second hierarchical level (e.g., DS3 and above; E-3 and above). (From Figure 1, p. 16, CCIR Rec. 556-1, 1994 F Series, Part 1; Ref. 15.)

comprises, in each direction, nine sets of digital multiplexing equipment at the ITU-T recommended hierarchical levels.

The HRDP contains nine identical digital radio sections of equal length, as shown in Figure 3.13.

3.6.4 Error Performance Objectives for Real Digital Radiolinks Forming Part of a High-Grade Circuit in an ISDN Network

When designing digital radiolinks in a high-grade circuit forming part of an ISDN, the following error performance should be incorporated in the requirements section of a RFP for a link of length L, between 280 and 2500 km:

A. BER $> 1 \times 10^{-3}$ for no more than $(L/2500) \times 0.054\%$ of any month. Integration time shall be 1 second. (See Note 1.)

B. BER $> 1 \times 10^{-6}$ for more than $(L/2500) \times 0.4\%$ of any month. Integration time shall be 1 minute. (See Notes 1, 5, and 6.)

C. Errored seconds for no more than $(L/2500) \times 0.32\%$ of any month. (See Notes 1, 2, and 6.)

D. RBER (residual BER) is given as

$$\text{RBER} \leq (L \times 5 \times 10^{-9})/2500$$

The BER performance criteria are to be complied with at the system bit rate. The ES criterion should be complied with at the 64-kbps level.

Note 1. The term "any month" used above is defined in ITU-R Rec. PN.581, where measurements are used to ensure compliance with the error performance values given above. Then the propagation conditions also need to be assessed and related to propagation data representative of "any month" conditions.

Note 2. The relationship between the errored seconds of a 64-kbps channel and the corresponding parameters, which may be measured directly at the bit rate of the radio-relay system, is still under study by the ITU-R organization. At this time, the errored seconds are to be measured only at the 64-kbps interface. However, it should be noted that if the objectives given in A, B, and D above are met, then the objective given in C above is usually satisfied, taking into account the typical cumulative error ratio distribution for high-grade systems.

Note 3. The provisional method of measurement of the RBER involves taking the BER measurements over a period of 1 month using a 15-minute integration time, discarding 50% of 15-minute intervals that contain the worst BER measurements, and taking the worst of the remaining measurements. The RBER limits and the method of measurement are under study by the ITU-R organization.

Note 4. Seconds when the BER $\geq 1 \times 10^{-3}$ should be excluded, when measuring degraded minute performance.

Note 5. Measurements of 1×10^{-6} BER criterion at the system bit rate, using different integration times, indicate that the errors occurring within a 1-minute period may be clustered. The DM objective may therefore become a more stringent requirement to be complied with than the SES objective and possibly more stringent than the DM objective at 64 kbps.

Note 6. The ES allowance includes all performance degradations other than unavailability.

3.6.4.1 *Error Burst Objectives.* CCIR Rec. 594-3 (Ref. 16) deals with error burst objectives. The principal cause of burst errors is fading on radio systems, typically LOS microwave. The BER floor for telecommunication systems is 1×10^{-3}. This BER value is based on supervisory signaling for speech telephony. Even loss of supervisory signaling for very short time durations (e.g., < 1 second) may cause telephone call dropout. In other words, active telephone calls on the circuit during such fade conditions may experience loss of each call and the return of a dial tone. Underlying multiplex alignment may also be lost, depending on the BCI* span.

Figure 3.14 shows a possible limit that would serve to control the incidence of multiple call dropout and is derived from material supplied by ITU-R from the United Kingdom.

*BCI = bit count integrity. This implies how long a digital bit stream remains in synchronization after loss of a time synchronized source. This means that the bit stream must now rely on its free-running clock.

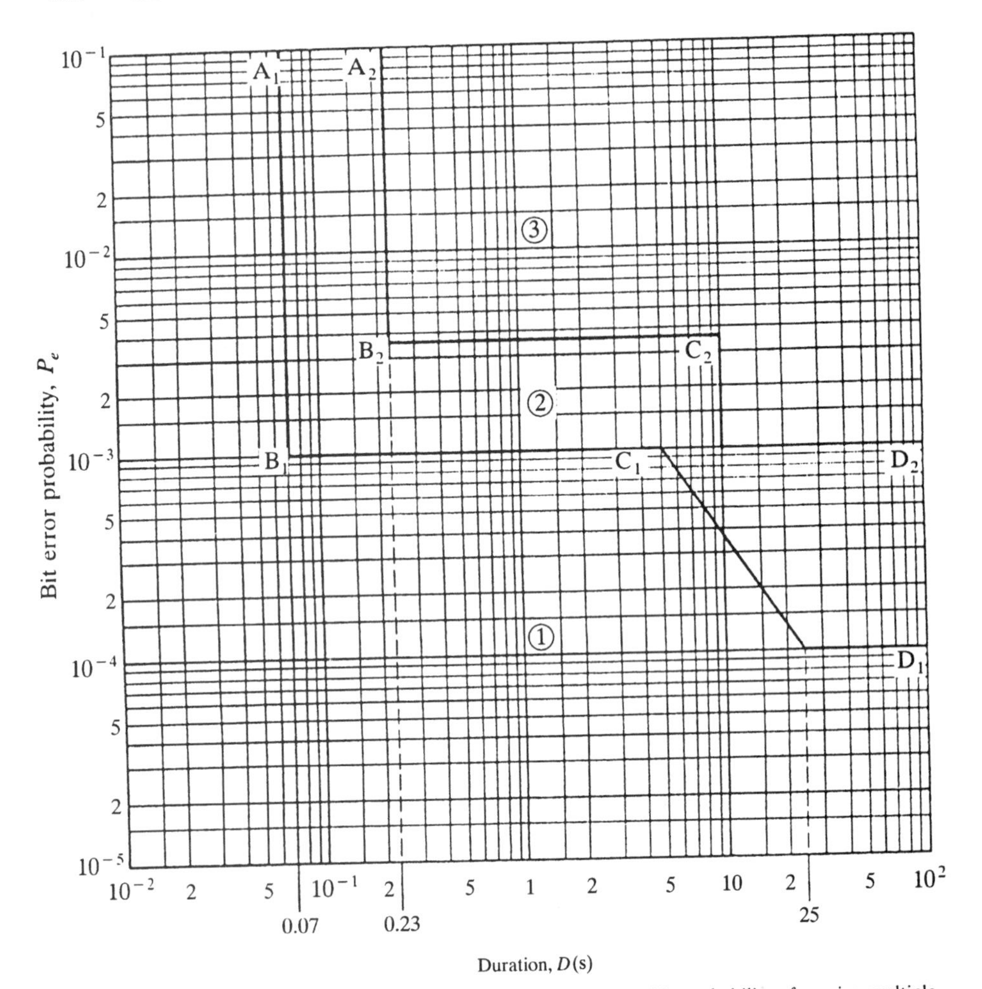

Area ①: error bursts confined within this area will have a negligible probability of causing multiple call dropout.

Area ②: error bursts extending into this area will have an increased probability of causing multiple call dropout and would probably be unacceptable.

Area ③: error bursts extending into this area are likely to have an unacceptably high probability of causing multiple call dropout.

$A_1 B_1 C_1$, $A_2 B_2 C_2$ } boundaries related to signalling performance criteria.

$C_1 D_1$, $C_2 D_2$ } boundaries related to subjective performance criteria.

Figure 3.14. Possible error burst objective. (From Figure 1, p. 20, CCIR Rec. 594-3, ITU-R Recommendations, 1994 F Series, Part 1; Ref. 16.)

3.6.5 Error Performance Objectives of a 27,500-km Hypothetical Reference Path

The ITU-R organization (Ref. 17) recommends that LOS microwave paths forming part of the international portion of a 27,500-km Hypothetical Reference Path (HRP) should be based on both distance-based and country-based allocations as specified in ITU-T Rec. G.826 (Ref. 18). (Also see Section 7.6.4.)

It is further stipulated that for each direction of a LOS microwave radio path forming part of an international portion of a constant bit rate path at or above the primary rate,* which extends (1) from country border to country border for an intermediate country or (2) from the international gateway (IG) to country border for a terminating country, the allocated error performance objectives are composed of a distance-based allocation of 1%/500 km and at least part or all of the country-based allocation of 2% per intermediate country or 1% per terminating country, of the total allocations for the 27,500-km HRP.

The following use is recommended for the country-based allocation:

- A constant value (up to the maximum value given in the paragraph above) will be applied to LOS microwave paths with a length L greater than a reference length L_{ref}.
- A fraction of the constant value, proportional to the length L, will be applied to LOS microwave paths with a length equal to or less than the reference length L_{ref}.

The value of the reference length L_{ref} is provisionally taken to be 1000 km.

Table 3.3 gives the error performance objectives applicable to each direction of a LOS microwave path of length L in the international portion of a constant bit rate path at or above the primary rate.

Note: Readers are encouraged to review ITU-T Rec. G.826 (Ref. 18) for concise definitions of parameters. Some ITU-T Rec. G.826 definitions are as follows:

Block. A set of consecutive bits associated with the path; each bit belongs to one and only one block.

Errored Block (EB). A block in which one or more bits are in error.

Errored Second (ES). A 1-second period with one or more errored blocks.

Severely Errored Second (SES). A 1-second period that contains $\geq 30\%$ errored blocks or at least one *severely disturbed period* (SDP).

Background Block Error (BBE). An errored block not occurring as part of a SES.

*Primary rate may refer to the North American DS1 (1.544 Mbps) or the European E-1 (2.048 Mbps).

TABLE 3.3 Error Performance Objectives for a 27,500-km HRP

Rate (Mbit/s)	1.5 to 5	> 5 to 15	> 15 to 55	> 55 to 160	> 160 to 3500
Errored second ratio	$0.04 \times (F_L + B_L)$	$0.05 \times (F_L + B_L)$	$0.075 \times (F_L + B_L)$	$0.16 \times (F_L + B_L)$	
Severely errored second ratio	$0.002 \times (F_L + B_L)$				
Background block error ratio	$3 \times 10^{-4} \times (F_L + B_L)$	$2 \times 10^{-4} \times (F_L + B_L)$			

Note: VC-11 and VC-12 (ITU-T Rec. G.709) paths are defined with a number of bits/block of 832 and 1120, respectively, that is, outside the bit/block range recommended for 1.5 to 5 Mbit/s paths in ITU-T Rec. G.826. For these block sizes, the BBER objective for VC-11 and VC-12 is $2 \times 10^{-4} \times (F_L + B_L)$, where:

Distance allocation factor	$F_L = 0.01 \times L/500$	L(km)
Block allowance factor, B_L		
For intermediate countries	$B_L = B_R \times 0.22 \times (L/L_{ref})$	for $L_{min} < L \leq L_{ref}$
	$B_R \times 0.02$	for $L > L_{ref}$
For terminating countries	$B_L = B_R \times 0.01 \times (L/L_{ref})$	for $L_{min} < L \leq L_{ref}$
	$B_R \times 0.01$	for $L > L_{ref}$
Block allowance ratio, B_R	$(0 < B_R \leq 1)$	
Reference length, L_{ref}	$L_{ref} = 1000$ km (provisionally)	

Source: Table 1, p. 49, ITU-R Rec. F.1092, 1994 F Series, Part 1 (Ref. 17).

Errored Second Ratio (ESR). The ratio of ES to total seconds in available time during a fixed measurement interval.

Severely Errored Second Ratio (SESR). The ratio of SES to total seconds in available time during a fixed measurement interval.

Background Block Error Ratio (BBER). The ratio of errored blocks to total blocks during a fixed measurement interval, excluding all blocks during SES and unavailable time.

Unavailable Time. A period of unavailable time begins at the onset of ten consecutive SES events. These 10 seconds are considered to be part of unavailable time. A new period of available time begins at the onset of ten consecutive non-SES events. These 10 seconds are considered to be part of available time.

Also see Section 3.6.2.

3.6.6 Jitter and Wander

Jitter and wander requirements remain under study by the ITU-R organization. Jitter and wander are equipment-related impairments and thus guidance may be taken from applicable ITU-T recommendations such as G.783 and G.958 (Refs. 19 and 20).

3.6.7 Error Performance from a Bellcore Perspective*

The requirements given below are for all one-way system options and apply at the maximum short-haul design length. For the requirements given below, a measurement period consists of a series of 1-second intervals.

> The BER at the interface levels DSX-1, DSX-1C, DSX-2, and DSX-3 shall be less than 2×10^{-10}, excluding all burst errored seconds in the measurement period. During a burst errored second,† neither the number of bit errors nor the number of bits is counted.
>
> The frequency of burst errored seconds, other than those caused by protection switching induced by hard equipment failures, shall average no more than four per day at each of the interface levels DSX-1, DSX-1C, DSX-2, and DSX-3. This is a long-term average over many days. Due to day-to-day variation, the number of burst errored seconds occurring on a particular day may be greater than the average.
>
> For systems interfacing at the DS1 level, the long-term percentage of errored seconds (measured at the DS1 rate) shall not exceed 0.04%. This is equivalent to 99.96% error-free seconds (EFS).

3.7 APPLICATION OF HIGH-LEVEL *M*-QAM TO HIGH-CAPACITY SDH/SONET FORMATS

Digital LOS radio systems are often required to transport SDH or SONET high-capacity formats. SDH, we will remember, is the synchronous digital hierarchy, a digital transmission format espoused by European administrations, whereas SONET (synchronous optical network) is a similar format used in North America. Table 3.4 shows the appropriate equivalence between SONET and SDH.

Table 3.5 gives examples of possible channel arrangements that allow transmission of the basic STM-1 (SONET STS-3) rate and multiple STM rates within existing radio-frequency channel plans.

*Section 3.6.7 is based on Bellcore TR-NWT-000499, Sec. 4.3 (Ref. 21).
†A burst errored second is any errored second containing at least 100 errors.

TABLE 3.4 SDH Bit Rates with SONET Equivalents

SDH Level	SDH Bit Rate (kbps)	SONET Equivalent Line Rate
1	155,520 (STM-1)	STS-3/OC-3
4	622,080 (STM-4)	STS-12/OC-12
16	2,488,320 (STM-16)	STS-48/OC-48

TABLE 3.5 Application of SDH/SONET Formats to Existing Radio-Frequency Channel Arrangements

Channel Spacing (MHz)	Capacity	Examples of Modulation Method[a]
20	1 × STM-1	256-QAM, 512-QAM
28, 29, 29.65, 30	1 × STM-1	64-QAM, 128-QAM, 256-QAM
28, 29, 29.65, 30	2 × STM-1	128-QAM(CC), 256-QAM(CC)
40	1 × STM-1	32-QAM, 64-QAM
40	2 × STM-1	32-QAM(CC), 64-QAM(CC), 512-QAM
55, 56, 60	1 × STM-1	16-QAM, 32-QAM
55, 56, 60	2 × STM-1	16-QAM(CC), 32-QAM(CC), 64-QAM(CC), 256-QAM
80	2 × STM-1	64-QAM
80	4 × STM-1, 1 × STM-4	64-QAM(CC)
110, 112	1 × STM-1	QPSK(4-QAM)
110, 112	2 × STM-1	16-QAM, 32-QAM
110, 112	4 × STM-1, 1 × STM-4	16-QAM(CC), 32-QAM(CC)
220	4 × STM-1, 1 × STM-4	16-QAM, 32-QAM

[a]The term QAM is intended to also encompass forward error correction or coded modulation techniques (like TCM). In this table CC is used as the abbreviation for "band re-use in the co-channel mode."

Source: Table 1, p. 227, ITU-R Rec. F.751-1, 1994 F Series Volume, Part 1 (Ref. 22).

3.8 CONSIDERATIONS OF FADING ON LOS DIGITAL MICROWAVE SYSTEMS

3.8.1 Introduction*

For a well-designed LOS microwave path that is not subject to diffraction fading or surface reflections, multipath conditions provide the dominant factor for fading below 10 GHz. Above about 10 GHz, the effects of rainfall (Chapter 9) determine maximum path lengths through system availability criteria. The necessary reduction in path length due to rainfall loss limitations tends to reduce the severity of multipath fading. These two principal causes of fading are mutually exclusive.† Given the split between availability and error performance objectives (see Appendix 1), rainfall effects contribute mainly to unavailability, and multipath propagation effects contribute mainly

*Section 3.8.1 is based on ITU-R Rec. F.1093 (Ref. 11).

†This means that we do not add fade margins (i.e., multipath and rainfall). The fade margin to be implemented is the larger of the two, not their sum.

to error performance. Another result of rainfall is backscatter from the raindrops. This may influence the selection of RF channel arrangements.

Propagation effects from rainfall tend not to be frequency dispersive, while multipath propagation caused by tropospheric layers can be, which may cause severe distortion and ISI on digital signals. Rainfall fading is covered in Chapter 9.

3.8.2 Other Views of Calculation of Fade Margins on Digital LOS Microwave*

Digital radio systems react to fading differently than their analog counterparts (see Chapter 2). In Chapter 2 we were concerned with a thermal fade margin (Section 2.7); here we are more concerned with the dispersiveness of a path. The concept of "net" or "effective" fade margin is used for digital systems. The "net" fade margin is defined as a single-frequency fade depth in dB that is exceeded for the same number of seconds as that selected for the BER threshold, usually 1×10^{-3}.

A composite fade margin accounts for the dispersiveness of the fading on a hop by using dispersion ratios, which can be used as a parameter to compare the dispersiveness of different hops in relation to single-frequency fading. This net fade margin is considered as the composite of the effects of thermal noise, ISI due to multipath dispersion, and interference from other radio systems. At the detector of a radio receiver during fading these three sources will give three voltage components, which add on a power basis since each is independent. Thus the total outage time is the sum of the contributions due to single-frequency fading, dispersion, and interference.

The dispersive fade margin, which we call DFM, may be determined from the measured net fade margin by correcting it for any thermal noise or interference contributions as necessary. Because the dispersive fade margin reflects the impact of multipath dispersion on a radio system, its value must depend on the fading and on the radio equipment. The first step is to determine the dispersive fade margin of a radio system on a path with a known dispersion ratio of DR_0. This value (in dB) is taken as a reference dispersive fade margin, DFMR. Then the dispersive fade margin that would be measured or predicted on a path with a dispersion ratio of DR is given by

$$\mathrm{DFM} = \mathrm{DFMR} - 10\log(\mathrm{DR}/\mathrm{DR}_0) \tag{3.14}$$

The referenced ITU-R recommendation states that calculations based on this procedure have shown good agreement with measured radio performance in the field in the presence of interference, as well as detailed estimates based on propagation models.

*Section 3.8.2 is based on ITU-R Rec. F.1093 (Ref. 11).

DR is given by the following relationship:

$$\mathrm{DR} = \frac{T_{\mathrm{IBPD}}}{T_{\mathrm{SFF}} \cdot (\mathrm{BF})^2} \tag{3.15}$$

where T_{IBPD} = amount of time that a chosen in-band power difference (IBPD) (i.e., the amount of dispersion on a hop) value is exceeded

T_{SFF} = amount of time that a chosen single-frequency fade (SFF) value is exceeded

BF = bandwidth correction factor, which is the ratio of 22 MHz to the measurement bandwidth

Modern digital radio systems (e.g., 64-QAM), equipped with adaptive time domain equalizers, experience outage time (i.e., BER $> 10^{-3}$) due to IBPD distortion in the region of 10–15 dB. Thus a suitable threshold for comparing dispersion would be 10 dB. The values of dispersion ratio measured on number of hops in North America and Europe are in the range 0.09–8.1 for hop lengths in the range 38–112 km. This is based on a value of 10 dB and 30 dB for IBPD (In-Band Power Difference) and single-frequency fade, respectively.

3.8.3 Multipath Fading Calculation Based on TIA TSB 10-F*

On digital microwave radiolinks, the fade margin consists of four factors that are power added and constitute the composite fade margin (CFM). These four factors are defined below:

TFM. Thermal fade margin (dB) (sometimes called the flat fade margin) is the fade margin discussed in Chapter 2, Section 2.7.2. TFM is the algebraic difference between the nominal RSL and the 1×10^{-3} BER outage threshold for flat (i.e., nondispersive) fades. Since interference affects unfaded baseband noise, TFM is the only fade margin that needs to be considered on analog LOS links.

DFM. Dispersive fade margin (dB), also to the 1×10^{-3} BER, is defined by the radio equipment manufacturer. It is determined by the type of modulation, the effectiveness of equalization employed in the receive path, and the multipath signal's delay time. This is standardized on manufacturer's datasheets as 6.3 ns. DFM characterizes the digital radio's robustness to dispersive (spectrum-distorting) fades. One means to improve DFM on some paths is to increase the antenna discrimination to reduce the level of longer-delay multipath signals, which can

*Section 3.8.3 was extracted from Section 4.2.3 of TIA TSB 10-F (Ref. 23).

unacceptably degrade a link's DFM. According to TIA (Ref. 23), a DFM greater than 50 dB is a good baseline criterion. *(Note the difference between DFM defined here and in Section 3.8.2.)*

EIFM. External interference fade margin (dB) is a receiver threshold degradation due to interference from a total of the three (MEA factor)* external systems (usually 1 dB, but depends on CFM objective). In the absence of adjacent channel interference (AIFM), EIFM is simply IFM.

AIFM. Adjacent-channel interference fade margin (dB). Receiver threshold degradation is due to interference from adjacent channel transmitters on the same path due to transmitters in one's own system. This is normally a negligible parameter except in cases of frequency diversity and multiline hot-standby systems.

These four fade margins are power added to derive the composite fade margin (CFM) as follows:

$$\mathrm{CFM} = 10\log(10^{-\mathrm{TFM}/10} + 10^{-\mathrm{DFM}/10} + 10^{-\mathrm{EIFM}/10} + 10^{-\mathrm{AIFM}/10}) \tag{3.16}$$

The outage time due to multipath fading in a nondiversity link is calculated by

$$T = (rT_0 \times 10^{-(\mathrm{CFM}/10)})/I_0 \tag{3.17}$$

where T = outage time in seconds
r = fade occurrence factor
$T_0 = (t/50)(8 \times 10^6)$ = length of fade season in seconds
t = average annual temperature in degrees Fahrenheit
CFM = composite fade margin
I_0 = space diversity improvement factor: factor = 1 for nondiversity; ≥ 1 for space diversity

The fade occurrence factor, r, is calculated from the basic outage equation for atmospheric multipath fading:

$$r = c\left(\frac{f}{4}\right)\left(\frac{D}{1.6}\right)^3 \times 10^{-5}$$

$$= x\left(\frac{15}{w}\right)^{1.3}\left(\frac{f}{4}\right)\left(\frac{D}{1.6}\right)^3 \times 10^{-5} \quad \text{(metric)} \tag{3.18a}$$

*MEA = multiple exposure allowance (dB). See Chapter 13, Section 13.3 for more discussion of MEA.

or

$$r = c\left(\frac{f}{4}\right)D^3 \times 10^{-5}$$

$$= x\left(\frac{50}{w}\right)^{1.3}\left(\frac{f}{4}\right)D^3 \times 10^{-5} \quad \text{(English)} \qquad (3.18b)$$

where c = climate–terrain factor (see Figure 3.15)
x = climate factor (see Figure 3.16)
w = terrain roughness: $6 \leq w \leq 42$ m for average 15 m or $20 \leq w \leq 140$ ft for average 50 ft
f = frequency (GHz)
D = path length (km or mi)

The space diversity improvement factor I_0 may be calculated by:

$$I_0 = 1.2 \times 10^{-3} s^2\left(\frac{f}{D}\right) \times 10^{\mathrm{CFM}/10}, \quad s \leq 15\ \mathrm{m} \quad \text{(metric)}$$

$$= 7 \times 10^{-5} s^2\left(\frac{f}{D}\right) \times 10^{\mathrm{CFM}/10}, \quad s \leq 50\ \mathrm{ft} \quad \text{(English)} \qquad (3.19)$$

where fade margins on both antennas are about equal, and s is the vertical antenna separation in meters (feet), center to center.

The space diversity improvement factor (I_0) may underestimate diversity improvements for small antenna spacings and overestimate diversity improvement for large antenna spacings on "flat land" microwave links.

For the purposes of this text, average climate ($x = 1$), temperature [10°C (50°F)], and terrain roughness [15 m (50 ft)] conditions may usually be

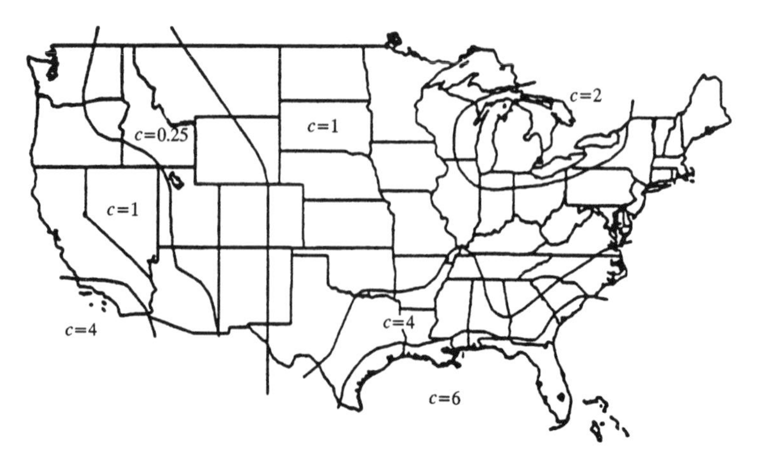

Figure 3.15. Values of climate–terrain factor, c.

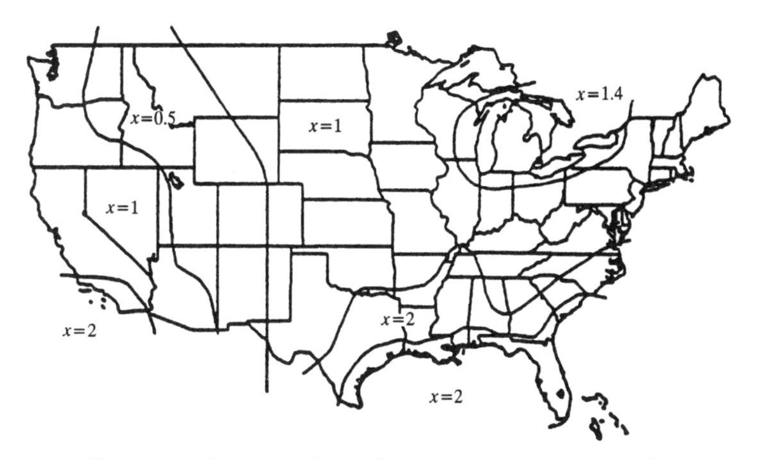

Figure 3.16. Values of climate factor, x (also see table below).

Hawaii/Caribbean	**Alaska**
$c = 4/x = 2$	$c = 0.25/x = 0.5$: coastal and mountainous areas $c = 1/x = 1$: flat permafrost tundra areas in west and north Alaska

assumed. This simplifies the outage time equation to

$$T = \frac{5fD^3 \times 10^{-\mathrm{CFM}/10}}{I_0} \quad \text{(metric)}$$

$$= \frac{20fD^3 \times 10^{-\mathrm{CFM}/10}}{I_0} \quad \text{(English)} \tag{3.20}$$

It is seen from the above equations that nondiversity multipath outage increases directly as a function of the path length cubed (D^3). Therefore short digital paths can usually meet outage objectives with less composite fade margin (more interference) since the outage probability of fading is low.

Since the total number of seconds in one year equals 31.5×10^6, the annual path reliability is computed from

$$\text{Path reliability (\%)} = \left(\frac{31.5 \times 10^6 - T}{31.5 \times 10^6} \right) \times 100 \tag{3.21}$$

The nondiversity outage equations can be rearranged to derive the analog radio fade margin (FM) or digital radio composite fade margin (CFM) required for a given outage time:

$$\text{FM or CFM (nondiversity)} = -10\log\left(\frac{T}{5fD^3}\right) \quad \text{(metric)}$$

$$= -10\log\left(\frac{T}{20fD^3}\right) \quad \text{(English)} \quad (3.22)$$

where T = outage time objective (s/yr)
f = frequency (GHz)
D = path distance (km or mi)

Space diversity improvement plays such a significant role in increasing path reliability, that it often allows higher interference levels that degrade (reduce) the composite fade margin of many digital links. By combining the nondiversity outage equation and the space diversity improvement factor equations, we arrive at the following equation for the annual outage in a space diversity path:

$$T_{\text{SD}} = \frac{4 \times 10^3 D^4 \times 10^{-\text{CFM}/5}}{s^2} \quad \text{(metric)}$$

$$= \frac{3 \times 10^5 D^4 \times 10^{-\text{CFM}/5}}{s^2} \quad \text{(English)} \quad (3.23)$$

Note that the frequency term has disappeared from the space diversity outage equation and the annual outage now varies as a function of D^4. Rearranging this equation to solve for the required fade margin or composite fade margin for a given outage time with space diversity gives:

$$\text{FM or CFM (space diversity)} = -5\log\left(\frac{2.5 \times 10^{-4} Ts^2}{D^4}\right) \quad \text{(metric)}$$

$$= -5\log\left(\frac{3.5 \times 10^{-6} Ts^2}{D^4}\right) \quad \text{(English)}$$

$$(3.24)$$

Calculation of the required fade margins for nondiversity or space diversity links with the above equations may provide improved spectrum utilization (efficiency) by permitting higher interference levels without overly degrading the required reliability for many short and diversity links. For example, if the required fade margin (above) is 25 dB, and the path calculations with no interference show 33 dB, an interference level 7 dB above the value calculated on the basis of threshold degradation [by equation (13.10), for instance] would probably not cause the hop outage to exceed objectives.

Since analog radios are nonregenerative, the baseband noise is additive on N tandem hops (typically per-hop noise plus $13 \log N$). Fading on different hops is noncorrelated, so the outage time (probability of outage) of a digital or analog radio system is equivalent to the sum of the outage times (probabilities of outage) of the individual hops. While the above outage and fade margin calculations are applicable to both analog and digital radio hops, analog radio noise buildup poses a more complex problem. With analog systems, one must consider the overall system noise objectives in parallel with the system reliability (outage) objectives. Most analog systems require significant increases in RSL above FM improvement threshold just to achieve acceptable baseband S/N.

3.8.4 Simple Calculations of Path Dispersiveness

Multipath delay for LOS microwave paths can be as high as 10 μs. Ideally this dispersion should be less than half a symbol period to avoid destructive ISI. For STM-1/STS-3 with a bit rate of 155 Mbps, using a binary waveform, this value should be < 0.003226 μs. If we use 64-QAM, the symbol period is six times as long or 0.01935 μs. Thus a path with dispersion on the order of 10 μs would be highly destructive for such symbol rates. Many paths, however, even with hop lengths on the order of 50 mi, display median dispersion in the very low nanosecond range with maxima in the range of 20 or 30 ns, which is still in the destructive range.

Reference 24 suggests two formulas to calculate maximum dispersion. The first depends entirely on path length:

$$\tau_m = 3.7(D/20)^3 \quad \text{(ns)} \tag{3.25}$$

where D is the path length in miles.

The second formula is based on path length in kilometers and half-power beamwidth of the antenna in degrees:

$$\tau_m = 1668 D \tan^2(\theta/2) \quad \text{(ns)} \tag{3.26}$$

Suppose a path is 60 km (37.5 mi) long and the half-power beamwidth is 1°. By the first formula the maximum dispersion is 24.39 ns; by the second formula it is 7.62 ns.

3.9 PATH ANALYSES OR LINK BUDGETS ON DIGITAL LOS MICROWAVE PATHS

The procedure for digital radiolink analysis is very similar to its analog counterpart described in Chapter 2, Section 2.6, or simply:

- Calculate EIRP.
- Algebraically add FSL and other losses due to the medium (P_L) such as gaseous absorption loss, if applicable.
- Add receiving antenna gain (G_r).
- Algebraically add the line losses (L_{Lr}).

Inserting minus signs for losses, we have the following familiar equation to calculate RSL (receive signal level):

$$\mathrm{RSL}_{\mathrm{dBW}} = \mathrm{EIRP}_{\mathrm{dBW}} - \mathrm{FSL}_{\mathrm{dB}} - P_L + G_r - L_{Lr} \tag{3.27}$$

In Section 3.1.1 we introduced an expression for E_b/N_0. We will use equipment manufacturer's error performance curves to determine a value for E_b/N_0 given a certain required BER. The manufacturer should also provide a modulation implementation loss or that loss is included in the E_b/N_0 value. A typical curve is shown in Figure 3.12 or the values in the W column may be used from Table 3.1. The following example may be useful.

Example 4. Assume a 15-mi (24-km) path with an operating frequency at 6 GHz on a link designed to transmit STS-3 (155.520 Mbps). A 64-QAM modulation will be employed requiring a 30-MHz bandwidth. The required BER is 1×10^{-6}; thus the theoretical E_b/N_0 is 21.4 dB. The modulation implementation loss is 2.0 dB, resulting in a practical E_b/N_0 of 23.4 dB. Assume no fading and a zero margin; the receiver noise figure is 5.0 dB, while the waveguide losses at each end are 1.5 dB. Find a reasonable transmitter output power and antenna aperture to meet these conditions.

Calculate N_0 using equation (3.2):

$$\begin{aligned} N_0 &= -204\ \mathrm{dBW} + 5\ \mathrm{dB} \\ &= -199\ \mathrm{dBW} \end{aligned}$$

We can now calculate E_b, because we know that its level must be 23.4 dB higher than the N_0 level (i.e., $E_b = 23.4 + N_0$). Then

$$E_b = -175.6\ \mathrm{dBW}$$

Calculate $\mathrm{RSL}_{\mathrm{dBW}}$:

$$\begin{aligned} \mathrm{RSL}_{\mathrm{dBW}} &= E_b + 10\log(\text{bit rate}) \\ &= E_b + 10\log(155{,}520{,}000) \\ &= -175.6 + 81.92\ \mathrm{dB} \\ &= -93.68\ \mathrm{dBW} \end{aligned}$$

Calculate FSL:

$$\text{FSL} = 36.58 + 20\log 15 + 20\log 6000$$
$$= 135.66\ \text{dB}$$

Apply equation (3.18):

$$\text{RSL}_{\text{dBW}} = \text{EIRP}_{\text{dBW}} - 135.66\ \text{dB} + G_r - 1.5\ \text{dB}$$

Simplifying this expression:

$$\text{RSL}_{\text{dBW}} = \text{EIRP}_{\text{dBW}} - 137.36 + G_r$$

G_r is the gain of the receive antenna; let G_t be the gain of the transmit antenna. Let the aperture of each antenna be 2 ft [equation (2.27a)]:

$$G_t = G_r = 20\log 2 + 20\log 6 + 7.5$$
$$= 29.08\ \text{dB}$$

Calculate the EIRP:

$$\text{EIRP}_{\text{dBW}} = P_0 + L_L + G_t$$

Let the transmitter output be 1 watt or 0 dBW; then

$$\text{EIRP} = 0 - 1.5\ \text{dB} + 29.08\ \text{dB}$$
$$= +27.58\ \text{dBW}$$

Now make a "trial" run for RSL [equation (3.13)]

$$\text{RSL}_{\text{dBW}} = +27.58\ \text{dBW} - 137.36\ \text{dB} - 1.5\ \text{dB} + 29.08\ \text{dB}$$
$$= -82.20\ \text{dBW}$$

The required RSL, with no margin, was −93.68 dBW. We end up with a 11.48-dB margin. To bring the link budget down by 11.48 dB, we can reduce antenna sizes and/or reduce transmit power. Reducing the antennas to 1-ft dishes at each end would allow us to subtract 6.02 × 2 or 12.04 dB (we multiply by 2 because there are two antennas—a transmit and a receive antenna) and we are left with 0.56 dB below what we wanted. We could increase the transmitter output power by 0.56 dB. The resulting output power would be 1.14 watts.

Of course, in practice, we would want a margin, depending on the time availability desired.

PROBLEMS AND EXERCISES

1 Compare spectrum occupancy for one voice channel (or equivalent) for LOS microwave: analog SSB, analog FM (conventional), and digital. Assume 1 bit/Hz.

2 Calculate N_0 for a low-noise receiver with a 1-dB noise figure.

3 Calculate E_b for a RSL of -85 dBW and a bit rate of 45 Mbps.

4 Calculate E_b/N_0 given 155 Mbps, a RSL of -82.5 dBW, and a receiver noise figure of 2.7 dB.

5 Describe bit packing. Why is it important in digital transmission, particularly when U.S. FCC licensing is required?

6 What is another term for *baud rate*?

7 Differentiate between theoretical bit packing (bits/Hz) and practical.

8 How many bits may be packed into 1-Hz bandwidth with 256-QAM—theoretical?

9 What type of digital modulation allows use of the Nyquist bandwidth and no more?

10 Using raised cosine filtering with $\alpha = 1.3$, what bandwidth would be required for 90 Mbps using 64-QAM?

11 As M increases for M-QAM modulation, what happens to power amplifier distortion?

12 We deal with theoretical and practical E_b/N_0 values. The practical values are always greater. By how much? Answer: By the value of _____ _____ _____ (3 words). This is brought about by certain degradations. Name at least three.

13 Why is digital modulation so much more sensitive to multipath fading than its analog counterpart?

14 Name at least three mitigation techniques to counteract the effects of multipath fading on a digital microwave link.

15 Define dispersion and its mechanics that cause ISI.

16 What are typical maximum dispersion values on LOS microwave links?

17 Name two types of diversity combining used on digital microwave.

18 Quantitatively (in general terms), what is the total improvement when space diversity is used with an adaptive equalization scheme?

19 What are the three principal degradation effects on digital LOS radiolinks?

20 What is the bottom-out BER threshold for LOS microwave digital radios? Why that particular value?

21 Two views were presented in the text on dispersive fade margin (DFM), one by ITU-R and the other by TIA. Discuss the difference(s).

22 Composite fade margin (CFM) consists of four components. What are they?

23 Two approaches were presented to calculate path dispersiveness (in ns). One was simply based on path length. The other was also based on path length and what else?

24 Carry out the following exercise: A path is 12 mi long. Connectivity will be by digital microwave, using a STS-3/STM-1 format/bit rate.* The desired BER is 1×10^{-7}; the modulation is 258-QAM. Equalization will be such that the thermal fade margin +2 dB will cover the necessary margin for dispersiveness. Time availability is 99.99%. Transmitter output is 1 watt, operating frequency is 11 GHz, and the receiver noise figure (a low-noise down converter) is 2 dB. Allow 2.5 dB for modulation implementation loss. Space diversity is employed with proper vertical antenna spacing. Total transmission line losses at each end are 2.2 dB. Disregard rainfall losses. What antenna apertures will be needed?

REFERENCES

1. Roger L. Freeman, *Telecommunication Transmission Handbook*, 3rd ed., Wiley, New York, 1991.
2. *Transmission Systems for Communications*, 5th ed., Bell Telephone Laboratories, Holmdel, NJ, 1982.
3. *FCC Rules and Regulations*, Part 21, Vol. VII, Federal Communications Commission, Washington, DC, Sept. 1982.
4. T. Noguchi, Y. Daido, and J. A. Nossek, "Modulation Techniques for Digital Microwave Radio," *IEEE Communications Magazine*, Vol. 24, No. 10, Oct. 1986.
5. *Principles of Digital Transmission*, Raytheon Company, Equipment Division, ER79-4307, Contract MDA-904-79-C-0470, Sudbury, MA, Nov. 1979.
6. *Digital Radio—Theory and Measurements*, Application Note 355A, Hewlett-Packard Company, San Carlos, CA, 1992.
7. *Characteristics of Digital Radio-Relay Systems Below About 17 GHz*, ITU-R Rec. F.1101, 1994 F Series Volume, Part 1, ITU, Geneva, 1994.
8. *Linear Modulation Techniques for Digital Microwave*, Harris Corp., RADC-TR-79-56, USAF Rome Laboratories, Rome, NY, 1979.
9. Private communication, W. P. Norris, Harris Electronic Systems Sector, Palm Bay, FL, Sept. 13, 1995.
10. K. Feher, *Digital Communications Microwave Applications*, Prentice-Hall, Englewood Cliffs, NJ, 1981.
11. *Effects of Multipath Propagation on the Design and Operation of Line-of-Sight Digital Radio-Relay Systems*, ITU-R Rec. F.1093, 1994 F Series Volume, Part 1, ITU, Geneva, 1994.

*The bit rates are the same; the formats are not exactly.

12. E. W. Allen, "The Multipath Phenomenon in Line-of-Sight Digital Transmission Systems," *Microwave Journal*, May 1984.
13. *Terminology Used for Radio-Relay Systems*, CCIR Rec. 592-2, 1994 F Series Volume, Part 1, ITU, Geneva, 1994.
14. *Error Performance Objectives for Real Digital Radio-Relay Links Forming Part of a High-Grade Circuit Within an Integrated Services Digital Network*, ITU-R Rec. F.634-3, 1994 F Series Volume, Part 1, ITU, Geneva, 1994.
15. *Hypothetical Reference Digital Path for Radio-Relay Systems Which May Form Part of an Integrated Services Digital Network with a Capacity Above the Second Hierarchical Level*, CCIR Rec. 556-1, 1994 F Series Volume, Part 1, ITU, Geneva, 1994.
16. *Allowable Bit Error Ratios at the Output of a Hypothetical Reference Digital Path for Radio-Relay Systems Which May Form Part of an Integrated Services Digital Network*, CCIR Rec. 594-3, 1994 F Series Volume, Part 1, ITU, Geneva, 1994.
17. *Error Performance Objectives for Constant Bit Rate Digital Path at or Above the Primary Rate Carried by Digital Radio-Relay Systems Which May Form Part of the International Portion of a 27,500 km Hypothetical Reference Path*, ITU-R Rec. F.1092, 1994 F Series Volume, Part 1, ITU, Geneva, 1994.
18. *Error Performance Parameters and Objectives for International, Constant Bit Rate Digital Paths at or Above the Primary Rate*, ITU-T Rec. G.826, ITU, Geneva, 1993.
19. *Characteristics of Synchronous Digital Hierarchy (SDH) Equipment Functional Blocks*, ITU-T Rec. G.783, ITU, Geneva, 1994.
20. *Digital Line Systems Based on the Synchronous Digital Hierarchy for Use on Optical Fibre Cables*, ITU-T Rec. G.958, ITU, Geneva, 1994.
21. *Transport Systems Generic Requirements (TSGR): Common Requirements*, Bellcore TR-NWT-000499, Issue 5, Bellcore, Piscataway, NJ, Dec. 1993.
22. *Transmission and Performance Requirements of Radio-Relay Systems for SDH-Based Networks*, ITU-R Rec. F.751-1, 1994 F Series Volume, Part 1, ITU, Geneva, 1994.
23. *Interference Criteria for Microwave Systems*, TIA Telecommunications Systems Bulletin, TSB 10-F, Telecommunications Industries Association, Washington, DC, 1994.
24. Eli Kolton, *Results and Analysis of Static and Dynamic Multipath in a Severe Atmospheric Environment*, NTIA Contractor Report 86-37, Boulder, CO, Sept. 1986.

4

FORWARD ERROR CORRECTION AND ADVANCED DIGITAL WAVEFORMS

4.1 OBJECTIVE

Forward error correction (FEC) and its related *coding gain* provide yet another technique to improve digital performance on radio systems. It does not give dramatic performance improvements such as, say, doubling the diameter of an antenna at both ends (some 12 dB). But it does provide some 2–6-dB improvement as a gain in a link budget. It also has some hidden advantages, such as with scintillation and rainfall fading. Ordinarily we have to pay some price for the added redundancy required for FEC with additional bandwidth. Some of the advanced digital modulation schemes described in the second section of this chapter can mitigate bandwidth expansion requirements.

In the second part of this chapter we discuss four advanced digital modulation/coding schemes that are considerably more robust when dealing with transmission impairments because of improved Euclidean distances (i.e., the geometrical distance between decision points in a signal constellation), reduced equivalent bandwidth requirements, and better performance under multipath conditions (Ref. 1).

4.2 FORWARD ERROR CORRECTION

4.2.1 Background and Objective

Error rate is a principal design factor for digital transmission systems. The digital public switched telecommunication network (PSTN) is based on 8-bit pulse code modulation (PCM), which, for speech transmission, can tolerate a BER degraded to 1×10^{-2} and still be intelligible. Continuous variable slope delta modulation (CVSD) remains intelligible with a BER of nearly 1×10^{-1}. For conventional telephony on the PSTN the grating error rate is determined by supervisory signaling.* This value of BER should be $\leq 1 \times 10^{-3}$, the threshold for the PSTN. Computer data users of the network drive the BER value to at least 1×10^{-6} at the receive end as recommended in CCITT Rec. G.821 (Ref. 2). For digital LOS microwave a single hop should display a BER of 1×10^{-9}.

To design a transmission system to meet a specified error performance, we must first consider the cause of errors. For this initial discussion, let us assume that intersymbol interference (ISI) is negligible. The errors derive from insufficient E_b/N_0, which results in bit mutilation by thermal noise peaks (additive white Gaussian noise or AWGN). Error rate is a function of signal-to-noise ratio or E_b/N_0 (see Section 3.1.1). As we have seen in Chapter 3, we can achieve a desired BER on a link by specifying the requisite E_b/N_0. On many satellite and tropospheric scatter links, there may be a more economically feasible way.

Errors derive from insufficient E_b/N_0 (for this initial discussion) because of deficient system design, equipment deterioration, or fading. On unfaded links or during unfaded conditions, these errors are random in nature, and during fading, errors are predominantly bursty in nature. The length of a burst can be related to fade duration.

There are several tools available to the design engineer to achieve a desired link error rate. Obviously, the first is to specify sufficient E_b/N_0 for the waveform selected, adding margin for link deterioration due to equipment aging and a margin allowance for fading as described in Chapters 2 and 3.

Another approach is to specify a lower E_b/N_0 for a BER, perhaps in the range of 1×10^{-5}, and implement an automatic repeat request (ARQ) regime. ARQ requires a return channel. It is usually implemented on an end-to-end basis or a section-by-section basis depending on the protocol used and whether it is a network layer or link-layer protocol. With ARQ, data messages are broken down into blocks, frames, or packets at the originating end. Each block, frame, or packet has appended a parity tail, often referred to as a frame check sequence (FCS). This tail is generated at the originating end of the link by a processor that determines the parity characteristics of the message or uses a cyclic redundancy check (CRC), and the tail is the remainder of that check, often 16 bits in length. At the receiving end a similar processing technique is used, and the locally derived remainder is compared to the remainder received from the distant end. If they are the same, the message is said to be error-free, and if not, the block or packet is in error.

There are several ARQ methods. One is called stop-and-wait ARQ. In this case, if the frame is error-free, the receiver transmits an ACK (acknowl-

*Supervisory signaling informs a circuit switch if a line is idle or busy.

edgment) to the transmitting end, which in turn sends the next frame or packet. If the frame is in error, the receiver sends a NACK (negative acknowledgment) to the transmitter, which now repeats the frame just sent. It continues to repeat it until the ACK signal is received.

A second type of ARQ is variously called "continuous ARQ" or "selective ARQ." At the transmitting end, in this case, sending of frames is continuous. There is accounting information, which involves some sort of sequential frame or packet numbering in each frame or packet header. Similar parity checking is carried out as before. When the receive end encounters a frame in error, it identifies the errored frame to the transmitter, which intersperses the repeated frames, each with a proper sequence number, with its regular continuous frame transmissions. Obviously, continuous ARQ is a more complex implementation than stop-and-wait ARQ, but no valuable circuit time is lost stopping and waiting.

Go-back-*n* ARQ is similar to continuous ARQ but rather than intersperse a repeated frame with the normal sequence of frames, the receiver tells the transmit end to "go-back-*n*" frames. The transmitter then repeats all frames from the errored frame forward. One can imagine that ARQ is completely unfeasible on voice circuits and other connectivities requiring a constant bit rate, such as conference TV. The far-end voice listener will not tolerate the ARQ delays or interspersed frames, which also imply delay.

On satellite circuits the delay problem is magnified because of the long propagation delays in addition to ARQ delay, on the order of 0.5 second. It is also bad on data circuits with connectivity through a geostationary satellite, particularly with stop-and-wait ARQ. Here the wasted circuit time is multiplied. Continuous ARQ mitigates much of that time waste.

Another approach is to use error-correction coding, often called channel coding. If offers a number of advantages and at least one major disadvantage. To meet BER requirements in a most cost-effective manner, it is then up to the design engineer to trade off system overbuild with error-correction techniques, of which there are two types: ARQ and FEC. A brief description of the former was given above. The latter is described in Section 4.2.2.

4.2.2 Basic Forward Error Correction

The IEEE (Ref. 3) defines a forward error-correcting system "as a system employing an error-correcting code and so arranged that some or all of the signals detected as being in error are automatically corrected at the receiving terminal before delivery to the data sink or to the telegraph receiver."

Our definition may help clarify the matter. FEC is a method of error control that employs the adding of systematic redundancy at the transmit end of a link such that many or all of the errors caused by the medium can be corrected at the receiver by means of a decoding algorithm (Ref. 4). The price we pay for this is the redundancy.

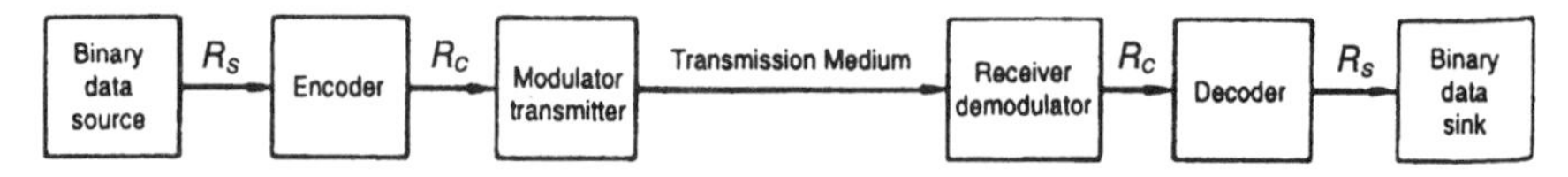

Figure 4.1. Simplified functional block diagram of a binary communication system employing FEC.

Figure 4.1 shows a digital communication system with FEC. The binary data source generates information bits at R_s bits per second. An encoder encodes these information bits for FEC at a code rate R. The output of the encoder is a binary sequence of R_c symbols per second. The coder output is related to the input bit rate by the following expression:

$$R_c = R_s/R \tag{4.1}$$

R, the code rate, is the ratio of the number of information bits to the number of encoded symbols for binary transmission. For example, if the information bit rate were 2400 bps and the code rate (R) were $1/2$, then the symbol rate (R_c) would be 4800 coded symbols per second. As we can see, the transmission rate is doubled without any increase in the information rate.

The encoder output sequence is then modulated and transmitted over the transmission medium or channel. Demodulation is then performed at the receive end. The output of the demodulator is R_c symbols per second, which is fed to the decoder. The decoder output to the data sink is the recovered 2400 bps (R_s). The entire system is assumed to be synchronous.

The major advantages of a FEC system are:

- No feedback channel required as with ARQ systems
- Constant information throughput (e.g., no stop-and-wait gaps)
- Decoding delay generally small and constant
- Significant coding gain achieved for an AWGN channel

There are two disadvantages to a FEC system. To effect FEC, for a fixed information rate, the bandwidth must be increased (refer to Section 4.3 for some exceptions) because, by definition, the symbol rate on the transmission channel is greater than the information bit rate. There is also the cost of added complexity of the coder and decoder.

4.2.2.1 Coding Gain. For a given information bit rate and modulation waveform (e.g., QPSK, 8-ary FSK, 16-QAM), the required E_b/N_0 for a specified BER with FEC is less than the E_b/N_0 required without FEC. The coding gain is the difference in E_b/N_0 values.

FEC can be used on any digital transmission system. There is a difference in application depending on the medium employed. Wire-pair, coaxial cable, and satellite radio systems (operating below 10 GHz) generally do not suffer from fading. LOS microwave, tropospheric scatter/diffraction, HF, and satellite systems operating above 10 GHz can be prone to fading.

Let's consider a satellite system operating below 10 GHz, where we can generally neglect rainfall losses. Here we can effect major savings by simply adding FEC processors at each end of a link. Suppose we can achieve 3 dB of coding gain. Thus, keeping the link BER unchanged from the uncoded to coded condition, the 3-dB coding gain can be used to reduce satellite EIRP by 3 dB without affecting error performance. This can provide many economic savings. First, we can reduce the size of the high-power amplifier (HPA) stage on the satellite transponder because only half the power output is required when compared to the uncoded link. Second, it also reduces the weight of the transponder payload, possibly by 75%. The principal reason for this is that the transmitter power supply size can be decreased. Third, battery weight can also be reduced, possibly up to 50% [batteries are used to power the transponder during eclipse (darkness)], with the concurrent reduction in solar cells. It is not only the savings in the direct cost of these items, but also the savings in lifting weight of the satellite to place it in orbit, whether by space shuttle or rocket booster. With unfaded conditions, assuming random errors, coding gains of 2–7 dB can be realized. The amount of gain achievable under these conditions is a function of the modulation type (waveform), the code employed, the coding rate [equation (4.1)], the constraint length, the type of decoder, soft or hard decision decoding, and the demodulation approach. Coding gain values are provided later in this section.

A FEC system can use one of two broad classes of error-correcting codes: block and convolutional. These are briefly described below. However, we first discuss demodulators using hard decisions or soft decisions as these could provide a form of synergism in the decoding process.

4.2.2.2 Hard and Soft Decision Demodulators. In this section, the outputs of a demodulator are defined on a continuum. Before these demodulator outputs can be processed with digital circuits, some form of amplitude quantization* is required. We consider the quantizer as part of the demodulator.

With binary modulation and no coding, a demodulator produces an output defined over a continuum for each bit transmitted. It makes a hard (irrevocable) decision as to which information bit was transmitted (i.e., a 1 or a 0) by determining the polarity of the demodulator output. We could say a 1-bit quantizer is used. Such a 1-bit quantization is also referred to as *hard quantization.* Without coding, providing additional amplitude information about the demodulator is of no help in the improvement of error performance.

*Quantization in this context means converting from an output defined by a continuum to an output with discrete values, usually based on the power of 2.

With coding, a decoding decision on a particular bit can be based on several demodulator outputs. Retaining some amplitude information rather than just the sign of the demodulator outputs can be very helpful. If, for example, the output of the demodulator is very large, there is confidence that a polarity decision on that demodulator output is correct. On the other hand, if the output is very small, there is a high probability that the output would be in error. We can design a decoder that uses this amplitude information, which in effect weighs the contributions of demodulator outputs to the decoding decision. Such decoders can perform better than similar decoders that only use polarity information. A quantizer that retains some amplitude information (i.e., more than 1 bit is retained) is called a *soft quantizer.*

No quantization refers to the ideal situation where no quantizer is used at the demodulator output: that is, all of the amplitude information is retained.

An interesting example is 8-ary PSK, where several methods of quantizing demodulator outputs are suggested. One method is to quantize the in-phase and quadrature outputs so that the signal space (see Chapter 3), consisting of signal components every 45° on a circle of radius $\sqrt{E_s}$, is divided into small squares as shown in Figure 4.2a, where E_s is the energy per symbol. Another method divides the receive signal space into pie-shaped wedges depending on the angle of the received signal component, as shown in Figure 4.2b. The particular quantization technique depends largely on implementation considerations (Ref. 5).

4.2.3 FEC Codes*

4.2.3.1 Block Codes. Block codes were the first error-correction coding techniques to be used. For this class of codes, the data are transmitted in blocks of symbols. In the binary case, for every block of K information bits input to the coder, $N - K$ redundant parity-check bits are generated as linear (modulo-2) combinations of the information bits and transmitted along with the information bits at a code rate of K/N bits per binary channel symbol. N represents the total number of binary symbols transmitted. Thus the code rate (R) is given by

$$R = K/N \tag{4.2}$$

As we see, the code rate is the ratio of the information bits to the total bits (binary symbols) transmitted: this is also the inverse of the bandwidth expansion factor. The more successful block-coding techniques have centered about finite-field algebraic concepts, culminating in various classes of codes that can be generated by means of a linear feedback register.

Linear block codes can be described by a $K \times N$ generator matrix **G**. If the K-symbol encoder input is represented by a K-dimensional column

*Section 4.2.3 has been abstracted from Ref. 5.

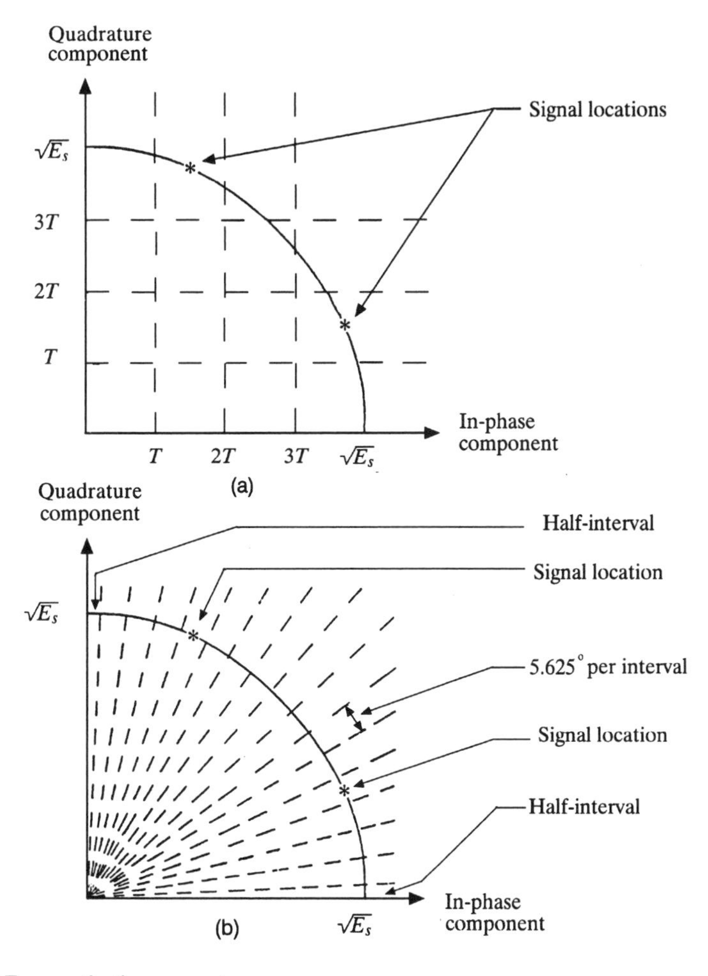

Figure 4.2. Diagrams of the first quadrant signal space quantization intervals for two possible 6-bit quantization techniques for 8-ary PSK. (From Ref. 5.)

vector **x**, and the encoder output by a K-dimensional column vector **v**, the encoder input–output relationship is given by

$$\mathbf{v} = \mathbf{x}\mathbf{G} \tag{4.3}$$

So the N-symbol encoder output blocks are linear algebraic combinations of the rows of the generator matrix. In the binary symbol case, the output blocks are bit-by-bit modulo-2* sums of the appropriate rows of **G**.

*Modulo-2 addition is binary addition without carries. The sign of modulo-2 addition is $\oplus$. For example, $0 \oplus 0 = 0$, $0 \oplus 1 = 1$, $1 \oplus 0 = 1$, and $1 \oplus 1 = 0$.

Usually block codes are decoded using algebraic techniques, which require the demodulator to make a hard decision on each received symbol. Hard quantization reduces the potential performance of a coding system. For BPSK and QPSK modulation, based only on an AWGN channel, the potential coding gain of a finely quantized coding system is about 2 dB more than that of a hard quantized system. However, block codes are now starting to use soft quantization and some of the 2 dB that was lost is being recovered. The implementation complexity of such systems, however, is usually greater than that of a hard-quantized system.

Another disadvantage of block codes as compared to convolutional codes is that with block codes the receiver must resolve an n-way ambiguity to determine the start of a block whereas with Viterbi- or feedback-decoded convolutional codes, a much smaller ambiguity needs to be resolved.

Block codes are often described with notation such as $(7, 4)$ meaning $N = 7$ and $K = 4$. In this case the information bits are stored in $K = 4$ storage devices and the device is made to shift $N = 7$ times. The first K symbols of the block output are the information symbols, and the last $N - K$ symbols are a set of check symbols that form the whole N-symbol word. A block code may also be identified with the notation (N, K, t), where t corresponds to the number of errors in a block of N symbols that the code will correct.

In the following subsections several specific block codes are reviewed.

4.2.3.1.1 Hamming Codes. Hamming codes are the simplest nontrivial class of codes with $N = 2^m - 1$ $(m = 2, 3, \ldots)$ encoder output symbols for each block of $K = N - m$ input symbols. These codes have a minimum distance of 3 and thus are able to correct all single errors or detect all combinations of 2 or fewer errors. Although Hamming codes are not very powerful, they belong to a class of block codes called perfect codes. An e-error correcting code $\{e = (d - 1)/2]_I\}$ is called a perfect code if every n-symbol sequence is at a distance of at most e from some N-symbol encoder output sequence.

Hamming codes are often described in terms of $N \times (N - K)$ dimensional parity-check matrix, $\mathbf{H}$, with the property that for each N-dimensional encoded output word $\mathbf{y}$

$$\mathbf{yH} = 0 \tag{4.4}$$

For Hamming codes, the N rows of the parity-check matrix are equal to all positive nonzero m-bit sequences. Given a parity-check matrix, a generator matrix can be determined.

If the binary additive noise sequence is represented by an N-dimensional vector $\mathbf{z}$, then the received signal is

$$\mathbf{y} = \mathbf{xG} \oplus \mathbf{z} \tag{4.5}$$

where $\oplus$ denotes the bit-by-bit modulo-2 addition.

Decoding is accomplished by multiplying this binary vector by the parity-check matrix to form an $N - K = m$-dimensional syndrome vector $\mathbf{S}$. Using equation (4.4), we then have

$$\mathbf{S} = \mathbf{yGH} \oplus \mathbf{zH} = \mathbf{zH} \tag{4.6}$$

Because of the form of $\mathbf{H}$, this m-bit syndrome specifies the locations of any single error, which can then be corrected. If the syndrome is zero, the decoder assumes no errors occurred.

Figure 4.3 gives bit error probability versus E_b/N_0 for block length $N = 2^m - 1$ Hamming codes with $m = 3$, 4, and 5 on an AWGN channel.

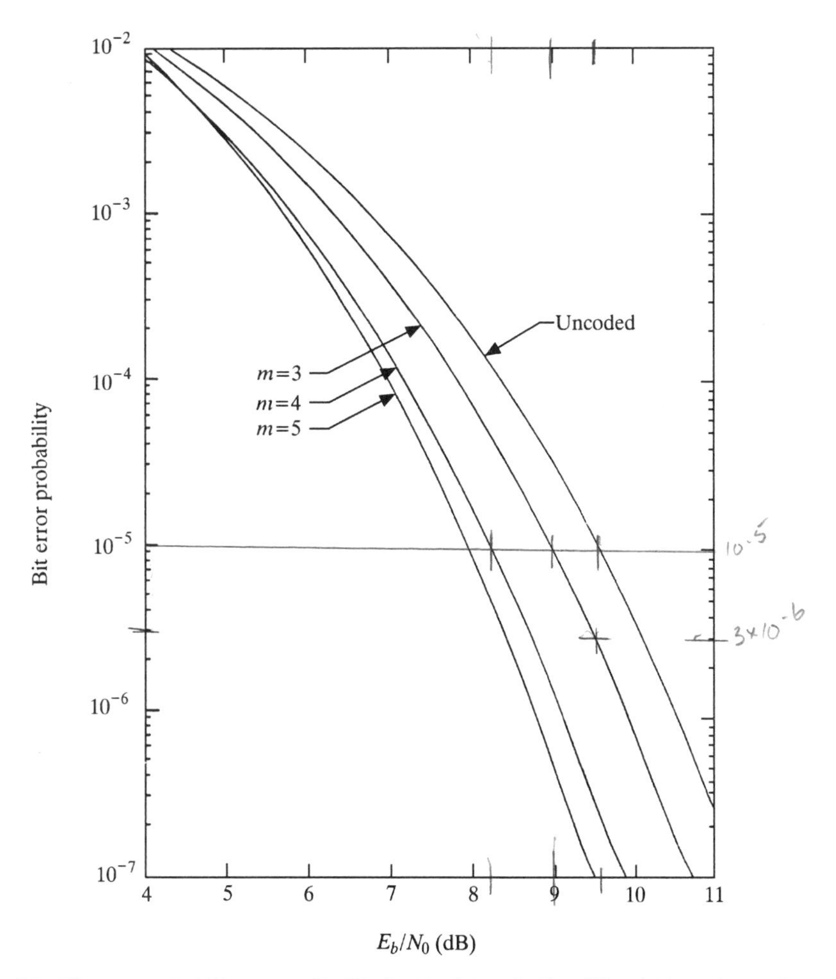

Figure 4.3. Bit error probability versus E_b/N_0 for block length $N = 2^m - 1$ Hamming codes with $m = 3$, 4, and 5 on an AWGN channel. (From Ref. 5.)

TABLE 4.1 E_b/N_0 Ratio Required to Achieve a BER of 1×10^{-5} with Hamming Coding for Several Modulation/Demodulation Techniques

Type of Interference	Modulation/Demodulation Technique	E_b/N_0 (in dB) Required to Achieve a 1×10^{-5} Bit Error Probability		
		$m = 3$	$m = 4$	$m = 5$
AWGN	BPSK or QPSK	9.0	8.3	8.0
AWGN	Octal-PSK with interleaving	12.2	11.5	11.2
AWGN	DBPSK with interleaving	10.2	9.4	9.1
AWGN	DQPSK with interleaving	11.4	10.7	10.4
Independent Rayleigh fading	Noncoherently demodulated binary FSK with optimum diversity	17.5 ($L = 8$ diversity)	16.7 ($L = 16$ diversity)	16.4 ($L = 16$ diversity)

Source: Reference 5.

Table 4.1 provides a summary of E_b/N_0 ratios required to achieve a 1×10^{-5} BER using Hamming codes for several modulation/demodulation with AWGN and Rayleigh fading.

4.2.3.1.2 Extended Golay Code. A widely used block code is the extended Golay code, where $N = 24$, $K = 12$ [i.e., (24, 12)]. It is formed by adding an overall parity bit to the perfect (23, 12) Golay code. The parity bit increases the minimum distance of the code from 7 to 8, producing a rate 1/2 code that is easier to work with than the rate 12/23 of the (23, 12) code.

Extended Golay codes are considerably more powerful than the Hamming codes described above. To achieve the improved performance, a more complex decoder at a lower rate is required involving a larger bandwidth expansion. Decoding algorithms that make use of soft decision demodulator outputs are available for these codes. When such soft decision decoding algorithms are used, the performance of an extended Golay code is similar to that of a simple Viterbi-decoded convolutional coding system of constraint length 5 ($K = 5$). While it is difficult to compare the implementation complexity of two different coding systems, it can be concluded that only when hard decision demodulator outputs are available, are extended Golay coding systems of the same approximate complexity as similar performance convolutional coding schemes. However, when soft decisions are available, convolutional coding is superior.

Figure 4.4 gives block, bit, and undetected error probabilities versus E_b/N_0 for BPSK or QPSK modulation, an AWGN channel, and extended Golay coding. Table 4.2 provides E_b/N_0 ratios required to obtain a 1×10^{-5} BER with extended Golay coding and several different modulation/demodulation techniques for AWGN and Rayleigh fading channels.

4.2.3.1.3 Bose–Chaudhuri–Hocquenghem (BCH) Codes. Binary BCH codes are a large class of block codes with a wide range of code parameters.

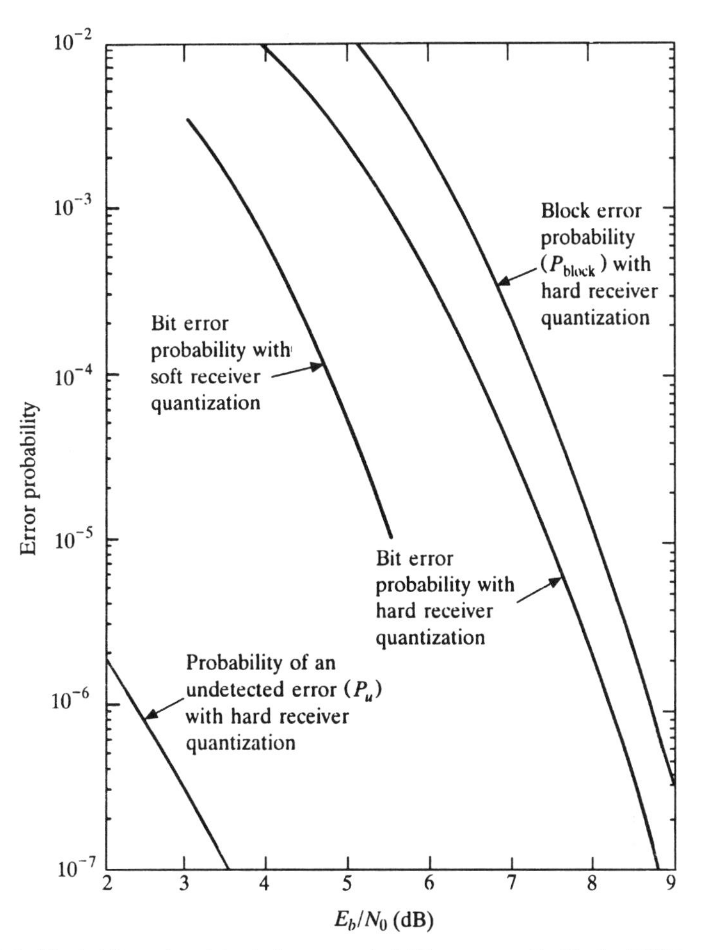

Figure 4.4. Block, bit, and undetected error probabilities versus E_b/N_0 for BPSK and QPSK modulation, an AWGN channel, and extended Golay coding. (From Ref. 5.)

TABLE 4.2 E_b/N_0 Required to Achieve a 1×10^{-5} BER with Extended Golay Coding and Several Modulation/Demodulation Techniques

Type of Interference	Modulation/Demodulation Technique	E_b / N_0 (in dB) Required to Achieve a 1×10^{-5} Bit Error Rate
AWGN	BPSK or QPSK	7.5 (hard decision), 5.6 (soft decision)
AWGN	Octal-PSK with interleaving	10.4 (hard decision)
AWGN	DBPSK with interleaving	9.0 (hard decision)
AWGN	DQPSK with interleaving	10.0 (hard decision)
Independent Rayleigh fading	Noncoherent binary PSK with optimum diversity (i.e., $L = 8$ or 16 channel bits per information bit with rate 1/2 coding)	15.9 (hard decision)

Source: Reference 5.

The so-called primitive BCH codes, which are the most common, have codeword lengths of the form $2^m - 1$, $m \geq 3$, where m describes the degree of the generating polynomial (i.e., the highest exponent value of the polynomial). For BCH codes, there is no simple expression for the N, K, t parameters. Table 4.3 gives these parameters for all binary BCH codes of length 255 and less. In general, for any m and t, there is a BCH code of length $2^m - 1$ that corrects any combination of t errors and that requires no more than m/t parity-check symbols.

BCH codes are cyclic and are characterized by a generating polynomial. A selected group of primitive generating polynomials is given in Table 4.4. The encoding of a BCH code can be performed with a feedback shift register of length K or $N - K$ (Ref. 1).

TABLE 4.3 BCH Code Parameters for Primitive Codes of Length 255 and Less

N	*K*	*t*	*N*	*K*	*t*	*N*	*K*	*t*
7	4	1	127	64	10	255	87	26
				57	11		79	27
15	11	1		50	13		71	29
	7	2		43	14		63	30
				36	15		55	31
31	26	1		29	21		47	42
	21	2		22	23		45	43
	16	3		15	27		37	45
	11	5		8	31		29	47
	6	7					21	55
			255	247	1		13	59
63	57	1		239	2		9	63
	51	2		231	3			
	45	3		223	4			
	39	4		215	5			
	36	5		207	6			
	30	6		199	7			
	24	7		191	8			
	18	10		187	9			
	16	11		179	10			
	10	13		171	11			
	7	15		163	12			
				155	13			
127	120	1		147	14			
	113	2		139	15			
	106	3		131	18			
	99	4		123	19			
	92	5		115	21			
	85	6		107	22			
	78	7		99	23			
	71	9		91	25			

Source: Reference 1.

TABLE 4.4 Selected Primitive Binary Polynomials

m	$p(x)$
3	$1 + x + x^3$
4	$1 + x + x^4$
5	$1 + x^2 + x^5$
6	$1 + x + x^6$
7	$1 + x^3 + x^7$
8	$1 + x^2 + x^3 + x^4 + x^8$
9	$1 + x^4 + x^9$
10	$1 + x^3 + x^{10}$
11	$1 + x^2 + x^{11}$
12	$1 + x + x^4 + x^6 + x^{12}$
13	$1 + x + x^3 + x^4 + x^{13}$
14	$1 + x + x^6 + x^{10} + x^{14}$
15	$1 + x + x^{15}$

Source: Reference 1.

Figure 4.5 gives the BER versus channel error rate performance of several block length 127 BCH codes. Channel error rate performance can be taken as the uncoded error performance of the modulation type given.

4.2.4 Binary Convolutional Codes*

Viterbi (Ref. 6) defines a convolutional coder as the following: "a linear finite state machine consisting of a K-stage shift register and n linear algebraic function generators. The input data which are usually, but not necessarily always, binary, are shifted along the register b bits at a time."

Convolutional codes are defined by a code rate, R, and a constraint length, K. The code rate is the information bit rate into the coder divided by the coder's output symbol rate. The constraint length K is defined as the total number of binary register stages in the coder.

Let's consider a convolutional encoder with $R = 1/2$ and $K = 3$. This is shown in Figure 4.6, which indicates the outputs for a particular binary input sequence assuming the state (i.e., the previous two data bits into the shift register) were 0. Modulo-2 addition is used. With the input and output sequences defined from right-to-left, the first three input bits—0, 1, and 1—generate the code outputs 00, 11, and 01, respectively. The outputs are shown multiplexed into a single code sequence. Of course, the coded sequence has twice the bit rate as the data sequence. We will return to this simple model again.

A tree diagram is often used instructively to show a convolutional code. A typical tree diagram is illustrated in Figure 4.7.

*Section 4.2.4 has been abstracted from Ref. 5.

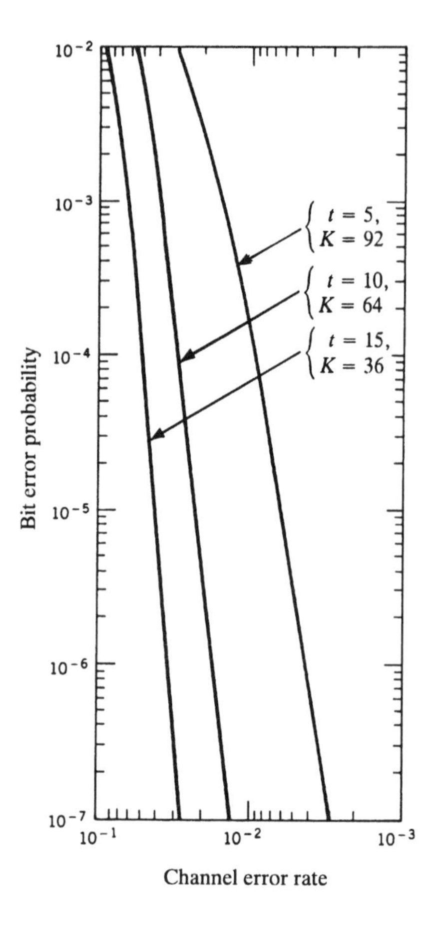

Figure 4.5. Bit error rate probability versus channel error rate performance of several BCH codes with block length of 127. (From Ref. 5.)

In Figure 4.7, if the first input bit is a zero, the code symbols are shown on the first upper branch, while if it is a one, the output symbols are shown on the first lower branch. Similarly, if the second input bit is a zero, we can trace the tree diagram to the next upper branch, while if it is a one, the diagram is traced downward. In such a way, all 32 possible outputs may be traced for the first five inputs.

From the tree diagram in Figure 4.7 it also becomes clear that after the first three branches the structure becomes repetitive. In fact, we readily recognize that beyond the third branch the code symbols on branches emanating from the two nodes labeled "*a*" are identical, and so on, for all the similarly labeled pairs of nodes. The reason for this is obvious from examination of the encoder. As the fourth bit enters the coder at the right, the first data bit falls off on the left and no longer influences the output code symbols. Consequently, the data sequences $100xy\ldots$ and $000xy$ generate the

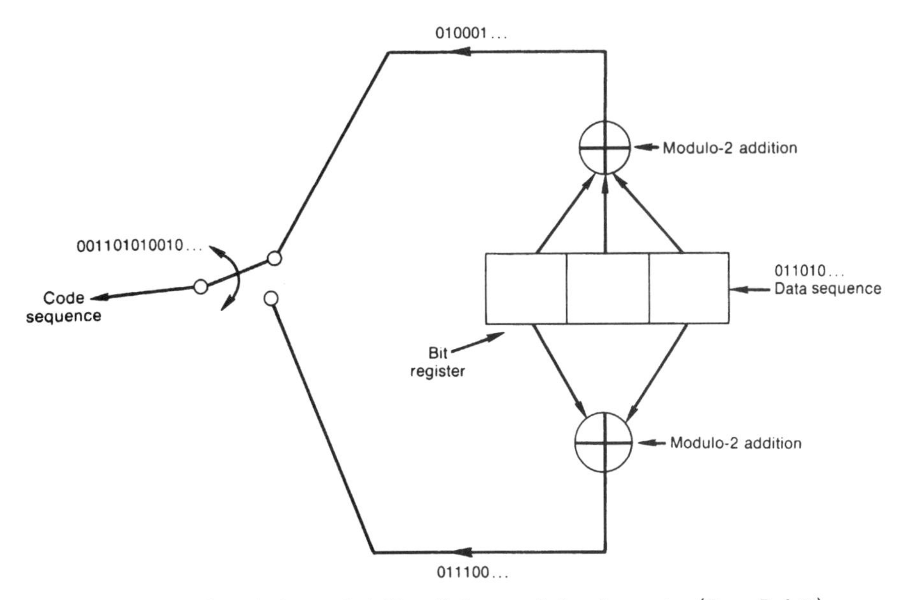

Figure 4.6. Rate 1/2, constraint length 3, convolutional encoder. (From Ref. 5.)

same code symbols after the third branch and, as is shown in the tree diagram, both nodes labeled "a" can be joined together.

This leads to redrawing the tree diagram as shown in Figure 4.8. This is called a trellis diagram, since a trellis is a treelike structure with remerging branches. A convention is adopted here, where the code branches produced by a zero input bit are shown as solid lines and code branches produced by a one input bit are shown as dashed lines.

The completely repetitive structure of the trellis diagram suggests a further reduction in the representation of the code to the state diagram in Figure 4.9. The "states" of the state diagram are labeled according to the nodes of the trellis diagram. However, since the states correspond merely to the last two input bits to the coder, we may use these bits to denote the nodes of states of this diagram.

It can be observed that the state diagram can be drawn directly observing the finite-state machine properties of the encoder and, particularly, the fact that a four-state directed graph can be used to represent uniquely the input–output relation of the finite-state machine, since the nodes represent the previous two bits, while the present bit is indicated by the transition branch. For example, if the encoder (synonymous with finite-state machine) contains the sequence 011, this is represented in the diagram by the transition from state $b = 01$ to state $d = 11$ and the corresponding branch indicates the code symbol outputs 01.

This example will be used when we describe the Viterbi decoder.

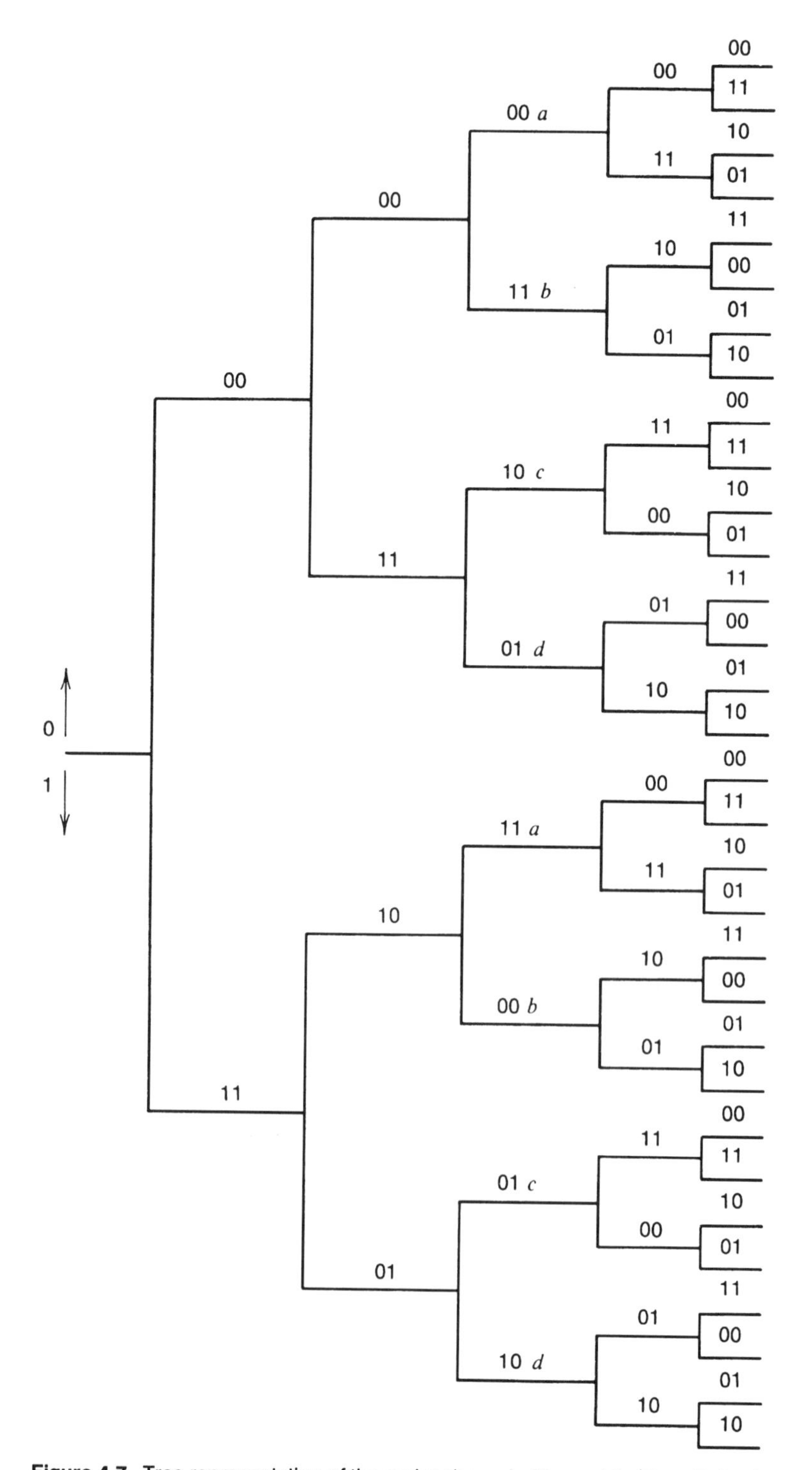

Figure 4.7. Tree representation of the coder shown in Figure 4.6. (From Ref. 5.)

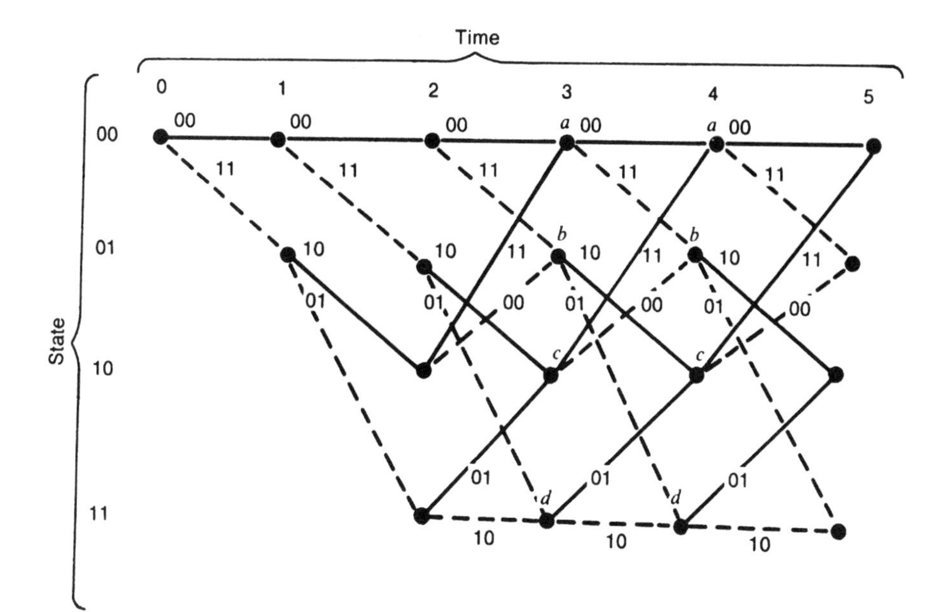

Figure 4.8. Trellis code representation for the encoder shown in Figure 4.6. (From Ref. 5.)

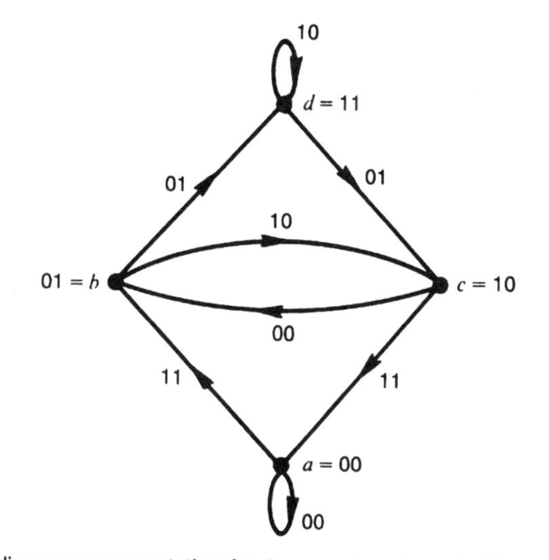

Figure 4.9. State diagram representation for the encoder shown in Figure 4.6. (From Ref. 5.)

4.2.4.1 Convolutional Decoding. Decoding algorithms for block and convolutional codes are quite different. Because a block code has a formal structure, advantage can be taken of the known structural properties of the words or algebraic nature of the constraints among the symbols used to represent an information sequence. One class of powerful codes with well-defined algorithms is the BCH codes described in Section 4.2.3.1.3.

The decoding of convolutional codes can be carried out by a number of different techniques. Among these are the simpler threshold and feedback decoders and those more complex decoders with improved performance (coding gain) such as the Viterbi decoder and sequential decoder. These techniques depend on the ability to home in on the correct sequence by designing efficient search procedures that discard unlikely sequences quickly. The sequential decoder differs from most other types of decoders in that when it finds itself on the wrong path in the tree, it has the ability to search back and forth, changing previously decoded information bits until it finds the correct tree path. The frequency with which the decoder has to search back and the depth of the backward searches are dependent on the value of the channel BER.

An important property of the sequential decoder is that, if the constraint length is large enough, the probability that the decoder will make an error approaches zero (i.e., a BER better than 1×10^{-9}). One cause of error is overflow, being defined as the situation in which the decoder is unable to perform the necessary number of computations in the performance of the tree search. If we define a computation as a complete examination of a path through the decoding tree, a decoder has a limit on the number of computations that it can make per unit time. The number of searches and computations is a function of the number of errors arriving at the decoder input, and the number of computations that must be made to decode one information bit is a random variable. An important parameter for a decoder is the average number of computations per decoded information bit. As long as the probability of bit error is not too high, the chances of decoder overflow will be low, and satisfactory performance will result.

For the previous discussion it has been assumed that the output of a demodulator has been a hard decision. By "hard" decision we mean a firm, irrevocable decision. If these were soft decisions instead of hard decisions, additional improvement in error performance (or coding gain) on the order of several decibels can be obtained. By a "soft" decision we mean that the output of a demodulator is quantized into four or eight levels (e.g., 2- or 3-bit quantization, respectively), and then certain decoding algorithms can use this additional information to improve the output BER. Sequential and Viterbi decoding algorithms can use this soft decision information effectively, giving them an advantage over algebraic decoding techniques, which are not designed to handle the additional information provided by the soft decision.

The soft decision level of quantization is indicated conventionally by the letter Q, which indicates the number of bits in the quantized decision sample. If $Q = 1$, we are dealing with a hard decision demodulator output; $Q = 2$ indicates a quantization level of 4; $Q = 3$ a quantization level of 8; and so on.

4.2.4.1.1 Viterbi Decoding. The Viterbi decoder is one of the more common decoders on links using convolutional codes. The Viterbi decoding algorithm is a path maximum-likelihood algorithm that takes advantage of the remerging path structure (see Figure 4.8) of convolutional codes. By path maximum-likelihood decoding we mean that of all the possible paths through the trellis, a Viterbi decoder chooses the path, or one of the paths, most likely in the probabilistic sense to have been transmitted. A brief description of the operation of a Viterbi decoder using a demodulator giving hard decisions is now given.

For this description our model will be a binary symmetric channel (i.e., BPSK, BFSK). Errors that transform a channel code symbol 0 to 1 or 1 to 0 are assumed to occur independently from symbols with a probability of p. If all input (message) sequences are equally likely, the decoder that minimizes the overall path error probability for any code, block, or convolution is one that examines the error-corrupted received sequence, which we may call $y_1, y_2, \ldots, y_j, \ldots,$ and chooses the data sequence corresponding to the sequence that was transmitted or $x_1, x_2, \ldots, x_j, \ldots,$ which is closest to the received sequence as measured by the Hamming distance. The Hamming distance can be defined as the transmitted sequence that differs from the received sequence by the minimum number of symbols.

Consider the tree diagram (typically Figure 4.7). The preceding statement tells us that the path to be selected in the tree is the one whose code sequence differs in the minimum number of symbols from the received sequence. In the derived trellis diagram (Figure 4.8) it was shown that the transmitted code branches remerge continually. Thus the choice of possible paths can be limited in the trellis diagram. It is also unnecessary to consider the entire received sequence at any one time to decide on the most likely transmitted sequence or minimum distance. In particular, immediately after the third branch (Figure 4.8) we may determine which of the two paths leading to node or state "a" is more likely to have been sent. For example, if 010001 is received, it is clear that this is a Hamming distance 2 from 000000, while it is a distance 3 from 111011. As a consequence, we may exclude the lower path into node "a." No matter what the subsequent received symbols will be, they will affect the Hamming distances only over subsequent branches after these two paths have remerged and, consequently, in exactly the same way. The same can be said for pairs of paths merging at the other three nodes after the third branch. Often, in the literature, the minimum-distance path of the two paths merging at a given node is called the "survivor." Only two things have to be remembered: the minimum-distance path from the received sequence (or survivor) at each node and the value of that minimum distance. This is necessary because at the next node level we must compare two branches merging at each node level, which were survivors at the previous level for different nodes. This can be seen in Figure 4.8 where the comparison at node "a" after the fourth branch is among the survivors of the comparison of nodes "a" and "c" after the third branch. For example, if received sequence over the first four branches is 01000111, the survivor at the third node level for node "a" is 000000 with distance 2 and at node "c" it is 110101, also with distance 2. In going from the third node level to the fourth, the received sequence agrees precisely with the survivor from "c" but has distance 2 from the survivor from "a." Hence the survivor at node "a" of the fourth level is the data sequence 1100 that produced the code sequence 11010111, which is at minimum distance 2 from the received sequence.

In this way we may proceed through the received sequence and at each step preserve one surviving path and its distance from the received sequence, which is more generally called a "metric." The only difficulty that may arise is the possibility that, in a given comparison between merging paths, the distances or metrics are identical. In this case we may simply flip a coin, as is done for block code words at equal distances from the received sequence. Even if both equally valid contenders were preserved, further received symbols would affect both metrics in exactly the same way and thus not further influence our choice.

Another approach to the description of the algorithm can be obtained from the state diagram representation given in Figure 4.9. Suppose we sought that path around the directed state diagram arriving at node "a" after the kth transition, whose code symbols are at a minimum distance from the received sequence. Clearly, this minimum-distance path to node "a" at time k can only be one of two candidates: the minimum-distance path to node "a" at time $k - 1$ and the minimum-distance path to node "c" at time $k - 1$. The comparison is performed by adding the new distance accumulated in the kth transition by each of these paths to their minimum distances (metrics) at time $k - 1$.

Thus it appears that the state diagram also represents a system diagram for this decoder. With each node or state, we associate a storage register that remembers the minimum distance path into the state after each transition as well as a metric register that remembers its (minimum) distance from the received sequence. Furthermore, comparisons are made at each step between the two paths that lead into each node. Thus four comparators must also be provided.

We will expand somewhat on the distance properties of convolutional codes following the example given in Figure 4.6. It should be noted that, as with linear block codes, there is no loss in generality in computing the distance from the all-zeros codeword to all other codewords, for this set of distances is the same as the set of distances from any specific codeword to all the others.

For this purpose we may again use either the trellis diagram or state diagram. First, we redraw the trellis diagram in Figure 4.8, labeling the branches according to their distances from the all-zeros path. Now consider all the paths that merge with the all-zeros path for the first time at some arbitrary node "j." From the redrawn trellis diagram (Figure 4.10), it can be seen that of these paths there will be just one path at distance 5 from the all-zeros path and this diverged from it three branches back. Similarly, there are two at distance 6 from it, one that diverged four branches back and the other that diverged five branches back, and so forth. It should be noted that

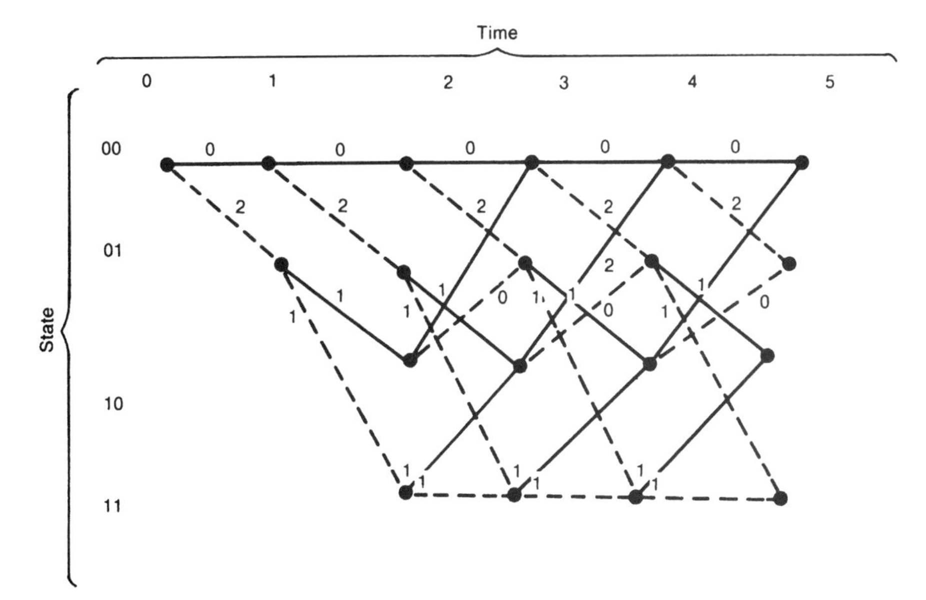

Figure 4.10. Trellis diagram labeled with distances from the all-zeros path. (From Ref. 5.)

the input bits for the distance 5 path are $00 \cdots 01000$ and thus differ in only one input bit from the all-zero path. The minimum distance, sometimes called the "minimum free distance," among all paths is thus seen to be 5. This implies that any pair of channel errors can be corrected, for two errors will cause the received sequence to be at a distance 2 from the transmitted (correct) sequence, but it will be at least at distance 3 from any other possible code sequence. In this way the distances of all paths from all-zeros (or any arbitrary) path can be determined from the trellis diagram.

4.2.4.2 *Systematic and Nonsystematic Convolutional Codes.* The term "systematic" convolutional code refers to a code on each of whose branches the uncoded information bits are included in the encoder output bits generated by that branch. Figure 4.11 shows an encoder for rate $1/2$ and $K = 2$ that is systematic.

For linear block codes, any nonsystematic code can be transformed into a systematic code with the same block distance properties. This is not the case for convolutional codes because the performance of a code on any channel depends largely on the relative distance between codewords and, particularly, on the minimum free distance. Making the code systematic, in general, reduces the maximum possible free distance for a given constraint length and code rate. For example, the maximum minimum-free-distance systematic code for $K = 3$ is that of Figure 4.11 and this has $d = 4$, while the nonsys-

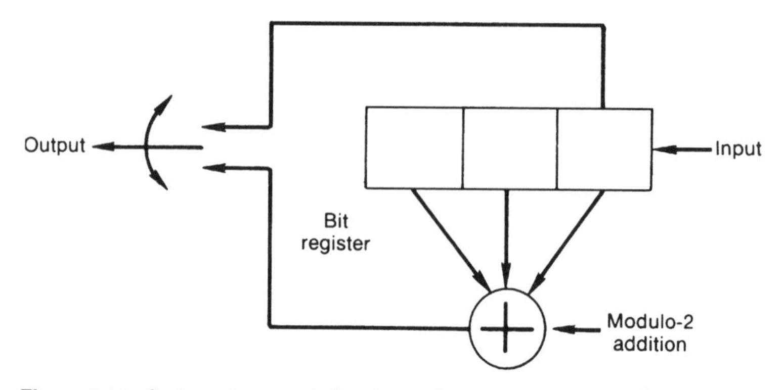

Figure 4.11. Systematic convolutional encoder, $K = 3$, $R = 1/2$. (From Ref. 5.)

TABLE 4.5 Comparison of Systematic and Nonsystematic $R = 1/2$ Code Distances

	Maximum, Minimum Free Distance	
K	Systematic	Nonsystematic
2	3	3
3	4	5
4	4	6
5	5	7

Source: Reference 5.

tematic $K = 3$ code of Figure 4.6 has a minimum free distance of $d = 5$. Table 4.5 shows the maximum free distance for $R = 1/2$ systematic and nonsystematic codes for $K = 2$ through 5. It should be noted that for large constraint lengths the results are even more widely separated.

4.2.5 Channel Performance of Uncoded and Coded Systems

4.2.5.1 Uncoded Performance. For uncoded systems a number of modulation implementations are reviewed in the presence of additive white Gaussian noise (AWGN) and with Rayleigh fading. The AWGN performance of BPSK, QPSK, and 8-ary PSK is shown in Figure 4.12. AWGN is typified by thermal noise or wideband white noise jamming. The demodulator for this system requires a coherent phase reference.

Another similar implementation is differentially coherent phase-shift keying. This is a method of obtaining a phase reference by using the previously received channel symbol. The demodulator makes its decision based on the change in phase from the previous to the present received channel symbol. Figure 4.13 gives the performance of DBPSK and DQPSK with values of BER versus E_b/N_0.

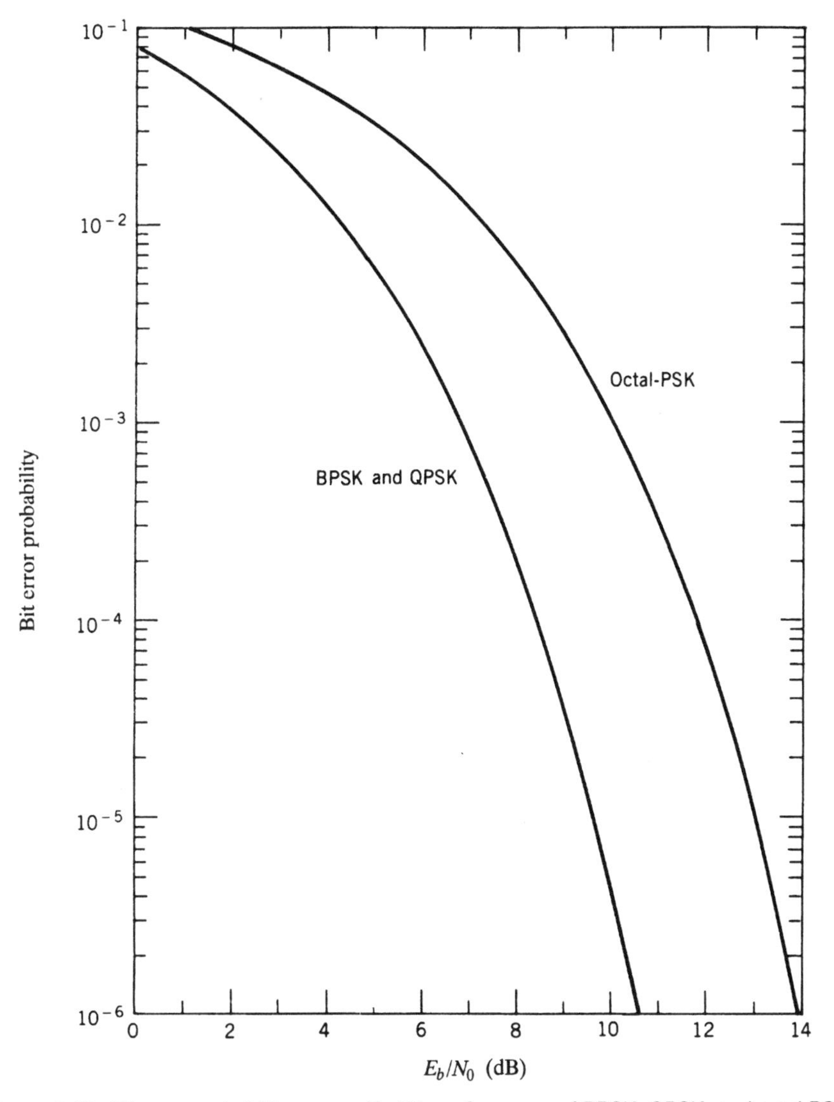

Figure 4.12. Bit error probability versus E_b/N_0 performance of BPSK, QPSK, and octal-PSK (8-ary PSK). (From Ref. 5.)

Figure 4.14 gives performance for M-ary FSK.

Independent Rayleigh fading can be assumed during periods of heavy rainfall on satellite links operating above about 10 GHz (see Chapter 9). Such fading can severely degrade error rate performance. The performance with this type of channel can be greatly improved by providing some type of diversity. Here we mean providing several independent transmissions for each information symbol. In this case we will restrict the meaning to some

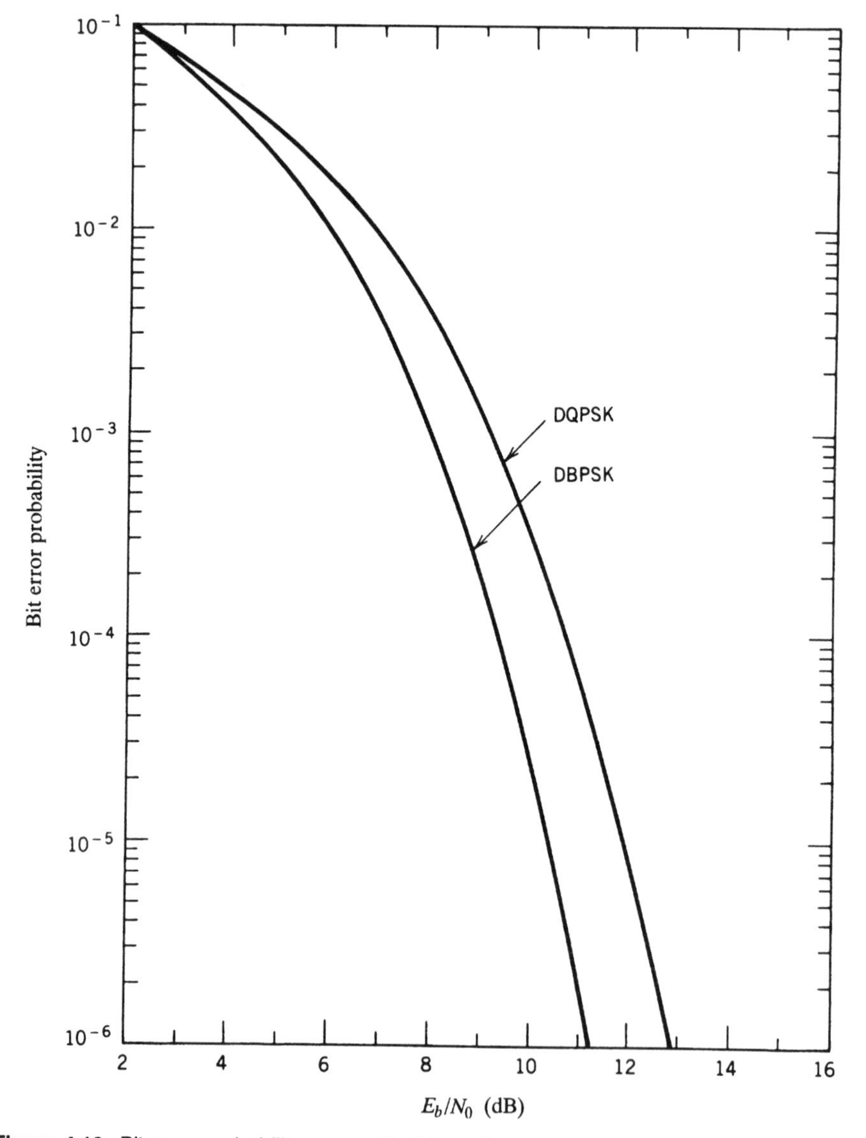

Figure 4.13. Bit error probability versus E_b/N_0 performance of DBPSK and DQPSK. (From Ref. 5.)

form of time diversity that can be achieved by repeating each information symbol several times and using interleaving/deinterleaving for the channel symbols. Figure 4.15 gives binary bit error probability for several orders of diversity (L = order of diversity; $L = 1$, no diversity) for the mean bit energy-to-noise ratio ($\overline{E}_b/N_0$). This figure shows that for a particular error rate there is an optimum amount of diversity. The modulation is binary FSK.

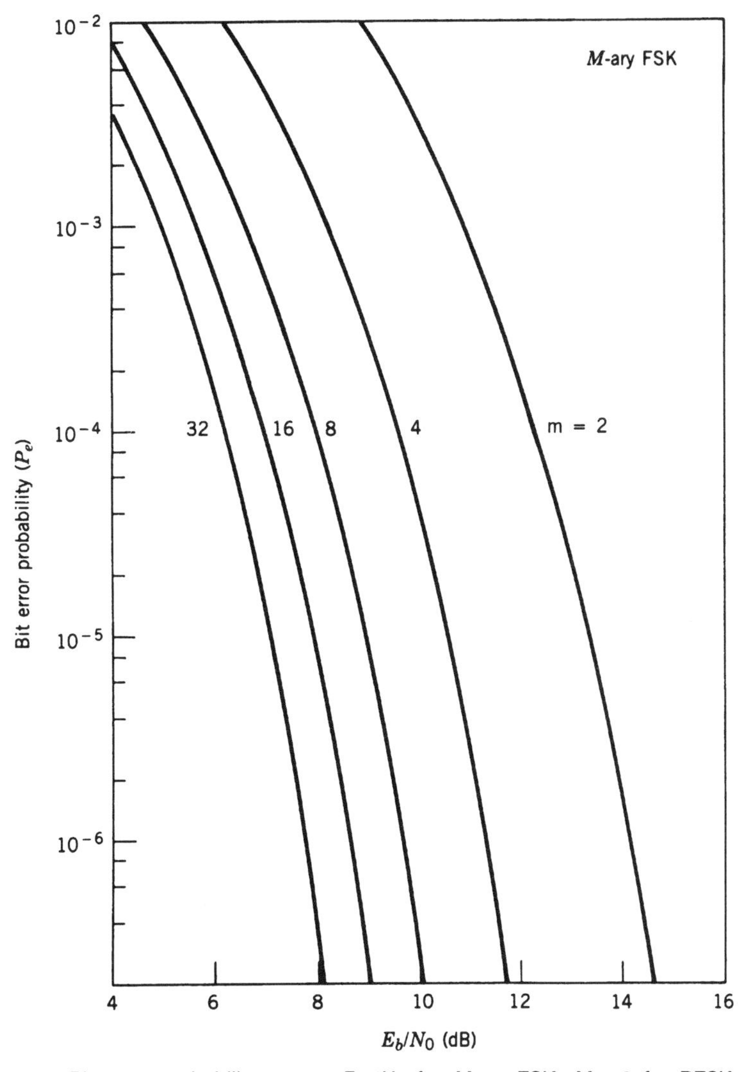

Figure 4.14. Bit error probability versus E_b/N_0 for M-ary FSK: $M = 2$ for BFSK. (From Ref. 5.)

Table 4.6 recaps error performance versus E_b/N_0 for the several modulation types considered. The reader should keep in mind that the values for E_b/N_0 are theoretical values. A certain modulation implementation loss should be added for each case to derive practical values. The modulation implementation loss value in each case is equipment driven.

4.2.5.2 Coded Performance. Figures 4.16, 4.17, and 4.18 show the BER performance $K = 7$, $R = 1/2$; $K = 7$, $R = 1/3$; and $K = 9$, $R = 3/4$ convo-

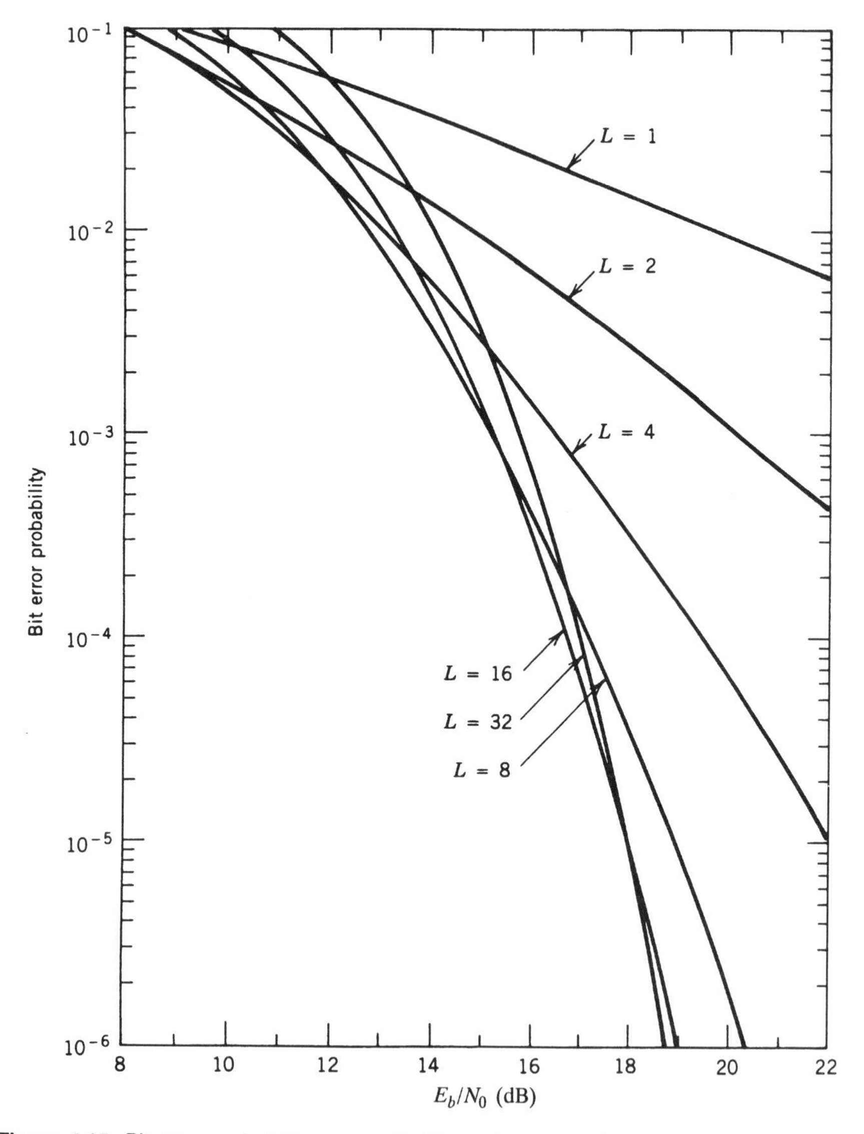

Figure 4.15. Bit error probability versus E_b/N_0 performance of binary FSK on a Rayleigh fading channel for several orders of diversity. L = order of diversity. (From Ref. 5.)

lutional coding systems on an AWGN channel with hard and 3-bit soft quantization. Figures 4.16 and 4.18 again illustrate the advantages of soft quantization discussed in Section 4.2.2.2.

Reference 5 reports that simulation results have been run for many other codes (from those discussed) on an AWGN channel. The results show that for rate 1/2 codes, each increment increase in the constraint length in the

TABLE 4.6 Summary of Uncoded System Performance

Channel	Modulation / Demodulation	E_b/N_0 (dB) Required for Given Bit Error Rate						
		10^{-1}	10^{-2}	10^{-3}	10^{-4}	10^{-5}	10^{-6}	10^{-7}
Additive white Gaussian noise	BPSK and QPSK	−0.8	4.3	6.8	8.4	9.6	10.5	11.3
	Octal-PSK	1.0	7.3	10.0	11.7	13.0	13.9	14.7
	DPBSK	2.1	5.9	7.9	9.3	10.3	11.2	11.9
	DQPSK	2.1	6.8	9.2	10.8	12.0	12.9	13.6
	Noncoherently demodulated binary FSK	5.1	8.9	10.9	12.3	13.4	14.2	14.9
	Noncoherently demodulated 8-ary MFSK	2.0	5.2	7.0	8.2	9.1	9.9	10.5
Independent Rayleigh fading	Binary FSK, $L = 1$	9.0	19.9	30.0	40.0	50.0	60.0	70.0
	Binary FSK, $L = 2$	7.9	14.8	20.2	25.3	30.4	35.4	40.4
	Binary FSK, $L = 4$	8.1	13.0	16.5	19.4	22.1	24.8	27.3
	Binary FSK, $L = 8$	8.7	12.8	15.3	17.2	18.9	20.5	22.0
	Binary FSK, $L = 16$	9.7	13.2	15.3	16.7	18.0	19.1	20.0
	Binary FSK, $L = 32$	10.9	14.1	15.8	17.1	18.1	18.9	19.7

Source: Reference 5.

range of $K = 3$ to $K = 8$ provides an approximate 0.4–0.5-dB improvement in E_b/N_0 at a BER of 1×10^{-5}.

Figure 4.19 shows the 3-bit quantization coding gain for the codes shown in Figures 4.16, 4.17, and 4.18.

Figure 4.20 shows bit error probability versus E_b/N_0 performance of a $K = 7$, $R = 1/2$ convolutional coding system with DBPSK modulation and an AWGN channel. Note that the performance is notably degraded compared to a coherent PSK system as shown in Figure 4.16.

Table 4.7 recapitulates coding gain information for several modulation types using $K = 7$, $R = 1/2$ convolutional coding and a Viterbi decoder assuming a BER of 1×10^{-5}. Table 4.8 summarizes E_b/N_0 requirements of several coded communication systems with BPSK modulation and a BER of 1×10^{-5}.

4.2.6 Coding with Bursty Errors

In our entire discussion so far we have limited our communication systems to channels with AWGN and *only* AWGN. As we mentioned, this would suffice for well-behaved propagation media. Typically by "well-behaved" media we mean wire-pair, coaxial cable, fiberoptic cable, very short line-of-sight (LOS) microwave links [e.g., < 1 mi (1.6 km)], and satellite links above a 5° elevation angle and operating below 10 GHz.*

* Purists will note certain exceptions in the low-UHF and VHF bands operating in the vicinity of the magnetic equator.

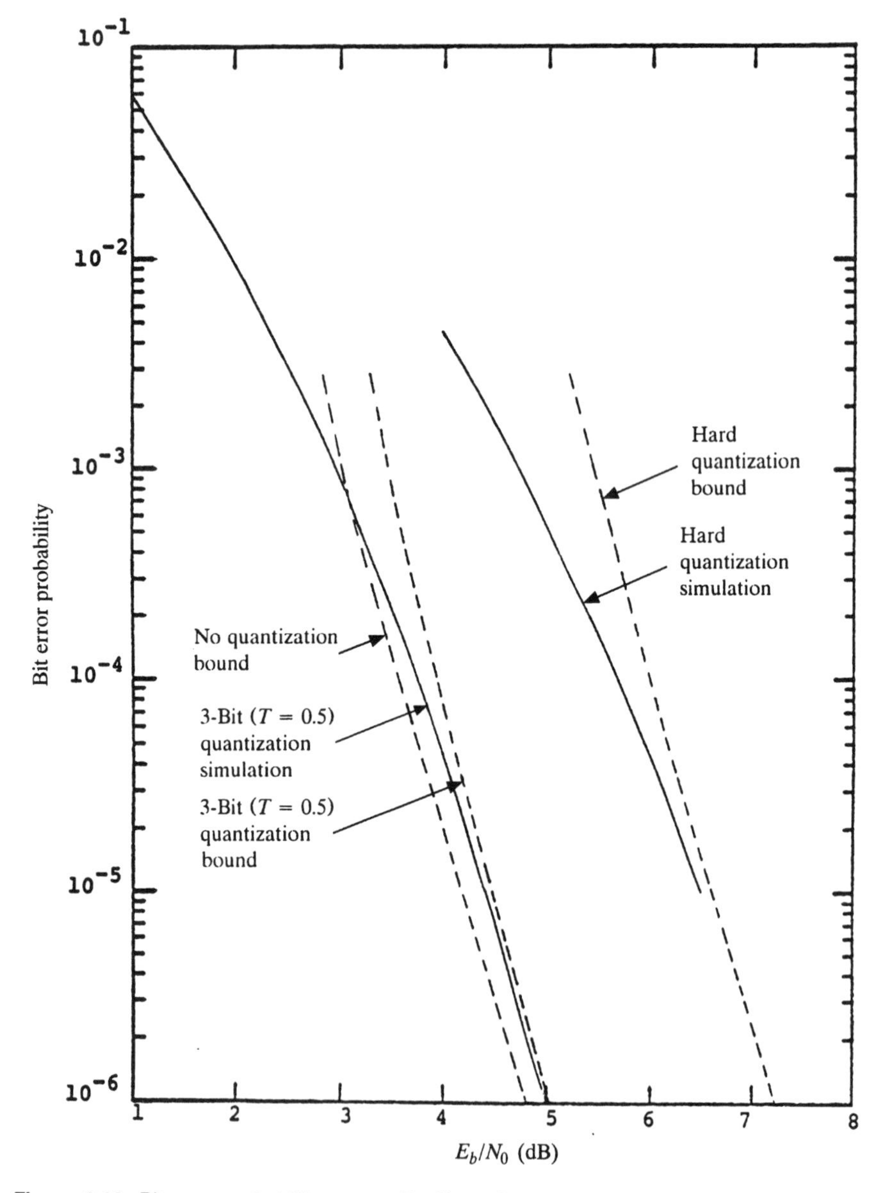

Figure 4.16. Bit error probability versus E_b/N_0 performance of a $K = 7$, $R = 1/2$ convolutional coding system with BPSK modulation and an AWGN channel. (From Ref. 5.)

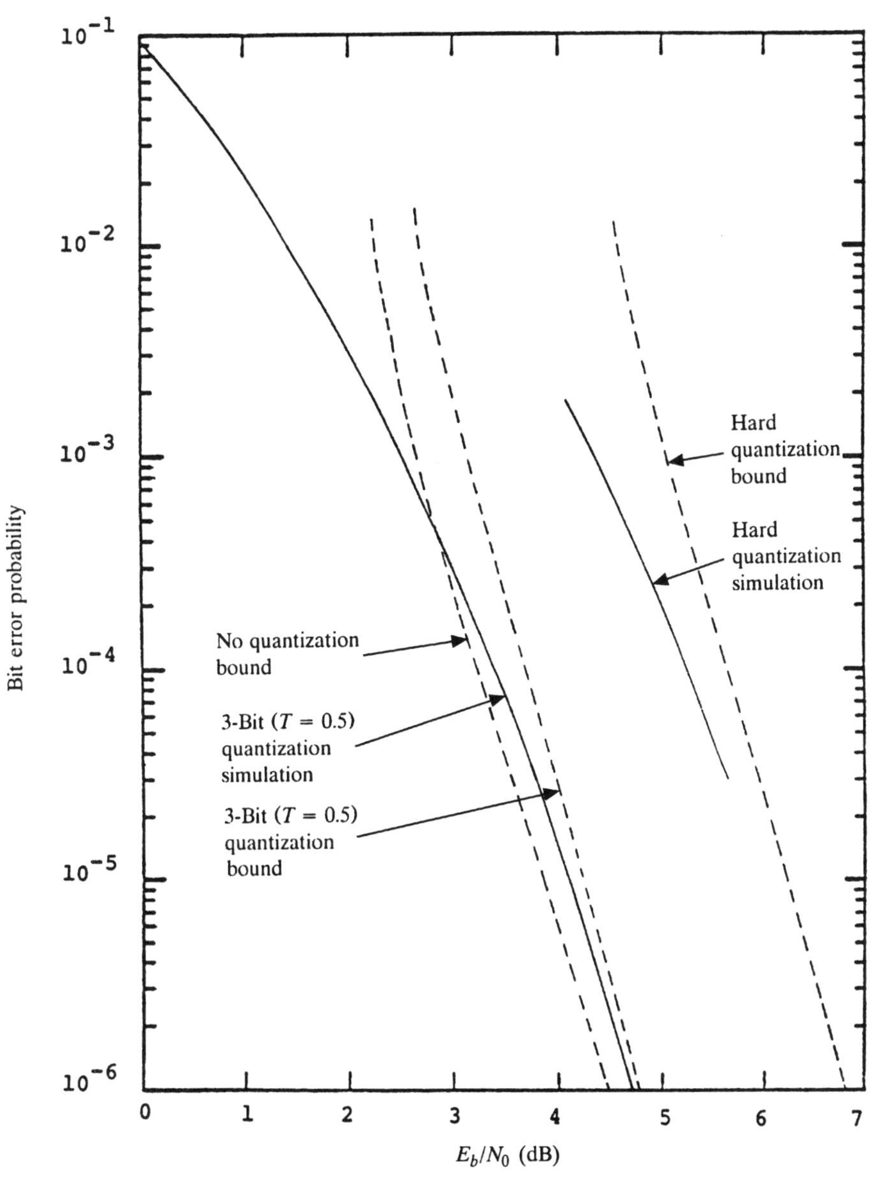

Figure 4.17. Bit error probability versus E_b/N_0 performance of $K = 7$, $R = 1/3$ convolutional coding system with BPSK modulation and an AWGN channel. (From Ref. 5.)

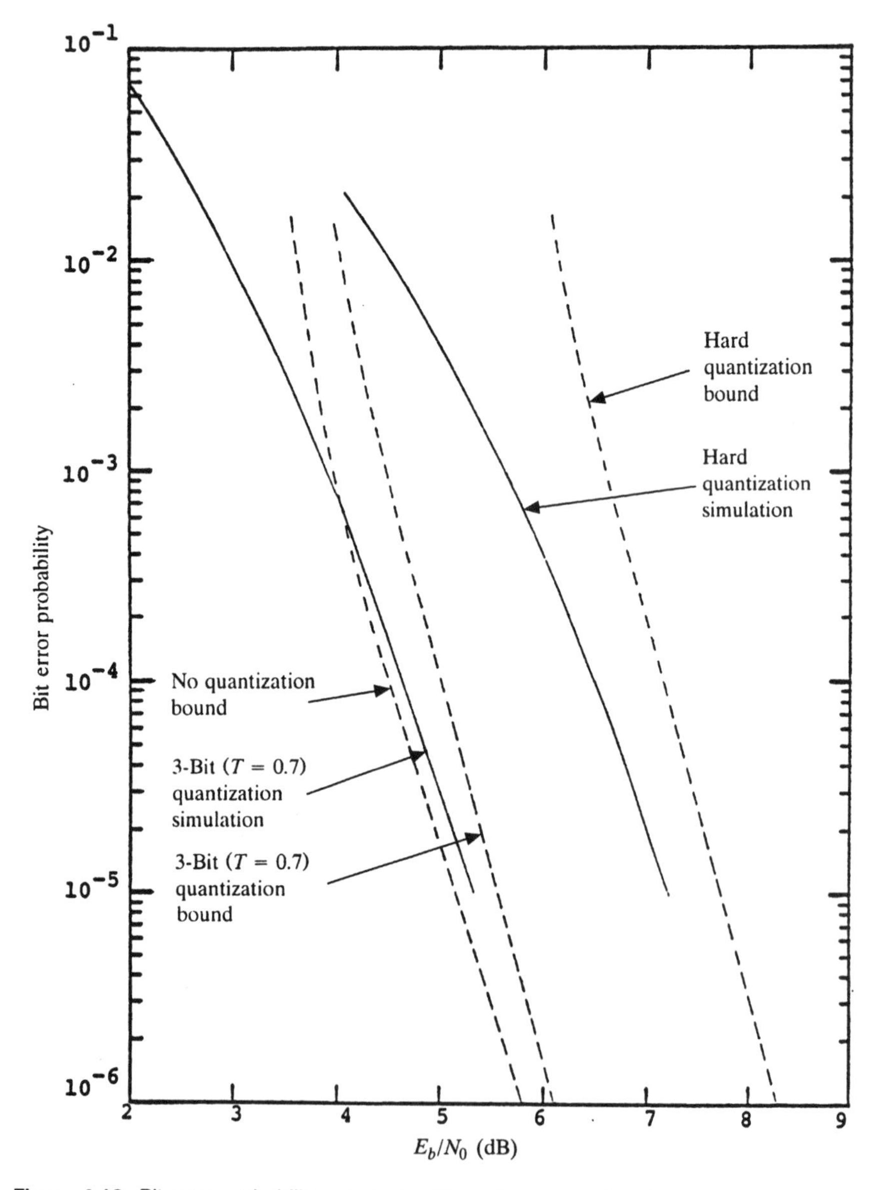

Figure 4.18. Bit error probability versus E_b/N_0 performance of a $K = 9$, $R = 3/4$ convolutional coding system with BPSK modulation and an AWGN channel. (From Ref. 5.)

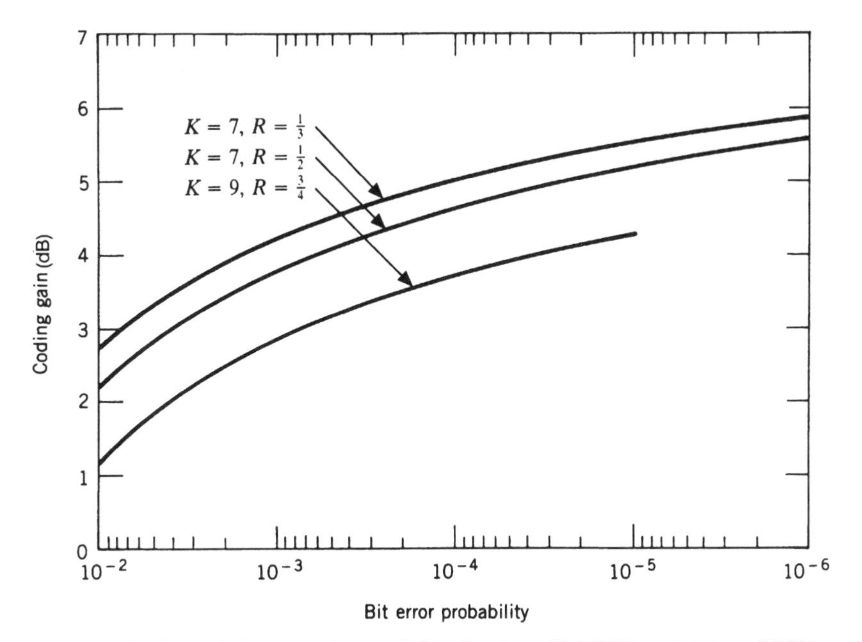

Figure 4.19. Coding gain for several convolutional codes with BPSK modulation, AWGN, and 3-bit receiver quantization. (From Ref. 5.)

Poorly behaved media are MF/HF radio, all LOS microwave on links over 1 mi (1.6 km) range, tropospheric scatter and diffraction circuits, and satellite communication paths above 10 GHz. There is an easy "fix," but beware of added delay.

There are basically two causes of error bursts (vis-à-vis random errors): fading and impulse noise. The characteristics of each are quite different, but the result is the same—error bursts. We will make an assumption here that during the period when there are no error bursts, the signal is benign with only a very few random errors.

It should be noted that impulse noise could be found on any of the media. However, the chances are very slight that impulse noise occurred on fiberoptic circuits. It could "invade" these circuits when still in the electrical domain.

The objective here is to randomize the bit stream. To do this we use a device called an *interleaver.* An interleaver takes blocks of bits and pseudo-randomizes the bits in each block. It is the interleaver that "blocks" the data, and we will call each block a *span*—because it is a time span. This span must be greater than the duration of a fade.

Suppose we found that, statistically, fades seldom had durations greater than 0.9 second. We might make the span of 1-second duration. What would be 1 second of data? At 9600 bps, it would be 9600 bits or binary coded symbols. So the interleaver would break the binary symbol stream up into

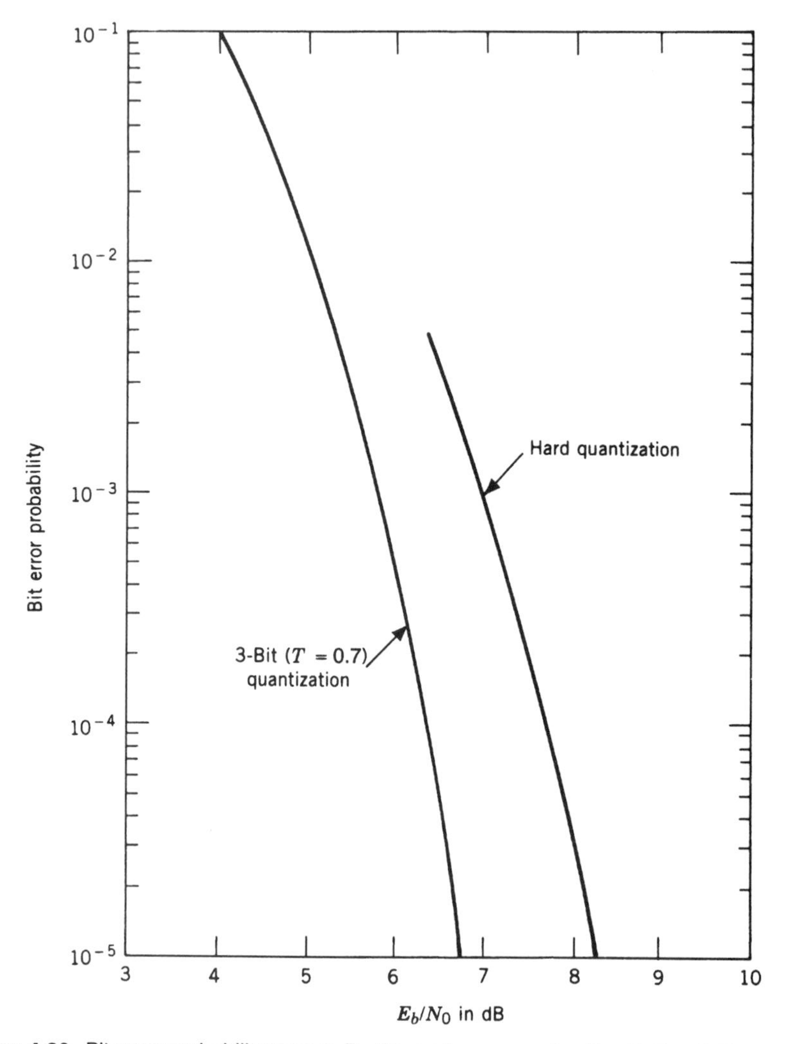

Figure 4.20. Bit error probability versus E_b/N_0 performance of a $K = 7$, $R = 1/2$ convolutional coding system with DBPSK modulation and an AWGN channel. (From Ref. 5.)

9600 symbol spans. Suppose it was a bit stream with DS1 (T1) at 1.544 Mbps and the coding rate was 1/2. Thus the symbol rate out of the coder would be 2×1.544 Mbps or 3.088 Mbps, which would be the length of the span or 1-second worth of binary symbols.

The danger here is the added delay of the interleaver. The span time is directly additive to the total delay. This might be quite acceptable on circuits carrying data, but it may be highly unacceptable on circuits carrying voice or even conference television. If the span time is only a few milliseconds, it might not alter the delay budget enough to affect speech telephony circuits.

TABLE 4.7 Summary of E_b/N_0 Requirements and Coding Gains of $K = 7$, $R = 1/2$ Viterbi-decoded Convolutional Coding Systems with Several Modulation Types, BER = 1×10^{-5}

Modulation	Number of Bits of Receiver Quantization per Binary Channel Symbol	E_b / N_0 (dB) Required for $P_b = 10^{-5}$	Coding Gain (dB)
Coherent biphase			
BPSK or QPSK	3	4.4	5.2
BPSK or QPSK	2	4.8	4.8
BPSK or QPSK	1	6.5	3.1
Octal-PSK[a]	1	9.3	3.7
DBPSK[a]	3	6.7	3.6
DBPSK[a]	1	8.2	2.1
Differentially[a] coherent QPSK	1	9.0	3.0
Noncoherently demodulated binary FSK	1	11.2	2.1

[a]Interleaving/deinterleaving assumed.

Source: Reference 5.

A block diagram of a typical interleaver implementation is shown in Figure 4.21. One important aspect of interleaving is that the interleaver and companion deinterleaver must be time synchronized.

4.3 ADVANCED SIGNAL WAVEFORMS*

In this section, four coded modulation techniques are discussed: block-coded modulation (BCM), trellis-coded modulation (TCM), multilevel-coded modulation (MLCM), and partial response with soft decoder. Tabular comparisons of the four coded modulation schemes are provided. Of interest to the system designer is coding gain, bandwidth expansion, and implementation complexity.

4.3.1 Block-Coded Modulation (BCM)

Here error correction is combined with a quadrature modulation scheme. With the BCM technique, multidimensional signal constellations are generated that have both large distances (i.e., good error performance) and also regular structures, allowing an efficient parallel demodulation architecture

*Section 4.3 is based on ITU-R Rec. F.1101, Appendix 2 to Annex 1 (Ref. 7).

TABLE 4.8 Summary of E_b/N_0 Requirements of Several Coded Communication Systems for a BER = 1×10^{-5} with BPSK Modulation

	Coding Type	Number of Bits Receiver Quantization	Coding[a] Gain (dB)
$K=7$, $R=\frac{1}{2}$	Viterbi-decoded convolutional	1	3.1
$K=7$, $R=\frac{1}{2}$	Viterbi-decoded convolutional	3	5.2
$K=7$, $R=\frac{1}{3}$	Viterbi-decoded convolutional	1	3.6
$K=7$, $R=\frac{1}{3}$	Viterbi-decoded convolutional	3	5.5
$K=9$, $R=\frac{3}{4}$	Viterbi-decoded convolutional	1	2.4
$K=9$, $R=\frac{2}{4}$	Viterbi-decoded convolutional	3	4.3
$K=24$, $R=\frac{1}{2}$	Sequential-decoded convolutional 20 kbps,[b] 1000-bit blocks	1	4.2
$K=24$, $R=\frac{1}{2}$	Sequential-decoded convolutional 20 kbps,[b] 1000-bit blocks	3	6.2
$K=10$, $L=11$, $R=\frac{1}{2}$	Feedback-decoded convolutional	1	2.1
$K=8$, $L=8$, $R=\frac{2}{3}$	Feedback-decoded convolutional	1	1.8
$K=8$, IL $=9$, $R=\frac{3}{4}$	Feedback-decoded convolutional	1	2.0
$K=3$, $L=3$, $R=\frac{3}{4}$	Feedback-decoded convolutional	1	1.1
(24, 12) Golay		3	4.0
(24, 12) Golay		1	2.1
(127, 92) BCH		1	3.3
(127, 64) BCH		1	3.5
(127, 36) BCH		1	2.3
(7, 4) Hamming		1	0.6
(15, 11) Hamming		1	1.3
(31, 26) Hamming		1	1.6

[a] Requirement for uncoded system is 9.6 dB.
[b] The same system at a data rate of 100 kbps has 0.5 dB less coding gain.

Notation:

K Constraint length of a convolutional code defined as the number of binary register stages in the encoder for such a code. With the Viterbi-decoding algorithm, increasing the constraint length increases the coding gain but also the implementation complexity of the system. To a much lesser extent the same is also true with sequential and feedback decoding algorithms.

L Look-ahead length of a feedback-decoded convolutional coding system defined as the number of received symbols, expressed in terms of the corresponding number of encoder input bits, that are used to decode an information bit. Increasing the look-ahead length increases the coding gain but also the decoder implementation complexity.

(n, k) Denotes a block code (Golay, BCH, or Hamming here) with n decoder output bits for each block of k encoder input bits.

Receiver Quantization describes the degree of quantization of the demodulator outputs. Without coding and biphase (0° or 180°) modulation the demodulator output (or intermediate output if the quantizer is considered as part of the demodulator) is quantized to 1 bit (i.e., the sign if provided). With coding, a decoding decision is based on several demodulator outputs, and the performance can be improved if in addition to the sign the demodulator provides some magnitude information.

Source: Reference 5.

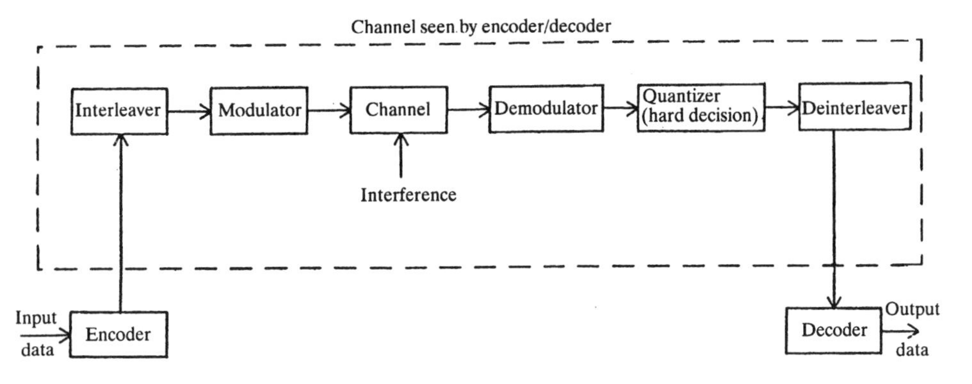

Figure 4.21. Block diagram of a coding system with interleaving/deinterleaving.

called staged decoding. It does so by obtaining a subset of the Cartesian product of a number of elementary (i.e., low-dimensional) signal sets by itself. The staged construction allows demodulation algorithms based on projection of the signal set into lower-dimensional, lower-size constellations. Such algorithms lend themselves naturally to a pipelined architecture. By combining algebraic codes of increasing Hamming distance with nested signal constellations of decreasing Euclidean distance, multidimensional signals with large distances can be generated.

Step partitioning of the signal constellation is the basis of construction of practical BCM schemes. Table 4.9 provides coding gains of BCM systems over uncoded QAM reference systems corresponding to the same number of net information bits per transmitted symbol. The table shows BCM families based on one-step partitioning (or "B-partition") and two-step partitioning (or "C-partition"). The data in the table are relevant only to AWGN channels.

TABLE 4.9 Coding Gains (dB) Over Corresponding Uncoded QAM Systems on AWGN Channels

Uncoded Signal Constellation	BCM Signal Constellation	Kind of Partition	Block-code Length (n)	Number of Dimensions ($2n$)	Asymptotic Coding Gain (dB)
16-QAM	24-QAM	B	2	4	1.5
	24-QAM	B	4	8	2.6
64-QAM	96-QAM	B	2	4	1.8
	80-QAM	B	4	8	2.6
64-QAM	128-QAM	C	4	8	3.1
256-QAM	368-QAM	B	2	4	1.7
	308-QAM	B	4	8	2.5

Source: Appendix 2 to Annex 1, Table 2, p. 243, ITU-R Rec. F.1101, 1994 F Series, Part 1 (Ref. 7).

BCM detection is based on choice of the codeword that is nearest to the received sequence in the Euclidean distance sense. BCM improves coding gain in a BER range from 10^{-3} to 10^{-4} without greatly increasing the number of redundant bits, by introducing block codes into multistate modulation schemes. For large QAM schemes, such as 256-QAM, the introduction of block codes into only 2 bits of the 8-bit baseband signal streams can improve error performance. This is because this method enables the addition of four times the number of redundant bits than conventional error correction schemes.

The two code bits just mentioned are used as "subset" signals and the remaining uncoded bits (6 bits in the case of 256-QAM) are mapped into signal space to maximize the Euclidean distance based on the Ungerboeck set partitioning method. These subset signals are decoded using a conventional error-correcting algorithm. The error correction of uncoded bits is performed only if the subset signal is corrected. At the specific timeslot, the uncoded bits are decoded by selecting a signal point nearest to the received signal point from the coded subset signals using soft decision information.

BCM can provide coding gains around 5 dB at a BER of 1×10^{-4}.

4.3.2 Trellis-Coded Modulation (TCM)

TCM is a generalized convolutional coding scheme with nonbinary signals optimized to achieve large "free Euclidean distance," d_E, among sequences of transmitted symbols. This results in a lower E_b/N_0 value or a smaller bandwidth to transmit binary digital signals at a given rate and BER.

TCM uses a redundant alphabet to achieve this advantage. It is obtained by convolutionally encoding k out of n information bits to be transmitted at a certain time. One redundant bit is added where the convolutional code rate is $k(k + 1)$. There is a symbol-mapping procedure that follows the convolutional encoding. The encoder bits determine the subset (or "submodulation") to which the transmitted symbol belongs. The encoded bits determine a particular signal point in that subset. Another name for the mapping procedures is *set partitioning*. Its purpose is to increase the minimum distance d_E among the symbols.

The cost of TCM coding gain is not an increase in the necessary transmission bandwidth but a higher modulation complexity. The reason for this is that the redundancy of coding is not in the time domain, as used in serial FEC, but is a "spatial" redundancy. A maximum-likelihood sequence estimation (MLSE), implemented as a Viterbi algorithm, is required by an optimum receiver for TCM sequence.

Figure 4.22 shows TCM performance versus the number of net information bits per symbol compared with the corresponding uncoded system.

The performance of TCM on digital LOS microwave systems, when used in conjunction with adaptive equalizers of medium complexity, uncoded and TCM systems, have nearly the same performance over multipath fading

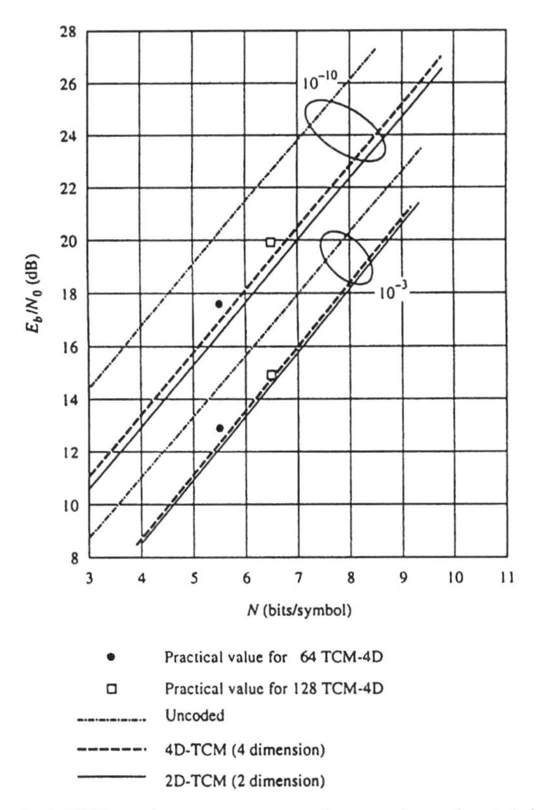

Figure 4.22. Expected TCM performance versus the number of net information bits per symbol compared with the corresponding uncoded system. (From Figure 4, p. 245, ITU-R Rec. F.1101; Ref. 7.)

channels. However, the improvements for lower BER values increase as the BER decreases. It also should be pointed out that the coding gain of a TCM system on a nonlinear channel is greater than on a linear channel. This advantage of TCM is of crucial importance to reduce residual BER in the case of high-complexity modulation schemes.

4.3.3 Multilevel-Coded Modulation (MLCM)

MLCM is based on set partitioning. Here, each level is regarded as an independent transmission path with different minimum square distance. Also, each level uses a different code with a different strength. Figure 4.23 illustrates an example of set partitioning of 16-QAM, where, of course, there are four levels and 16 states. These 16 states (A) are divided into subsets B0 and B1. In these subsets, the minimum square distance among each subset is

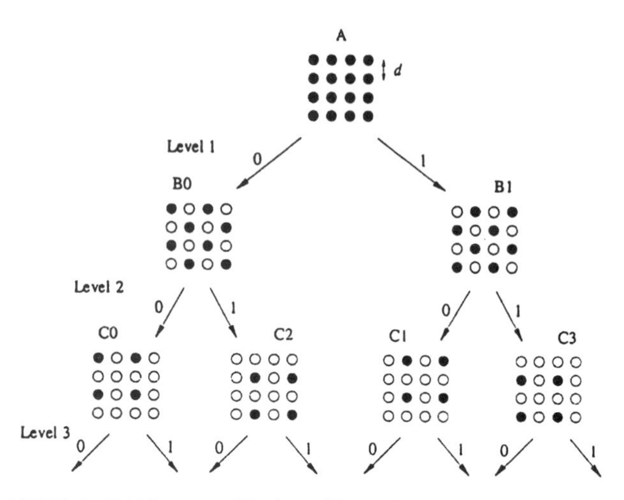

Figure 4.23. MCLM 16-QAM set partitioning. (From Figure 5, p. 246, ITU-R Rec. F.1101; Ref. 7.)

twice that of set A. Subsets B0 and B1 are divided into subsets C0, C2 and C1, C3, respectively.

In the subsets $Ci(i = 0\text{–}3)$, the minimum square distance is $4d^2$. This same partitioning is done until the number of states becomes one in each subset. Thus the 16 states are divided into sets of subsets with increased minimum square distance. In this stage, however, level 1 error performance is determined by the minimum square distance of the A states set. Thus, in order to increase the free Euclidean distance, d_E, coding is performed at the lower level. In this manner the total error performance is improved. MCLM codes are not necessarily restricted to convolutional codes. Nevertheless, convolutional codes are preferred for the lower levels, while other codes, such as block codes, may be used for other levels. The coding rates at each level are chosen separately. Thus the coding rate for MLCM can be selected rather freely. In the case of 16-QAM, for example, if the coding rate is $\frac{1}{2}$ for level 1, $\frac{3}{4}$ for level 2, $\frac{23}{24}$ for level 3, and no coding for level 4, then the total coding rate R becomes:

$$R = \left(\tfrac{1}{2} + \tfrac{3}{4} + \tfrac{23}{24} + 1\right) = 3.2/4$$

The parallel output of each encoder is converted to serial and inputs a mapping circuit. Therefore the results of one coding step correspond to a number of symbols. As a consequence, the coding rate is at least half of the modulation rate. MCLM coding gain is a function of its coding rate and coding methods.

MCLM decoding is carried out by a method called *multistage decoding.* The lowest level is decoded first and, based on the result, the next level is

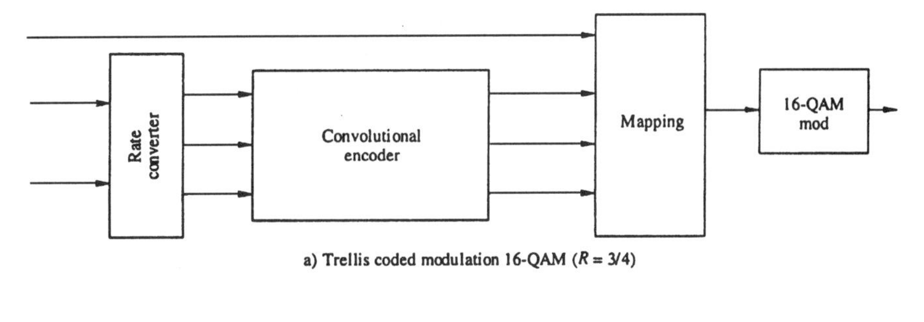

a) Trellis coded modulation 16-QAM ($R = 3/4$)

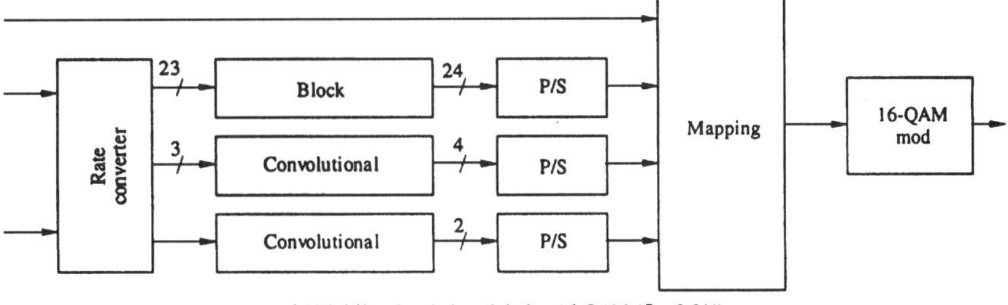

b) Multilevel coded modulation 16-QAM ($R = 3.2/4$)

Figure 4.24. Simplified functional block diagrams of the transmitter side for coded modulation. (From Figure 6, p. 247, ITU-T Rec. F.1101; Ref. 7.)

decoded. Higher levels use a similar scheme. Figure 4.24 is a simplified functional block diagram of TCM and MLCM.

Figure 4.25 illustrates a comparison between BER performance for 128-QAM systems with different redundancies based on simulation.

4.3.4 Partial Response with a Soft Decoder

Quadrature partial response (QPR) technology is a candidate for application to digital LOS microwave systems. It may be combined with other coding techniques to improve BER performance. One of the QPR techniques is ambiguity zone detection or AZD. This is a simple form of a maximum-likelihood (soft) decoder. A 9-QPR eye diagram and a block diagram of AZD correction are shown in Figures 4.26 and 4.27, respectively.

Certain sequences of symbols are forbidden in partial response coding. If we let M be the number of baseband levels from partial response coding, then the coded signal cannot transverse more than N levels between consecutive symbols, where $N = (M + 1)/2$. The only way such forbidden sequences can occur is when a symbol has been received in error. Again, the assumption is that errors are caused by AWGN and are displaced only one level from the correct level into adjacent levels. Under normal E_b/N_0 ratios of interest, this is a valid assumption. When such a forbidden sequence occurs, it is called a *partial response violation* (PRV). Given that a symbol

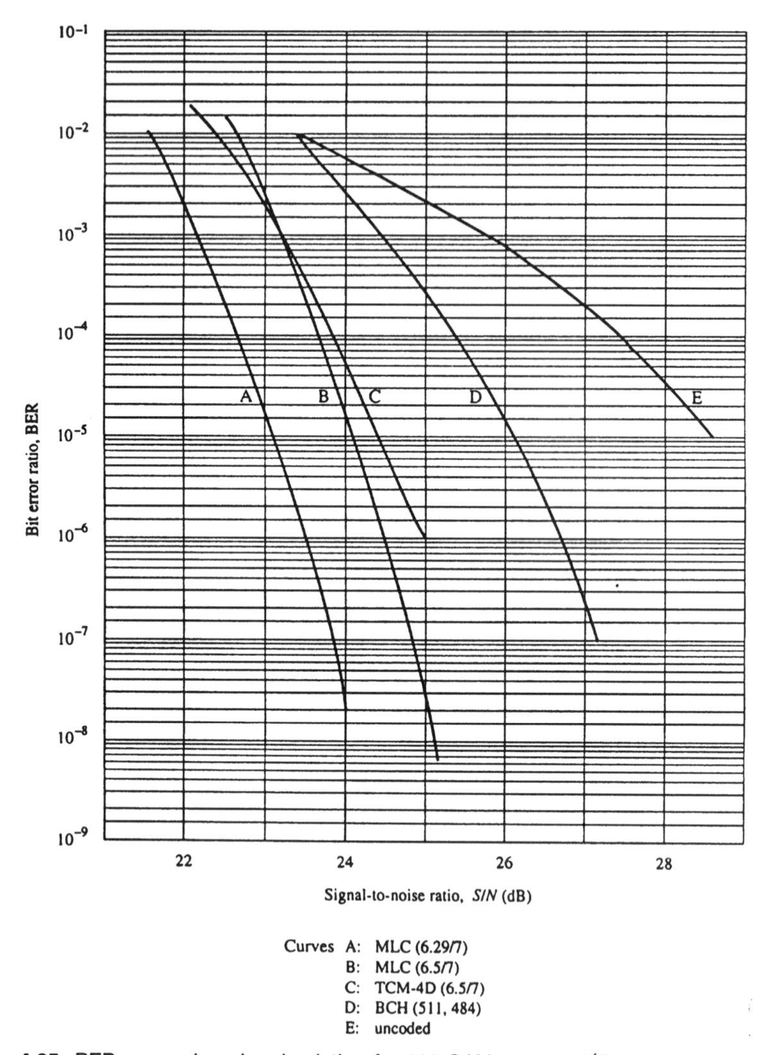

Figure 4.25. BER comparison by simulation for 128-QAM systems. (From Figure 7, p. 248, ITU-R Rec. F.1101; Ref. 7.)

error has occurred in the past D symbols, the probability of detecting a PRV is

$$P = 1 - (1 - 1/N)^D \tag{4.7}$$

With an ambiguity zone detector, the eye is divided vertically between the pinpoints (i.e., ideal locations of the eye centers) into two types of regions. An ambiguity weight (AW) of 0 is given to the symbols sampled in the regions

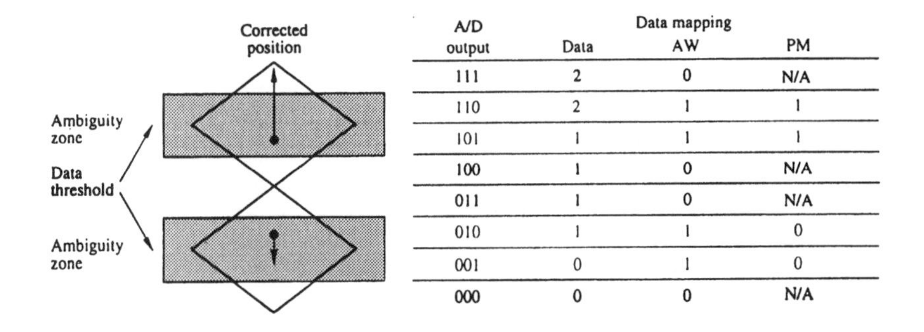

A/D output	Data	AW	PM
	Data mapping		
111	2	0	N/A
110	2	1	1
101	1	1	1
100	1	0	N/A
011	1	0	N/A
010	1	1	0
001	0	1	0
000	0	0	N/A

AW: ambiguity weight
PM: position marker

Figure 4.26. A9-QPR eye diagram showing ambiguity zones. (From Figure 8, p. 249, ITU-R Rec. F.1101; Ref. 7.)

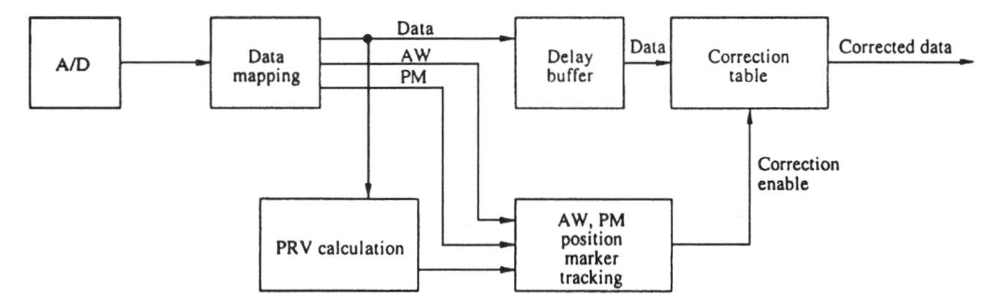

Figure 4.27. AZD correction block diagram. (From Figure 9, p. 249, ITU-R Rec. F.1101; Ref. 7.)

closest to the pinpoints. Their relative ambiguity positions above or below the pinpoints are also marked for later error correction. The further symbols are measured from the pinpoints, the more likely they will be in error. Thus these upper and lower ambiguity zones are set up and symbols in these zones are considered suspect. The decoded symbols and their ambiguity weights and position markers are fed into the error (PRV) detection and correction circuitry.

The decoder will look back D symbols, when a PRV is detected, to see if any decision is made in the ambiguity zones that has AW = 1. A correction is made if and only if both conditions are met. Using the position marker, the correction is made by pushing the ambiguity decision up or down.

Table 4.10 compares performance of various coded (advanced) modulation waveforms.

TABLE 4.10 Comparison of Advanced/Coded Modulation Waveforms

System[a]	Variants	W (dB)	S/N (dB)	Nyquist Bandwidth[b] (b_n)
	Coded Modulation Schemes			
BCM[c]	96 BCM-4D (QAM one-step partition)	24.4	29.0	B/6
	88 BCM-6D (QAM one-step partition)	23.8	28.8	B/6
	16 BCM-8D (QAM one-step partition)	15.3	18.5	B/3, 75
	80 BCM-8D (QAM one-step partition)	23.5	28.4	B/6
	128 BCM-8D (QAM two-step partition)	23.6	28.2	B/6
TCM[d]	16 TCM-2D	12.1	14.3	B/3
	32 TCM-2D	13.9	17.6	B/4
	128 TCM-2D	19.0	23.6	B/6
	512 TCM-2D	23.8	29.8	B/8
	64 TCM-4D	18.3	21.9	B/5.5
	128 TCM-4D	20.0	24.9	B/6.5
	512 TCM-4D	24.8	31.1	B/8.5
MLCM[e]	32 MLCM	14.1	18.3	B/4.5
	64 MLCM	18.1	21.7	B/5.5
	128 MLCM	19.6	24.5	B/6.5
QPR with AZD	9-QPR with coherent detection	11.5	14.5	B/2
	25-QPR with coherent detection	14.0	18.8	B/3
	49-QPR with coherent detection	15.5	21.5	B/4

[a]BCM = block-coded modulation; TCM = trellis-coded modulation; MLCM = multilevel-coded modulation; and AZD = ambiguity zone detection.
[b]The bit rate B does not include code redundancy.
[c]The block-code length is half the number of the BCM signal dimensions.
[d]The performances depend on the implemented decoding algorithm. In this example, an optimum decoder is used.
[e]In this example convolutional code is used for the lower two levels and block codes are used for the third level to give overall redundancies as for those of TCM-4D. Specifically, the redundancies are 3/2 and 8/7 on the two convolutional-coded levels and 24/23 on the block-coded third level.
Note: See Section 3.4.3 for explanations of terminology and notations.
Source: Table 1b, p. 242, ITU-R Rec. F.1101 (Ref. 7).

PROBLEMS AND EXERCISES

1 What is an objective BER on a digital LOS microwave link?

2 There are two basic causes of error (for this discussion) on digital microwave links. What are they?

3 Why is ARQ not feasible on digital links carrying telephony?

4 FEC is a great step forward on digital links to improve performance or achieve "coding gain" comparatively inexpensively. Give the other side of the coin. In other words, what must we pay for this advantage?

5 Define coding rate.

6 Define coding gain (simply).

7 Show what coding gain buys us on a satellite circuit.

8 Give a short narrative description of how soft decision modulators work.

9 Define a block code.

10 What is a major difference between block codes and convolutional codes?

11 Block codes can be described by the notation (N, K, t). What does each symbol mean?

12 Convolutional codes can be described by the notation r and K. Give the meaning of each.

13 What are the three basic convolutional decoders? List in order of power (coding gain) and complexity.

14 Channel performance on a digital radio channel using FEC depends on what? Name at least four items.

15 Carefully describe how interleaving/deinterleaving works. What bodes against its use?

16 With convolutional coding, how can we increase coding gain?

17 We describe four families of advanced digital modulation schemes. What advantages are provided by these modulation schemes? Name at least two.

18 Why would we wish to increase Euclidean distance? (Think signal space constellations and the mechanics of a fading "hit.")

19 Describe the set partitioning of MLCM.

20 Discuss how AZD works.

21 FEC requires additional bandwidth due to added redundancy. Describe how TCM can ameliorate some of this problem.

22 Of the two major classes of errors, FEC (unless we fix it somehow) works on only one class. What is it?

REFERENCES

1. *Error Protection Manual, with Summary and Supplement*, Computer Sciences Corporation, Falls Church, VA, 1973, NTIS AD-759 836.
2. *Error Performance of an International Digital Connection Forming Part of an Integrated Services Digital Network*, CCITT Rec. G.821, Fascicle III.5, IXth Plenary Assembly, Melbourne, 1988.

3. *The New IEEE Standard Dictionary of Electrical and Electronic Terms*, 5th ed., IEEE Std 100-1992, IEEE, New York, 1992.
4. *Satellite Communications Reference Data Handbook*, Vol. I, Computer Science Corporation, Falls Church, VA, Apr. 1983, under DCA contract DCA100-81-C-0044.
5. *Error Control Handbook*, Linkabit Corporation, San Diego, CA, July 1976, under USAF contract F44620-76-C-0056.
6. A. J. Viterbi, "Convolutional Codes and Their Performance in Communication Systems," *IEEE Transactions on Communication Technology*, Vol. Com-19, pp. 751–792, Oct. 1971.
7. *Characteristics of Digital Radio-Relay Systems Below About* 17 *GHz*, ITU-R Rec. F.1101, 1994 F Series Volume, Part 1, ITU, Geneva, 1994, Appendix 2 to Annex 1, "Coded Modulation Techniques."

5

OVER-THE-HORIZON RADIOLINKS

5.1 OBJECTIVES AND SCOPE

This chapter deals with the design of radiolinks that operate beyond line-of-sight (BLOS). Two transmission modes achieving over-the-horizon communications are examined: *diffraction* and *scattering.*

Diffraction over one or more obstacles can be the predominant transmission mode on most shorter paths (i.e., less than 100 mi, 160 km) displaying long-term median transmission loss (path loss) on the order of 170–190 dB. On longer paths (i.e., from 100 to about 500 mi (160–800 km) or more) scattering off a nonhomogeneous atmosphere will be the predominant transmission mode with long-term median transmission loss of 180–260 dB.

Emphasis will be placed on the design of radiolinks operating in the troposcatter mode. A brief review of the design for knife-edge and smooth earth diffraction links is also presented.

5.2 APPLICATION

Over-the-horizon radiolinks, whether based on the diffraction or scatter transmission modes, use larger installations and are more expensive than their line-of-sight (LOS) counterparts. The scattering and diffraction phenomena limit the frequency bands of operation from about 250 MHz to under 6 GHz. Table 5.1 provides a useful comparison of LOS and scatter/diffraction systems.

A summary of some advantages of tropospheric scatter and diffraction links is presented below:

- Military/tactical for transportability where the total number of installations is minimized for point-to-point area coverage.
- Reduces the number of relay stations required to cover a given large distance when compared to LOS radiolinks. Tropospheric scatter/diffraction systems may require from one-tenth to one-third the number of relay stations as a LOS radiolink system over the same path.

TABLE 5.1 Comparison of LOS Microwave Versus Tropo/Diffraction

Item	LOS	Tropo/Diffraction
VF channels	Up to 1800/ 2700 per carrier	Up to 240 per carrier
Bit rate	90 Mbps or more per carrier	2400 bps to 4 Mbps per carrier
Path length	1–50 mi	50–500 mi
Transmit power	0.1–10 W	100–50,000 W
Receiver noise figure	4–12 dB	Less than 4 dB
Diversity	None or dual	Dual or quadruple
Antenna aperture size	1–12 ft	6–120 ft

- Provides reliable multichannel communication across large stretches of water (e.g., over inland lakes, to offshore islands or oil rigs, between islands) or between areas separated by inaccessible terrain.
- May be ideally suited to meet toll-connecting requirements in areas of low population density.
- Useful for multichannel connectivity crossing territories of another political administration.
- Requires less maintenance staff per route-kilometer than comparable LOS radiolinks over the same route.
- Allows multichannel communication with isolated areas, especially when intervening territory limits or prevents the use of repeaters.

5.3 INTRODUCTION TO TROPOSPHERIC SCATTER PROPAGATION

The picture of the tropospheric scatter mechanism is still pretty speculative. One reasonable explanation is expressed in USAF Technical Order 31Z-10-13 (Ref. 1). It states that the ability of the troposphere to act as a refractive medium is based on the variations of refractivity caused by heating and cooling of the atmosphere's water content. The variation is inversely proportional to altitude, with the greatest variation taking place nearest the earth's surface. The atmosphere is constantly in motion with respect to a point on earth, and this motion causes small irregularities, in refractivity, colloquially called "blobs." Such blobs are large with respect to the wavelengths used in tropospheric scatter radio systems and present a slightly different index of refraction than the surrounding medium. These relatively abrupt changes in the index of refraction produce a scattering effect of the incident radio beam. However, nearly all of the energy of the ray beam passes through these irregularities continuing onward out through free space, and only a small amount of energy is scattered back toward earth.

Figure 5.1 illustrates the concept of *common volume*, the common area subtended by the transmit and receive ray beams of a tropospheric scatter

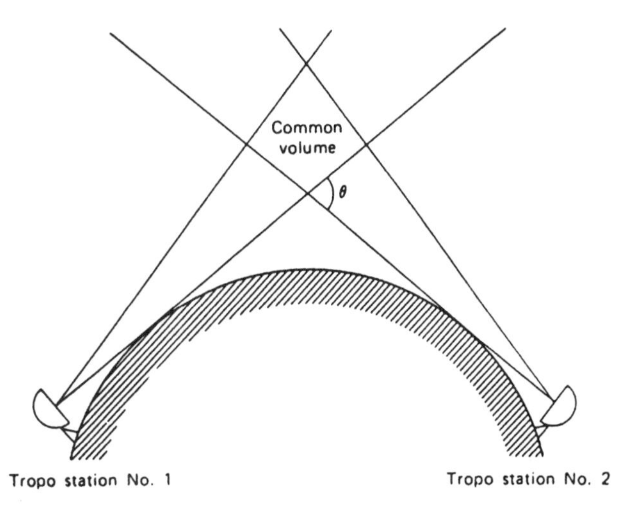

Figure 5.1. Tropospheric scatter model; θ is the scatter angle.

line. It is in this common volume where, for a particular link, the useful scattering takes place to effect the desired communications.

The scatter angle θ governs the received power level in an inverse relationship, with the receive level falling off rapidly as the scatter angle increases. The scatter angle is the angle between a ray from the receiving antenna and a corresponding ray from the transmit antenna, as shown in Figure 5.1.

Fading is common on scatter/diffraction paths. Two general types of fading are identified: short-term and long-term fading.

Short-term fading is characterized by changes in the signal level around the hourly median value. Short-term fading or fast fading is more prevalent on short hops where the fade rate may be as many as 20 fades per second.

The amplitude of these short-term fades follows a Rayleigh distribution (Figure 5.2). From the figure it will be seen that the median (50 percentile) is exceeded by about 5.5 dB 10% of the time, and 90% of the time the signal does not drop more than 8 dB below the median value. The difference between the 10% and 90% values is called the fading range, which is a measure of signal swing about the median. Typically, on a scatter hop this range is 13.5 dB if the signal follows completely a Rayleigh distribution. Values of less than 13.5 dB indicate the presence of a nonfading component.

Diversity is universally used on troposcatter links. Diversity tends to keep short-term fading within reasonable bounds, to no greater than 10 dB below the mean value of a single Rayleigh fading signal (Ref. 1), which is usually accommodated in the noise performance objectives of the system.

Long-term fading is another matter, which must be accommodated in the overall link design. Here we are dealing with long-term variations in signal

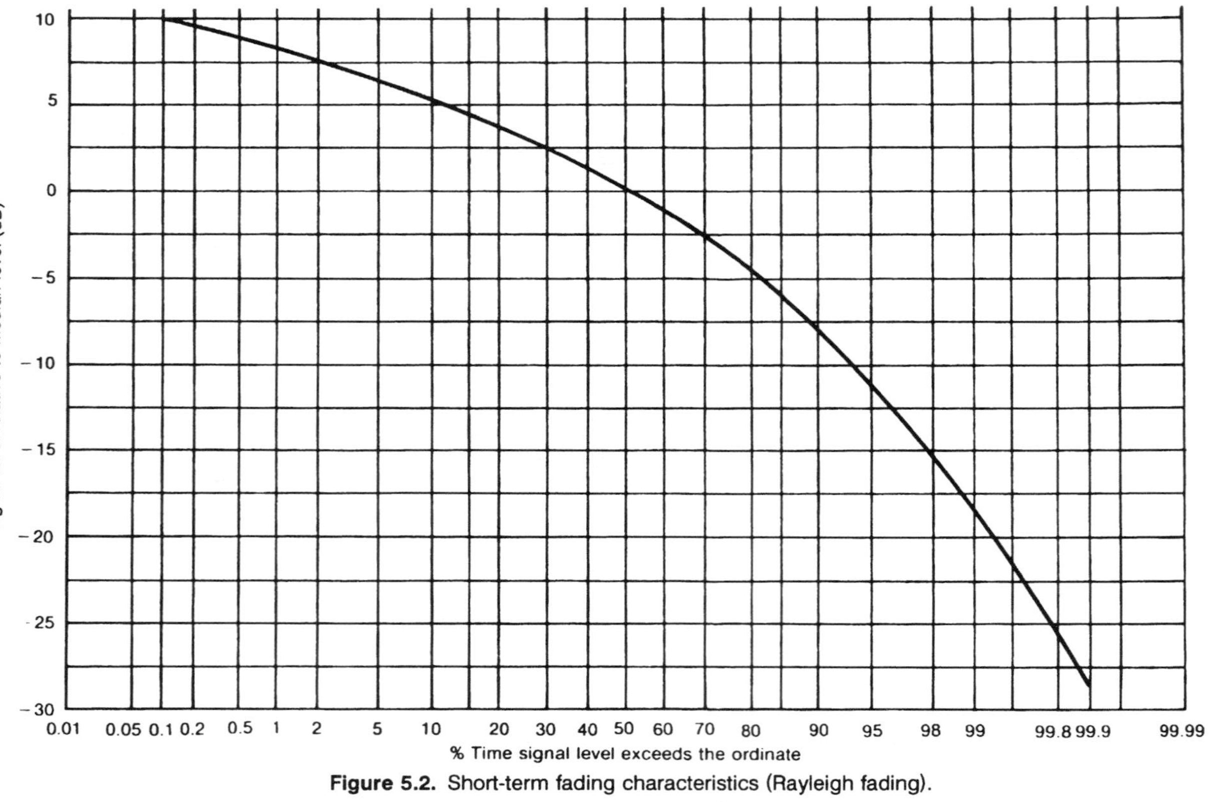

Figure 5.2. Short-term fading characteristics (Rayleigh fading).

level, which can be defined as variations in the hourly median values of transmission loss (path loss) over periods of hours, days, months, or years. Such long-term fading is caused by slow changes in the general condition of the transmission medium—the troposphere.

Long-term fades are characterized by day-to-day fades and seasonal fades. The day-to-day variations are caused by changes in various properties of the scatter volume such as temperature, density, moisture content, and altitude of the scattering layers. Seasonal variations are due to similar causes but can be traced to seasonal cycles. The best propagation conditions are usually summer days and the worst case is during winter nights.

Long-term fading generally follows a log-normal distribution, and the fading *decreases* as hop length increases. Whereas the fade range on an 80-km circuit is around 30 dB, for a 500-km circuit it is about 11 dB. Increasing hop length generally requires an increase in scatter angle, raising the altitude of the common volume. The scatter properties of the atmosphere tend to change more at lower elevations and become more uniform at higher elevations.

In the path analysis phase of link design (Section 5.4.2) we will deal only with long-term fading.

5.4 TROPOSPHERIC SCATTER LINK DESIGN

5.4.1 Site Selection, Route Selection, Path Profile, and Field Survey

5.4.1.1 Introduction. Site selection, route selection, path profile, and field survey are carried out in a similar manner as outlined and discussed in Sections 2.2, 2.3, and 2.5. Here, then, we will point out the differences and areas of particular emphasis.

5.4.1.2 Site Selection. Consider these important factors. Whereas on LOS systems several decibels of calculation error may impact hop cost by one or several thousand dollars, impact on over-the-horizon hops may be on the order of hundreds of thousands of dollars or more. Thus special attention must be paid to accuracy in site position, altitudes and horizon angles, and bearings.

Tropospheric scatter/diffraction sites will be larger, often requiring greater site improvement including fresh water, sanitary systems, living quarters, and more prime power and larger backup power plants. Radiated electromagnetic interference (EMI) is of greater concern. Takeoff angle (θ_{et}, θ_{er}) is critical. For each degree reduction of takeoff angle there is a 12-dB reduction (approximately) in median long-term transmission loss.

5.4.1.3 Route Selection. The route should be selected with first choice to those sites with the most *negative* and last choice to those with the largest positive takeoff angle. The effect of slight variations in path length is negligible for constant takeoff angles. The transmission loss on an over-the-horizon link will vary only slightly for changes in path length of less than about 10 mi (16 km). In a given area, therefore, it is usually best to select the highest feasible site, which also provides adequate shielding from potential interference, even though this may result in a slightly longer path than some location at a lower elevation.

Reference 2 suggests the following formula to estimate takeoff angle, which is valid only for smooth earth (or over water paths):

$$\theta_{et,er} = -0.000686(h_{ts,rs})^{1/2} \tag{5.1}$$

where $\theta_{et,er}$ is in radians, h_{ts} is the height in meters of the transmitting antenna above mean sea level (MSL), and h_{rs} is the height in meters of the receiving antenna above MSL. Formula (5.1) is based on an effective earth radius of 4250 km, which is representative of a worst-case condition.

5.4.1.4 *Path Profile.* A path profile of a proposed tropo/diffraction route is carried out in a similar manner as in Section 2.3. Tropo engineers prefer the use of $\frac{4}{3}$ paper. Takeoff angle is a more important parameter than K-factor. Thus $\frac{4}{3}$ paper may prove more convenient in the long run. Key obstacles to be plotted are the horizons from each site. The horizon is the first obstacle that the ray beam will graze.

Basic tropospheric scatter path geometry is shown in Figure 5.3. From the path profile we will derive the following:

d = great circle distance between sites (km)
d_{Lt} = distance from transmitter horizon (km)
d_{Lr} = distance from receiver to receiver horizon (km)
h_{ts} = elevation above MSL of the center of the transmitting antenna (km)
h_{rs} = elevation above MSL of the center of the receiving antenna (km)
h_{Lt} = elevation above MSL of the transmitter horizon point (km)
h_{Lr} = elevation above MSL of the receiver horizon point (km)

5.4.2 Link Performance Calculations

5.4.2.1 *Introduction.* Free-space loss (FSL) is based on theory and on unfaded LOS paths; the calculated receive signal level (RSL) and the measured level will turn out to be within 0.1 and 0.2 dB or less of each other. Transmission loss equations and curves for diffraction and troposcatter paths are empirical, based on hundreds of paths in many parts of the world. The

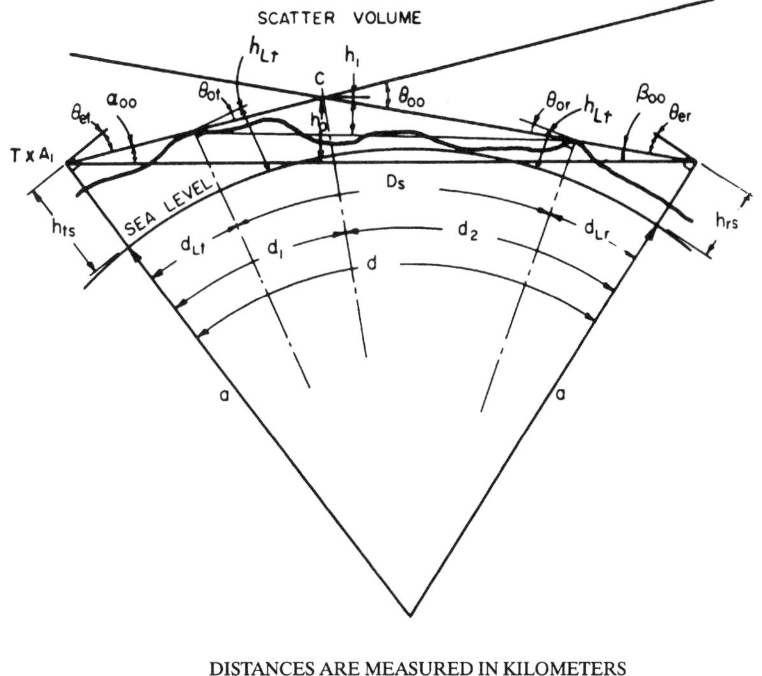

DISTANCES ARE MEASURED IN KILOMETERS
ALONG A GREAT CIRCLE ARC.

$$\theta_{00} = \frac{D_s}{a} + \theta_{0t} + \theta_{0r} = \frac{d}{a} + \theta_{et} + \theta_{er}$$

Figure 5.3. Tropospheric scatter path geometry.

cited references contain empirical methods to calculate tropo transmission loss:

1. NBS (NBS Tech. Note 101) (Ref. 3).
2. CCIR (CCIR Rec. 617-1) (Ref. 4).
3. Yeh (Ref. 5).
4. Rider (Ref. 6).
5. Collins (Ref. 7).
6. Longley–Rice (computer model) (Ref. 8).

In our opinion, methods following NBS Tech. Note 101 are the most well accepted worldwide. The method to be described is based on NBS 101 as set forth in USAF Technical Order 31Z-10-13 (Ref. 1), which we have simplified. However, the reader is warned that there can be considerable variation between calculated median loss values and measured loss, as much as 6 dB in some cases. It is for this reason that the term service probability must be dealt with, as will be described.

5.4.2.2 *Two Definitions.* The terms *time availability* and *service probability* must be distinguished and understood. We will first calculate the long-term median basic transmission loss L_{cr}. This is then corrected for a particular climatic region to derive the basic median transmission loss L_n. If the RSL on a link were calculated using this value (i.e., L_n), the RSL would reach or exceed this value only 50% of the time. This is the "time availability" of the link. Of course, we would wish an improved time availability of a link, usually 99% or better. If we design the link for a time availability of 99%, then 1% of the time the RSL will be less than the objective. In Section 2.7 we called this the link availability or propagation reliability.

The basic median transmission loss is described by the notation $L_n(0.5, 50)$, or more generally $L_n(Q, q)$. Here the q refers to the time availability and Q refers to the *service probability*.

The service probability concept is used to obtain a measure of prediction uncertainty due to our lack of complete knowledge regarding the propagation mechanism, the semiempirical nature of the prediction formulas, and the uncertainties in equipment performance.

A service probability of 0.5 or 50% tells us that only half the links in a large population given identical input conditions will meet the time availability value. Often we engineer links for a service probability of 0.95 or 95%. Then only 5% of the links will fail to meet the time availability. The last step in calculating transmission loss is to extend the transmission loss to take into account prediction uncertainty, which we call here service probability.

5.4.2.3 *Propagation Mode.* There are two possible modes for over-the-horizon transmission: diffraction or troposcatter. In most cases the path profile will tell us the mode. For paths just over LOS, the diffraction mode will predominate. For long paths, the troposcatter mode will predominate. If the basic median transmission loss of one mode is 15 dB more than the other mode, that with the higher loss can be neglected, and the one with the lower loss predominates.

To aid in determining the mode of propagation, the following criteria (Ref. 1) may be used:

- The distance (in kilometers) at which diffraction and forward scatter losses are approximately equal is $65(100/f)^{1/3}$, where f is the operating frequency in megahertz. For distances less than this value, diffraction will generally be predominant; for those distances greater than this value, the tropo mode predominates.
- For most paths having an angular distance (see Section 5.4.2.6) of at least 20 mrad, the diffraction effects may be neglected and the path can be considered to be operating in the troposcatter mode.

5.4.2.4 *Method of Approach.* First, we will discuss the equation for calculation of basic long-term tropospheric scatter transmission loss L_{bsr}, then the geometry and substeps to carry out the calculation. The next step will be to calculate the basic long-term diffraction transmission loss L_{bd} for two selected diffraction modes. Then we discuss the mixed-mode case, tropo/diffraction. The next operation is to calculate the reference value of basic transmission loss L_{cr} and extend to the basic transmission loss value $L_n(0.5, 50)$ for a region of interest. This value is then extended for the desired time availability, for the 50% service probability. The last step, if desired, is to again extend the value for an improved service probability, usually 95%.

5.4.2.5 Basic Long-Term Median Tropospheric Scatter Loss, L_{bsr}

$$L_{bsr}(\text{dB}) = 30 \log F - 20 \log d + F(\theta d) - F_0 + H_0 + A_a \quad (5.2)$$

where f = the operating frequency in megahertz
d = greater circle path length in kilometers
$F(\theta d)$ = attenuation function
F_0 = scattering efficiency correction factor
H_0 = frequency gain function
A_a = atmospheric absorption
θd = product of the angular distance and path distance (angular distance in radians, path distance distance in kilometers)

We disregard the F_0 and H_0 terms. It should be pointed out that the CCIR method (Ref. 4) is similar to the NBS method, and it too disregards the scattering efficiency correction factor and the frequency gain function.

A_a, the atmospheric absorption term, is calculated using Figure 5.4 and employing distance d and operating frequency f.

The attenuation function $F(\theta d)$ is one of the most significant terms in calculating L_{bsr}. Figures 5.11 through 5.15 plot values of decibel loss versus the product θd for values of surface refractivity and the path asymmetry factor s. Section 5.4.2.6 describes the steps necessary to calculate values of θ, the scatter angle in radians, sometimes called the angular distance.

5.4.2.6 Calculation of θ, the Scatter Angle. Turn to Section 5.4.1.4 and annotate the values taken from the path profile. These are used as the basis for calculating θ or angular distance. Compute values of $h_{st,sr}$, the average height of the surface above MSL (km).

Compute h_{st} and h_{sr}, where h_{st} and h_{sr} are the average heights above MSL (km) for the transmit site and receive site:

$$h_{st} = h_{Lt} \quad \text{if } h_{Lt} < h_{ts} + 0.15 \text{ km} \quad (5.3)$$

$$h_{sr} = h_{Lr} \quad \text{if } h_{Lr} < h_{rs} + 0.15 \text{ km} \quad (5.4)$$

or

$$h_{st} = h_{ts} \quad \text{if } h_{Lt} > h_{ts} + 0.15 \text{ km} \quad (5.5)$$

$$h_{sr} = h_{rs} \quad \text{if } h_{Lr} > h_{rs} + 0.15 \text{ km} \quad (5.6)$$

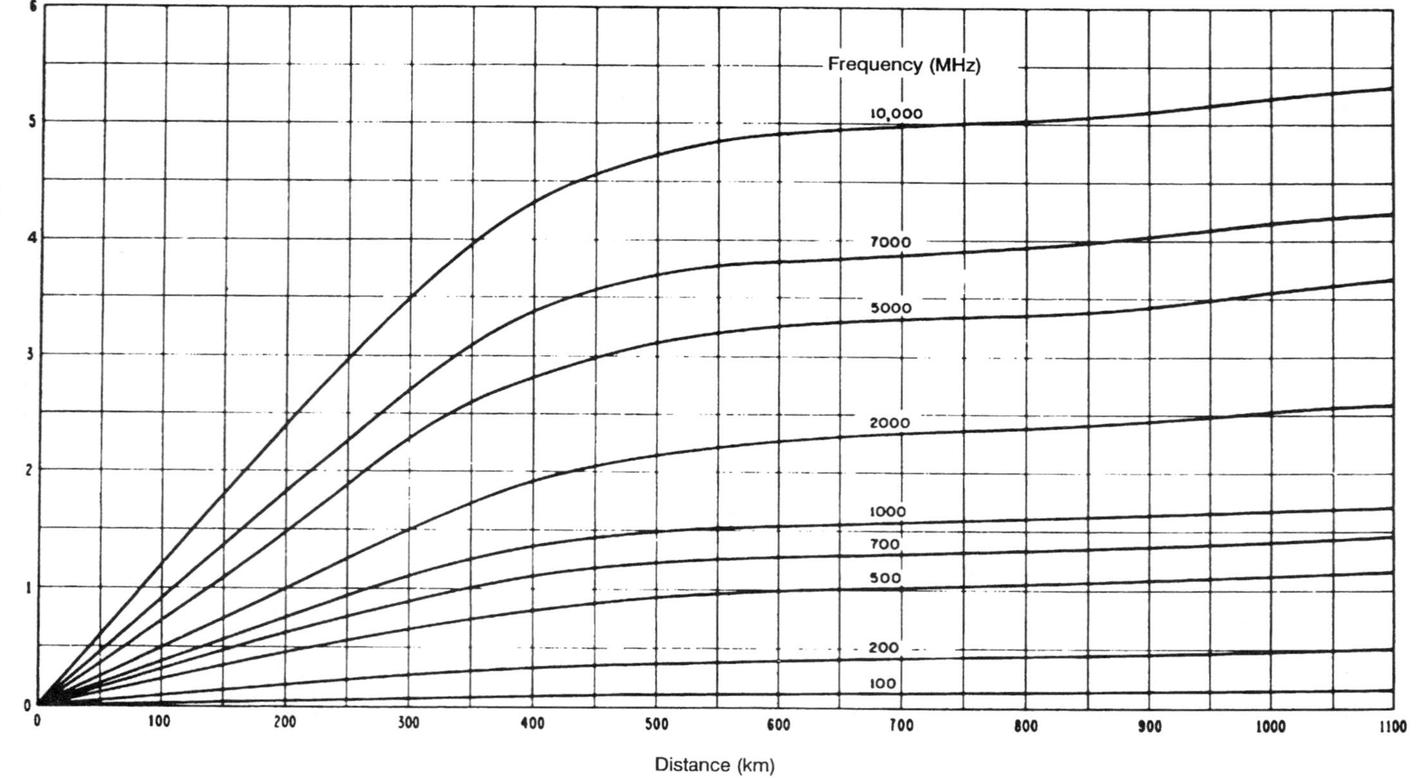

Figure 5.4. Estimate of median oxygen and water vapor absorption in decibels. (Ref. 1).

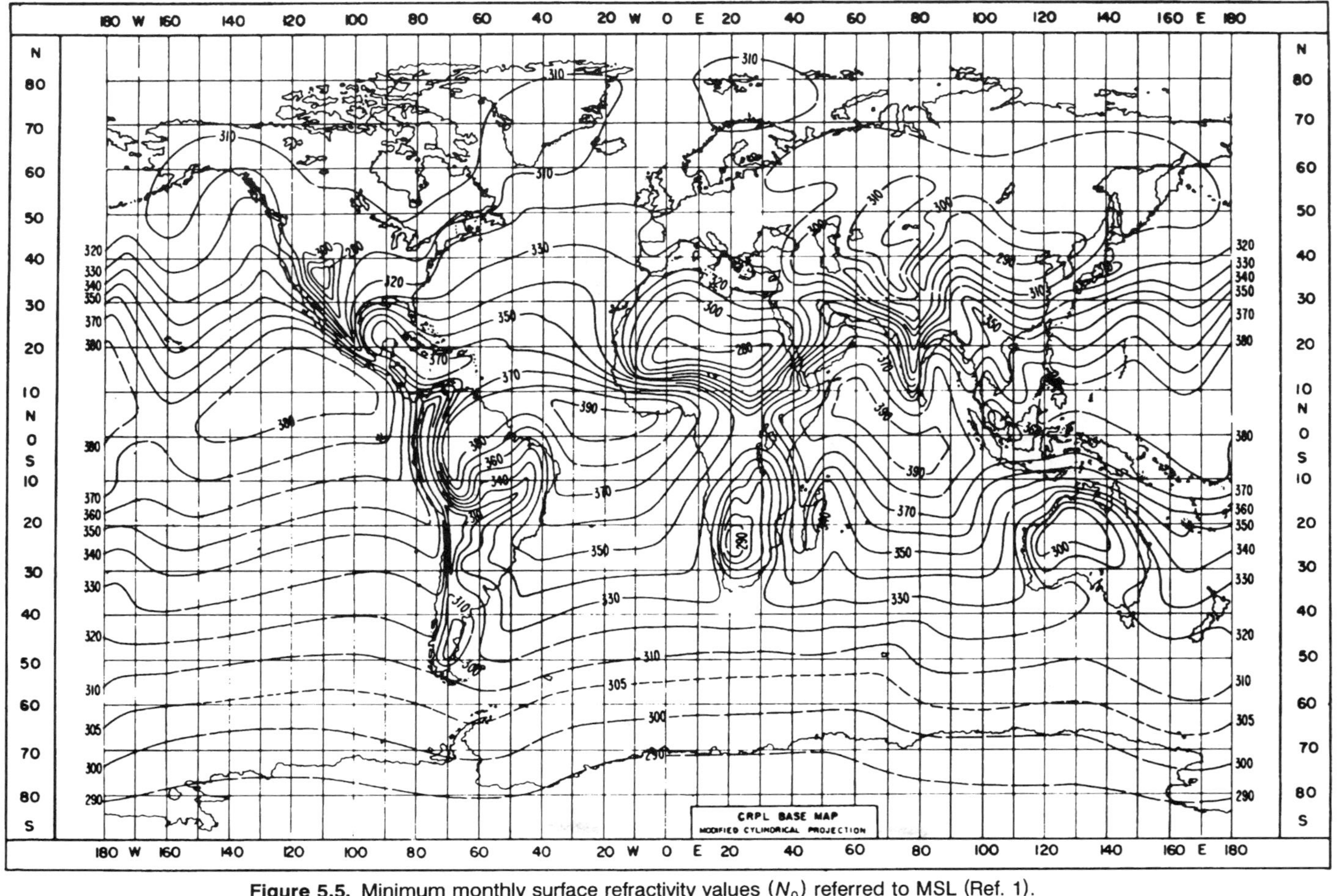

Figure 5.5. Minimum monthly surface refractivity values (N_0) referred to MSL (Ref. 1).

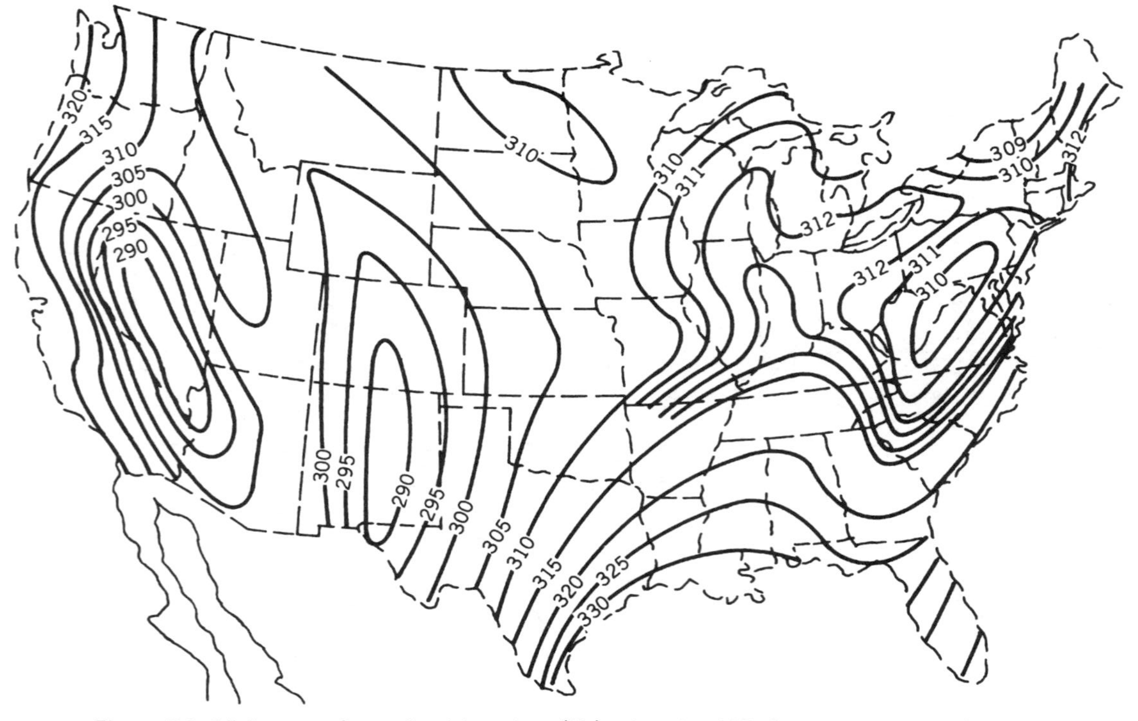

Figure 5.6. Minimum surface refractivity values (N_0) referred to MSL for average winter afternoon, continental United States (CONUS) (Ref. 1).

If there are two or more values of h_{Lt} or h_{Lr}, average the values before using the preceding equations.

Determine N_0 for the locations of h_{st} and h_{sr}. N_0 is the surface refractivity reduced to sea level. Use Figure 5.5 to obtain values for N_0 for locations outside CONUS. Use Figure 5.6 for continental United States (CONUS) locations.

Compute values of N_{st} and N_{sr}, which are values of refractivity (N) at the surface of the earth for the transmit and receive sites, respectively:

$$N_{st,\,sr} = N_0 \exp(-0.1057 h_{st,\,sr}) \tag{5.7}$$

Compute N_s, which is the value of N for the total path:

$$N_s = \tfrac{1}{2}(N_{st} + N_{sr}) \tag{5.8}$$

Obtain the value for a, the effective earth radius, from Figure 5.7.

Compute values of the takeoff angles (horizon angles) θ_{et} for transmit site and θ_{er} for the receiving site:

$$\theta_{et} = \frac{h_{Lt} - h_{ts}}{d_{Lt}} - \frac{d_{Lt}}{2a} \tag{5.9}$$

$$\theta_{er} = \frac{h_{Lr} - h_{rs}}{d_{Lr}} - \frac{d_{Lr}}{2a} \tag{5.10}$$

Compute D_s, the distance between horizon points:

$$D_s = d - (d_{Lt} + d_{Lr}) \tag{5.11}$$

The following procedures are used to compute the effective height of the transmitting and receiving antennas, respectively—h_{te} and h_{re}:

1. Compute the average height above MSL, $\bar{h}_t$, $\bar{h}_r$, of the central 80% of the horizon between each antenna and its respective point, h_{Lt}, h_{Lr}.

$$\bar{h}_{t,r} = \frac{1}{25} \sum_{i=3}^{27} h_{ti,ri} \quad \text{(km)} \tag{5.12}$$

All elevations are in kilometers. Divide the distance between each antenna location and its respective horizon into 31 evenly spaced intervals, $d_{ti,ri}$, where $i = 0, 1, 2, 3, \ldots, 30$. From the path profile, write the elevation $h_{ti,ri}$ corresponding to each. The heights above MSL of the ground below the transmitting and receiving antenna, are h_{t0} and h_{r0}, respectively, and $h_{t30,\,r30}$ is the height of the transmitter and receiver horizons, respectively.

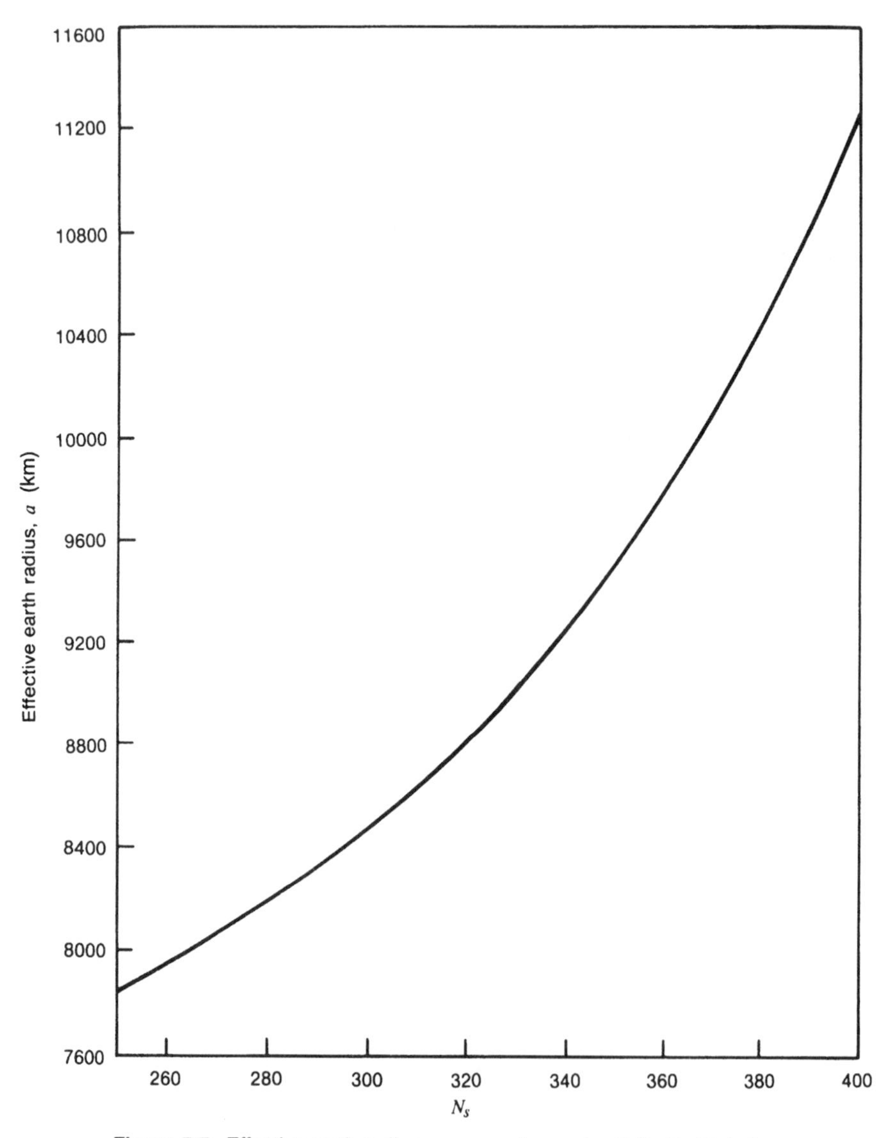

Figure 5.7. Effective earth radius versus surface refractivity N_s (Ref. 1).

2. Compute h_t and h_r, which are the heights of the transmit and receive antennas above intervening terrain between the transmit site and its horizon and the receiving site and its horizon. If $h_{t0,\,r0} > \bar{h}_{t,\,r}$, then

$$h_t = h_{ts} - \bar{h}_t \tag{5.13}$$

$$h_r = h_{rs} - \bar{h}_r \tag{5.14}$$

If $h_{t0, r0} < \bar{h}_{t,r}$, then

$$h_t = h_{ts} - h_{t0} \tag{5.15}$$

$$h_r = h_{rs} - h_{r0} \tag{5.16}$$

3. Compute h_{te} and h_{re}, which are the effective heights of the transmitting and receiving antennas, respectively. If $h_{t,r} < 1$ km, then

$$h_{te, re} = h_{t,r} \tag{5.17}$$

If $h_{t,r} > 1$ km, then

$$h_{te, re} = h_{t,r} - \Delta h_e \tag{5.18}$$

where Δh_e is an antenna height correction factor. To find this value, use Figure 5.8.

We now have the necessary inputs to calculate the angular distance θ, also called the scatter angle. The reader should refer to Figure 5.3 for clarification of geometrical references and relationships. All angles are in radians; heights and distances are in kilometers. There are 12 numbered steps in the calculation, and each step number is followed by the letter t (for theta) to distinguish from other step series.

1t. Compute θ_{0t} and θ_{0r}, which are the angular elevations of the horizon ray at the transmitting and receiving points, respectively:

$$\theta_{0t} = \theta_{et} + \frac{d_{Lt}}{a} \tag{5.19}$$

$$\theta_{0r} = \theta_{er} + \frac{d_{Lr}}{a} \tag{5.20}$$

where a is the effective earth radius.

2t. Calculate α_{00} and β_{00} (these angles are defined in Figure 5.3):

$$\alpha_{00} = \frac{d}{2a} + \theta_{et} + \frac{h_{ts} - h_{rs}}{d} \tag{5.21}$$

$$\beta_{00} = \frac{d}{2a} + \theta_{er} + \frac{h_{rs} - h_{ts}}{d} \tag{5.22}$$

3t. Compute θ_{00}, which is the uncorrected scatter angle (radians):

$$\theta_{00} = \alpha_{00} + \beta_{00} \tag{5.23}$$

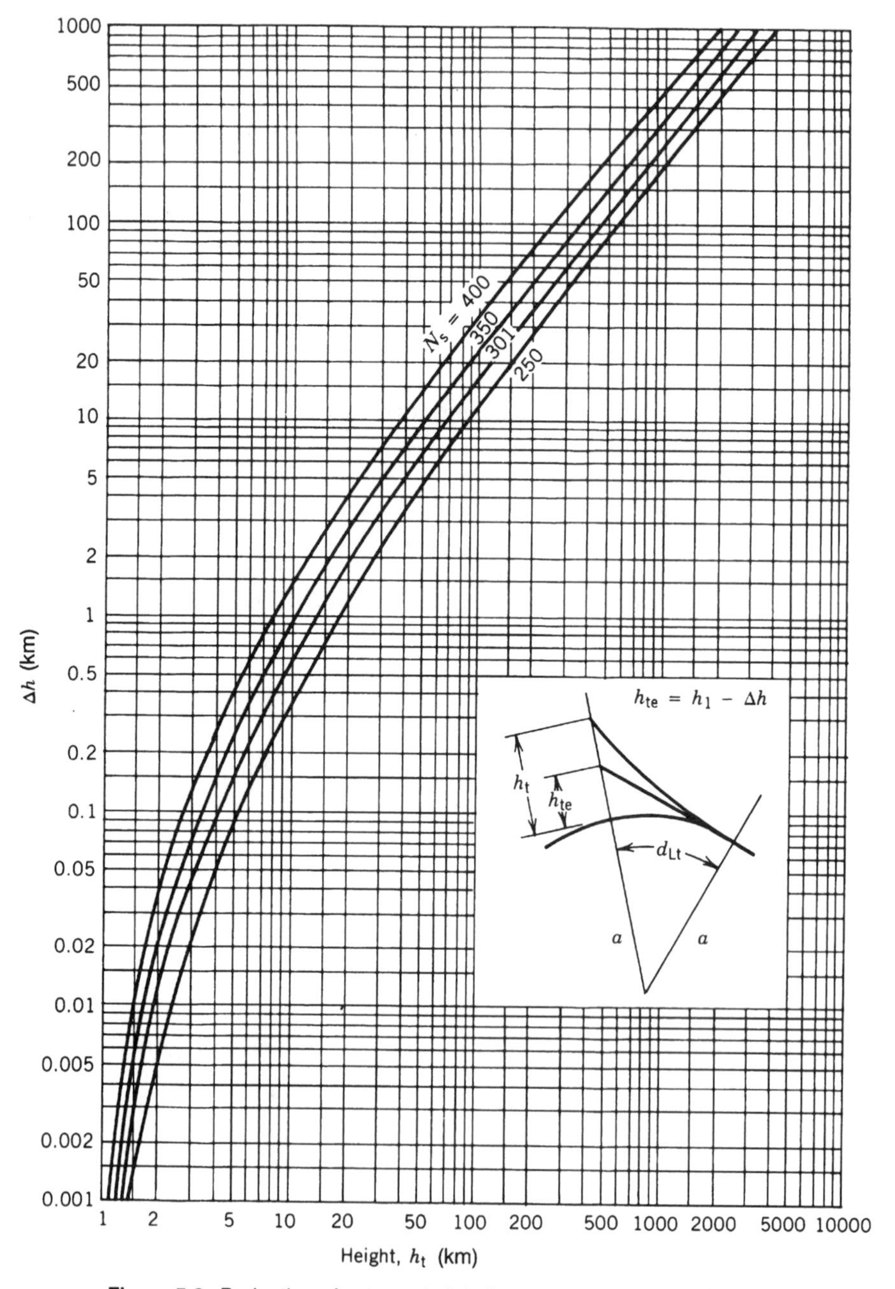

Figure 5.8. Reduction of antenna height for a very high antenna (Ref. 1).

4t. Compute d_{st} and d_{sr}, which are the distances from the crossover of the horizon rays to the transmitter and receiver points, respectively:

$$d_{st} = \frac{d\alpha_{00}}{\theta_{00}} - d_{Lt} \tag{5.24}$$

$$d_{sr} = \frac{d\beta_{00}}{\theta_{00}} - d_{Lr} \tag{5.25}$$

If θ_{0t} or θ_{0r} is negative, compute

$$d'_{st} = d_{st} = |a\theta_{0t}| \quad \text{for negative } \theta_{0t} \tag{5.26}$$

$$d'_{sr} = d_{sr} - |a\theta_{0r}| \quad \text{for negative } \theta_{0r} \tag{5.27}$$

where $d'_{st,\,sr}$ are the adjusted values of d_{st} and d_{sr} for negative values of θ_{0t} and θ_{0r}, respectively.

5t. Determine $\Delta\alpha_0$ and $\Delta\beta_0$ as a function of θ_{0t} and θ_{0r}; $\Delta\alpha_0$ relates to θ_{0t} and $\Delta\beta_0$ relates to θ_{0r}. If $0 < \theta_{0t,0r} < 0.1$, use steps 6 and 7 and then proceed to step 10.
If $0.1 < \theta_{0t,0r} < 0.9$, use step 8 and proceed to step 10.
If $\theta_{0t,0r}$ is negative, use step 9 and then proceed to step 10. For $\theta_{0t,0r}$ greater than 0.9, the corresponding value of $\Delta\alpha_0$ or $\Delta\beta_0$ is negligible. $\Delta\alpha_0$ and $\Delta\beta_0$ are parameters used to correct α_{00} and β_{00} for a refractivity N_s of 301.

6t. From Figure 5.9 determine $\Delta\alpha_0$ and $\Delta\beta_0$ for $N_s = 301$. Note in Figure 5.9 that θ_{0t} and θ_{0r} are in milliradians and that the x-axis is valid for both. Also note that the derived values of $\Delta\alpha_0$ and $\Delta\beta_0$ are in milliradians and must be converted to radians.

7t. Determine $\Delta\alpha_0$ and $\Delta\beta_0$ for given values of N_s. Read $C(N_s)$ from Figure 5.10 as a function of N_s. The parameter $C(N_s)$ is used to correct α_{00} and β_{00} for the effects of a nonlinear refractivity index.

$$\Delta\alpha_0(N_s) = C(N_s)\,\Delta\alpha_0(301) \tag{5.28}$$

$$\Delta\beta_0(N_s) = C(N_s)\,\Delta\beta_0(301) \tag{5.29}$$

8t. Repeat steps 6 and 7 for $\theta_{0t} = 0.1$ and $\theta_{0r} = 0.1$ and to each add the following factor to the values of step 7. Use Figure 5.9 for $\theta_{0t,0r} = 0.1$. For $\Delta\alpha_0$ use θ_{0t} and d_{st} and for $\Delta\beta_0$ use θ_{0r} and d_{sr}.

$$\text{Factor (8t)} = N_s(9.97 - \cot\theta_{0t,0r})[1 - \exp(-0.05d_{st,sr})](10^{-6}) \tag{5.30}$$

9t. Repeat steps 6 and 7 for $\theta_{0t} = 0$ and $\theta_{0r} = 0$ using Figure 5.9.

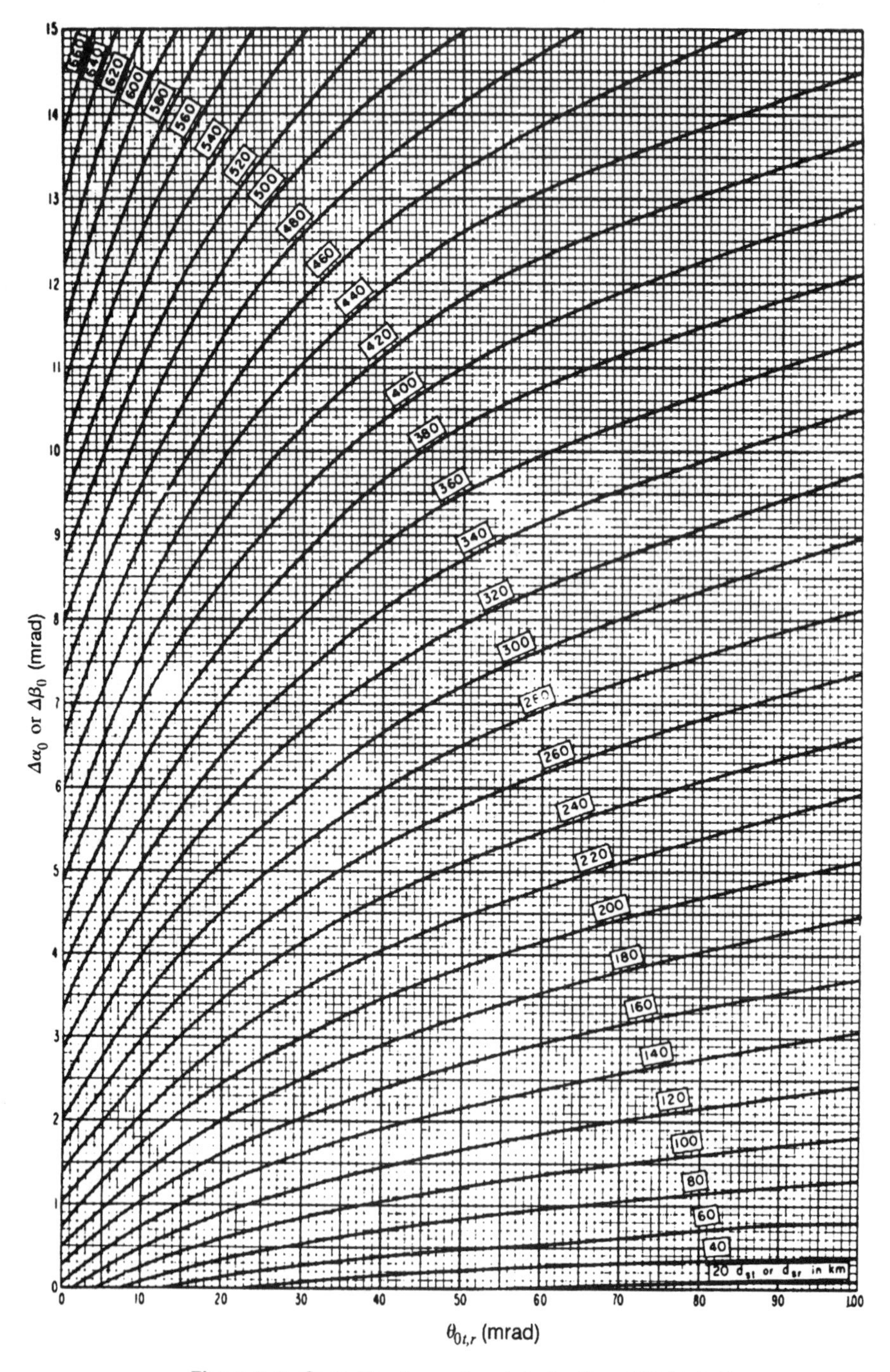

Figure 5.9. Correction terms $\Delta\alpha_0\,\Delta\beta_0$ for $N_s = 301$ (Ref. 1).

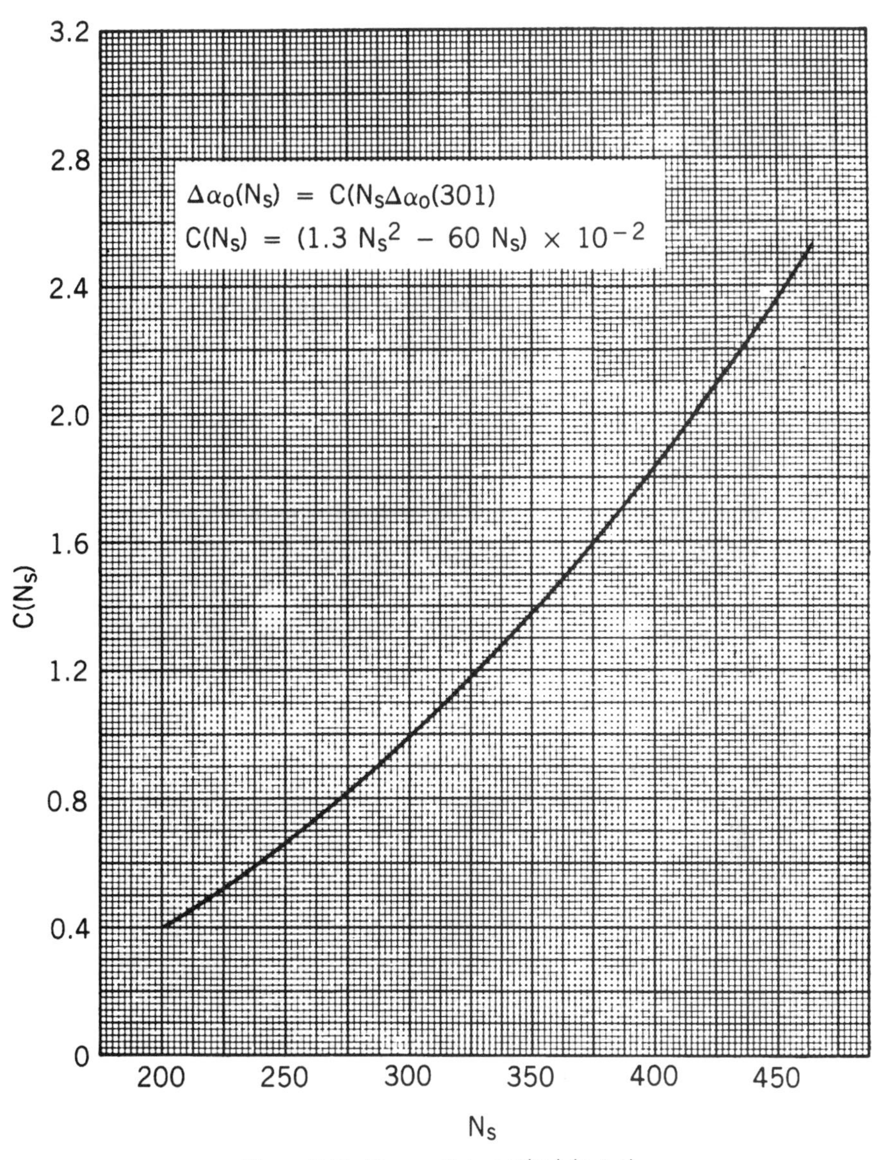

Figure 5.10. The coefficient $C(N_s)$ (Ref. 1).

10t. Calculate α_0 and β_0, which are the corrected values of α_{00} and β_{00}:

$$\alpha_0 = \alpha_{00} + \Delta\alpha_0 \tag{5.31}$$

$$\beta_0 = \beta_{00} + \Delta\beta_0 \tag{5.32}$$

11t. Calculate the scatter angle (angular distance) θ:

$$\theta = \alpha_0 + \beta_0 \tag{5.33}$$

12t. Calculate the path asymmetry factor s:

$$s = \frac{\alpha_0}{\beta_0} \tag{5.34}$$

5.4.2.7 Calculation of Basic Long-Term Median Tropospheric Scatter Loss, L_{bsr}.

Calculate the product θd (i.e., multiply θ times d). θ is the scatter angle (in radians) and d is the great circle distance of the tropo hop in kilometers.

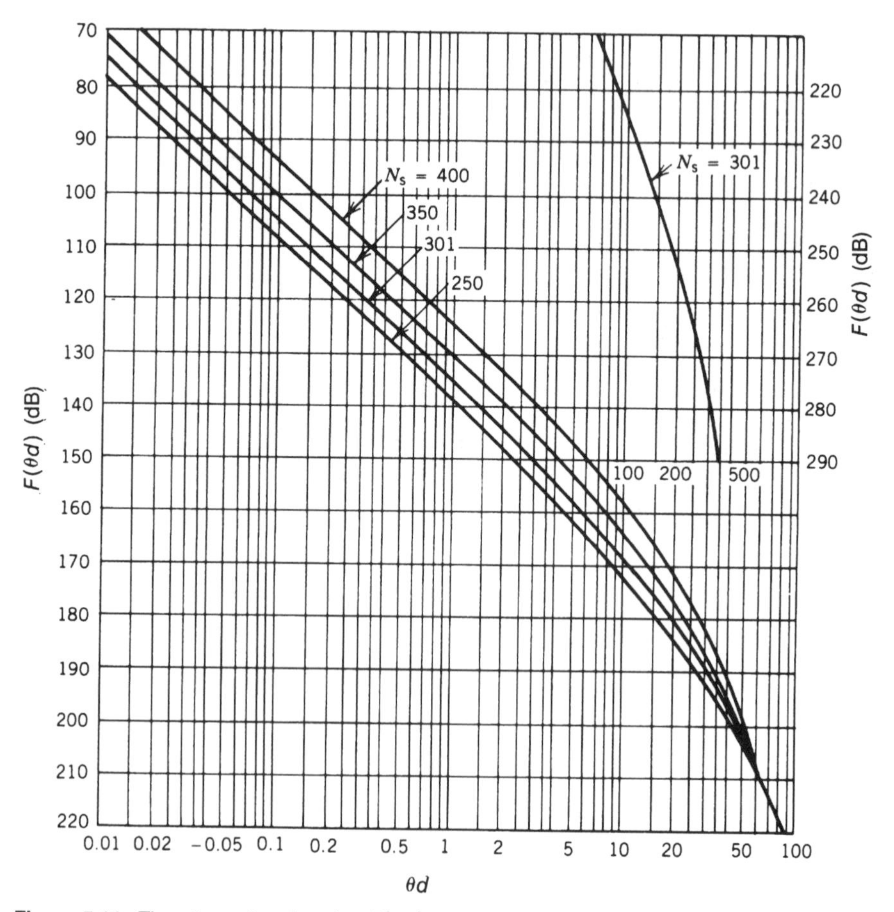

Figure 5.11. The attenuation function $F(\theta d)$; d is measured in kilometers and θ is the scatter angle in radians (Ref. 1).

Compute the basic median tropospheric scatter transmission loss L_{bsr}:

$$L_{bsr} = 30 \log f - 20 \log d + F(\theta d) + A_a \tag{5.35}$$

Remember, in this simplified method, we omit the terms F_0 and H_0. Obtain values for $F(\theta d)$ as follows: if $\theta d \leq 10$, read from Figure 5.11; if $\theta d > 10$, read from Figures 5.12–5.15 for given values of N_s; if $s > 1$, read curves s_1, where $s_1 = 1/s$.

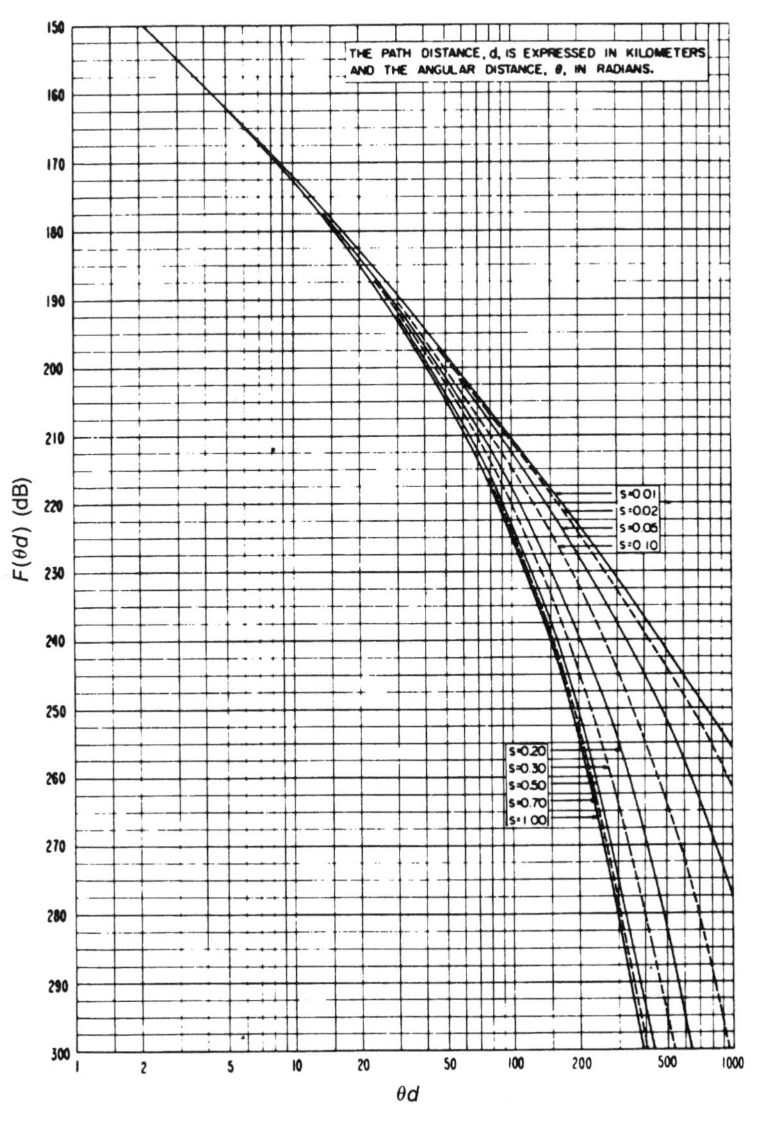

Figure 5.12. The function $F(\theta d)$ for $N_s = 250$ (Ref. 1).

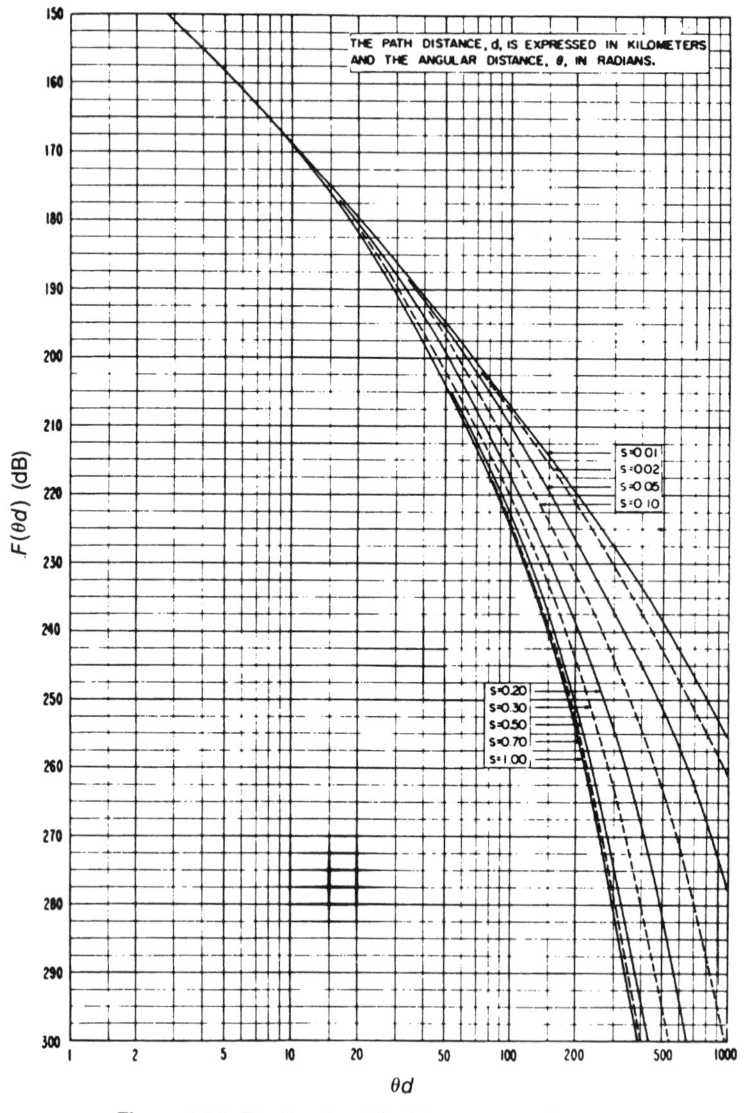

Figure 5.13. The function $F(\theta d)$ for $N_s = 301$ (Ref. 1).

5.4.2.8 Basic Long-Term Median Diffraction Loss, L_{bd}

5.4.2.8.1 Types of Diffraction Paths. Figure 5.16 illustrates six types of diffraction paths. Three of the more common paths are described below: knife-edge diffraction over a single, isolated obstacle (1) with and (2) without reflections (Section 5.4.2.8.2). Step numbers use the letter k; (3) smooth earth diffraction (such as over comparatively flat desert or over water where

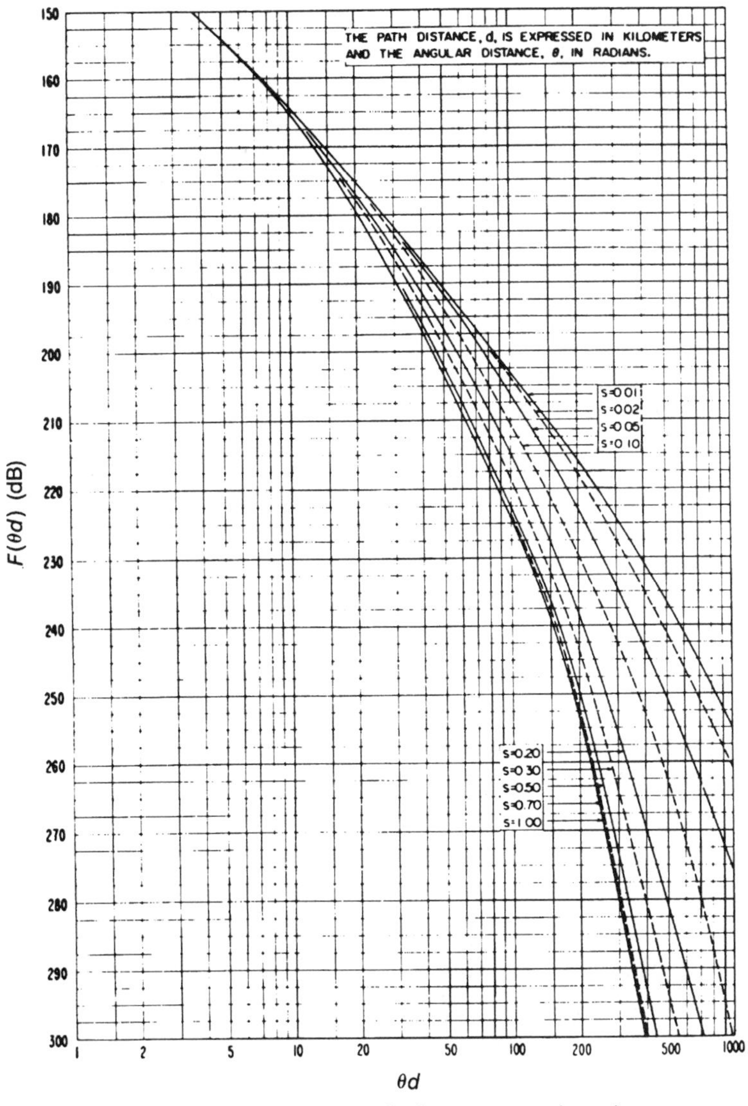

Figure 5.14. The function $F(\theta d)$ for $N_s = 350$ (Ref. 1).

diffraction is the predominant mode) (Section 5.4.2.8.3). Step numbers use the letter *s*.

Note: All the following distances and elevations given are in kilometers; all angles are in radians.

5.4.2.8.2 Knife-Edge Diffraction Over a Single, Isolated Obstacle. For this diffraction mode a common horizon is required between the transmit and

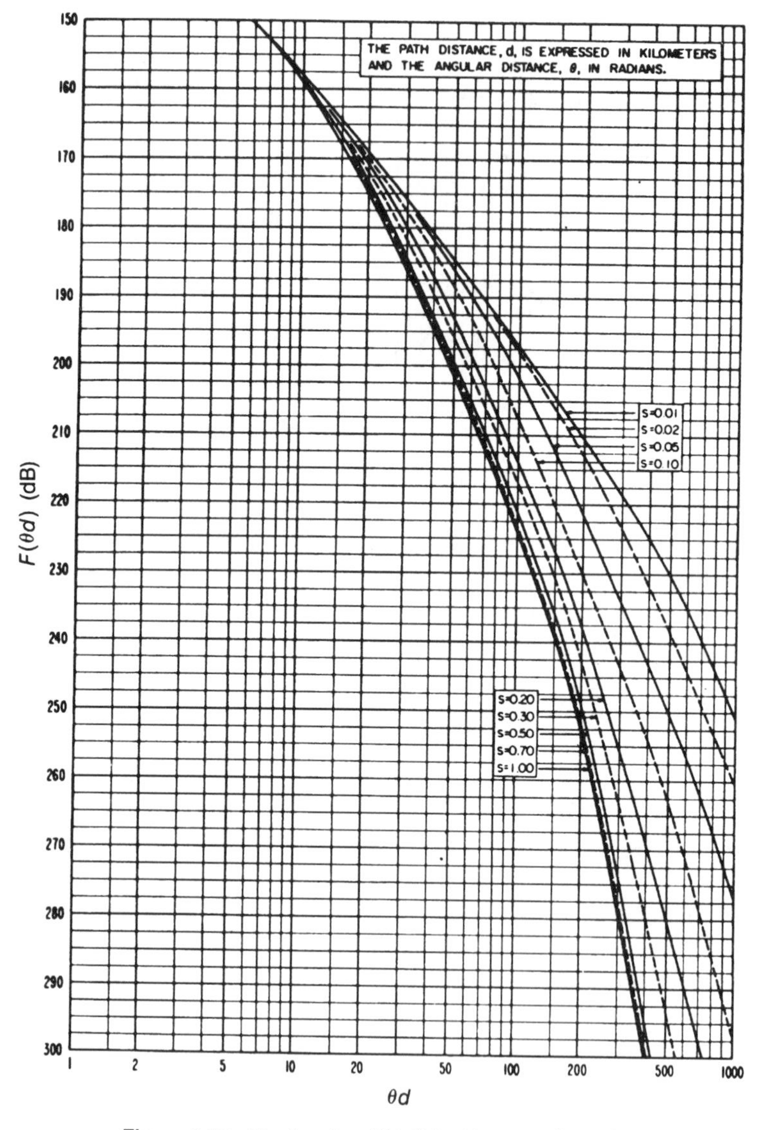

Figure 5.15. The function $F(\theta d)$ for $N_s = 400$ (Ref. 1).

receive antennas. The common horizon is usually a sharp mountain top or sharp ridge.

1k. For an ideal knife-edge without ground reflections, calculate L_{bd} using step 11k and employ the terms $A(v, 0)$, L_{bf}, and A_a.

2k. For an ideal knife-edge with ground reflections again use step 11k and employ terms $A(v, 0)$, L_{bf}, A_a, $G(h_1)$, and $G(h_2)$.

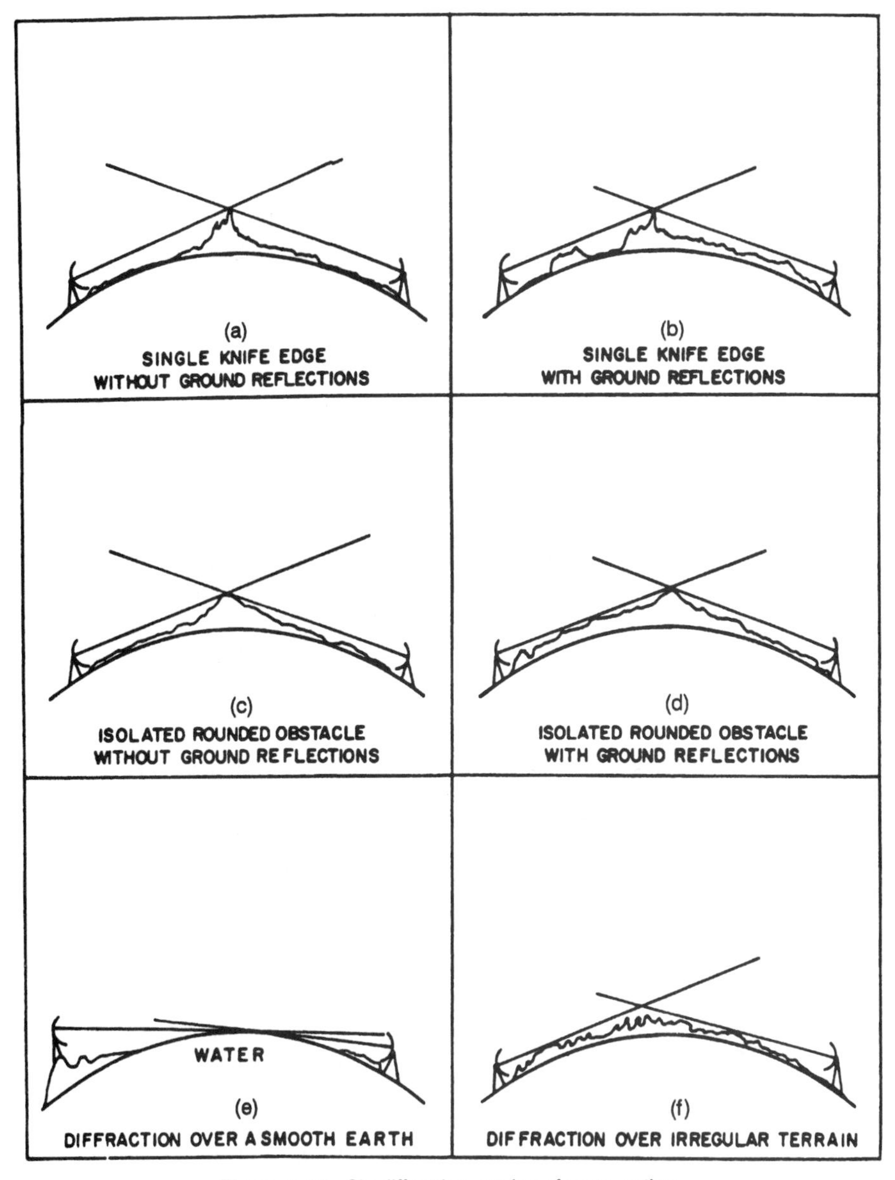

Figure 5.16. Six diffraction modes of propagation.

3k. Calculate υ:

$$\upsilon = 2.583\theta\sqrt{\frac{fd_{Lt}d_{Lr}}{d}} \tag{5.36}$$

where θ is the scatter angle as calculated in the previous section, step 11t; f is the operating frequency; d_{Lt} and d_{Lr} are from the path profile; and d is the great circle distance taken from the path profile.

4k. If $\upsilon < 2.4$, go to step 5k. If $\upsilon > 2.4$, go to step 6k.

5k. Determine $A(\upsilon, 0)$ from Figure 5.17 or by step 6k.

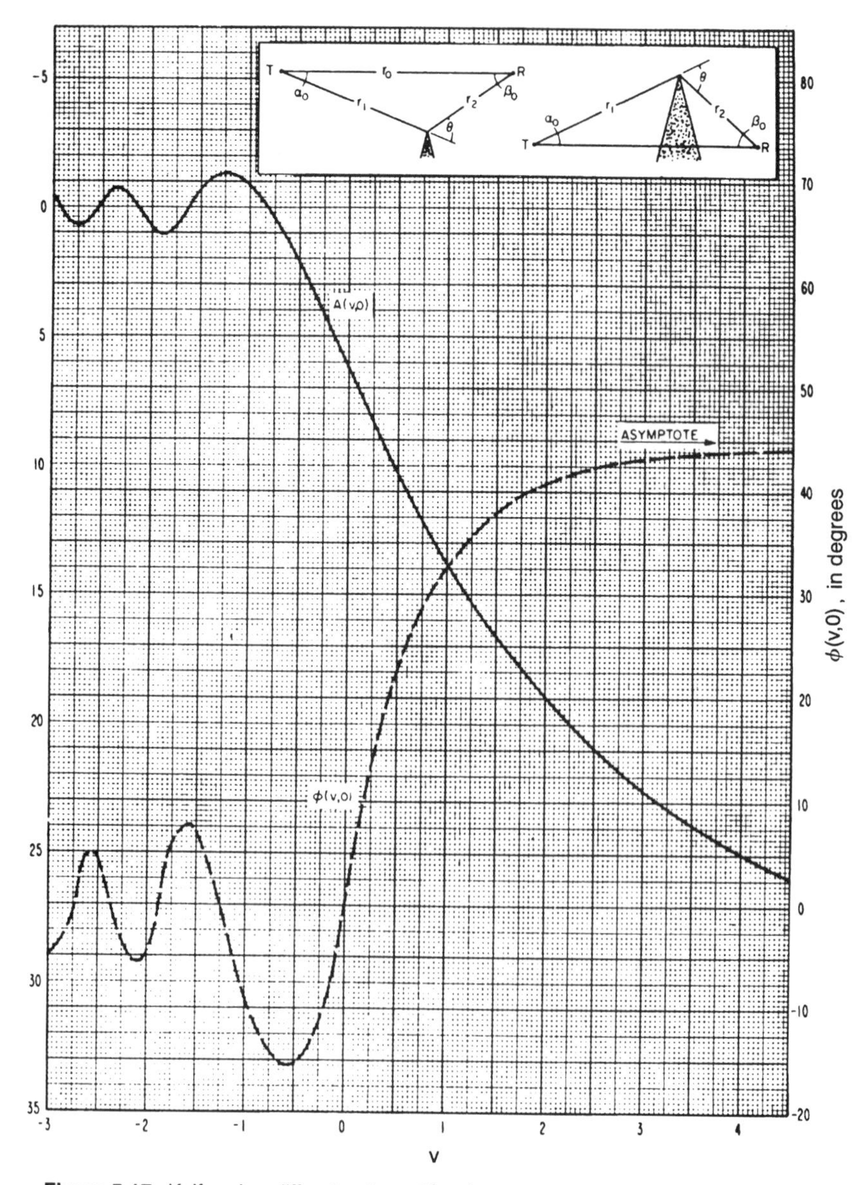

Figure 5.17. Knife-edge diffraction loss $A(v, 0)$ in decibels below free space (Ref. 1).

6k. Calculate $A(\upsilon, 0)$, which is the diffraction attenuation relative to free space as a function of υ:

$$A(\upsilon, 0) = 12.953 + 20 \log \upsilon \tag{5.37}$$

7k. Calculate L_{bf}, which is the free-space loss (dB):

$$L_{bf} = 32.45 + 20 \log f + 20 \log d \tag{5.38}$$

where f is the operating frequency in megahertz and d is the great circle distance of the hop in kilometers.

8k. Steps 9k and 10k are used to compute the additional loss introduced by ground reflections. Reflections may occur in either or both the transmitting or receiving portions of the path. For the applicable portion compute $\bar{h}$ and $G(\bar{h})$ in steps 9k and 10k. If reflections may be neglected in one portion of the path, its corresponding value of $G(\bar{h})$ is then zero.

9k. Calculate $\bar{h}_1$ and $\bar{h}_2$:

$$\bar{h}_{1,2} = 7.23 \sqrt[3]{\frac{f^2}{d_{Lt}^2, d_{Lr}^2}} \left(h_{te,re}\right)^{4/3} \tag{5.39}$$

For $\bar{h}_1$ use d_{Lt} and h_{te} and for $\bar{h}_2$ use d_{Lr} and h_{re}, where h_{te} and h_{re} are the effective antenna heights from the previous section.

10k. Read the value of $G(\bar{h}_{1,2})$ for corresponding values of $\bar{h}_1, \bar{h}_2$ from Figure 5.18, where $G(\bar{h}_{1,2})$ is the reflection loss(es) in decibels in the transmitting and receiving segments, respectively.

11k. Compute the basic long-term median diffraction loss L_{bd}:

$$L_{bd} = L_{bf} + A(\upsilon, 0) + A_a - G(\bar{h}_1) - G(\bar{h}_2) \tag{5.40}$$

where A_a is obtained from Figure 5.4.

5.4.2.8.3 *Diffraction Over Smooth Earth*

1s. Calculate C_0:

$$C_0 = \left(\frac{8497}{a}\right)^{1/3} \tag{5.41}$$

where a = effective earth radius from Figure 5.7 given N_s (Section 5.4.2.6)

C_0 = a parameter for computing $K(a)$

2s. From Figure 5.19 determine K for an earth radius of 8497 km [$K(a = 8497)$], where K is a function of the operating frequency.

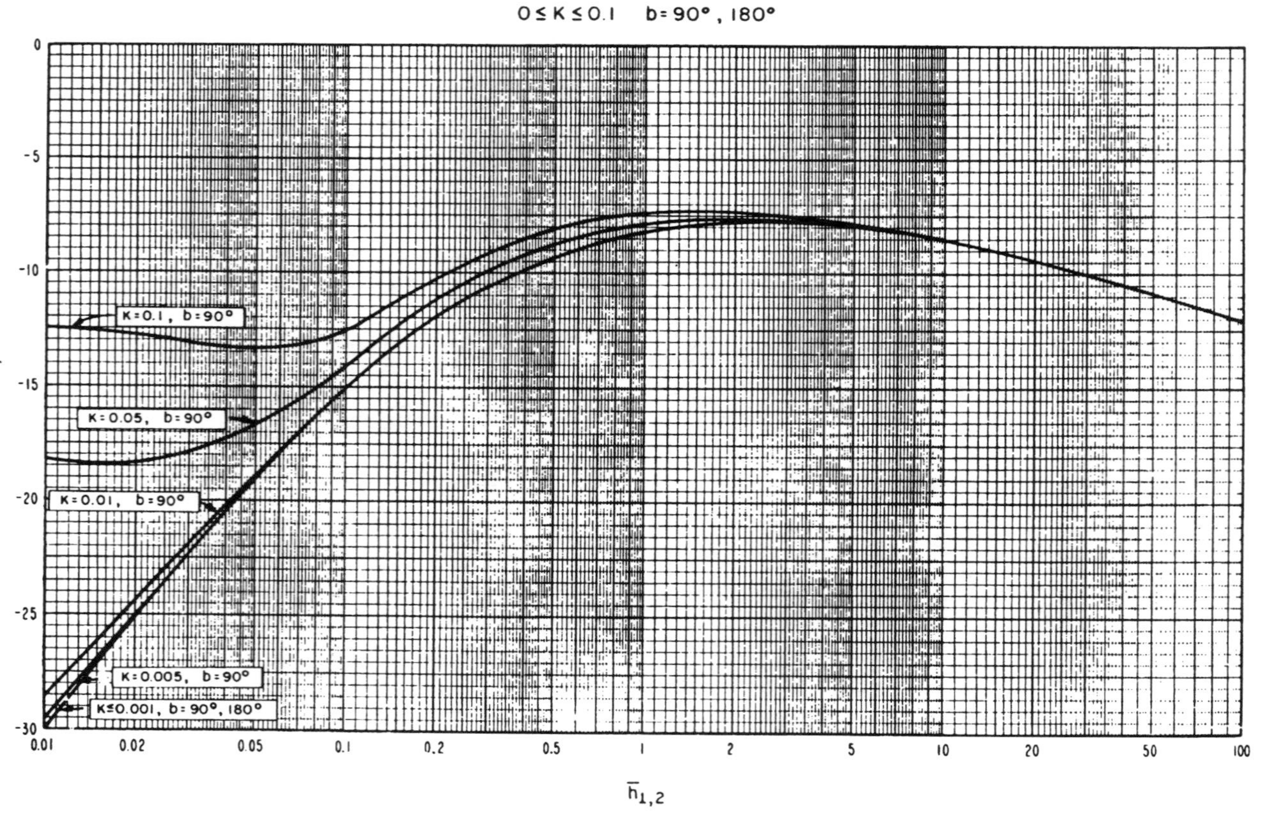

Figure 5.18. The residual height gain function, $G(\overline{h}_{1,2})$.

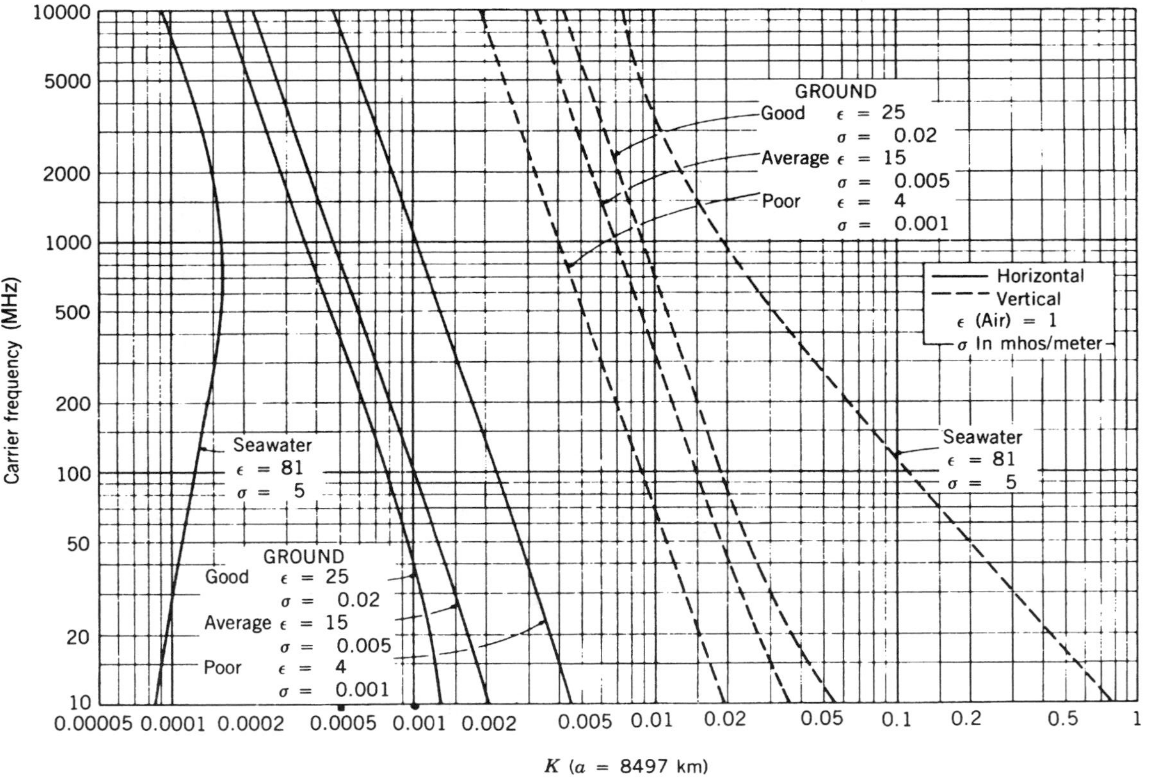

Figure 5.19. The parameter K for an effective earth radius $a = 8497$ km (Ref. 1).

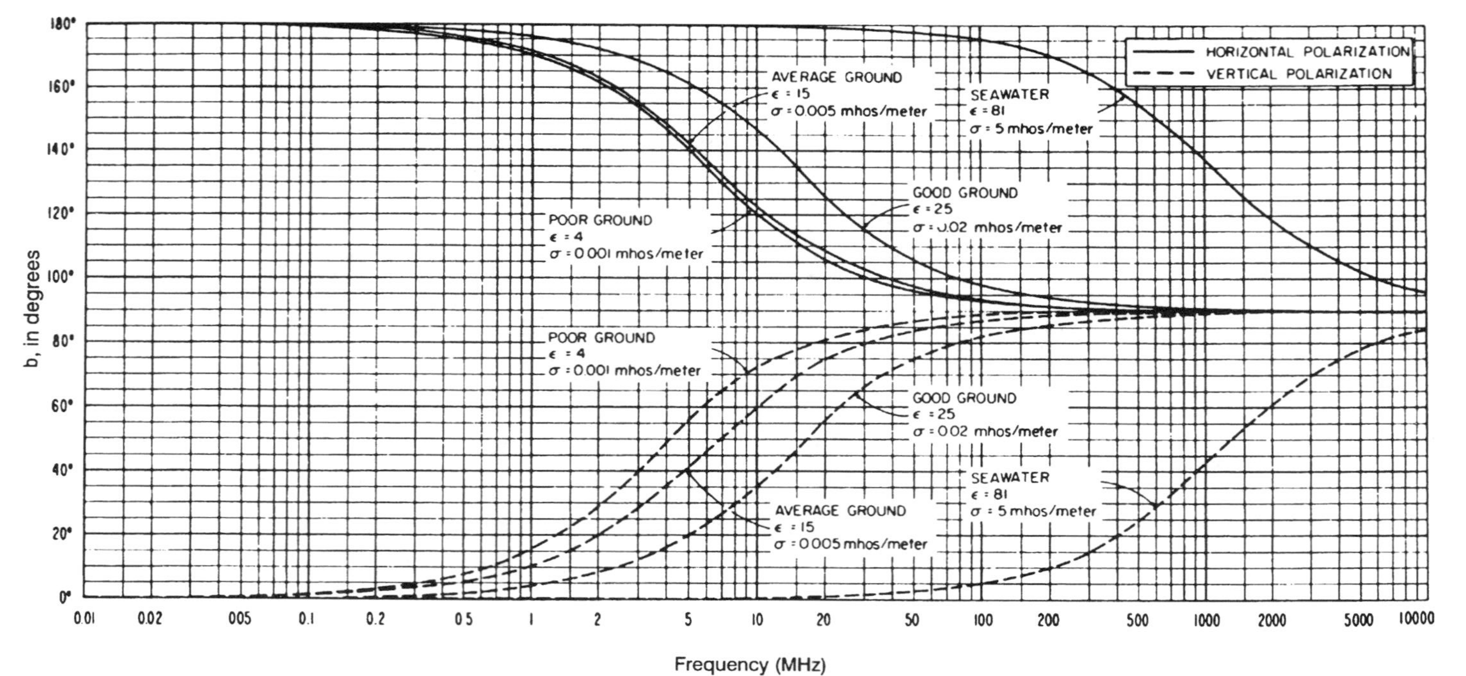

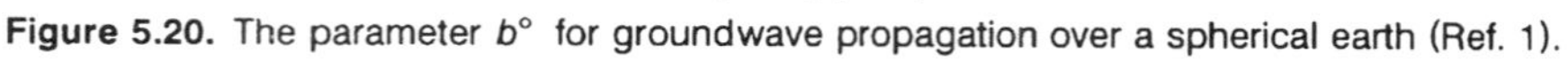
Figure 5.20. The parameter b° for groundwave propagation over a spherical earth (Ref. 1).

3s. Determine $b°$ from Figure 5.20, where $b°$ also is a function of the operating frequency.

4s. Calculate $K(a)$, which is a value of K for an effective earth radius other than 8497 km:

$$K(a) = C_0 K(a = 8497 \text{ km}) \tag{5.42}$$

5s. From Figure 5.21 determine $B(K, b°)$, which is a function of $K(a)$ and $b°$.

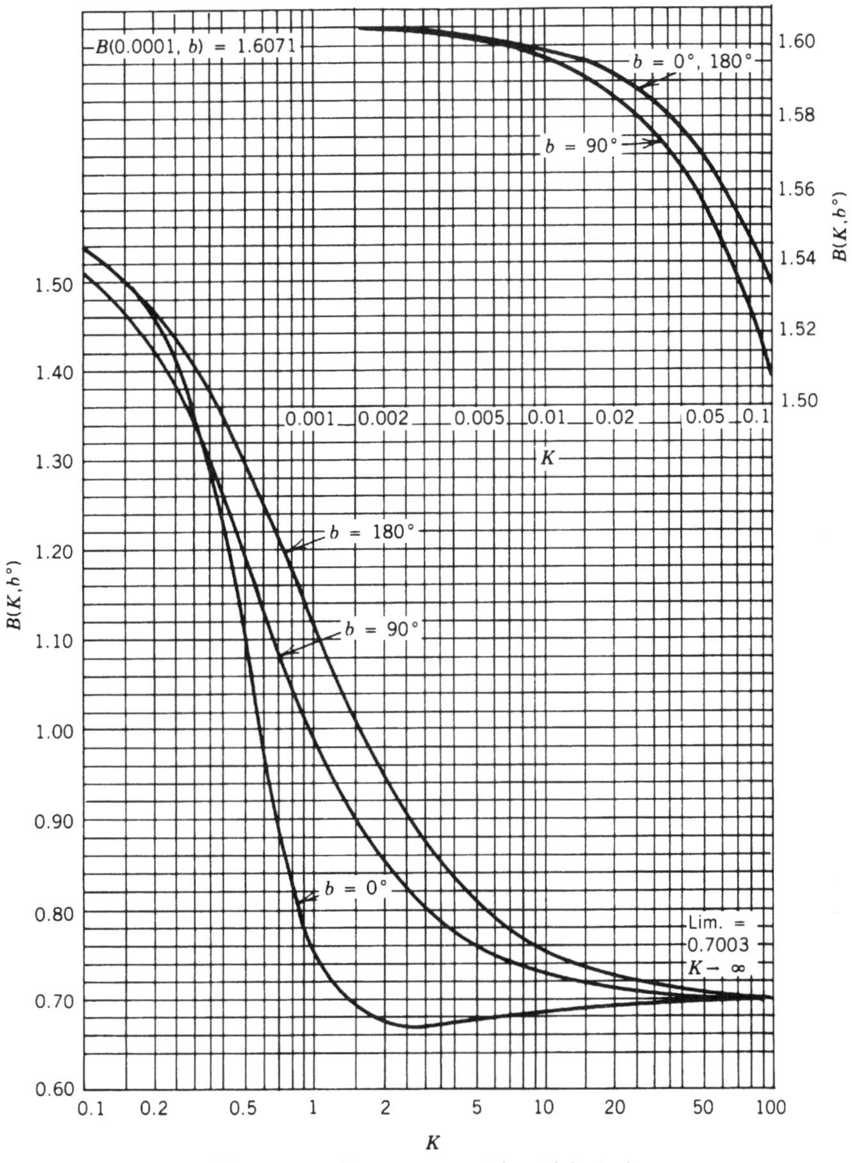

Figure 5.21. The parameter $B(K, b°)$ (Ref. 1).

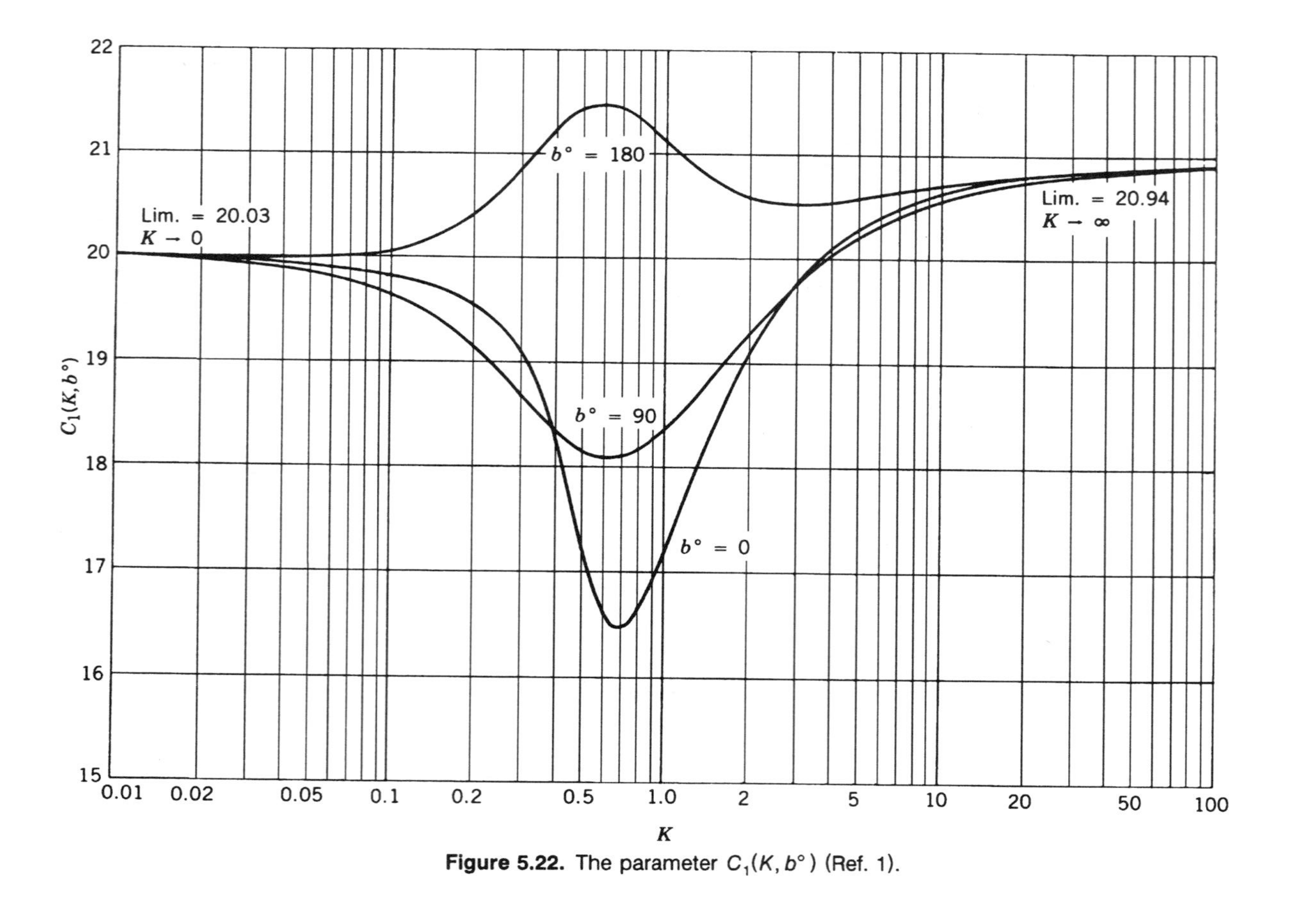

Figure 5.22. The parameter $C_1(K, b^\circ)$ (Ref. 1).

6s. Determine $C_1(K, b°)$ from Figure 5.22.

7s. Calculate B_0:

$$B_0 = f^{1/3} C_0^2 B(K, b°) \quad (5.43)$$

8s. Calculate the following values:

$$x_0 = dB_0 \quad (5.44a)$$

$$x_1 = d_{Lt} B_0 \quad (5.44b)$$

$$x_2 = d_{Lr} B_0 \quad (5.44c)$$

where d is the great circle distance for the path and d_{Lt} and d_{Lr} are the distances from the transmitter and receiver terminals to their respective horizons, derived from path profile.

9s. Determine $G(x_0)$, $F(x_1)$, and $F(x_2)$ from Figure 5.23. These values consider the effect of distance and antenna height on the diffraction loss.

10s. Calculate A, which is the diffraction attenuation relative to free space:

$$A_{\text{dB}} = G(x_0) - F(x_1) - F(x_2) - C_1(K, b°) \quad (5.45)$$

11s. Calculate L_{bf}, the free-space loss:

$$L_{bf} = 32.45 + 20 \log f + 20 \log d \quad (5.46)$$

where f is the operating frequency in megahertz and d is the great circle distance in kilometers.

12s. Calculate L_{bd}, which is the long-term median value of transmission loss (A_a is the atmospheric gaseous loss from Figure 5.4):

$$L_{bd} = L_{bf} + A + A_a \quad (5.47)$$

5.4.2.9 Calculation of the Combined Reference Value of Long-Term Basic Transmission Loss, L_{cr}. It is assumed here that a positive identification of the predominant transmission mode has not been made. If the tropo mode displays a transmission loss 15 dB or more greater than the diffraction mode, the diffraction mode predominates and

$$L_{cr} = L_{bd} \quad (5.48a)$$

If the contrary occurs, where the diffraction mode displays a transmission loss 15 dB or more greater than the tropo mode, then the tropo mode

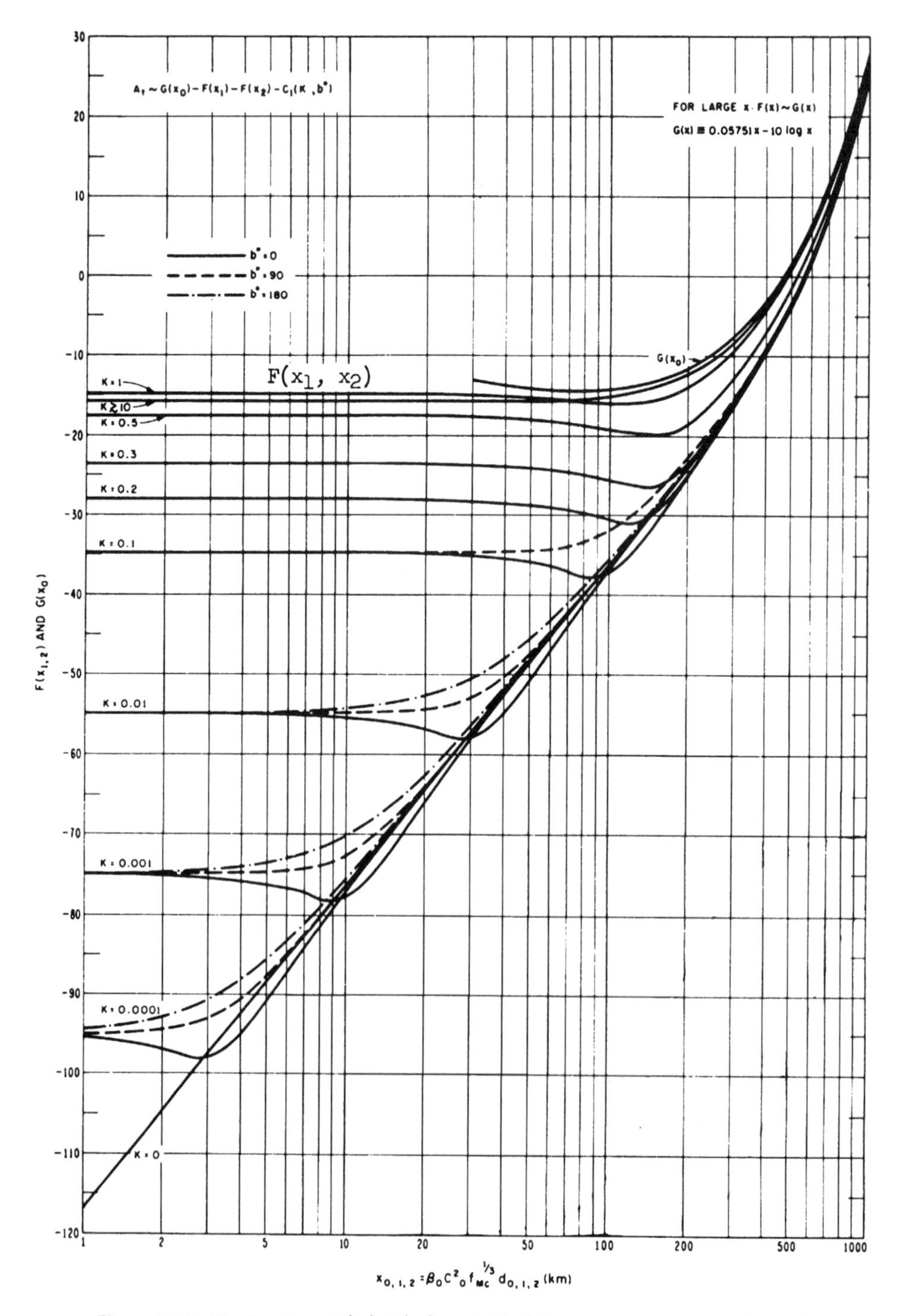

Figure 5.23. The functions $F(x_1)$, $F(x_2)$, and $G(x_0)$ for the range $0 \leq K \leq 1$ (Ref. 1).

predominates and

$$L_{cr} = L_{bsr} \tag{5.48b}$$

If the two modes have transmission losses within 15 dB of each other, then

$$L_{cr} = L_{bd} - R(0.5) \tag{5.48c}$$

The value of $R(0.5)$ is determined from Figure 5.24. The value of L_{cr} will be used in the next section.

5.4.2.10 Calculation of Effective Distance, d_e. Calculate d_{sl}, which is the distance where the diffracted and scatter fields are approximately equal over smooth earth:

$$d_{sl} = 65\left(\frac{100}{f}\right)^{1/3} \tag{5.49}$$

where f is the operating frequency in megahertz. Compute d_L, which is the sum of d_{Lt} and d_{Lr} for smooth earth with a radius of 9000 km:

$$d_L = 3\sqrt{2h_{te} \times 10^3} + 3\sqrt{2h_{re} \times 10^3} \tag{5.50a}$$

where h_{te} and h_{re} are the effective antenna heights from step 3 of Section 5.4.2.6.

If $d \leq d_L + d_{sl}$, then

$$d_e = \frac{130d}{d_L + d_{sl}} \tag{5.50b}$$

If $d > d_L + d_{sl}$, then

$$d_e = 130 + d - d_L - d_{sl} \tag{5.50c}$$

5.4.2.11 Calculation of the Basic Long-Term Median Transmission Loss, $L_n(0.5, 50)$. $L_n(0.5, 50)$ is the long-term loss adjusted for climatic region and effective distance providing a path time availability of 50% and a service probability of 50%. There are eight climatic regions given in Table 5.2. Select the climatic region from the table appropriate for the path in question. If two climatic regions appear applicable, calculate V_n for both regions and average the value. Note that the small n after the V is indicative of the climatic region. After selecting the climatic region (n), next obtain the appropriate value (in dB) for $V_n(0.5, d_e)$ from Figure 5.25 for the region and

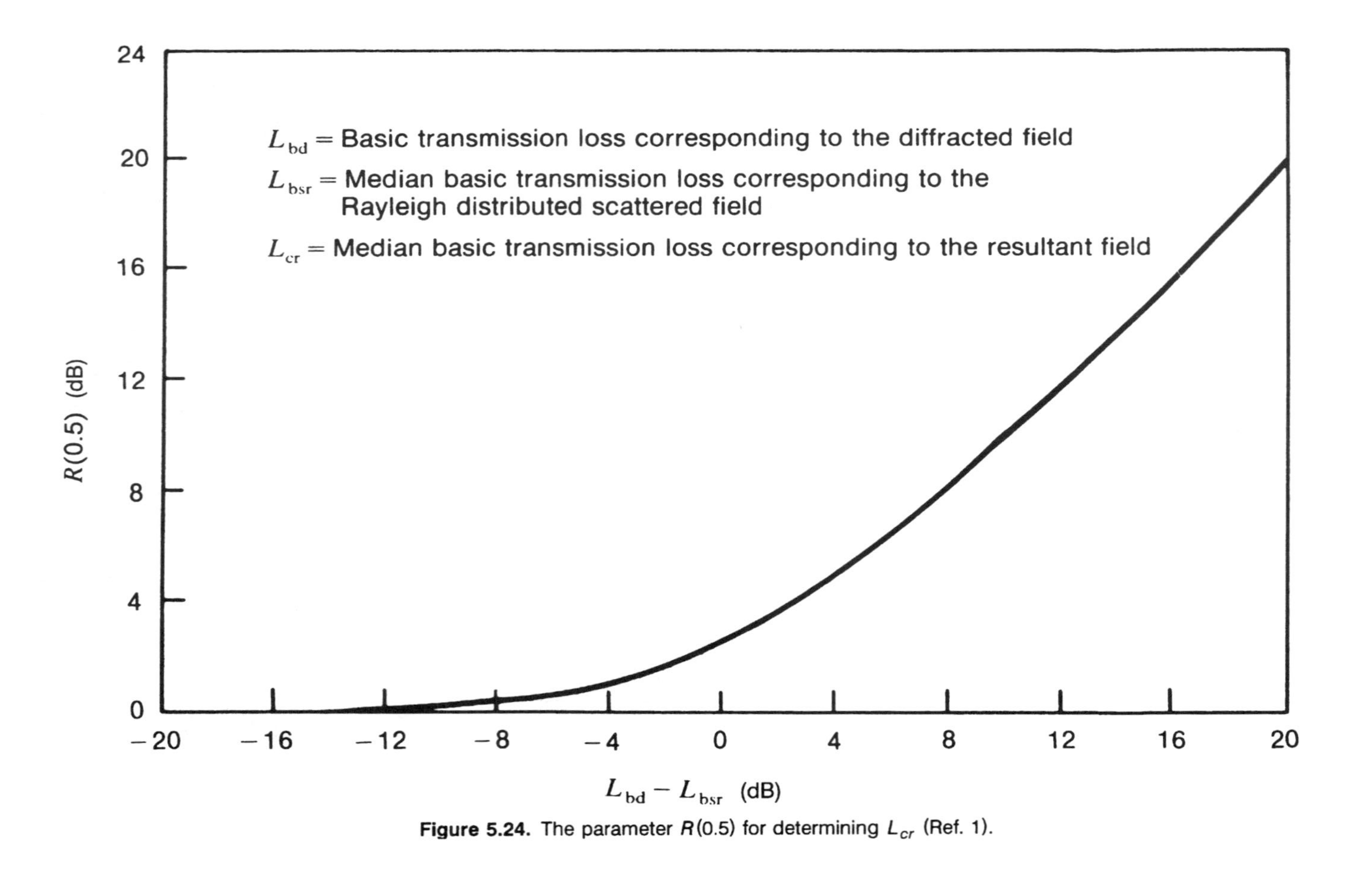

Figure 5.24. The parameter $R(0.5)$ for determining L_{cr} (Ref. 1).

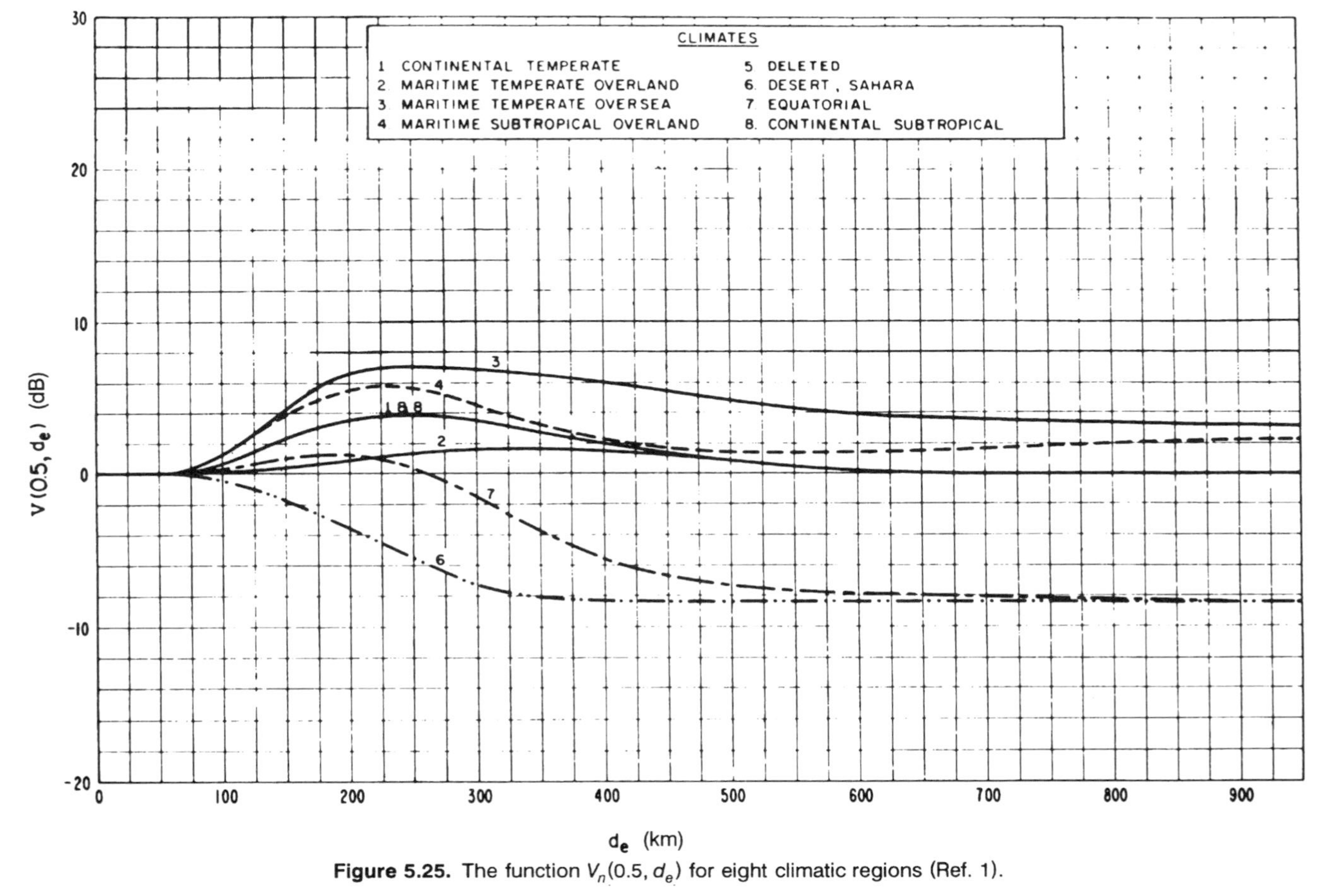

Figure 5.25. The function $V_n(0.5, d_e)$ for eight climatic regions (Ref. 1).

TABLE 5.2 Characteristics of Climatic Regions

Climate 1. Continental Temperature

This region is characterized by an annual mean N_s of about 320 N-units with an annual range of monthly mean N_s of 20–40 N-units. A continental climate in a large land mass shows extremes of temperature in a "temperate" zone, such as 30°–60° north or south latitude. Pronounced diurnal and seasonal changes in propagation are expected to occur. On the east coast of the United States the annual range of N_s may be as much as 40–50 N-units owing to contrasting effects of arctic or tropical maritime air masses that may move into the area from the north or from the south.

Climate 2. Maritime Temperature Overland

This region is characterized by an annual mean N_s of about 320 N-units with a rather small annual range of monthly mean N_s of 20–30 N-units. Such climatic regions are usually located from 20° to 50° north or south latitude, near the sea, where prevailing winds, unobstructed by mountains, carry moist maritime air inland. These conditions are typical of the United Kingdom, the west coasts of North America and Europe, and the northwestern coastal areas of Africa.

Although the islands of Japan lie within this range of latitude, the climate differs in showing a much greater annual range of monthly mean N_s, about 60 N-units, the prevailing winds have traversed a large land mass and the terrain is rugged. One would therefore not expect to find ratio propagation conditions similar to those in the United Kingdom, although the annual mean N_s is 310–320 N-units in each location. Climate 1 is probably more appropriate than Climate 2 in this area, but ducting may be important in coastal and oversea areas of Japan as much as 5% of the time in summer.

Climate 3. Maritime Temperature Oversea

This region is characterized by coastal and oversea areas with the same general characteristics as those for Climate 2. The distinction made is that a radio path with both horizons on the sea is considered to be an oversea path; otherwise Climate 2 is used. Ducting is rather common for a small fraction of time between the United Kingdom and the European Continent and along the west coast of the United States and Mexico.

Climate 4. Maritime Subtropical Overland

This region is characterized by an annual mean N_s of about 370 N-units with an annual range of monthly mean N_s of 30–60 N-units. Such climates may be found from about 10° to 30° north and south latitude, usually on lowlands near the sea with definite rainy and dry seasons. Where the land area is dry, radio ducts may be present for a considerable part of the year.

Climate 5. Maritime Subtropical Oversea

This region is characterized by conditions observed in coastal areas with the same range of latitude as Climate 4. The curves for this climate were based on an inadequate amount of data and have been deleted. It is suggested that the curves for Climates 3 or 4 be used, selecting whichever seems more applicable to each specific case.

Climate 6. Desert, Sahara

This region is characterized by an annual mean N_s of about 280 N-units with year-round semiarid conditions. The annual range of monthly mean N_s may be from 20 to 80 N-units.

Climate 7. Equatorial

This region is characterized by a maritime climate with an annual mean N_s of about 360 N-units and annual range of 0–30 N-units. Such climates may be observed from 20° N to 20° S latitude and are characterized by monotonous heavy rains and high average summer temperatures. Typical equatorial climates occur along the Ivory Coast and in the Congo of Africa.

TABLE 5.2 (*continued*)

Climate 8. Continental Subtropical

This region is typified by the Sudan and monsoon climates, with an annual mean N_s of about 320 N-units and an annual range of 60–100 N-units. This is a hot climate with seasonal extremes of winter drought and summer rainfall, usually located from 20° to 40° N latitude.

A continental polar climate, for which no curves are shown, may also be defined. Temperatures are low to moderate all year round. The annual mean N_s is about 310 N-units with an annual range of monthly mean N_s of 10–40 N-units. Under polar conditions, which may occur in middle latitudes as well as in polar regions, radio propagation would be expected to show somewhat less variability than in a continental temperature climate. Long-term median values of transmission loss are expected to agree with the reference value L_{cr}.

High mountain areas or plateaus in a continental climate are characterized by low values of N_s and year-round semiarid conditions. The central part of Australia with its hot dry desert climate and an annual range of N_s as much as 50–70 N-units may be intermediate between Climates 1 and 6.

Source: Reference 3.

effective path distance (d_e) calculated in the previous section. Then calculate $L_n(0.5, 50)$:

$$L_n(0.5, 50) = L_{cr} - V_n(0.5, 50) \tag{5.51}$$

The transmission loss $L_n(0.5, 50)$ now must be extended from its present time availability of 50% to the desired time availability.

5.4.2.12 *Extending to the Desired Time Availability, q.* To extend the time availability from its median or 50% value, we must first calculate an additional loss, $Y_n(q, d_e, f)$, to be added to the value of L_n calculated above; q is the desired time availability, d_e is the effective distance from Section 5.4.2.10, and f is the operating frequency. The value for Y_n deals with the fading characteristics of the path. Fading is a function of operating frequency, path length, and climatic region (n). If more than one climatic region is indicated (i.e., a particular path has characteristics of two regions), then we calculate Y_n as follows for regions called i and j:

$$Y_n(q, d_e) = \sqrt{0.5Y_i^2(q, d_e) + 0.5Y_j^2(q, d_e)} \tag{5.52}$$

Climatic Regions 1, 6, and 8 require special treatment to calculate or derive values for $Y_n(q, d_e)$.

We first treat Climatic Region 1 (continental temperature climate). For this case there are two approaches to calculate values for $Y_1(q, d_e)$. The first approach is to use the curves directly, which are found in Figures 5.28, 5.29, and 5.30. For the second, more refined approach, we derive a value for

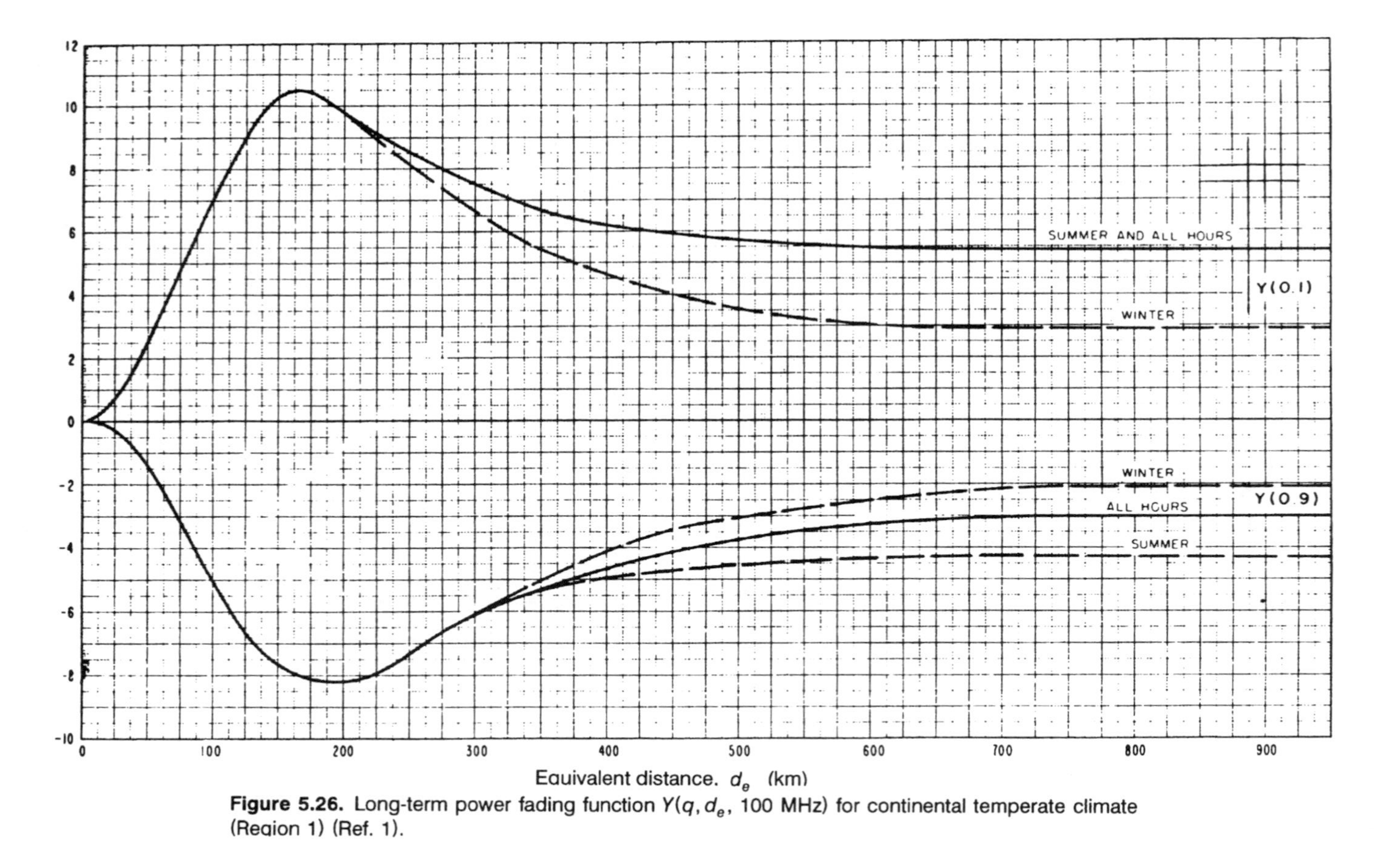

Figure 5.26. Long-term power fading function $Y(q, d_e, 100\text{ MHz})$ for continental temperate climate (Region 1) (Ref. 1).

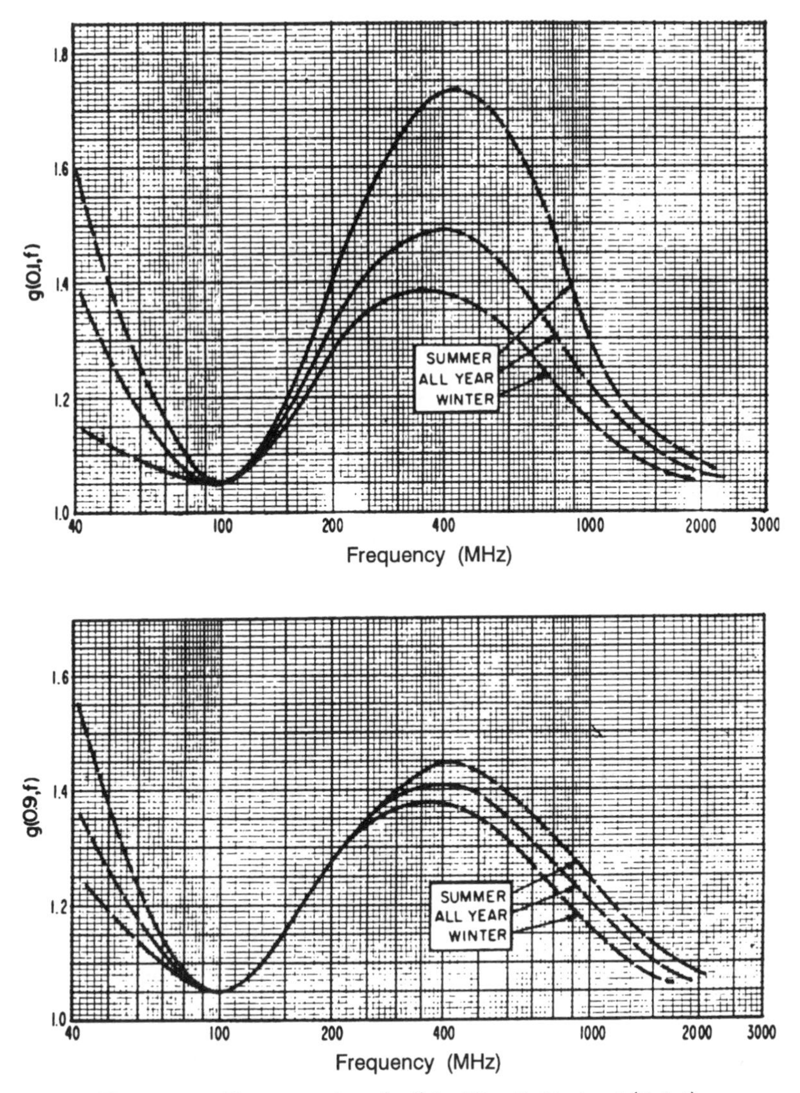

Figure 5.27. The parameter $g(q, f)$ for Climatic Region 1 (Ref. 1).

$Y_1(0.9, d_e)$ for 100 MHz from Figure 5.26 and then multiply the value thus derived by a frequency correction factor $g(0.9, f)$ for the desired frequency f. This factor is taken from Figure 5.27 and is used for operating frequencies 2 GHz and below. The product so derived is then applied to the formulas given in the legend box in Figures 5.28, 5.29, and 5.30.

Example 1. Let $d_e = 200$ km, $f = 900$ MHz. Calculate the value of $Y_1(0.99, 200, 900)$.

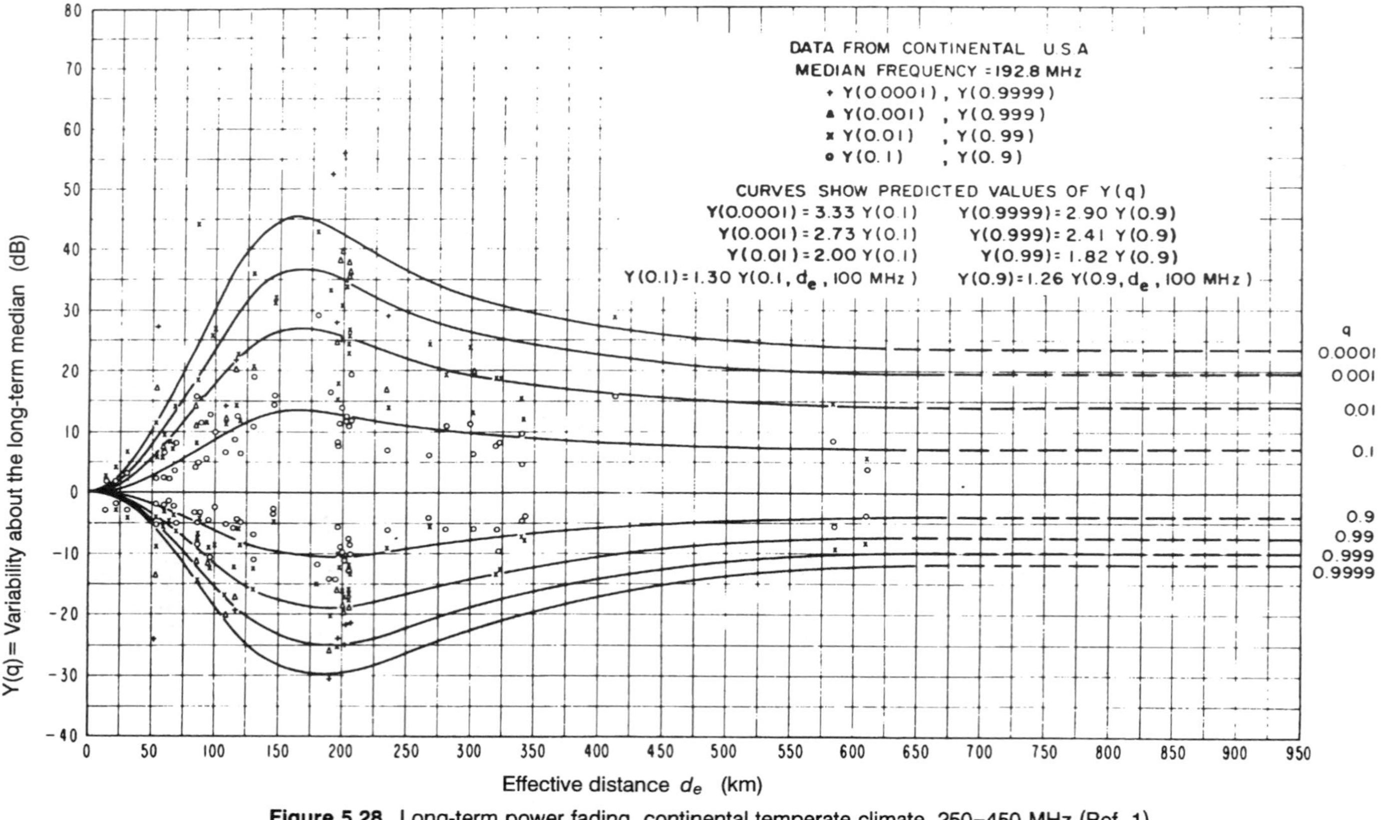

Figure 5.28. Long-term power fading, continental temperate climate, 250–450 MHz (Ref. 1).

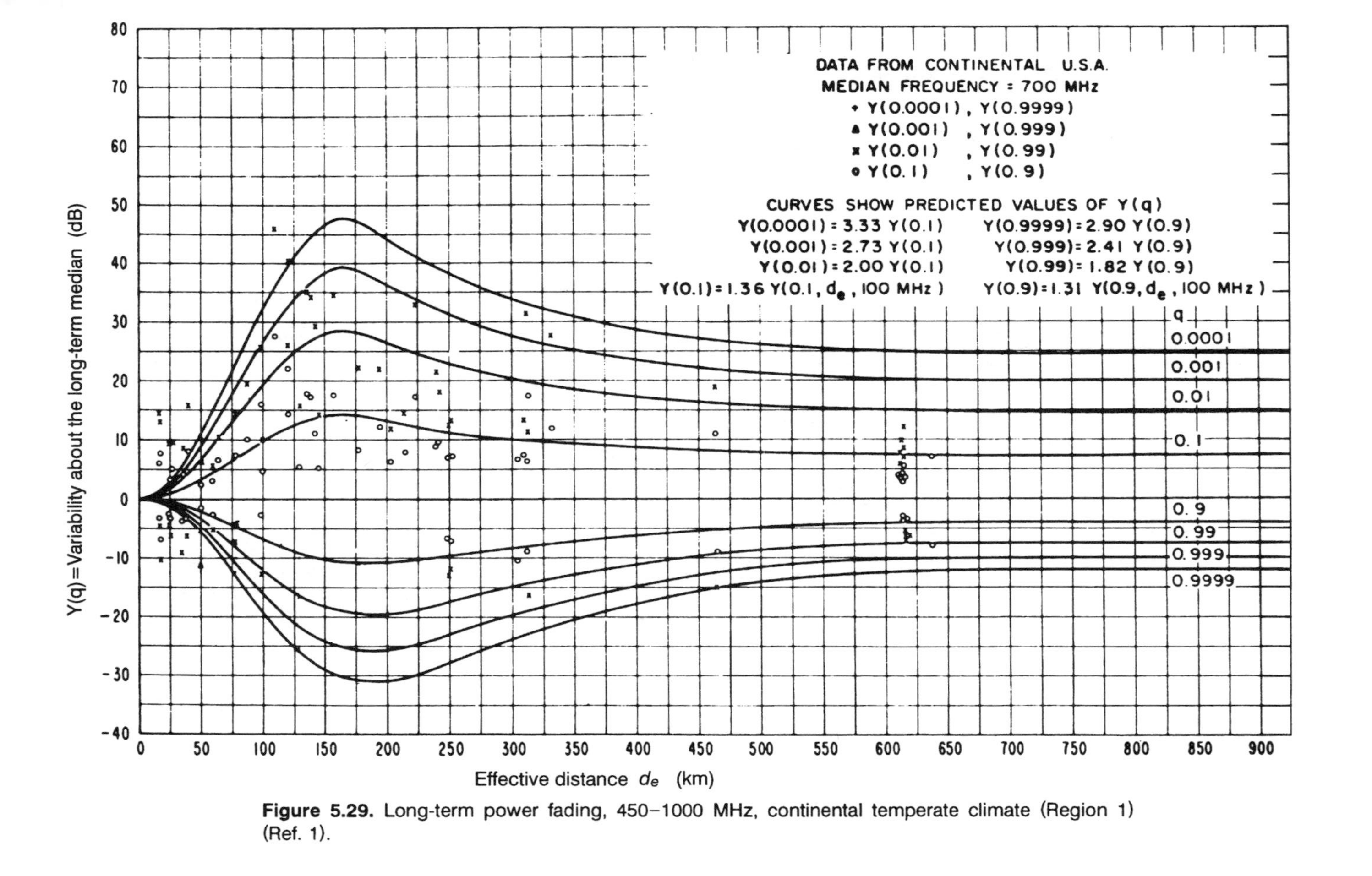

Figure 5.29. Long-term power fading, 450–1000 MHz, continental temperate climate (Region 1) (Ref. 1).

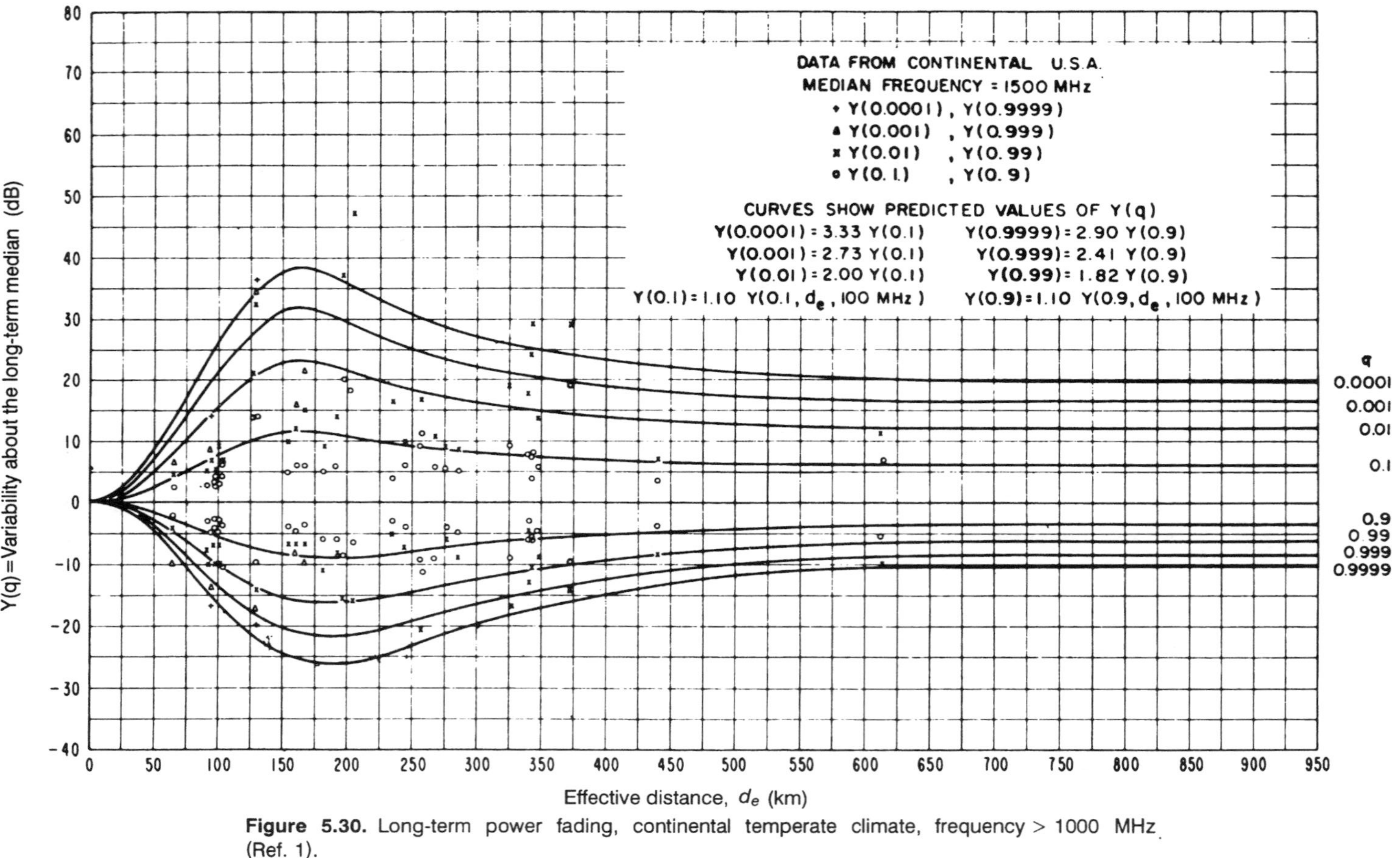

Figure 5.30. Long-term power fading, continental temperate climate, frequency > 1000 MHz (Ref. 1).

From Figure 5.29 we derive −19.9 dB directly, using the first approach.

For the second approach, the function $Y_1(0.9, 200, 100)$ from Figure 5.26 is −8.1 dB. This is the value for Y_1 for a 0.90 time availability, effective distance of 200 km, and frequency of 100 MHz. The frequency correction factor is taken from Figure 5.27, and we see that for 900 MHz (worst case—summer) the value is 1.275. Multiplying these two values together (1.275 × 8.1 dB), we get −10.33 dB. Apply the formula from the legend in Figure 5.29 for $q = 0.99$ and $Y(0.99) = -10.33 \times 1.82 = -18.8$ dB or 1.1 dB less (loss) than by the first approach.

Reference 2 reports that the more refined approach can improve accuracies by more than 2 dB.

A similar approach is used for Climatic Regions 6 and 8. A frequency correction factor $g(q, f)$ is derived from Figure 5.31 for the specified frequency. Figures 5.32 and 5.33 gives values of Y_n for Climatic Regions 6 and 8, respectively, for 1000 MHz. The next step is to determine values for long-term power fading from Figures 5.32 and 5.33 for the appropriate effective distance d_e. This value is then multiplied by the frequency correction factor $g(q, f)$ to determine $Y_n(q, d_e, f)$.

Example 2. Determine $Y_n(0.999, 350, 400)$ (i.e., Climatic Region 6, path time availability 0.999, effective distance of 350 km, and operating frequency of 400 MHz). First, derive the frequency correction factor from Figure 5.31. Here $g(q, 400) = 1.2$. Next, determine the value for $Y_6(0.999, 350, 1000)$ from Figure 5.32. This value is −17 dB. $Y_6(0.999, 350, 400) = -17 \text{ dB} \times 1.2 = -20.4$ dB.

For the remaining climatic regions—2, 3, 4, and 7—we determine $Y_n(q, d_e, f)$ for the applicable climatic region n, path time availability q, effective distance d_e, and applicable frequency f. Read values of Y_n for the appropriate climatic regions from Figures 5.33–5.39.

To calculate $L_n(q)$ the transmission loss for any time availibility q, use the following formula:

$$L_n(q) = L_n(0.5) - Y_n(q) \tag{5.53}$$

where $L_n(q)$ for a service probability of 0.5 (i.e., 50%) is taken from formula (5.51).

5.4.2.13 *Extending to the Desired Service Probability.*

The service probability concept is used to obtain a measure of the prediction uncertainty due to our lack of complete knowledge of the propagation mechanism. The formulas we use are of a semiempirical nature and are based on a finite data sample.

The processes in tropospheric scatter propagation are extremely complex, and it is neither possible nor practical to provide numerical values of all possible parameters and their effects on the time distribution of transmission

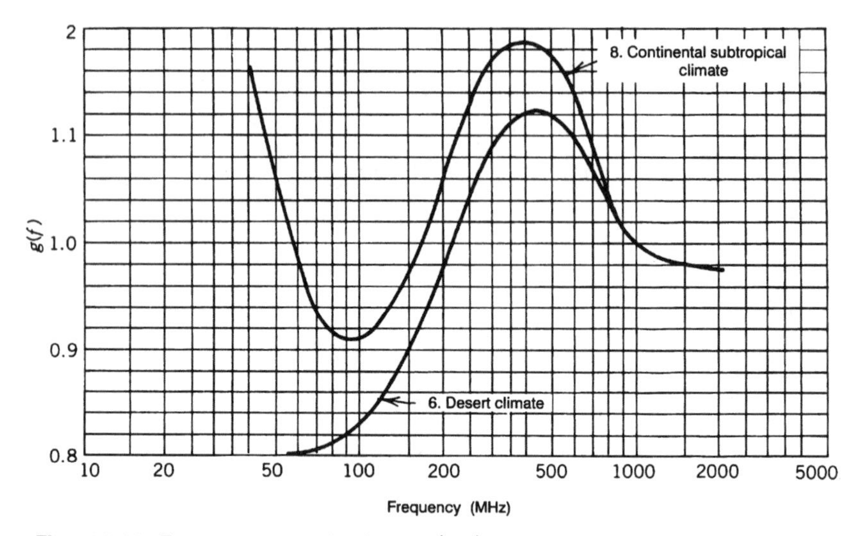

Figure 5.31. Frequency correction factor $g(q, f)$ for Climatic Regions 6 and 8 (Ref. 1).

loss. Consequently, the specific values calculated in accordance with the material in the previous sections must be considered as mean values resulting from an ensemble of propagation paths for which the parameters used in the transmission loss calculations are exactly identical, but which differ from each other in additional respects that cannot be included in the formulation of the models and methods used. It is reasonable to expect that long-term measurements over such an ensemble of paths or links would produce a random (or Gaussian) distribution of transmission loss values for each percentile of time, with the mean of such a distribution identical to the calculated value. The standard deviation of this distribution would then characterize the uncertainty inherent in the prediction or modeling process. The service probability (Q) is the parameter used to specify prediction uncertainty. A common service probability is specified at $Q = 0.95$. This means if we built 100 paths using these same parameters, that, once installed, 95 would equal or exceed specifications and 5 would fail. To calculate the service probability factor, $z_{m0}(Q)\sigma_{rc}(q)$, we first compute $\sigma_c^2(q)$ and $\sigma_{rc}(q)$ for the required values of q, the path time availability:

$$\sigma_c^2(q) = 12.73 + 0.12Y_n^2(q, d_e) \tag{5.54}$$

$$\sigma_{rc}(q) = \sqrt{\sigma_c^2(q) + \sigma_r^2} \tag{5.55}$$

where $\sigma_{rc}(q)$ is the prediction error between the predicted and the observed values of transmission loss for a fraction q of all hours. $\sigma_c^2(q)$ is the path-to-path variance of the difference between observed and predicted long-term median values of transmission loss. σ_r^2 is the variance assigned to

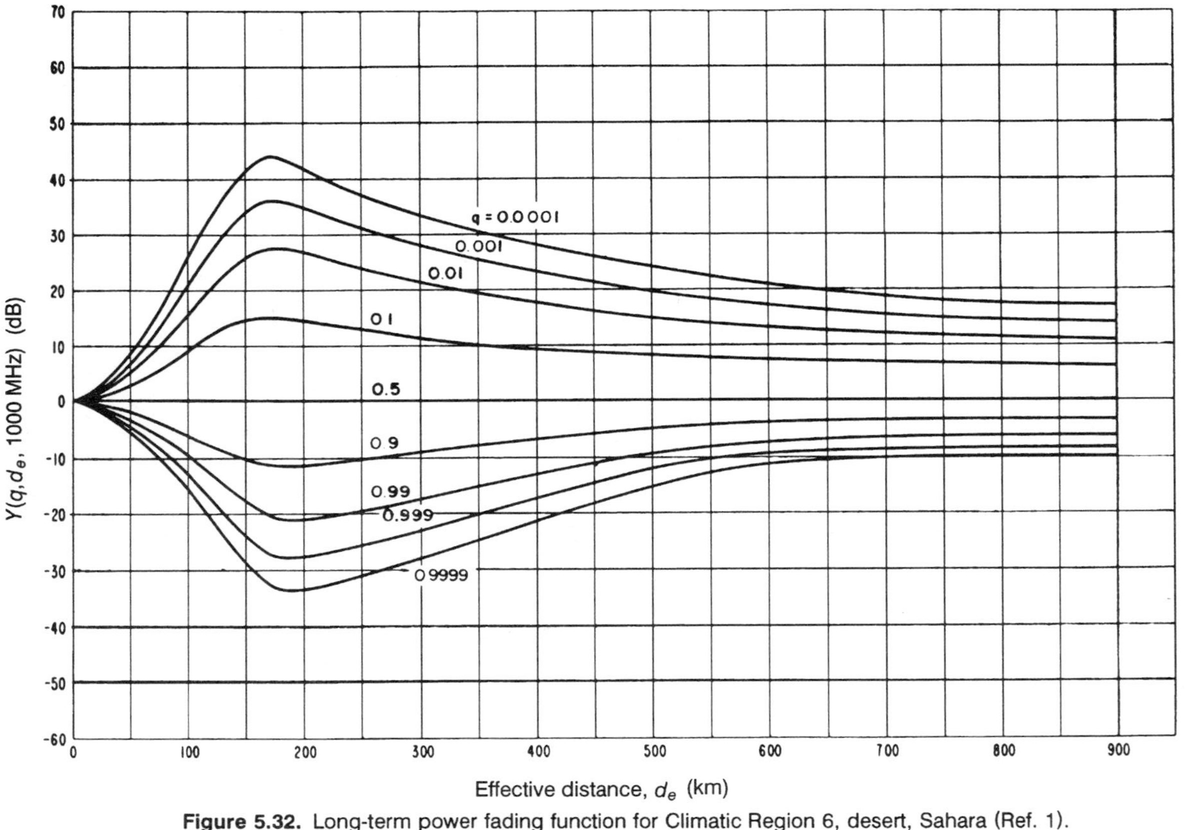

Figure 5.32. Long-term power fading function for Climatic Region 6, desert, Sahara (Ref. 1).

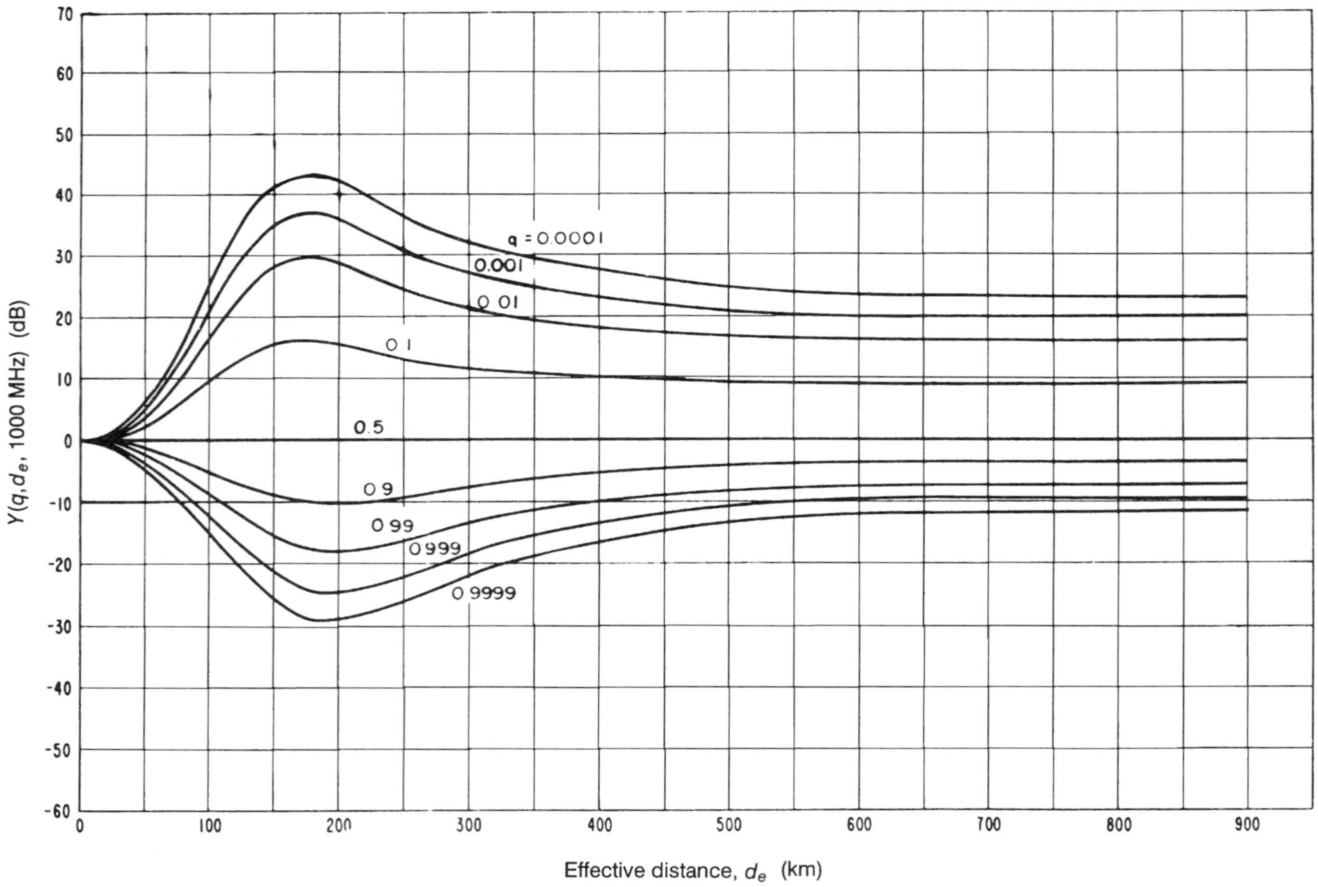

70
60
50
40
30
20
10
0
-10
-20
-30
-40
-50
-60
0
100
200
300
400
500
600
700
800
900
q = 0.0001
0.001
0.01
0.1
0.5
0.9
0.99
0.999
0.9999
$Y(q, d_e, 1000\ \text{MHz})$ (dB)
Effective distance, d_e (km)

Figure 5.33. Long-term power fading function for Climate Region 8, continental subtropical (Ref. 1).

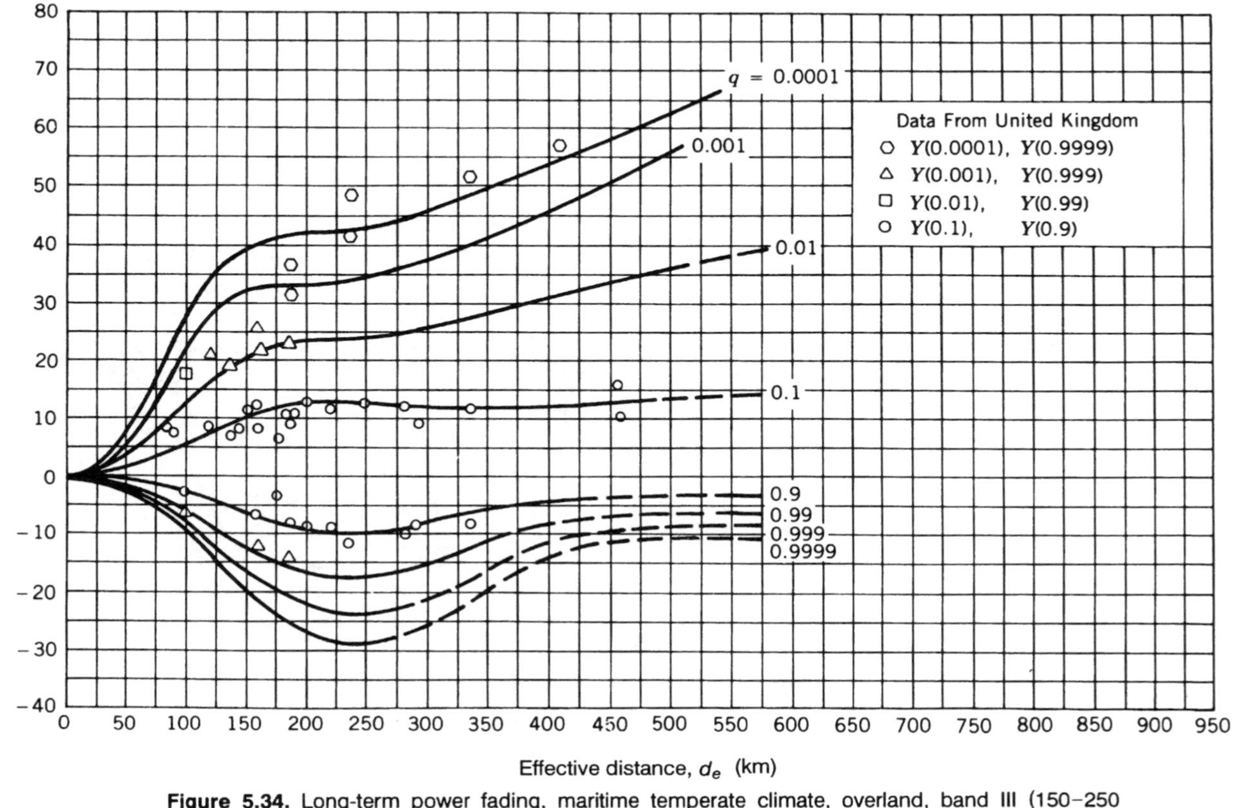

Figure 5.34. Long-term power fading, maritime temperate climate, overland, band III (150–250 MHz) (Ref. 1).

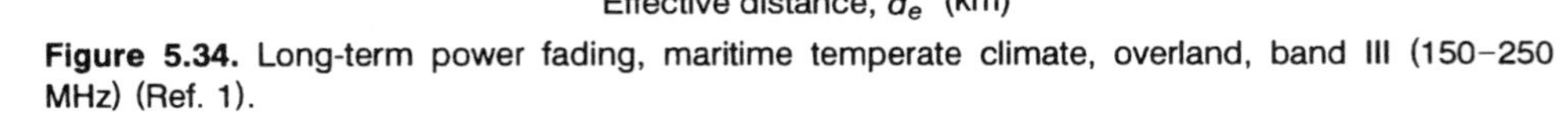

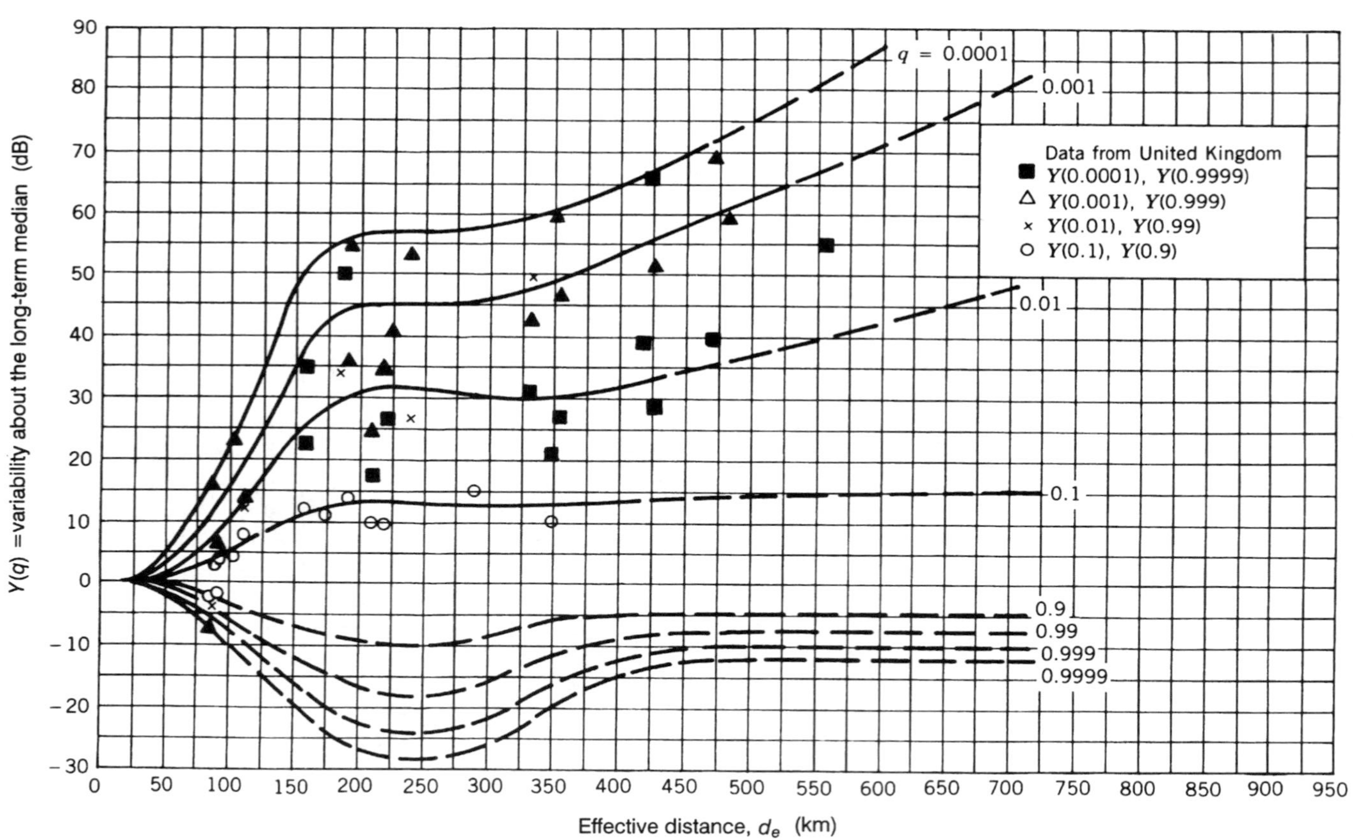

Figure 5.35. Long-term power fading, maritime temperate climate, overland, bands IV and V (450–1000 MHz), Climatic Region 2 (Ref. 1).

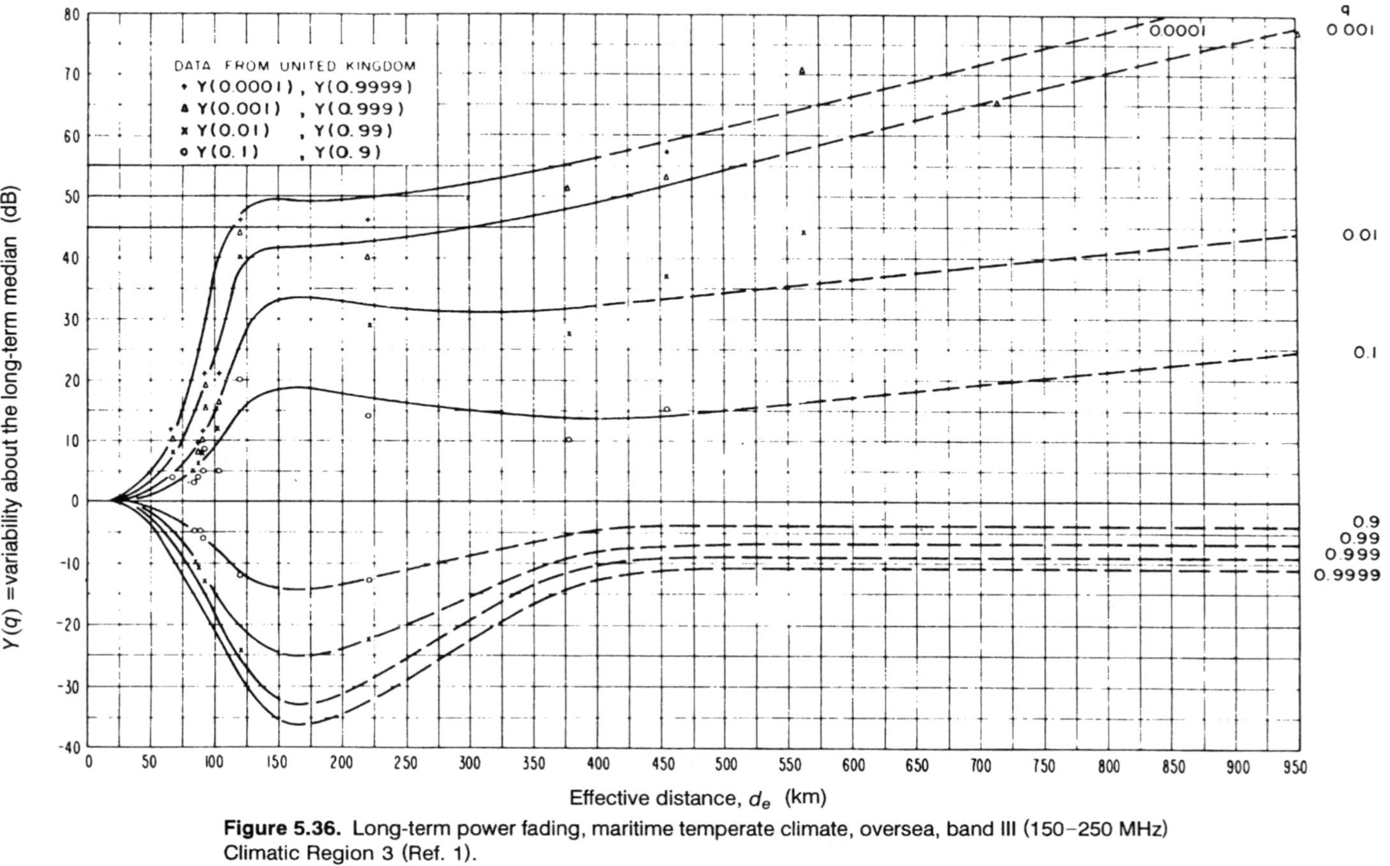

Figure 5.36. Long-term power fading, maritime temperate climate, oversea, band III (150–250 MHz) Climatic Region 3 (Ref. 1).

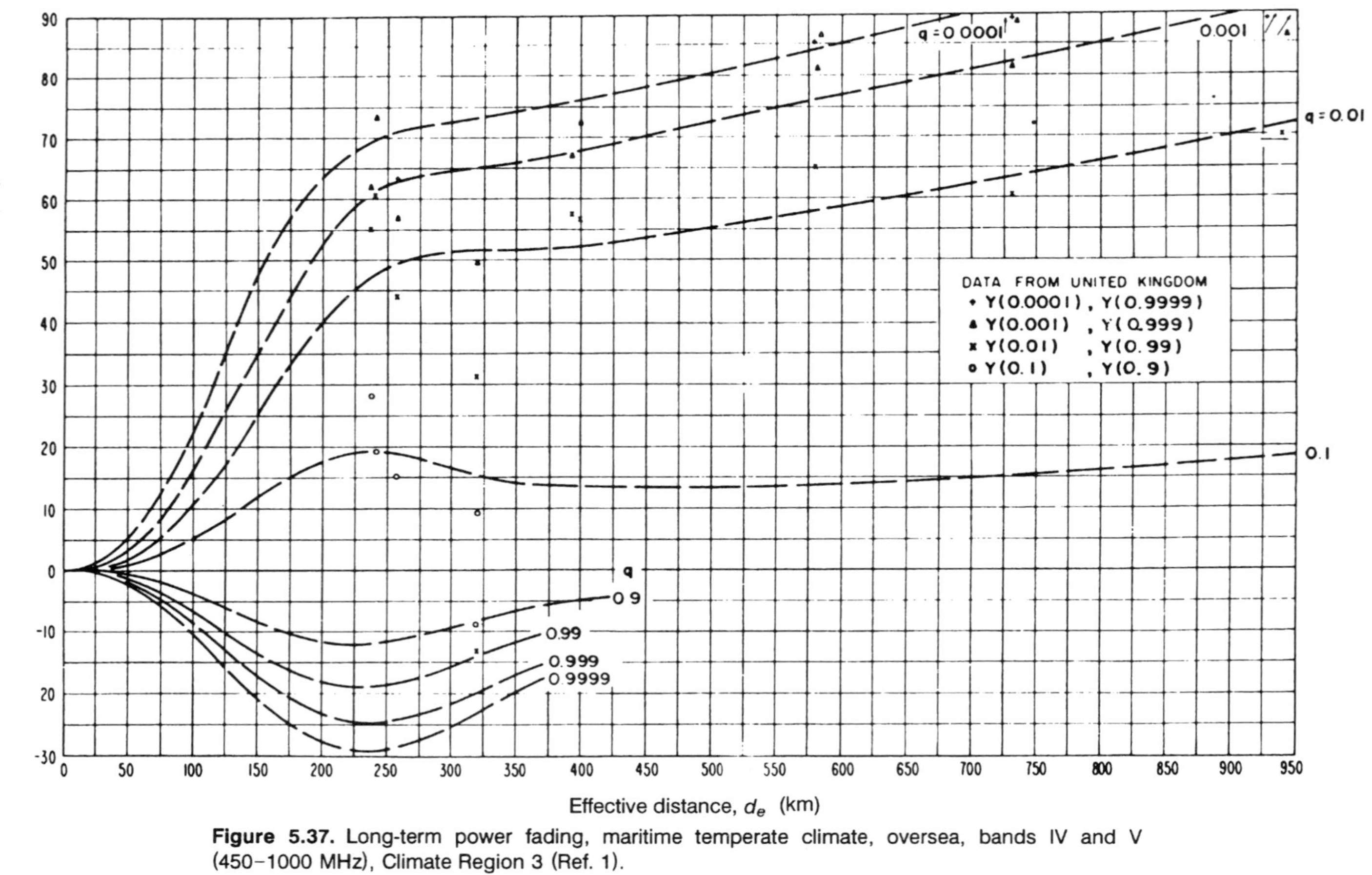

Figure 5.37. Long-term power fading, maritime temperate climate, oversea, bands IV and V (450–1000 MHz), Climate Region 3 (Ref. 1).

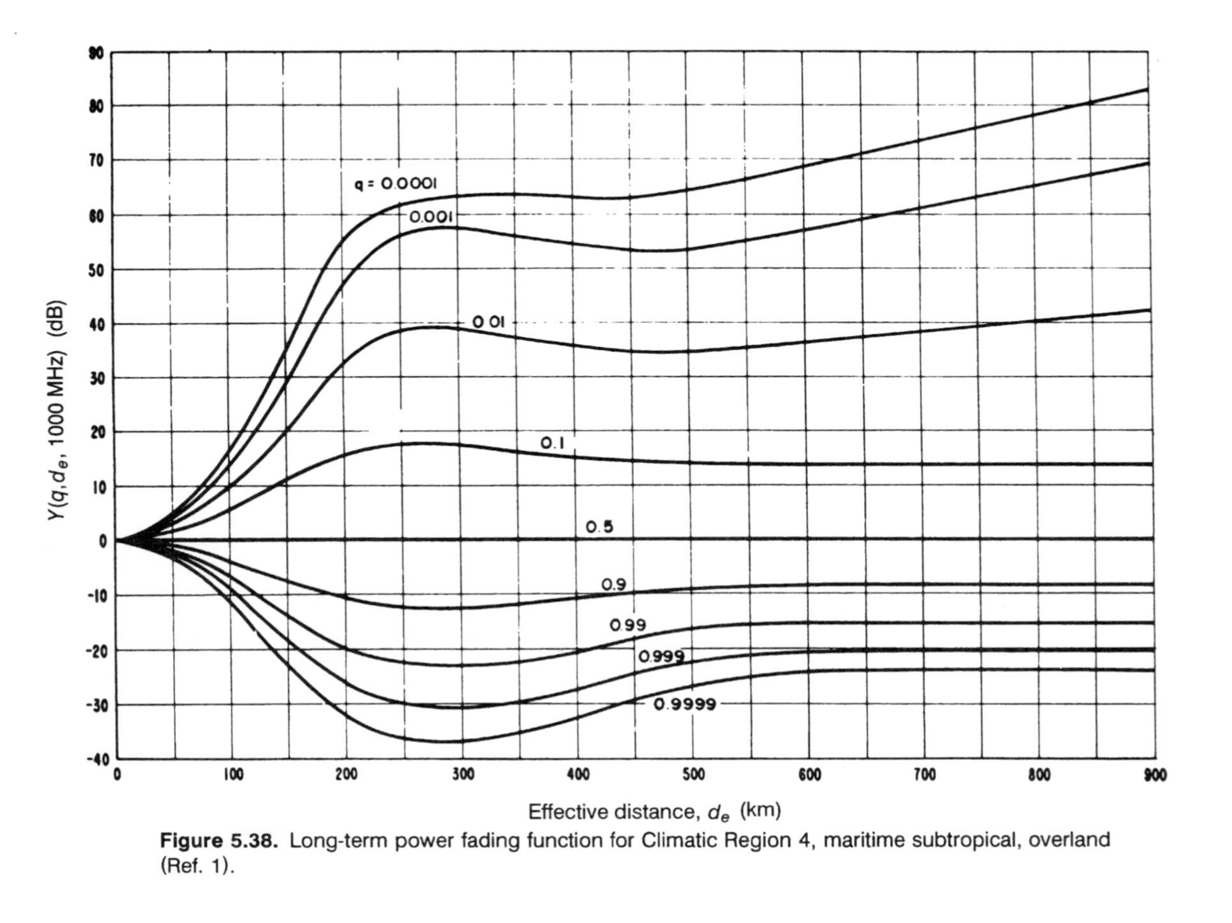

Figure 5.38. Long-term power fading function for Climatic Region 4, maritime subtropical, overland (Ref. 1).

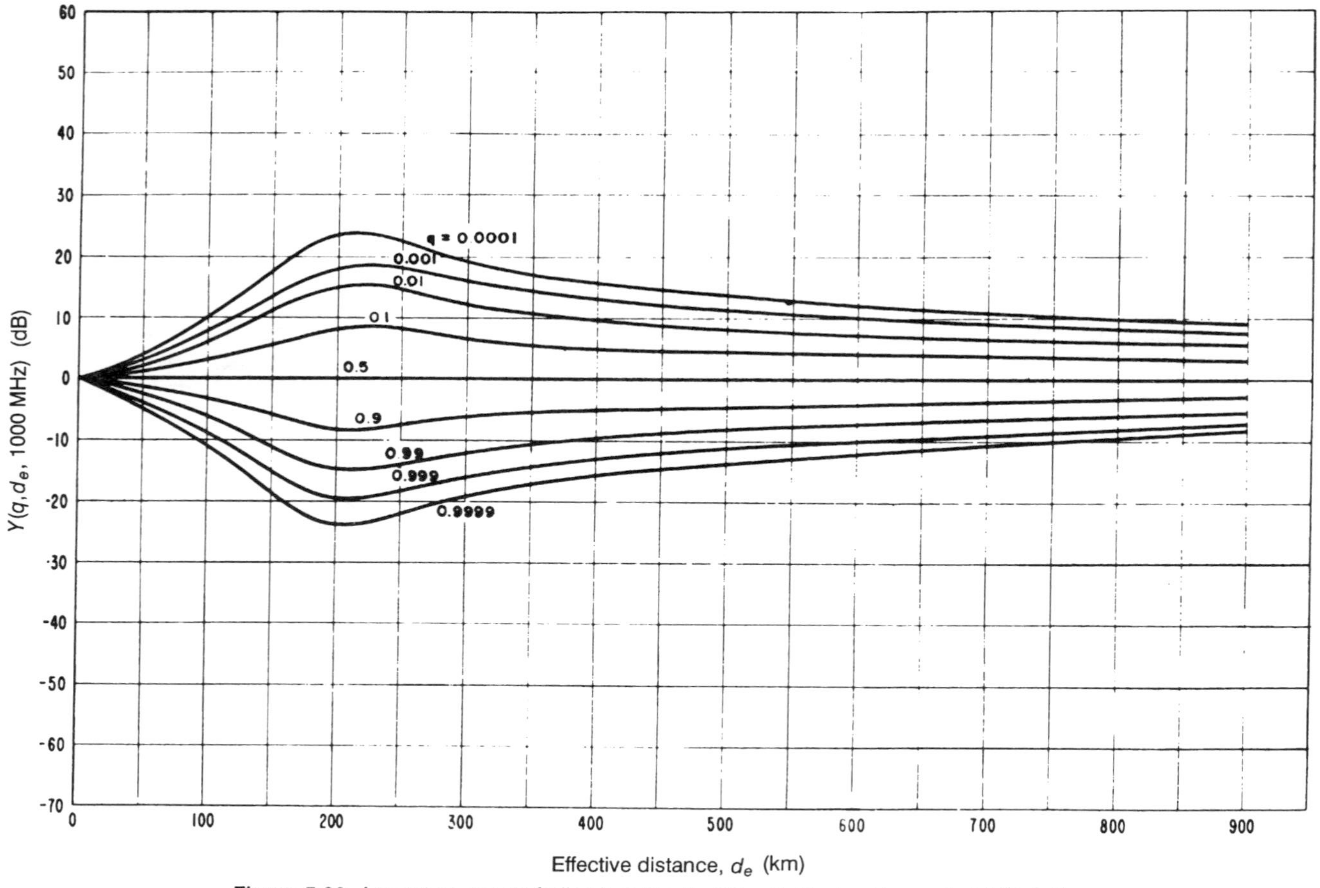

Figure 5.39. Long-term power fading function for Climatic Region 7, equatorial (Ref. 1).

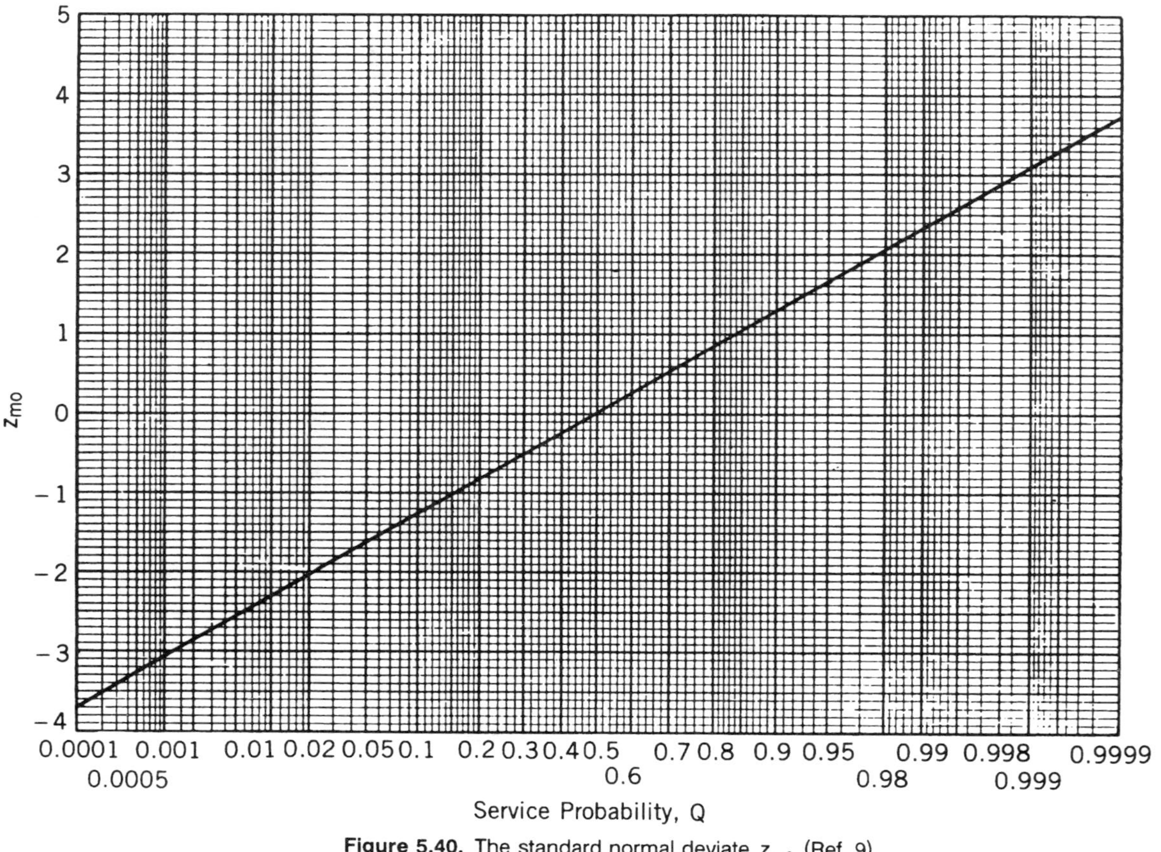

Figure 5.40. The standard normal deviate z_{m0} (Ref. 9).

errors in estimating equipment parameters; use a value of 4 dB^2 for new circuits and 2 dB^2 for established circuits. Y_n was derived in Section 5.4.2.12.

The next step is to determine $z_{m0}(Q)$ from Figure 5.40. Often 0.95 is specified. $z_{m0}(Q)$ is the standard normal deviate.

When calculating the product $z_{m0}(Q)\sigma_{rc}(q)$ for the values of q, it has been found through experience that tables are useful and may be constructed as follows. First, for the 50% service probability case:

Time Availability (q) (%)	$Y_n(q)$ (dB)	Transmission Loss (dB)
50		
90		
99		
99.9		
99.99		

A second table is then constructed to extend the 50% service probability case to 95% (i.e., $Q = 0.95$) or to some other desired service probability:

Time Availability (q) (%)	$z_{m0}(Q)\sigma_{rc}(q)$ (dB)	Transmission Loss (dB)
50		
90		
99		
99.9		
99.99		

Example 3. In Climatic Region 1 a tropo path has an effective distance of 325 km and $L_n(0.5) = 218$ dB. Prepare the two tables in accordance with this information. Assume the operating frequency to be 4000 MHz and thus the frequency correction factor equals 1 and may be neglected. Use Figure 5.30 directly to determine $Y_1(q)$.

The calculations are based on formula (5.53):

$$L_n(q) = L_n(0.5) - Y_n(q)$$

50% Service Probability Case

Time Availability (%)	$Y_1(q)$ (dB)	Transmission Loss (dB)
50	0	218
90	−6	224
99	−12	230
99.9	−15	233
99.99	−17.5	235.5

Calculate $\sigma_c^2(0.5)$ (i.e., where $q = 0.50$ or 50%):

$$\sigma_c^2(0.5) = 12.73 + 0$$

$$\sigma_{rc}(0.5) = \sqrt{12.73 + 4}$$

$$= 4.09$$

$$z_{m0}(0.95) = 1.65 \quad \text{(from Figure 5.40)}$$

$$z_{m0}(0.95) \times \sigma_{rc}(0.5) = 1.65 \times 4.09$$

$$= 6.75 \text{ dB}$$

Calculate σ_c^2 (0.90):

$$\sigma_c^2(0.90) = 12.73 + 0.12 \times 6^2$$

$$= 12.73 + 4.32$$

$$= 17.05$$

$$\sigma_{rc}(0.90) = \sqrt{17.05 + 4}$$

$$= 4.59$$

$$z_{m0}(0.95) \times \sigma_{rc}(0.90) = 1.65 \times 4.59$$

$$= 7.57 \text{ dB}$$

and so on for the other values of q to complete the table.

EXTENDING TO A 95% SERVICE PROBABILITY

Time Availability (%)	**$z_{m0}(0.95) \times \sigma_{rc}(q)$ (dB)**	**Transmission Loss (dB)**
50	6.75	224.75
90	7.57	231.57
99	9.62	239.62
99.9	10.91	243.91
99.99	12.06	248.41

5.4.2.14 Aperture-to-Medium Coupling Loss

5.4.2.14.1 Definition. In Chapter 2 antenna gain was treated as free-space antenna gain or the ratio of the maximum radiated power density at a point in space to the theoretical maximum radiated power density of an isotropic antenna at the same relative point in space. No degradation factors were included. In tropospheric scatter and diffraction links such degradations cannot be neglected. Whereas LOS microwave antennas are seldom mounted near ground and antenna apertures are relatively small, in tropo and diffraction links antennas are mounted near ground level and their apertures are comparatively large. This gives rise to two forms of gain degradation:

- Distortion of wavefront due to antenna position
- Radiation pattern distortion due to ground reflections

These results in a loss of antenna gain when compared to free-space gain. We call this *aperture-to-medium coupling loss*. Reference 1 calls this multipath coupling loss and Ref. 2 calls it *loss in path antenna gain.*

In Chapter 2 we could define path antenna gain in decibels as the sum of the transmit and receive antenna gains in decibels or

$$G_p = G_t + G_r \tag{5.56}$$

For the transhorizon systems described in this chapter, we must take into account aperture-to-medium coupling loss, and path antenna gain is defined as

$$G_p = G_t + G_r - L_{gp} \tag{5.57}$$

where L_{gp} is the aperture-to-medium coupling loss or loss in path antenna gain.

There are two ways of treating this parameter:

1. As a loss added to the transmission loss.
2. As a degradation to antenna gains.

We prefer the former. Now we can treat antenna gains as we had previously in Chapter 2. But once we do this, we must add the value of L_{gp} to the path transmission loss.

5.4.2.14.2 Calculation of Aperture-to-Medium Coupling Loss. One common method of calculation is given in CCIR Rec. 617-1 (Ref. 4) and assumes each antenna gain does not exceed 55 dB and that there is no significant difference in gains between G_t and G_r. The following formula applies to the CCIR approach:

$$L_{gp} = 0.07 \exp[0.055(G_t + G_r)] \tag{5.58}$$

where G_t and G_r, the gains of the transmitting and receiving antennas, respectively, are given in decibels.

Another method of calculation is given in Ref. 2 and is shown in Figure 5.41. L_{gp} is the aperture-to-medium coupling loss. (Reference 1 can be consulted for still a third method of calculation.)

Example 4. Assume 15-ft parabolic reflector antennas at both ends of a 4-GHz transhorizon link. At 55% efficiency the antenna gains are 43 dB. What is the aperture-to-medium coupling loss?

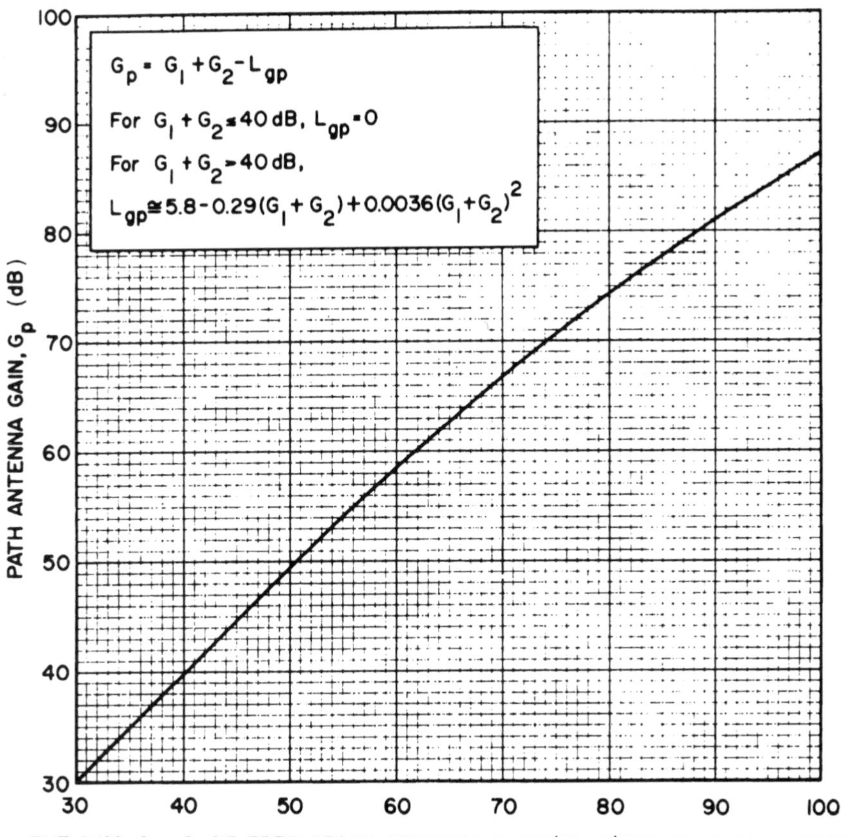

Figure 5.41. Path antenna gain G_p for transhorizon links (Ref. 2).

Method 1 (*CCIR*):

$$\begin{aligned} L_{gp} &= 0.07 \exp[0.055(43 + 43)] \\ &= 0.07 \exp(4.73) \\ &= 7.93 \text{ dB} \end{aligned}$$

Method 2 (*Ref*. 2):

$$\begin{aligned} L_{gp} &= 5.8 - 0.29(43 + 43) + 0.0036(43 + 43)^2 \\ &= 7.49 \text{ dB} \end{aligned}$$

5.4.2.15 *Sample Problem—Tropospheric Scatter Transmission Loss.* A tropospheric scatter path is 283.1 km long; the transmitting frequency is 5000 MHz; the radio refractivity N_0 is 315; the heights above sea level of the antennas are $h_{ts} = 0.2804$ km, $h_{rs} = 0.2439$ km; the elevations above MSL of the horizon points are $h_{Lt} = 0.2195$ km, $h_{Lr} = 0.2743$ km; distances from transmitter and receiver sites to their respective horizons are $d_{Lt} = 39.6$ km,

$d_{Lr} = 8.8$ km. Prepare tables of transmission loss values for both 50% and 95% service probabilities and path time availabilities from 90% to 99.99%. Assume Climatic Region 1.

Calculation of basic long-term median tropospheric scatter loss L_{bsr} (Section 5.4.2.5): Compute h_{st} and h_{sr}. Equations (5.3) and (5.4) are valid for this case and then $h_{st} = h_{Lt} = 0.2195$ km and $h_{sr} = h_{Lr} = 0.2743$ km.

Compute values of N_{st} and N_{sr} for h_{st} and h_{sr} [equation (5.7)]:

$$N_{st} = 315 \exp(-0.1057 \times 0.2195)$$
$$N_{st} = 307.8$$
$$N_{sr} = 315 \exp(-0.1057 \times 0.2743)$$
$$N_{sr} = 306$$

Compute N_s [equation (5.8)]:

$$N_s = 306.9$$

Compute effective earth radius a from Figure 5.7 given $N_s = 306.9$: $a = 8580$ km.

Compute the horizon takeoff angles θ_{et} and θ_{er} [equations (5.9) and (5.10)]:

$$\begin{aligned}\theta_{et} &= \frac{0.2195 - 0.2804}{39.6} - \frac{39.6}{2 \times 8580} \\ &= -0.00154 - 0.00231 \\ &= -0.00385 \text{ radian} \\ \theta_{er} &= \frac{0.2743 - 0.2439}{8.8} - \frac{8.8}{2 \times 8580} \\ &= 0.00345 - 0.00051 \\ &= 0.00294 \text{ radian}\end{aligned}$$

Compute the distance between horizon points [equation (5.11)]:

$$\begin{aligned}D_s &= 283.1 - 39.6 - 8.8 \\ &= 234.7 \text{ km}\end{aligned}$$

Compute the effective antenna heights h_{te} and h_{re}. Divide the distance to the radio horizon for the transmitter and receiver sites, respectively, into 31 uniform segments. Then compute h_t and h_r [equations (5.13)–(5.16)]:

$$d_{Lt} = 39.6 \text{ km} \quad \text{and} \quad d_{Lr} = 8.8 \text{ km}$$

$$\frac{39.6}{31} = \text{(approx.) } 1.3 \text{ km}$$

$$\frac{8.8}{31} = \text{(approx.) } 0.28 \text{ km}$$

Build two tables as shown below, starting at segment 3 and finishing at segment 27 showing altitude at each segment. (Normally this is taken from the topo maps used in the path profile.)

Table for Transmitter				Table for Receiver			
i		d_{tt} (km)	h_{ti} (m)	i		d_{ri} (km)	h_{ri} (m)
3	3 × 1.3	3.9	270	3	3 × 0.28	0.84	240
4	4 × 1.3	5.2	260	4	4 × 0.28	1.12	250
5	5 × 1.3	6.5	260	5	5 × 0.28	1.4	250
6	6 × 1.3	7.8	250	6	6 × 0.28	1.68	250
7	7 × 1.3	9.1	250	7	7 × 0.28	1.96	250
8	8 × 1.3	10.4	250	8	8 × 0.28	2.24	250
9	9 × 1.3	11.7	250	9	9 × 0.28	2.52	250
10	10 × 1.3	13	250	10	10 × 0.28	2.8	250
11	11 × 1.3	14.3	240	11	11 × 0.28	3.08	260
12	12 × 1.3	15.6	240	12	12 × 0.28	3.36	260
13	13 × 1.3	16.9	240	13	13 × 0.28	3.64	260
14	12 × 1.3	18.2	240	14	14 × 0.28	3.92	260
15	15 × 1.3	19.5	240	15	15 × 0.28	4.2	260
16	16 × 1.3	20.8	240	16	16 × 0.28	4.48	260
17	17 × 1.3	22.1	240	17	17 × 0.28	4.76	260
18	18 × 1.3	23.4	240	18	18 × 0.28	5.04	260
19	19 × 1.3	24.7	250	19	19 × 0.28	5.32	260
20	20 × 1.3	26	250	20	20 × 0.28	5.6	260
21	21 × 1.3	27.3	240	21	21 × 0.28	5.88	260
22	22 × 1.3	28.6	240	22	22 × 0.28	6.16	260
23	23 × 1.3	29.9	240	23	23 × 0.28	6.44	260
24	24 × 1.3	31.2	230	24	24 × 0.28	6.72	260
25	25 × 1.3	32.5	230	25	25 × 0.28	7.0	260
26	26 × 1.3	33.8	230	26	26 × 0.28	7.28	260
27	27 × 1.3	35.1	230	27	27 × 0.28	7.56	270
		Total	6100			Total	6420

$$\overline{h}_{ti} = \frac{6100}{25} = 244 \text{ m or } 0.244 \text{ km} \qquad \overline{h}_{r} = \frac{6420}{25} = 256.8 \text{ m or } 0.2568 \text{ km}$$

In the case of the transmit site, h_{t0} is larger than $\overline{h}_{ti}$; then

$$h_t = h_{ts} - \overline{h}_t \quad \text{[equation (5.13)]}$$

$$h_t = 0.2804 - 0.244 = 0.0364 \text{ km}$$

In the case of the receiver site, the reverse is true, and

$$h_r = h_{rs} - h_{r0}$$
$$= 0.2439 - 0.2429$$
$$= 10 \text{ m} \quad \text{or} \quad 0.010 \text{ km}$$

(where h_{r0} is the ground elevation at the receiver site, about 10 m below antenna center, where h_{rs} elevation is measured).

Because h_{t0} and h_{r0} are less than 1 km, use equation (5.17) and $h_{te} = h_t$ and $h_{re} = h_r$:

$$h_{te} = 0.0364 \text{ km}$$
$$h_{re} = 0.010 \text{ km}$$

Calculate the angular distance θ (steps 1t through 11t):

1t. Compute $\theta_{0t,0r}$ [equations (5.19) and (5.20)]:

$$\theta_{0t} = -0.00385 + 39.5/8580$$
$$= -0.00385 + 0.0046$$
$$= 0.00075 \text{ radian}$$
$$\theta_{0r} = 0.00294 + 8.8/8580$$
$$= 0.00396 \text{ radian}$$

2t. Calculate α_{00} and β_{00} [equations (5.21) and (5.22)]:

$$\alpha_{00} = 0.0165 - 0.00385 + \frac{0.2804 - 0.2439}{283.1}$$
$$= 0.01278 \text{ radian}$$
$$\beta_{00} = 0.0165 + 0.00294 + \frac{0.2439 - 0.2804}{283.1}$$
$$= 0.0193 \text{ radian}$$

3t. Compute θ_{00} [equation (5.23)]:

$$\begin{aligned}\theta_{00} &= \alpha_{00} + \beta_{00} \\ &= 0.01278 + 0.0193 \\ &= 0.03208 \text{ radian}\end{aligned}$$

4t. Compute d_{st} and d_{sr} [equations (5.24) and (5.25)]:

$$\begin{aligned}d_{st} &= \frac{283.1 \times 0.01278}{0.03208} - 39.6 \\ &= 73.2 \text{ km}\end{aligned}$$

$$\begin{aligned}d_{sr} &= \frac{283.1 \times 0.0193}{0.03208} - 8.8 \\ &= 161.5 \text{ km}\end{aligned}$$

5t. Determine $\Delta\alpha_0$ and $\Delta\beta_0$ as a function of θ_{0t} and θ_{0r}. θ_{0t} and θ_{0r} are greater than 0 but less than 0.1; thus proceed to steps 6t, 7t, and 10t.

6t. From Figure 5.9 determine $\Delta\alpha_0$ and $\Delta\beta_0$ for $N_s = 301$:

$$\Delta\alpha_0 = 0$$

$$\Delta\beta_0 = 0.5 \text{ milliradian} \quad \text{or} \quad 0.0005 \text{ radian}$$

7t. Read $C(N_s)$ from Figure 5.10; $N_s = 306.9$ from above:

$$\begin{aligned}C(N_s) &= 1.05 \\ \Delta\alpha_0(N_s) &= 0 \\ \Delta\beta_0(N_s) &= 1.05 \times 0.0005 \\ &= 0.000525\end{aligned}$$

(Note: Steps 8t and 9t are not required in this example.)

10t. Calculate α_0 and β_0 using equations (5.31) and (5.32) (correction factors are from step 7t):

$$\alpha_0 = \alpha_{00} + 0 = 0.01278 \text{ radian}$$

$$\beta_0 = \beta_{00} + 0.000525 = 0.01982 \text{ radian}$$

11t. Calculate the scatter angle (angular distance) θ:

$$\theta = 0.01278 + 0.01982$$
$$= 0.0326 \text{ radian}$$

12t. Calculate the path asymmetry factor s:

$$s = \frac{0.01278}{0.01982}$$
$$= 0.645$$

Calculate the product θd, where θ is the scatter angle or angular distance and d is the path length. The units must be in radians and kilometers, respectively.

$$\theta d = 0.0326 \times 283.1$$
$$= 9.23 \text{ km-radian}$$

Read $F(\theta d)$ from Figure 5.11 ($N_s = 306.9$):

$$F(\theta d) = 167.5 \text{ dB}$$

Compute the basic median tropospheric scatter transmission loss L_{bsr}:

$$L_{bsr} = 30 \log 5000 - 20 \log 283.5 + 167.5 \text{ dB} + 2.7 \text{ dB}$$

$$L_{bsr} = 232.12 \text{ dB}$$

(The value for A_a is taken from Figure 5.4.)

Section 5.4.2.3 tells us that if the angular distance exceeds 20 milliradians, the path is predominantly tropospheric scatter. This path is 32.6 milliradians, significantly greater than 20 milliradians. Now we can say that $L_{cr} = L_{bsr} =$ 232.12 dB.

Calculate the effective distance d_e (Section 5.4.2.10):

$$d_{sl} = 65\left(\frac{100}{5000}\right)^{1/3} \quad \text{[equation (5.49)]}$$

$$= 17.67 \text{ km}$$

$$d_L = 3\sqrt{2 \times 0.0364 \times 10^3} + 3\sqrt{2 \times 0.010 \times 10^3} \quad \text{[equation (3.50)]}$$

$$= 25.6 + 13.42$$

$$= 39 \text{ km}$$

Because d is greater than $d_{sl} + d_L$,

$$d_e = 130 + 283.1 - 17.67 - 39 \quad \text{[equation (5.50b)]}$$

$$= 356 \text{ km}$$

Calculate the basic long-term median transmission loss $L_n(0.5, 50)$ [equation (5.51)] for Climatic Region 1:

$$L_n(0.5, 50) = 232.12 \text{ dB} - 3 \text{ dB} \quad \text{(Figure 5.25)}$$

$$= 229.12 \text{ dB}$$

Calculate values for $Y_1(q, 356)$ and arrange them in tabular form for values of q. Use Figure 5.30 directly.

TRANSMISSION LOSS TABLE: 50% SERVICE PROBABILITY

q (%)	Y_1 (dB)	Transmission Loss (dB)
50	0	229.12
90	5.5	234.62
99	10	239.12
99.9	13.5	242.62
99.99	16.5	245.62

Prepare a second table for the 95% service probability case. From Figure 5.40, z_{m0} is 1.65. Use equations (5.54) and (5.55).

TRANSMISSION LOSS TABLE: 95% SERVICE PROBABILITY

q (%)	$z_{m0}(95)\sigma_{r0}(q)$ (dB)	Transmission Loss (dB)
50	6.75	235.8
90	7.44	242.1
99	8.84	248.0
99.9	10.25	252.9
99.99	11.6	257.22

5.5 PATH CALCULATION / LINK ANALYSIS

5.5.1 Introduction

We have shown how to calculate transmission loss for over-the-horizon paths, both tropospheric scatter and diffraction. In either case, the calculation is complex. For LOS paths the calculation is rather simple.

Once we have the transmission loss that includes fade margin (long term), the approach to link analysis follows the same methodology as LOS. There is one exception: multipath.

5.5.2 Path Intermodulation Noise—Analog Systems

For an analog LOS system there were two analog noise components: equipment noise and feeder distortion (waveguide distortion). With tropo and diffraction paths we must consider one more intermodulation (IM) noise component, namely, path IM noise.

Compute which is the maximum delay of the multiecho as compared to the main beam, expressed in seconds:

$$\Delta = 5.21d(\Omega_t + \Omega_r)(4\alpha_0 + 4\beta_0 + \Omega_t + \Omega_r) \times 10^{-8} \tag{5.59}$$

where d = path length in kilometers

α_0, β_0 = main beam angles (radians) of the transmitting and receiving antennas, respectively; these were defined in Section 5.4.2.5

Ω_t, Ω_r = 3-dB beamwidths (radians) of the transmitting and receiving antennas, respectively; read beamwidths from Figure 5.42.

Calculate D, the rms composite deviation, for values of M = 2, 3, 4, 6, 8, and 9; M is the peak deviation ratio:

$$D = \frac{Mf_m}{4.46} \tag{5.60}$$

where f_m is the maximum baseband frequency. This value may be taken from Table 5.3.

Compute $\gamma_{0.5}$ and $\gamma_{0.01}$, the phase distortion (radians) corresponding to values of D (taken from the previous step) not exceeded more than 50% and 1% of the time:

$$\gamma_{0.5} = 8(\Delta D)^2 \tag{5.61}$$

$$\gamma_{0.01} = 2600(\Delta D)^2 \tag{5.62}$$

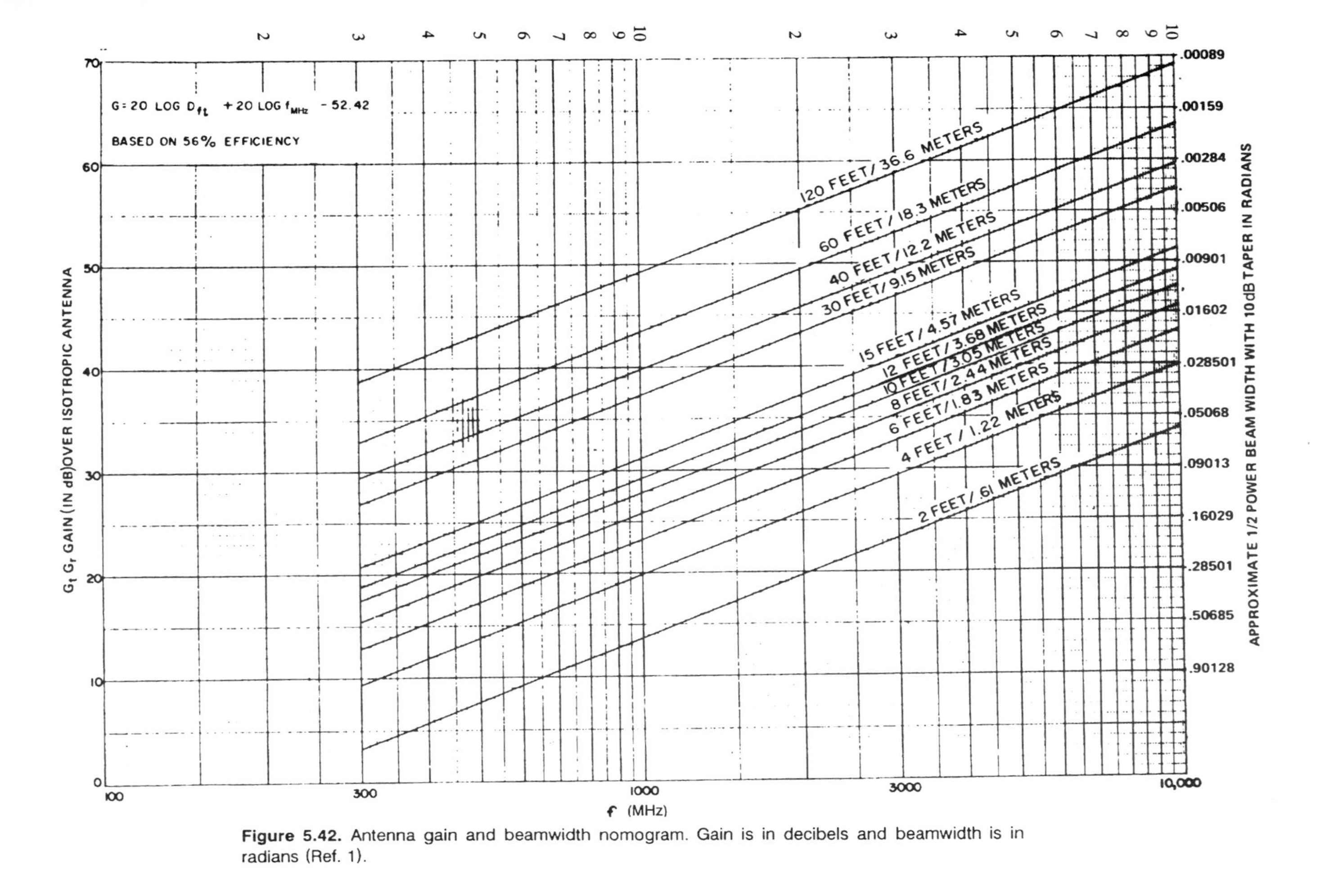

Figure 5.42. Antenna gain and beamwidth nomogram. Gain is in decibels and beamwidth is in radians (Ref. 1).

TABLE 5.3 Various System Parameters Calculated as a Function of N, the Number of FDM 4-kHz Voice Channels (Based on CCIR Standards)

N^a	Baseband Spectrum (Δf_b) (kHz)^b	Baseband Spectrum (Δf_b) (kHz)^c	BWR (dB)^d	NLR (dBm0)^e	$I(N)$ rms^f	$L(N)$ (dBm0)^g	$I(N)$ Peak^h	C (dBp)^i
12	12–60		11.90	3.3	1.46	16.3	6.53	11.10
24	12–108		14.91	4.5	1.68	17.5	7.50	12.91
36	12–156	60–204	16.67	5.2	1.82	18.2	8.13	13.97
48	12–204	60–252	17.92	5.72	1.93	18.72	8.63	14.70
60	12–252	60–300	18.89	6.11	2.02	19.11	9.03	15.30
72	12–300	60–360	19.86	6.43	2.09	19.42	9.46	15.93
84		60–408	20.50	6.7	2.16	19.7	9.66	16.30
96		60–456	21.06	6.92	2.22	19.92	9.91	16.86
108		60–504	21.60	7.1	2.27	20.1	10.11	17.00
120		60–552	22.01	7.3	2.32	20.3	10.35	17.21
180		60–804	23.80	8.02	2.57	21.0	11.2	18.28
240		60–1052	25.05	8.80	2.75	21.80	12.3	18.75
300		60–1300	26.02	9.77	3.08	22.77	13.0	18.75
600		60–2540	29.03	12.78	4.36	25.78	12.46	18.75

[a] N = number of 4-kHz channels.
[b] Baseband spectrum beginning at 12 kHz.
[c] Baseband spectrum beginning at 60 kHz.
[d] BWR (dB) = bandwidth ratio = 10 log(occupied baseband bandwidth/3.1 kHz).
[e] NLR (dBm0) = Noise loading ratio
= $-1 + 4 \log N$ for $12 < N < 240$ (CCIR Standard)
= $-15 + 10 \log N$ for $N \geq 240$ (CCIR Standard)
= $-10 + 10 \log N$ for all N (DCA Standard)
[f] $I(N)$ rms = antilog(NLR/20).
[g] $L(N)$ (dBm0) = NLR + 13(peak loading factor).
[h] $I(N)$ peak = antilog $[L(N)/20]$.
[i] C(dBp) = BWR − NLR + I_w(2.5 dB) (multichannel to single-channel conversion factor).

Determine $H(\gamma_{0.5})$ and $H(\gamma_{0.01})$ for values of D. Read values from Figure 5.43. Note that for $\gamma < 0.05$, $H(\gamma) = \gamma^2$.

Compute R for values of D, where R is a factor for computing $(S/N)_p$:

$$R + 0.288\left(\frac{f_m}{D}\right)^2 \tag{5.63}$$

Compute $(S/N)_p$ for various values of $H(\gamma)$ and D. $(S/N)_p$ is the signal-to-path IM noise ratio in dBp:

$$\left(\frac{\mathrm{S}}{\mathrm{N}}\right)_{\mathrm{p}} = 10 \log[RH(\gamma)] + C + I_d \tag{5.64}$$

where C is the multichannel to single-channel conversion factor, which is taken from Table 5.3 and I_d is the diversity improvement factor. For I_d, use one of the following values applicable for the equipment to be used on the link: postdetection combining, dual diversity—3.0 dB; quadruple diversity—6.0 dB; predetection combining, dual diversity—2.5 dB; quadruple diversity—5.0 dB.

Convert $(S/N)_p$ to N_p (pWp).

$$N_p = \text{antilog}\frac{1}{10}\left[90 - \left(\frac{S}{N}\right)_p\right] \tag{5.65}$$

For the value of N_p for worst-hour performance, use the equivalent value of $H(\gamma_{0.01})$.

Example 5. Assume the previous example in Section 5.4.2.15 and that the antennas have 30-ft apertures; α_0 and β_0 are 0.01278 and 0.01982 radian, respectively; 60 FDM voice channels occupying a baseband of 12–252 kHz (Table 5.3); and let $M = 3$ for simplification.

From Figure 5.42, Ω_t and Ω_r are 0.008 radian each. Then

$$\begin{aligned}\Delta &= 5.2(283.1)(0.008 + 0.008)\\ &\quad \times (4 \times 0.01278 + 4 \times 0.01982 + 2 \times 0.008) \times 10^{-8}\\ &= 5.2(283.1)(0.016)(0.146) \times 10^{-8} \quad \text{[equation (5.59)]}\\ &= 3.44 \times 10^{-8}\ \text{s}\end{aligned}$$

Calculate D, the rms composite deviation [equation (5.60)]:

$$D = \frac{3 \times 252 \times 10^3}{4.46}$$

$$D = 169{,}507\ \text{Hz}$$

Compute

$$\begin{aligned}\gamma_{0.01} &= 2600(3.44 \times 10^{-8} \times 169{,}507)^2\\ &= 0.0884\ \text{radian}\end{aligned}$$

Determine $H(\gamma_{0.01})$ from Figure 5.43. The value is 7×10^{-3}.

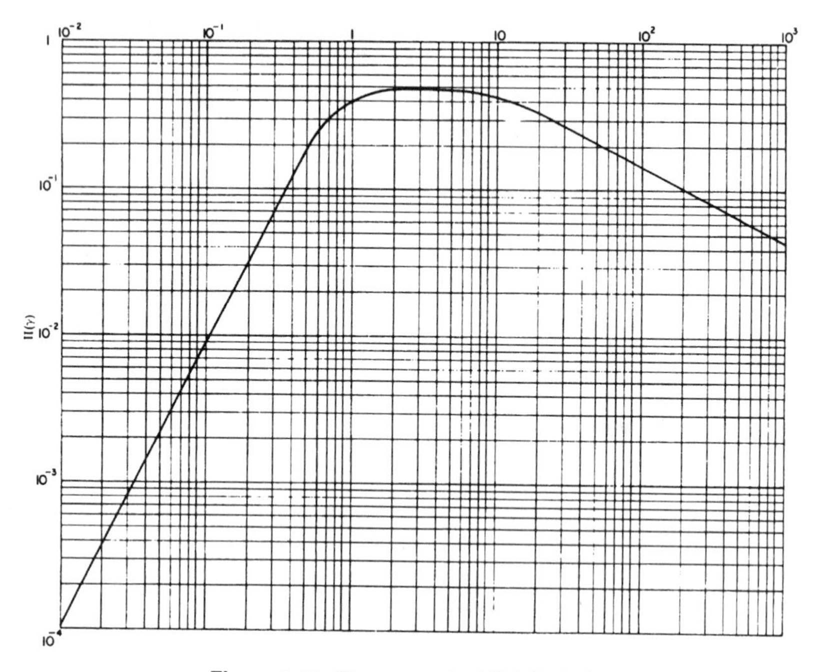

Figure 5.43. The parameter $H(\gamma)$ (Ref. 1).

Compute R for the value of D (D = 169,507 Hz); use equation (5.63):

$$R = 0.288\left(\frac{252 \times 10^3}{169{,}507}\right)^2$$

$$= 0.636$$

Compute $(S/N)_p$ [equation (5.64)]:

$$\left(\frac{S}{N}\right)_p = 10\log(0.636)(7 \times 10^{-3}) + 15.3 + 6$$

(The value for I_d assumes postdetection combining, quadruple diversity; therefore I_d = 6.0 dB.)

$$\left(\frac{S}{N}\right)_p = +235 + 15.3 + 6$$

$$= 44.8 \text{ dB}$$

$$N_p = 33{,}113 \text{ pWp } 1\% \text{ of the time}$$

If we calculate it for the median (50% of the time) [equation (5.61)], we arrive at a value of 0.346 pWp. This latter value would be used for total noise calculation.

5.5.3 Sample Link Analysis

Let us treat the link given in Section 5.2.4.15 for two cases using 95% service probability: (1) 50% path availability and (2) 99% path availability. Transmitters are 1 kW, antennas have 30-ft apertures, line losses are 3 dB at each end, quadruple diversity is used, there is a 6-dB diversity improvement, operating frequency is 5000 MHz, there are 60 voice channels FDM/FM, and receiver noise figure is 3 dB.

Calculate EIRP with an antenna gain of 51 dB:

$$\begin{aligned}\text{EIRP} &= +30\ \text{dBW} - 3\ \text{dB} + 51\ \text{dB}\\ &= +78\ \text{dBW}\end{aligned}$$

Calculate the aperture-to-medium coupling loss using Method 2 in Section 5.4.2.14:

$$\begin{aligned}L_{gp} &= 5.8 - 0.29(51 + 51) + 0.0036(51 + 51)^2\\ &= 5.8 - 29.58 + 37.45\\ &= 13.67\ \text{dB}\end{aligned}$$

Add this value to the transmission losses for each case.

Compute the total transmission loss for each case:

Path time availability 50%: 235.8 dB + 13.67 dB = 249.47 dB

Path time availability 99%: 248.0 dB + 13.67 dB = 261.67 dB

Calculate the isotropic receive level (IRL) for each case:

$$\text{IRL}_{\text{dBW}} = +78\ \text{dBW} - 249.47\ \text{dB} = -171.47\ \text{dBW} \quad (50\%)$$

$$\text{IRL}_{\text{dBW}} = +78\ \text{dBW} - 261.67\ \text{dB} = -183.67\ \text{dBW} \quad (99\%)$$

Calculate the RSL for each case:

$$\mathrm{RSL}_{\mathrm{dBW}} = -171.47\ \mathrm{dBW} - 3\ \mathrm{dB} + 51\ \mathrm{dB}$$
$$= -123.47\ \mathrm{dBW} \quad (50\%)$$
$$\mathrm{RSL}_{\mathrm{dBW}} = -183.67\ \mathrm{dBW} - 3\ \mathrm{dB} + 51\ \mathrm{dB}$$
$$= -135.67\ \mathrm{dBW} \quad (99\%)$$

[*Note:* RSL = IRL − line losses + antenna gain (see Section 2.6.2).]

Calculate the FM improvement threshold of the far-end receiver (see Section 2.8.3):

$$P_{\mathrm{FM(dBW)}} = -204\ \mathrm{dBW} + 3\ \mathrm{dB} + 10\ \mathrm{dB} + 10 \log B_{\mathrm{IF}}$$

Calculate B_{IF} in hertz using Carson's rule. (Follow the procedures given in Section 2.6.6.) Table 5.3 gives the highest modulating frequency for a 60-channel configuration as 252 kHz. We must now calculate peak deviation ΔF_p [equation (2.24)].

Calculate NLR; equation (2.20) is used for 60-channel operation:

$$\mathrm{NLR}_{\mathrm{dB}} = -1 + 4 \log N \quad \text{where } N = 60$$
$$\mathrm{NLR} = -1 + 7.11$$
$$= 6.11\ \mathrm{dB}$$

Calculate peak deviation:

$$\Delta F_p = 4.47d\left[\log^{-1}(6.11/20)\right]$$

Use 200 kHz per channel from Table 2.4.

$$\Delta F_p = 4.47 \times 200 \times 2.02\ \mathrm{kHz}$$
$$= 1806\ \mathrm{kHz}$$

Apply Carson's rule:

$$B_{\mathrm{IF}} = 2(1806 + 252)\ \mathrm{kHz}$$
$$= 4116\ \mathrm{kHz}$$
$$P_{\mathrm{FM(dBW)}} = -204\ \mathrm{dBW} + 3\ \mathrm{dB} + 10\ \mathrm{dB} + 10 \log(4116 \times 10^3)$$
$$= -124.85\ \mathrm{dBW}$$

The link RSL should be equal to or greater than this value for both the 50% and 99% time availability cases. Compare the values:

		FM Improvement Threshold (dBW)	Margin (dB)
RSL (50%)	−123.47	−124.85	+1.38
RSL (99%)	−135.67	−124.85	−10.82

The preceding argument presented is one of the classic methods of carrying out a link analysis. A baseline system has been established. It does not meet requirements as shown in the preceding table: 50% of the time FM improvement threshold is achieved; 1% of the time we are short 10.82 dB. What alternatives are open to us?

1. Accept a reduced availability. This is probably unacceptable.
2. Increase transmitter output power. Increasing the output power to 10 kW would give an across-the-board link improvement of 10 dB.
3. Add threshold extension. (See Section 5.6 for a description of threshold extension.) With a modulation index of at least 3 ($M = 3$), a 7-dB link improvement can be achieved.
4. Increase antenna aperture. We must remember that as aperture increases, so does aperture-to-medium coupling loss. Thus there is no decibel-for-decibel improvement.
5. Reduce the number of voice channels, thus reducing bandwidth, which will improve threshold.
6. Resite or reconfigure links to reduce transmission loss.
7. Some marginal improvements can be achieved by lowering receiver noise temperature, using lower-loss transmission lines or higher-efficiency antennas. A thermal noise improvement can be achieved by implementing quadruple diversity rather than dual diversity.

The link design engineer can parametrically compare these alternatives using present worth of annual charges technique for commercial systems or life cycle costs technique for military systems. Present worth of annual charges is described in Ref. 9.

5.6 THRESHOLD EXTENSION

Threshold extension is a method of lowering the FM improvement threshold on a FM link by replacing a conventional FM modulator with a threshold extension demodulator. Assuming a modulation index of 3 ($M = 3$), the amount of improvement that can be expected over a conventional FM demodulator is about 7 dB. In the example given in Section 5.5.3, where the

FM improvement threshold was given as −124.85 dBW for the conventional demodulator, implementing a threshold extension demodulator would lower or extend the threshold to −131.85 dBW.

Threshold extension works on a FM feedback principle, which reduces the equivalent instantaneous deviation, thereby reducing the required bandwidth B_{IF}. This, in turn, effectively lowers the receiver noise threshold. A typical receiver with a threshold extension module may employ a tracking filter, which instantaneously tracks the deviation with a steerable bandpass filter having a 3-dB bandwidth of approximately four times the top baseband frequency. The control voltage for the filter is derived by making a phase comparison between the feedback signal and the IF input signal.

5.7 DIGITAL TRANSHORIZON RADIOLINKS*

5.7.1 Introduction

Digital tropospheric scatter and diffraction links are being implemented on all new military construction and will be attractive for commercial telephony as we approach the era of an all-digital network. The advantages are similar to those described in the previous chapter. However, on tropo and diffraction paths the problem of dispersion due to multipath can become acute. It can be handled—in fact, it can be turned to advantage with the proper equipment.

The design of a digital tropo or diffraction link in most respects is similar to the design of its analog counterpart. Siting, path profiles, and calculation of transmission loss, including aperture-to-medium coupling loss, use identical procedures as those previously described. However, the approach to link analysis and the selection of a modulation scheme differ. These issues will be discussed below.

5.7.2 Digital Link Analysis

Digital tropo/diffraction link analysis can be carried out by a method that is very similar to that described in Section 3.9:

- Calculate the EIRP.
- Algebraically add the transmission loss (Section 5.4.2).
- Add the receiving antenna gain (G_r).
- Algebraically add the line losses incurred up to the input of the LNA (low-noise amplifier).

We must account for the aperture-to-medium coupling loss. As was previously suggested, we can add this value to the transmission loss. Another

*Section 5.7 is based on ITU-R Rec. F.1106 Annex 1, 1994 F Series Volume, Part 1 (Ref. 10).

method is to subtract this loss from the receiving antenna gain (G_r), but we must account for aperture-to-medium coupling loss.

One point of guidance: the analysis will follow a single string on a link. We mean here that the EIRP is calculated for a single transmitter and its associated antenna and then down through a single diversity branch on the receiving side as though it were operating alone.

The objective of this exercise is to calculate E_b/N_0. From equation (3.4):

$$\frac{E_b}{N_0} = \text{RSL}_{\text{dBW}} - 10\log(\text{BR}) - (-204\ \text{dBW} + \text{NF}_{\text{dW}}) \tag{5.66}$$

[i.e., the first two terms in equation (5.66) represent E_b and the last two terms represent N_0, where the subtraction sign implies logarithmic division]. In equation (5.66) BR is the bit rate in bits per second and NF_{dB} is the noise figure of the receive chain or the noise figure of the LNA, the latter being sufficient in most cases.

Calculate RSL [similar to equation (3.18)]

$$\text{RSL}_{\text{dBW}} = \text{EIRP}_{\text{dBW}} + T_L + G_r + L_{Lr} \tag{5.67}$$

where T_L is the transmission loss including aperture-to-medium coupling loss, G_r is the gain of the receiving antenna, and L_{Lr} is the total line losses from the antenna feed to the input of the LNA.

In tropo/diffraction receiving systems, in addition to waveguide loss, there are duplexer losses, preselector filter loss, transition losses, and possibly others.

Example 6. Refer to the example link analysis discussed in Section 5.5.3. Assume a digital link with a bit rate of 1.544 Mbps operating at 5000 MHz requiring a 99% time availability and 95% service probability; the transmission loss is 248.0 dB, the aperture-to-medium coupling loss = 13.67 dB, and, therefore, T_L = 261.67 dB; 30-ft dishes are used on each end of the link; the receiver noise figure is 3 dB; line losses are 3 dB at each end; modulation is QPSK; and transmitter output is 10 kW. Calculate E_b/N_0 and BER using Figure 5.44.

First calculate EIRP [equation (2.6)] and then RSL [equation (3.18)]:

$$\begin{aligned}\text{EIRP}_{\text{dDW}} &= 10\log 10{,}000 - 3\ \text{dB} + 51\ \text{dB} \\ &= +88\ \text{dBW} \\ \text{RSL} &= +88\ \text{dBW} - 261.67\ \text{dB} + 51\ \text{dB} - 3\ \text{dB} \\ &= -125.67\ \text{dBW}\end{aligned}$$

Use equation (3.14) to calculate E_b/N_0:

$$\begin{aligned}E_b/N_0 &= -125.67\ \text{dBW} - 10\log 1.544 \times 10^6 + 204\ \text{dBW} - 3\ \text{dB} \\ &= 13.44\ \text{dB}\end{aligned}$$

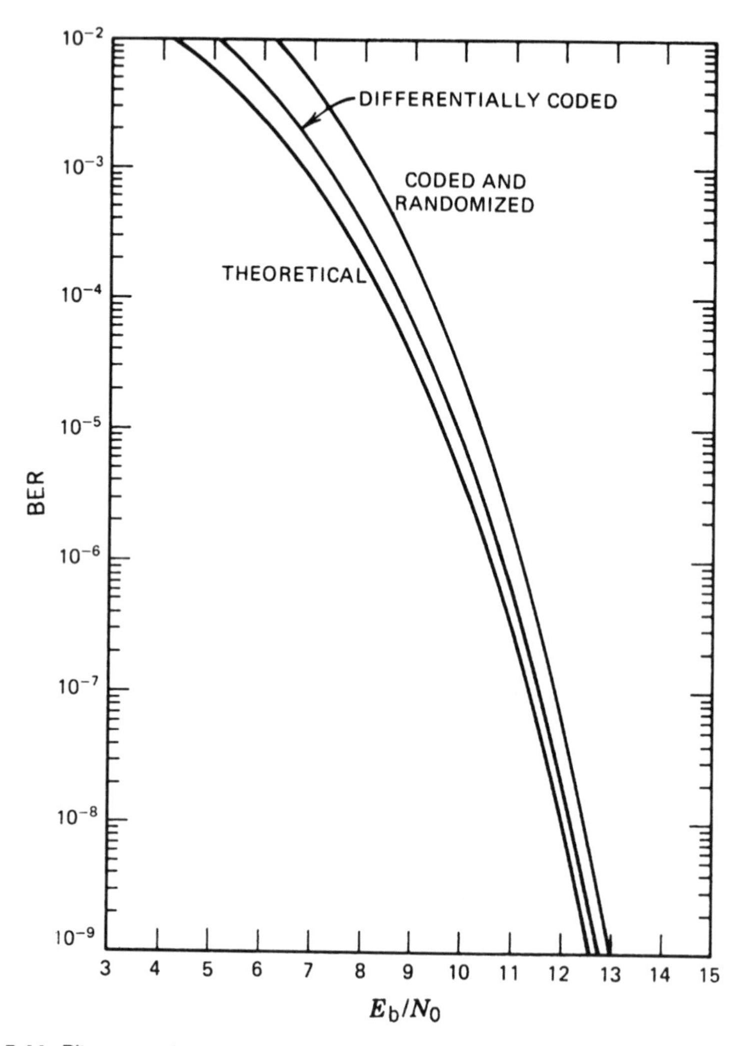

Figure 5.44. Bit error probability versus E_b/N_0 for coherent BPSK and QPSK (Ref. 9, Section 9-10.2).

If we allow a 3-dB modulation implementation loss, we are left with a net E_b/N_0 of 10.44 dB. Turning to Figure 5.44 we find the equivalent BER on the theoretical curve to be 1×10^{-6}. We would expect to have this value BER 99% of the time.

5.7.3 Dispersion

Dispersion is the principal cause of degradation of BER on digital transhorizon links. With conventional waveforms such as BPSK, MPSK, BFSK, and

MFSK, dispersion may be such, on some links, that BER performance is unacceptable.

Dispersion is simply the result of some signal power from an emitted pulse that is delayed, with that power arriving later at the receiver than other power components. The received pulse appears widened or smeared or what we call dispersed. These late arrival components spill over into the time slots of subsequent pulses. The result is intersymbol interference (ISI), which deteriorates bit error rate.

Expected values of dispersion on transhorizon paths vary from 30 to 350 ns. The cause is multipath. The delay can be calculated from equation (5.59). This equation shows that delay is a function of path length, antenna beamwidth, and the scatter angle components α_0 and β_0.

5.7.4 Some Methods of Overcoming the Effects of Dispersion

One simple method to avoid overlapping pulse energy is to time-gate the transmitted energy, which allows a resting time after each pulse. Suppose we were transmitting a megabit per second and we let the resting time be half a pulse width. Then we would be transmitting pulses of 500 ns of pulse width, and there would be a 500-ns resting time after each pulse, time enough to allow the residual delayed energy to subside. The cost in this case is a 3-dB loss of emitted power.

A two-frequency approach taken to reach the same objective in the design of the Raytheon AN/TRC-170 DAR modem, which is the heart of this digital troposcatter radio terminal, is to transmit on two separate frequencies alternatively gating each. The two-frequency pulse waveform is simply the time interleaving of two half-duty cycle pulse waveforms, each on a separate frequency. This technique offers two significant advantages over the one-frequency waveform. First, the two signals (subcarriers) are interleaved in time and are added to produce a composite transmitted signal with nearly constant amplitude, thereby nearly recovering the 3 dB of power lost due to time-gating. The operation of this technique is shown in Figure 5.45.

The second advantage is what is called intrinsic or implicit diversity. This can be seen as achieved in two ways. First, the residual energy of the "smear" can be utilized, whereas in conventional systems it is destructive (i.e., causes intersymbol interference). Second, on lower bit rate transmission, where the bit rate is R, R is placed on each subcarrier, rather than $R/2$ for the higher bit rates. The redundancy at the lower bit rates gives an order of in-band diversity. The modulation on the AN/TRC-170 is QPSK on each subcarrier. The maximum data rate is 4.608 Mbps, which includes a digital orderwire and service channel.

The AN/TRC-170 operates in the 4.4–5.0-GHz band with a transmitter output power of 2 kW. The receiver noise figure is 3.1 dB. In its quadruple-diversity configuration with 9.5-ft antennas and when operating at a trunk bit rate of 1.024 Mbps, the terminal can support a path loss typically of 240 dB

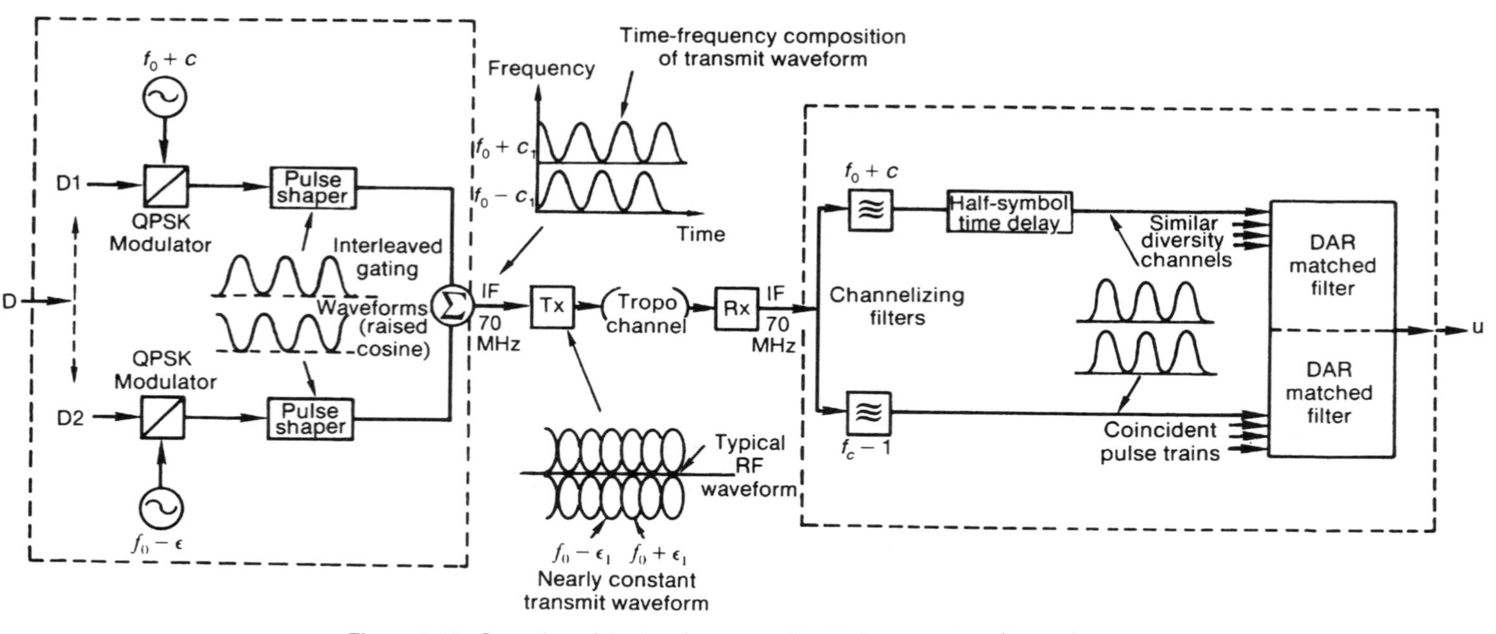

Figure 5.45. Operation of the two-frequency AN/TRC-170 modem (Ref. 11).

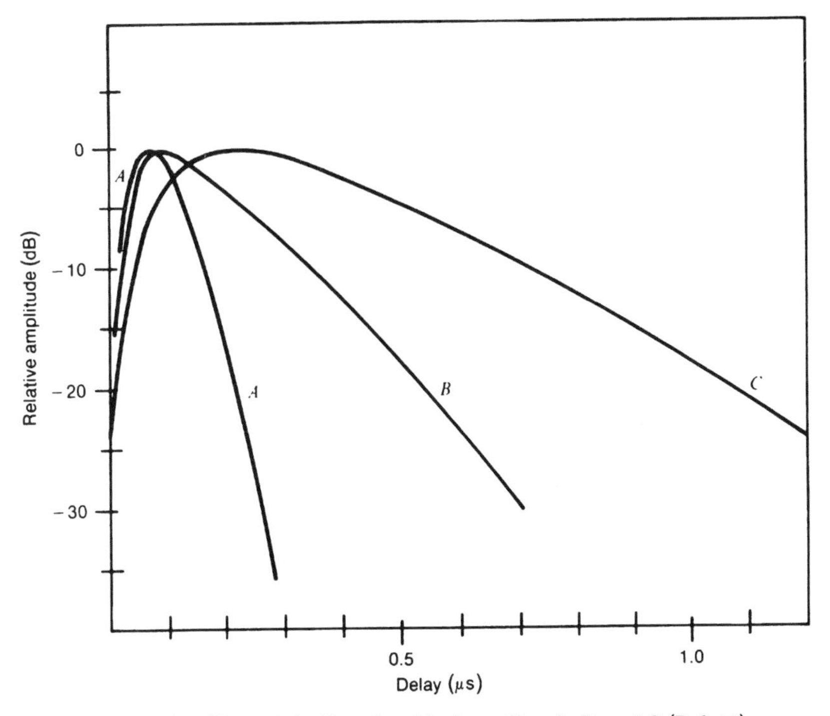

Figure 5.46. Characterization of multipath profiles *A*, *B*, and *C* (Ref. 12).

(BER = 1×10^{-5}). This value is based on an implicit diversity advantage for a multipath delay spread typical of profile *B* (see Figure 5.46). On a less dispersive path based on profile *A*, the performance would be degraded by about 2.6 dB or a transmission loss of 237.4 dB. With a more dispersive path based on profile *C*, we would expect a 1.1-dB improvement over that of profile *B* (Ref. 12).

The more dispersive the path, the better the equipment operates up to about 1 μs of rms dispersion. This maximum value would be shifted upward or downward depending on the data rate.

Three multipath profiles are shown in Figure 5.46; rms values for the multipath delay spread of each profile are as follows:

Profile *A*: 65 ns
Profile *B*: 190 ns
Profile *C*: 380 ns

5.7.5 Some ITU-R Perspectives on Transhorizon Radio Systems

5.7.5.1 Space Diversity in Troposcatter Systems. The theoretical diversity distances have been examined in terms of horizontal correlation distance D_h normal to the path, the horizontal correlation distance D_a parallel to (along) the path, and the vertical correlation distance D_v. D_h is the parameter most often used:

$$D_h = 3\lambda a/4d \tag{5.68}$$

where d = path distance
a = effective earth radius (see Section 1.3.3 and Figure 1.8)
λ = wavelength

Many troposcatter systems use the value 100λ for D_h.

5.7.5.2 *Angle Diversity in Troposcatter Systems.* Angle diversity is a useful method that enables maximum use of a limited frequency band. Some United Kingdom test results between 1.7 and 2.7 GHz over paths of 250–350 km show that with antennas in the range 18–25 m in diameter, correlation coefficients of 0.2 and 0.4 are obtained with two radio beams separated by approximately one beamwidth, with the lower figure resulting from vertical angle diversity. These results agree closely with theory. A significant medium-term decorrelation was apparent from the trials, as was a marked tendency for the short-term cross-correlation coefficient between beams to fall with decreasing signal levels. The former effect is unique to angle diversity. The overall performance of a correctly engineered angle diversity link can be expected to approach that of frequency diversity despite the small increase in transmission loss attributable to necessary engineering compromises.

Theoretical considerations predict that angle diversity achieves most effective results at large scatter angles.

Further confirmation of the efficiency of angle diversity was obtained in trials in Japan at 1.8 GHz over a 256-km path with 19-m diameter antennas (gain = 47.5 dB), where correlation coefficients of less than 0.4 were obtained for beam separations more than about 6 mrad.

5.7.5.3 *Effects of Multipath Dispersion on the Performance of Digital Transhorizon Systems*

5.7.5.3.1 Introduction to Time Dispersion Due to Multipath. The multipath dispersion over a digital transhorizon link is considerably worse than the corresponding dispersion over a LOS link for a given frequency band. Consequently, the onset of multipath dispersion, rather than signal-to-noise ratio as the limiting factor in communications performance, occurs at a lower transmission rate for transhorizon links than for LOS links.

A significant improvement in the performance can be obtained by intrinsic diversity improvement using adaptive equalization.

Predictions of (1) the increase in error probability due to multipath dispersion for a given set of equipment and path parameters, and (2) the improvements in diversity performance and/or the reduction of intersymbol interference that can be achieved by adaptive equalization, are expressed with respect to σ, the rms standard deviation of the delay time, which is obtained by expressing the power impulse response of the transhorizon path as a probability density function.

5.7.5.3.2 Performance Predictions and Measurements. The average bit error probability P_e in a digital transhorizon link subject to multipath dispersion is dependent on the following parameters:

$2\sigma/T$ Normalized multipath dispersion (where T is symbol period)
W (Energy per bit/noise spectral density) per diversity channel

For "weak multipath dispersion" ($2\sigma \ll T$), no special protection against intersymbol interference is required as normal multiple diversity reception techniques guarantee the requisite probability level of irreducible errors.

In the case of "medium multipath dispersion" ($2\sigma < T$), passive measures against antisymbol interference are used in addition to diversity reception. A combination of various passive methods with matched filtering of the multipath signal characterizes the active methods used to combat intersymbol interference when the elimination of intersymbol interference is accompanied by the effect of implicit diversity in multipath dispersion.

With "strong multipath dispersion" ($2\sigma > T$) the intersymbol interference can only be combated by using special adaptive methods of reception.

In practice, both short-term and long-term variations are observed in the multipath dispersion. Measurements performed in the United Kingdom indicate that for a median 2σ value of 106 ns, the standard deviations of the long-term variation in 2σ taken from 92-s averages of 10-ms samples was 15 ns. The standard deviation of the short-term variation in 2σ taken from the 10-ms samples was 50 ns. For the 124-km test link transmitting 2048 kbps with 4-CPSK, the median value of $2\sigma/T$ was 0.1.

5.7.5.3.3 Adaptive Equalization. The use of adaptive equalization reduces intersymbol interferences and thereby increases the transmission capacity of digital transhorizon systems. Adaptive equalization permits the diversity inherent within multipath dispersion to be accommodated in receiver design, thereby resulting in a predictable improvement in error performance for a given value of σ.

Ideally, adaptive equalization (linear equalization) in a multipath radio channel presupposes a cascade connection of the filter matched with the incoming signal and the transversal filter. However, in real conditions the input filter is matched with the transmitted signal so that, with the elimination of the influence of multisymbol interference, the implicit diversity effect cannot be adequately achieved.

The use of decision feedback is a nonlinear method of signal processing and it can be used to compensate for the intersymbol interference caused by precursor signal elements.

Reception with evaluation of a discrete sequence by means of the Viterbi algorithm is considered to be a method for solving, with a maximum of *a posteriori* probability, the problem of evaluating the sequence of a time-discrete Markov process with a finite number of states. When account is taken of the matched filtering of the multipath signal, the Viterbi algorithm is considered to be the method that offers optimum signal reception in a multipath communication channel.

Spectral processing of a multipath signal involves extracting a number of signal frequency bands at the receiving end when the signal spectrum is wider than the frequency correlation bandwidth of the communication channel. Combining signal samples in the frequency range with specific weighting coefficients makes it possible not only to eliminate intersymbol interference but also—with matched filtering of the output component signal—to achieve an implicit diversity effect.

Successful transmission of information rates of up to 12.6 Mbps at 4.6 GHz using adaptive modems over transhorizon links has been reported.

5.8 TROPOSCATTER FREQUENCY BANDS AND THE SHARING WITH SPACE RADIO-COMMUNICATION SYSTEMS*

Radio Regulations (RR) Article 27 is not only applicable to LOS microwave but also to transhorizon radio systems. Its main points are that EIRP shall not exceed +55 dBW and that the power delivered by a transmitter to an antenna shall not exceed +13 dBW in the fixed or mobile service in the frequency bands between 1 and 10 GHz. Most transhorizon systems exceed these limits.

RR Article 27 recognizes that transhorizon systems in the ranges 1700–1710 MHz, 1970–2010 MHz, 2025–2110 MHz, and 2200–2290 MHz may exceed the EIRP limit and the transmitter power limit.

5.8.1 Frequency Bands Shared with Space Services (Space-to-Earth)

Frequency bands of main interest lie in the ranges 1525–2500 MHz, 2500–2690 MHz, and 3400–7750 MHz. According to RR Article 28, the

*Section 5.8 is based on ITU-R Rec. F.698-2, 1994 F Series Volume, Part 1 (Ref. 13).

relevant frequency bands below 5 GHz are more specifically:

- Between 1525 and 2500 MHz:
  1525–1530 MHz (for Regions 1 and 3)
  1670–1690 MHz
  1690–1700 MHz (for certain countries)
  1700–1710 MHz
  2025–2110 MHz
  2200–2300 MHz
- Between 2500 and 2690 MHz:
  2500–2690 MHz
- Above 3400 MHz:
  3400–4200 MHz
  4500–4800 MHz

The relevant space radiocommunication services are the fixed-satellite, broadcasting-satellite, meteorological-satellite, space operation, and space research services. It is necessary to consider both interference to transhorizon radio-relay systems caused by space stations in space services and interference to earth stations in space services caused by transhorizon radio-relay systems. It should also be taken into account that not all satellites are in the geostationary-satellite orbit.

It should be further noted that, according to RR Article 8, the bands 1525–1559 MHz, 1626.5–1660.5 MHz, 2160–2200 MHz, 2483.5–2500 MHz, and 2500–2520 MHz are allocated to the mobile-satellite services (space-to-earth) and the coordination and notification procedures for these bands are laid down, but the absolute maximum limits of power flux density produced by space stations are not defined.

PROBLEMS AND EXERCISES

1 Show diagrammatically the "common volume" and "scatter angle."

2 There are two types of fading encountered on tropospheric scatter links. Define and name the types of distributions (statistical distributions of fade events) encountered.

3 Given smooth earth conditions and site elevations of transmitter and receiver of 150 and 250 m above MSL, respectively, what are the takeoff angles at the transmitter and receiver sites?

4 Given sea level refractivity of 310, calculate surface refractivity for a tropo link where the transmitter and receiver sites are 1221 and 1875 m above MSL, respectively.

5 A radiolink 200 km long is to be installed in the eastern United States; the sea level refractivity is 320. The transmitter site is 158 m above MSL and the receiver site is 315 m above MSL. Allow 11 m to antenna centers at each end. The altitude of the transmitter horizon is 330 m and the receiver horizon is 490 m, and the distances to each horizon are 60 and 30 km, respectively. Assume $h_{st} = h_{ts} = 169$ m and $h_{sr} = h_{rs} = 326$ m. Calculate the effective earth radius.

6 Using the information given in question 5, calculate the horizon takeoff angles.

7 As in question 6, calculate the effective antenna heights assuming linear increments in altitude.

8 As in question 6, calculate the angular distance (or scatter angle).

9 Using information developed in questions 5–8, with an operational frequency of 1800 MHz, calculate the basic transmission loss L_{bsr}.

10 From question 9, calculate the long-term median transmission loss L_n.

11 From question 10, the service probability factor is 95%; compute the path loss for a time availability of 99.9%.

12 Assuming 60-ft antennas, calculate the aperture-to-medium coupling loss using both methods given in this chapter.

13 Using information from questions 11 and 12, define the parameters of a baseline system that uses quadruple diversity. Separate the diversity antennas by 400 wavelengths. Select waveguide or coaxial cable transmission line using best judgment. What other losses will be encountered? Calculate values of RSL.

14 What is the FM improvement threshold if the IF bandwidth is 4260 kHz (60 channels of FDM telephony)?

15 Scale the system in question 5 allowing 3 pWp/km. (How much bigger or smaller than baseline?) Allow NPR = 55 dB.

16 For a digital modulation scheme, an E_b/N_0 of 13 dB is required 99.9% of the time to achieve an error rate of 10^{-5}. Rescale the system from baseline (question 5).

17 What sort of dispersion will be found on the sample path in question 16? Give the answer in nanoseconds.

18 If the scatter angle on a certain path can be reduced by 1.5°, will the transmission loss be increased or decreased and by approximately how much?

REFERENCES

1. *General Engineering Beyond-Horizon Radio Communications*, USAF T.O. 31Z-10-13, U.S. Department of Defense, Washington, DC, Oct. 1971.
2. *Military Handbook Facility Design for Tropospheric Scatter (Transhorizon Microwave System Design)*, MIL-HDBK-417, U.S. Department of Defense, Washington, DC, Nov. 1977.
3. P. L. Rice, A. C. Longley, K. A. Norton, and A. P. Barsis, *Transmission Loss Prediction for Tropospheric Communication Circuits*, NBS Tech. Note 101, U.S. National Bureau of Standards, Boulder, CO, May 1965 (revised Jan. 1967).
4. *Propagation Data and Prediction Methods Required for Trans-Horizon Radio Relay Systems*, CCIR Rec. 617-1, 1994 PN Series Volume, ITU, Geneva, 1994.
5. L. P. Yeh, "Simple Methods for Designing Troposcatter Circuits," *IRE Transactions on Communications Systems*, Sept. 1960.
6. C. C. Rider, "Median Signal Level Prediction for Tropospheric Scatter," *Marconi Review*, Third quarter, 1962.
7. *Instruction Manual for Tropospheric Scatter—Principles and Applications*, US-AEPG-SIG 960-67, U.S. Army Electronic Proving Ground, Ft. Huachuca, AZ, Mar. 1960.
8. (Longley–Rice): G. A. Hufford, A. G. Longley, and W. A. Kissick, *A Guide to the Use of the ITS Irregular Terrain Model in the Areas Prediction Mode*, U.S. Department of Commerce, Washington, DC, Apr. 1982. NTIA Report, 82–100.
9. R. L. Freeman, *Reference Manual for Telecommunications Engineering*, 2nd Ed., Wiley, New York, 1993.
10. *Effects of Propagation on the Design and Operation of Trans-horizon Radio-Relay Systems, Annex 1—Factors Concerning Effects of Propagation on the Design and Operation of Trans-horizon Radio-Relay Systems*, ITU-R Rec. F.1106, 1994 F Series Volume, Part 1, ITU, Geneva, 1994.
11. W. J. Connor, "AN/TRC-170—A New Digital Troposcatter Communication System," IEEE ICC '78 Conference Record.
12. T. E. Brand, W. J. Connor, and A. J. Sherwood, "AN/TRC-170—Troposcatter Communication System," NATO Conference on Digital Troposcatter, Brussels, Mar. 1980.
13. *Preferred Frequency Bands for Trans-horizon Radio-Relay System, Annex 1—Factors Affecting the Choice of Frequency Bands for Trans-horizon Radio-Relay Systems*, ITU-R Rec. F.698-2, 1994 F Series Volume, Part 1, ITU, Geneva, 1994.

6

BASIC PRINCIPLES OF SATELLITE COMMUNICATIONS

6.1 INTRODUCTION, SCOPE, AND APPLICATIONS

Satellite communications was in its infancy in 1962. It has seen remarkable growth since then. This growth is tempered by orbital and frequency congestion and the accelerating implementation of fiberoptic links.

Initially, commercial satellite radiolinks were used on long transoceanic circuits almost exclusively. The largest satellite network, with operation dating back to 1965, is INTELSAT* with satellite series from INTELSAT I through INTELSAT VII, and INTELSAT VIII now in development. Domestic satellite communications was pioneered by Canada with the ANIK/TeleSat series of satellites. ANIK/TeleSat essentially provides three services: high-capacity trunking among major Canadian population centers, thin-line connectivity to sparsely populated rural areas such as the Canadian Arctic, and TV programming relay.

Ships at sea are now being served by INMARSAT† using INMARSAT satellites where user uplinks and downlinks operate around 1.5 GHz. Earth stations serving INMARSAT are found in nearly all maritime countries. The United States and Russia have a joint venture for search and rescue (SAR) alert and location by satellite.

There are regional and domestic satellite systems. Regions with operational satellites serving them are Europe and the Arab states. Large countries have national domestic satellite systems. Among these we find Mexico, Canada, Brazil, Japan, and Indonesia. The United States is probably the leader in this area. It is said that three-quarters of the satellite transponder bandwidth over North America is dedicated to TV programming relay where the C-band orbit is full, and the Ku-band orbit is nearly full. In this business

*INTELSAT = International Telecommunication Satellite (consortium) based in Washington, DC.
†INMARSAT = International Marine Satellite (consortium) based in London, UK.

area we include:

- Broadcasters
- Cable TV (CATV)
- Hotel/motel service
- Direct broadcast service

Very small aperture terminal (VSAT) service is another broad application area. This is discussed in Chapter 8.

The United States armed forces have a number of their own specialized satellite systems such as Defense Satellite Communication System (DSCS), Fleet Satellite (FLTSAT), and MILSTAR. MILSTAR is an advanced digital-processing satellite system operating in the 44/20-GHz bands. Great Britain, Russia, and NATO have their own dedicated systems.

The use of satellite systems for telephone trunking for the PSTN is notably diminishing in favor of fiberoptic links, which display much less propagation delay. However, a number of specialized systems are coming on-line in the near future (i.e., 1997–2000). These are designed for PCS (personal communication system) and cellular services (see Chapter 12).

This chapter starts our tutorial discussion on satellite communications and is slanted more for analog connectivity. Chapter 7 covers digital communications by satellite and Chapter 8 deals with VSAT systems, which are entirely digital. Chapter 9 covers system design above 10 GHz, where rainfall loss and gaseous absorption are covered.

6.2 SATELLITE SYSTEMS—AN INTRODUCTION

This text deals with two broad categories of communication satellites. The first is the RF repeater satellite, affectionately called the *bent-pipe* satellite. The second is the processing satellite, which is used exclusively on digital circuits, where, as a minimum, the satellite demodulates the uplink signal to baseband and regenerates that signal for retransmission back to earth. Analog circuits exclusively use "bent-pipe" techniques; digital circuits may use either variety. The bent-pipe satellite is simply a frequency-translating RF repeater. Figure 6.1 is a simplified functional block diagram of the communication payload of such a satellite. Processing satellites are described in Chapter 7.

6.2.1 Satellite Orbits

Satellites can be categorized in two ways. The first deals with their altitude above the earth's surface. There are LEOs, MEOs, and GEOs. These abbreviations stand for low earth orbit, medium earth orbit, and geostationary earth orbit, respectively.

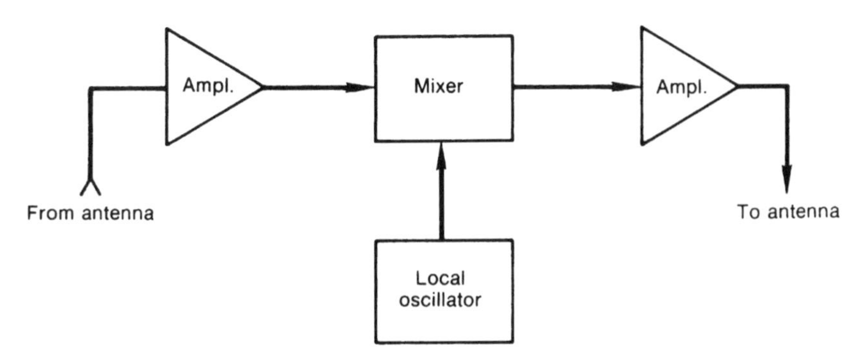

Figure 6.1. Simplified functional block diagram of a satellite communications transponder. This is the conventional translating RF repeater or "bent-pipe" satellite configuration.

Satellites can also be characterized by their orbits:

- Polar
- Equatorial
- Inclined

The figure a satellite defines in orbit is an ellipse. Of course, a circle is a particular class of ellipse. The Molniya, so popular with the Russians, is a highly inclined elliptical orbit.

The discussion in this chapter will dwell almost entirely on geostationary satellites. A geostationary satellite has a circular orbit and is classified as equatorial. Its orbital period is one sidereal day (23 h, 56 min, 4.091 s) or nominally 24 hours. Its inclination is 0°, which means that the satellite is always directly over the equator. A geostationary satellite appears stationary over any location on earth, that is within optical view.

Geostationary satellites are conventionally located with respect to the equator (0° latitude) and a subsatellite point, which is given in degrees longitude at the earth's surface. The satellite's range at this point, and only at this point, is 35,784 km (22,235 statue miles) above the earth's surface (sea level). Table 6.1 gives details and parameters of the geostationary satellite.

Table 6.1 also outlines several of the advantages and disadvantages of a geostationary satellite. Most of these points are self-explanatory. For satellites not at geosynchronous altitude and not over the equator, there is the appearance of movement. The movement with relation to a point on earth will require some form of automatic tracking by the earth station antenna to always keep it pointed at the satellite. If a satellite system is to have full earth coverage using a constellation of geostationary satellites, a minimum of three satellites would be required, separated by 120°. As an earth station moves northward or southward from the equator, the elevation angle to a geosta-

TABLE 6.1 The Geostationary Satellite Orbit

For the special case of a synchronous orbit—satellite in prograde circular orbit over the equator:	
Altitude	19,322 nautical miles, 22,235 statute miles, 35,784 km
Period	23 h, 56 min, 4.091 s (one sidereal day)
Orbit inclination	0°
Velocity	6879 statute miles/h
Coverage	42.5% of earth's surface (0° elevation)
Number of satellites	Three for global coverage with some areas of overlap (120° apart)
Subsatellite point	On the equator
Area of no coverage	Above 81° north and south latitude
Advantages	Simpler ground station tracking
	No handover problem
	Nearly constant range
	Very small Doppler shift
Disadvantages	Transmission delay
	Range loss (free-space loss)
	No polar coverage

Source: Reference 1.

tionary satellite decreases (see Section 6.2.3). Elevation angles below 5° are generally undesirable, as will be discussed subsequently. This is the rationale in Table 6.1 for "area of no coverage." Handover refers to the action taken by a satellite earth terminal antenna when a nongeostationary (often misnamed "orbiting satellite") disappears below the horizon (or below 5° elevation angle) and its antenna slews to a companion satellite of the system that is just appearing above the opposite horizon. It should be pointed out here that geostationary satellites do have small residual relative motions. Over its subsatellite point, a geostationary satellite carries a small apparent orbit in the form of a figure eight because of higher space harmonics of the earth's gravitation and tidal forces from the sun and moon. The satellite also tends to drift off station because of the gravitational attraction of the sun and moon as well as solar winds. Without correction, the inclination plane drifts roughly 0.86° per year (Ref. 1, Section 13.4.1.9).

6.2.2 Elevation Angle

The elevation angle or "look angle" of a satellite terminal antenna is the angle measured from the horizontal to the point on the center of the main beam of the antenna when the antenna is pointed directly at the satellite. This concept is shown in Figure 6.2. Given the elevation angle of a geostationary satellite, we can define the range. We will need the range, d in Figure 6.2, to calculate the free-space loss or spreading loss for the satellite radiolink.

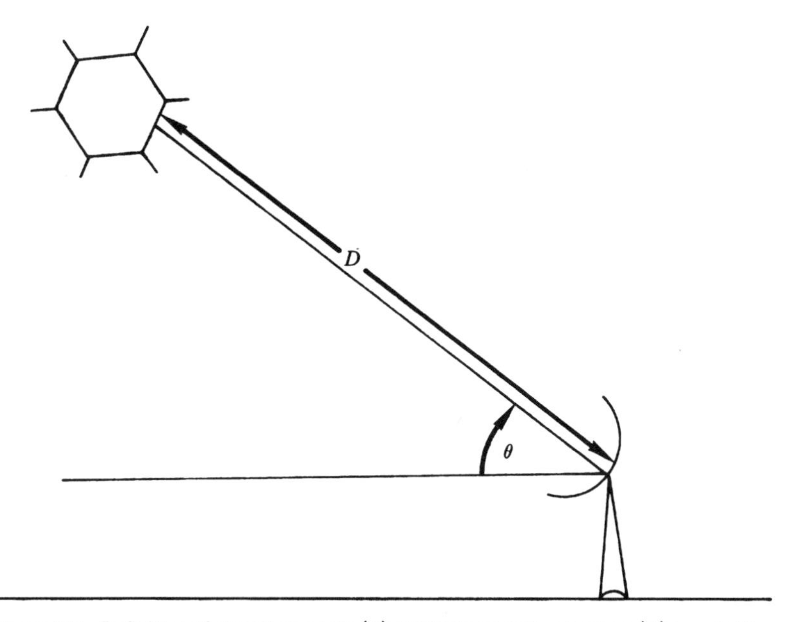

Figure 6.2. Definition of elevation angle (θ) or "look angle"and range (d) to satellite.

6.2.3 Determination of Range and Elevation Angle of a Geostationary Satellite

Geostationary satellites operate at an altitude of about 35,785 km above sea level. Unless an earth station is directly under a satellite, however, the distance d of Figure 6.2 will be greater than 35,785 km. The value of d can be established by use of the law of cosines of plane trigonometry. Consider first that the earth station is on the same longitude as the subsatellite point, taken to be at 0° latitude. The subsatellite point is located where a straight line from the satellite to the center of the earth intersects the earth's surface. See Figure 6.3. From the figure we can now state that

$$D^2 = r_0^2 + (h + r_0)^2 - 2r_0(h + r_0)\cos\theta' \tag{6.1}$$

where θ' is latitude. The equatorial radius of the earth is 6378.16 km, the polar radius is 6356.78 km, and the mean radius is 6371.03 km (Ref. 2). To obtain the most accurate value of d, it would be necessary to take into account the departure of the earth from sphericity, but an approximate value of d can be obtained by taking r_0, the earth's radius, to be 6378 km and h, the height of the satellite above the earth's surface, to be 35,785 km in equation (6.1). Divide all terms in equation (6.1) by $(h + r_0)^2$ or $(42{,}163)^2$,

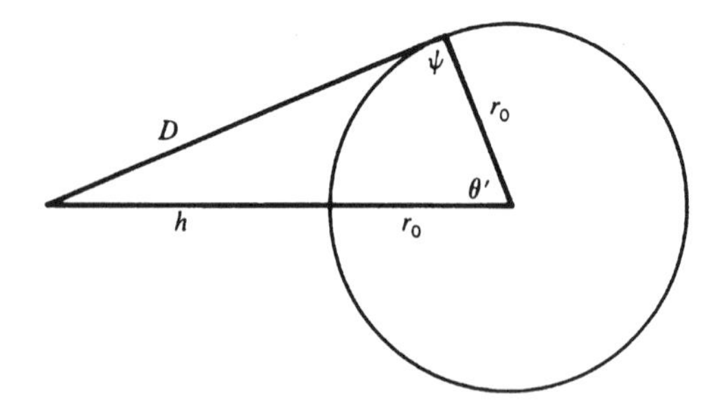

Figure 6.3. Geometry for calculation of distance *D* of a geostationary satellite from an earth terminal on earth's surface.

which gives

$$\left(\frac{d}{h + r_0}\right)^2 = f^2 + 1 - 2f \cos \theta' \tag{6.2}$$

where $f = r_0/(h + r_0) = 0.1513$. Once d is known, then all three sides of the triangle in Figure 6.3 are known and the angle ψ can be determined by applying the law of cosines again. The applicable equation is

$$(h + r_0)^2 = d^2 + r_0^2 - 2r_0 d \cos \psi \tag{6.3}$$

The elevation angle θ measured from the horizontal at the earth terminal is equal to $\psi - 90°$.

For an earth terminal not on the same meridian as the subsatellite point, we can use the equation (from the spherical law of cosines)

$$\cos z = \cos \theta' \cos \phi' \tag{6.4}$$

in equation (6.1) in place of $\cos \theta'$, where ϕ' is the difference in longitude between the subsatellite point and the earth terminal; $\cos Z$ is the angular distance of a great circle path for the special case that one of the end points is at 0° latitude (Figure 6.4). Also, the expression follows from the law of cosines for sides from spherical trigonometry (Ref. 3, Section 44) The azimuth angle α of an earth–space path can be determined by using (Ref. 3, Section 44)

$$\cos \alpha = \tan \theta' \cot Z \tag{6.5}$$

The angle α is shown in Figure 6.4a for an earth terminal located to the east of the subsatellite point. The azimuth angle measured from the north in this case would then be $180° + \alpha$. For an earth terminal located to the west of the subsatellite point (Figure 6.4b), the angle from true north is $180° - \alpha$.

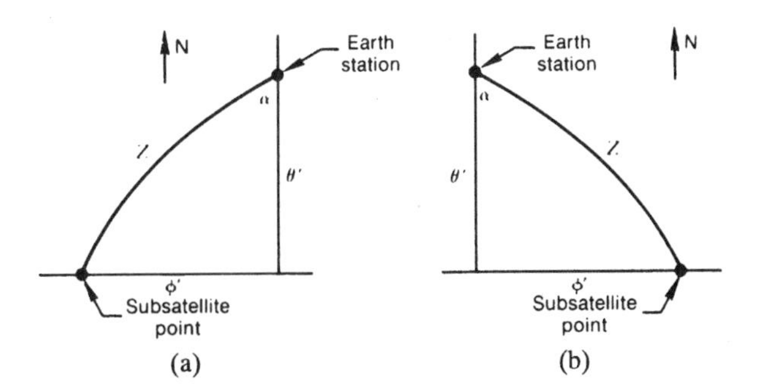

Figure 6.4. Projection of right spherical triangles on earth's surface.

Example 1. Calculate azimuth and elevation angles and distance from a point 40° N, 105° W for a satellite where the subsatellite point is located at 119° W.

$$\cos Z = \cos 40^\circ \cos 14^\circ = 0.743$$

From equation (6.1) we find that $d = 37{,}666$ km and the elevation angle $\theta = 43.73°$; the azimuth angle is 201.2°.

Range, elevation, and azimuth angles may also be determined by nomogram with sufficient accuracy for link analyses (link budget). See Figure 6.5

6.3 INTRODUCTION TO LINK ANALYSIS OR LINK BUDGET

6.3.1 Rationale

To size or dimension a satellite terminal correctly, we will want to calculate the receive signal level (RSL) at the terminal. The methodology is very similar to what we used in Chapters 2, 3, 4, and 5. However, there are certain legal constraints of which we should be aware.

6.3.2 Frequency Bands Available for Satellite Communications

The frequency bands assigned for satellite communications are given in Table 6.2 as corrected by recent Radio Regulations (Ref. 4). Generally, these frequency bands are referred to in band pairs. One of the band pairs, usually of higher frequency than the other, is used for the uplink path (i.e., terminal to satellite), and the other is assigned the downlink path. In this text we will be dealing with the following commonly used frequency band pairs (uncorrected for recent Radio Regulations, but commonly accepted in the industry):

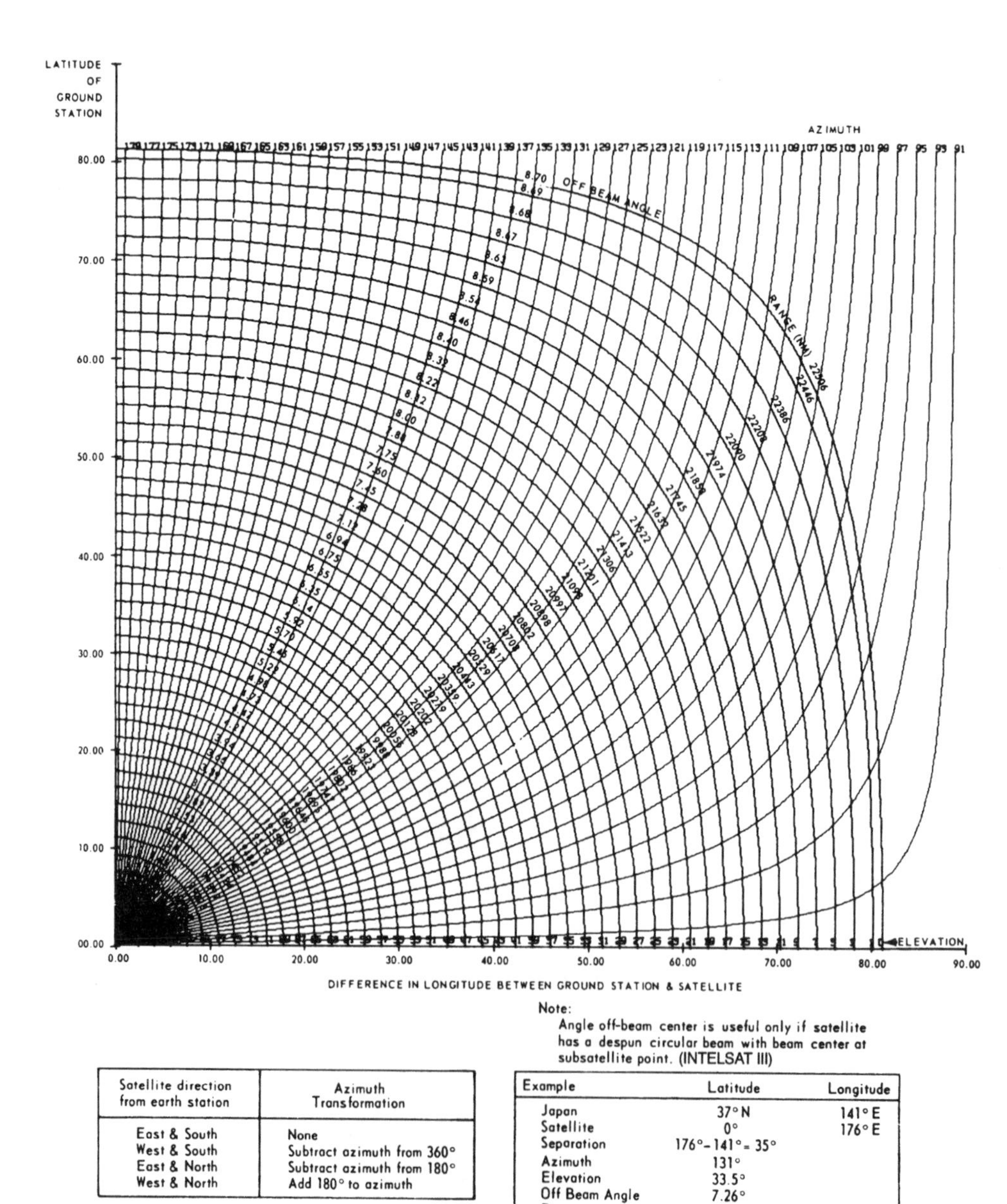

Note:
Angle off-beam center is useful only if satellite has a despun circular beam with beam center at subsatellite point. (INTELSAT III)

Satellite direction from earth station	Azimuth Transformation
East & South	None
West & South	Subtract azimuth from 360°
East & North	Subtract azimuth from 180°
West & North	Add 180° to azimuth

Example	Latitude	Longitude
Japan	37° N	141° E
Satellite	0°	176° E
Separation	176° − 141° = 35°	
Azimuth	131°	
Elevation	33.5°	
Off Beam Angle	7.26°	
Range	20,680 NM	

Figure 6.5. Determination of range to a geostationary satellite, azimuth, and elevation angles. (From Ref. 1, courtesy of COMSAT, Washington, DC.)

TABLE 6.2 Fixed Satellite Service Frequency Allocations[a]

Uplink (MHz)	Region	Bandwidth (MHz)	Applicable Radio Regulation Article
5,725–5,850	1	125	RR8-135
5,850–7,025	1, 2, 3	1225	RR8-135
7,900–8,400	1, 2, 3	500	RR8-137, RR8-139
12,500–12,750	1	250	RR8-146
12,750–13,250	1, 2, 3	500	RR8-150
13,750–14,800	1, 2, 3	1050	RR8-150
17,300–18,400	1, 2, 3	1100	RR8-155, RR8-158
27,000–27,500	2, 3	500	RR8-156
27,500–31,000	1, 2, 3	3500	RR8-166, 167
42,500–43,500	1, 2, 3	1000	RR8-173
47,200–50,200	1, 2, 3	3000	RR8-173
50,400–52,400	1, 2, 3	2000	RR8-173
71,000–74,000	1, 2, 3	3000	RR8-176
Downlink	**Region**	**Bandwidth (MHz)**	**Applicable Radio Regulation Article**
3,400–4,200	1, 2, 3	500	RR8-128
4,500–4,800	1, 2, 3	300	RR8-128
7,250–7,750	1, 2, 3	500	RR8-135, RR8-137
10,700–11,700	1, 2, 3	1000	RR8-146
11,700–12,200	2	500	RR8-146
12,500–12,750	1, 3	250	RR8-146
17,700–21,200	1, 2, 3	3500	RR8-155, 158
37,500–40,500	1, 2, 3	3000	RR8-172
81,000–84,000	1, 2, 3	3000	RR8-178

[a]Based on *Radio Regulations*, edition 1990, revised 1994 (Ref. 4).

6/4 GHz	5925–6425 MHz	Uplink	Commercial
	3700–4200 MHz	Downlink	
8/7 GHz	7900–8400 MHz	Uplink	Military
	7250–7750 MHz	Downlink	
14/11 GHz	14.0–14.5 GHz	Uplink	Commercial
	11.7–12.2 GHz	Downlink	
30/20 GHz	27.5–30.5 GHz	Uplink	Commercial
	17.7–20.2 GHz	Downlink	
30/20 GHz	30.0–31.0 GHz	Uplink	Military
	20.2–21.2 GHz	Downlink	
44/20 GHz	43.5–45.5 GHz	Uplink	Military
	20.2–21.2 GHz	Downlink	

Tables 6.3–6.5 are based on WARC-79 (Ref. 5). The reader is urged to consult the latest edition of the Radio Regulations, especially the footnotes, when using these tables.

TABLE 6.3 WARC 79 Intersatellite Service Allocations

Band (GHz)	Bandwidth (GHz)
22.55–23.55	1
32.00–33.00	1
54.25–58.20	3.95
59.0–64.0	5
116.0–134.0	18
170.0–182.0	2
185.0–190.0	5

Source: Reference 5, reprinted with permission.

TABLE 6.4 Mobile Satellite Services Allocations

Earth-to-Space (MHz)	Region	Bandwidth (MHz)	Space-to-Earth (MHz)	Region	Footnote
121.45–121.55[a,b,h]	1, 2, 3		—		3572A
242.95–243.05[a,b,h]	1, 2, 3		—		3572A
235.0–322.0[a,h]	1, 2, 3	87			3618
335.4–399.9[a,h]	1, 2, 3	64.4			3618
405.5–406.0[c]	Canada	0.5	—		3533A
406.0–406.1[b]	1, 2, 3	0.1	—		3633A
406.1–410.0[c]	Canada	3.9	—		3634
608.0–614.0[d,e]	2	6.0	—		
806.0–890.0[c,h]	2, 3	3.0			3662C
	Norway				3662CA
	Sweden				3670B
942.0–960.0[c,h]	3, Norway	18			3662C
	Sweden				3662CA
1645.5–1646.5[f]	1, 2, 3	1.0	1554.0–1545.0[f]	1, 2, 3	3695A
(GHz)			(GHz)		
7.90–8.025[a]	1, 2, 3	125	7.250–7.375[a]	1, 2, 3	3764B
14.00–14.50[e,g]	1, 2, 3	500	—		
29.50–30.0[e]	1, 2, 3	500	19.70–20.20[e]	1, 2, 3	
30.00–31.00	1, 2, 3	1000	20.20–21.20	1, 2, 3	
43.50–47.00[h]	1, 2, 3	3500	39.50–40.50	1, 2, 3	3814C
50.40–51.40[e]	1, 2, 3	1000	—		—
66.00–71.00[h]	1, 2, 3	5000			3814C
71.00–74.00	1, 2, 3	3000	81.00–84.00	1, 2, 3	—
95.00–100.00[h]	1, 2, 3	5000			3814C
135.00–142.00[h]		7000			3814C
190.00–200.00[h]		10000			3814C
252.00–265.00[h]		13000			3814C

[a]Footnote allocation.
[b]Emergency position indicating radio beacons only.
[c]Footnote allocation excludes aeronautical mobile satellite services.
[d]Excludes aeronautical mobile satellite services.
[e]Secondary allocation.
[f]Distress and safety operations only.
[g]Footnote allocation to land mobile satellite service only.
[h]No direction specified.

Source: Reference 6, reprinted with permission.

TABLE 6.5 Broadcasting Satellite Service Allocations

Earth-to-Space (GHz)	Region	Bandwidth (MHz)	Space-to-Earth (GHz)	Region	Bandwidth (MHz)
Feeder links for the broadcast satellite service may, in principle, use any of the fixed satellite service (earth-to-space) bands listed in Table 6.2 with appropriate coordination. However, the following bands were set aside for exclusive or preferential use by such feeder links:					
			0.62–0.79[a]	1, 2, 3	170
			2.50–2.69[b]	1, 2, 3	190
10.70–11.70	1	1000	11.70–12.1	1, 3	400
14.50–14.80	1,[c] 2, 3	300	12.10–12.2	1, 2, 3	100
17.30–18.1	1, 2, 3	800	12.20–12.5	1, 2	300
			12.50–12.7	2, 3[b]	200
			12.70–12.75	3[b]	50
27.00–27.50	2, 3	500	22.50–23.00	2, 3	500
47.20–49.20	1, 2, 3	2000	40.50–42.50	1, 2, 3	2000
			84.00–86.00	1, 2, 3	2000

[a]Limited to TV.
[b]Limited to community reception.
[c]Excluding Europe.

Source: Reference 6, reprinted with permission.

6.3.3 Free-Space Loss or Spreading Loss

Section 1.2 gave a method of calculation of free-space loss (FSL) or, what some call, spreading loss. Equations (1.7), (1.9a), and (1.9b) apply. We just plug in frequency and distance (range), being sure to use the proper units.

Example 2. The elevation angle to a geostationary satellite is 21° and the transmitting frequency is 3.941 GHz. What is the free-space loss in decibels?

Use equation (1.9c), where the distance unit is the nautical mile (nm):

$$L_{\mathrm{dB}} = 37.80 + 20\log 21{,}201\ \mathrm{nm} + 20\log 3941\ \mathrm{MHz}$$

$$= 196.24\ \mathrm{dB}$$

6.3.4 Isotropic Receive Level—Simplified Model

Consider a downlink at 3941 MHz from a geostationary satellite with a free-space loss of 196.24 dB. Let the satellite have an EIRP of +29 dBW. What would be the isotropic receive level (IRL) for an earth station with a 21° elevation angle? All other losses are disregarded. The path is line-of-sight

(LOS) by definition. The approach is the same as in Chapter 2 (see Section 2.6.2). Equation (2.7) applies:

$$\begin{aligned} IRL_{dBW} &= EIRP + FSL_{dB} \\ &= +29\text{ dBW} + (-196.24)\text{ dB} \\ &= -167.24\text{ dBW} \end{aligned}$$

The equation has left out a number of small, but when added together, not insignificant losses. Some of these losses are radome loss, pointing losses, polarization loss, gaseous absorption loss, and excess attenuation due to rainfall.

6.3.5 Limitation of Flux Density on Earth's Surface

The frequency bands shown in Section 6.3.2 are shared with other services, such as terrestrial point-to-point radiolink microwave. The flux density of satellite signals on the earth's surface must be limited so as not to interfere with terrestrial radio services band-sharing with satellite systems. CCIR recommends the following flux density limits [excerpted from ITU-R Rec. SF.358-4 (Ref. 7)].

Maximum Power Flux Density on Earth Surface

Unanimously recommends:

1. that, in frequency bands in the range 2.5 to 27.5 GHz shared between systems in the Fixed Satellite Service and LOS radio-relay systems, the maximum power flux density produced at the surface of the earth by emissions from a satellite, for all conditions and methods of modulation, should not exceed:

   1.1. in the band 2.5 to 2.690 GHz, in any 4-kHz band:

$$\begin{array}{llll} -152 & \text{dB(W/m}^2) & \text{for} & \theta \leq 5^\circ \\ -152 + 0.75(\theta - 5) & \text{dB(W/m}^2) & \text{for} & 5^\circ < \theta \leq 25^\circ \\ -137 & \text{dB(W/m}^2) & \text{for} & 25^\circ < \theta \leq 90^\circ \end{array}$$

   1.2 In the band 3.4 to 7.750 GHz, in any 4-kHz band:

$$\begin{array}{llll} -152 & \text{dB(W/m}^2) & \text{for} & \theta \leq 5^\circ \\ -152 + 0.5(\theta - 5) & \text{dB(W/m}^2) & \text{for} & 5^\circ < \theta \leq 25^\circ \\ -142 & \text{dB(W/m}^2) & \text{for} & 25^\circ < \theta \leq 90^\circ \end{array}$$

   1.3 in the band 8.025 to 11.7 GHz, in any 4-kHz band:

$$\begin{array}{llll} -150 & \text{dB(W/m}^2) & \text{for} & \theta \leq 5^\circ \\ -150 + 0.5(\theta - 5) & \text{dB(W/m}^2) & \text{for} & 5^\circ < \theta \leq 25^\circ \\ -140 & \text{dB(W/m}^2) & \text{for} & 25^\circ < \theta \leq 90^\circ \end{array}$$

1.4 in the band 12.2 to 12.75 GHz, in any 4-kHz band:

-148	dB(W/m^2)	for	$\theta \leq 5°$
$-148 + 0.5(\theta - 5)$	dB(W/m^2)	for	$5° < \theta \leq 25°$
-138	dB(W/m^2)	for	$25° < \theta \leq 90°$

1.5 in the bands 17.7 to 19.7 GHz, 22.55 to 23.55 GHz, 24.45 to 24.75 GHz, and 25.25 to 27.5 GHz, in any 1-MHz band:

-115	dB(W/m^2)	for	$\theta \leq 5°$
$-115 + 0.5(\theta - 5)$	dB(W/m^2)	for	$5° < \theta \leq 25°$
-105	dB(W/m^2)	for	$25° < \theta \leq 90°$

where θ is the angle of arrival of the radio-frequency wave (degrees above the horizontal);

2. that the aforementioned limits relate to the power flux density and angles of arrival, which would be obtained under free-space propagation conditions.

Note 1. Definitive limits applicable in shared frequency bands are laid down in Nos. 2561 to 2580.1 of Article 28 of the Radio Regulations (RR). Study of these problems is continuing, which may lead to changes in the recommended limits.

Note 2. Under RR Nos. 2581 to 2585, the power flux density limits in bands between 17.7 and 27.5 GHz shall apply provisionally to the band 31.0–40.5 GHz until such time as definitive values have been recommended, endorsed by a competent world radio communication conference (RR No. 2582.1).

6.3.5.1 Calculation of Power Flux Density Levels. The signal level of a wave incident on a satellite antenna is measured by the power flux density S (expressed in watts per square meter) of the approaching wavefront. If we consider the emitter as a point source, the emitted energy spreads uniformly in all directions. The satellite antenna can be considered a point on the surface of a sphere, where the emitter is at the center of the sphere. The radius R_s of the sphere is the distance from the emitter to the satellite antenna. The surface area of the sphere in square meters is given by

$$A_s = 4\pi R_s^2 \tag{6.6}$$

If the distance to the satellite from the emitter is 37,750 km, then

$$A_s = 1.791 \times 10^{16} \text{ m}^2$$

The flux density at the satellite is calculated as if the EIRP of the emitter uniformly covered the total surface A_s. The resulting flux density is given by

$$S = \frac{(k_A)\text{EIRP}}{4\pi R_s^2} = \frac{(k_A)\text{EIRP}}{A_s} \tag{6.7}$$

where k_A is the atmospheric attenuation factor (e.g., gaseous attenuation), which is less than unity. Converting to decibel form gives

$$S(\text{dBW/m}^2) = \text{EIRP}_{\text{dBW}} - 10\log A_S - L_A \tag{6.8}$$

where $L_A = -10\log k_A$, which is the atmospheric attenuation factor in decibels. For a bent-pipe satellite we are often given a flux density value (S) to saturate the satellite transponder traveling wave tube (TWT) final amplifier. The earth station EIRP required from a satellite terminal emitter antenna is

$$\text{EIRP} = 10\log A_s + S + L_A \tag{6.9}$$

Example 3. What earth station EIRP is required to saturate a TWT HPA of satellite transponder if the transponder requires -88 dBW/m^2 power flux density for TWT saturation? The distance to the satellite is 23,000 statute miles, and 1 dB is allotted for atmospheric gas attenuation.

First convert 23,000 statute miles to kilometers:

$$23{,}000 \times 1.609 = 37{,}007 \text{ km}$$

Then

$$\begin{aligned} A_s &= 4\pi(37{,}007) \times 37{,}007 \times 10^6 \\ &= 1.3695 \times 10^{15} \times 4 \times 3.14159 \\ &= 1.72098 \times 10^{16} \end{aligned}$$

and

$$\begin{aligned} \text{EIRP} &= 162.36 - 88 - 1 \\ &= +73.36 \text{ dBW} \end{aligned}$$

6.3.6 Thermal Noise Aspects of Low-Noise Systems

We deal with very low signal levels in space communication systems. Downlink signal levels are in the approximate range of -154 to -188 dBW. The objective is to achieve sufficient S/N or E_b/N_0 at demodulator outputs. There are two ways of accomplishing this:

1. By increasing system gain, usually with antenna gain.
2. By reducing system noise.

In this section we will give an introductory treatment of thermal noise analytically, and later the term G/T (figure of merit) will be introduced.

Around noise threshold, thermal noise predominates. To set the stage, we quote from Ref. 8:

> The equipartition law of Boltzmann and Maxwell (and the works of Johnson and Nyquist) states that the available power per unit bandwidth of a thermal noise source is
>
> $$p_n(f) = kT \quad \text{watts/Hz} \tag{6.10}$$
>
> where k is Boltzmann's constant (1.3806×10^{-23} joule/K) and T is the absolute temperature of the source in kelvins.

Looking at a receiving system, all active components and all components displaying ohmic loss generate noise. In LOS radiolinks, system noise temperatures are in the range of 1500–4000 K, and the noise of the receiver front end is by far the major contributor. In the case of space communications, the receiver front end may contribute less than one-half the total system noise. Total receiving system noise temperatures range from as low as 70 K up to 1000 K (for those types of systems considered in this chapter).

In Chapters 2–5, receiving system noise was characterized by noise figure expressed in decibels. Here, where we deal often with system noise temperatures of less than 290 K, the conventional reference of basing noise at room temperature is awkward. Therefore noise figure is not useful at such low noise levels. Instead, it has become common to use effective noise temperature T_e [equation (6.13)].

It can be shown that the available noise power at the output of a network in a band B_w is (Ref. 8)

$$p_n = g_a(f)(T + T_e)B_w \tag{6.11}$$

where g_a is the network power gain at frequency f, T is the noise temperature of the input source, and T_e is the effective input temperature of the network. For an antenna–receiver system, the total effective system noise temperature T_{sys}, conventionally referred to the input of the receiver, is

$$T_{sys} = T_{ant} + T_r \tag{6.12}$$

where T_{ant} is the effective input noise temperature of the antenna subsystem and T_r is the effective input noise temperature of the receiver subsystem. The ohmic loss components from the antenna feed to the receiver input also generate noise. Such components include waveguide or other transmission line, directional couplers, circulators, isolators, and waveguide switches.

It can be shown that the effective input noise temperature of an attenuator is (Ref. 8)

$$T_e = \frac{p_a}{kB_w g_a} - T_s = \frac{T(1 - g_a)}{g_a} \tag{6.13}$$

where T_s is the effective noise temperature of the source, the lossy elements have a noise temperature T, k is Boltzmann's constant, g_a is the gain (available loss), and p_a is the noise power at the output of the network.

The loss of the attenuator l_a is the inverse of the gain or

$$l_a = \frac{1}{g_a} \tag{6.14}$$

Substituting into equation (6.13) gives

$$T_e = T(l_a - 1) \tag{6.15}$$

It is accepted practice (Ref. 8) that

$$n_f = 1 + \frac{T_e}{T_0} \tag{6.16}$$

where in Ref. 8 n_f is called the noise figure. Other texts call it the noise factor and

$$\mathrm{NF_{dB}} = 10 \log_{10} n_f \tag{6.17}$$

and T_0 is standard temperature or 290 K. $\mathrm{NF_{dB}}$ is the conventional noise figure discussed in Section 2.6.3.1.

From equation (6.15) the noise figure (factor) is

$$n_f = 1 + \frac{(l_a - 1)T}{T_0} \tag{6.18}$$

If the attenuator lossy elements are at standard temperature (e.g., 290 K), the noise figure equals the loss (the noise factor equals the numeric of the loss):

$$n_f = l_a \tag{6.19}$$

or expressed in decibels,

$$\mathrm{NF_{dB}} = 10 \log l_a = L_{a\,(\mathrm{dB})}$$

For low-loss (i.e., ohmic-loss) devices whose loss is less than about 0.5 dB, such as short waveguide runs, which are at standard temperature, equation (6.15) reduces to a helpful approximation

$$T_e \approx 66.8L \tag{6.20}$$

where L is the loss of the device in decibels.

The noise figure in decibels may be converted to effective noise temperature by

$$\mathrm{NF_{dB}} = 10\log_{10}\left(1 + \frac{T_e}{290}\right) \tag{6.21}$$

Example 4. If a noise figure were given as 1.1 dB, what is the effective noise temperature?

$$1.1\ \mathrm{dB} = 10\log\left(1 + \frac{T_e}{290}\right)$$

$$0.11 = \log\left(1 + \frac{T_e}{290}\right)$$

$$1 + \frac{T_e}{290} = \log^{-1}(0.11)$$

$$1 + \frac{T_e}{290} = 1.29$$

$$T_e = 84.1\ \mathrm{K}$$

6.3.7 Calculation of C/N_0

We present two methods to carry out this calculation. The first method follows the rationale given in Sections 3.9 and 5.7.2. C/N_0 is measured at the input of the first active stage of the receiving system. For space receiving systems this is the low-noise amplifier (LNA) or other device carrying out a similar function. Figure 6.6 is a simplified functional block diagram of such a receiving system.

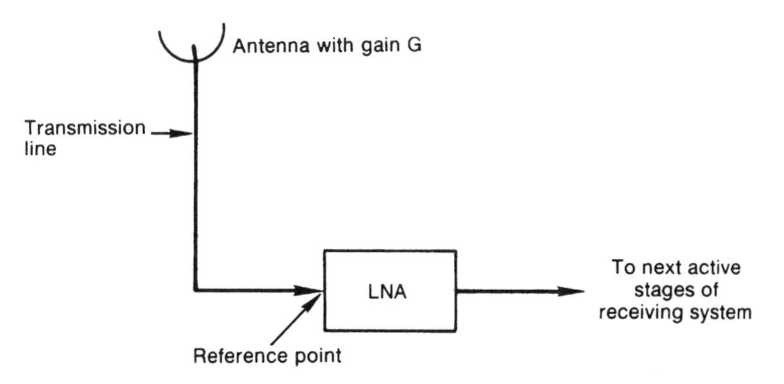

Figure 6.6. Simplified block diagram of space receiving system.

C/N_0 is simply the carrier to noise ratio, where N_0 is the noise density in 1 Hz of bandwidth. C is the receive signal level (RSL). Restating equation (6.10):

$$N_0 = kT \tag{6.22}$$

where k is Boltzmann's constant and T is the effective noise temperature, in this case of the space receiving system. We can now state this identity:

$$C/N_0 = C/kT \tag{6.23}$$

Turning to Figure 6.6, we see that if we are given the signal level impinging on the antenna, which we call the IRL, the receive signal level (RSL or C) at the input to the LNA is the IRL plus the antenna gain minus the line losses, or stated in equation form:

$$C_{\mathrm{dBW}} = \mathrm{IRL}_{\mathrm{dBW}} + G_{\mathrm{ant}} - L_{L\,(\mathrm{dB})} \tag{6.24}$$

where L_L represents the line losses in decibels. These losses will be the sum of the waveguide or other transmission line losses, antenna feed losses, and, if used, directional coupler loss, waveguide switch loss, power split loss, bandpass filter loss (if not incorporated in LNA), circulator/isolator losses, and so forth.

To calculate N_0, equation (6.22) can be restated as

$$N_0 = -228.6\ \mathrm{dBW} + 10 \log T_{\mathrm{sys}} \tag{6.25}$$

where -228.6 dBW is the theoretical value of the noise level in dBW for a perfect receiver (noise factor of 1) at absolute zero in 1 Hz of bandwidth. T_{sys} is the receiving system effective noise temperature, often called just system noise temperature.

Example 5. Given a system (effective) noise temperature of 84.1 K, what is N_0?

$$\begin{aligned} N_0 &= -228.6\ \mathrm{dBW/Hz} + 10 \log 84.1 \\ &= -228.6 + 19.25 \\ &= -209.35\ \mathrm{dBW/Hz} \end{aligned}$$

To calculate C or RSL, consider the following example.

Example 6. The IRL from a satellite is -155 dBW; the earth station receiving system (space receiving system) has an antenna gain of 47 dB, an antenna feed loss of 0.1 dB, a waveguide loss of 1.5 dB, a directional coupler insertion loss of 0.2 dB, and a bandpass filter loss of 0.3 dB; the system noise temperature (T_{sys}) is 117 K. What is C/N_0?

Calculate C (or RSL):

$$\begin{aligned} \mathrm{C} &= -155\ \mathrm{dBW} + 47\ \mathrm{dB} - 0.1\ \mathrm{dB} - 1.5\ \mathrm{dB} - 0.2\ \mathrm{dB} - 0.3\ \mathrm{dB} \\ &= -110.1\ \mathrm{dBW} \end{aligned}$$

Calculate N_0:

$$\begin{aligned} \mathrm{N}_0 &= -228.6\ \mathrm{dBW/Hz} + 10\log 117\ \mathrm{K} \\ &= -207.92\ \mathrm{dBW/Hz} \end{aligned}$$

Thus

$$\mathrm{C/N}_0 = \mathrm{C}_{\mathrm{dBW}} - \mathrm{N}_{0\,(\mathrm{dBW})} \tag{6.26}$$

In this example, substituting:

$$\begin{aligned} \mathrm{C/N}_0 &= -110.1\ \mathrm{dBW} - (-207.92\ \mathrm{dBW}) \\ &= 97.82\ \mathrm{dB} \end{aligned}$$

The second method of calculating $\mathrm{C/N}_0$ involves G/T, which is discussed in the next section.

6.3.8 Gain-to-Noise Temperature Ratio, *G/T*

G/T can be called the "figure of merit" of a radio receiving system. It is most commonly used in space communications. It not only gives an experienced engineer a "feel" of a receiving system's capability to receive low-level signals effectively, it is also used quite neatly as an algebraically additive factor in space system link budget analysis.

G/T can be expressed by the following identity:

$$\frac{G}{T} = G_{\mathrm{dB}} - 10\log T \tag{6.27}$$

where G is the receiving system antenna gain and T (better expressed as T_{sys}) is the receiving system noise temperature. Now we offer a word of caution. When calculating G/T for a particular receiving system, we must stipulate where the reference plane is. In Figure 6.6 it was called the "reference point." It is at the reference plane where the system gain is measured. In other words, we take the gross antenna gain and subtract all losses (ohmic and others) up to that plane or point. This is the net gain at that plane.

System noise is treated in the same fashion. Equation (6.12) stated

$$T_{\mathrm{sys}} = T_{\mathrm{ant}} + T_r$$

The antenna noise temperature T_{ant}, coming inward in the system, includes all noise contributors, including sky noise, up to the reference plane. Receiver noise T_r includes all noise contributors from the reference plane to the baseband output of the demodulator.

In most commercial space receiving systems, the reference plane is taken at the input to the LNA, as shown in Figure 6.6. In many military systems it is taken at the foot of the antenna pedestal. In one system, it was required to be taken at the feed. It can be shown that G/T will remain constant as long as we are consistent regarding the reference plane.

Calculation of the net gain (G_{net}) to the reference plane is straightforward. It is the gross gain of the antenna minus the sum of all losses up to the reference plane. Calculation of T_{sys} is somewhat more involved. We use equation (6.12). The calculation of T_{ant} is described in Section 6.3.8.1 and of T_r in Section 6.3.8.2.

6.3.8.1 *Calculation of Antenna Noise Temperature, T_{ant}.* The term T_{ant}, or antenna noise, includes all noise contributions up to the reference plane. Let us assume for all further discussion in this chapter that the reference plane coincides with the input to the LNA (Figure 6.6). There are two "basic" contributors of noise: sky noise and noise from ohmic losses.

Sky noise is a catchall for all external noise sources that enter through the antenna, through its main lobe, and through its sidelobes. External noise is largely due to extraterrestrial sources and thermal radiation from the atmosphere and the earth. Cosmic noise is a low level of extraterrestrial radiation that seems to come from all directions.

The sun is an extremely strong source of noise, which can interrupt satellite communications when it passes behind the satellite being used and thus lies in the main lobe of an earth station's antenna receiving pattern. The moon is a much weaker source, which is relatively innocuous to satellite communications. Its radiation is due to its own temperature and reflected radiation from the sun.

The atmosphere affects external noise in two ways. It attenuates noise passing through it, and it generates noise because of the energy of its constituents. Ground radiation, which includes radiation of objects of all kinds in the vicinity of the antenna, is also thermal in nature.

For our discussion we will say that sky noise (T_{sky}) varies with frequency, elevation angle, and surface water-vapor concentration. Figures 6.7, 6.8, and 6.9 give values of sky noise for elevation angles (θ) of 0°, 5°, 10°, and 90°, and for water-vapor concentrations of 3, 10, and 17 g/m^3. These figures do not include ground radiation contributions.

Antenna noise T_{ant} is the total noise contributed to the receiving system by the antenna up to the reference plane. It is calculated by the formula (Ref. 10)

$$T_{\text{ant}} = \frac{(l_a - 1)290 + T_{\text{sky}}}{l_a} \tag{6.28}$$

where l_a is the numeric equivalent of the system ohmic losses (in decibels) up

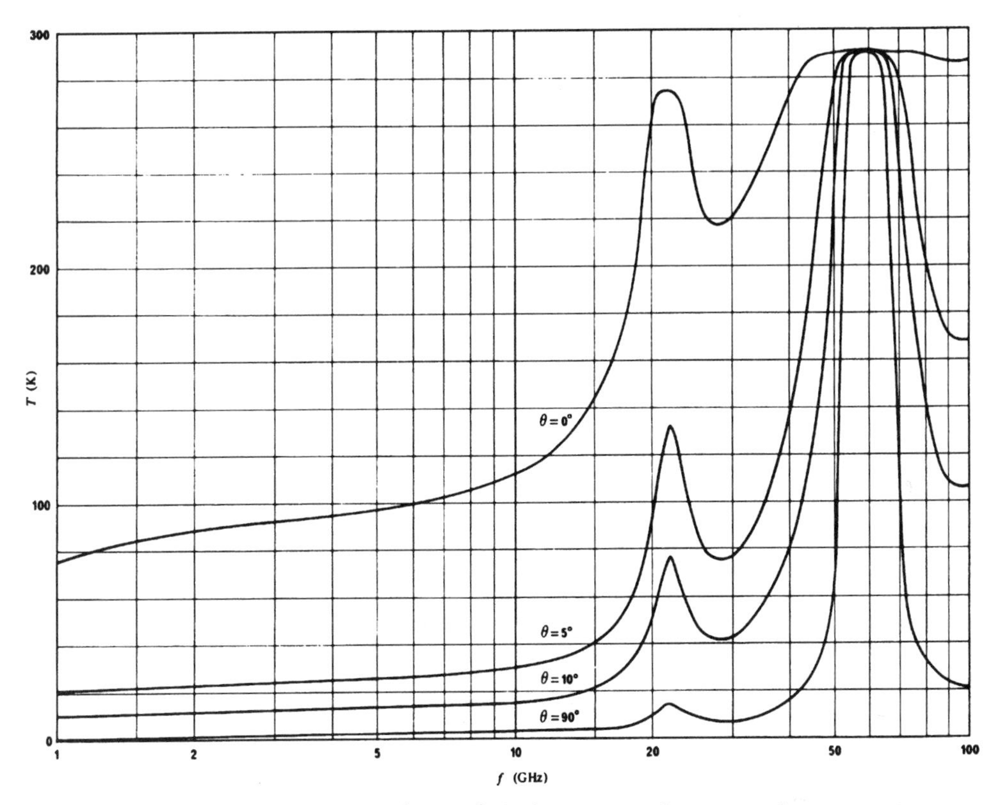

Figure 6.7. Sky noise temperature (clear air). Surface pressure is 1 atm; surface temperature is 20°C; surface water-vapor concentration is 3 g/m^3; θ is the elevation angle. (From CCIR Rep. 720; Ref. 9. Courtesy of ITU-CCIR, Geneva.)

to the reference plane. Then, l_a may be expressed as

$$l_a = \log_{10}^{-1} \frac{L_a}{10} \tag{6.29}$$

where L_a is the sum of the losses to the reference plane.

Example 7. Assume an earth station with an antenna at an elevation angle of 10°, clear sky, 3 g/m^3 water-vapor concentration, and ohmic losses as follows: waveguide loss of 2 dB, feed loss of 0.1 dB, directional coupler insertion loss of 0.2 dB, and bandpass filter insertion loss of 0.4 dB. These are the losses up to the reference plane, which is taken as the input to the LNA (Figure 6.10). What is the antenna noise temperature T_{ant}? The operating frequency is 12 GHz.

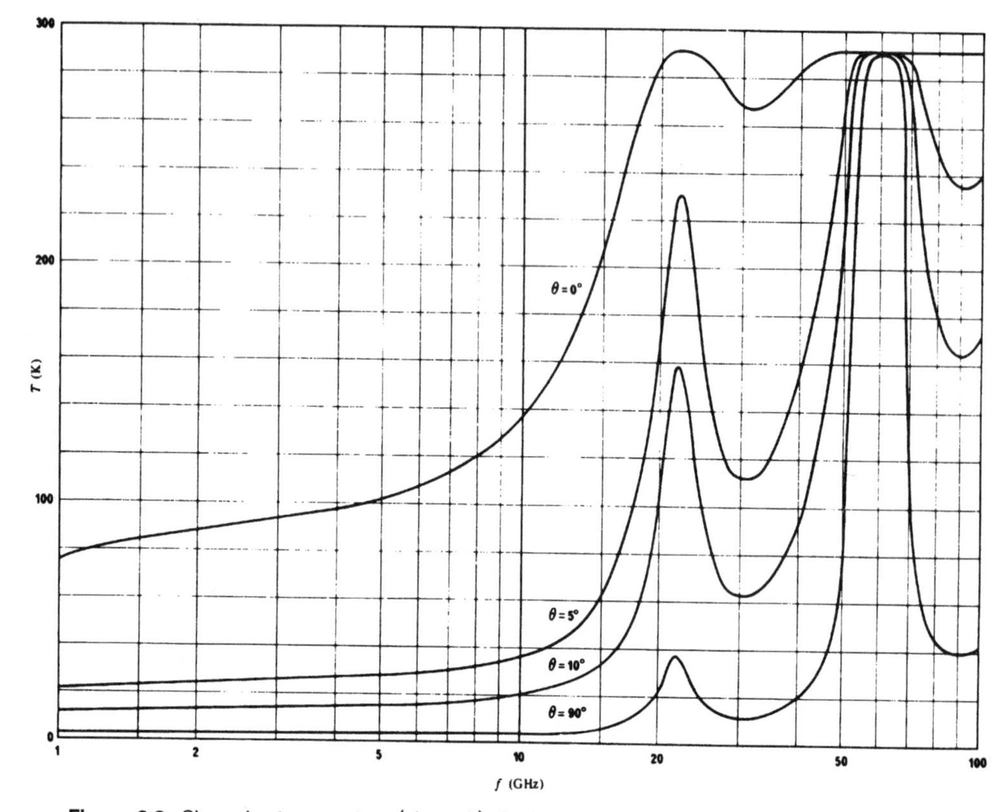

Figure 6.8. Sky noise temperature (clear air). Surface pressure is 1 atm; surface temperature is 20°C; surface water-vapor concentration is 10 g/m³; θ is the elevation angle. (From CCIR Rep. 720; Ref. 9. Courtesy of ITU-CCIR, Geneva.)

Determine the sky noise from Figure 6.7. For an elevation angle of 10° and a frequency of 12 GHz, the value is 19 K.

Sum the ohmic losses up to the reference plane:

$$L_a = 0.1 \text{ dB} + 2 \text{ dB} + 0.2 \text{ dB} + 0.4 \text{ dB} = 2.7 \text{ dB}$$

$$l_a = \log^{-1}(2.7/10)$$

$$= 1.86$$

Substitute into equation (6.28):

$$T_{\text{ant}} = \frac{(1.86 - 1)290 + 19}{1.86}$$

$$= 144.3 \text{ K}$$

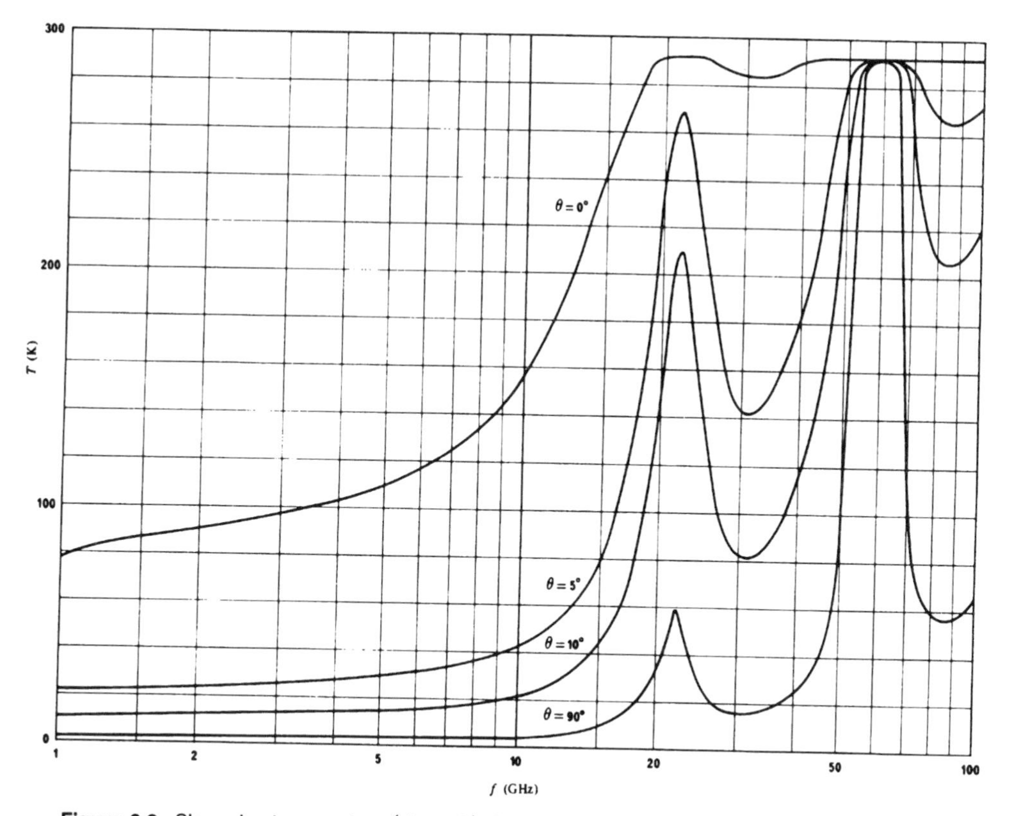

Figure 6.9. Sky noise temperature (clear air). Surface pressure is 1 atm; surface temperature is 20°C; surface water-vapor concentration is 17 g/m^3; θ = elevation angle. (From CCIR Rep. 720; Ref. 9. Courtesy of ITU-CCIR, Geneva.)

6.3.8.2 Calculation of Receiver Noise, T_r. A receiver will probably consist of a number of stages in cascade, as shown in Figure 6.11. The effective noise temperature of the receiving system, which we will call T_r, is calculated from the traditional cascade formula:

$$T_r = y_1 + \frac{y_2}{G_1} + \frac{y_3}{G_1 G_2} + \cdots + \frac{y_n}{G_1 G_2 \cdots G_{n-1}} \tag{6.30}$$

where y is the effective noise temperature of each amplifier or device and G is the numeric equivalent of the gain (or loss) of the device.

Example 8. Compute T_r for the first three stages of a receiving system. The first stage is a LNA with a noise figure of 1.1 dB and a gain of 25 dB. The second stage is a lossy transmission line with 2.2-dB loss. The third and final stage is a postamplifier with a 6-dB noise figure and a gain of 30 dB.

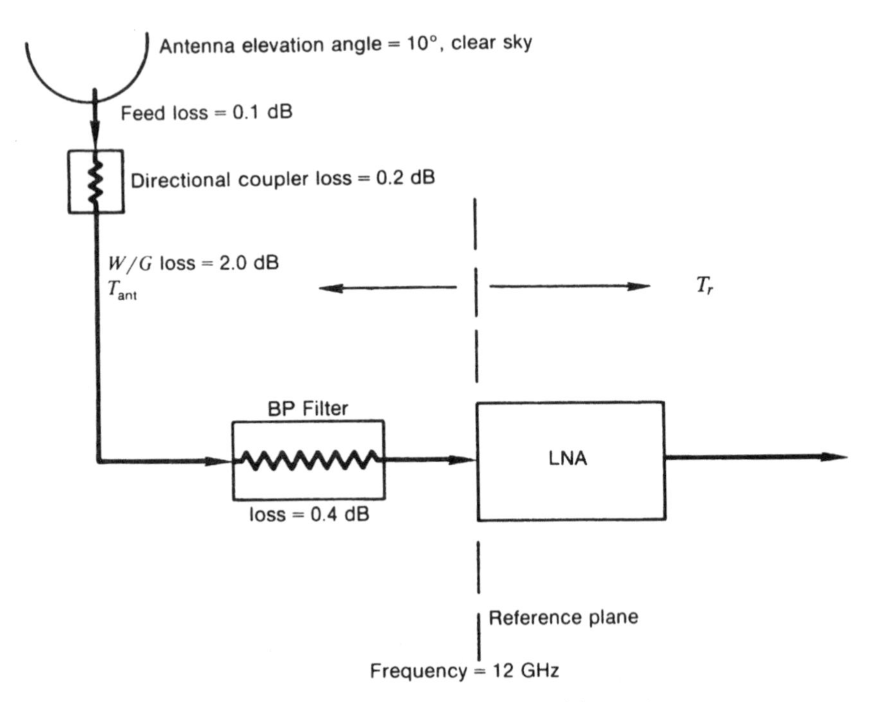

Figure 6.10. Example earth station receiving system.

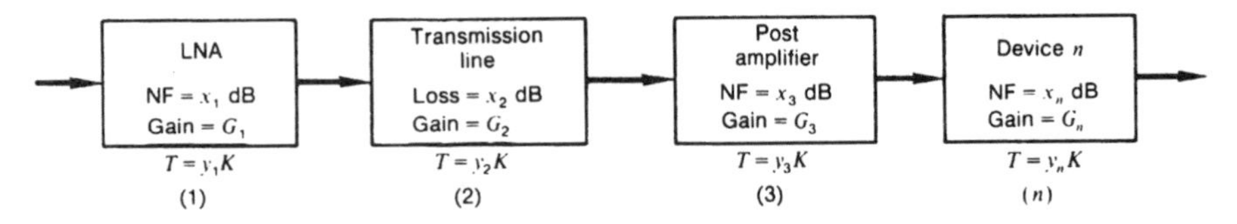

Figure 6.11. Generalized cascaded amplifiers/attenuators for noise temperature calculation.

Convert the noise figures to equivalent noise temperatures, using equation (6.21):

$$1.1 \text{ dB} = 10 \log\left(1 + \frac{T_e}{290}\right)$$

$$T_e = 83.6 \text{ K}$$

$$6.0 = 10 \log\left(1 + \frac{T_e}{290}\right)$$

$$T_e = 864.5 \text{ K}$$

Calculate the noise temperature of the lossy transmission line, using equation (6.11). First calculate l_a:

$$l_a = \log^{-1}\left(\frac{L_a}{10}\right)$$
$$= 1.66$$
$$T_e = (1.66 - 1)290$$
$$= 191.3 \text{ K}$$

Calculate T_r, using equation (6.30):

$$T_r = 83.6 + \frac{191.3}{316.2} + \frac{864.5}{316.2 \times 1/1.66}$$
$$= 83.6 + 0.605 + 4.53$$
$$= 88.735 \text{ K}$$

It should be noted that in the second and third terms, we divided by the numeric equivalent of the gain, not the gain in decibels. In the third term, of course, it was not a gain, but a loss (e.g., 1/1.66), where 1.66 is the numeric equivalent of a 2.2-dB loss. It will also be found that in cascaded systems the loss of a lossy device is equivalent to its noise figure.

6.3.8.3 Example Calculation of G/T. A satellite downlink operates at 21.5 GHz. Calculate the G/T of a terminal operating with this satellite. The reference plane is taken at the input to the LNA. The antenna has a 3-ft aperture displaying a 44-dB gross gain. There is 2 ft of waveguide with 0.2 dB/ft of loss. There is a feed loss of 0.1 dB; a bandpass filter has 0.4-dB insertion loss; a radome has a loss of 1.0 dB. The LNA has a noise figure of 3.0 dB and a 30-dB gain. The LNA connects directly to a downconverter/IF (intermediate frequency) amplifier combination with a single sideband noise figure of 13 dB.

Calculate the net gain of the antenna to the reference plane.

$$G_{\text{net}} = 44 \text{ dB} - 1.0 \text{ dB} - 0.1 \text{ dB} - 0.4 \text{ dB} - 0.4 \text{ dB}$$
$$= 42.1 \text{ dB}$$

This will be the value for G in the G/T expression.

Determine the sky noise temperature value at the 10° elevation angle, clear sky with dry conditions at 21.5 GHz. Use Figure 6.7. The value is 63 K.

Calculate L_A, the sum of the losses to the reference plane. This will include, of course, the radome loss:

$$L_A = 1.0 \text{ dB} + 0.1 \text{ dB} + 0.4 \text{ dB} + 0.4 \text{ dB} = 19 \text{ dB}$$

Calculate l_a, the numeric equivalent of L_A from the 1.9-dB value [equation (6.30)]:

$$l_a = \log^{-1}(1.9/10)$$
$$= 1.55$$

Calculate T_{ant}, using equation (6.28):

$$T_{\text{ant}} = \frac{(1.55 - 1)290 + 63}{1.55}$$
$$= 143.55 \text{ K}$$

Calculate T_r, using equation (6.30). First convert the noise figures to equivalent noise temperatures using equation (6.21). The LNA has a 3.0-dB noise figure, and its equivalent noise temperature is 290 K; the down-converter/IF amplifier has a noise figure of 13 dB and an equivalent noise temperature of 5496 K:

$$T_r = 290 + 5496/1000$$
$$= 295.5 \text{ K}$$

Calculate T_{sys} using equation (6.12):

$$T_{\text{sys}} = 143.55 + 295.5$$
$$= 439.05 \text{ K}$$

Calculate G/T using equation (6.27):

$$\frac{G}{T} = 42.1 \text{ dB} - 10\log(439.05)$$
$$= +15.67 \text{ dB/K}$$

The following discussion, taken from Ref. 6, further clarifies G/T analysis.

Figure 6.12 shows a satellite terminal receiving system and its gain/noise analysis. The notation used is that of the reference document.

The following are some observations of Figure 6.12 (using notation from the reference):

a. The value of T_S is different at every junction.
b. The value of G/T (where $T = T_S$) is the same at every junction.
c. The system noise temperature at the input to the LNA is influenced largely by the noise temperature of the components that precede the LNA and the LNA itself. The components that follow the LNA have a negligible contribution to the system noise temperature at the LNA input junction (reference plane) if the LNA gain is sufficiently high.

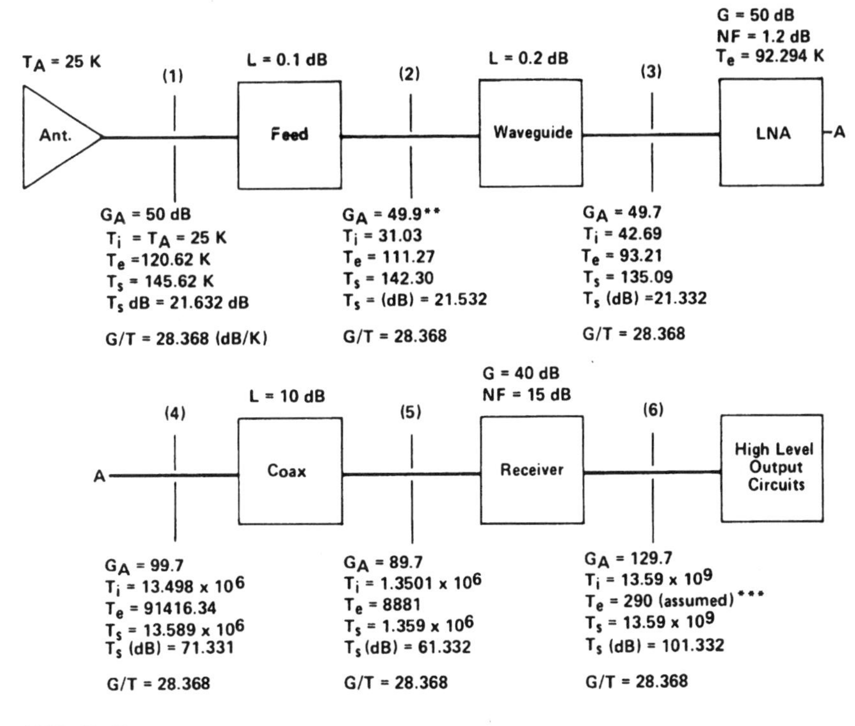

*G/T = G_A/T_s

**Dimensions are same as at position (1)

***T_e for the output circuits is likely to be inconsequential in comparison with T_i at (6) and can usually be neglected.

Figure 6.12. Example of noise temperature and G/T calculations for cascade of two ports. Note that G/T is independent of position in cascade. (From Scientific-Atlanta Satellite Communications Symposium 1982, Paper 2A, "Noise Temperature and G/T of Satellite Receiving Systems"; Ref. 6. Reprinted with permission.)

The parameters that have significant influence on G/T are the following:

a. The antenna gain and the antenna noise temperature.
b. The antenna elevation angle. The lower the angle, the higher the sky noise, thus, the higher the antenna noise, and hence the lower the G/T for a given antenna gain.
c. Feed and waveguide insertion losses. The lower the insertion loss of these devices, the higher the G/T.
d. LNA. The lower the noise temperature of the LNA, the higher the G/T. The higher the gain of the LNA, the less the noise contribution of the stages following the LNA. For instance, in Figure 6.12, if the gain of the LNA were reduced to 40 dB, the value of T_S would increase to

144.1 K. This means that the value of G/T would be reduced by about 0.26 dB. For a LNA with a gain of only 30 dB, the G/T would then drop by an additional 1.96 dB.

6.3.9 Calculation of C/N_0 Using the Link Budget Technique

The link budget is a tabular method of calculating space communication system parameters. It is similar to the approach used in Chapters 2, 3, and 5, where it was called link analysis. In the method presented here the starting point of a link budget is the platform EIRP. The platform can be a terminal or a satellite. In an equation, it would be expressed as follows:

$$\frac{C}{N_0} = \text{EIRP} - \text{FSL}_{\text{dB}} - (\text{other losses}) + G/T_{\text{dB/K}} - k \qquad (6.31)$$

where FSL is the free-space loss, k is Boltzmann's constant expressed in dBW or dBm, and the "other losses" may include the following:

- Polarization loss
- Pointing losses (terminal and satellite)
- Off-contour loss
- Gaseous absorption losses
- Excess attenuation due to rainfall (as applicable)

The off-contour loss refers to spacecraft antennas that are not earth coverage, such as spot beams, zone beams, and MBA (multiple beam antenna), and the contours are flux density contours on the earth's surface. This loss expresses in equivalent decibels the distance from the terminal platform to a contour line. Satellite pointing loss, in this case, expresses contour line error or that the contours are not exactly as specified.

6.3.9.1 ***Some Link Loss Guidelines.*** Free-space loss is calculated using equations (1.7), (1.9a), and (1.9b). Care should be taken in the use of units for distance (range) and frequency.

When no other information is available, use 0.5 dB as an estimate for polarization loss and pointing losses (e.g., 0.5 dB for satellite pointing loss and 0.5 dB for terminal pointing loss).

For off-contour loss, the applicable contour map should be used, placing the prospective satellite terminal in its proposed location on the map and estimating the loss. Figure 6.13 is a typical contour map. Also see Figures 6.23 and 6.24.

Atmospheric absorption losses are comparatively low for systems operating below 10 GHz. These losses vary with frequency, elevation angle, and altitude. For the 7-GHz downlink, 0.8 dB is appropriate for a 5° elevation angle, dropping to 0.5 dB for 10° and 0.25 dB for 15°; all values are for sea level. For 4 GHz, recommended values are 0.5 dB for 5° elevation angle and

0.25 dB for 10°; all values are for sea level. Atmospheric absorption losses are treated more extensively in Chapter 9.

Excess attenuation due to rainfall is also rigorously treated in Chapter 9. This attenuation also varies with elevation angle and altitude. Suggested estimates are 0.5 dB at 5° elevation angle for 4-GHz downlink band, 0.25 dB at 10°, and 0.15 dB at 15°, with similar values for the 6-GHz uplink; for the 7-GHz military band, 3 dB for 5°, 1.5 dB at 10°, and 0.75 dB at 15°, all values at sea level. Use similar values for the 8-GHz uplink.

6.3.9.2 *Link Budget Examples.* The link budget is used to calculate C/N_0 when other system parameters are given. It is also used when C/N_0 is given when it is desired to calculate one other parameter such as G/T or EIRP of either platform in the link.

Example 9. We have 4-GHz downlink, FDM/FM, 5° elevation angle. Satellite EIRP is +30 dBW. Range to satellite (geostationary) is 22,208 nautical miles or 25,573 statute miles from Figure 6.5. Terminal $G/T = +20$ dB/K. Calculate C/N_0.

Calculate free-space loss from equation (1.9b):

$$\begin{aligned} \text{FSL}_{\text{dB}} &= 36.58 + 20\log(4 \times 10^3) + 20\log(25{,}573) \\ &= 196.78 \text{ dB} \end{aligned}$$

EIRP of satellite	+30 dBW
Free-space loss	−196.78 dB
Satellite pointing loss	0.5 dB
Off-contour loss	0.5 dB
Atmospheric absorption loss	0.5 dB
Rainfall loss	0.5 dB
Polarization loss	0.5 dB
Terminal pointing loss	0.5 dB
Isotropic receive level	−169.78 dBW
Terminal G/T	+20 dB/K
Sum	−149.78 dBW
Boltzmann's constant	−(−228.6 dBW)
C/N_0	78.82 dB

Example 10. Calculate the required satellite G/T, where the uplink frequency is 6.0 GHz and the terminal EIRP is +70 dBW, and with a 5° terminal elevation angle. The required C/N_0 at the satellite is 102.16 dB (typical for an uplink video link). The free-space loss is calculated as in Example 9 but with a frequency of 6.0 GHz. This loss is 200.3 dB. Call the satellite G/T value X.

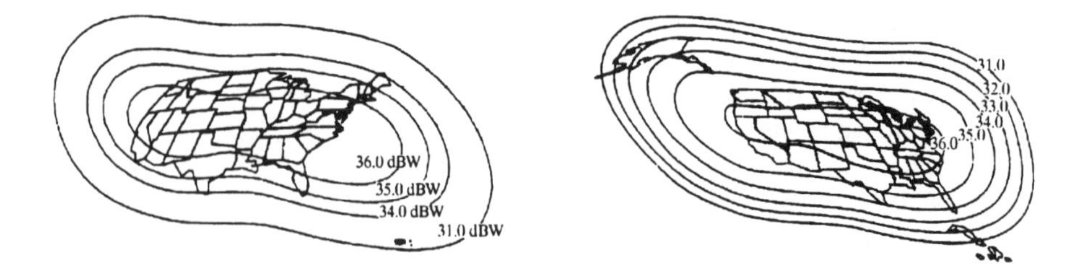

Figure 6.13. Typical 4-GHz EIRP contours for Spacenet II, 69° W (left) and ASC I, 128° W (right), 36-MHz transponders. Copyright 1997 GE Capital Services Corp., McLean, VA.

Terminal EIRP	+70 dBW	
Terminal pointing loss	0.5 dB	
Free-space loss	200.3 dB	
Polarization loss	0.5 dB	
Satellite pointing loss	0.5 dB	
Atmospheric absorption loss	0.5 dB	
Off-contour loss	0.5 dB	
Rainfall loss	0.5 dB	
Isotropic receive level at satellite	−133.3 dBW	
G/T of satellite	X dB/K	(initially let $X = 0$ dB/K)
Sum	−133.3 dBW	
Boltzmann's constant	−(−228.6 dBW)	
C/N_0 (as calculated)	95.3 dB	
C/N_0 (required)	102.16 dB	
G/T	+6.86 dB/K	(difference)

This G/T value, when substituted for X, will derive a C/N_0 of 102.16 dB. It is not advisable to design a link without some margin. Margin will compensate for link degradation as well as errors of link budget estimation. The more margin (in decibels) that is added, the more secure we are that the link will work. On the other hand, each decibel of margin costs money. Some compromise between conservation and economic realism should be met. For this link, 4 dB might be such a compromise. If it were all allotted to satellite G/T, then the new G/T value would be +10.86 dB/K. Other alternatives to build in a margin would be to increase transmitter power output, thereby increasing EIRP, and increasing terminal antenna size. The power of using the link budget can now be seen easily. The decibels flow through on a one-for-one basis, and it is fairly easy to carry out trade-offs.

It should be noted that when the space platform (satellite) employs an earth coverage (EC) antenna, satellite pointing loss and off-contour loss are disregarded. The beamwidth of an EC antenna is usually accepted as 17° or 18° when the satellite is in geostationary orbit.

6.3.9.3 Calculating System C/N_0. The final C/N_0 value we wish to know is that at the terminal receiver. This must include, as a minimum, the C/N_0 for the uplink and C/N_0 for the downlink. If the satellite transponder is shared (e.g., simultaneous multicarrier operation on one transponder), a value for C/N_0 for satellite intermodulation noise must also be included. The basic equation to calculate C/N_0 for the system is given as

$$\left(\frac{C}{N_0}\right)_{(s)} = \frac{1}{1/(C/N_0)_{(u)} + 1/(C/N_0)_{(d)}} \tag{6.32}$$

Example 11. Consider a bent-pipe satellite system where the uplink $C/N_0 =$ 105 dB and the downlink $C/N_0 = 95$ dB. What is the system C/N_0? Convert each value of C/N_0 to its numeric equivalent:

$$\log^{-1}(105/10) = 3.16 \times 10^{10}; \qquad \log^{-1}(95/10) = 0.316 \times 10^{10}$$

Invert each value and add. Invert this value and take 10 log.

$$\left(\frac{C}{N_0}\right)_{(s)} = 94.6 \text{ dB/Hz}$$

Many satellite transponders simultaneously permit multiple-carrier access, particularly when operated in the FDMA or SCPC regimes. FDMA (frequency division multiple access) and SCPC (single-channel-per-carrier) systems are described subsequently in this chapter.

When transponders are operated in the multiple-carrier mode, the downlink signal is rich in intermodulation (IM) products. Cumulatively, this is IM noise, which must be tightly controlled, but cannot be eliminated. Such noise must be considered when calculating $(C/N_0)_{(s)}$ when two or more carriers are put through the same transponder simultaneously. The following equation now applies to calculate $(C/N_0)_{(s)}$:

$$\left(\frac{C}{N_0}\right)_{(s)} = \frac{1}{(C/N_0)_{(u)}^{-1} + (C/N_0)_{(d)}^{-1} + (C/N_0)_{(i)}^{-1}} \tag{6.33}$$

where the subscripts s, u, d, and i refer to system, uplink, downlink, and intermodulation, respectively.

The principal source of IM noise (products) in a satellite transponder is the final amplifier, often called HPA (high-power amplifier). With present

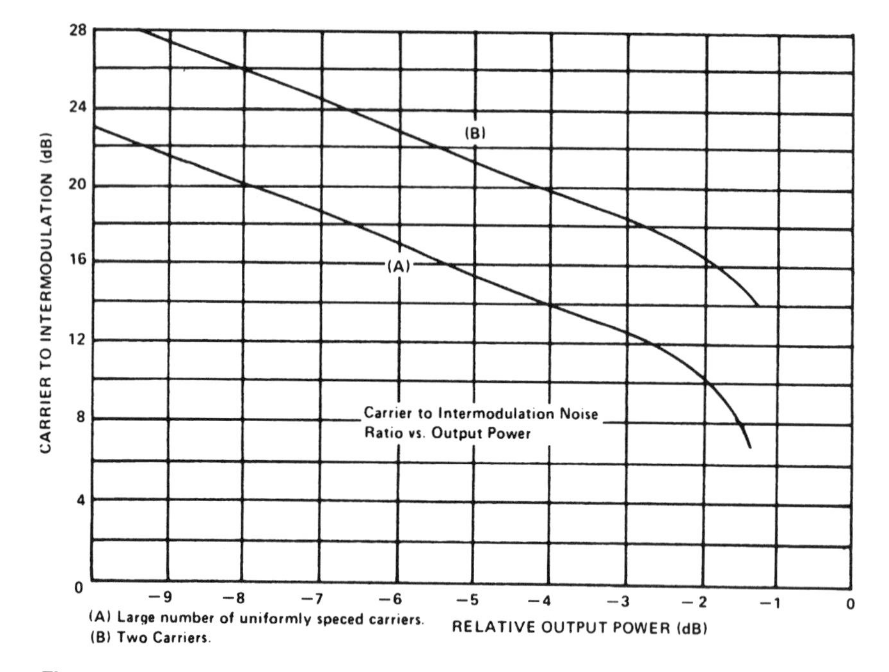

Figure 6.14. TWT power transfer characteristics. (Courtesy of Scientific-Atlanta; Ref. 6.)

technology, this is usually a TWT (traveling-wave tube) amplifier, although SSAs (solid-state amplifiers) are showing increasing implementation on space platforms.

Figure 6.14 shows typical IM curves for a bent-pipe type transponder using a TWT transmitter. The lower curve in the figure (curve A) is the IM performance for a large number of equally spaced carriers within a single transponder. The upper curve (curve B) shows the IM characteristics of two carriers.

In order to bring the IM products to an acceptable level and thus not degrade the system C/N_0 due to poor C/IM performance, the total power of the uplink must be "backed off" or reduced, commonly to a C/IM ratio of 20 dB or better. The result, of course, is a reduction in downlink EIRP. Figure 6.15 shows the effect of input (drive) power reduction versus output power of a TWT. To increase the C/IM ratio to 24 dB for the two-carrier case, we can see from Figure 6.14 that the input power must be backed off by 7 dB. Figure 6.15 shows that this results in a total downlink power reduction of 3 dB. It should also be noted that the resulting downlink power must be shared among the carriers actually being transmitted. For example, if two equal-level carriers share a transponder, the power in each carrier is 3 dB lower than the backed-off value.

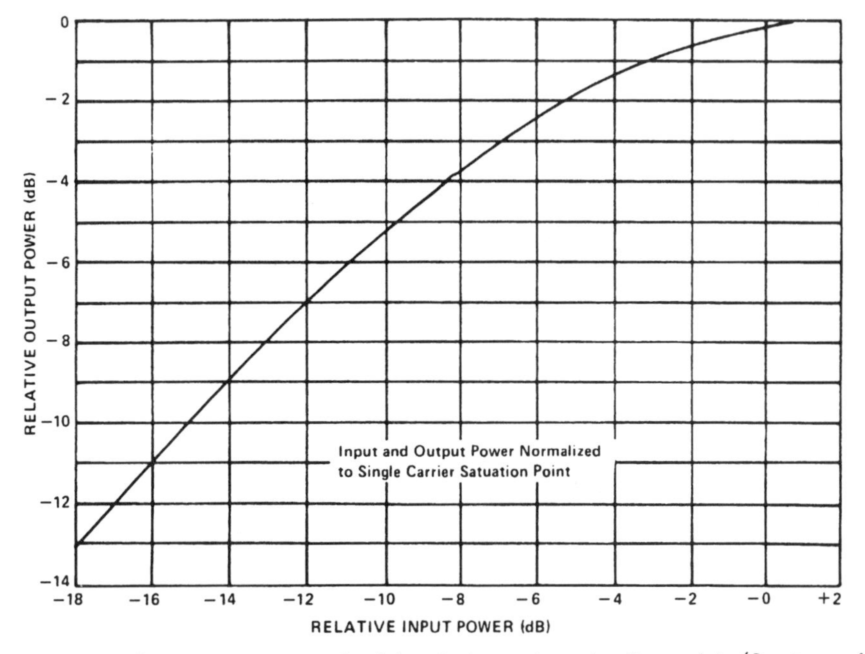

Figure 6.15. Output power normalized to single carrier saturation point. (Courtesy of Scientific-Atlanta; Ref. 6.)

6.3.10 Calculating S/N

6.3.10.1 In a VF Channel of a FDM/FM Configuration. To calculate S/N and psophometrically weighted noise (pWp0) in a voice channel when C/N in the IF is given, use the following formula (Ref. 6):

$$\frac{\mathrm{S}}{\mathrm{N}} = \frac{\mathrm{C}}{\mathrm{N}} + 20\log\left(\frac{\Delta F_{\mathrm{TT}}}{f_{\mathrm{ch}}}\right) + 10\log\left(\frac{B_{\mathrm{IF}}}{B_{\mathrm{ch}}}\right) + P + W \tag{6.34}$$

where ΔF_{TT} = rms test tone deviation
f_{ch} = highest baseband frequency
B_{ch} = voice channel bandwidth (3.1 kHz)
P = top VF channel emphasis improvement factor
W = psophometric weighting improvement factor (2.5 dB)

Once the voice channel S/N has been calculated, the noise in the voice channel in picowatts may be determined from

$$\text{Noise} = \log^{-1}\{[90 - (\mathrm{S/N})]/10\} \quad (\mathrm{pWp0}) \tag{6.35}$$

Table 6.6 lists the transmission parameters of INTELSAT V through VIIA, which we will use below in an example problem. These are typical bent-pipe satellite system parameters for communication links.

TABLE 6.6 INTELSAT V, VA, VA(IBS) VI, VII, VIIA, and K Transmission Parameters: Regular FDM/FM Carriers

Carrier Capacity (Number of Channels)	Top Baseband Frequency (kHz)	Allocated Satellite BW Unit (MHz)	Occupied Bandwidth (MHz)	Deviation (rms) for 0 dBm0 Test Tone (kHz)	Multichannel rms Deviation (kHz)	Carrier-to-Total Noise Temperature Ratio at Operating Point 8000 ± 200 pWOp from RF Sources (dBW / K)	Carrier-to-Noise Ratio in Occupied BW (dB)	Ratio of Unmodulated Carrier Power to Maximum Carrier Power Density Under Full-Load Conditions
n	f_m	B_a	B_o	f_r	f_{mc}	C/T	C/N	(dB/4 kHz)
12	60	1.25	1.125	109	159	−154.7	13.4	20.0
24	108	2.5	2.00	164	275	−153.0	12.7	22.3
36	156	2.5	2.25	168	307	−150.0	15.1	22.8
48	204	2.5	2.25	151	292	−146.7	18.4	22.6
60	252	2.5	2.25	136	276	−144.0	21.1	22.4
60	252	5.0	4.0	270	546	−149.9	12.7	25.3
72	300	5.0	4.5	294	616	−149.1	13.0	25.8
96	408	5.0	4.5	263	584	−145.5	16.6	25.6
132	552	5.0	4.4	223	529	−141.4	20.7	24.2[a] ($X = 1$)
96	408	7.5	5.9	360	799	−148.2	12.7	27.0
132	552	7.5	6.75	376	891	−145.9	14.4	27.5
192	804	7.5	6.4	297	758	−140.6	19.9	25.8[a] ($X = 1$)
132	552	10.0	7.5	430	1020	−147.1	12.7	28.0
192	804	10.0	9.0	457	1167	−144.4	14.7	28.6
252	1052	10.0	8.5	358	1009	−139.9	19.4	27.0[a] ($X = 1$)

252	1052	15.0	12.4	577	1627	−144.1	13.6	30.0
312	1300	15.0	13.5	546	1716	−141.7	15.6	30.2
372	1548	15.0	13.5	480	1646	−138.9	18.4	30.1
432	1796	15.0	13.0	401	1479	−136.2	21.2	27.6[a] ($X = 1$)
312	1300	17.5	15.75	663	2081	−143.4	13.2	31.2
372	1548	17.5	15.75	583	1999	−140.8	15.9	31.0
432	1796	17.5	15.75	517	1919	−138.5	18.2	30.8
432	1796	20.0	18.0	616	2276	−139.9	16.1	31.5
492	2044	20.0	18.0	558	2200	−137.8	18.2	31.4
552	2292	20.0	18.0	508	2121	−136.0	20.0	30.2[a] ($X = 1$)
432	1796	25.0	20.7	729	2688	−141.4	14.1	32.2
492	2044	25.0	22.5	738	2911	−140.3	14.8	32.6
552	2292	25.0	22.5	678	2833	−138.5	16.6	32.5
612	2540	25.0	22.5	626	2755	−136.9	18.1	32.4
612	2540	36.0	32.4	983	4325	−141.0	12.5	34.3
792	3284	36.0	32.4	816	4085	−137.0	16.5	34.1
972	4028	36.0	32.4	694	3849	−133.8	19.7	32.8[a] ($X = 1$)
792	3284	36.0	36.0	930	4653	−138.3	14.8	34.7
972	4028	36.0	36.0	802	4417	−135.2	17.8	34.5

[a]This value is X dB lower than the value calculated according to the normal formula used to derive this ratio:

$$10 \log_{10}(f_{mc}\sqrt{2\pi}/4)$$

where X is the value in brackets in the last column and f_{mc} is the rms multichannel deviation in kilohertz. This factor is necessary in order to compensate for low modulation index carriers, which are not considered to have a Gaussian power density distribution.

Source: INTELSAT, IESS-301, Rev. 3 (Ref. 11).

Example 12. Using Table 6.6, calculate S/N and psophometrically weighted noise in a FDM/FM derived voice channel for a 972-channel system. First use equation (6.34) to calculate S/N and then equation (6.35) to calculate the noise in pWp. Use the value f_r for rms test tone deviation, f_m for the maximum baseband frequency, and B_{IF} for the allocated satellite bandwidth. The emphasis improvement may be taken from Figure 2.15. C/N is 17.8 dB in this case:

$$\begin{aligned} \mathrm{S/N} &= 17.8\ \mathrm{dB} + 20\log(802 \times 10^3/4028 \times 10^3) \\ &\quad + 10\log(36 \times 10^6/3.1 \times 10^3) + 2.5\ \mathrm{dB} + 4.5\ \mathrm{dB} \\ &= 51.43\ \mathrm{dB} \\ \mathrm{Noise} &= \log^{-1}[(90 - 51.43)/10] \\ &= 7194.5\ \mathrm{pWp} \end{aligned}$$

6.3.10.2 *For a Typical Video Channel.*

As suggested by Ref. 6,

$$\frac{\mathrm{S}}{\mathrm{N_v}} = \frac{\mathrm{C}}{\mathrm{N}} + 10\log 3\left(\frac{\Delta f}{f_m}\right)^2 + 10\log\left(\frac{B_{\mathrm{IF}}}{2B_v}\right) + W + CF \qquad (6.36)$$

where $\mathrm{S/N_v}$ = peak-to-peak luminance signal-to-noise ratio
Δf = peak composite deviation of the video
f_m = highest baseband frequency
B_v = video noise bandwidth (for NTSC systems this is 4.2 MHz)
B_{IF} = IF noise bandwidth
W = emphasis plus weighting improvement factor (12.8 dB for NTSC North American systems)
CF = rms to peak-to-peak luminance signal conversion factor (6.0 dB)

For many satellite systems, without frequency reuse, a 500-MHz assigned bandwidth is broken down into twelve 36-MHz segments.* Each segment is then assigned to a transponder. For video transmission either a half or full transponder is assigned.

Example 13. A video link is relayed through a 36-MHz (full transponder) transponder bent-pipe satellite, where the peak composite deviation is 11 MHz and the C/N is 14.6 dB. What is the weighted signal-to-noise ratio? Assume NTSC standards.

*Many new systems coming on-line use wider bandwidths.

Using equation (6.36), we find

$$\frac{\mathrm{S}}{\mathrm{N}} = 14.6\ \mathrm{dB} + 10\log\left[3(11/4.2)^2\right] + 10\log(36/8.4) + 12.8 + 6$$

$$= 52.9\ \mathrm{dB}$$

In this case the video noise bandwidth is 4.2 MHz, which is also the highest baseband frequency. European systems would use 5 MHz. The reader should consult ITU-R Rec. BT.470-3 (Table 1 of Ref. 12) or latest revision.

Equation (6.36) is only useful for C/N above 11 dB. Below that C/N value, impulse noise becomes apparent and equation (6.36) is not valid.

6.3.10.3 *S/N of the Video-Related Audio or Aural Channel.* Conventionally, the aural subcarrier is placed above the video in the baseband. To calculate the signal-to-noise ratio of the derived audio, we first calculate the $\mathrm{C/N_{sc}}$ or carrier-to-noise ratio of the audio subcarrier when the C/N of the main carrier is given. $\mathrm{C/N_{sc}}$ may be calculated from

$$\left(\frac{\mathrm{C}}{\mathrm{N}}\right)_{\mathrm{sc}} = \frac{\mathrm{C}}{\mathrm{N}} + 10\log\left(\frac{B_{\mathrm{IF}}}{2B_{\mathrm{sc}}}\right) + 10\log\left[\left(\frac{\Delta F_c}{f_{\mathrm{sc}}}\right)^2\right] \quad (6.37)$$

where ΔF_c = peak deviation of the main carrier by the subcarrier
I_{sc} = subcarrier frequency
B_{sc} = subcarrier filter noise bandwidth
B_{IF} = IF noise bandwidth

Now we can calculate the signal-to-noise ratio of the aural channel by the following equation:

$$\left(\frac{\mathrm{S}}{\mathrm{N}}\right)_a = \left(\frac{\mathrm{C}}{\mathrm{N}}\right)_{\mathrm{sc}} + 10\log\left[3\left(\frac{\Delta F_{\mathrm{sc}}}{f_m}\right)^2\right] + 10\log\left(\frac{B_{\mathrm{sc}}}{2B_a}\right) + E \quad (6.38)$$

where f_m = maximum audio frequency
B_a = audio noise bandwidth
ΔF_{sc} = peak subcarrier deviation
E = audio pre-/de-emphasis advantage, which usually is given the value of 12 dB

Example 14. Calculate the S/N of the aural channel where the aural subcarrier frequency is 6.8 MHz; the peak carrier deviation is 2 MHz; the peak subcarrier deviation is 75 kHz; and the aural channel bandwidth is 15 kHz. Assume C/N is 14 dB, B_{sc} is 600 kHz, and the audio noise bandwidth is 30 kHz.

First calculate C/N_{sc} using equation (6.37):

$$\left(\frac{C}{N}\right)_{sc} = 14\text{ dB} + 10\log[36 \times 10^6/(2)(600 \times 10^3)]$$

$$+ 10\log\left[(2 \times 10^6/6.8 \times 10^6)^2\right]$$

$$= 14 + 14.77 - 10.63$$

$$= 18.14\text{ dB}$$

Calculate S/N_a by equation (6.38):

$$\left(\frac{S}{N}\right)_a = 18.14\text{ dB} + 10\log\left[3(75 \times 10^3/15 \times 10^3)^2\right]$$

$$+ 10\log(600 \times 10^3/30 \times 10^3) + 12\text{ dB}$$

$$= 18.14 + 18.75 + 13 + 12$$

$$= 61.89\text{ dB}$$

6.3.10.4 Calculation of S/N for Analog Single-Channel-per-Carrier (SCPC) Systems. SCPC is a method of accommodating multiple carriers on a single satellite transponder. It is a satellite-access method, similar to FDMA, where each carrier is modulated by a single channel, either a voice channel or a program channel. Access methods are discussed in Section 6.4. SCPC modulation techniques may be analog (FM) or digital. In this subsection only the analog (FM) technique is discussed. Companding, weighting, and emphasis are used.

Channel signal-to-noise ratio may be calculated when C/N is given by

$$\frac{S}{N} = \frac{C}{N} + 10\log\left[3\left(\frac{\Delta F}{f_m}\right)^2\right] + 10\log\left(\frac{B_{IF}}{2\,B_a}\right) + W + C \qquad (6.39)$$

where ΔF = peak deviation
f_m = highest modulating frequency (baseband frequency)
B_a = audio noise bandwidth
B_{IF} = IF bandwidth
W = emphasis plus weighting improvement factor
C = companding advantage

Example 15. Calculate the signal-to-noise ratio of a FM SCPC VF channel, where the C/N is 10 dB; the peak deviation is 7.3 kHz; the IF noise bandwidth is 25 kHz; the weighting/emphasis advantage is 7 dB; and the companding advantage is 17 dB.

Use formula (6.39):

$$\frac{S}{N} = 10\text{ dB} + 10\log\left[3(7.3 \times 10^3/3.4 \times 10^3)^2\right]$$

$$+ 10\log(25 \times 10^3/6.8 \times 10^3) + 7\text{ dB} + 17\text{ dB}$$

$$= 51.1\text{ dB}$$

The voice channel bandwidth in this example has a 3.4-kHz bandwidth.

6.3.10.5 System Performance Parameters. Table 6.7 gives some typical performance parameters for video, aural channel, and FDM/FM for a 1200-channel configuration.

6.4 ACCESS TECHNIQUES

6.4.1 Introduction

Access refers to the way in which a communication system uses a satellite transponder. There are three basic access techniques:

1. FDMA (frequency division multiple access)
2. TDMA (time division multiple access)
3. CDMA (code division multiple access)

With FDMA a satellite transponder is divided into frequency band segments, where each segment is assigned to a user. The number of segments can vary from one, where an entire transponder is assigned to a single user, to literally hundreds of segments, which is typical of SCPC operation. For analog telephony operation, each segment is operated in FDM/FM mode. In this case FDM group(s) and/or supergroup(s) are assigned for distinct distant location connectivity.

TDMA works in the time domain. Only one user appears on the transponder at any given time. Each user is assigned a time slot to access the satellite. System timing is crucial with TDMA. It lends itself only to digital transmission, typically PCM.

CDMA is particularly attractive to military users due to its antijam and low probability of intercept (LPI) properties. With CDMA, the transmitted signal is spread over part or all of the available transponder bandwidth in a time–frequency relationship by a code transformation. Typically, the modulated RF bandwidth is ten to hundreds of times greater than the information bandwidth.

TABLE 6.7 Some Typical Satellite Link Performance Parameters (Analog Operation)

System Parameters	Units	FDM / FM	Video
TV video			
C/N	dB	—	14.6
Maximum video frequency	MHz	—	4.2
Overdeviation	dB	—	—
Peak operating deviation	MHz	—	10.7
FM improvement	dB	—	13.2
BW improvement	dB	—	6.3
Weighting/emphasis improvement	dB	—	12.8
P-rms Conversion Factor	dB	—	6.0
Total improvement	dB	—	38.3
S/N (peak-to-peak/rms-luminance signal)	dB	—	52.9
TV program channel (subcarrier)			
Peak carrier deviation	MHz	—	2.0
Subcarrier frequency	MHz	—	7.5
FM improvement	dB	—	11.5
BW improvement	dB	—	14.0
Total improvement	dB	—	2.5
C/N_{sc} (subcarrier)	dB	—	16.9
Peak subcarrier deviation	kHz	—	75
Maximum audio frequency	kHz	—	15
FM improvement	dB	—	18.8
BW improvement	dB	—	13.8
Emphasis improvement	dB	—	12.0
Total improvement	dB	—	44.6
S/N (audio)	dB	—	59.2
FDM/FM			
Number of channels	—	1200	—
Test tone deviation (rms)	kHz	650	—
Top baseband frequency	kHz	5260	—
FM improvement	dB	18.2	—
BW improvement	dB	40.7	—
Weighting improvement	dB	2.5	—
Emphasis (top slot) improvement	dB	4.0	—
Total improvement	dB	29.0	—
TT/N (test tone to noise ratio)	dB	49.0	—
Noise	pWp0	12,589	—

Source: Scientific-Atlanta Satellite Communications Symposium 1982. Courtesy of Scientific-Atlanta (Ref. 6).

CDMA had previously only been attractive to the military user because of its antijam properties. Since 1980 some interest in CDMA has been shown by the commercial sector for demand access for large populations of data circuit/network users with bursty requirements in order to improve spectral utilization. For further discussion of spread spectrum systems refer to Chapter 13.

Figure 6.16 depicts the three basic types of satellite multiple access techniques. The horizontal axis represents time and the vertical axis represents spectral bandwidth.

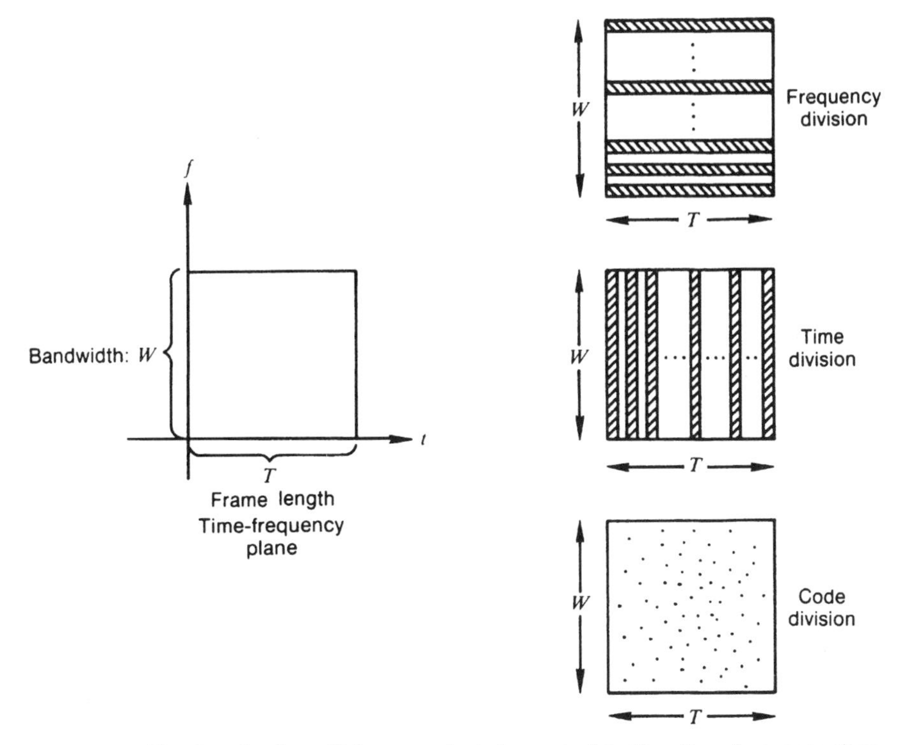

Figure 6.16. The three basic multiple access techniques depicted in a time–frequency plane.

6.4.2 Frequency Division Multiple Access (FDMA)

FDMA has been the primary method of multiple access for satellite communication systems for the past 20 years and will probably remain so for quite some time. In the telephone service, it is most attractive for permanent direct or high-usage (HU) routes, where, during the busy hour, trunk groups to a particular distinct location require 5 or more Erlangs (e.g., one or more FDM groups).

For this most basic form of FDMA, a single earth station transmits a carrier to a bent-pipe satellite. The carrier contains a FDM configuration in its modulation consisting of groups and supergroups for distinct distant locations. Each distant location receives and demodulates the carrier but demultiplexes only those FDM groups and supergroups pertaining to it. For full duplex service, a particular earth station receives, in return, one carrier from each distant location with which it provides connectivity. So, on its downlink, it must receive, and select by filtering, carriers from each distant location and demultiplex only that portion of each derived baseband that contains FDM channelization destined for it.

Another form of FDMA is SCPC, where each individual telephone channel independently modulates a separate radio-frequency carrier. Each carrier may be frequency modulated or digitally modulated, often PSK. SCPC is useful on low-traffic-intensity routes (e.g., less than 5 Erlangs) or for overflow from FDM/FDMA. SCPC will be discussed in Section 6.4.2.1.

FDMA is a mature technology. Its implementation is fairly straightforward. Several constraints must be considered in system design. Many of these constraints center around the use of TWT as the HPA in satellite transponders. Depending on the method of modulation employed, amplitude and phase nonlinearities must be considered to minimize IM products, intelligible crosstalk, and other interfering effects by taking into account the number and size (bandwidth) of carriers expected to access a transponder. These impairments are maintained at acceptable levels by operating the transponder TWT at the minimum input backoff necessary to ensure that performance objectives are met. However, this method of operation results in less available channel capacity when compared to a single access mode.

TWT amplifiers operate most efficiently when they are driven to an operating point at or near saturation. When two or more carriers drive the TWT near its saturation point, excessive IM products are developed. These can be reduced to acceptable levels by reducing the drive, which, in turn, reduces the amplifier efficiency. This reduction in drive power is called backoff.

In early satellite communication systems, IM products created by TWT amplitude nonlinearity were the dominant factor in limiting system operation. However, as greater power became available and narrow bandwidth transponders capable of operation with only a few carriers near saturation became practical, maximum capacity became dependent on a carefully evaluated trade-off analysis of a number of parameters, which include the following:

1. Satellite TWT impairments including in-band IM products produced by both amplitude and phase nonlinearities and intelligible crosstalk caused by FM-AM-PM conversion during multicarrier operation.
2. FM transmission impairments not directly associated with the satellite transponder TWT such as adjacent carrier interference caused by frequency spectrum overlap between adjacent carriers, which gives rise to convolution and impulse noise in the baseband; dual path distortion between transponders; interference due to adjacent transponder IM earth station out-of-band emission; and frequency reuse co-channel interference.
3. General constraints including available power and allocated bandwidth; uplink power control; frequency coordination; and general vulnerability to interference.

Backoff was discussed briefly in Section 6.3.9.3. From Figures 6.14 and 6.15 we can derive a rough rule-of-thumb that tells us that for approximately every decibel of backoff, IM products drop ~ 2 dB for the multicarrier case (i.e., more than three carriers on a transponder). Also, for every decibel of backoff on TWT driving power, TWT output power drops 1 dB. Of course, this causes inefficiency in the use of the TWT. As the number of carriers is increased on a transponder, the utilization of the available bandwidth becomes less efficient. If we assume 100% efficient utilization of a transponder with only one carrier, then with two carriers the efficiency drops to about 90%, with four to 60%, with eight to about 50%, and with fourteen carriers to about 40%.

Crosstalk is a significant impairment in FDMA systems. It can result from a sequence of two phenomena:

1. An amplitude response that varies with frequency, producing amplitude modulation coherent with the original frequency modulation of a RF carrier or FM/AM transfer.
2. A coherent amplitude modulation that phase-modulates all carriers occupying a common TWT amplifier due to AM/PM conversion.

As a carrier passes through a TWT amplifier, it may produce amplitude modulation from gain–slope anomalies in the transmission path. Another carrier passing through the same TWT will vary in phase at the same rate as the AM component and thereby pick up intelligible crosstalk from any carrier sharing the transponder. Provisions should be made in specifying TWT amplifier characteristics to ensure that AM/PM conversion and gain–slope variation meet system requirements. Intelligible crosstalk should be 58 dB down or better, and, with modern equipment, this goal can be met quite easily.

Out-of-band RF emission from an earth station is another issue; 500 pWp0 is commonly assigned in a communication satellite system voice channel noise budget for earth station RF out-of-band emission. The problem centers on the earth station HPA TWT. When a high-power, wideband TWT is operated at saturation, its full output power can be realized, but it can also produce severe IM RF products to the up-path of other carriers in a FDMA system. To limit such unwanted RF emission, we must again turn to the backoff technique of the RF drive to the TWT. Some systems use as much as 7-dB backoff. For example, a 12-kW HPA with 7-dB backoff is operated at about 2.4 kW. This is one good reason for overbuilding earth station TWTs and accepting the inefficiency of use.

Uplink power control is an important requirement for FDM/FM/FDMA systems. Sufficient power levels must be maintained on the uplink to meet signal-to-noise ratio requirements on the derived downlinks for bent-pipe transponders. On the other hand, uplink power on each carrier accessing a particular transponder must be limited to maintain IM product generation (IM noise) within specifications. This, of course, is the backoff discussed above. Close control is required of the power level of each carrier to keep transmission impairments within the total noise budget.

One method of meeting these objectives is to study each transponder configuration on a case-by-case basis before the system is actually implemented on the satellite. Once the proper operating values have been determined, each earth station is requested to provide those values of uplink carrier levels. These values can be further refined by monitoring stations that precisely measure resulting downlink power of each carrier. The theoretical power levels are then compared to both the reported uplink and the measured downlink levels.

Energy dispersal is yet another factor required in the design of a bent-pipe satellite system. Energy dispersal is used to reduce the spectral energy density of satellite signals during periods of light loading (e.g., off busy hour). The reduction of maximum energy density will also facilitate efficient use of the geostationary satellite orbit by minimizing the orbital separation needed between satellites using the same frequency band and multiple-carrier operation of broadband transponders. The objective is to maintain spectral energy density the same for conditions of light loading as for busy hour loading. Several methods of implementing energy dispersal are described in ITU-R Rec. S.446-4 (Ref. 13).

One method of increasing satellite transponder capacity is by *frequency reuse*. As the term implies, an assigned frequency band is used more than once. The problem is to minimize mutual interference between carriers operating on the same frequency but accessing distinct transponders. There are two ways of avoiding interference in a channel that is used more than once:

1. By orthogonal polarization.
2. By multiple-exclusive spot beam antennas.

Opposite-hand circular or crossed-linear polarizations may be used to effect an increase in bandwidth by a factor of 2. Whether the potential increase in capacity can be realized depends on the amount of cross-polarization discrimination that can be achieved for the co-channel operation. Cross-polarization discrimination is a function of the quality of the antenna systems and the effect of the propagation medium on polarization of the transmitted signal. The amount of polarization "twisting" or distortion is a function of the elevation angle for a given earth station. The lower the angle, the more the twist.

Isolation between co-channel transponders should be greater than 25 dB. INTELSAT VII specifies 29-dB minimum isolation between polarizations.

Whereas a satellite operating in a 500-MHz bandwidth might have only 12 transponders (36- or 40-MHz bandwidth each), with frequency reuse, 24 transponders can be accommodated with the same bandwidth. Thus the capacity has been doubled.

Table 6.6 shows a typical transponder allocation for FDMA operation.

6.4.2.1 *Single-Channel-per-Carrier (SCPC) FDMA.* A number of SCPC systems have been implemented and are now in operation, and many more are coming on-line. There are essentially two types of SCPC systems: preassignment and demand assignment (DAMA). The latter requires some form of control system. Some systems occupy an entire transponder, whereas others share a transponder with video service, leaving a fairly large guardband between the video portion of the transponder passband and the SCPC portion. Such transponder sharing is illustrated in Figure 6.17.

Channel spacings on a transponder vary depending on the system. INTELSAT systems commonly use 45-kHz spacing. Others use 22.5, 30, and 36 kHz as well as 45 kHz. Modulation is FM or BPSK/QPSK. The latter lends itself well to digital systems, whether PCM or CVSD (continuous variable slope delta modulation).

If we were to divide a 36-MHz bandwidth on a transponder into uniform 45-kHz segments, the total voice channel capacity of the transponder is 800 VF channels. These are better termed half-channels, because, for telephony, we always measure channel capacity as full duplex channels. Thus 400 of the 800 channels would be used in the "go" direction, and the other 400 would be used in the reverse direction. A SCPC system is generally designed for a 40% activity factor. Thus, statistically, only 320 of the channels can be expected to be active at any instant during the busy hour. This has no effect on the bandwidth, only on the loading. Most systems use voice-activated service. In this case, carriers appear on the transponder for the activated channels only. This provides probably the worst case for IM noise for all conventional bent-pipe systems during periods of full activation. A simplified drawing of a SCPC system is shown in Figure 6.18.

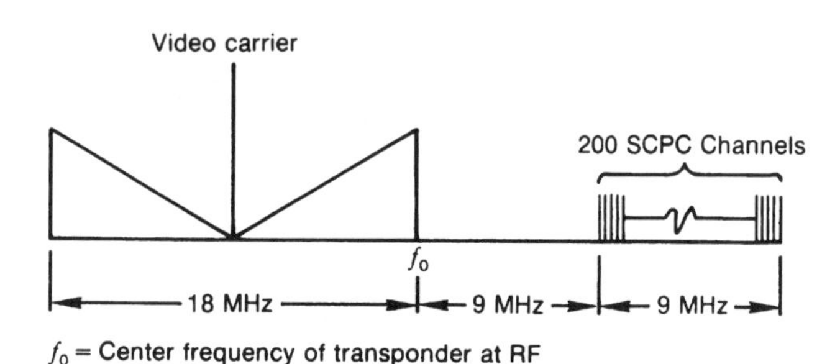

f_0 = Center frequency of transponder at RF

Figure 6.17. Satellite transponder frequency plan, video plus SCPC.

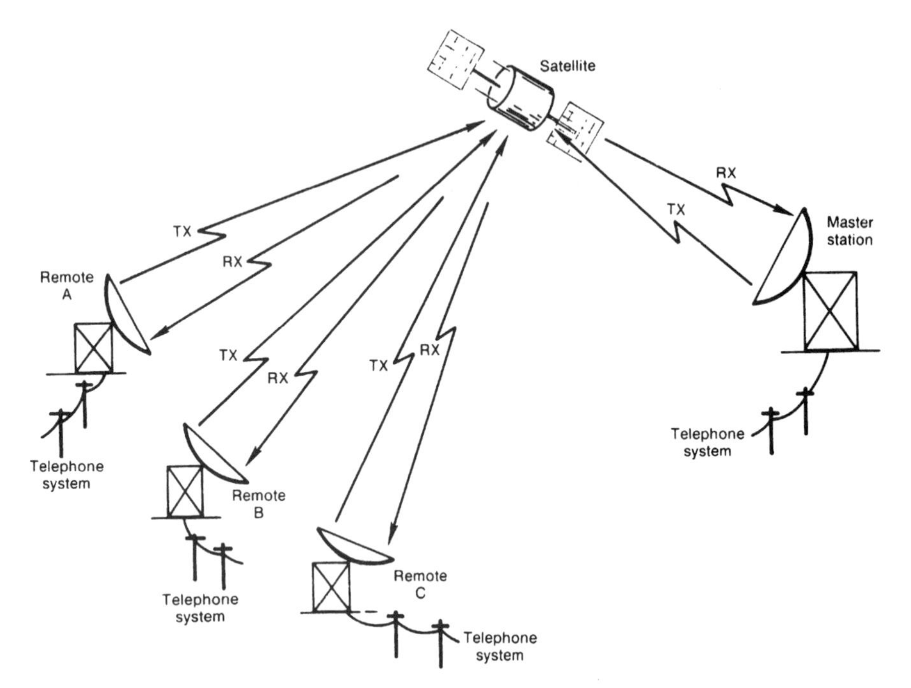

Figure 6.18. A typical FM SCPC system.

SCPC channels can be preassigned or DAMA. The preassigned technique is only economically feasible in situations where source/sink locations have very low traffic density during the busy hour (e.g., less than several Erlangs).

DAMA SCPC systems are efficient for source/sink* locations of comparatively low traffic intensity, especially under situations with multiple-location community of interest (e.g., trunks to many distinct locations) but under 12 channels to any one location. This demarcation line at 12 channels, of course, is where the designer should look seriously at establishing fixed-assignment FDMA/FDM channel group allocation.

The potential improvement in communication system capacity and efficiency is the primary motivation for DAMA systems. Our primary consideration in this section is voice traffic. Call durations average several minutes. The overhead time to connect and disconnect is 1 or 2 s in an efficient channel assignment system. Some of the issues of importance in demand assignment systems are:

- User requirements such as traffic intensity in Erlangs or ccs, number of destinations, and grade of service (probability of blocked calls)
- Capacity improvements due to implementation of DAMA scheme

*Sink also called destination.

- Assignment algorithms (centralized versus distributed control)
- Equipment cost and complexity

Consider an example to demonstrate the potential improvement due to demand assignment. (See Table 6.8.) An earth station is required to communicate with 40 destinations, and the traffic intensity to each destination is 0.5 Erlang with a blocking probability of $P_B = 0.01$. Assume that the call arrivals have a Poisson distribution and the call holding times have an exponential distribution. Then for a trunk group with n channels in which blocked calls are cleared, based on an Erlang B distribution (Ref. 1, Section 1-2), A, in Table 6.8, gives the traffic intensity in Erlangs. If the system is designed for preassignment, each destination will require four channels to achieve $P_B = 0.01$. For 40 destinations, 160 channels will be required for the preassigned case. In the DAMA case, the total traffic is considered since any channel can be assigned to any destination. The total traffic load is 20 Erlangs (40×0.5) and the Erlang B formula gives a requirement of only 30 channels. The efficiency of the system with demand assignment is improved over the preassigned by a factor of 5.3 (160/30). Table 6.8 further expands on this comparison using various traffic loadings and numbers of destinations. Reference 15, Chapter 1, provides good introductory information on traffic engineering.

There are essentially three methods for controlling a DAMA system:

1. Polling.
2. Random-access central control.
3. Random-access distributed control.

The polling method is fairly self-explanatory. A master station (Figure 6.18) "polls" all other stations in the system sequentially. When a positive reply is

TABLE 6.8 Comparison of Preassigned Versus DAMA Channel Requirements Based on a Grade of Service of $P_B = 0.01$

Number of Destinations	Channel Requirements[a]					
	$A = 0.1$		$A = 0.5$		$A = 1.0$	
	Preassigned	DAMA	Preassigned	DAMA	Preassigned	DAMA
1	2	2	4	4	5	5
2	4	3	8	5	10	7
4	8	3	16	7	20	10
8	16	4	32	10	40	15
10	20	5	40	11	50	18
20	40	7	80	18	100	30
40	80	10	160	30	200	53

[a] A = Erlang traffic intensity.

Source: "Demand Assignment," Subsection 3.6.5 in H. L. Van Trees, *Satellite Communications,* IEEE Press, 1979 (Ref. 14). Reprinted with permission of IEEE Press.

received, a channel is assigned accordingly. As the number of earth stations in the system increases, the polling interval becomes longer, and the system tends to become unwieldy because of excessive postdial delay as a call attempt waits for the polling interval to run its course.

With random-access central control, the status of channels is coordinated by a central computer located at the master station. Call requests (called *call attempts* in switching) are passed to the central computer via a digital orderwire (i.e., digitally over a radio service channel), and a channel is assigned, if available. Once the call is completed and the subscriber goes on-hook, the speech path is taken down and the channel used is returned to the demand-access pool of idle channels. According to system design, there are a number of methods to handle blocked calls [all trunks busy (ATB) in telephone switching], such as queueing and second attempts.

The distributed control random-access method utilizes a processor controller at each earth station accessing the system. All earth stations in the network monitor the status of all channels via continuous updating of channel status information by means of the digital orderwire circuit. When an idle channel is seized, all users are informed of the fact, and the circuit is removed from the pool. Similar information is transmitted to all users when circuits are returned to the idle condition. One problem, of course, is the possibility of double seizure. Also the same problems arise regarding blockage (ATB) as in the central-control system. Distributed control is more costly and complex, particularly in large systems with many users. It is attractive in the international environment because it eliminates the "politics" of a master station.

6.4.3 Brief Overview of Time Division Multiple Access (TDMA)

TDMA operates in the time domain and is applicable only to digital systems because information storage is required. To make Section 6.4 complete, a brief overview of TDMA is presented. Chapter 7 gives a more rigorous description of TDMA.

With TDMA, use of a satellite transponder is on a time-sharing basis. At any given moment in time, only one earth station accesses the satellite. Individual time slots are assigned to earth stations operating with that transponder in a sequential order. Each earth station has full and exclusive use of the transponder bandwidth during its time-assigned segment. Depending on bandwidth of the transponder and type of modulation used, bit rates of 10–200 Mbps are used.

If only one carrier appears on the transponder at any time, then intercarrier IM products cannot be developed and the TWT-based HPA of the transponder may be operated at full power. This means that the TWT requires no backoff in drive and may be run at or near saturation or maximum efficiency. With multicarrier FDMA systems, backoffs of 5–10 dB are not uncommon.

TDMA utilizes the transponder bandwidth more efficiently too. Reference 16 compares approximate channel capacities of an INTELSAT IV global beam transponder operating with normal INTELSAT Standard A (30-m antenna) earth stations using FDMA and TDMA, respectively. Assuming 10 accesses, typical capacity using FM/FDMA is approximately 450 one-way voice channels. With TDMA, using standard 64-kbps PCM voice channels, the capacity of the same transponder is 900 voice channels. If digital speech interpolation (DSI) is now implemented on the TDMA system, the voice channel capacity is increased to approximately 1800 channels. DSI is discussed in Chapter 7.

Still another advantage of TDMA is flexibility. Flexibility is not only a significant benefit to large systems but is often the key to system viability in smaller systems. Nonuniform accesses pose no problem in TDMA because time-slot assignments are easy to adjust. This applies to initial network configuration, assignments, reassignments, and demand assignments. This is ideal for a long-haul system where traffic adjustments can be made dynamically as the busy hour shifts from east to west. Changes can also be made for growth or additional services.

Disadvantages of TDMA are timing requirements and complexity. Accesses may not overlap. Obviously, overlapping causes interference between sequential accesses. Guard times between accesses must be made minimum for efficient operation. The longer the guard times, the shorter the burst

TABLE 6.9 FDMA Versus TDMA

Advantages	Disadvantages
FDMA	
Mature technology	IM in satellite transponder output
No network timing	Requires careful uplink power control
Easy FDM interface	Inflexible to traffic load
TDMA	Still emerging technology
Maximum use of transponder power	Network timing
No careful uplink power control	Complex control
Flexible to dynamic traffic loading	Major digital buffer considerations
Straightforward interface with digital network	Difficult to interface with FDM
More efficient transponder bandwidth usage	
Digital format compatible with	
Forward error control	
Source coding	
Demand access algorithms	
Applicable to switched satellite service	

length and/or number of accesses. Typically 5–20 accesses can be accommodated per transponder with guard times on the order of 50–200 ns. Frame durations vary from 1 to 20 ms (Ref. 17).

As the world's telecommunication network evolves from analog to digital and as frequency congestion increases, there will be more and more demand to shift to TDMA operation in satellite communications.

Table 6.9 compares FDMA with TDMA.

6.5 INTELSAT SYSTEMS

6.5.1 Introduction

INTELSAT is probably, overall, the best known of the families of communication satellite systems. The INTELSAT family, starting with INTELSAT III of the mid-1960s, has been in operation for the longest period. The key word is international. Its original concept was to provide international trunk connectivity, particularly transoceanic, for telephone administrations. INTELSAT satellites are bent-pipe FDMA/FDM/FM. Provision has been made for TDMA, and in mid-1985 TDMA was finally implemented. Current operational satellites are INTELSAT V, VI, and VII. Frequency reuse is common practice. Transponders operate in the 6/4- and 14/11/12-GHz bands. The INTELSAT VII transponder frequency plan is shown in Figure 6.19a and the INTELSAT VIII transponder frequency plan is shown in Figure 6.19b. Table 6.10 summarizes some of the key characteristics of the INTELSAT space segment for INTELSAT V and VI.

INTELSAT specifies certain requirements for earth stations of member countries to operate with INTELSAT satellites. Earth stations are defined by class: INTELSAT Standard A, B, C, D, E, and F. Table 6.11 summarizes several key characteristics on a comparison basis. The following subsections provide discussion of each INTELSAT class.

6.5.2 INTELSAT Type A Standard Earth Stations*

The Type A earth station is used primarily for intercontinental direct trunk groups and television. The driving parameter is the required G/T of +35.0 dB/K at 4 GHz. The receiving system must accommodate the entire 500-MHz satellite bandwidth. To achieve the G/T, generally 16-m antennas are required and receiving system noise temperatures are in the range of 80 K. To reach the nearly +90 dBW EIRP, 5-kW transmitters are required. If TV is to be transmitted simultaneously with high-density FDM, often two 5-kW transmitters are used, one for TV and one for FDM, and the outputs of the two transmitters are power combined.

*Also see Section 6.5.8.

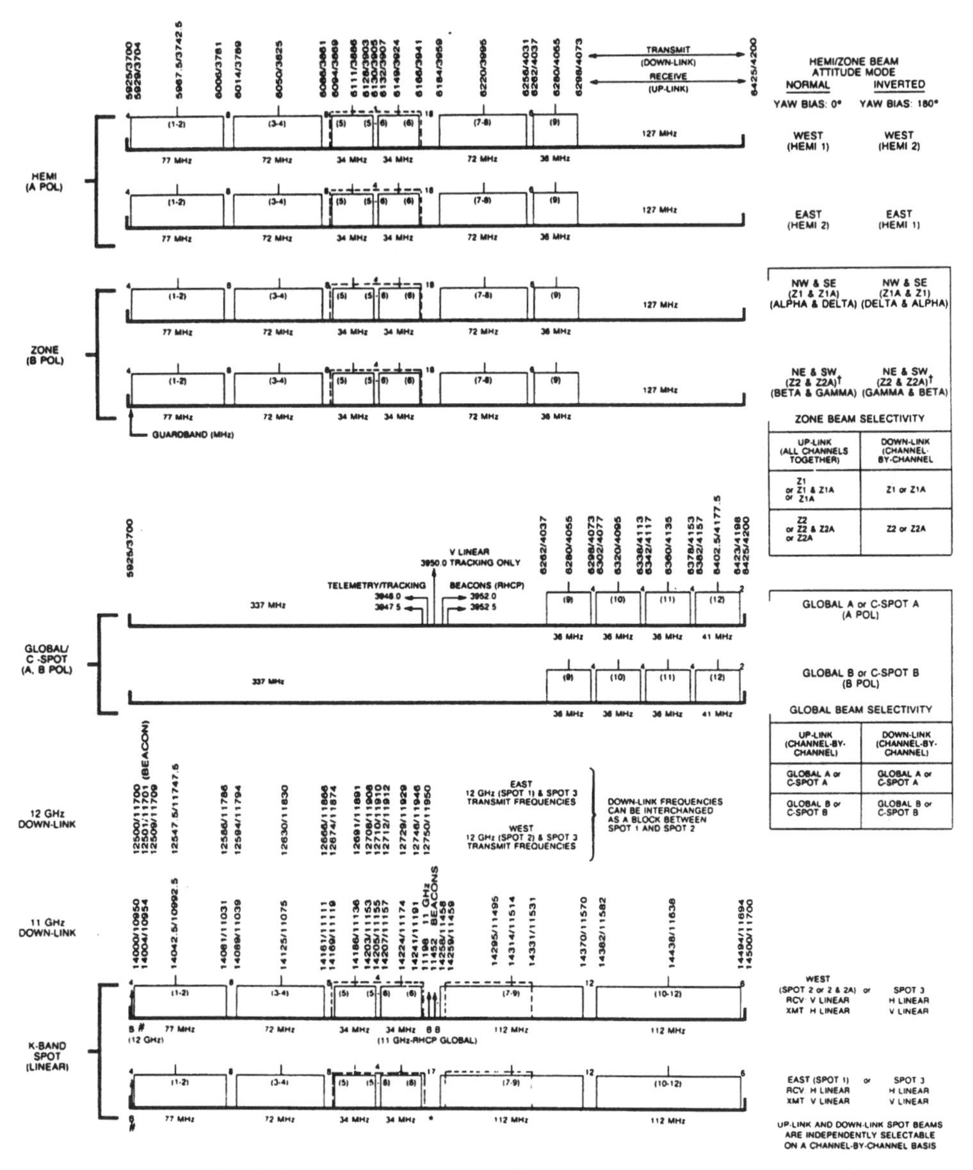

Figure 6.19a. INTELSAT VII transponder layout. (From INTELSAT IESS-409, Ref. 18.)

NOTES

1. On spot 11-GHz downlink there is a 250-MHz gap between (5-6) and (7-9) (11200-11450 MHz).
2. On spot 12-GHz downlink there is a 9-MHz guardband before (1-2) to accomodate the beacon.
3. The 11-GHz or 12-GHz spot frequencies may be selected independently for each beam and each channel.
4. Northeast and southwest zones are always labeled Z2 and Z2A respectively regardless of attitude mode.
5. At 342°E in the inverted mode, Z1 is the northwest and Z1A is in the southeast.
6. The 12-GHz beacons are transmitted through the communications antennas and are of the same transmit polarization.

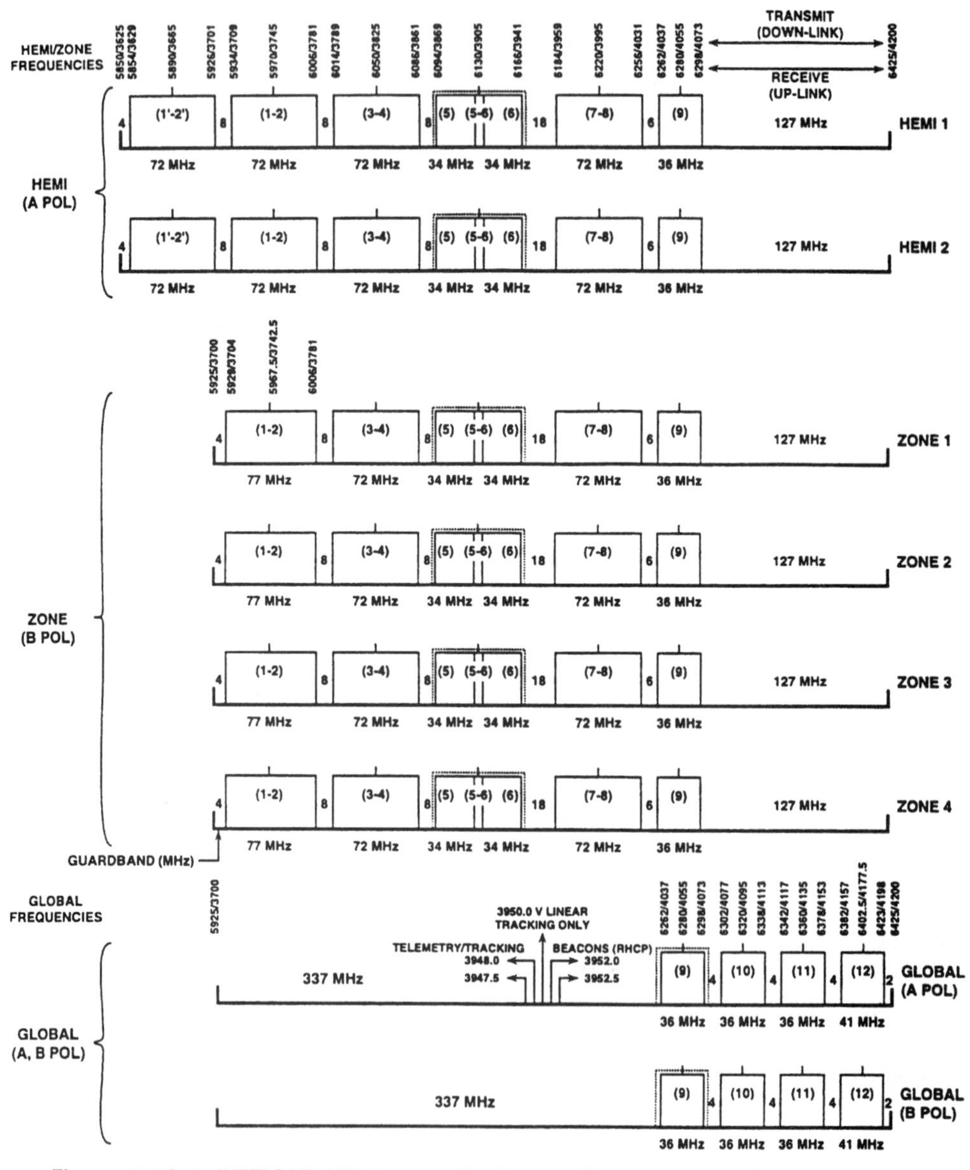

Figure 6.19b-1. INTELSAT VIII transponder layout: C-band portion. (From INTELSAT IESS-417; Ref. 19.)

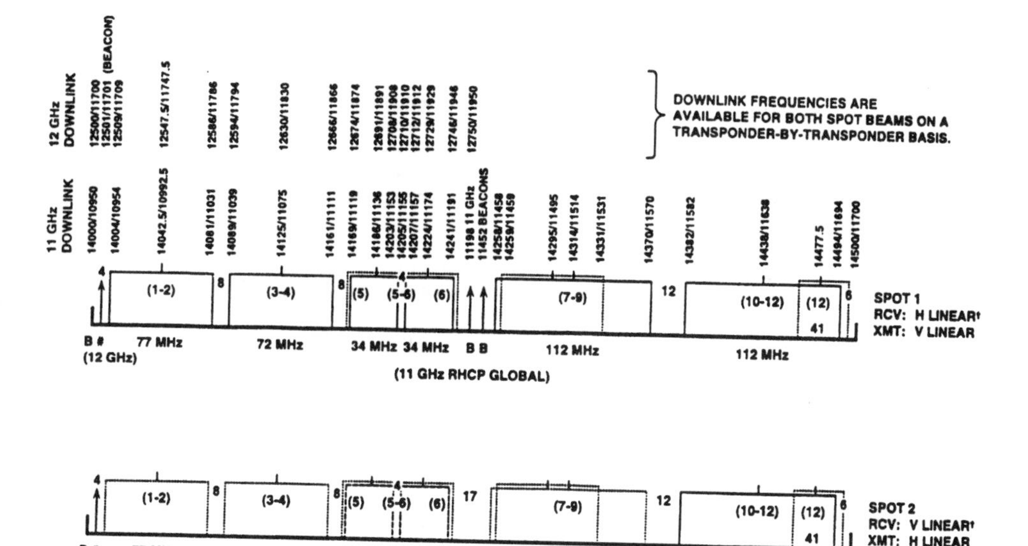

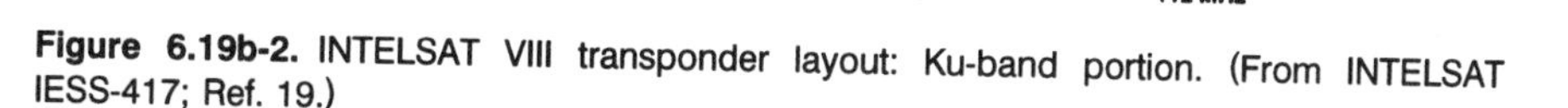

Figure 6.19b-2. INTELSAT VIII transponder layout: Ku-band portion. (From INTELSAT IESS-417; Ref. 19.)

NOTES

1. See Tables 1, 2, and 3 of IESS-413 for further information.
2. On spot 11-GHz downlink there is a 250-MHz gap between (5-6) and (7-8)/(7-9) (11200-11450 MHz).
3. On spot 12-GHz downlink there is a 9-MHz guardband before (1-2) to accomodate the beacon.
4. The polarization senses of each spot beam can be changed by ground command independent of the other spot beam.
5. The 11-GHz or 12-GHz spot frequencies may be selected independently for each beam and each transponder.
6. The 12-GHz beacons are transmitted through the communications antennas and are of the same transit polarization.
7. Both spot 1 and 2 may operate with the 41-MHz portion of the uplink in transponder (10-12) corresponding to the bandwidth defined by the downlink A or B pol global transponder 12.
8. Up to the maximum of six, out of ten, KU-band transpnders can be operated simultaneously with some limtations on the selection of specific transponders.
9. Uplink and downlink spot beams are independently selectable on a transponder-by-transponder basis.

SCPC/DAMA is cost effective on these large installations for overflow during busy hours and for low-density traffic relations, less than about 5 Erlangs.

The large antennas used on INTELSAT Standard A earth stations produce very narrow beamwidths, on the order of 0.3°. Because of satellite suborbital motion and drift, and with such narrow beamwidths, automatic

TABLE 6.10 Summary of Some Key Characteristics of INTELSAT VI and VII

Characteristic[a]		INTELSAT VII		INTELSAT VI
G/T (dB/K)				
6 GHz				
	Global coverage	−11.6		−14.0
	HC	−8.5		−9.2
				−9.5(1)
	ZC	−5.5	Z2, Z4	−7.0/−7.5[b]
			Z1, Z3	−2.0/−3.0[b]
14 GHz				
	East spot	+4.8		+1.0
	West spot	+2.5		+5.0
EIRP (dBW)				
4 GHz				
	Global coverage	+29(7-9)		+23.5 (9)
		+26		+26.5
	EH, WH			+28.0 (9)
	EH, WH			+31.0
	Hemi (1)	+33		
	Hemi (2)	+33		
	Z1-4	+33.0		+28.0 (9)
	Z1-4	+33.0		+31.0
11 GHz				
	East spot	+45.5		+41.1
	West spot	+47.5		+44.4
Polarization				
4 GHz				
	Global coverage	RHC		
	Global coverage (A)	LHC		RHC
	Global coverage (B)	RHC		LHC
	HC	LHC		RHC
	Z1, Z2	RHC		
	Z1-4			LHC
14 GHz	(east orthogonal to west)	Linear		Linear
11 GHz	(east orthogonal to west)	Linear		Linear
6 GHz	(opposite orthogonality to 14-GHz counterpart)			
	Global coverage (A)	RHC		LHC
	Global coverage (B)	LHC		RHC
	HC	RHC		LHC
	Z1, Z2	LHC		
	Z1-4			RHC

[a]HC = hemispherical coverage; ZC = zone coverage; EH = east hemispherical; WH = west hemispherical; RHC, LHC: right-hand circular polarization and left-hand circular polarization, respectively. Spot refers to spot-beam antennas.

[b]Denotes high and low levels.

Source: INTELSAT IESS-409 (Ref. 18).

TABLE 6.11 Comparison of Some Key Parameters of INTELSAT Standard Earth Stations

	INTELSAT Earth Stations by Standard Class					
Parameter	A	B	C	D	E	F
Service	FDM/FM, TV/FM, SCPC/QPSK (PCM)	TV/FM FDM/FM for TV aural SCPC/QPSK (PCM)	FDM/FM only	LDTS[a] CFM[b] (SCPC)	Business SVC IBS[c]	Business SVC (data with FEC)(CQPSK)
Frequency bands (GHz)	6/4	6/4	14/11	6/4	14/11	6/4
Minimum G/T (dB/K)	35.0	31.7	37.0	D1 = 22.7 D2 = 31.7	E1 = 25.0 E2 = 29.0 E3 = 34.0	F1 = 22.7 F2 = 27.0 F3 = 29.0
EIRP TV (max)(dBW)	+88	+85				
EIRP FDM (max)(dBW)	+90.2	+74.7 (TV aural)	+90.3[d]	D1 = +55.3 D2 = +49.6		F1 = +91.3 F2 = +88.9 F3 = +87.1 E1 = +87.7 E2 = +85.2 E3 = +79.5
EIRP SCPC (max)(dBW)		+69.8				
Minimum elevation angle (deg)	5	5	10	5	10	5
Tracking	Automatic open and closed loop	Same	Same	Same D2; manual D1	Same E3; manual E1, E2	Same F3; automatic or manual F1, F2

[a]LDTS = low-density telephone service.
[b]CFM = companded FM.
[c]IBS = international business service.
[d] + 85.8 dBW for INTELSAT VII.

Source: INTELSAT IESS-101 and cataloged documents (Ref. 20).

tracking is required, usually active tracking, monopulse of conscan. Further discussion of earth station design may be found in Chapter 14.

6.5.2.1 Noise on a Telephone Channel for an INTELSAT Link. INTELSAT allows 10,000 pWp of noise power in the receive telephone channel. It will be noted that this value corresponds to the noise power value on a 2500-km CCIR hypothetical reference circuit. This 10,000-pWp noise power value is budgeted by INTELSAT as follows:

Space segment	8,000 pWp
Earth stations	1,000 pWp
Terrestrial interference	1,000 pWp
Total	10,000 pWp

Earth station noise contribution is broken down as follows:

Earth station transmitter noise excluding multicarrier IM noise and group delay noise	250 pWp
Noise due to total group delay after necessary equalization	500 pWp
Other earth station receiver noise	250 pWp
Total	1000 pWp

The INTELSAT space segment allocates 8000 pWp to uplink and downlink thermal noise, transponder IM noise, earth station out-of-band emission, co-channel interference within the operating satellite, and interference from adjacent satellite networks. Within the 8000-pWp allocation, an allowance of 500 pWp is reserved for earth station RF out-of-band emission caused by multicarrier IM from other earth stations in the system.

All noise power values shown are referenced to the zero-test-level point in the overall system (Ref. 11).

6.5.3 INTELSAT Standard B Earth Stations

Standard B earth stations are smaller than their Standard A counterpart. They operate in the 6/4-GHz band and the G/T is +31.7 dB/K. They transmit/receive TV and related aural channels. The latter are in a FDM/FM configuration. Standard B also operates with companded FDM/FM and SCPC using QPSK. It operates with E1/DS1 bit streams in the International Business Service (IBS) (Ref. 21) and Intermediate Data Rate (IDR) carriers. Maximum data rates in such cases are E-3/DS3 rates up to the nominal 45 Mbps.

EIRP requirements of Standard B earth stations vary with the service required. For example, an EIRP of +93.7 dBW is required for IDR operation at the DS3 data rate, low-gain antenna, operating with INTELSAT V. This EIRP value is reduced to +84.5 dBW for INTELSAT VII operation. The EIRP values for Standard A and Standard B earth stations are for the minimum elevation angles of 5°.

SCPC (single channel per transponder) is one of the services offered by Standard B earth stations. Figure 6.20 shows the SCPC frequency plan using a 36-MHz transponder fully dedicated to this service. It has a maximum capacity of 399 full-duplex voice channels.

On digital configurations, INTELSAT often requires the use of FEC (see Chapter 4) with a rate 3/4 or 7/8 to meet link BER requirements. INTELSAT digital offerings are described further in Chapter 7.

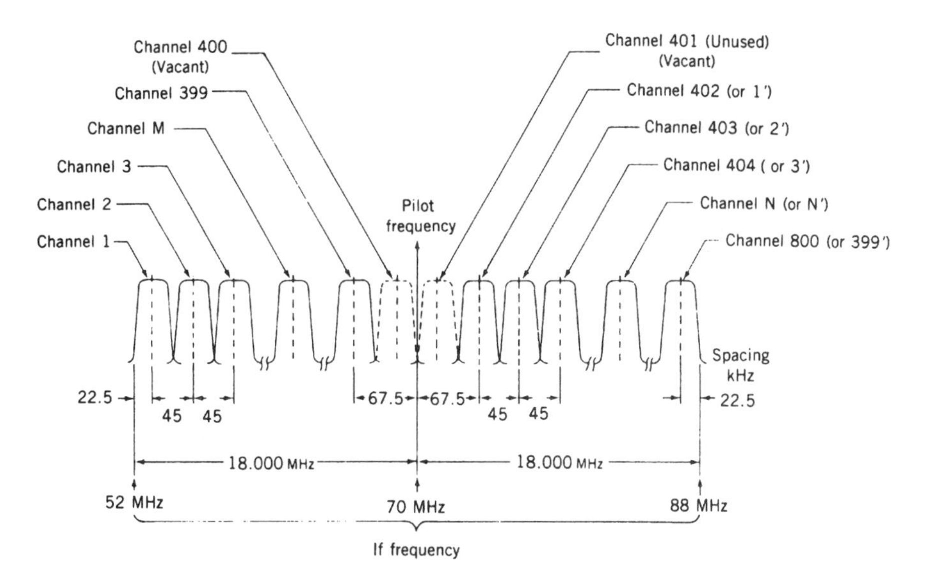

Figure 6.20. SCPC frequency plan at 70-MHz IF for full transponder operation. (From INTELSAT IESS-305, Ref. 22.)

6.5.4 INTELSAT Standard C Earth Stations

INTELSAT Standard C earth stations operate in the 14/11-GHz frequency band with INTELSAT V–VII satellites. These earth stations can provide FDM/FM service, companded FDM/FM, IDR digital carriers, and IBS.

Because INTELSAT C earth stations operate in the 14/11-GHz band, a rainfall margin is provided between 11 and 13 dB for G/T and between 5 and 7 dB for uplink margins. These margins should maintain a time availability of 99.96%.

Minimum G/T for a Standard C earth station is 37.0 dB/K. The maximum EIRP value is +90.3 dBW while operating with INTELSAT VI and lower. The EIRP is reduced to +85 dBW for those facilities operating with INTELSAT VII and VIII.

The minimum elevation angle for the C-type earth station is not given in the applicable INTELSAT IESS documents. One industry guideline is 10°. It should be appreciated that excess attenuation due to rainfall is tightly tied to elevation angle.

6.5.5 INTELSAT Standard D Earth Stations*

INTELSAT Standard D earth stations provide more modest capabilities than earth stations based on Standards A, B, and C. Standard D-1 has the

*Section 6.5.5 is based on information provided in INTELSAT IESS-305 (Ref. 22).

TABLE 6.12 SCPC/CFM Characteristics and Transmission Parameters

Parameter	Requirement
Audio channel input bandwidth	300–3400 Hz
Modulation	FM with companding
Companding	2 : 1 Syllabic
Carrier control	Voice activated
Channel spacing	30.0 kHz[a]
Channel bandwidth	30.0 kHz
IF noise bandwidth	25.0 kHz
rms Test tone deviation for 0 dBm0 at 1 kHz	5.1 kHz
C/N per channel at nominal operating point	10.2 dB
C/N_0 at nominal operating point	54.2 dB-Hz
C/N per channel at threshold	6.2 dB
C/N_0 in IF bandwidth at threshold	50.2 dB-Hz

[a]The spacing between the two carrier slots adjacent to the center of the frequency band will be 45 kHz.

Source: INTELSAT IESS-305, Table 1 (Ref. 22).

capability of only one voice channel, whereas D-2 supports a greater number of voice channels.

INTELSAT calls the service provided by D-type earth stations the *Vista service*. It is companded FM using SCPC techniques with a RF slot of 30 kHz per voice channel. The SCPC/CFM characteristics and transmission parameters are shown in Table 6.12.

The maximum global beam EIRP values for SCPC/CFM carriers transmitted by a Standard D earth station are provided below. The EIRP values for INTELSAT V, VA, VA(IBS), and VI apply to earth stations with a 30° elevation angle. The EIRP values for INTELSAT VII and VIIA apply to earth stations with a 10° elevation angle. The appropriate G/T values for D-1 and D-2 earth stations are also shown. For operation with INTELSAT V through VI at elevation angles below 30°, IESS-402 (Ref. 23) should be consulted.

Maximum Uplink EIRP per Carrier (dBW)			G/T of Receiving Station (dB/K)
V, VA, VA(IBS), VI[a]	VII[b]	VIIA[b]	
55.6	55.3	53.2	22.7 (D-1)
51.7	49.6	48.8	31.7 (D-2)

[a]The transponder gain step is Extra High for INTELSAT V, VA, and VA(IBS) and High Gain for INTELSAT VI.

[b]The transponder saturation flux density is -86.0 dBW/m^2.

6.5.6 INTELSAT Standard E Earth Stations

The Standard E earth station is designed primarily for the "business service" and operates in the nominal 14/11-GHz band pair utilizing INTELSAT V, VA, VI, VII, and VIII satellites. The business service basically provides users with data connectivity using standard E-1 and DS1 formats up to and including E-3/DS3 rates. This type of transmission is covered in Chapter 7.

Two grades of IBS are offered on the INTELSAT V through VI satellites: Basic and Super. Basic IBS is designed to maximize channel capacity in both the C-band (6/4 GHz) and Ku-band (14/11 GHz) transponders. Maximum capacity is achieved by providing a better yearly channel availability for C-band uplinks than for Ku-band uplinks, due to the excess attenuation because of rainfall experienced at Ku band. As an option, Super IBS is offered at Ku-band to provide an availability equivalent to C-band through an increase in uplink EIRP.

Maximum EIRPs for the FEC = 3/4 rate for 8.448-Mbps service are:

E-1 = +86.3 dBW
E-2 = +80.7 dBW
E-3 = +76.3 dBW

Subtract 1.5 dB from these EIRP values for the rate 1/2 service. Lower standard bit rates use appropriately low EIRP values.

G/T values for Standard E earth station receiving systems are given below:

E-1 $G/T \geq 25$ (but < 29) $+ 20 \log_{10} f/11.0$ dB/K
E-2 $G/T \geq 29$ (but < 34) $+ 20 \log_{10} f/11.0$ dB/K
E-3 $G/T \geq 34$ $+ 20 \log_{10} f/11.0$ dB/K

An additional value of X dB is added to provide a rainfall margin, which can vary from 11 to 15 dB. Reference Tables 1–6 in INTELSAT IESS-208 (Ref. 24).

The minimum elevation angle for E-type earth stations is not provided in the appropriate IESS documents. One industry guideline is a minimum angle of 10° for Ku-band operation. It must be appreciated, as discussed in Chapter 9, that excess attenuation due to rainfall is strongly tied to elevation angle.

E-1 earth stations may operate with reduced bandwidths as follows:

Receive: 10.95–11.2 GHz; *Transmit*: 14.0–14.25 GHz or
Receive: 11.45–11.7 GHz; *Transmit*: 14.25–14.5 GHz

Earth station feed elements must be capable of operating across the entire Ku transmit and receive bands (i.e., 10.95–11.7 GHz and 14.0–14.5 GHz).

E-2 and E-3 earth stations must operate across the entire Ku transmit and receive frequency bands.

6.5.7 INTELSAT Standard F Earth Stations

The INTELSAT Standard F earth station operates in INTELSAT's IBS and is similar in most respects to the Standard E earth stations, except that it operates in the 6/4-GHz frequency pair. These earth stations may also be used for the IDR digital services. F-type earth stations may also provide companded frequency division multiplex/frequency modulation (CFDM/FM) telephony service.

The specified G/T values for INTELSAT Standard F earth stations are as follows:

F-1 $G/T = 22.7 + 20\log_{10} f/4 \quad \mathrm{dB/K}$
F-2 $G/T = 27.0 + 20\log_{10} f/4 \quad \mathrm{dB/K}$
F-3 $G/T = 29.0 + 20\log_{10} f/4 \quad \mathrm{dB/K}$

where f is the frequency in gigahertz.

Maximum EIRP values vary with the service and satellite used. For example, F-3 using IDR has a maximum of +87.7 dBW employing a hemispheric antenna in "high gain" with a traffic bit rate of 44.736 Mbps (DS3). F-2, transmitting E-2 (8.448 Mbps), has a maximum EIRP of +82.4 dBW. In both cases the satellite series is INTELSAT V, VA, VA(IBS), and VI.

When operating with INTELSAT VII, the F-1 maximum EIRP value is +86.4 dBW, rate 3/4 coding, 6/4-GHz band for E-1 (2.048 Mbps) traffic rate. For F-2, based on the same parameters as F-1, the maximum EIRP is +87.6 dBW for the E-3 rate (i.e., 34.368 Mbps). For F-3, based on the same parameters as F-1, the EIRP value is +86.4 dBW for the DS3 traffic rate.

6.5.8 Basic INTELSAT Space Segment Data Common to All Families of Standard Earth Stations

Tables 6.13 and 6.14 give earth station polarization requirements to operate with INTELSAT V through VIIIA and K.

Table 6.15 provides minimum tracking requirements for INTELSAT earth stations operating in the 6/4-GHz band. Table 6.16 provides similar information for INTELSAT Standard C and E earth stations that operate in the 14/11- and 14/12-GHz bands.

6.5.9 Television Operation Over INTELSAT

This section briefly describes INTELSAT "occasional TV" operation in the 6/4-GHz band for INTELSAT Standard A and B earth stations. It is based on INTELSAT IESS-306 (Rev. 3) (Ref. 26).

Two methods of transmission can be used to access global beam transponders of occasional-use TV and its associated sound program channel(s). One method employs open-network transmission parameters, which are briefly

TABLE 6.13 Earth Station Polarization Requirements to Operate with INTELSAT V Through VIIIA and K Satellites (14/11- and 14/12-GHz Bands)

		Linear Polarization[a]	
Satellite	Coverage	Earth Station Transmit	Earth Station Receive
V, VA, VA(IBS), and VI	East Spot	Horizontal	Vertical
	West Spot	Vertical	Horizontal
VII	Spot 1 and Spot 3[b]	Horizontal	Vertical
	Spot 2	Vertical	Horizontal
VIIA	S1, S2X, and S3[b]	Horizontal	Vertical
	S1X, S2	Vertical	Horizontal
VIII[c]	Spot 1	Horizontal	Vertical
	Spot 2	Vertical	Horizontal
VIIIA	Spot 1		
	805	Horizontal	Vertical
	806	Vertical	Horizontal
K	EUH and NAH	Horizontal	Horizontal
	EUV and NAV	Vertical	Vertical
	SAV	[d]	Vertical

[a]Users are referred to the IESS-400 series modules for the definition of horizontal and vertical linear polarization and the dependence of the polarization orientation on the geographic location of the earth station.
[b]On INTELSAT VII (F-3, F-4, F-5, and F-9) and INTELSAT VIIA (F-6, F-7, and F-8), Spot 3 receive and transmit antenna polarization senses can be switched in orbit by ground command.
[c]The polarization sense of either INTELSAT VIII Spot beam can be changed independently by ground command. Users are urged to confirm with INTELSAT the polarization sense of the Spot beam that will be utilized.
[d]Earth stations located in the South American beam are only required to receive in the vertical polarization.
EUH = European horizontal
EUV = European vertical
NAH = North American horizontal
NAV = North American vertical
SAV = South American vertical
Source: Table 7, INTELSAT IESS-208 (Rev. 1) (Ref. 24).

discussed below. The other method recognizes closed-network transmission parameters, provided certain conditions are met.

6.5.9.1 *Open-Network Transmission Parameters.* Occasional-use TV/FM carriers in the INTELSAT system generally utilize the 6/4-GHz transponder slot 12 of the global beam. It should be noted that other global beam slots may also be used to meet specific traffic demands. Two basic modes of operation are available:

1. Full-transponder TV in which a single TV/FM carrier occupies the transponder.
2. Half-transponder TV in which either two TV/FM carriers are simultaneously transmitted through a single transponder or a single TV/FM carrier is transmitted through one-half of a transponder.

TABLE 6.14 Earth Station Polarization Requirements to Operate with INTELSAT V Through VIIA Satellites (6/4-GHz Band)

	INTELSAT V		INTELSAT VA/VA(IBS)		INTELSAT VI		INTELSAT VII/VIIA	
Coverage	Earth Station Transmit	Earth Station Receive	Earth Station Transmit	Earth Station Receive	Earth Station Transmit	Earth Station Receive	Earth Station Transmit	Earth Station Receive
1. Global A	LHCP	RHCP	LHCP	RHCP	LHCP	RHCP	LHCP	RHCP
2. Global B	N/A	N/A	RHCP	LHCP	RHCP	LHCP	RHCP	LHCP
3. West Hemisphere (Hemi 1)[a,b]	LHCP	RHCP	LHCP	RHCP	LHCP	RHCP	LHCP	RHCP
4. East Hemisphere (Hemi 2)[a,b]	LHCP	RHCP	LHCP	RHCP	LHCP	RHCP	LHCP	RHCP
5. NW Zone (Z1)[c] ZA[b]	RHCP	LHCP	RHCP	LHCP	RHCP	LHCP	RHCP	LHCP
6. NE Zone (Z3)[c] ZB[b]	RHCP	LHCP	RHCP	LHCP	RHCP	LHCP	RHCP	LHCP
7. SW Zone (Z2)[c] ZC[b]	N/A	N/A	N/A	N/A	RHCP	LHCP	RHCP	LHCP
8. SE Zone (Z4)[c] ZD[b]	N/A	N/A	N/A	N/A	RHCP	LHCP	RHCP	LHCP
9. C-Spot A	N/A	N/A	N/A	RHCP	N/A	N/A	LHCP	RHCP
10. C-Spot B	N/A	N/A	N/A	LHCP	N/A	N/A	RHCP	LHCP

[a]Hemi 1, Hemi 2, ZA, ZB, ZC, ZD nomenclature applies to INTELSAT VII and VIIA only.

[b]This indicates the normal mode of operation for INTELSAT VII and VIIA; the inverted mode implies different beams in the East and West, as illustrated in IESS-409.

[c]Z1, Z2, Z3, Z4 nomenclature applies to INTELSAT VI only.

Notes: LHCP = left-hand circularly polarized; RHCP = right-hand circularly polarized, N/A = not applicable to this spacecraft.

Source: INTELSAT IESS-207 (Rev. 1), p. 22 (Ref. 25).

TABLE 6.15 Minimum Tracking Requirements for INTELSAT Earth Stations Operating in the 6/4-GHz Band

Earth Station Standard	V, VA, and VA(IBS)	VI, VII, VIIA, VIII, and VIIIA[a]
A[b]	Manual and autotrack	Manual and autotrack
B[b]	Manual and autotrack	Manual and autotrack
D-2[b,c,d]	Manual and autotrack	Manual and autotrack
D-1[e]	"Fixed" antenna[f]	"Fixed" antenna
F-3[g]	Autotrack	"Fixed" antenna[e,h,i]
F-2[g]	Manual E/W only[f] (weekly peaking)	"Fixed" antenna[e]
F-1[e,g]	"Fixed" antenna[f]	"Fixed" antenna

[a]Antenna tracking requirements are based on provisional INTELSAT VIIA, VIII, and VIIIA station-keeping tolerances. These tolerances will be reviewed after operational experience is gained with these new series of spacecraft.
[b]Users are urged to include in their designs a provision to add program steering.
[c]The use of autotracking is recommended for Standard D-2.
[d]The tracking requirements shown for Standard D-2 earth stations is only mandatory for earth stations with a RFP issued after 1994. (Prior to the release of IESS-207 only autotrack was recommended.)
[e]"Fixed" antenna mounts will still require the capability to be steered from one satellite position to another, as dictated by operational requirements (typically once or twice every 2–3 years). These antennas should also be capable of being steered at least over a range of $\pm 5^\circ$ from beam center for the purpose of verifying that the antenna pointing is correctly set toward the satellite, and for providing a means of verifying the sidelobe characteristics in the range.
[f]Standard D-1, F-1, and F-2 users should consider the possible need to upgrade the earth station with manual or autotrack systems in the event it becomes necessary to operate with satellites having higher than nominal inclination.
[g]These minimum requirements apply to Standard F earth stations that either transmit or transmit and receive.
[h]Standard F-3 users are encouraged to consider autotrack designs. Earth stations using fixed antennas must meet all specifications of this document irrespective of the satellite position within the box defined by nominal station-keeping limits.
[i]Standard F-3 users should take into consideration their tracking requirements under contingency operation with another satellite series.
Note: The minimum tracking requirements are subject to the earth station meeting the axial ratio requirements of paragraph 1.3.2 of the reference publication.
Source: Table 3, INTELSAT IESS-207 (Rev. 1) (Ref. 25).

Full-transponder TV uses a 30-MHz TV/FM carrier. Half-transponder TV was converted from 17.5-MHz to 20.0-MHz TV/FM carriers in 1987.* The 30-MHz and 17.5-MHz TV/FM carriers were originally designed for transmission through transponder slot 12 of the INTELSAT IVA global beam (+22.5-dBW EIRP and 36-MHz allocated bandwidth). The increased EIRP (+23.5 dBW for INTELSAT V, VA, and VA(IBS); +26.5 dBW for INTELSAT VI; and +29 dBW for INTELSAT VII and VIIA) and allocated bandwidth (41 MHz of the INTELSAT V, VA, VA(IBS), VI, VII, and VIIA global beam transponder slot 12) allows improved half-transponder TV performance via 20-MHz TV/FM carriers.

*Although the INTELSAT network has been converted from 17.5-MHz to 20-MHz TV parameters, 17.5-MHz parameters are still shown in the reference document for special applications and as a design reference for services such as leased transponders.

TABLE 6.16 Minimum Tracking Requirements for INTELSAT Ku-Band Earth Stations

Earth Station	INTELSAT V, VA, and VA(IBS) (±0.1° N/S and ±0.1° E/W)	INTELSAT VI, VII, VIIA, VIII, VIIIA, and K[a] (±0.05° N/S and ±0.05° E/W)
C	Autotrack[b]	Autotrack[b]
E-1	Manual, E/W only (weekly peaking)	Fixed antenna[c]
E-2	Manual, E/W and N/S (peaking every 3–4 hours)	Fixed antenna[c]
E-3	Autotrack[b]	Autotrack[b]

[a]Antenna tracking requirements are based on provisional INTELSAT VIIA, VIII, and VIIIA station-keeping tolerances. These tolerances will be reviewed after operational experience is gained with these new series of spacecraft.
[b]Step-track operation can experience difficulties in a Ku-band environment due to severe fading or scintillations. Users may wish to consider systems that utilize step-tracking in conjunction with program tracking during periods of adverse atmospheric conditions.
[c]"Fixed" antenna mounts will still require the capability to be steered from one satellite position to another, as dictated by operational requirements (typically once or twice every 2–3 years). These antennas should also be capable of being steered at least over a range of ±5° from beam center for the purpose of verifying that the antenna pointing is correctly set toward the satellite and for providing a means of verifying the sidelobe characteristics in this range.
Source: Table 8, INTELSAT IESS-208 (Rev. 1) (Ref. 24).

TV-associated sound program channels are available in two basic modes of operation:

1. For both full-transponder and half-transponder TV operation, the TV-associated sound program signal frequency modulates a subcarrier, which is then combined with the TV video baseband before final frequency modulation is effected.
2. For full-transponder TV operation, the TV-associated sound program channels comprising sound program circuits, coordination circuits, and cue and commentary channels are transmitted via two 24-channel telephony carriers within the same transponder.

6.5.9.2 Closed-Network Transmission. Closed-network TV and TV-associated sound program channel(s) are assigned the same RF carrier frequency assignments and satellite EIRP resources as 20-MHz and 30-MHz open-network carriers. Closed-network transmission parameters are only approved to access the INTELSAT space segment if the conditions of Section 4 of the reference document (IESS-306) are met.

6.5.9.3 Some Basic System Parameters–Pre-emphasis/De-emphasis. For TV/FM carriers, the pre- and de-emphasis characteristics are in accordance with ITU-R Rec. S.405-1. Interchangeable networks for both 525- and 625-line transmission may be provided.

TABLE 6.17 INTELSAT V (F5–F9), VA, VA(IBS), and VI Maximum Earth Station EIRP Requirements for TV/FM Carriers[a]

		EIRP for 10° Elevation Angle[b] (dBW)			
		Earth Station			
Carrier Size	Gain Step[c]	"A" to "A"	"B" to "A"	"A" to "B"	"B" to "B"
Video (17.5 MHz)	Low	85.4	85.0	85.4	85.0
Video (20 MHz)	Low	85.4	85.0	85.4	85.0
Video (30 MHz)	High	85.4	80.8	85.4	85.0

[a]The EIRP values shown in the table do not include 2-dB margin for possible satellite saturation flux density variation for INTELSAT V, VA, VA(IBS), and VI. For INTELSAT VII, changes in transponder saturation flux density are compensated for by changing the receiver gain step, since this transponder has several small gain step attenuators.
[b]The earth station types shown for various connections are: "A"= revised or previous Standard A and "B"= Standard B.
[c]Gain step = Transponder gain step, where Low = TWT attenuator "in" and High = TWT attenuator "out."
Source: INTELSAT IESS-306 (Rev. 3), Annex B (Ref. 26).

Table 6.17 gives examples of maximum EIRP values for TV/FM carriers on INTELSAT V, VA, VA(IBS), and VI. Table 6.18 provides maximum EIRP values for INTELSAT VII and VIIA.

Tables 6.19 and 6.20 provide television transmission parameters for half- and full-transponder operation.

Figure 6.21 shows a typical INTELSAT VII coverage diagram.

TABLE 6.18 INTELSAT VII and VIIA Maximum Earth Station EIRP Requirements for TV/FM Carriers[a]

		EIRP for 10° Elevation Angle[b] (dBW)			
		Earth Station			
Carrier Size	Saturation Flux Density (dBW/m^2)[c]	"A" to "A"	"B" to "A"	"A" to "B"	"B" to "B"
Video (20 MHz)	−73.0	85.0	85.0	85.0	85.0
Video (30 MHz)	−77.0[d]	85.0	80.8	85.0	85.0

[a]The EIRP values shown in the table do not include 2-dB margin for possible satellite saturation flux density variation for INTELSAT V, VA, VA(IBS), and VI. For INTELSAT VII, changes in transponder saturation flux density are compensated for by changing the receiver gain step, since this transponder has several small gain step attenuators.
[b]The earth station types shown for various connections are: "A"= revised or previous Standard A and "B"= Standard B.
[c]Saturation flux density (= the transponder gain step attenuation setting) will be adjusted to provide a nominal saturation flux density corresponding to the value shown in this column.
[d]The transponder saturation flux density is −81.0 dBW/m^2 for Standard "B" to "A."
Source: INTELSAT IESS-306 (Rev. 3), Annex B (Ref. 26).

TABLE 6.19 INTELSAT VII and VIIA Half-Transponder 20-MHz TV Operation

General Characteristics

Allocated satellite bandwidth (MHz)	20.0	
Receiver noise bandwidth (MHz)	18.0	
Television standard	525/60	625/50
Maximum video bandwidth[a] (MHz)	4.2	6.0
Peak-to-peak low-frequency (15-kHz) deviation of a pre-emphasized video signal (MHz)	5.95[b]	5.30[b]
Peak deviation due to 1.0-V peak-to-peak test tone (MHz)	9.4[c]	9.4[c]
Differential gain	10%	10%
Differential phase	4°	4°

Receive Characteristics

Television standard	525/60 or 625/50							
Earth station[d]	"A" to "A(P)"	"B" to "A(P)"	"A" to "A(R)"	"B" to "A(R)"	"A" to "B"		"B" to "B"	
Elevation angle (receive)	10°	10°	10°	10°	10°	55°	50°	90°
C/T total at operating point (dBW/K)	−135.1	−135.1	−138.9	−138.9	−141.8	−141.1	−141.1	−140.8
Video signal-to-weighted noise ratio[a] (dB)	54.6 52.9 52.9	54.6 52.9 52.9	50.8 49.0 49.0	50.8 49.0 49.0	48.0 46.2 46.2	48.8 47.0 47.0	48.8 47.0 47.0	49.0 47.2 47.2
C/N in IF noise bandwidth (dB)	20.9	20.9	17.1	17.1	14.3	15.0	15.0	15.3
Transmit and receive IF noise bandwidth (MHz)	18.0							
Amount of over-deviation in IF bandwidth[e] (dB) 525/60	5.8							
Amount of over-deviation in IF bandwidth[e] (dB) 625/50	9.9							

[a] $|X|\ \ |Y|\ \ |Z|$ format should be interpreted as follows: X values refer to System M (NTSC); Y values refer to Systems B, C, G, and H (PAL); and Z values refer to Systems D, K, and L (SECAM). All video signal-to-weighted noise ratios are calculated with weighting plus de-emphasis factors based on the unified weighting network (Rec. ITU-R CMTT.567-3), which has a noise bandwidth of 5 MHz for all TV systems.

[b] Excluding the maximum peak-to-peak deviation of 1 MHz due to the application of energy dispersal waveform.

[c] Peak frequency deviation of TV/FM carrier due to a 1.0-V peak-to-peak test tone at the crossover frequency of the pre-emphasis characteristic (excluding the maximum peak-to-peak deviation of 1 MHz due to the application of energy dispersal waveform).

[d] "A(P)" denotes previous Standard A earth station ($G/T \geq 40.7$ dB/K), and "A(R)" denotes revised Standard A earth station ($G/T \geq 35.0$ dB/K). "A" denotes either "A(P)" or "A(R)."

[e] The amount of overdeviation (OD) is defined as:

$$\text{OD} = 20 \log_{10}(\Delta f_{tt}/\Delta f_{cr}) \quad \text{dB}$$

where Δf_{tt} = peak frequency deviation (in MHz) of TV/FM carrier due to 1.0-V peak-to-peak test tone at the crossover frequency of pre-emphasis characteristic, and Δf_{cr} = peak frequency deviation (in MHz) of the TV/FM carrier if it were designed to exactly match the receive filter bandwidth based on Carson's rule, $\Delta f_{cr} = \frac{1}{2}B_{IF} - f_m$. Note that B_{IF} is the receive IF bandwidth (in MHz) and f_m is the maximum video bandwidth (in MHz).

Source: INTELSAT IESS-306 (Rev. 3), Annex C (Ref. 26).

TABLE 6.20 INTELSAT VII and VIIA Full-Transponder 30-MHz TV Operation

General Characteristics

Allocated satellite bandwidth (MHz)	30	
Receiver noise bandwidth (MHz)	22.5 to 30	
Television standard	525/60	625/50
Maximum video bandwidth[a] (MHz)	4.2	6.0
Peak-to-peak low-frequency (15-kHz) deviation of a pre-emphasized video signal (MHz)	6.8[b]	5.1[b]
Peak deviation due to 1.0-V peak-to-peak test tone (MHz)	10.75[c]	9.0[c]
Differential gain	10%	10%
Differential phase	+3°	+3°

Receive Characteristics

Television standard	525.60 or 625/50					
Earth station[d]	"A" to "A(P)"	"B" to "A(P)"	"A" to "A(R)"	"B" to "A(R)"	"A" to "B"	"B" to "B"
Elevation angle (receive)	10°	10°	10°	10°	10°	10°
C/T total at operating point (dBW/K)	−132.3	−134.7	−135.0	−136.4	−137.3	−137.3
Video signal-to-weighted noise ratio[a] (dB)	58.6 55.3 55.3	56.3 52.9 52.9	56.0 52.6 52.6	54.5 51.2 51.2	53.6 50.3 50.3	53.6 50.3 50.3
C/N in IF noise bandwidth (dB)	21.5	19.2	18.8	17.4	16.5	16.5
Receive IF noise bandwidth (MHz)	30.0[e]					
Amount of over-deviation in IF bandwidth[f] (dB) 525/60	—					
Amount of over-deviation in IF bandwidth[f] (dB) 625/50	—					

[a] |X| |Y| |Z| format should be interpreted as follows: X values refer to System M (NTSC); Y values refer to Systems B, C, G, and H (PAL); and Z values refer to Systems D, K, and L (SECAM). All video signal-to-weighted noise ratios are calculated with weighting plus de-emphasis factors based on the unified weighting network (Rec. ITU-R CMTT.567-3), which has a noise bandwidth of 5 MHz for all TV systems.

[b] Excluding the maximum peak-to-peak deviation of 1 MHz due to the application of energy dispersal waveform.

[c] Peak frequency deviation of TV/FM carrier due to a 1.0-V peak-to-peak test tone at the crossover frequency of the pre-emphasis characteristic (excluding the maximum peak-to-peak deviation of 1 MHz due to the application of energy dispersal waveform).

[d] "A(P)" denotes previous Standard A earth station ($G/T \geq 40.7$ dB/K), and "A(R)" denotes revised Standard A earth station ($G/T \geq 35.0$ dB/K). "A" denotes either "A(P)" or "A(R)."

[e] If 22.5-MHz IF noise bandwidth is used in place of 30.0-MHz filters, the C/N value will be improved by about 1.2 dB.

[f] The amount of overdeviation (OD) is defined as:

$$OD = 20 \log_{10}(\Delta f_{tt}/\Delta f_{cr}) \quad \text{dB}$$

where Δf_{tt} = peak frequency deviation (in MHz) of TV/FM carrier due to 1.0-V peak-to-peak test tone at the crossover frequency of pre-emphasis characteristic, and Δf_{cr} = peak frequency deviation (in MHz) of the TV/FM carrier if it were designed to exactly match the receive filter bandwidth based on Carson's rule, $\Delta f_{cr} = \frac{1}{2}B_{IF} - f_m$. Note that B_{IF} is the receive IF bandwidth (in MHz) and f_m is the maximum video bandwidth (in MHz).

Source: INTELSAT IESS-306 (Rev. 3), Annex C (Ref. 26).

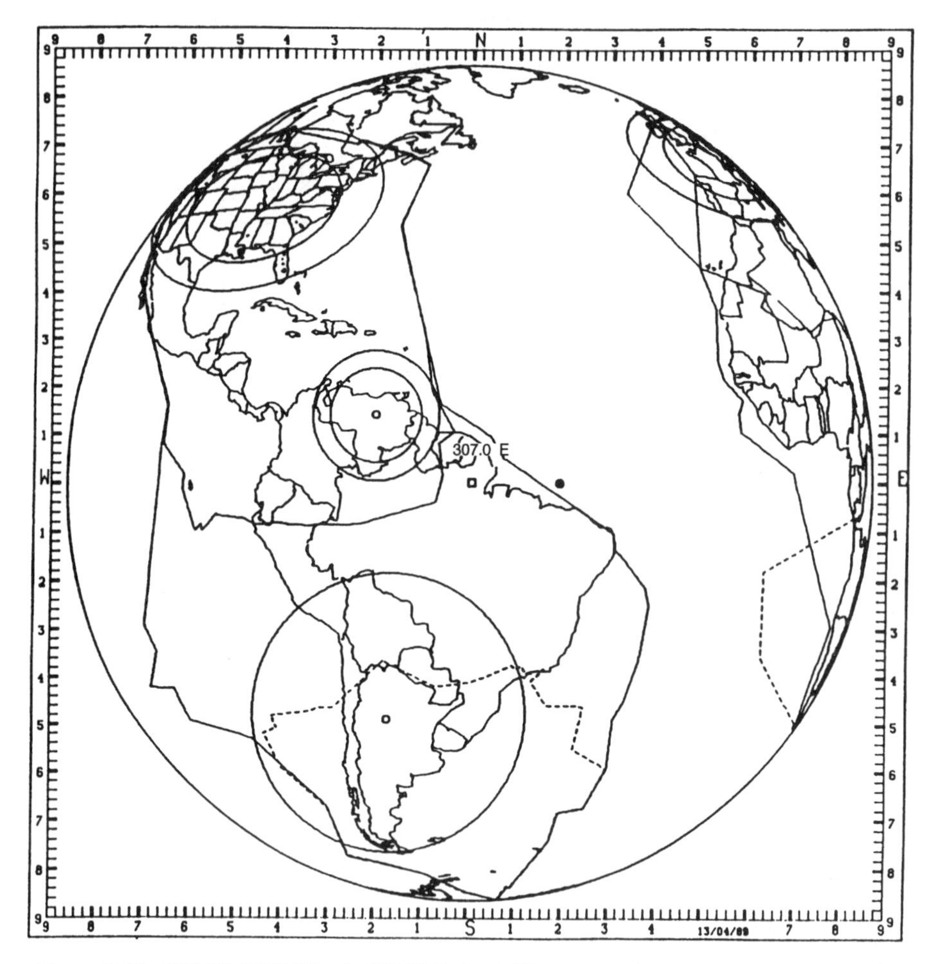

Figure 6.21. INTELSAT VII Atlantic 307.0° East satellite, expected antenna beam coverage for hemispheric, zone, and spot beams. *Note:* This satellite antenna coverage diagram illustrates the preliminary spacecraft platform pointing bias and spot beam pointing for this location. After the final pointing is selected, it or an appropriate pointing range will be used when determining the satellite antenna pattern advantage (Beta Factor) for each earth station as described in IESS-402. (The C-band spot beams are congruent.) The nominal beam edge contours are illustrated: □, subsatellite point; and ⊕, satellite antenna platform. Pointing bias = 1.90° East; yaw bias = 0°. It should be noted that INTELSAT uses a nonconventional 360° longitude scheme for identifying subsatellite points. (Figure and note are from INTELSAT IESS-409, Figure 7; Ref. 18.)

6.6 DOMESTIC AND REGIONAL SATELLITE SYSTEMS

6.6.1 Introduction

Domestic and regional satellite communication systems are attractive to regions of the world with a large community of interest. These regions may be just one country with a comparatively large geographical expanse, such as the United States, India, Brazil, Indonesia, Australia, Japan, Mexico, and Russia, or a group of countries with a common interest, such as Europe, or a common culture, such as the Arabic-speaking countries, or a common language, such as Hispanic America. Many of these systems also serve areas that are sparsely populated to deliver quality telephone/data service and TV programming. Canada's ANIK/TeleSat is a good example. Still another family of systems serves the business community, providing basically a digital offering. One important group of this type provides long-distance enterprise networking. In this case, the "enterprise" leases all or part of one or more satellite transponders. VSAT (very small aperture terminal) systems are in this category. They are really burgeoning in many parts of the world. VSAT systems are described in Chapter 8.

About three-quarters of the transponder space, particularly in the Ku-band, over North America supply TV programming to CATV headends and for hotels and motels.

6.6.2 Rationale

The basic, underlying guideline for any telecommunication system is cost effectiveness. Consider a very large enterprise network system with, say, 1000 terminals. Here we may want to invest more in the space segment to allow as much economy as possible on that large terminal segment. One thousand is a big multiplier. This idea is a driving factor in VSAT networks. If the terminal segment is small, say, ten terminals, we'd invest more in these terminals so that space segment charges may be reduced.

6.6.3 Approaches to Cost Reduction

Terminal cost reduction can be achieved by the following means:

- Reducing performance requirements
- Eliminating connecting links
- Reducing/optimizing bandwidth
- Augmenting the space segment
- Mass production of terminals
- Increasing the minimum elevation angle

We consider here a network where satellite terminals are right on customers' premises or on enterprise-owned buildings where connectivity to the end user may be measured in less than 100 m (330 ft).

Such a network has no need for the tight and strict performance requirements of INTELSAT described in Section 6.5. INTELSAT provides long-haul, international connectivity, where there may be long tails on each end. Each tail may have several switches in tandem.

As a consequence, the enterprise network can notably reduce performance of the intervening satellite links without sacrificing end-user performance. In other words, the enterprise satellite network can be much more economical than its INTELSAT counterpart. For example, if we desire an end user to have a BER of better than 1×10^{-6} more than 80% of the time (see CCITT Rec. G.821), then certainly the network need not provide a BER better than 5×10^{-7}.

If the enterprise network in question had many terminals, as we suggested above, it might be wise to spend more on the space segment so that the cost of the terminal segment can be decreased. It would go without saying that we'd do all in our power to optimize transponder bandwidth usage to reduce space segment costs. The network cost per terminal can notably be reduced because of the quantity of terminals involved, compared, say, to a ten-terminal network.

A rigid cost–benefit analysis may show that the use of satellite communications may be the best approach for long-distance connectivity for a large, geographically dispersed enterprise network. Keep in mind such a network requires good engineering design, contractor selection, contract monitoring, acceptance, and cutover. Downstream it must be maintained. Here we recommend another cost–benefit analysis: Do we do it ourselves or out-source a portion or all of the job?

6.6.4 A Typical Satellite Series That Can Provide Transponder Space for Enterprise Networks*

Hughes Communications, Inc. is the sponsor of the Galaxy Satellite System with more than four satellites in the North American arc. The system provides full and partial transponder point-to-point and point-to-multipoint C-band and Ku-band services for domestic (U.S.) and transborder telecommunication connectivity. Services are available to users in the contiguous 48 states of the United States, Alaska, Hawaii, parts of Canada, Mexico, and the Caribbean basin. In-orbit protection is available for both C-band and Ku-band satellites.

The primary usage of C-band services is to support the cable TV and TV broadcast industries as well as audio and data services. Ku-band services include VSAT communications, broadcast video, compressed video, video time sharing, and audio and data services.†

Figure 6.22 shows the Galaxy VII Ku-band frequency/polarization plan. Figure 6.23 gives Ku-band Galaxy VII EIRP contours for the satellite at 91° W longitude, vertical polarization. Figure 6.24 gives saturation flux

*Material in this section was kindly provided by Hughes Communications, Inc., Los Angeles, CA.

†It should be appreciated that the trend for PSTN long-distance digital connectivity is away from satellites and more and more on fiber optic links, transcontinental and around the world.

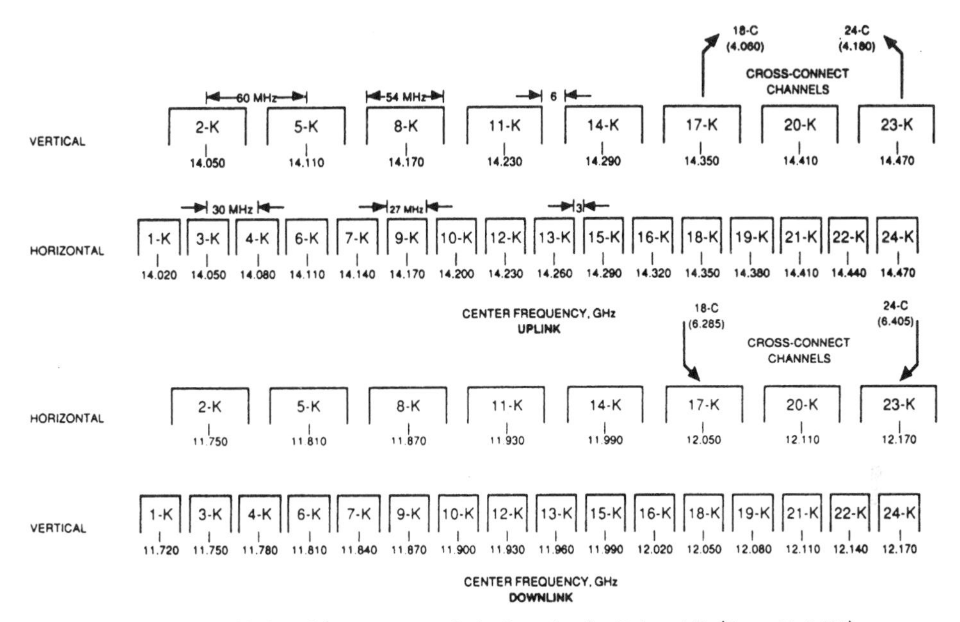

Figure 6.22. Ku-band frequency/polarization plan for Galaxy VII. (From Ref. 27.)

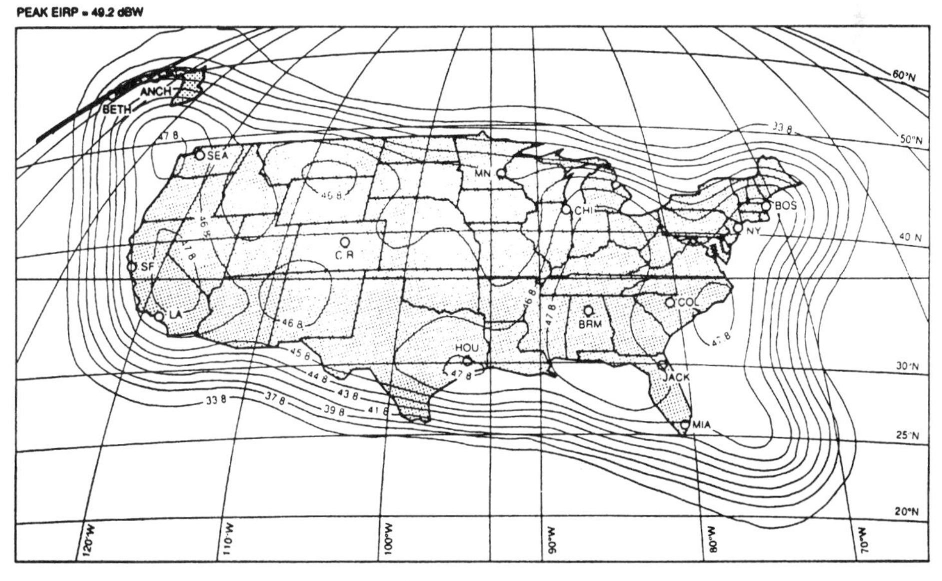

Figure 6.23. Ku-band EIRP contours (in +dBW) for Galaxy VII, 91° W longitude for vertical polarization.

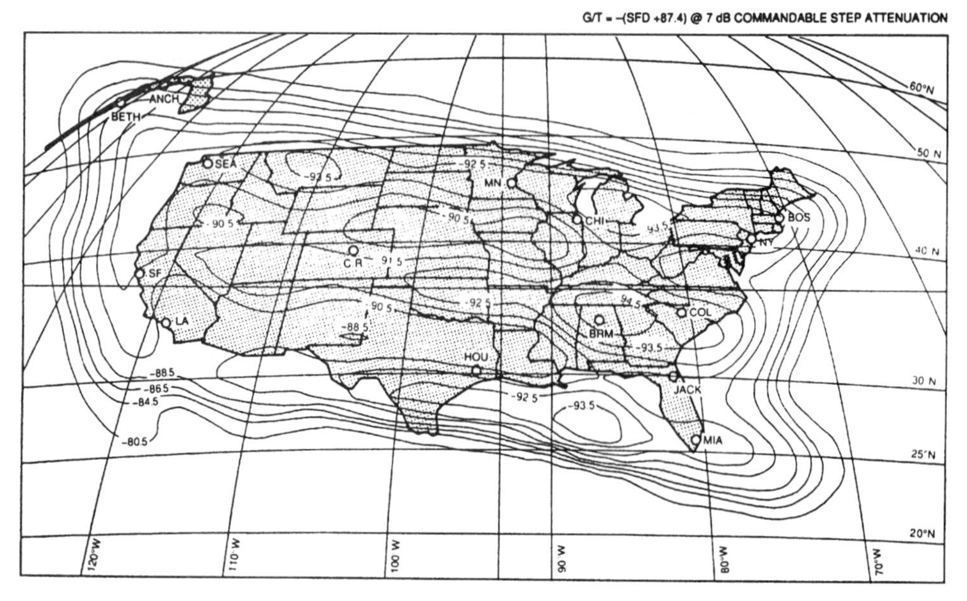

Figure 6.24. Ku-band saturation flux density (SFD) contours for Galaxy VII, 91° W longitude for horizontal polarization.

density (SFD) contours for Galaxy VII at 91° W longitude, horizontal polarization.

PROBLEMS AND EXERCISES

1 Draw a simplified block diagram of a "bent-pipe" satellite. Show the local oscillator mixing frequency to derive the current 4-GHz band from the 6-GHz uplink band.

2 List four advantages and four disadvantages of geostationary satellites used for telecommunications relay.

3 What is the range (distance) to a geostationary satellite if the elevation angle for a particular earth station is 27°?

4 Calculate the free-space (spreading) loss to a geostationary satellite at 6105 MHz when the elevation angle is 23°.

5 Calculate the isotropic receive level at an earth station if a satellite transponder radiates +31 dBW on 7.305 GHz and the elevation angle is 15°. Consider only free-space loss.

6 Calculate the flux density (in dBW/m) impinging on a satellite antenna if the EIRP of an earth station is +65 dBW and the earth station is directly under the satellite (i.e., at the subsatellite point).

7 Downlink signals are generally of low level at the earth's surface.

a. Give at least two reasons why.

b. Give at least two general ways in which we can achieve sufficient "sensitivity" at an earth station to satisfactorily utilize these weak signals.

8 What is the thermal noise level (in dBW) in a 1-Hz bandwidth of the theoretically perfect receiver operating at absolute zero.

9 What is the receiving system noise temperature (in kelvins) when the antenna noise is 105 K and the receiver noise is 163 K.

10 As one lowers the elevation angle of an earth station antenna toward the horizon, what happens to T_{sys}? Give at least two reasons why.

11 The noise figure of a certain LNA is 1.25 dB. What is the effective noise temperature of the LNA (in kelvins)?

12 A section of waveguide has an ohmic loss of 0.3 dB. What is its (approximate) equivalent noise temperature when inserted in the transmission line system?

13 What is the spectral noise density in 1-Hz bandwidth (N_0) of an earth station receiving system where T_{sys} is 97 K?

14 If the receive signal level (RSL) at the input of a certain LNA is −131 dBW where T_{sys} is 141 K at the same reference point, what, then, is the value of C/N_0?

15 G/T for a particular earth station is given as 35.0 dB/K where G at the reference plane is 55 dB. What is T_{sys}?

16 Name at least three components of sky noise.

17 Calculate the antenna noise of an earth station where the elevation angle is 10°, the operating frequency is 7300 MHz, feed loss is 0.1 dB, waveguide losses are 1.9 dB, and the bandpass filter has an insertion loss of 0.5 dB at 7300 MHz. The reference plane is at the input of the LNA.

18 Calculate T_{sys} from question 17 where the LNA has a noise figure of 1.19 dB. Disregard noise contributions of subsequent receiver stages after the LNA.

19 Calculate receiver noise temperature where a LNA has a noise figure of 1.05 dB and a gain of 30 dB, and the noise temperature of a subsequent postamplifier is 450 K.

20 Calculate the value of G/T of a satellite earth terminal given the following parameters: sky noise is 53 K, total ohmic losses at the reference plane of the antenna subsystem is 1.94 dB, receiver noise figure is 1.65 dB, gross antenna gain is 47 dB. Disregard nonohmic losses.

21 Calculate downlink C/N_0 for an earth coverage antenna of a satellite where the EIRP of that beam is +34 dBW, the earth station G/T is 21.5 dB/K, and the total downlink losses are 202 dB.

22 Calculate the required G/T of a satellite where the required uplink C/N_0 at the satellite transponder is 85 dB/Hz, the terminal EIRP is +62 dBW, the operating frequency is 6250 MHz, and the elevation angle is 21°; the satellite uses a spot beam and the earth terminal is somewhere inside the 3-dB contour footprint. Use reasonable loss values.

23 What is the system C/N_0 where the uplink C/N_0 is 85 dB; the downlink C/N_0 is 71 dB; and the transponder IM C/N_0 is 96 dB?

24 A satellite FDM/FM/FDMA system with 972 VF channels on one carrier is designed for a noise power value on the downlink VF channel derived output of 10,000 pW0p. Calculate S/N in the derived voice channel where the emphasis improvement factor is 4 dB. Use INTELSAT values.

25 Compare FDMA with TDMA.

26 Using the rule of thumb, how much backoff is required to achieve a C/IM of 24 dB when initially the C/IM value is 16 dB?

27 Where would SCPC DAMA have application?

28 Name the three methods used to control DAMA operation.

29 Why do INTELSAT Standard A earth stations require such a large G/T?

30 On an INTELSAT Standard A earth station system, 80% of the noise in a derived voice channel is allocated to the space segment. Give at least two reasons why, explaining the sources of the noise.

31 We are to design a domestic satellite from scratch. Using INTELSAT Standard A as a model for the starting point, name six ways to reduce system cost.

32 What is the minimum aperture antenna assuming 60% antenna efficiency for a SCPC system to receive just one 45-kHz channel and allow 20,000 pWp noise in the derived voice channel. The system operates in the 6/4-GHz band pair. Assume satellite EIRP of +10 dBW and an elevation angle of 25°. Use reasonable loss values.

REFERENCES

1. Roger L. Freeman, *Reference Manual for Telecommunications Engineering*, 2nd ed., Wiley, New York, 1994.
2. Warren L. Flock, *Propagation Effects on Satellite Systems at Frequencies below* 10 *GHz*, University of Colorado, 1983. Prepared for NASA, NASA Ref. Pub. 1108.
3. *Reference Data for Radio Engineering*, 5th ed., ITT–Howard W. Sams, Indianapolis, 1968.
4. *Radio Regulations*, ed. 1990, revised 1994, ITU, Geneva, 1994.
5. *World Administrative Radio Congress—1979 (WARC-79)*, ITU, Geneva, 1980.
6. *Satellite Communications Symposium—'82*, Scientific-Atlanta, Norcross, GA, 1982.
7. *Maximum Permissible Values of Power Flux-Density at the Surface of the Earth Produced by Satellites in the Fixed-Satellite Service Using the Same Frequency Bands Above 1 GHz as Line-of-Sight Radio-Relay Systems*, ITU-R Rec. SF.358-4, ITU-R SF Series, ITU, Geneva, 1994.
8. *Transmission Systems for Communications*, 5th ed., Bell Telephone Laboratories, Holmdel, NJ, 1982.
9. *Radio Emission from Natural Sources in the Frequency Range Above About 50 MHz*, CCIR Rep. 720, Recommendations and Reports of the CCIR 1978, XIVth Plenary Assembly, Kyoto, 1978.
10. Raul Pettai, *Noise in Receiving Systems*, Wiley, New York, 1984.
11. *Performance Characteristics for Frequency Division Multiplex / Frequency Modulation (FDM / FM) Telephony Carriers*, INTELSAT IESS-301 (Rev. 3), INTELSAT, Washington, DC, May 1994.
12. *Television Systems*, ITU-R Rec. BT.470-3, 1994 BT Series Volume, ITU, Geneva, 1994.
13. *Carrier Energy Dispersal for Systems Employing Angle Modulation by Analogue Signals or Digital Modulation in the Fixed-Satellite Service*, ITU-R Rec. S.446-4, 1994 S Series Volume, ITU, Geneva, 1994.
14. Harry L. Van Trees, *Satellite Communications*, IEEE Press, New York, 1979, Sec. 3.6.5, "Demand Assignment."
15. Roger L. Freeman, *Telecommunication System Engineering*, 3rd ed., Wiley, New York, 1996.
16. Harry L. Van Trees, *Satellite Communications*, IEEE Press, New York, 1979, Sec. 3.6.3.1, "Time Division Multiple Access."
17. G. Maral and M. Bosquet, *Satellite Communication Systems*, 2nd ed., Wiley, Chichester, UK, 1993.
18. *INTELSAT VII Satellite Characteristics*, INTELSAT IESS-409, INTELSAT, Washington, DC, June 1989.
19. *INTELSAT VIII Satellite Characteristics*, INTELSAT IESS-417, INTELSAT, Washington, DC, Aug. 1994.
20. *Introduction and Approved IESS Document List* (and cataloged documents therein), INTELSAT IESS-101, Rev. 30, INTELSAT, Washington, DC, May 1995.
21. *QPSK/FDMA Performance Characteristics for INTELSAT Business Service (IBS)*, INTELSAT IESS-309, Rev. 4, INTELSAT, Washington, DC, May 1994.

22. *SCPC/CFM Performance Characteristics for the INTELSAT VISTA Service*, INTELSAT IESS-305, Rev. 2, INTELSAT, Washington, DC, May 1994.

23. *Earth Station EIRP Adjustment Factors to Account for Satellite Antenna Pattern Advantage and Path Loss Differential with Elevation Angle*, INTELSAT IESS-402, Rev. 4, INTELSAT, Washington, DC, May 1994.

24. *Standards C and E Antenna and Wideband RF Performance of Ku-Band Earth Stations Accessing the INTELSAT Space Segment for Standard Service*, INTELSAT IESS-208, Rev. 1, INTELSAT, Washington, DC, Nov. 1994.

25. *Standards A, B, D and F Antenna and Wideband RF Performance Characteristics of C-Band Earth Stations Accessing the INTELSAT Space Segment*, INTELSAT IESS-207, Rev. 1, INTELSAT, Washington, DC, Aug. 1994.

26. *Performance Characteristics for Television / Frequency Modulation (TV / FM) Carriers with TV-Associated Sound Program Transmission (FM Subcarriers), Operation with C-Band Global Beam Transponders*, INTELSAT IESS-306, Rev. 3, INTELSAT, Washington, DC, May 1994.

27. *Galaxy/Weststar Brochure*, Hughes Communications Corp. El Segundo, CA, 1989.

7

DIGITAL COMMUNICATIONS BY SATELLITE

7.1 INTRODUCTION

The North American PSTN (public switched telecommunications network) is nearly entirely digital, especially in its long-distance (toll) networks. Other parts of the world are moving in that direction rapidly. International satellite routes still have numerous analog links. We expect by the year 2007 that nearly all satellite communications will be digital, including television.

INTELSAT has a wide variety of digital offerings; nearly all VSAT (very small aperture terminal) networks are digital (Chapter 8); U.S. and NATO satellite circuits are entirely digital, such as DSCS, FLTSAT, and MILSTAR.

In this chapter we will discuss two basic classes of digital satellites. The first is the familiar bent-pipe satellite (Chapter 6), where "what comes down is a reasonable replica of what goes up," but with some degradation. The second class of digital satellite is the processing satellite, where "what comes down is not necessarily an exact replica of what went up" and, if it is the same, there is notably less degradation of the signal. In this latter case the satellite demodulates and regenerates the uplink signal and may carry out decoding/recoding (Chapter 4) and various levels of switching and routing. One example of this type of satellite is NASA's ACTS (Advanced Communications Technology Satellite), operating in the 30/20-GHz band, and the U.S. Department of Defense's MILSTAR.

Two digital access methods are described: FDMA and TDMA. FDMA will include INTELSAT FDMA schemes, coding, and performance. We will discuss TDMA in some detail and DSI (digital speech interpolation).

Digital transmission offers many advantages (see Ref. 17, Chapter 8). Regeneration at all digital network nodes and repeaters prevents noise accumulation, a primary concern with analog systems. The principal disadvantages are error accumulation and critical timing requirements.

7.2 DIGITAL OPERATION OF A BENT-PIPE SATELLITE SYSTEM

7.2.1 General

There are three approaches to digital bent-pipe operation and access: (1) FDMA mode, (2) TDMA mode, and (3) CDMA mode.

We will discuss the first two in this chapter. CDMA (code division multiple access or spread spectrum) is covered in Chapter 10. In neither of the three cases is the signal regenerated in the satellite.

7.2.2 Digital FDMA Operation

Digital FDMA operation is very similar to analog FDMA operation discussed in Section 6.4.2. Rather than place a FDM/FM signal in a preassigned transponder frequency slot, a digital waveform is transmitted in that segment using modulation techniques covered in Sections 3.3 and 4.2. Quadrature phase-shift keying (QPSK) modulation is commonly employed.

Because there is no signal regeneration in the satellite transponder, the downlink signals suffer a certain amount of distortion due to the satellite transponder, which essentially is a downconverting mixer. For one thing, there is additive thermal noise, and when there is multichannel activity on the same transponder, IM (intermodulation) noise can be a major impairment. A primary source of IM noise is the transponder final amplifier, which is usually a TWT (traveling wave tube). Solid-state amplifiers (SSAs), which are now being implemented in some new satellite designs, tend to reduce IM products because of improved linearity over their TWT counterparts. Both types of amplifiers are inefficient. Noise and distortion result in degraded error performance.

7.2.2.1 INTELSAT Digital Operation. In this section, INTELSAT's Intermediate Data Rate (IDR) digital carrier system is briefly reviewed followed by some highlights of the INTELSAT Business Service (IBS).

7.2.2.1.1 Intermediate Data Rate Digital Carrier System Requirements. IDR digital carriers in the INTELSAT system utilize coherent QPSK modulation operating at information rates ranging from 64 kbps to 44.736 Mbps. The information rate is defined as the bit rate entering the channel unit prior to the application of any overhead or FEC. Any information rate from 64 kbps to 44.736 Mbps inclusive may be transmitted. INTELSAT has, however, defined a set of recommended information rates that are based on ITU-T hierarchical rates (see ITU-T Rec. G.703) as well as ITU-T ISDN rates covered in the I-series recommendations. These recommended bit rates are given in Table 7.1. Bit error rate performance is provided in Table 7.2. Figure 7.1 shows an IDR channel unit.

For certain information rates—namely, 1.544, 2.048, 6.312, 8.448, 32.064, 34.368, and 44.736 Mbps—an overhead structure has also been defined to

TABLE 7.1 Recommended IDR Information Rates and Associated Overhead

Number of 64-kbps Bearer Channels (n)	Information Rate ($n \times 64$ kbps)	Type of Overhead		
		No Overhead[a]	With 96 kbps IDR Overhead[b]	With 6.7% IBS Overhead[b,c]
1	64	X[d]		X
2	128			X
3	192	X		
4	256			X
6	384	X		X
8	512			X
12	768			X
16	1,024			X
24	1,536			X
24	1,544[e]		X	
30 (31)	2,048[e]		X	
90 (93)	6,312[e]		X	
120 (124)	8,448[e]		X	
480	32,064[e]		X	
480 (496)	34,368[e]		X	
630 (651) or 672	44,736[e]		X	

[a]For rates less than 1544 kbps, it is possible to use any $n \times 64$ kbps information rate without overhead, but the only INTELSAT-recommended rates are 64, 192, and 384 kbps. The use of the optional Reed–Solomon outer coding is not defined for any information rate less than 1.544 Mbps, which does not use overhead.

[b]The optional Reed–Solomon outer coding can be used with the information rates shown with an "X."

[c]The carriers in this column are small IDR carriers that can be used with the circuit multiplication concept described in Appendix B of the reference publication. [For a definition of the IBS overhead framing, see IESS-309 (IBS).]

[d]X = Recommended rate corresponding to the type of overhead.

[e]These are standard ITU-T hierarchical bit rates. Other $n \times 64$ kbps information rates above 1.544 Mbps are also possible and must have an ESC overhead of at least 96 kbps (the overhead framing will be defined on a case-by-case basis by INTELSAT).

Source: Table 1, INTELSAT IESS-308, Rev. 7 (Ref. 1).

provide engineering service circuits (ESCs) and maintenance alarms. This overhead increases the data rate of these carrier sizes by 96 kbps. The overhead adds its own frame alignment signal and thus passes the information data stream transparently. For the provision of ESC and alarms on IDR (intermediate data rate) carrier sizes larger than 1.544 Mbps, and overhead of 96 kbps must be used. For other information rates larger than 1.544 Mbps that are not listed above, INTELSAT will determine the overhead framing structure on a case-by-case basis.

The occupied satellite bandwidth unit for IDR carriers is approximately equal to 0.6 times the transmission rate.* To provide a guardband between

*The transmission rate (R) is defined as the bit rate entering the QPSK modulator at the earth station (i.e., after the addition of any overhead or FEC) and is equal to twice the symbol rate at the output of the QPSK modulator.

TABLE 7.2 IDR Performance[a]

	INTELSAT V, VA, VA(IBS), and VI
Weather Condition	Minimum BER Performance[b,c] (% of year)
Clear sky	$\leq 10^{-7}$ for $\geq$ 95.90%
Degraded	$\leq 10^{-6}$ for $\geq$ 99.36%
Degraded	$\leq 10^{-3}$ for $\geq$ 99.96%

	INTELSAT VII, VIIA, and K	
Weather Condition	Minimum BER Performance[d] (% of year)	Typical BER Performance[c] (% of year)
Clear sky	$\leq 2 \times 10^{-8}$ for $\geq$ 95.90%	$\leq 10^{-10}$ for $\geq$ 95.90%
Degraded	$\leq 2 \times 10^{-7}$ for $\geq$ 99.36%	$\leq 10^{-9}$ for $\geq$ 99.36%
Degraded	$\leq 7 \times 10^{-5}$ for $\geq$ 99.96%	$\leq 10^{-6}$ for $\geq$ 99.96%
Degraded	[e]	$\leq 10^{-3}$ for $\geq$ 99.98%

[a]These values account for propagation-related effects only. They do not include the effect of earth station equipment problems or misoperation, such as improper tracking. For a complete listing of the uppath and downpath margins used in the reference link budgets for all beam connections (C-to-C, Ku-to-Ku and cross-strapped), refer to the Appendices of Ref. 1. Refer to Appendix H of Ref. 1 for the IDR performance expected for earth stations using equipment that exceeds the minimum IESS performance requirements.

[b]The minimum BER performance conforms with Rec. ITU-R S.614-2. For INTELSAT VI satellites, INTELSAT will endeavor to provide the minimum performance shown for INTELSAT VII / VIIA, whenever transponder capacity and earth station uplink EIRP constraints permit.

[c]The typical clear-sky performance values shown are based on specified modem performance characteristics and earth station G/T and are the basis of carrier lineups to ensure that the minimum performance requirements can be met under operational conditions. In practice, the BER performance and link availability can be even better, depending on the type of equipment and additional G/T performance selected by the user.

[d]The minimum BER performance conforms with the bit-rate-dependent criteria given in Note 1 of Rec. ITU-R S.1062. Recommendation ITU-R S.1062 has been developed by the ITU Radiocommunication Sector for satellite links, operating at or above the primary rate. Performance that meets the criteria of Note 1 of the Recommendation will fully comply with the requirements of Recommendation ITU-T G.826. The performance shown in Table 7.2 applies to 2.048 Mbps IDR carriers where the value of α (average number of errors per burst) is equal to 10. For the minimum performance applicable to other information rates, refer to Rec. ITU-R S.1062. The typical IDR performance will either meet or exceed the minimum BER performance recommended for the information rates shown in Note 1 of Rec. ITU-R S.1062. For IDR information rates of 8.448, 32.064, 34.368, and 44.736 Mbps, which are not listed in Note 1, the minimum performance requirements in the ITU-R Recommendation for 51.0 Mbps should be assumed. In the case of small IDR carriers (information rates less than 1.544 Mbps), the minimum BER performance requirements for 2.0 Mbps in the ITU-R Recommendation should be assumed.

[e]Not specified in Recommendation ITU-R S.1062.

Source: Tables 2 and 3, INTELSAT IESS-308, Rev. 7A (Ref. 1).

adjacent carriers, the nominal satellite bandwidth unit is equal to 0.7 times the transmission rate. The actual carrier spacing may be larger and is determined by INTELSAT, based on the particular transponder frequency plan. In particular, for the case of two nominal 45-Mbps rate 3/4 FEC IDR carriers within a 72-MHz C-band transponder, the allocated RF bandwidth is 36 MHz for each carrier.

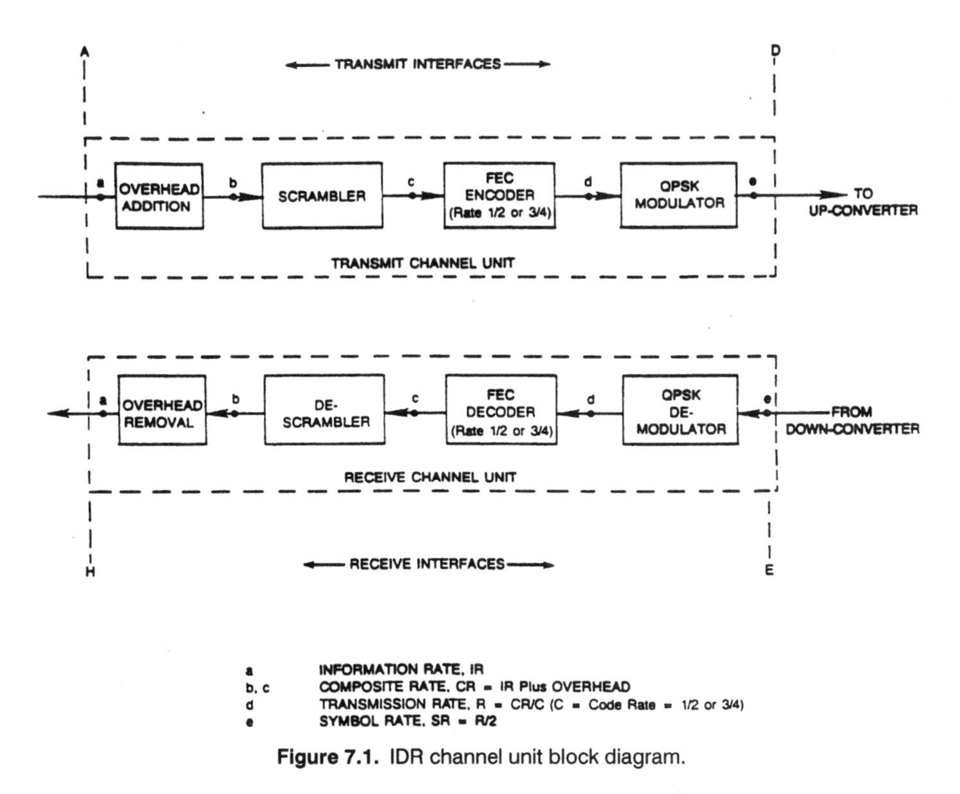

Figure 7.1. IDR channel unit block diagram.

IDR carriers may share transponders with carriers employing other approved modulation techniques and/or with other IDR carriers. Receiving earth stations should be designed so that a carrier can be received in the presence of adjacent carriers, which may use any approved modulation technique.

7.2.2.1.2 EIRP Values. The maximum 6-GHz and 14-GHz example EIRP values and associated transmission parameters are given in Tables 7.3 and 7.4 for INTELSAT VII and VIIA, respectively.

The EIRP values given in these tables are for a 10° elevation angle and located at beam edge. For other angles (i.e., > 10°) and for earth station locations other than at satellite antenna beam edge, see IESS-402 for correction factors K_1 and K_2. The term "case" refers to specific satellite scenarios, applicable only to INTELSAT VII and VIIA (Tables 7.3 and 7.4). The applicable case is described below the appropriate table footnotes. "High gain" and "low gain" refer to the power attenuator setting.

Table 7.5 gives QPSK characteristics and transmission parameters for IDR carriers using rate 3/4 FEC for INTELSAT VII, VIIA, and K.

Table 7.6 provides one example of transmission parameters, for INTELSAT VII, VIIA, and K recommended IDR carriers, FEC rate 3/4.

TABLE 7.3 Maximum Earth Station EIRP[a] for Rate $\frac{3}{4}$ FEC[b] (6/4 GHz)[c] INTELSAT VII

Rx Station	EIRP (dBW)						
	A (Previous)	A (Revised)	B	F-3	F-2	F-1	
Off-Axis EIRP	[6.6	4.9	3.1	0.9	−1.5	−6.3]	(Rate $\frac{3}{4}$)
Indicator[d] (dB)	[9.5	7.8	6.0	3.8	1.4	−3.4]	(Rate $\frac{1}{2}$)
Information Rate (kbps)							
64	52.2	53.9	55.7	57.9	60.3	65.1	
192	57.0	58.7	60.5	62.7	65.1	69.9	
384	60.0	61.7	63.5	65.7	68.1	72.9	
512	61.5	63.2	65.0	67.2	69.6	74.4	
1,024	64.5	66.2	68.0	70.2	72.6	77.4	
1,544	66.3	68.0	69.8	72.0	74.4	79.2	
2,048	67.5	69.2	71.0	73.2	75.6	80.4	
6,312	72.2	73.9	75.7	77.9	80.3	85.1	
8,448	73.5	75.2	77.0	79.2	81.6	86.4	
32,064	79.2	80.9	82.7	84.9	87.3	[e]	
34,368	79.5	81.2	83.0	85.2	87.6	[e]	
44,736	80.7	82.4	84.2	86.4	[e]	[e]	
X (for other information rates)	4.1	5.8	7.6	9.8	12.2	17.0	

[a]These maximum EIRP values have been computed using a saturation flux density of −75.0 dBW/m^2 for full-transponder loading conditions. Other saturation flux densities may be used depending on the actual traffic loading. For 36-MHz Hemi/Zone Transponder 9, a saturation flux density of −79 dBW/m^2 is used and the EIRP values are 1.5 dB lower for Revised Standard A, 2.5 dB lower for Standards B and F-3, and 3.5 dB lower for Standards F-2 and F-1.
[b]To obtain the maximum EIRP for rate $\frac{1}{2}$ FEC, a correction factor of 1.1 dB must be subtracted from the EIRP values shown for rate $\frac{3}{4}$ FEC.
[c]The above maximum EIRP values are applicable for the following uplink/downlink beam connections, ocean regions, and spacecraft locations:

Uplink		Downlink			D/L EIRP	U/L Margin	D/L Margin	S/C Location
Beam	Direction	Beam	Direction	Region	(dBW)	(dB/%)	(dB/%)	(°E)
H/Z	All	H/Z	All	All	33.0	3/0.02	4/0.02	All
Global	All	Hemi	All	All	33.0	3/0.02	4/0.02	All

[d]Off-axis EIRP density indicator (dB) applies to all values in the column below the number.
[e]Denotes carriers not intended for this connection.
Source: Table E.5, Appendix E, INTELSAT IESS-308, Rev. 7 (Ref. 1).

7.2.2.1.3 INTELSAT Business Service (IBS). IBS utilizes QPSK and FDMA to provide digital communications between Standard A, B, C, E, and F earth stations to facilitate the use of national gateway, urban gateway, and customer premise type earth stations but is not intended to be used for public switched telephony. This is what primarily sets it apart from IDR services described earlier.

TABLE 7.4 Maximum Earth Station EIRP[a] for Rate $\frac{3}{4}$ FEC IDR[b] (dBW), 14/11 or 14/12 GHz[c]: INTELSAT VIIA

Rx Station	EIRP (dBW)					
	C (Previous)	C (Revised)	E-3	E-2	E-1	
Off-Axis EIRP Indicator[d] (dB)	[9.7	8.8	7.6	2.1	−4.1]	(Rate $\frac{3}{4}$)
	[13.0	12.1	10.9	5.4	−0.8]	(Rate $\frac{1}{2}$)
Information Rate (kbps)						
64	50.4	51.3	52.5	58.0	64.2	
192	55.2	56.1	57.3	62.8	69.0	
384	58.2	59.1	60.3	65.8	72.0	
512	59.7	60.6	61.8	67.3	73.5	
1,024	62.7	63.6	64.8	70.3	76.5	
1,544	64.5	65.4	66.6	72.1	78.3	
2,048	65.7	66.6	67.8	73.3	79.5	
6,312	70.4	71.3	72.5	78.0	84.2	
8,448	71.7	72.6	73.8	79.3	85.5	
32,064	77.4	78.3	79.5	85.0	[e]	
34,368	77.7	78.6	79.8	85.3	[e]	
44,736	78.9	79.8	81.0	[e]	[e]	
X (for other information rates)	2.3	3.2	4.4	9.9	16.1	

[a]These maximum EIRP values have been computed using a saturation flux density of −78.0 dBW/m^2 for full-transponder loading conditions and applies to both 72- and 112-MHz transponders. Other saturation flux densities may be used depending on the actual traffic loading. The EIRP values shown in the table apply to the uplink inner contour and the downlink outer contour (see IESS-402 for correction factors).
[b]To obtain the maximum EIRP for rate $\frac{1}{2}$ FEC, subtract 1.5 dB from the EIRP values shown for rate $\frac{3}{4}$ FEC.
[c]The above maximum EIRP values are applicable for the following uplink/downlink beam connections, ocean regions, and spacecraft locations:

Uplink		Downlink		Outer Contour EIRP	U/L Margin	D/L Margin	
Beam	Direction	Beam	Direction	(dBW)	(dB/%)	(dB/%)	Region
Spot 1	Any	Spot 1	Any	44.7	7/0.02	13/0.02	All
Spot 1	Any	Spot 2	Any	43.7	7/0.02	13/0.02	All
Spot 1X	Any	Spot 1X	Any	44.8	7/0.02	13/0.02	All
Spot 1X	Any	Spot 2X	Any	43.4	7/0.02	13/0.02	All

[d]Off-axis EIRP density indicator (dB) applies to all values in the column below the number.
[e]Denotes carriers not intended for this connection.

Source: Table E.5, Appendix E, INTELSAT IESS-308, Rev. 7 (Ref. 1).

TABLE 7.5 QPSK Characteristics and Transmission Parameters for IDR Carriers Using Rate $\frac{3}{4}$ FEC (INTELSAT VII, VIIA, and K)

Parameter	Requirement				
1. Information rate, IR (bits/s)	64 kbps to 44.736 Mbps				
2. Overhead data rate for carriers with IR ≥ 1.544 Mbps	96 kbps				
3. Forward error correction encoding	Rate $\frac{3}{4}$ convolutional encoding/Viterbi decoding				
4. Energy dispersal (scrambling)	As per Figures 15 and 16 (main text of Ref. 1)				
5. Modulation	Four-phase coherent PSK				
6. Ambiguity resolution	Combination of differential encoding (180°) and FEC (90°)				
7. Clock recovery	Clock timing must be recovered from the received data stream				
8. Minimum carrier bandwidth (allocated)[a, b]	$0.7R^c$ Hz or [0.933 (IR + Overhead)]				
9. Noise bandwidth (and occupied bandwidth)	$0.6R$ Hz or [0.8 (IR + Overhead)]				
10. Composite rate[d] E_b/N_0 at BER (rate $\frac{3}{4}$ FEC)	10^{-3}	10^{-6}	10^{-7}	10^{-8}	10^{-10}
a. Modems back-to-back	5.3 dB	7.6 dB	8.3 dB	8.8 dB	10.3 dB
b. Through satellite channel[a]	5.7 dB	8.0 dB	8.7 dB	9.2 dB	11.0 dB
11. Transmission rate E_b/N_0 at BER (rate $\frac{3}{4}$ FEC)	10^{-3}	10^{-6}	10^{-17}	10^{-8}	10^{-10}
a. Modems back-to-back	4.05 dB	6.35 dB	7.05 dB	7.55 dB	9.05 dB
b. Through satellite channel[a]	4.45 dB	6.75 dB	7.45 dB	7.75 dB	9.75 dB
12. C/T[a] at typical operating point ($< 10^{-10}$ BER)	$-217.6 + 10\log_{10}(\text{IR} + \text{OH})$, dBW/K				
13. C/N[a] in noise bandwidth at typical operating point	12.0 dB				
14. Typical bit error rate at operating point	Less than 1×10^{-10}				
15. C/T[a] at threshold (BER = 10^{-6})	$-220.6 + 10\log_{10}(\text{IR} + \text{OH})$, dBW/K				
16. C/N[a] in noise bandwidth at threshold (BER = 10^{-6})	9.0 dB				
17. C/T[a] at BER = 10^{-3}	$-222.9 + 10\log_{10}(\text{IR} + \text{OH})$, dBW/K				
18. C/N[a] in noise bandwidth at BER = 10^{-3}	6.7 dB				

[a]In the special case of two 45-Mbps carriers operating in an INTELSAT VII/VIIA 72-MHz transponder, the E_b/N_0, C/T, and C/N values shown are increased by 0.4 dB (10^{-3}), 0.7 dB (10^{-6}), 1.0 dB (10^{-7}), 1.3 dB (10^{-8}), and 2.0 dB (10^{-10}). The allocated bandwidth is 36.0 MHz.

[b]The allocated bandwidth will be equal to 0.7 times the transmission rate, rounded up to the next higher odd integer multiple of 22.5 kHz (for information rates less than or equal to 10 Mbps) or integer multiple of 125 kHz (for information rates greater than 10 Mbps).

[c]R is the transmission rate in bits per second and equals (IR + OH) times $\frac{4}{3}$ for carriers employing rate $\frac{3}{4}$ FEC.

[d]Composite data rate = (information data rate + overhead).

[e]A BER of 10^{-6} is used as the threshold point for link availability with the fade margin allocation tables.

Note: IR = information rate; OH = overhead.

Source: Table D.2, Appendix D, INTELSAT IESS-308, Rev. 7 (Ref. 1).

TABLE 7.6 Transmission Parameters for INTELSAT VII, VIIA, and K Recommended IDR Carriers (Rate $\frac{3}{4}$ FEC)[a]

Information Rate[b] (bits/s)	Overhead Rate (kbps)	Data Rate (IR + OH) (bits/s)	Transmission Rate (bits/s)	Occupied Bandwidth (Hz)	Allocated Bandwidth (Hz)	C/T^c (dBW/K)	C/N_0^c (dB-Hz)	C/N^c (dB)
						10^{-10}	10^{-10}	10^{-10}
64 k	0	64 k	85.33 k	51.2 k	67.5 k	−169.5	59.1	12.0
192 k	0	192 k	256.00 k	153.6 k	202.5 k	−164.8	63.8	12.0
384 k	0	384 k	512.00 k	307.2 k	382.5 k	−161.8	66.8	12.0
512 k[d]	34.1[d]	546.1 k[d]	728.18 k[d]	436.9 k[d]	517.5 k[d]	−160.2[d]	68.4[d]	12.0[d]
1.024 M[d]	68.3[d]	1.092 k[d]	1.456 M[d]	873.8 k[d]	1057.5 k[d]	−157.2[d]	71.4[d]	12.0[d]
1.544 M	96	1.640 M	2.187 M	1.31 M	1552.5 k	−155.5	73.1	12.0
2.048 M	96	2.144 M	2.859 M	1.72 M	2002.5 k	−154.3	74.3	12.0
6.312 M	96	6.408 M	8.544 M	5.13 M	6007.5 k	−149.5	79.1	12.0
8.448 M	96	8.544 M	11.392 M	6.84 M	7987.5 k	−148.3	80.3	12.0
32.064 M	96	32.160 M	42.880 M	25.73 M	30125.0 k	−142.5	86.1	12.0
34.368 M	96	34.464 M	45.952 M	27.57 M	32250.0 k	−142.2	86.4	12.0
44.736 M	96	44.832 M	59.776 M	35.87 M	41875.0 k	−141.1	87.5	12.0
44.736 M[e]	96	44.832 M	59.776 M	35.87 M	36000.0 k	−142.4	86.2	10.7

[a]The table illustrates parameters for recommended carrier sizes. However, any other information rate between 64 kbps and 44.736 Mbps may be used.
[b]For carrier information rates of 10 Mbps and below, carrier frequency spacings will be odd integer multiples of 22.5 kHz. For rates greater than 10 Mbps, they will be any integer multiple of 125 kHz.
[c]C/T, C/N_0, and C/N values have been calculated to provide a clear-sky link BER of better than 10^{-10} and assume the use of rate $\frac{3}{4}$ FEC.
[d]See Appendix B of Ref. 1. The approach described in Appendix B for 512 kbps may also be applied to other IDR carrier sizes ($n \times 64$ kbps) equal to or smaller than 1.536 Mbps, where n may be equal to 1, 2, 4, 6, 8, 12, 16, or 24.
[e]In the special case of transmissions of two 45-Mbps carriers in an INTELSAT VII/VIIA 72-MHz transponder with Standard A or B earth stations, the maximum uplink EIRP values can be maintained at the present levels (however, the nominal clear-sky BER = 10^{-7}). Operation of two 45-Mbps carriers with the Reed–Solomon outer code is provisional. The uplink EIRP values are the same with and without the Reed–Solomon outer coding.
Source: Table D.3, Appendix D, INTELSAT IESS-308, Rev. 7 (Ref. 1).

A brief overview of IBS was provided in Chapter 6. The objective here is to provide the reader with additional information, stressing more the digital aspects of IBS and its performance.

NOTES AND EIRP. EIRP stability is to be maintained within ±0.5 dB for Standard A and B earth stations, and to ±1.5 dB for E- and F-type earth stations.

Tropospheric scintillation can occur in C-band or Ku-band under both adverse weather and clear-sky conditions. The effects of scintillation may be significant on links having elevation angles less than 20°. On links having elevation angles near 5°, scintillation effects can be severe. As a consequence of scintillation, antennas employing active tracking on low-elevation-angle paths may experience antenna mispointing or may transmit excessive EIRP level when uplink power control is employed. The use of program tracking is therefore highly recommended on links operating with elevation angles less than 20° for those periods when tropospheric scintillation is severe and is recommended as the primary tracking method for antennas with elevation angles below 10°.

ADVERSE WEATHER CONDITIONS. (A) For 6-GHz uplinks (closed and open networks). In the event of severely adverse weather conditions, the 6-GHz power flux density at the satellite may be permitted to drop 2 dB below the nominal value. It should be recognized, however, that this will result in degraded channel performance at cooperating receiving earth stations.

(B) For 14-GHz uplinks (open network with basic or super IBS). At 14 GHz, in order to meet the required performance objective, it is mandatory that means be provided to prevent the power flux density at the satellite from falling more than M dB below the nominal clear-sky value for more than K percent of the time in a year. The values for M and K may be found in the appendices to IESS-309 (Ref. 2). However, the values of M, in most cases, are 7 dB and, in several cases, 5 dB. The value for K is 0.2% (i.e., time availability 99.98%) in most cases.

These requirements can be met by either using diversity earth stations (discussed in Chapter 9) or uplink power control. When using the latter, INTELSAT recommends that when the uplink excess attenuation is greater than 1.5 dB, control of transmitter power should be applied to restore the flux density at the satellite to -1.5 dB, ± 1.5 dB of the nominal, to the extent that it is possible with the total power control range available.

SCRAMBLING. Scrambling must be provided at IBS earth stations to ensure that a uniform spreading is applied to the transmitted carrier at all times. INTELSAT recommends that a particular type of scrambler be used. It is described in the reference specification.

BIT ERROR PERFORMANCE CHARACTERISTICS. The channel unit will have the BER performance requirement given below in an IF back-to-back mode. These values apply with the scrambler enabled, with the specified FEC, and under standard operating conditions as specified in IESS-309. The BER performance includes the effects of carrier slips.

	Composite Data Rate E_b/N_0 (dB)	
BER Better Than	**Rate $\frac{1}{2}$ FEC**	**Rate $\frac{3}{4}$ FEC**
10^{-3}	4.2	5.3
10^{-4}	4.7	6.0
10^{-6}	6.1	7.6
10^{-8}	7.2	8.8
10^{-10}	9.0	10.3

The E_b/N_0 is referred to the modulated carrier power and to the composite data rate (information rate plus overhead) entering the FEC coder.

SLIP CONTROL. The incoming buffer is reset whenever the channel loses service. It is also reset when it is full or when it becomes empty. Resets are done at frame boundaries. The time interval between frame slips is at least 40 days.

TIMING. The timing of the digital signals at the earth station in both directions of transmission is assumed to be derived in one of three ways:

1. From a national clock with an accuracy of 1×10^{-11} as recommended in ITU-T Rec. G.811.
2. From a local earth station clock with an accuracy of at least 1×10^{-9} over the 40-day interval between frame slips (refer to Ref. 3).
3. From an incoming clock received from a remote earth station by satellite.

IBS SYSTEM FRAMING. A frame structure is defined to support the transmission of overhead information such as signaling, alarms, unique words for scrambling and encryption synchronization, an earth-station-to-earth-station communication channel, and encryption control. In addition, spare bits are available for possible future use in providing features such as station identification and multipoint control.

The framing is based on the structure of ITU-T Rec. G.732. Instead of using the nonsymmetric structure of that recommendation in which odd and even frames are defined differently, a double ITU-T Rec. G.732 frame structure was selected by INTELSAT to provide frame-to-frame symmetry without regard to odd or even frame number.

Data entering the encryptor and scrambler must be framed up as follows: The transmission frame consists of 64 bytes (512 bits) of which 60 bytes (480 bits) contain payload and 4 bytes (32 bits) are overhead. The overhead bits are assigned as follows: an 8-bit frame alignment field (byte 0), an 8-bit message field (byte 32), and 16 bits for signaling (bytes 16 and 48). This frame structure is shown in Figure 7.2.

FRAME ALIGNMENT. Frame alignment is carried out using the frame alignment signal, comprising a 7-bit code in byte 0. Bit 2 in byte 32 is set to 1 in order to avoid possible duplication of the frame alignment signal. Frame alignment is assumed to be lost when three or four consecutive frame alignment signals have each been received with one or more errors.

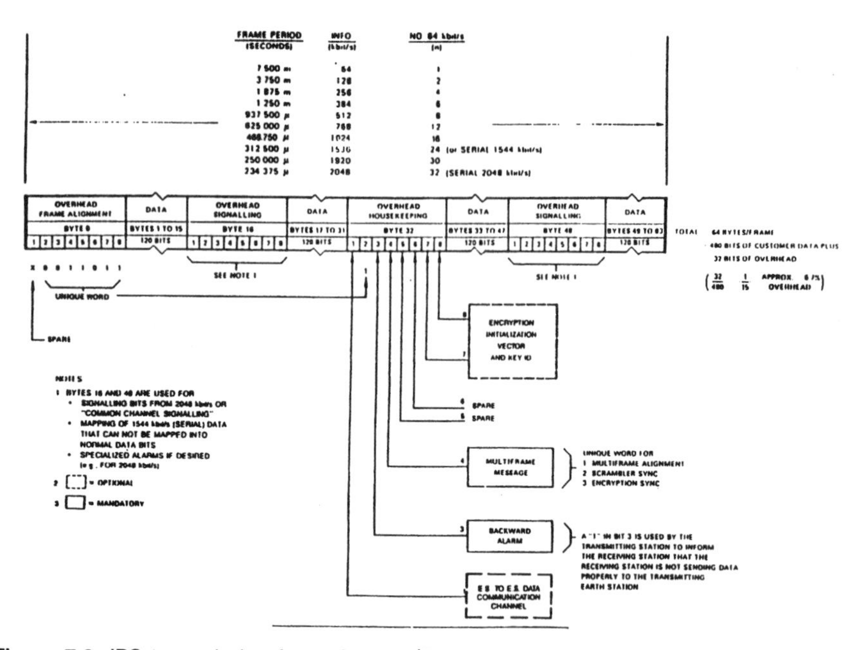

Figure 7.2. IBS transmission frame format. (From Figure 10, INTELSAT IESS-309, Rev. 4; Ref. 2.)

Frame alignment is assumed to be recovered when the following is detected:

- For the first time, the presence of the correct frame alignment signal in byte 0
- The absence of the frame alignment signal in byte 32 by verifying that bit 2 is a "1"
- For the second time, the presence of the correct frame alignment signal in byte 0 of the next frame

In the event of loss of alignment, a continuous frame alignment signal search is initiated. On correct receipt of a frame alignment signal, the recovery sequence given above in the previous paragraph is initiated.

The loss and recovery sequence is consistent with ITU-T Rec. G.732.

IBS TRANSMISSION PARAMETERS. Table 7.7 lists IBS QPSK characteristics and transmission parameters for rate $\frac{3}{4}$ FEC for INTELSAT V through VI. Table 7.8 gives open-network transmission parameters for rate $\frac{3}{4}$ FEC with INTELSAT V through VI. Table 7.9 lists QPSK characteristics and transmission parameters for Basic and Super IBS carriers using rate $\frac{3}{4}$ FEC and working with INTELSAT VII and VIIA. Table 7.10 provides an example of

TABLE 7.7 QPSK Characteristics and Transmission Parameters for Basic and Super IBS Carriers with Rate $\frac{3}{4}$ FEC (INTELSAT V Through VI)

Parameter	Requirement			
1. Information rate, IR (bits/s)	64 kbps to 8.448 Mbps			
2. Overhead data rate for IBS carriers				
(a) Closed network	10% × information rate			
(b) Open network	1/15 × information rate (≃ 6.7%)			
3. Forward error correction encoding	Rate $\frac{3}{4}$ convolutional encoding/Viterbi decoding			
4. Energy dispersal (scrambling)	Synchronous scrambler/descrambler			
5. Modulation	Four-phase coherent PSK			
6. Ambiguity resolution	Combination of differential encoding (180°) and FEC (90°)			
7. Clock recovery	Clock timing must be recovered from the received data stream			
8. Minimum carrier bandwidth (allocated)[a]	$0.7R^b$ Hz or [0.933 (IR + Overhead)]			
9. Noise bandwidth (and occupied bandwidth)	$0.6R$ Hz or [0.8 (IR + Overhead)]			
10. Composite rate[c] E_b/N_0 at BER (rate $\frac{3}{4}$ FEC)	10^{-3}	10^{-4}	10^{-6}	10^{-8}
a. Modems back-to-back	5.3 dB	6.1 dB	7.6 dB	8.8 dB
b. Through satellite channel	5.7 dB	6.5 dB	8.0 dB	9.2 dB
11. Transmission rate E_b/N_0 at BER (rate $\frac{3}{4}$ FEC)	10^{-3}	10^{-4}	10^{-6}	10^{-8}
a. Modems back-to-back	4.05 dB	4.85 dB	6.45 dB	7.55 dB
b. Through satellite channel	4.45 dB	5.25 dB	6.85 dB	7.95 dB
12. C/T at clear sky				
(a) C-Band: Basic IBS (10^{-8} BER)	$-219.4 + 10\log_{10}(\text{IR} + \text{OH})$, dBW/K			
(b) Ku-Band: Basic IBS ($< 10^{-8}$ BER)	$-218.1 + 10\log_{10}(\text{IR} + \text{OH})$, dBW/K			
Super IBS ($< 10^{-8}$ BER)	$-215.9 + 10\log_{10}(\text{IR} + \text{OH})$, dBW/K			
13. C/N in noise bandwidth at clear sky				
(a) C-Band: Basic IBS (10^{-8} BER)	10.1 dB			
(b) Ku-Band: Basic IBS ($< 10^{-8}$ BER)	11.5 dB			
Super IBS ($< 10^{-8}$ BER)	13.7 dB			
14. C/T at threshold				
(a) C-Band: Basic IBS (10^{-3} BER)	$-222.9 + 10\log_{10}(\text{IR} + \text{OH})$, dBW/K			
(b) Ku-Band: Basic IBS (10^{-6} BER)	$-220.6 + 10\log_{10}(\text{IR} + \text{OH})$, dBW/K			
Super IBS (10^{-3} BER)	$-222.9 + 10\log_{10}(\text{IR} + \text{OH})$, dBW/K			
15. C/N in noise bandwidth at threshold BER				
(a) C-Band: Basic IBS (10^{-3} BER)	6.7 dB			
(b) Ku-Band: Basic IBS (10^{-6} BER)	9.0 dB			
Super IBS (10^{-3} BER)	6.7 dB			

[a]The allocated bandwidth will be equal to 0.7 times the transmission rate, rounded up to the next higher odd integer multiple of 22.5 kHz.

[b]R is the transmission rate in bits per second and equals (IR + OH) times 4/3 for carriers employing rate $\frac{3}{4}$ FEC.

[c]Composite data rate = (information data rate + overhead).

Note: IR = information rate in bps. OH = overhead.

Source: Table F.2, Appendix F, INTELSAT IESS-309, Rev. 4 (Ref. 2).

TABLE 7.8 Example C-Band IBS Open-Network Transmission Parameters for Rate $\frac{3}{4}$ FEC, INTELSAT V Through VI (About 6.7% Overhead)[a]

Information Rate (kbps)	Data Rate Including Overhead[b] (kbps)	Transmission Rate[c] (kbps)	Occupied Bandwidth Unit (Hz)	Allocated Bandwidth Unit (Hz)	Number of 22.5-kHz Slots for Allocated Bandwidth	C/T^e (dBW/K)	C/N_0^e (dB-Hz)	C/N^e (dB)
						10^{-8}	10^{-8}	10^{-8}
64	68.3	91	55 k	67.5 k	3	−170.9	57.7	
128	136.5	182	110 k	157.5 k	7	−167.9	60.7	
256	273.1	365	219 k	292.5 k	13	−164.9	63.7	
384	409.6	547	328 k	382.5 k	17	−163.1	65.5	10.1
512	546.1	729	437 k	517.5 k	23	−161.9	66.7	
768	819.2	1093	656 k	787.5 k	35	−160.1	68.5	
1024	1092.3	1457	874 k	1.058 M	47	−158.9	69.7	
1536	1638.4	2185	1.311 M	1.553 M	69	−157.1	71.5	
1544	1638.4	2185	1.311 M	1.553 M	69	−157.1	71.5	
1920	2048.0	2731	1.639 M	1.913 M	85	−156.3	72.3	
2048	2184.5	2913	1.748 M	2.048 M	91	−155.9	72.7	

[a]Depending on the actual transponder and link conditions, INTELSAT may establish the clear-sky setting of the link at a C/N better than or equal to 10.1 dB in order to ensure the margins identified in Table F.7 of Ref. 2 are provided. The C/N, C/T, and C/N_0 values for 10^{-3} and 10^{-6} are 3.5 dB and 1.1 dB, respectively, less than those shown for clear sky ($\leq 10^{-8}$). In the case of Ku-band operation where the clear-sky C/N includes rain margin, the C/T, C/N_0, and C/N values shown in the above table can be corrected by the following factors:

Up to +1.5 dB for 5-dB system margin above 10^{-3} BER

Up to +3.5 dB for 7-dB system margin above 10^{-3} BER

For example, for a Ku-band 64-kbps carrier with 5 dB of allocated system margin, the clear-sky C/T would be equal to −170.9 dBW/K + 1.5 dB = −169.4 dBW/K. Similarly, the clear-sky C/N_0 would be equal to 57.7 + 1.5 dB = 59.2 dB/Hz and the clear-sky C/N = 10.1 + 1.5 dB = 11.6 dB.

[b]The assumed composite data rate (information rate plus overhead) E_b/N_0 is 9.2 dB for a BER of 10^{-8} at C-band.

[c]Transmission rate = (information rate + 1/15 overhead) × 4/3.

[d]The bandwidth allocated to the carrier in the satellite transponder is an odd multiple of 22.5 kHz.

[e]The C/T and C/N_0 values have been chosen to correspond with those of the rate $\frac{3}{4}$ FEC IBS closed-network parameters in order to ensure that the same lineup procedures and values are used for both the open and closed networks. This will, in general, result in a C/N value slightly higher than 10.1 dB.

Source: Table F.6, Appendix F, INTELSAT IESS-309, Rev. 4 (Ref. 2).

Basic IBS open-network transmission parameters for rate $\frac{3}{4}$ FEC and working with INTELSAT VII and VIIA.

7.2.3 TDMA Operation on a Bent-Pipe Satellite

7.2.3.1 Introduction. With a time division multiple access (TDMA) arrangement on a bent-pipe satellite, each earth station accessing a transponder is assigned a time slot for its transmission, and all uplinks use the same carrier frequency on a particular transponder. (See Section 6.4.3.) We recall

TABLE 7.9 QPSK Characteristics and Transmission Parameters for Basic and Super IBS Carriers Using FEC Rate $\frac{3}{4}$ (INTELSAT VII, VIIA, and K)

Parameter	Requirement				
1. Information rate, IR (bits/s)	64 kbps to 8.448 Mbps				
2. Overhead data rate for IBS carriers					
(a) Closed network	10% × information rate				
(b) Open network	1/15 × information rate (≃ 6.7%)				
3. Forward error correction encoding	Rate $\frac{3}{4}$ convolutional encoding/Viterbi decoding				
4. Energy dispersal (scrambling)	Synchronous scrambler/descrambler, as per Figure B (of Ref. 2.)				
5. Modulation	Four-phase coherent PSK				
6. Ambiguity resolution	Combination of differential encoding (180°) and FEC (90°)				
7. Clock recovery	Clock timing must be recovered from the received data stream				
8. Minimum carrier bandwidth (allocated)[a]	$0.7R^b$ Hz or [0.933 (IR + Overhead)]				
9. Noise bandwidth (and occupied bandwidth)	$0.6R$ Hz or [0.8 (IR + Overhead)]				
10. Composite rate[c] E_b/N_0 at BER (rate $\frac{3}{4}$ FEC)	10^{-3}	10^{-6}	10^{-7}	10^{-8}	10^{-10}
a. Modems back-to-back	5.3 dB	7.6 dB	8.3 dB	8.8 dB	10.3 dB
b. Through satellite channel	5.7 dB	8.0 dB	8.7 dB	9.2 dB	11.0 dB
11. Transmission rate E_b/N_0 at BER (rate $\frac{3}{4}$ FEC)	10^{-3}	10^{-6}	10^{-7}	10^{-8}	10^{-10}
a. Modems back-to-back	4.05 dB	6.35 dB	7.05 dB	7.55 dB	9.05 dB
b. Through satellite channel	4.45 dB	6.75 dB	7.45 dB	7.95 dB	9.75 dB
12. C/T at clear sky					
(a) C-Band: Basic IBS (10^{-8} BER)	$-219.4 + 10\log_{10}(IR + OH)$, dBW/K				
Super IBS ($< 10^{-10}$ BER)	$-217.6 + 10\log_{10}(IR + OH)$, dBW/K				
(b) Ku-Band: Basic IBS ($< 10^{-8}$ BER)[d]	$-217.9 + 10\log_{10}(IR + OH)$, dBW/K				
Basic IBS ($< 10^{-8}$ BER)[e]	$-215.9 + 10\log_{10}(IR + OH)$, dBW/K				
Super IBS ($< 10^{-10}$ BER)[d]	$-215.6 + 10\log_{10}(IR + OH)$, dBW/K				
Super IBS ($< 10^{-10}$ BER)[e]	$-213.6 + 10\log_{10}(IR + OH)$, dBW/K				
13. C/N in noise bandwidth at clear sky					
(a) C-Band: Basic IBS (10^{-8} BER)	10.1 dB				
Super IBS ($< 10^{-10}$ BER)	12.0 dB				
(b) Ku-Band: Basic IBS ($< 10^{-8}$ BER)[d]	11.7 dB				
Basic IBS ($< 10^{-8}$ BER)[e]	13.7 dB				
Super IBS ($< 10^{-10}$ BER)[d]	14.0 dB				
Super IBS ($< 10^{-10}$ BER)[e]	16.0 dB				
14. C/T at threshold					
(a) C-Band: Basic IBS (10^{-3} BER)	$-222.9 + 10\log_{10}(IR + OH)$, dBW/K				
Super IBS (10^{-6} BER)	$-220.6 + 10\log_{10}(IR + OH)$, dBW/K				
(b) Ku-Band: Basic IBS (10^{-3} BER)	$-222.9 + 10\log_{10}(IR + OH)$, dBW/K				
Super IBS (10^{-6} BER)	$-220.6 + 10\log_{10}(IR + OH)$, dBW/K				
15. C/N in noise bandwidth at threshold BER					
(a) C-Band: Basic IBS (10^{-3} BER)	6.7 dB				
Super IBS (10^{-6} BER)	9.0 dB				
(b) Ku-Band: Basic IBS (10^{-3} BER)	6.7 dB				
Super IBS (10^{-6} BER)	9.0 dB				

[a] The allocated bandwidth will be equal to 0.7 times the transmission rate, rounded up to the next higher odd integer multiple of 22.5 kHz.

[b] R is the transmission rate (in bits/s) and equals (IR + OH) times 2 for carriers employing rate $\frac{1}{2}$ FEC or times 4/3 for those employing rate $\frac{3}{4}$ FEC.

[c] Composite data rate = (information data rate + overhead).

[d] With 5-dB rain fade margin to threshold BER.

[e] With 7-dB rain fade margin to threshold BER.

Note: IR = information rate; OH = overhead.

Source: Table G.2, Appendix G, INTELSAT IESS-309, Rev. 4 (Ref. 2).

TABLE 7.10 Example Basic IBS Open Network Transmission Parameters for FEC Rate $\frac{3}{4}$, INTELSAT VII, VIIA, and K (1/15 or about 6.7% Overhead)[a]

Information Rate (kbps)	Data Rate Including Overhead[b] (kbps)	Transmission Rate[c] (kbps)	Occupied Bandwidth Unit (Hz)	Allocated Bandwidth[d] Unit (Hz)	Number of 22.5-kHz Slots for Allocated Bandwidth	C/T^e (dBW/K) $< 10^{-8}$	C/N_0^e (dB/Hz) $< 10^{-8}$	C/N^e (dB) $< 10^{-8}$
64	68.3	91	55 k	67.5 k	3	−170.9	57.7	
128	136.5	182	110 k	157.5 k	7	−167.9	60.7	
256	273.1	365	219 k	292.5 k	13	−164.9	63.7	
384	409.6	547	328 k	382.5 k	17	−163.1	65.5	10.1
512	546.1	729	437 k	517.5 k	23	−161.9	66.7	
768	819.2	1093	656 k	787.5 k	35	−160.1	68.5	
1024	1092.3	1457	874 k	1.058 M	47	−158.9	69.7	
1536	1638.4	2185	1.311 M	1.553 M	69	−157.1	71.5	
1544	1638.4	2185	1.311 M	1.553 M	69	−157.1	71.5	
1920	2048.0	2731	1.639 M	1.913 M	85	−156.3	72.3	
2048	2184.5	2913	1.748 M	2.048 M	91	−155.9	72.7	

[a]Depending on the actual transponder and link conditions, INTELSAT may establish the clear-sky setting of the C-band link at a C/N better than or equal to 10.1 dB in order to ensure the margins identified in Table G.7 of Ref. 2 are provided. The C/N, C/T, and C/N_0 values for 10^{-3} and 10^{-6} are 3.5 dB and 1.1 dB, respectively, less than those shown for clear sky ($< 10^{-8}$). In the case of Ku-band operation where the clear-sky C/N includes rain margin, the C/T, C/N_0, and C/N values shown in the above table can be corrected by the following factors:

Up to +1.5 dB for 5-dB system margin above 10^{-3} BER

Up to +3.5 dB for 7-dB system margin above 10^{-3} BER

For example, for a Ku-band 64-kbps carrier with 5 dB of allocated system margin, the clear-sky C/T would be equal to −170.9 dBW/K + 1.5 dB = −169.4 dBW/K. Similarly, the clear-sky C/N_0 would be equal to 57.7 + 1.5 dB = 59.2 dB/Hz and the clear-sky C/N = 10.1 + 1.5 dB = 11.6 dB.

[b]The assumed composite data rate (information rate plus overhead) E_b/N_0 is 9.2 dB for a clear-sky BER of typically better than 10^{-8} at C-band.

[c]Transmission rate = (information rate plus 1/15 overhead) × 4/3.

[d]The bandwidth allocated to the carrier in the satellite transponder is an odd multiple of 22.5 kHz.

[e]The C/T and C/N_0 values have been chosen to correspond with those of the rate $\frac{3}{4}$ FEC IBS closed-network parameters in order to ensure that the same lineup procedures and values are used for both the open and closed networks. This will, in general, result in a C/N value slightly higher than 10.1 dB.

Source: Table G.6(a), Appendix G, INTELSAT IESS-309, Rev. 4 (Ref. 2).

that a major limitation of FDMA systems is the required backoff of drive in a transponder to reduce IM products developed in the TWT final amplifier owing to simultaneous multicarrier operation. With TDMA, on the other hand, only one carrier appears at the transponder input at any one time, and, as a result, the TWT can be run to saturation minus a small fixed backoff to reduce waveform spreading. This results in more efficient use of a transponder and permits greater system capacity. In some cases capacity can be doubled when compared to an equivalent FDMA counterpart. Another advantage of a TDMA system is that the traffic capacity of each access can be modified on a nearly instantaneous basis. The loading of a long-haul system can be varied as the busy hour moves across it, assuming that accesses are located in different time zones. This is difficult to achieve on a conventional FDMA system.

7.2.3.2 Description of TDMA Operation. Figure 7.3 shows a typical TDMA frame. As we mentioned, each user is assigned a timeslot in which the user transmits a traffic burst for the duration of the slot. An important requirement of TDMA is that transmission bursts do not overlap. To ensure nonoverlap, bursts are separated by a guard time, which is analogous to a guard band in FDMA. However, this guard band is nonuseful time except for protection against overlap. As a result, guard time reduces system efficiency. The amount of guard time, of course, is a function of system timing. The better the timing and synchronization systemwide, the shorter we can make the guard times. Typical guard times for operating systems are on the order of 100–1000 ns.

A TDMA frame, as shown in Figure 7.3, is a complete cycle of bursts, with one burst per access. The burst length per access, in the context of this chapter, need not be of uniform duration; in fact, it usually is not. Burst length can be controlled dynamically in quasi-real time. It can be made a function of the traffic load of a particular access at a particular time. This

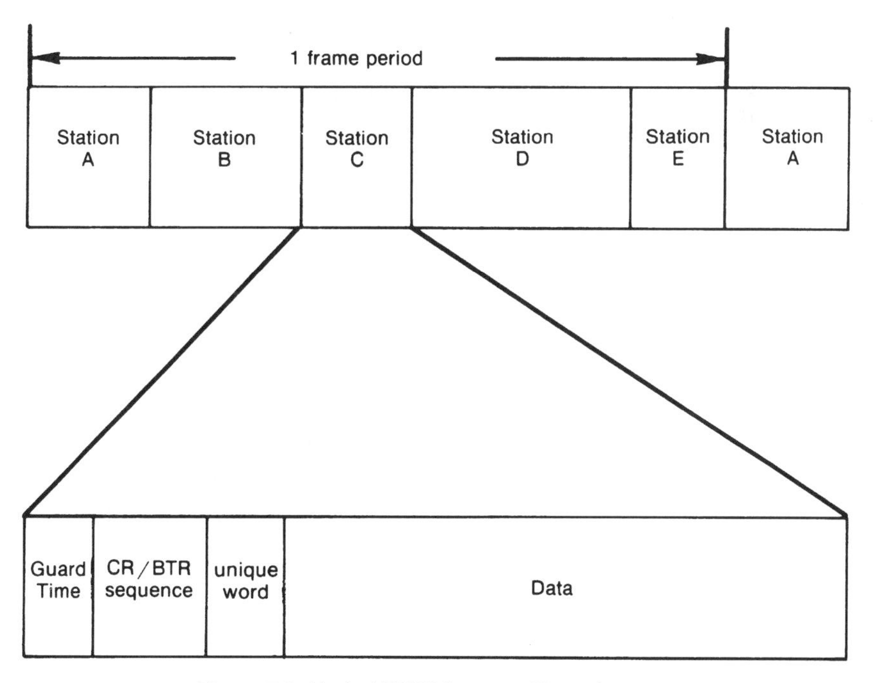

Figure 7.3. Typical TDMA frame and burst formats.

nonuniformity is purposefully exaggerated in the figure. The frame period is the time required to sequence one cycle of bursts through a frame.

The number of accesses on a transponder can vary from 3 to over 100. Obviously, the number of accesses is a function of the traffic intensity of each access, assuming a full-capacity system. It is also a function of the transponder bandwidth, its group delay over the bandwidth, and the digital modulation employed (e.g., packing ratio per unit bandwidth and sideband characteristics). For example, more bits can be packed per unit bandwidth with an 8-ary PSK signal of fixed duration than a BPSK signal of equal duration. For high-capacity TDMA systems, frame periods vary from 100 μs to 2 ms. As an example, the INTELSAT TDMA systems has a frame period of 2 ms.

Figure 7.3 also shows a typical burst format, which we will call an access subframe. The first data field of the subframe is the CR/BTR, which stands for carrier recovery (CR), bit timing recovery (BTR). This symbol sequence is particularly necessary on a coherent PSK system, where the CR is used by the PSK demodulator in each receiver to recover local carrier (corrects internal reference phase) and the BTR to synchronize (sync) the local clock. In the INTELSAT system CR/BTR is 176 symbols long. Other systems may use as few as 30 symbols. Of course, CR/BTR is overhead, and thus it would be desirable to shorten its duration as much as possible.

Generally, the minimum number of bits (or symbols—in QPSK systems we remember that one symbol or baud carries 2 bits of information) required in a CR/BTR sequence is only roughly a function of system bit or symbol rate. Carrier recovery and bit timing recovery at the receiver must be accomplished by realizable phase-lock loops and/or filters that have a sufficiently narrow bandwidth to provide satisfactory output signal-to-noise ratio (SNR). There is a trade-off in system design between acquisition time (implying a wider bandwidth) and SNR (implying a narrower bandwidth). Reference 3 suggests CR/BTR bandwidths of 0.5–2% of the symbol rate providing a good compromise between acquisition time and BER performance resulting from a finite-loop-output SNR. Adaptive phase-lock loops that acquire in a wideband mode and track in a narrower bandwidth can be used to reduce the CR/BTR overhead.

The next field in the burst subframe shown in Figure 7.3 is the unique word (UW), which establishes an accurate time reference in the received burst. The primary purpose of the UW is to perform the local clock alignment function. It can also be used as a transmit station identifier. Alternatively, the UW can be followed by a transmit station identifier sequence (SIC—station identification code).

The loss of either the BR/CTR or UW is fatal to the receipt of a burst. For voice traffic, a lost burst causes clicks and sounds like impulse noise to the listener. In the case of a data bit stream, large blocks of data can be lost due to a "skew" or slip of alignment. TDMA systems are designed for a probability of miss or false detection of 1×10^{-8} or better per burst to maintain a required threshold BER of 1×10^{-4}. A major guideline we should not lose sight of is the point where supervisory signaling (on telephone speech channels) will be lost. This value is a BER of approximately 1×10^{-3}. The design value of 1×10^{-8} threshold will provide a mean time to miss or false detection of several hours with a frame length on the order of 1 ms (Ref. 3).

7.2.3.3 TDMA Channel Capacity. Satellite communication systems may be bandwidth limited or power limited. For the bandwidth-limited case, the nominal capacity of a satellite transponder using TDMA may be approximated by the following expression (Ref. 3):

$$R_b = W + B - C_w \tag{7.1}$$

where R_b = link bit rate expressed in decibels (i.e., 100 bps is equivalent to 20 dB, 1000 bps to 30 dB, 1 Mbps to 60 dB, 2 Mbps to 63 dB, and so forth)

W = bandwidth of the satellite transponder expressed in decibels

B = bit rate to symbol rate ratio expressed in decibels

C_w = ratio of the transponder bandwidth to the possible band-limited symbol rate through the transponder (if no other value available, use 0.8 dB)

Example 1. A typical transponder has a bandwidth of 36 MHz, and QPSK modulation is employed. What is the satellite link transmission bit rate?

$$\begin{aligned} R_b &= 75.6 \text{ dB} + 3 \text{ dB} - 0.8 \text{ dB} \\ &= 77.8 \text{ dB} \end{aligned}$$

The bit rate is

$$R = \log^{-1}(77.8/10) = 60.26 \text{ Mbps}$$

If a satellite channel is power limited on the downlink, the following expression may be used to determine R_p, (Ref. 3):

$$R_p = \text{EIRP}_{\text{dBW}} - P_L + \frac{G}{T} - K - \frac{E_b}{N_0} - M \tag{7.2}$$

where R_p = satellite transmission link bit rate expressed in decibels for the power-limited case

EIRP = effective isotropic radiated power of the transponder (in dBW)

P_L = path loss of the downlink in decibels (for the 4-GHz case, use 197 dB)

K = Boltzmann's constant (-228.6 dBW/Hz/K)

E_b/N_0 = value for the required BER
M = total system link margin in decibels
G/T = the earth station in question G/T

Example 2. Given an EIRP from a satellite transponder of +22.5 dBW, G/T of the earth terminal of 40.7 dB/K, coherent QPSK modulation, an 8-dB margin, which includes modulation implementation loss, and an E_b/N_0 of 9.6 dB for a BER of 1×10^{-5}, what is the bit rate for the power-limited case?

$$R_p = 22.5 \text{ dBW} - 197 \text{ dB} + 40.7 \text{ dB/K} + 228.6 \text{ dBW} - 9.6 \text{ dB} - 8 \text{ dB}$$

$$= 77.2 \text{ dB}$$

which is equivalent to 52.48 Mbps.

We note that if the satellite transponder EIRP is increased 10 dB, the bit rate would increase by an equivalent 10 dB and force us into the band-limited regime, where we will then use equation (7.1). Obviously, we are not going to get 524 Mbps through a 36-MHz transponder utilizing QPSK or even 16-ary PSK.

7.2.3.3.1 Elementary Analysis of Burst Rate and Duration. Assume for argument's sake that a frame is 1 s long, the frame burst rate is 100 Mbps, and there are ten timeslots of uniform length. If there were no guard times, each slot (or user) would burst 10 Mbps. If we want a net of 100 Mbps across the frame, then somehow we must compensate for the guard times. If the guard time were 1 ms, each user must burst 1.01×10 or 10.1 Mbps.

Reference 4 defines TDMA throughput as the number of bits per second that are revenue bearing, which we have called the payload. If we are working with a given burst rate R_b measured in bits per second, we can relate guard time to a certain number of bits.

For example, let the burst rate of a frame be 100 Mbps, the frame is 1-s duration, the guard time is 1 ms, and there are 10 users. Thus there is a total of 10×1 or 10 ms total of guard time so that the total guard time is equivalent to losing 1 Mbps of data throughput or 0.1 Mbps in 1 ms (100 Mbps/1000) and 10 ms guard time is 10×0.1 Mbps or 1 Mbps.

Overhead bits and reference bursts also count against throughput (Ref. 4). Let throughput efficiency be η; then

$$\eta = 1 - (P + 2)(p + g)/RT_f \quad (7.3)$$

where P = number of bursts in a frame
R = bit rate (in bps)
p = number of bits in the subburst header
g = equivalent number of bits corresponding to the guard time
T_f = frame duration (Ref. 4)

Example 3. Suppose the burst rate (per second) is 100 Mbps, the frame duration is 2 ms, guard time is 500 ns, there are 10 bursts per frame (P), and there are 250 bits in the header. The guard time equivalent in bits is $500 \times 10^{-9} \times 100 \times 10^6 = 10{,}000 \times 10^{-3} = 50$ bits.

$$\begin{aligned}\eta &= 1 - 12(300)/(100 \times 10^6 \times 2 \times 10^{-3}) \\ &= 1 - 3600/200{,}000 \\ &= 1 - 0.018 \\ &= 0.982 \text{ or } 98.2\%\end{aligned}$$

For a given frame period, throughput increases as guard time and header overhead decrease. This implies the following:

- A reduction in guard times. This is a function of the timing/synchronization system used. The better the system, the more guard times can be reduced.
- A reduction of header length. Reference 4 points out that receivers should have fast and dependable carrier and bit timing recovery. In this context, the advantage of differential demodulation* with respect to coherent demodulation can be seen. However, transmission throughput is not the only criterion and the degraded performance in terms of differential demodulation and BER must not be forgotten. Also, the UW length can be reduced. This then will involve an increase in the probability of false alarm in detection of the unique word.

System efficiency can be defined as the ratio of useful capacity (traffic that bears revenue, or simply the payload) measured in bits per second to the available capacity. Available capacity is the theoretical capacity of the channel limited only by thermal noise.

$$Ca = B \log_2\left(1 + \frac{E_s T_s}{N_0 B}\right)$$

*Differential demodulation depends on the phase of the previous symbol for present symbol decision. Coherent demodulation depends on a local, internal phase reference.

where Ca = available capacity
B = channel bandwidth
E_s = energy per symbol
T_s = symbol duration
N_0 = noise spectral density (thermal noise)

System efficiency also depends on traffic patterns, networking, method of multiple access, modulation, demodulation approach, coding, and type of decoder. The desired end product is net user throughput as in any other data transmission system.

7.2.3.4 TDMA System Clocking, Timing, and Synchronization. It was previously stressed that an efficient TDMA system must have no burst overlap, on the one hand, and as short a guard time as possible between bursts, on the other hand. We are looking at guard times in the nanosecond regime. The satellites under discussion here are geostationary. For a particular TDMA system, the range to a satellite can vary from 23,000 to 26,000 statute miles. We can express these range values in time equivalents by dividing by the velocity of propagation in free space or 186,000 mi/s. These values are 23,000/186,000 and 26,000/186,000 or 123.469 and 139.573 ms. The time difference for a signal to reach a geostationary satellite from a very-low-elevation-angle earth station and a very-high-elevation-angle earth station is 16.104 ms (e.g., 139.573 − 123.469) or 16,104 μs or 16,104,335 ns. Of course, this is a worst case, but still feasible. The TDMA system must be capable of handling these orders of time differences among the accessing earth stations. How do we do it and meet the guidelines set out previously? We must also keep in mind that geostationary satellites actually are in motion in a suborbit, causing an additional time difference and Doppler shift, both varying dynamically with time.

There are two generic methods used to handle the problem: "open loop" and "closed loop." Open-loop methods are characterized by the property that an earth station's transmitted burst is not received by that station. We mean here that an earth station does not monitor the downlink of its own signal for sync and timing purposes. By not using its delayed receiving signal for timing, the loop is not closed; hence it is open.

Closed-loop covers those synchronization techniques in which the transmitted signals are returned through the bent-pipe transponder repeater to the transmitting station. This permits nearly perfect synchronization and high-precision ranging. The term "closed loop" derives from the looping back through the satellite of the transmitted signal permitting the transmitting TDMA station to compare the time of the transmitted-burst leading edge to that of the same burst after passing through the satellite repeater. The TDMA transmitter is then controlled by the result, an early or late arrival relative to the transmitting station's time base. (*Note*: These definitions of open loop and closed loop should be taken in context and not confused with open-loop and closed-loop tracking discussed subsequently in Chapter 10.)

One open-loop method uses no active form of synchronization. It is possible to achieve accuracies from 5 μs to 1 ms (Ref. 5) through what can be termed "coarse sync." The system is based on very stable free-running clocks, and an approximation is made of the orbit parameters where burst positioning can be done to better than 200-μs accuracy. The method was used on some early TDMA trial systems and on some military systems and will probably be employed on satellite-based data networks, particularly with long frames.

One of the most common methods used to synchronize a family of TDMA accesses is by a reference burst. A reference burst is a special preamble only, and its purpose is to mark the start of frame with a burst codeword. The station transmitting the reference burst is called a reference station. The reference bursts are received by each member of the family of TDMA accesses, and all transmissions of the family are locked to the time base of the reference station. This, of course, is a form of open-loop operation. Generally, a reference burst is inserted at the beginning of frame. Since reference bursts pass through the bent-pipe repeater and usually occupy the same bandwidth as traffic bursts, they provide each station in the family with information on Doppler shift, time delay variations due to satellite motion, and channel characteristics. However, the reference burst technique has some drawbacks. It can only serve one repeater and a specific pair of uplinks and downlinks. There are difficulties with this technique in transferring such results accurately to other repeaters, other beams, or stations in different locations. The reference bursts also add to system overhead by using a bandwidth/time product not strictly devoted to the transfer of useful, revenue-bearing data/information. However, we must accept that some satellite capacity must be devoted to achieve synchronization.

7.3 DIGITAL SPEECH INTERPOLATION

DSI is designed for speech operation to increase system capacity. It is based on the fact that there is active speech on a full duplex voice circuit only a fraction of the time. For one thing, there is the talk–listen effect. In normal operation, while one end of a speech circuit talks, the other end listens. For this effect alone, there is only 50% usage. Also, there are many pauses in normal speech. DSI exploits these periods of nonusage and speech pauses.

A similar system was implemented on undersea cables operating in the analog mode. It was called TASI (time-assigned speech interpolation). With a significantly large number of voice channels, TASI could enhance transmission capacity by a factor of 2.

When describing TASI or DSI, we talk about the time occupied by a caller's speech as a speech spurt. With 100% of a TASI terminal connected to active circuits, speech is actively present on a busy channel only about 40% of the time. If, on the other hand, all circuits are not busy, the average speech activity on a TASI terminal is further decreased. The percentage of busy circuits is called the incoming channel activity, and the percentage of time that speech spurts occupy a channel is the speech spurt activity, or simply speech activity.

7.3.1 Freeze-out and Clipping

The operation of TASI and DSI exploits low speech spurt activity by assigning transmission channels only when a speech spurt is present. As the number of channels increases, the process becomes more efficient. If two speech users were to use one channel through interpolation, a large portion of the speech will be lost owing to competition for simultaneous occupancy. The spurt of one user will "freeze-out" any other attempt to use the channel by another user, and that freeze-out will continue until the spurt is terminated.

When a larger portion of users use a comparatively smaller number of available channels, there is always a finite probability that the number of conversations requiring service will exceed the number of channels providing that service. This competition causes an impairment called "competitive clipping," where the initial portion of a speech spurt is clipped. The percentage of time that speech is lost due to such competition is called percentage of freeze-out or freeze-out fraction. In the design of a TASI or DSI system, the fraction of speech lost must be acceptably small. Reference 6 gives a freeze-out fraction of 0.5% for TASI systems.

The most common freeze-outs clip initial portions of speech spurts from near zero to several hundred milliseconds. Clips longer than 50 ms cause perceptible mutilation, and the percentage of clips longer than 50 ms should be kept to less than 2%.

Another form of clipping is "connect clipping." This type of clipping is caused by the channel assignment process. The presence of speech on an incoming telephone channel on the transmit side of a TASI terminal is sensed by a speech detector, which initiates a request for a channel. A processor assigns an idle transmission channel to the incoming channel in response to that request and also informs the distant end specifying the outgoing channel to which the transmission channel is to be connected. During the time required to make the total connection, speech is clipped. This type of clipping only occurs when the demand for channels exceeds the operational transmit channels available. As the demand increases for service, connect clipping becomes more prevalent and, on a fully loaded TASI system, connect clipping may occur on every speech spurt. Thus a system design objective is to minimize transmit channel connect time.

7.3.2 TASI-Based DSI

One type of DSI is based on the TASI concept. It operates with 8-bit PCM words, 8000 samples per second. The incoming PCM signals in TDM format are processed by a transmit assignment processor. When speech activity is detected by the processor on an incoming PCM timeslot, it is assigned an available transmit slot. The distant-end processor is alerted, via a control channel, of the slot assignment and makes the corresponding connection on its terrestrial side. The control channel is carried on the same PCM TDM frame. Figure 7.4 shows a TASI-type DSI system.

Digital TASI has a number of advantages over its analog counterpart. The fact that it is all-digital lends itself more to digital processor control. Connect clipping is reduced because switching and control are faster. When the system becomes more loaded, competitive clipping can be reduced or eliminated by dropping the least significant bit on each PCM word. This increases quantizing distortion somewhat but is a lesser impairment than competition clipping. However, the time required when bit reduction is invoked is very low, and thus the impairment is hardly noticeable.

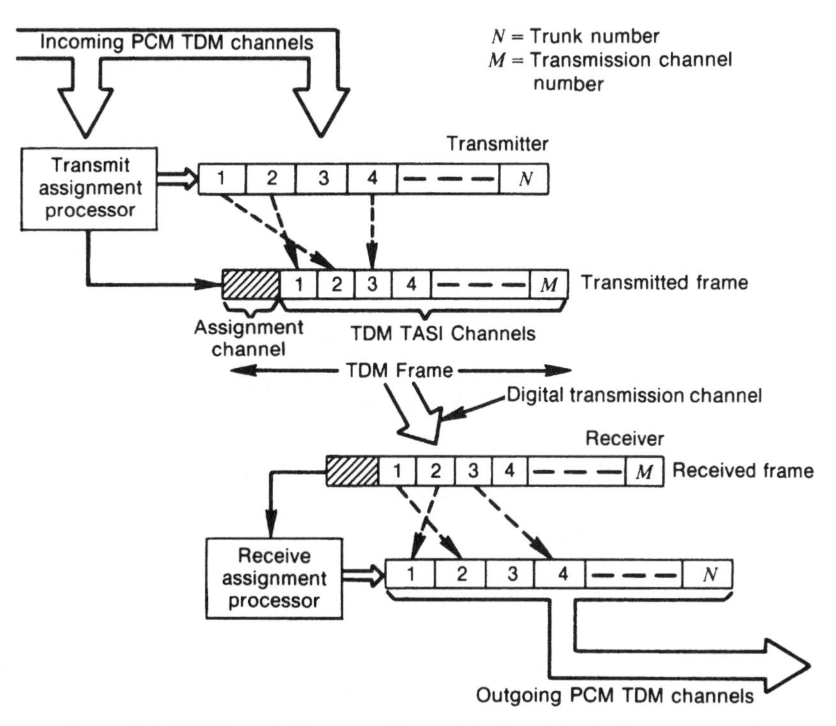

Figure 7.4. TASI-type DSI system. (From Ref. 6; reprinted with permission.)

7.3.3 Speech Predictive Encoding DSI

The basic idea of this type of DSI is that it eliminates PCM frame-to-frame redundancy that exists in ordinary PCM transmission. This redundancy includes that due to pauses and redundancy in the speech spurts themselves. It permits a two or more times increase in speech channel capacity compared to conventional, unprocessed PCM.

Redundancy is reduced by storing a frame and transmitting one frame behind. The two consecutive frames are compared in the transmit processor and only the nonredundant information is transmitted to the distant end. The redundant information is called predictable and the nonredundant information is called unpredictable. We need some way to tell the distant end how much of the information is predictable and where it is. This is done with a "sample assignment word" (SAW).

One system described in Ref. 6 can transmit up to 64 active speech channels in the frequency spectrum allotted to a 32-channel conventional PCM system corresponding to a bit rate of 2.048 Mbps. The PCM sample derived during each sample period from the 64 incoming channels is compared with samples previously sent to the receiver and stored in memory at the transmitter. Any samples that differ by an amount equal to or less than some given number of quantizing steps, called the aperture, are discarded and not sent to the receiver. These are the predictable samples. The remaining unpredictable samples are transmitted to the receiver and replace the values formerly stored in memories at both the transmitter and receiver. The aperture is adjusted automatically as a function of activity observed over the 64 incoming channels on each frame so that the number of samples transmitted is nearly constant.

The transmission frame of the predictive system is composed of an initial SAW slot followed by a number of 8-bit timeslots that carry the individual PCM samples that are unpredictable. The SAW contains one bit for each of the incoming telephone channels. Thus, for a 64-channel terrestrial system, the SAW contains 64 bits. The bit corresponding to a given channel is a "1" if the frame contains a sample for that particular channel and a "0" if it does not. Thus the SAW contains all the information needed to distribute the samples among the 64 outgoing channels at the distant receive end.

At the receiver the unpredictable samples received in the transmission channel frame replace previously stored samples in the receiver's 64-channel memory as directed by the SAW. The samples in memory, in the form of a conventional PCM frame, are reslotted into the outgoing terrestrial channels at the appropriate rate. The most recent frame thus contains new samples on the channels that have been updated by the most recent transmission channel frame and repetitions of the samples that have not been updated.

The term DSI advantage is the ratio of incoming terrestrial channels to the transmitter to the number of required transmit channels of the DSI system. If there were 120 terrestrial channels occupying only 54 satellite transmission channels, the DSI advantage would be 120/54 or 2.22.

Both the TASI and predictive methods of DSI offer significant enhancements in the capacity of digital transmission of speech. The two methods achieve interpolation advantages greater than 2, with the predictive methods achieving a slightly higher value than TASI with systems carrying a smaller number of channels (on the order of 120 PCM channels or less).

The TASI method degrades transmission by initial clips of speech spurts when approaching full-capacity loading. The frequency of occurrence of destructive initial clips can be kept low enough to produce little degradation by properly adjusting the DSI advantage. The probability of initial clips longer than 50 ms should be no greater than 2% to meet the degradation acceptability criteria. The technique of bit reduction during periods of heavy traffic loading (e.g., from 8-bit samples to 7-bit samples) can reduce the probability of initial clips. Just the 1-bit reduction reduces the probability of occurrence by more than an order of magnitude.

Prediction distortion is the major cause of degradation of the predictive methods of DSI. The amount of this distortion varies as the fraction of samples predicted varies in response to changes in the ensemble average of voice spurt activity. Again the DSI advantage is the controlling figure. The prediction noise produced is controlled by the DSI advantage. Reference 6 suggests designing the DSI advantage such that the probability that more than 25% of the samples are predicted during speech spurts is 0.25. This results, on the average, in a 0.5-dB degradation in the subjectively assessed speech-power-to-quantization-noise-power ratio. The predictive method is adaptive and yields to occasions of higher than average voice spurt activity by automatically increasing the fraction of samples predicted. It is this feature that eliminates the possibility of damaging initial clips.

7.4 INTELSAT TDMA / DSI SYSTEM

7.4.1 Overview

The INTELSAT TDMA/DSI system, as specified in IESS-307 (Rev. B), has traffic terminals that operate at 120 Mbps and incorporate the use of DSI. The TDMA system can be used with INTELSAT satellites V, VA, VA(IBS), VI, VII, and VIIA. The operation is with 80-MHz hemi and zone beam transponders at 6/4 MHz. For INTELSAT VI system, the equipment will be capable of operation with switched-satellite (SS) TDMA transponders at 6/4 GHz.

As shown in Figure 7.5, there is an east hemispheric (hemi) and west hemispheric beam, east zone beam, and west zone beam. Normally, zone beam coverage areas will also be contained within hemispheric beam coverage areas. Zone and hemispheric beams use opposite senses of polarization.

Figure 7.5 shows a satellite with typical east-to-west and west-to-east connectivities of both zone and hemispheric beams. Two dual-polarized reference stations located in each zone coverage area are thus able to monitor and control both zone and hemispheric beam transponders. Each reference station generates one reference burst per transponder, and each transponder is served by two reference stations. This provides redundancy by enabling traffic terminals to operate with either reference burst. The two pairs of reference stations provide network timing and control the operation of traffic terminals and other reference stations.

Reference stations include a TDMA system monitor (TSM), which is used to monitor system performance and diagnose system faults. In addition, the TSM is employed to assist users in carrying out their traffic terminal lineups.

The traffic terminals operate under control of a reference station and transmit and receive bursts containing traffic and system management information. Traffic terminals include interfaces that are used to connect termi-

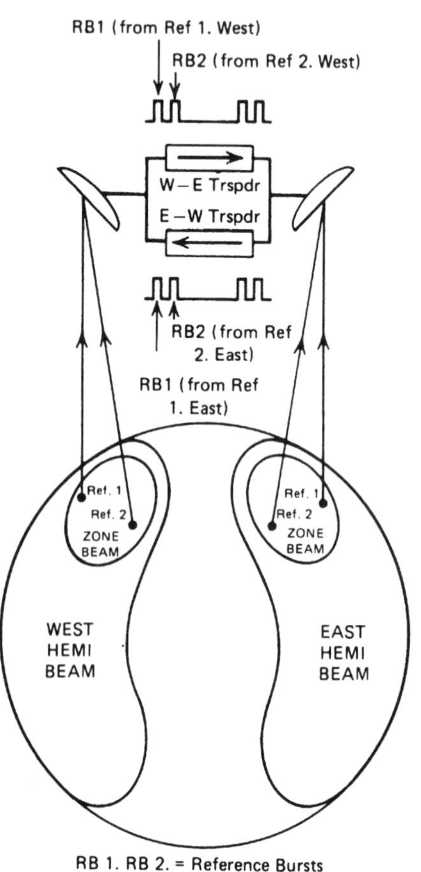

Figure 7.5. Overview of the INTELSAT/DSI system. (From Ref. 7.)

nals to the terrestrial networks. The INTELSAT TDMA/DSI system uses two types of interfaces: DSI for voice traffic (together with a limited amount of nonvoice traffic) and digital noninterpolated (DNI) traffic for data and noninterpolated voice traffic.

The TDMA system and terminal functions and capabilities review is given in Table 7.11. Terminal modulation parameters are given in Table 7.12.

7.4.2 Frame, Multiframe, and Burst Format

The INTELSAT TDMA frame contains traffic bursts and reference bursts RB1 and RB2 as shown in Figure 7.6. Each reference burst is transmitted by a separate reference station and, under normal conditions, both reference stations are active. As we mentioned, one reference station is designated as

TABLE 7.11 System Capabilities and Terminal Functions

System Capabilities	
Clear-sky BER	1×10^{-10}
Degraded sky BER: better than	1×10^{-6}
Nominal transmission bit rate	120 Mbps
TDMA frame length (nominal)	2 ms
The TDMA/DSI system is designed for multiple transponder operation via communities of synchronized transponders.	
FEC is applied to the traffic bursts.	
Features of the Reference Station	
Each transponder is served by two reference stations.	
Each reference station generates one reference burst per transponder to perform the following functions:	
• Provide open-loop acquisition information to traffic terminals and other reference stations	
• Provide synchronization information to traffic terminals and other reference stations	
• Provide burst time plan change control	
• Provide common synchronization across multiple-satellite transponders, which permits transponder hopping	
• Provide voice and teleprinter order wires	
Features of Traffic Terminals	
The traffic terminals perform the following functions necessary to provide traffic-carrying bursts synchronized to reference bursts:	
• Perform acquisition and synchronization	
• Generate bursts containing traffic and housekeeping information	
• Interface with terrestrial networks	
• Perform transponder hopping where necessary	
• Provide voice and teleprinter orderwires	
• Accept 64-kbps PCM traffic and apply digital speech interpolation selectively	
• Accept 32-kbps ADPCM traffic from the DCME system (see INTELSAT IESS-501)	

TABLE 7.12 Modulation Parameters

Nominal bit rate	120.832 Mbps
Nominal symbol rate	60.416 Mbaud
Mode of operation	Burst
Modulation	Four-phase PSK
Demodulation	Coherent
Encoding	Absolute (i.e., no differential encoding)
Carrier and bit timing	48 Symbols unmodulated
Recovery sequence	128 Symbols modulated
Unique word length	24 Symbols
Phase ambiguity resolution	By unique word detection with unique word

Source: Table 3.1, INTELSAT IESS-307, Rev. A and B (Ref. 8).

the primary reference station, and the other is designated as the secondary (backup) reference station. The traffic terminals respond to the secondary reference station only when the primary reference station fails. The nominal guard time between bursts is 64 symbols.

In Figure 7.6 we have two types of bursts: reference bursts and traffic bursts. The reference burst consists of a preamble and a control and delay channel (CDC). The traffic burst consists of the preamble and a traffic field consisting of one or more DSI and/or DNI subbursts. The preamble includes the CR/BTR, the UW, teleprinter orderwire channels, the service channel, and voice orderwire (VOW) channels.

As we discussed earlier, the CR/BTR sequence enables the TDMA modem to acquire and synchronize received bursts. The 24-symbol UW is used to differentiate reference bursts and traffic bursts, resolve the fourfold ambiguity inherent in QPSK modulation, and mark the beginning of a multiframe. Eight teleprinter orderwires and two voice orderwires are allocated 8 and 64 symbols, respectively, in each reference burst and traffic burst. Eight symbols form a service channel that is used to exchange control and housekeeping information throughout the TDMA network. Finally, in the

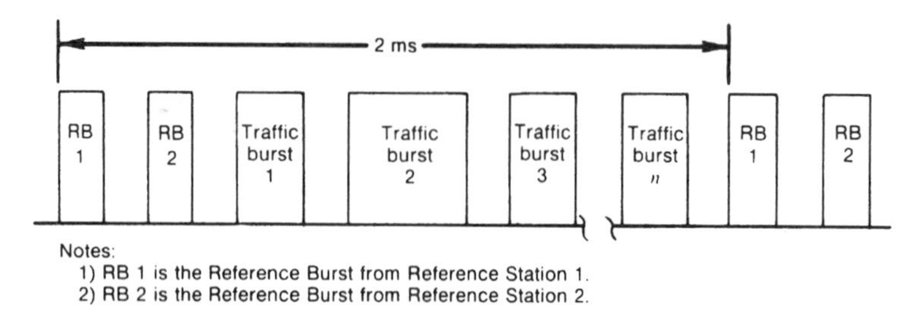

Figure 7.6. Structure of the INTELSAT TDMA frame.

reference burst, eight symbols are allocated to the CDC, which is used to control the traffic terminals' acquisition and synchronization.

A TDMA frame is a common 2-ms period within a community of synchronized transponders. To provide common timing within a community, a *start-of-TDMA-frame* (SOF) is defined by the reference bursts RB1 contained in one of those transponders as designated by INTELSAT. This transponder is called the timing reference transponder (TRT). The position of each burst within a community is referred to the SOF. Only RB1 in the TRT bridges the SOF. No other burst, whether reference or traffic, bridges the SOF at any time.

A terminal is controlled by a pair of reference bursts (RB1 and RB2) contained in a transponder designated by INTELSAT. The RB1 in this transponder relative to RB1 in the TRT is nominally offset by a predetermined amount, $T1_n$. The value of $T1_n$ for each transponder is provided as part of the burst time plan and is a multiple of 16 symbols. The time relationships in the TDMA frame of the TRT, SOF, and $T1_n$ for a community are shown in Figure 7.7.

A burst's nominal position in the frame is provided by INTELSAT as part of the burst time plan. Nominal position is defined by the last symbol of the UW with respect to the SOF instant. The burst time plan does not require a traffic burst to bridge the SOF instant. The nominal position and length of the traffic bursts are assigned according to a burst time plan established by INTELSAT. The nominal positions of a burst are assigned in multiples of 16 symbols.

The burst time plan does not require any traffic terminal to transmit or receive two consecutive bursts with a distance in symbols, between the nominal end of the first burst and the nominal start of the CR/BTR sequence of the second burst, smaller than 64 symbols, if the burst time plan does *not* require transponder hopping between the bursts, or smaller than 80 symbols if transponder hopping is required.

A terminal must be capable of transmitting up to 16 nonoverlapping bursts per frame containing in total up to 32 subbursts. A terminal must be capable of receiving up to 32 traffic bursts containing in total up to 32 subbursts. The TDMA burst format is shown in Figure 7.8.

FEC is applied to the entire traffic section of selected traffic bursts, as directed by INTELSAT, after the voice orderwires. The code used is the BCH code (128, 112) using a code rate of $\frac{7}{8}$.

7.4.2.1 TDMA Multiframe. Sixteen contiguous TDMA frames constitute a multiframe. Special unique words are used both in the reference bursts and in the traffic bursts as multiframe markers to designate the first frame of the multiframe, which is called "frame 0." It also designates the type of burst. All multiframe markers of all bursts in a community of synchronized transponders occur in the same frame.

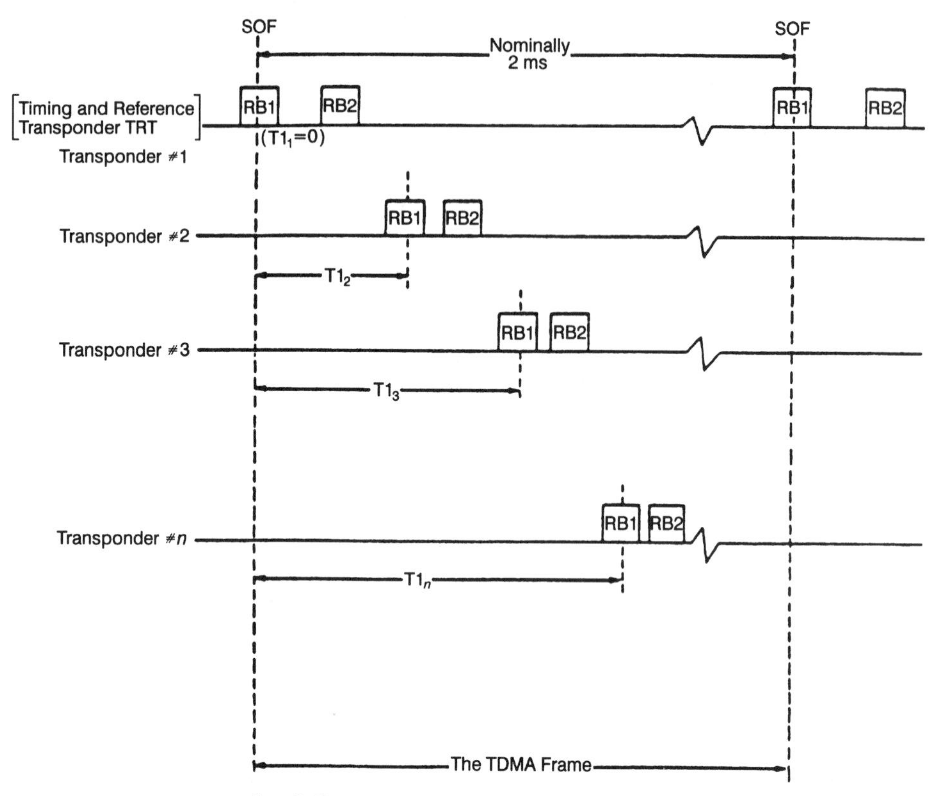

Figure 7.7. Timing relationships at the satellite between TRT, SOF, and T1, for a community of synchronized transponders ($i = 1, 2, \ldots, n$). (From Figure 4.1, INTELSAT IESS-307; Ref. 8.)

In the first frame of a multiframe, reference bursts RB1 and RB2 contain unique word No. 1 (UW1) and unique word No. 2 (UW2), respectively, whereas the traffic bursts contain unique word No. 3 (UW3). In the remaining 15 frames of the multiframe (frames 1 through 15), reference bursts RB1 and RB2 and the traffic bursts contain unique word No. 0 (UW0). Figure 7.9 illustrates the format of the TDMA multiframe.

In a given transponder of a community, the RB1 multiframe marker occurs nominally $T1_n$ symbols from the RB1 multiframe marker in the TRT.

7.4.2.2 *Control and Delay Channel.* Eight symbols of the reference burst are allocated for the transmission of the CDC. The CDC carries the following information:

- Identification number of terminal addressed
- Reference station status code

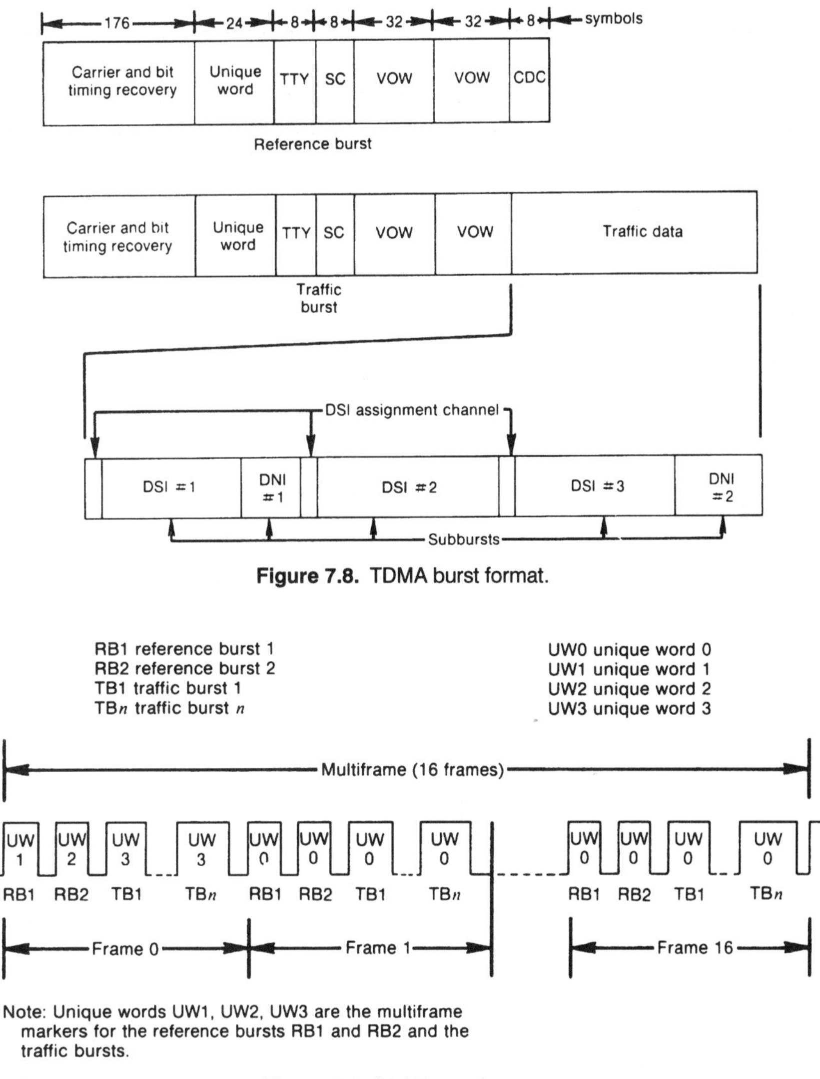

Figure 7.8. TDMA burst format.

Figure 7.9. Multiframe format.

- Control codes to be used by the controlled terminals for acquisition and synchronization
- Transmit delay information (D_n) to be used by the controlled terminals for acquisition and synchronization

The structure of the CDC is shown in Figure 7.10.

Transmission of the 32 bits of information of the CDC takes place over an interval of 16 frames at the rate of 2 bits per frame. These 2 bits are transmitted with eight-fold repetition in an eight-symbol slot. The CDC

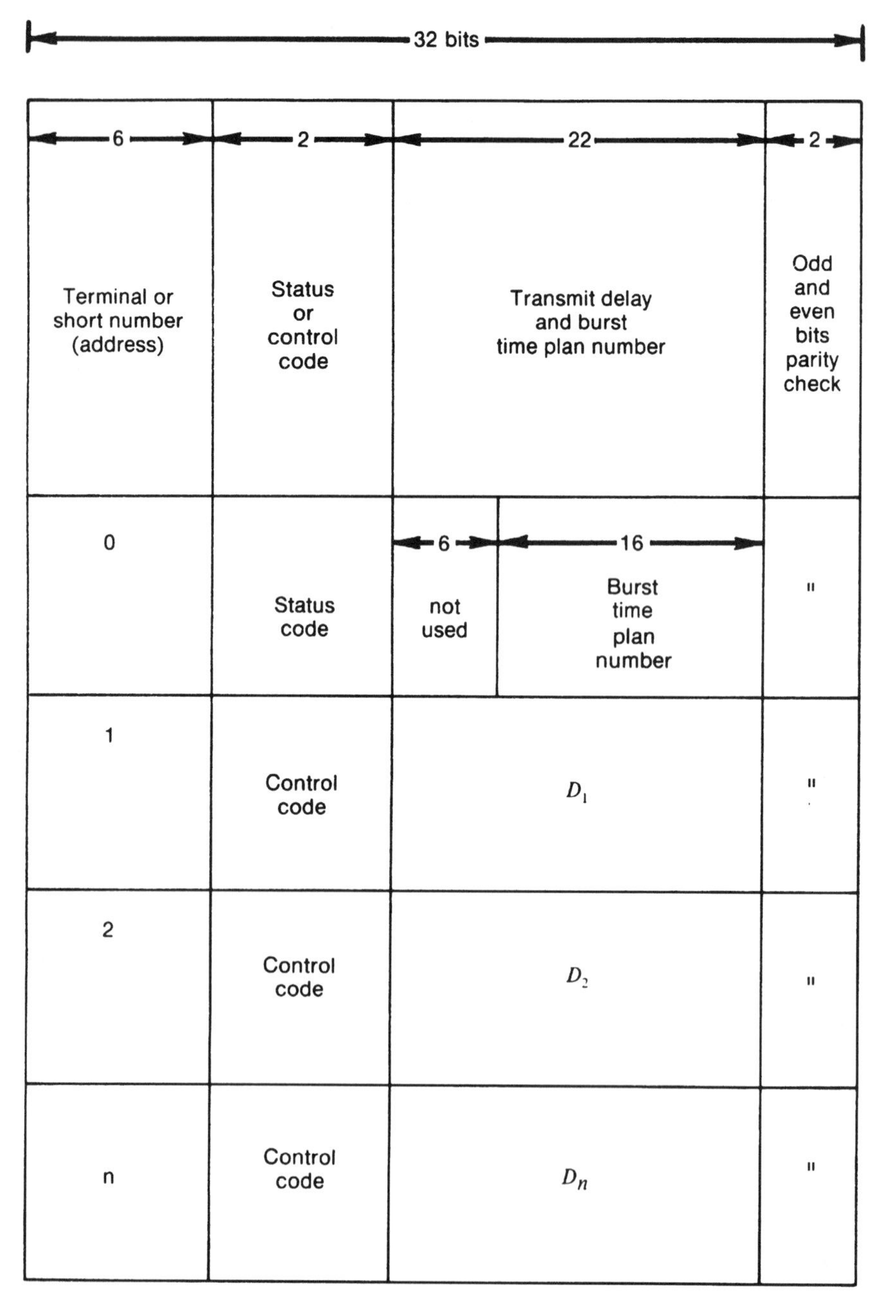

Figure 7.10. Reference burst control and delay channel (CDC). (From Figure 4.7, INTELSAT IESS-307; Ref. 8.)

messages transmitted in RB1 and RB2 in the same multiframe have the same terminal address number.

7.4.3 Acquisition and Synchronization

In this context acquisition is the process by which a TDMA terminal initially places its burst into the assigned position within a TDMA frame. To start the procedure, a terminal must acquire and maintain reception of the reference bursts. There are primary reference bursts (PRBs) and secondary reference bursts (SRBs). Normally the PRB is used by an acquiring station. If it fails, it may use the SRB to start the acquisition phase. A terminal may enter the acquisition phase if it receives, via the CDC, an acquisition control code from the reference station, a status code, and a value for transmit delay, D_n. The transmit delay is the time between the reception of a reference burst and the transmission of the acquiring terminal's own burst. The reference station calculates the value of the transmit delay based on knowledge of the satellite position. This method is referred to as "open-loop" acquisition.

The terminal derives and maintains receive frame and receive multiframe timing by using the UW detection timing and status information of the reference bursts. When the receive frame and receive multiframe timing become available, the terminal may start locating and receiving traffic bursts. The terminal transmit multiframe timing is derived by applying the delay D_n from the receive multiframe timing. The transmit frame is derived from the transmit multiframe by internal timing. When the transmit frame and transmit multiframe timing are established, the terminal may initiate acquisition under the control of a reference burst, normally the PRB. In this phase the terminal transmits the preamble of a designated burst. When the SYNC code is received, the transmit frame and multiframe synchronization is achieved and the acquisition is complete. When the acquisition phase successfully terminates, the terminal enters the synchronization phase and transmits all its traffic bursts.

7.4.4 Transponder Hopping

The traffic terminals are designed to hop across a maximum of four transponders, which can be separated in frequency and/or polarization. Since the TDMA/DSI system employs mutually synchronized reference bursts in all transponders, traffic bursts transmitted into different transponders will be separated by fixed time intervals. This allows reference stations to control the position of only one of the terminal's bursts, since the others are synchronized by fixed time offsets.

7.4.5 Digital Speech Interpolation Interface

In order to make the most efficient use of the satellite capacity, the INTELSAT TDMA system incorporates DSI on most telephone channels. This DSI capability increases the capacity of the TDMA transmission system by the interleaving of speech spurts from different terrestrial channels on the same satellite channel (interpolation).

DSI gain is defined as the ratio of the number of incoming terrestrial channels to the number of available normal satellite channels. The definition excludes assignment channels, preassigned noninterpolated satellite channels, and terrestrial channels assigned to preassigned noninterpolated satellite channels. The number of satellite channels and terrestrial channels served are established by INTELSAT.

Channels carried by the DSI system may be subject to competitive clipping, which is the time lost at the beginning of a speech spurt between the request for assignment and the availability of a satellite channel. Competitive clipping is a function of speech statistics, the DSI gain, and speech detector properties. INTELSAT follows the industry-accepted criterion in that the DSI gain will be established such that competitive clipping lasting more than 50 ms occurs on less than 2% of the voice spurts.

The INTELSAT DSI system operates with a 2-ms frame synchronized to the TDMA frame. A terminal comprises one or more DSI modules. One DSI subburst per TDMA frame is generated by each DSI module. Individual modules may be designed for either multidestination or single-destination operation. Terrestrial channels are preassigned to destinations.

The origin of a subburst is predetermined from its position in the TDMA frame. Two channels constituting a circuit are processed by the same DSI module.

A DSI module is expandable to a maximum capacity of 240 input terrestrial channels. The actual utilization of the module may vary from one terrestrial channel to its maximum capacity and is expandable in increments of one terrestrial channel.

Multidestination and single-destination DSI modules are capable of providing preassigned noninterpolated satellite channels to accommodate any 127 of the 240 terrestrial channels. The actual number of terrestrial channels carried on preassigned noninterpolated channels is expandable from 0 to 127. The use of preassigned noninterpolated satellite channels reduces the number of satellite channels in the interpolated pool. In this case, to avoid competitive clipping (due to excessive DSI gain), the number of terrestrial channels subject to interpolation must also be reduced (Ref. 8).

7.5 PROCESSING SATELLITES

Processing satellites, as distinguished from "bent-pipe" satellites, operate in the digital mode and, as a minimum, demodulate the uplink signal to baseband for regeneration. As a maximum, at least as we envision today, they operate as digital switches in the sky. In this section we will discuss satellite systems that demodulate and decode in the transponder and then we present some ideas on switching schemes suggested in the NASA 30/20-GHz system.

This will be followed by a section on coding gain and a section on link analysis for processing satellites.

7.5.1 Primitive Processing Satellite

The most primitive form of satellite processing is the implementation of on-board regenerative repeaters. This only requires that the uplink signal be demodulated and passed through a hard limiter or a decision circuit. The implementation of regenerative repeaters accrues the following advantages:

- Isolation of the uplink and downlink by on-board regeneration prevents the accumulation of thermal noise and interference. Co-channel interference is a predominant factor of signal degradation because of the measures taken to augment communication capacity by such means as frequency reuse.
- Isolating the uplink and downlink makes the optimization of each link possible. For example, the modulation format of the downlink need not be the same as that for the uplink.
- Regeneration on the satellite makes it possible to implement various kinds of signal processing on board the satellite. This can add to the communication capacity of the satellite and provide a more versatile set of conveniences for the user network.

Reference 9 points out that it can be shown on a PSK system that a regenerative repeater on board can save 6 dB on the uplink budget and 3 dB on the downlink budget over its bent-pipe counterpart, assuming the same BER on both systems.

Applying this technique to a digital TDMA system requires carrier recovery and bit timing recovery. Although the carrier frequencies and clocking are quite close among all bursts, coherency of the carrier and the clocking recovery may not be anticipated between bursts. To regenerate baseband signals on TDMA systems effectively, carrier recovery and bit timing recovery are done in the preamble of each burst and must be done very rapidly to maintain a high communication efficiency. There are two methods that can be implemented to resolve the correct phase of the recovered carrier. One method uses reference codewords in the transmitted bit stream at regular intervals. The other solution is to use differential encoding on the transmitted bit stream. Although this latter method is simpler, it does degrade BER considering equal C/N_0 of each approach.

An ideal regenerative repeater for a satellite transponder is shown in Figure 7.11 for PSK operation. It will carry out the following functions (Ref. 9):

- Carrier generation
- Carrier recovery

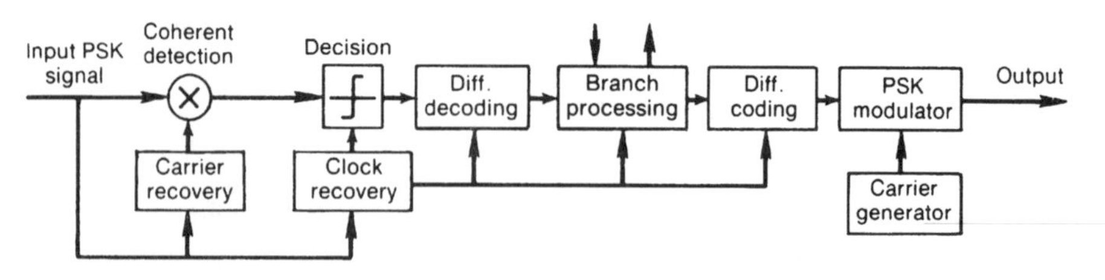

Figure 7.11. Configuration of an ideal regenerative repeater.

- Clock recovery
- Coherent detection
- Differential decoding
- Differential encoding
- Modulation
- Signal processing
- Symbol/bit decision

The addition of FEC coding/decoding on the uplink and on the downlink carries on-board processing one step further. FEC coding and decoding are discussed subsequently. In a fading environment such as one might expect with satellite communication systems operating above 10 GHz, during periods of heavy rainfall, an interleaver would be added after the coder and a deinterleaver before the decoder. Fading causes burst errors, and conventional FEC schemes handle random errors. Interleavers break up a digital bit stream by shuffling coded symbols such that symbols in error due to the burst appear to the decoder as random errors. Of course, interleaving intervals or spans should be significantly longer than the fade period expected.

7.5.2 Switched-Satellite TDMA (SS/TDMA)

We now carry satellite processing one step further by employing antenna beam switching. This technique provides bulk trunk routing, increasing satellite capacity by additional frequency reuse. Figure 7.12 depicts the concept. TDMA signals from a geographical zone are cyclically interconnected to other beams or zones so that a set of transponders appears to have beam-hopping capability. A sync window or reference window is usually required to synchronize the TDMA signals from earth terminals to the on-board switch sequence.

Sync window is a generic method to allow earth stations to synchronize to a switching sequence being followed in the satellite. A switching satellite, as described here, consists of a number of transmitters cross-connected to receivers by a high-speed time division switch matrix. The switch matrix connections are changed throughout the TDMA frame to produce the

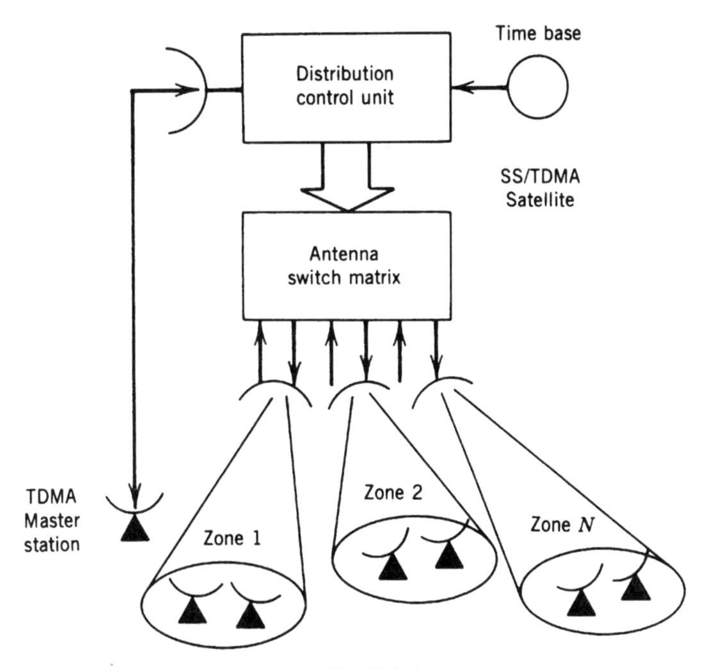

Figure 7.12. SS/TDMA concept.

required interconnections of earth terminals. A special connection at the beginning of the frame is the sync window, during which signals from each spot-beam zone are looped back to their originating spot-beam zone, thus forming the timing reference for all zones. This establishes closed-loop synchronization.

Figure 7.13 shows a scheme for locking and tracking a sync window in a satellite switching sequence. A burst of two tones, F_1 and F_2, is transmitted by a single access station. Only the portion that passes through the sync window is received back at that access station.

The basic concept is to measure and compare the received subbursts F_1 and F_2 as shown in the figure. Although a very narrow bandwidth and full RF power are used, digital averaging over many frames still is required. The difference is used to control the F_1/F_2 burst to a resolution of one symbol, and the process is continually repeated in closed loop. The sync bursts to the TDMA network are slaved, and the network is thus synchronized to the sync window and the switching sequence on the satellite.

With SS/TDMA the network connectivity and the traffic volume between zones can be adapted to changing needs by reprogramming the processor antennas. Also, of course, the narrow beams (e.g., higher-gain antennas) increase the uplink C/N and the downlink EIRP for a given transponder HPA power output. However, the applicability of the system must be analyzed carefully. Since there is a limit to the speed at which the antenna beam

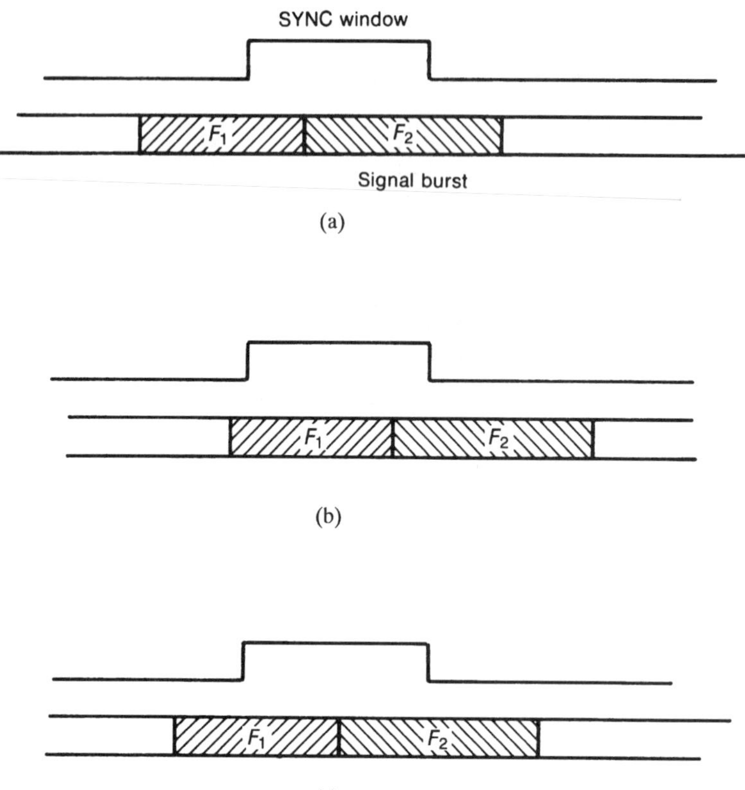

Figure 7.13. Sync window for SS/TDMA and use of bursts of two frequencies: (a) signal burst too early; (b) signal burst too late; and (c) synchronized to SYNC window. (From Ref. 5; reprinted with permission.)

can be switched, as the data symbol rate increases, guard times become an increasing fraction of the message frame, resulting in a loss of efficiency. This can be offset somewhat by the use of longer frames, but this will require increased buffering and will increase the transmission time.

The scheme becomes very inefficient if a large proportion of the traffic is to be broadcast to many zones. SS/TDMA is not applicable to channelized satellites where multiple transponders share common transmit and receive antennas and the signals in each transponder cannot be synchronized for transmission between common terminal zones (Ref. 5).

Figure 7.14 is a block diagram of a beam-switching processor in which narrow scanning beams are implemented by the use of a processor-controlled MBA (multiple-beam antenna) for both uplinks and downlinks. The output of the address selector is modified by the memory to provide control signals to the beam-switching network, which, in turn, controls the antenna-weight-

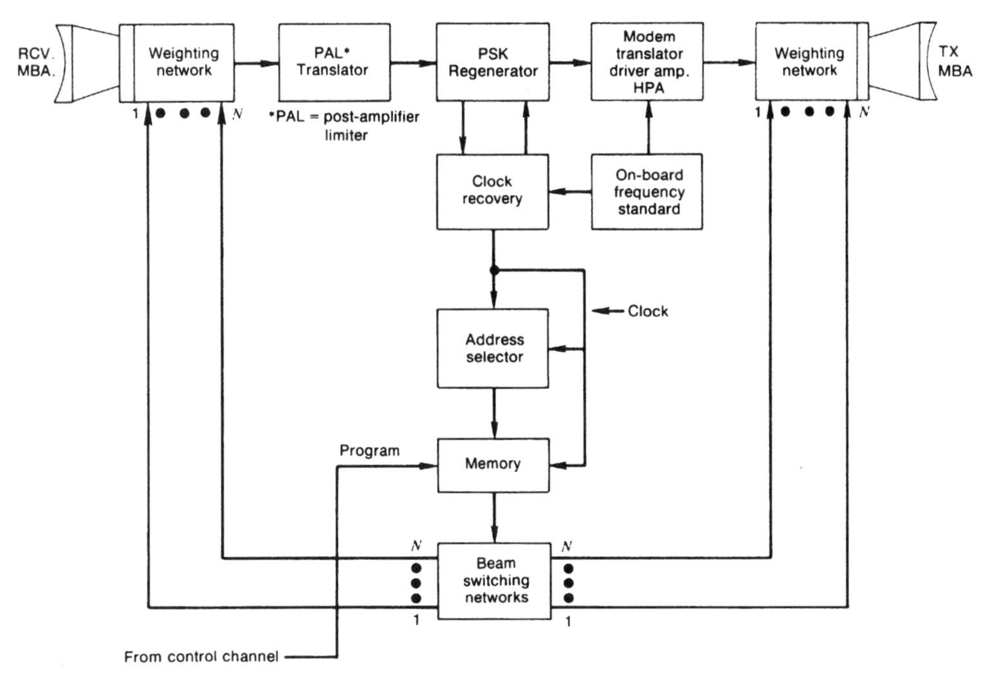

Figure 7.14. Block diagram of a beam-switching processor.

ing networks. This process steers the uplink and downlink antennas to the selected zones for a time interval that matches the time interval of a data frame to be exchanged between those selected zones. Initial synchronization of the system is obtained from a synchronizing signal transmitted to the processor from the TDMA master station. The memory can be updated at any time by signals from the master terminal via a control channel. Often, in practice, two memories are provided so that one can be updated while the other is on-line, thus preserving traffic continuity.

7.5.3 IF Switching

When the principal requirement for switching is for traffic routing rather than just for antenna selection, one method is to convert all uplink signals to a common IF, followed by a switching matrix and upconverting translators. This provides greater switching capability and flexibility. It allows for a wider choice of switching components in the design, since many more device types show good performance at the lower IF frequencies than at the higher uplink and downlink RF frequencies. Table 7.13 lists four switch matrix architectures with their advantages and disadvantages. Of these, Ref. 6 states that the coupler crossbar offers the best performance for a large switching matrix

TABLE 7.13 Comparison of Switch-Matrix Architectures

Description	Advantages	Disadvantages
Fan-out/ fan-in	Broadcast mode capability	High input VSWR High insertion loss Redundancy difficult to implement
Single-pole/ multiple-throw	Low insertion loss	Poor reliability
Rearrangeable switch	Low insertion loss	Poor reliability Random interruptions Control algorithm complicated
Coupler crossbar	Planar construction (minimum size, weight, and volume) Broadcast mode capability Redundancy easy to implement Good input/output VSWR[a] Signal output level independent of the path Enhanced reliability	Difficult feedthroughs Difficult broadbanding Isolation hard to maintain

[a]VSWR = voltage standing wave ratio.

Source: Reference 10.

(e.g., on the order of 20 × 20—20 inlets and 20 outlets). Figure 7.15 is a block diagram of such a crossbar matrix. In this case the crosspoints could be PIN diodes or dual-gate FETs.

7.5.4 Intersatellite Links

Carrying the on-board processing concept still further, we now consider intersatellite links that can greatly extend the coverage area of a network. An intersatellite link (ISL) is a full duplex link between two satellites, and other similar links can be added providing intersatellite connectivity among satellites in a large constellation. A number of military satellite constellations (e.g., MILSTAR) have cross-linking capability. The 58–62-GHz band has been assigned for this purpose by the ITU. Laser cross-links are actively under consideration by the U.S. Department of Defense.

An ISL capability can have a significant impact on system design. The following three system design characteristics are most directly affected:

1. *Connectivity.* The ISL can be used to provide connectivity among users served by regional satellites still retaining the required level of interconnec-

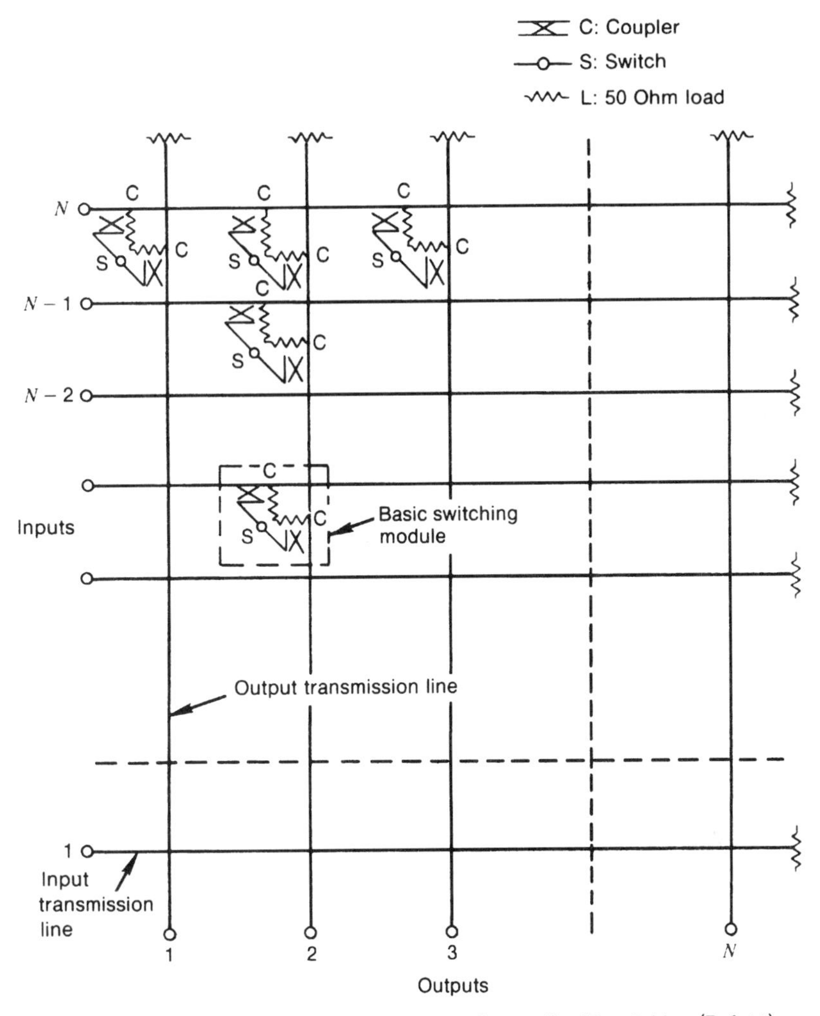

Figure 7.15. Coupler crossbar switching matrix for satellite IF switching (Ref. 10).

tivity within a community of users. It is assumed that each satellite in the system will serve a region with a high intraregion community of interest. The community of interest among regions will be lower. Thus each satellite will handle the high-traffic-intensity intraregion and cross-links will serve as tandem relays for the lower intensity traffic between regions.

2. *Capacity.* As described previously, uplinks and downlinks will have a high fill factor, and with cross-links implemented, low-traffic-intensity uplinks and downlinks will not be required to serve other regions with low interregional community of interest.

3. *Coverage.* Many satellite systems are designed for worldwide coverage where users of one satellite footprint require connectivity to users not in view of that satellite. The use of cross-links eliminates the need for earth relay at a dual-antenna earth station. The cross-link saves money and reduces propagation delay. Generally, an ISL is more economic than adding the additional uplink and downlink to the next satellite.

The cross-link concept incurs other advantages:

- Not affected by climatic conditions (e.g., the link does not pass through the atmosphere, assuming geostationary satellites)
- Low antenna noise temperature
- Low probability of earth-based intercept
- For military systems using 60-GHz band, a low probability of ground-based jamming owing to satellite antenna discrimination and the high atmospheric absorption in the 60-GHz band

An intersatellite link or cross-link system consists of four subsystems: receiver, transmitter, antenna or lens subsystem, and acquisition and tracking subsystem. The subsystems perform the following functions (Ref. 10):

1. *Receiver.* The receiver, in the generic case, detects, demodulates, and decodes the received signal. It interfaces the uplink/downlink system by cross-strapping at baseband. In some implementations this may also require format/protocol conversion. In a primitive cross-link system there may be no decoding and minimal format conversion.

2. *Transmitter.* In the general case, the transmitter selects the traffic for cross-linking. This may be accomplished by detecting a unique header word for routing. The signal for cross-linking is then encoded, which, in turn, modulates a carrier (or a laser); the signal is then upconverted and amplified. In a primitive system, there may not be any coding step.

3. *Antenna / Lens Subsystem.* For conventional RF transmission a suitable dish or lens antenna is required to radiate the transmitted signal and receive the incoming signal. For an optical system, a suitable lens or mirror, or combination, would be required to direct the laser signal.

4. *Tracking and Acquisition.* The two satellites involved, whether in a geostationary or nongeostationary orbit, require an antenna system that acquires and tracks to an appropriate accuracy, usually specified to some fraction of a beamwidth. Such techniques as raster scan and monopulse can be used for tracking. Initial pointing may be ground controlled from the TT & C* or master station/stations. It should be noted that it is not neces-

*TT & C = telemetry, tracking, and command.

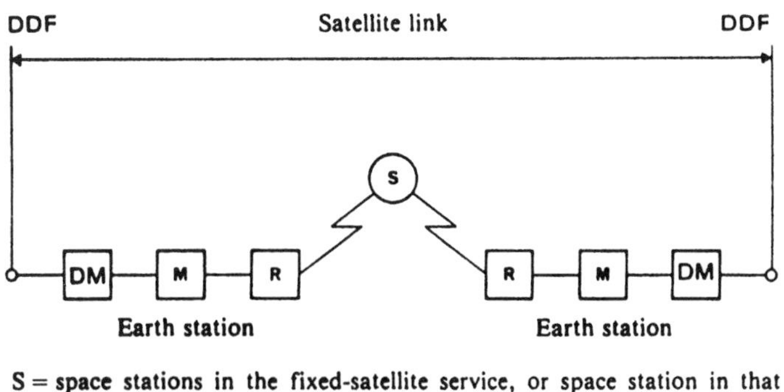

S = space stations in the fixed-satellite service, or space station in that service interconnected by satellite-to-satellite links

DM = digital multiplex equipment (including TDMA, DSI, and LRE equipment if used)

M = modem equipment

R = IF/RF equipment

Figure 7.16. Hypothetical reference digital path.

sary to orient the entire spacecraft to these accuracies, but only the ISL antenna reflector or feed (Ref. 10).

7.6 PERFORMANCE CONSIDERATIONS FOR DIGITAL SATELLITE COMMUNICATIONS

7.6.1 Hypothetical Reference Digital Path for Systems Using Digital Transmission in the Fixed-Satellite Service*

The hypothetical reference digital path (HRDP) consists of one earth–space–earth link, whose space portion may consist of one or more satellite-to-satellite links. The HDRP includes the equipment, as shown in Figure 7.16, and should interface the terrestrial network at either side through a DDF (digital distribution frame). The multiplex equipment (DM) illustrated in Figure 7.16 should also include TDMA, DSI, and LRE (low rate encoding) equipment, as appropriate, if used at the earth station. It should also include any equipment required for the purpose of compensating for the effects of satellite link transmission time variation, when it is used at an earth station.

*Section 7.6.1 is based on ITU-R Rec. S.521-2 (Ref. 11).

7.6.2 BERs at the Output of a HDRP for Systems Using PCM Telephony

The BER at the output of a HRDP should not exceed the following (from Ref. 12):

- 1×10^{-6}, 10 minutes mean value for more than 20% of any month
- 1×10^{-4}, 1 minute mean value for more than 0.3% of any month
- 1×10^{-3}, 1 second mean value for more than 0.05% of any month

"Any month" should be interpreted to mean "worst month" over the previous four years of statistics.

Note 1. The performance of a digital satellite system is generally much more sensitive to a variation in performance in the radio-frequency part of the network than is the case for its analog counterpart. It is therefore particularly important to allow adequate margins for degradation, which may occur during the life of the system if this recommended performance is to be maintained.

Note 2. The BER objectives given above include the effects due to interference noise and noise due to atmospheric absorption and rain, but exclude the unavailable time due to equipment degradation or failure.

These objectives have been set by ITU-R only for circuits involved in PCM telephony.

7.6.3 Allowable Error Performance for a HRDP in the Fixed-Satellite Service Operating Below 15 GHz When Forming Part of an International Connection in an ISDN*

The BER at the output (i.e., at either end of a two-way connection) of a satellite HRDP operating below 15 GHz and forming part of a 64-kbps ISDN connection should not exceed during the available time the values given below:

- 1×10^{-7} for more than 10% of any month
- 1×10^{-6} for more than 2% of any month
- 1×10^{-3} for more than 0.03% of any month (see Note 2)

Note 1. These HRDP BER performance objectives are sufficient to meet the requirements of ITU-T Rec. G.821 under all envisioned operating conditions. The ITU-T allocations for fixed satellite service (FSS) HRDP, which are considered to apply to the available time over the period of the order of any

*Section 7.6.3 is based on ITU-R Rec. S.614-3 (Ref. 13).

month, can be stated as follows:

- Fewer than 2% of the 1-minute intervals to have a BER worse than 1×10^{-6}
- Fewer than 0.03% of 1-second intervals to have a BER worse than 1×10^{-3}
- Fewer than 1.6% of 1-second intervals to have errors

Note 2. The value of 0.03% of any month relates to the measured BER during the available time. The objective could be met, for example, by designing the satellite system to an unavailability objective of 0.2% of the worst month (total time). By using the 10% availability factor (ratio of available to total time while the BER is worse than 1×10^{-3}), this would correspond to 0.02% of the available time of any month. Furthermore, it is necessary to include an allowance to accommodate contributions to those severely errored seconds, which occur when the BER is better than 1×10^{-3}. Taking as an example 0.01% of the worst month for this allowance, the total performance objective would be 0.3% of the available time of the worst month.

Table 7.14 gives end-to-end satellite HRDP error performance objectives for an international ISDN.

TABLE 7.14 Overall End-to-End HRDP Error Performance Objectives for International ISDN Connections

Performance Classification	Overall End-to-End Objectives[a]	Satellite HRDP Objectives[a]
Degraded minutes[b,c]	Fewer than 10% of 1-min intervals to have a bit error ratio worse than[d] 1×10^{-6}	Fewer than 2% of 1-min intervals to have a bit error ratio worse than[a] 1×10^{-6}
Severely errored seconds[b]	Fewer than 0.2% of 1-s intervals to have a bit error ratio worse than 1×10^{-3}	Fewer than 0.03% of 1-s intervals to have a bit error ratio worse than 1×10^{-3}
Errored seconds[b]	Fewer than 8% of 1-s intervals to have any errors (equivalent to 92% error-free seconds)	Fewer than 1.6% of 1-s intervals to have any errors (equivalent to 98.4% error-free seconds)

[a]Overall end-to-end and satellite HRDP performance objectives are expressed in terms of available time.

[b]The terms "degraded minutes," "severely errored seconds," and "errored seconds" are used as a convenient and concise performance objective "identifier." Their usage is not intended to imply the acceptability, or otherwise, of this level of performance.

[c]The 1-min intervals mentioned are derived by removing unavailable time and severely errored seconds from the total time and then consecutively grouping the remaining seconds into blocks of 60.

[d]For practical reasons, at 64 kbps, a minute containing four errors (equivalent to an error ratio of 1.04×10^{-6}) is not considered degraded. However, this does not imply relaxation of the error ratio objective of 1×10^{-6}.

Source: Table 1, ITU-R Rec. S.614-3 (Ref. 13).

7.6.3.1 Availability and Severely Errored Second Performance. In the derivation of performance models used to meet ITU-T Rec. G.821, it is necessary to consider the proportion of time a link is declared available. The generally accepted definition for unavailable time is as follows:

> A period of unavailable time begins when the BER in each second is worse than 1×10^{-3} for a period of 10 consecutive seconds. These 10 s are considered to be unavailable time. The period of unavailable time terminates when the BER in each second is better than 1×10^{-3} for a period of 10 consecutive seconds. These 10 s are considered to be available time and would contribute to the severely errored second performance objective. Excessive BER is only one of the factors contributing towards the total unavailable time. Definitions concerning availability can be found in ITU-T Recommendation G.106.
>
> Availability must be taken into account in the design of satellite transmission links which experience occasional periods of attenuation during precipitation which exceed the margins of the system. This is particularly true at frequencies above 10 GHz and propagation studies illustrate this fact.

In some operational TDMA systems the terminals make BER measurements on the UW of each received traffic burst over successive periods of less than 10 s. The period has a duration of 4 s (128 multiframes) in the case of the EUTELSAT TDMA system. When a BER threshold of 1×10^{-3} is exceeded during one measurement period, a set of high BER maintenance alarms are exchanged between the transmit and receive TDMA terminals. This causes the sending of particular signaling sequences toward the international switching center (ISC) from each of the two terminals. These sequences may be interpreted as call release messages and may cause the interruption of the calls concerned.

7.6.4 Allowable Error Performance for a HRDP Operating at or Above the Primary Rate*

(The Impact of ITU-T Rec. G.826)
Also see Sections 3.6.2 and 3.6.4.

7.6.4.1 Introduction. Consistent with ITU-T Recommendation G.821, the requirements of ITU-T Recommendation G.826 (Ref. 15) are given in terms of errored intervals (EIs). The terminology between the two recommendations is similar but the definitions of the parameters are different. For ITU-T Rec. G.826, the EIs are defined in terms of errored blocks (EBs) as opposed to individual bit errors. The purpose here is to allow the verification of adherence to the performance requirements of ITU-T Rec. G.826 on an in-service basis. The specification of performance in terms of block errors instead of bit errors has important consequences for systems where the errors tend to occur in groups (i.e., bursts), such as systems employing scrambling

*Section 7.6.4 is based on ITU-R Rec. S.1062 (Ref. 14).

and FEC. The block used in G.826 is that group of contiguous bits that normally makes up the inherent monitoring block or frame of the transmission system being employed.

7.6.4.2 Definitions

Events

Errored Block (EB). A block in which one or more bits are in error.

Errored Second (ES). A 1-second period with one or more errored blocks. SES (defined below) is a subset of ES.

Severely Errored Second (SES). A 1-second period that contains $\geq 30\%$ errored blocks (see Note 1) or at least one severely disturbed period (SDP) (see Note 2).

For out-of-service (OOS) measurements, a SDP occurs when, over a minimum period of time equivalent to four contiguous blocks or 1 ms, whichever is larger, either all the contiguous blocks are affected by a high binary error density of $\geq 1 \times 10^{-2}$, or a loss of signal information is observed. For in-service monitoring purposes, a SDP is estimated by the occurrence of a network defect. The term defect is defined in the relevant annexes of G.826 for the different network fabrics such as plesiochronous digital hierarchy (PDH) (typically E-1 series and DS1 series formats), synchronous digital hierarchy (SDH), or cell-based systems (such as ATM).

Note 1. The reference ITU-R recommendation cautions that, for historical reasons, SESs on some PDH systems are defined with a different percentage of errored blocks.

For maintenance purposes, the values from 30% may be used and these values may vary with transmission rate.

Note 2. SDP events may persist for several seconds and can be precursors to periods of unavailability especially when there are no restoration/protection procedures in use. SDPs persisting for T seconds, where $2 \leq T < 10$ (some network operators refer to these events as "failures"), can have a severe impact on service, for example, the disconnection of switched services. The only way to limit the frequency of these events is through the limit imposed for the severely errored seconds ratio (SESR).

Background Block Error (BBE). An errored block not occurring as part of an SES.

Parameters

Errored Second Ratio (ESR). The ratio of ES to total seconds in available time during a fixed measurement interval.

Severely Errored Seconds Ratio (SESR). The ratio of SES to total seconds in available time during a fixed measurement interval.

Background Block Error Ratio (BBER). The ratio of errored blocks to total blocks during a fixed measurement interval, excluding all blocks during SES and unavailable time.

7.6.4.3 *Performance Objectives.* Table 7.15 gives the end-to-end objectives of ITU-T Rec. G.826 (Ref. 15). These performance objectives are given as a function of transmission system bit rate. The ranges of block sizes accommodated at these bit rates are also given. It should be noted that the block size is associated with the frame structure of the particular transmission system. Ranges of block size are given by the recommendation so as not to prejudice the development of future transmission systems. These objectives are specified for available time.

Independent of the actual distance spanned, any satellite hop in the international or national portion receives 35% allocation of the end-to-end objectives. The performance objectives for a satellite HRDP are given in Table 7.16 for transmission rates between 1.5 and 3500 Mbps.

TABLE 7.15 End-to-End Objectives for a 27,500-km International Digital Connection at or Above the Primary Rate

Rate (Mbps)	1.5 to 5	> 5 to 15	> 15 to 55	> 55 to 160	> 160 to 3500	> 3500
Bits/block	2000–8000[a]	2000–8000	4000–20,000	6000–20,000	15,000–30,000[b]	For further study
ESR	0.04	0.05	0.075	0.16	[c]	For further study
SESR	0.002	0.002	0.002	0.002	0.002	For further study
BBER	3×10^{-4}	2×10^{-4}	2×10^{-4}	2×10^{-4}	10^{-4}	For further study

[a]VC-11 and VC-12 (ITU-T Recommendation G.709) paths are defined with a number of bits/block of 832 and 1120, respectively, that is, outside the recommended range for 1.5–5-Mbps paths. For these block sizes, the BBER objective for VC-11 and VC-12 is 2×10^{-4}.

[b]Because bit error ratios are not expected to decrease dramatically as the bit rates of transmission systems increase, the block sizes (bits) used in evaluating very high bit-rate paths should remain within the range of 15,000–30,000 bits/block. Preserving a constant block size for very high bit-rate paths results in relatively constant BBER and SESR objectives for these paths. As currently defined, VC-4-4c (ITU-T Recommendation G.709) is a 601-Mbps path with a block size of 75,168 bits/block. Since this exceeds the maximum recommended block size for a path of this rate, VC-4-4c paths should not be estimated in service using this table. The BBER objective for VC-4-4c using the 75,168 bit block size is taken to be 4×10^{-4}. There are currently no paths defined for bit rates greater than VC-4-4c (> 601 Mbps). Digital sections are defined for higher bit rates and guidance on evaluating the performance of digital sections can be found in the reference publication.

[c]Due to the lack of information on the performance of paths operating above 160 Mbps, no ESR objectives are recommended at this time. Nevertheless, ESR processing should be implemented within any error performance measuring devices operating at these rates for maintenance or monitoring purposes.

Source: Table 1, ITU-R Rec. S.1062 (Ref. 14).

TABLE 7.16 Satellite HRDP Performance Objectives for an International or National Digital Connection Operating at or Above the Primary Rate

Rate (Mbps)	1.5 to 5	> 5 to 15	> 15 to 55	> 55 to 160	> 160 to 3500
ESR	0.014	0.0175	0.0262	0.056	[a]
SESR	0.0007	0.0007	0.0007	0.0007	0.0007
BBER	1.05×10^{-4}	0.7×10^{-4}	0.7×10^{-4}	0.7×10^{-4}	0.35×10^{-4}

[a]Due to the lack of information on the performance of paths operating above 160 Mbps, no ESR objectives are recommended at this time. Nevertheless, ESR processing should be implemented within any error performance measuring devices operating at these rates for maintenance or monitoring purposes.

Source: Table 2, ITU-R Rec. S.1062 (Ref. 14).

TABLE 7.17 Relationship Between Bit Rate, Block Size, and Number of Blocks per Second

Bit Rate (Mbps)	Block Size (bits)	Number of Blocks per Second
1.544	4,632	$333\frac{1}{3}$
2.048	2,048	1,000
6.312	3,156	2,000
44.736	4,760	$9{,}398\frac{63}{119}$
51.84	6,480	8,000
155.52	19,440	8,000

Source: Table 3, ITU-R Rec. S.1062 (Ref. 14).

7.6.4.4 Monitoring Blocks. The block used in G.826 is the inherent monitoring block of the transmission system employed. Table 7.17 gives the block size and number of block(s) for various transmission rates.

7.7 LINK BUDGETS FOR DIGITAL SATELLITES

7.7.1 Commentary

The link budget, of course, is the primary tool for dimensioning both earth station terminals and their companion satellites. It is used for trade-offs of performance versus cost (cost because we are dimensioning equipment). The approach we use is almost identical to that described in Section 6.3.9. Rather than S/N, we use E_b/N_0, which was described in Chapter 3. Performance may be enhanced on digital circuits by the use of FEC (forward error correction), with or without interleaving. FEC and interleaving are described in Chapter 4. When FEC is used, we achieve *coding gain*, not available to us on equivalent analog links.

We made a transition from analog line-of-sight radiolinks to digital from Chapter 2 to Chapter 3. A similar transition is made here between Chapters 6 and 7. However, here we must distinguish between bent-pipe satellites operating in a FDMA/digital mode and processing satellites. In the case of the bent-pipe satellite, the link budget follows down just as we did in Section 6.3.9 to the calculation of C/N_0 for the uplink and C/N_0 for the downlink. Then we use formula (6.32) in Section 6.3.9.3 to calculate the $(C/N_0)_{(s)}$ (i.e., system C/N_0). From this value we can derive E_b/N_0. Given the modulation type and coding, we can then derive BER from the E_b/N_0 value. We can also do the reverse—start with a BER goal and work backward.

Example 4. A certain satellite system operates with 2.048 Mbps, the modulation is QPSK with coherent detection; it uses convolutional rate $\frac{1}{2}$ coding with Viterbi decoder (hard decision); modulation implementation loss is 2 dB; $(C/N_0)_{(s)} = 77$ dB; and we desire a 5-dB margin. Calculate the BER allowing for the margin.

Calculate net E_b/N_0:

$(C/N_0)_{(s)}$	77 dB
10 log (bit rate)	63.11 dB
E_b/N_0	13.89
Margin	5.0 dB
Modulation implementation	loss 2 dB
Net E_b/N_0	6.89 dB

Turn back to Chapter 4 for our parameters. Figure 4.16 shows that the BER is about 4×10^{-6} with a net E_b/N_0 of 6.89 dB.

In the case of a processing satellite (such as NASA ACTS), the uplink and downlink are isolated. We do not use formula (6.32). We directly calculate E_b/N_0 and BER for each link (up and down); or we plug in BER for each, derive the equivalent E_b/N_0, and work backward.

Consider the following two examples.

Example 5. A specific uplink at 6 GHz working into a processing satellite is to have a BER of 1×10^{-5}. The modulation is QPSK and the data rate is 10 Mbps. The terminal EIRP is +65 dBW, and the free-space loss to the satellite is 199.2 dB. What satellite G/T will be required without coding? What G/T will be required with FEC? The satellite uses an earth coverage antenna. See Table 7.18.

Discussion. Select the required E_b/N_0 first with no coding (Figure 4.12) and then with convolutional coding $R = \frac{1}{2}$ and $K = 7$ (Table 4.7). A modulation implementation loss of 2 dB is used in both cases. Allow a 4-dB link margin. The uncoded QPSK E_b/N_0 is 9.6 dB; the coded value is 6.5 dB. Initially, set the satellite $G/T = 0$ dB/K. Be sure to work from the "net" E_b/N_0.

TABLE 7.18 Example Uplink Power Budget

Terminal EIRP	+65 dBW	
Terminal pointing loss	0.5 dB	
Free-space loss	199.2 dB	
Satellite pointing loss	0.0 dB	(earth coverage)
Polarization loss	0.5 dB	
Atmospheric losses	0.5 dB	(Section 6.3.9)
Rainfall (excess attenuation)	0.25 dB	(Section 6.3.9) (10° elevation angle)
Isotropic receive level	−135.95 dBW	
G/T	0.0 dB	
Sum	−135.95 dBW	
Boltzmann's constant	−(−228.6 dBW)	
C/N_0	92.65 dB	
−10 log(bit rate)	−70.00 dB	
E_b/N_0	+22.65 dB	
Required E_b/N_0	−9.6 dB	
Implementation loss	−2.0 dB	
Margin	11.05 dB	
Allowable margin	−4 dB	
Excess margin	7.05 dB	

The satellite G/T can be −7.05 dB for the uncoded system.
For the coded system with a 3.1-dB coding gain, the satellite G/T can be degraded to −10.15 dB [−7.05 dB + (−3.1 dB)]

Example 6. A satellite downlink has a +30-dBW EIRP at 7.3 GHz. The desired BER is 1×10^{-6}; the modulation is BPSK with coherent detection; FEC is implemented with convolutional coding, rate = $\frac{1}{2}$, $K = 7$, and Viterbi decoding with 3-bit receiver quantization; the bit rate is 45 Mbs and the free-space loss is 202.0 dB. What is the terminal G/T assuming a 5-dB margin? Assume the satellite uses a spot beam. See Table 7.19.

Discussion. The required G/T is the sum of 26.18 + 4.9 + 2.0 + 5.0 or +38.08 dB/K. If that value is now substituted for the initial G/T of 0 dB/K, the last entry in the table or "sum" would drop to 0. The value for the required E_b/N_0 was derived first from Figure 4.12, left-hand curve, using the BER = 1×10^{-6} (10.5 dB), thence to Figure 4.19, where the coding gain was 5.6 dB for $K = 7$, $R = \frac{1}{2}$. We subtracted, leaving the required E_b/N_0 at 4.9 dB. The rainfall loss was, in a way, arbitrary. Generally, the link designer can neglect rainfall loss below 10 GHz. We added it as a safety factor. As we will show in Chapter 9, 3 dB in this frequency band, at a 10° elevation angle, would provide performance 99.9% of the time for central North America. However, interleaving using a sufficient interleaving interval (about 4 or 5 s*)

*Could we use full-duplex voice on circuits with delays of this magnitude?

TABLE 7.19 Example Downlink Power Budget

Satellite EIRP	+30 dBW
Satellite pointing loss	0.5 dB
Footprint error	0.25 dB
Off-footprint center loss	1.0 dB
Terminal pointing loss	0.5 dB
Polarization loss	0.5 dB
Atmospheric losses	0.5 dB
Free-space loss	202.0 dB
Rainfall loss	3.0 dB
Isotropic receive level	−178.25 dBW
Terminal G/T	0.0 dB/K
Sum	−178.25 dBW
Boltzmann's constant	−(−228.6) dBW
C/N_0	50.35 dB
10 log(BR)	−76.53 dB
Difference	−26.18 dB
Required	−4.9 dB
Modulation implementation loss	−2.0 dB
Margin	−5.0 dB
Sum	38.08 dB

and coding would permit us to mitigate the excess attenuation due to rainfall nearly completely. Note that the coding gain used was for AWGN conditions only. The incidental losses (i.e., polarization, pointing errors) are good estimates and probably can be improved upon with a real system where firm values can be used.

Table 7.20 is a typical digital link power budget provided by Scientific-Atlanta. It is an actual link working with INTELSAT IDR service. Similar values could be expected from INTELSAT IBS and EUTELSAT SMS satellite transponders.

PROBLEMS AND EXERCISES

1 Explain the basic difference between bent-pipe digital satellite systems and processing satellite systems.

2 For the most basic and elemental processing satellite systems, trace a typical signal as it enters the receive antenna through the satellite up the launching of that signal out the antenna on its downlink.

3 What is the principal advantage of TDMA operation from the point of view of satellite transmitter efficiency? What is the principal design factor in TDMA operation that is not a requirement in FDMA operation?

TABLE 7.20 Typical Digital Link Budget with INTELSAT IDR Service, 384 kbps

Satellite	VI F2	VI F2	
Beam type	Zone	Zone	
Type of service	QPSK	QPSK	
Transmit/receive connectivity	7/6	6/7	m/m
Information rate	384	384	kbps
Coding rate (1/2 = 0.5, 3/4 = 0.75)	0.5	0.5	—
PSK type (1 = BPSK, 2 = QPSK)	2	2	—
Occupied bandwidth	384	384	kHz
Carrier spacing	1.4	1.4	—
Allocated bandwidth	538	538	kHz
I. UPLINK			
Earth station EIRP per carrier	53.3	51.6	dBW
Pointing losses	1.3	1.0	dB
Path loss	199.8	199.8	dB
Atmosphere/rain attenuation	0.1	0.1	dB
Isotropic antenna area	37.3	37.3	dB/m^2
SFD at beam edge	−77.6	−77.6	dBW/m^2
G/T at beam edge	−7.0	−7.0	dB/K
Uplink footprint advantage	2.0	3.0	dB
Input backoff per carrier	31.1	31.4	dB
Uplink thermal C/N	19.8	19.5	dB
Uplink IM EIRP density	17.0	16.3	dBW/4 kHz
Uplink intermodulation C/N	16.5	15.5	dB
Total uplink C/(N + 1)	14.8	14.0	dB
II. TRANSPONDER INTERMODULATION NOISE			
IMP density at beam edge	−37.0	−37.0	dBW/4 kHz
C/IM	22.6	22.3	dB
III. DOWNLINK			
Transmit/receive connectivity	7/6	6/7	m
Beam edge saturated XPDR EIRP	31.0	31.0	dBW
Downlink footprint advantage	3.0	2.0	
Input to output backoff	5.5	5.5	dB
XPDR output backoff per carrier	25.6	25.9	dB
D/L EIRP per carrier at E/S	8.4	7.1	dBW
Path loss	195.9	195.9	dB
Atmosphere/rain attenuation	0.1	0.1	dB
Pointing losses	0.4	0.5	dB
Earth station G/T	25.9	27.7	dB/K
Rain G/T degradation	0.0	0.0	dB
Downlink thermal C/N	10.7	11.0	dB
IV. CO-CHANNEL INTERFERENCE	17.0	17.0	dB
V. TOTAL C/(N + 1) NOISE			
Phase noise degradation	0.4	0.4	dB
Total C/(N + 1)	8.0	8.0	dB
$C/(N_0 + I_0)$ total	63.9	63.8	dB-Hz
E_b/N_0 total	8.0	8.0	dB
Decoding (Viterbi or sequential)	SEQ	SEQ	—
BER (data) or S/N (FM) objective	10^{-7}	10^{-7}	—
E_b/N_0 (data) required	7.0	7.0	dB
Margin	1.0	1.0	dB

Source: Courtesy of Scientific-Atlanta (Ref. 16).

4 Argue the operational advantages of TDMA versus FDMA.

5 Name factors limiting TDMA digital throughput.

6 Explain the rationale for using CR/BTR and its location in a typical TDMA user frame.

7 There are three generic approaches to digital bent-pipe operation and access. Name them.

8 In which mode of operation/access would we expect to use INTELSAT IDR and IBS?

9 All INTELSAT IDR offerings use FEC coding. What would be their basic rationale for this?

10 For TDMA operation, explain the importance of the earth station buffer.

11 In a TDMA user frame, what part of the overhead establishes the time reference? (It establishes the local clock alignment function.)

12 A transponder operating in the TDMA mode has a bandwidth of 80 MHz and 8-ary PSK modulation is used. What maximum link transmission bit rate can be expected? Assume bandwidth-limited operation.

13 How do overhead bits and guard time durations affect TDMA system efficiency?

14 A satellite transponder is power limited, working in the TDMA mode with an EIRP of +19 dBW, operating with an earth terminal with a G/T of +33 dB/K using BPSK modulation with coherent detection; a 6-dB margin is provided; there is a 2-dB modulation implementation loss. The required E_b/N_0 for a BER of 1×10^{-5} is 9.6 dB (theoretical). What is the maximum bit rate for this power-limited case?

15 For TDMA operation, timing and synchronization of earth stations accessing a particular transponder are vitally important. There are two generic ways to handle this. What are they? Explain the difference between them.

16 Timing and synchronization of member TDMA earth stations due to their disparate distances and resulting time delays are bad enough. What other timing variable makes the situation even more difficult? Incorporate the word *dynamic* in the answer.

17 Explain how the *reference burst* method of TDMA timing operates.

18 Differentiate TASI-based DSI with speech-predictive DSI.

19 Describe the principal degradations of DSI.

20 Explain the operation of speech-predictive DSI.

21 What is the principal purpose of a control and delay channel in an operational TDMA system?

22 Give the three technical advantages of a primitive processing satellite.

23 Name at least five functions that are carried out by an on-board regenerative repeater.

24 Describe switched-satellite TDMA (SS/TDMA) operation. What is actually being switched in the satellite?

25 Explain a common application of IF switching in a satellite.

26 Give at least two reasons why the band 58–62 GHz was selected for intersatellite links (cross-links).

27 List three advantages of cross-links.

28 Name another transmission medium under consideration for cross-links.

29 A digital satellite link, based on the ITU-R HRDP, should achieve what BER more than 90% of the time?

30 What may happen to calls transiting a digital satellite system when the BER falls below 1×10^{-3}?

31 Define an errored second, a severely errored second, and an errored block based on the ITU-R HRDP.

32 Carry out the following exercise. Prepare a reasonable link budget for INTELSAT VII. The link employs FEC with convolutional coding rate $\frac{3}{4}$, $K = 7$, and Viterbi decoder with 3-bit quantization. The desired BER $= 1 \times 10^{-10}$. Assign earth terminal parameters for antenna aperture diameter using 65% efficiency and for transmitter output power and receiving system noise temperature. The bit rate is 44.736 Mbps using INTELSAT IDR offering. A 7-dB rainfall margin is desired. The elevation angle is 27°. Adjust the margin value, if required. Compare these parameters with those of the appropriate INTELSAT standard earth station found in Chapter 6.

REFERENCES

1. *Performance Characteristics for Intermediate Data Rate (IDR) Digital Carriers*, INTELSAT IESS-308 with Rev. 7 and 7A, INTELSAT, Washington, DC, Aug. 1994.
2. *QPSK / FDMA Performance Characteristics for INTELSAT Business Services (IBS)*, IESS-309 through Rev. 4, INTELSAT, Washington, DC, May 1994.
3. O. G. Gabbard and P. Kaul, "Time-Division Multiple Access," IEEE Electronics and Aerospace Systems Convention, Oct. 7–9, 1974, as reproduced in H. L. Van Trees, *Satellite Communications*, IEEE Press, New York, 1979.

4. G. Maral and M. Bousquet, *Satellite Communications Systems*, 2nd ed., Wiley, Chichester, UK, 1993.
5. V. K. Bhargava et al., *Digital Communications by Satellite*, Wiley, New York, 1991.
6. S. J. Campanella, "Digital Speech Interpolation," *COMSAT Tech. Review*, Spring 1976, as reproduced in H. L. Van Trees, *Satellite Communications*, IEEE Press, New York, 1979.
7. Benjamin Pontano and Jack Dicks, *INTELSAT TDMA/DSI System*, INTELSAT, Washington, DC, 1983.
8. *INTELSAT TDMA/DSI System Specifications*, INTELSAT IESS-307 with Rev. A and B, INTELSAT, Washington, DC, Mar. 1991.
9. K. Koga, T. Muratani, and A. Ogawa, "On-Board Regenerative Repeaters Applied to Digital Satellite Communications," *Proceedings of the IEEE*, Mar. 1977, as reproduced in H. L. Van Trees, *Satellite Communications*, IEEE Press, New York, 1979.
10. N. R. Edwards, *Satellite Communications Reference Data Handbook*, Computer Sciences Corp., Falls Church, VA, Mar. 1983 under DCA contract DCA100-81-0044.
11. *Hypothetical Reference Digital Path for Systems Using Digital Transmission in the Fixed-Satellite Service*, CCIR Rec. S.521-2, 1994 S Series Volume, ITU, Geneva, 1994.
12. *Allowable Bit Error Ratios at the Output of a Hypothetical Reference Digital Path for Systems in the Fixed-Satellite Service Using Pulse Code Modulation for Telephony*, ITU-R Rec. S.522-5, 1994 S Series Volume, ITU, Geneva, 1994.
13. *Allowable Error Performance for a Hypothetical Reference Digital Path in the Fixed-Satellite Service Operating Below 15 GHz when Forming Part of an International Connection in an Integrated Services Digital Network*, ITU-R Rec. S.614-3, 1994 S Series Volume, ITU, Geneva, 1994.
14. *Allowable Error Performance for a Hypothetical Reference Digital Path Operating at or Above the Primary Rate*, ITU-R Rec. S.1062, 1994 S Series Volume, ITU, Geneva, 1994.
15. *Error Performance Parameters and Objectives for International, Constant Bit Rate Digital Paths at or Above the Primary Rate*, ITU-T Rec. G.826, ITU, Helsinki, Nov. 1993.
16. Private communication between Gerry Eining of Scientific-Atlanta and Roger Freeman, 7 May 1996.
17. Roger L. Freeman, *Telecommunication System Engineering*, Third Edition, Wiley, New York, 1996.

8

VERY SMALL APERTURE TERMINALS

8.1 DEFINITIONS OF VSAT

A very small aperture terminal (VSAT) is a digital satellite terminal where *economy* is the key word. The term *very small aperture*, of course, refers to the size of the terminal antenna. The diameters of VSAT parabolic antennas vary from 0.6 m (2 ft) to 2.4 m (7.8 ft), depending a great deal on the capabilities desired from the terminal. These can vary from a data connectivity (inbound) of 1200 bps up to a full DS1 or E-1.

In most cases, the definition denotes a family of modest "out terminals" and a comparatively large "hub" terminal. This implies a wheel made up of a hub and spokes. In fact, most VSAT architectures can be seen as hub and spokes, the spokes being the connectivities to the VSAT outstations, most often in a star network configuration as shown in Figure 8.1. The larger hub, in theory, compensates for the smaller handicapped VSAT outstations.

With the VSAT star network, traffic can be one-way or two-way. VSAT networks have extended the definition to include mesh-connected networks with no hub, as shown in Figure 8.2. The star configuration does not lend itself well to VSAT-to-VSAT voice communication because of the added delay, whereas the mesh architecture does lend itself to voice interconnectivity among VSATs.

Some sources even call any network of small satellite terminals a VSAT network.

8.2 VSAT NETWORK APPLICATIONS

VSATs are commonly implemented in private networks. Why they are attractive depends largely on a country's telecommunication infrastructure. In the United States and Canada, economy is the driving factor. Such VSAT

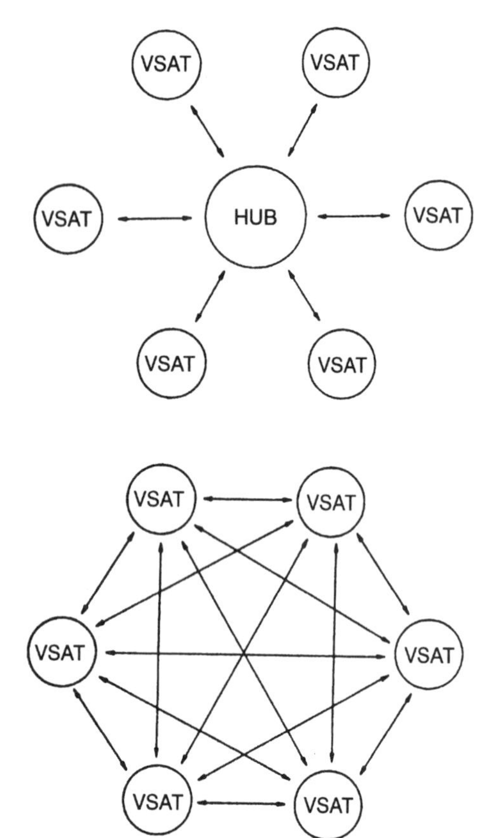

Figure 8.1. The conventional VSAT star network configuration.

Figure 8.2. A VSAT network based on a mesh architecture.

networks bypass the local and long-distance telephone companies and ostensibly save money by doing so.

In many other countries, national governments permitting, well-engineered VSAT networks can provide sterling quality service, whereas the local telecommunication administrations cannot. There is a third category that includes countries with a poor infrastructure and where many communities are afforded no electrical communications whatsoever.

8.2.1 One-Way Applications

This application basically involves data distribution from the hub outward to VSAT receive-only terminals. These data include the following:

- Press releases, news from press agencies, or the like
- Stocks, bonds, and commodity information
- Remote loading of computer programs

- Weather information from meteorological agencies, typically to airports
- Video distribution utilizing compressed video, typically 1.544 or 2.048 Mbps

Another data application is in the direction VSAT-to-hub for data collection purposes. This application may involve remote sensors on oil pipelines, environmental monitoring, and electric power utilities remote facilities. However, with many of these applications, some form of network control is necessary making two-way communication desirable.

8.2.2 Two-Way Applications

The most widely used application of VSAT communications is for diverse types of two-way data communications. Such a network provides complete flexibility for file transfer and all types of interactive data exchanges such as inquiry/response. In most configurations, the hub is co-located with corporate headquarters. Typical two-way applications are the following:

- Point-of-sale operations
- Financial, banking, and insurance information from field branches to central headquarters
- Credit card verification
- ATM (automatic teller machine) operations
- Hotel and motel reservations and all other types of reservations
- Support for shipping and freight handling facilities
- Inventory control and cash flow
- Technical support network, manufacturer-to-manufacturer representatives
- Supervisory control and data acquisition (SCADA), pipelines, railroads
- Extension of local area networks

If sufficient capacity is built into a VSAT network, telephone connectivities are feasible from outstation to headquarters and vice versa. It is not feasible because of added delay from outstation to other outstation via the hub. It would be feasible on VSAT mesh networks. With current video compression techniques, video conferencing may also be feasible (Ref. 1).

In some emerging nations, VSAT-like networks provide rural telephone connectivity.

8.3 TECHNICAL DESCRIPTION OF VSAT NETWORKS AND THEIR OPERATION

8.3.1 Introduction

The most common network topology of a VSAT network is the star configuration illustrated in Figure 8.1. The hub is the centerpiece and is almost always colocated with the corporate headquarters or state capital or national capital, for that matter. The hub may have an antenna with 5-m (16-ft) to 11-m (36-ft) diameter, whereas a VSAT may have an antenna diameter in the range of 0.6 (1 ft) to 2.4 m (8 ft). The RF output at the hub will vary from 100 to 1000 W, whereas a VSAT will be in the range of 1–10 W.

To reduce space segment recurring charges, it is incumbent on the system designer to use as small a bandwidth as possible on a satellite transponder. Outbound traffic is usually carried on a TDM bit stream, 56 or 64 kbps, 128, 256, 384 kbps (etc.). Inbound traffic, depending greatly on the traffic profile, will use some type of demand assignment process or contention, polling, or other protocol with bit rates ranging from 1200 bps to 64 kbps or greater.

Definition. Inbound means traffic or circuit(s) in the direction of VSAT to hub; outbound means traffic or circuit(s) in the direction of hub to VSAT.

VSATs commonly operate in the Ku-band because of the more favorable EIRP permitted on downlinks when compared to C-band. However, this does not mean to imply that C-band operation is ruled out. Let's examine a typical Ku-band two-way VSAT operation. Of course, excess attenuation due to rainfall must be contended with at Ku-band and can be minimal or neglected at C-band.

8.3.2 A Link Budget for a Typical VSAT Operation at Ku-Band

The model VSAT system in question is a two-way operation using Ku-band frequencies. The outbound link is 128 kbps in a TDM format employing QPSK modulation with coherent detection using rate $\frac{1}{2}$ convolutional coding, $K = 7$ and 3-bit quantization, and Viterbi decoder. The inbound (traffic) link has a transmission rate of 32 kbps using a HDLC*-type frame format, QPSK with similar FEC. The 128-kbps link with its rate $\frac{1}{2}$ coding has a coded symbol rate of 256 symbols/s. This link requires 200-kHz RF channel; the 32-kbps information rate and 64-kbps coded symbol rate require a 50-kHz RF channel. The BER, under clear-sky conditions, is 1×10^{-9}; for degraded conditions the BER may drop to 1×10^{-6}. There is a 2-dB modulation implementation loss so that the E_b/N_0 for clear-sky operation is 8.5 dB; for degraded operation, it is 6.7 dB for the 128-kbps outbound channel. The

*HDLC = high level data link control.

inbound 32-kbps channel also requires 8.5 dB for clear-sky and 6.7 dB for degraded operation. To counter rainfall loss at Ku-band, there is a 4-dB margin or both links. The elevation angle at both the hub and outstation VSAT is 10°. The range (distance) to the satellite (from Figure 6.5) is 25,220 sm.

The outbound uplink frequency is 14,100 MHz; its equivalent downlink operates at 11,800 MHz. The inbound uplink frequency is 14,300 MHz; its equivalent downlink frequency is 12,000 MHz. The satellite transponders in question each have an EIRP of +44 dBW over a 72-MHz bandwidth, assuming full loading. Transponder/satellite G/T in either case is 0.0 dB/K.

The inbound carrier downlink has an EIRP of +12.4 dBW; for the outbound downlink, the EIRP for the VSAT carrier is +18.4 dBW. These EIRP values were calculated assuming a uniform power density across the entire transponder bandwidth of 72 MHz. Therefore the EIRP = +44 dBW − 10 log(72,000/200) = +18.4 dBW.* The +12.4-dBW value is calculated in a similar fashion or $\text{EIRP}_{\text{dBW}} = +44 \text{ dBW} - 10\log(72{,}000/50)$.

The hub facility has the following terminal parameters: transmitter power output, 500 W or +27 dBW; line loss, 2 dB; antenna aperture, 5 m or 16.25 ft. Its gain at 14,100 MHz is 53.5 dB and at 11,800 MHz it is 52.0 dB; $T_{sys} = 200$ K, so the hub G/T is +29.0 dB/K. The EIRP = +78.5 dBW.

Postulated parameters of the VSAT terminal to operate in this system are as follows:

G/T = ? The antenna aperture is unknown. We will assume its efficiency is 65%.

The T_{sys} for the receiving system consists of the sum of T_{ant} and T_r. $T_r = 100$ K and $T_{ant} = 120$ K. Thus $T_{sys} = 220$ K. These are typical values.

The VSAT EIRP is unknown. The transmission line losses are 1 dB; the transmitter power output is unknown (in the range of 0.5–10 watts). The downlink (outbound) link budget will determine the antenna aperture.

The free-space loss values are:

$$
\begin{aligned}
(14{,}100 \text{ MHz}) \quad \text{FSL}_{\text{dB}} &= 36.58 + 20\log 14{,}100 + 20\log 25{,}220 \\
&= 36.58 + 82.98 + 88.03 \\
&= 207.59 \text{ dB} \\
(14{,}300 \text{ MHz}) \quad \text{FSL}_{\text{dB}} &= 36.58 + 83.11 + 88.03 \\
&= 207.71 \text{ dB} \\
(12{,}000 \text{ MHz}) \quad \text{FSL}_{\text{dB}} &= 36.58 + 81.58 + 88.03 \\
&= 206.19 \text{ dB} \\
(11{,}800 \text{ MHz}) \quad \text{FSL}_{\text{dB}} &= 36.58 + 81.44 + 88.03 \\
&= 206.05 \text{ dB}
\end{aligned}
$$

*This is the same as dividing 25,188 watts by 360 because there are 360 200-kHz segments in 72 MHz.

Outbound Link Budget

Uplink	
EIRP hub	+78.5 dBW
FSL	−207.59 dB
Polarization loss	−0.5 dB
Terminal pointing loss	−0.5 dB
Satellite pointing loss	−0.5 dB
Atmospheric loss	−0.3 dB
Isotropic receive level	−130.89 dBW
Satellite G/T	0.0 dB/K
Sum	−130.89 dBW/K
Boltzmann's constant	−(−228.6 dBW)
C/N_0	97.71 dB
Downlink	
EIRP satellite	+18.4 dBW
FSL	−206.05 dB
Polarization loss	−0.5 dB
Satellite pointing loss	−0.3 dB
Atmospheric loss	−0.2 dB
Terminal pointing loss	−0.5 dB
Isotropic receive level	−189.15 dBW
VSAT G/T	0.00 dB/K
Sum	−189.15 dBW
Boltzmann's constant	+228.6 dBW/Hz
C/N_0	39.95 dB

What net C/N_0 is required for an E_b/N_0 of 8.5 dB?

$$N_0 = -228.6 \text{ dBW} + 10 \log T_{sys}$$

$$T_{sys} = 220 \text{ K} \quad \text{(given above)}$$

Thus

$$N_0 = -228.6 \text{ dBW} + 10 \log 220$$
$$= -205.17 \text{ dBW}$$

E_b must have a level 8.5 dB higher than −205.17 dBW or −196.67 dBW. The bit rate on the channel is 128 kbps; thus C = RSL = −196.67 dBW + $10 \log(128 \times 10^3)$ or −196.67 + 51.07 dB = −145.6 dBW.

Then the objective

$$C/N_{0(t)} = -145.6 \text{ dBW} - (-)205.17 \text{ dBW}$$
$$= 59.57 \text{ dB}$$

Neglecting satellite generated noise (IM products);

$$C/N_{0(t)} = \frac{1}{1/(C/N_{0(u)}) + 1(C/N_{0(d)})} \qquad (6.32)$$

Convert decibel values to equivalent numerics.

$$59.57 \text{ dB} = 905{,}733 \text{ numeric} \qquad \text{(objective value)}$$

$$97.1 \text{ dB} = 5{,}128{,}613{,}840 \text{ numeric} \qquad \text{(calculated uplink value)}$$

$$905{,}733 = 1/\left[1/(5128 \times 10^6) + 1/(C/N_{0(d)})\right]$$

$$C/N_{0(d)} \approx 60 \text{ dB}$$

Placing that value in the downlink budget above, we can now calculate a value of G/T for the VSAT. The calculated C/N_0 was 39.45 dB; the required C/N_0 is 60 dB so there is a shortfall in 20.55 dB in C/N_0. Substitute 20.55 dB for the value of G/T. In other words, the G/T should be 20.55 dB/K rather than 0.00 dB/K. If we want a 4-dB margin, we would add 4 dB to this value, or the G/T would be 24.55 dB. We will use this latter value to calculate antenna aperture.

To calculate the required antenna aperture of the VSAT, we need the antenna gain by using the mathematical identity for G/T. T_{sys} was calculated to be 220 K.

$$G/T = G_{dB} - 10 \log 220$$

$$24.55 \text{ dB} = G_{dB} - 23.42 \text{ dB}$$

$$G_{dB} = 47.97 \text{ dB}$$

Antenna gain may be calculated from the formula

$$G_{dB} = 20 \log f_{MHz} + 20 \log D_{ft} + 10 \log \eta - 49.92 \qquad (8.1)$$

where η is the antenna efficiency, in this case 0.65 (65%).

$$G_{dB} = 20 \log D_{ft} + 20 \log(11{,}800 \text{ MHz}) + 10 \log 0.65 - 49.92$$

$$47.97 \text{ dB} = 20 \log D + 81.43 - 1.87 - 49.92$$

$$20 \log D = 18.33$$

$$D = 8.25 \text{ ft or } 2.5 \text{ m}$$

The next problem is to calculate the VSAT uplink transmit power. The EIRP of that uplink is based on an antenna with a 2.5-m (7.79-ft) aperture. Calculate the antenna gain at the uplink frequency of 14,300 MHz.

$$\begin{aligned} G_{dB} &= 20 \log(8.25) + 20 \log(14{,}300) + 10 \log(0.65) - 49.92 \quad \text{(dB)} \\ &= 18.32 + 83.11 - 1.87 - 49.92 \\ &= 49.65 \text{ dB} \end{aligned}$$

Make a trial run with a 1-watt transmitter (0 dBW). The EIRP of the VSAT uplink is then $= 0 - 1$ dB + 49.65 dBW = +48.65 dBW. Now run the inbound link budget.

Uplink	
EIRP VSAT hub	+48.65 dBW
FSL (14,300 MHz)	−207.71 dB
Polarization loss	−0.5 dB
Satellite pointing loss	−0.5 dB
Terminal pointing loss	−0.5 dB
Atmospheric loss	−0.4 dB
Isotropic receive level	−159.96 dBW
Satellite G/T	00.0 dB/K
Sum	−159.96 dBW/K
Boltzmann's constant (−)	−(−228.6 dBW/Hz)
C/N_0	68.63 dB
Downlink	
EIRP satellite	+12.4 dBW
FSL (12,000 MHz)	−206.19 dB
Polarization loss	−0.5 dB
Satellite pointing loss	−0.5 dB
Terminal pointing loss	−0.5 dB
Atmospheric loss	−0.3 dB
Isotropic receive level	−195.59 dBW
Hub G/T	+29.0 dB/K
Sum	−166.59 dBW
Boltzmann's constant (−)	−(−228.6 dB/Hz)
C/N_0	62.01 dB

Calculate the numeric equivalents of each C/N_0 value.

68.63 dB has a numeric value of 7,309,840.4
62.01 dB has a numeric value of 1,588,547

Calculate $C/N_{0(t)}$:

$$C/N_{0(t)} = 61.11 \text{ dB}$$

$$N_0 = -228.6 \text{ dBW/K} + 10\log(200) \text{ K}$$

$$= -205.59 \text{ dBW/Hz}$$

$$C/N_{0(t)} = C_{\text{dBW}} - N_0 \qquad (C = \text{RSL})$$

$$61.11 \text{ dB} = \text{RSL}_{\text{dBW}} - (-205.59 \text{ dBW/Hz})$$

$$-205.59 \text{ dBW} + 61.11 \text{ dB} = \text{RSL}$$

$$\text{RSL} = -144.48 \text{ dBW}$$

$$E_b = -144.48 \text{ dBW} - 10\log(32 \times 10^3)$$

$$= -144.51 \text{ dBW} - 45.05 \text{ dB}$$

$$= -189.53 \text{ dBW}$$

$$E_b/N_0 = -189.53 - (-205.39 \text{ dBW/Hz}) \qquad \text{(calculated)}$$

$$= 16.06 \text{ dB}$$

$$E_b/N_0 \text{ (required)} = 8.5 \text{ dB}$$

$$\text{Margin} = 7.56 \; dB$$

The VSAT transmitter power of 1 watt is sufficient. It will be noted that the C/N_0 on the inbound uplink is more than sufficient (i.e., 68 dB); it is the companion downlink that controls the total C/N_0 value (only 62 dB) (Ref. 2).

8.3.3 Summary of VSAT RF Characteristics

VSAT operation commonly uses either 6/4-GHz band (C-band) or 14/12-GHz band (Ku-band).* As operational frequencies increase, receiver noise performance degrades. At C-band we can expect a LNA with 50-K noise temperature; at Ku-band, 100 K. Antenna noise temperature (T_{ant}) at C-band (5° elevation angle) is 100 K and at Ku-band (10° elevation angle) it is 106 K. Thus typical T_{sys} for C-band VSAT operation is 150 K and for Ku-band it is 206 K. Line losses for both bands are taken at 1.5 dB for this particular model. In the case of Ku-band, the LNA is placed as close as practical to the feed to reduce line losses.

We now construct Table 8.1, which will give typical G/T values for several discrete antenna diameters for both C-band and Ku-band operation. The table is based on the T_{sys} figures given above. The antenna gains are based on 65% aperture efficiency (η). Formula (2.27c) was used to calculate parabolic antenna gains.

8.4 ACCESS TECHNIQUES

The most common VSAT architecture is the interactive network based on star topology (hub and spokes, Figure 8.1). Reference 3 describes a VSAT network with as many as 16,000 outstations. The author is familiar with

*There is nascent VSAT activity in Ka-band (30/20-GHz band) with the advent of the NASA ACTS. Surely other satellites will follow, and in time there will be equal or greater VSAT activity in the Ka-band.

TABLE 8.1 Typical VSAT *G/T* Values[a]

Antenna Aperture	Gain, C-Band	*G/T*, C-Band	Gain, Ku-Band	*G/T*, Ku-Band
0.5 m (1.625 ft)	23.46 dB	+1.7 dB/K	33 dB	+9.86 dB/K
0.75 m (2.44 ft)	27.0 dB	+5.23 dB/K	36.53 dB	+13.39 dB/K
1.0 m (3.25 ft)	29.49 dB	+7.73 dB/K	39.03 dB	+15.89 dB/K
1.5 m (4.875 ft)	33.01 dB	+11.24 dB/K	42.55 dB	+19.41 dB/K
2.0 m (6.5 ft)	35.51 dB	+13.75 dB/K	45.05 dB	+21.91 dB/K
2.5 m (8.125 ft)	37.45 dB	+15.69 dB/K	47.0 dB	+23.86 dB/K
3.0 m (9.75 ft)	39.03 dB	+17.54 dB/K	48.57 dB	+25.43 dB/K

[a]The reference frequencies used for antenna gain calculations are 4000 (C-band) and 12,000 MHz (Ku-band). The table includes 1-dB line loss for both bands.

networks with up to 2500 outstations interoperating with one hub. There are many access techniques, and the type selected will be fairly heavily driven by the traffic profile. The access technique will often determine the efficiency of usage of the space segment. For example, a completely assigned FDMA regime would prove to be very ineffective use of transponder bandwidth if there were hundreds of outstations, each interchanging short, interactive messages with the hub with a medium to low activity factor.

In selecting the type of channel assignment technique for a VSAT network, the following factors should be considered:

- Statistical properties of the traffic
- Permissible delay in transmission, including channel setup and propagation delay
- Efficiency of channel sharing, throughput performance
- Complexity, equipment, and implementation cost
- Operations and maintenance

For example, for credit card verifications and transactions, delay is probably the most important factor, throughput much less so. Whereas with file transfer and batch transactions, throughput is more important than delay (comparatively).

We will discuss three categories of access: random access, demand assigned, and fixed assigned. Here, of course, we refer to inbound channels. Outbound service is assumed to be a TDM bit stream and is discussed in Section 8.4.5.

8.4.1 Random Access

8.4.1.1 Pure Aloha*. Random access schemes lend themselves well to short and bursty traffic. In this case, the inbound channel is shared by several or many VSATs. This is really a contention scheme. When a VSAT has traffic, it bursts the traffic on the inbound channel, taking a chance that another VSAT is not transmitting at the same time. If there is a collision—in other words, another VSAT is transmitting traffic at the same time—the traffic is corrupted, and both VSATs must try again. Each VSAT has a random backoff algorithm. In theory, the backoff time will be different for each terminal, and the second attempt will be successful. A VSAT knows if an attempt is successful because it will receive an acknowledgment from its associated hub on the TDM outbound channel.

This scheme works out well when traffic volume is small. Delay is normally short because most transmissions only require one exchange with the hub.

As traffic volume picks up, the system becomes more and more unwieldy. The point where this occurs, according to Ref. 4, is when throughput approaches 25–30%.† As we increase loading above the 30% value for throughput, the probability of collision increases, as do transaction delays. When the traffic volume exceeds a further limiting value (some argue 50%), the system will tend to "crash," meaning that throughput begins to approach zero because of nearly continuous collisions, backoffs, and reattempts. Flow control mechanisms can help prevent this from happening.

The type of access described here is called pure Aloha. The principal advantage of pure Aloha is its simplicity. There is no time synchronization required, and the hub and VSATs do not need precise timing control (Refs. 3–6).

8.4.1.2 Slotted Aloha. Slotted Aloha is more complex than pure Aloha in that it requires time synchronization among VSATs. In this case users can transmit only in discrete timeslots. With such a scheme, two (or more) users can collide with each other only if they start transmitting exactly at the same time. One disadvantage of slotted Aloha is the wasted periods of time when a message or packet does not use up the total timeslot allowed. Slotted Aloha has about twice the efficiency of pure Aloha or about 34% throughput versus 18% for pure Aloha. These percentage values are points where throughput begins to level off or decrease because collisions begin to increase (Refs. 3 and 7).

*Aloha derives from Hawaii. The University of Hawaii developed a random access technique for a digital data radio system connecting island campuses.

†The CCIR (Ref. 3) places this point at an 18% throughput value. As further traffic is encountered, actual throughput drops.

8.4.1.3 Selective Reject (SREJ) Aloha. With SREJ Aloha, messages or packets are broken down into subpackets. These subpackets have fixed length and can be received independently. Each subpacket has its own acquisition preamble and header. Generally, there is no total collision. Some subpackets may collide, but not whole messages; some of the message or main packet gets through successfully. In SREJ Aloha, only the subpackets that are corrupted are retransmitted. This reduces retransmission because the only retransmission required is a smaller subpacket, not the whole message. SREJ Aloha does not need time synchronization for messages or subpackets. It can achieve higher throughput than pure Aloha and is well suited for variable-length messages (Refs. 3 and 4).

8.4.1.4 R-Aloha or Aloha with Capacity Reservation. This is still another version of Aloha. It is useful when there are a few high-traffic-intensity users and other low-intensity, sporadic users. The high-intensity users have reserved slots and the remainder of the slots are for low-intensity users. This latter group operates on a contention basis similar to pure Aloha. There are many variants of this reservation system (Ref. 6).

One efficient derivative is called slot reservation or demand assignment TDMA (DA/TDMA). When a VSAT has data packets to be transmitted, a request is sent to the hub, which then replies with TDMA slot assignments. The request specifies the length and number of data packets to be sent. Upon receipt of the slot assignment, the VSAT transmits the data packets in the assigned slots without any risk of collision because the hub informs all participating VSATs that particular slots have been reserved. Although requiring more response time since a delay time of two satellite round trips is spent before the actual packets are transmitted, the reservation mode is very effective for the case of longer messages. It is pointed out that this reservation scheme is different from conventional demand assignment multiple access (DAMA) schemes in that the reservations are made on a packet-by-packet (or for a group of packets) basis and not for a continuous channel.

The reservation request can be made on a dedicated request channel or on a traffic channel, which operates on a random access (pure Aloha) scheme. Of course, there is the added overhead for reservation requests. Figure 8.3 is a conceptual drawing of reservation mode transmission. Figure 8.4 compares throughput versus delay performance for three Aloha access schemes (Ref. 3).

8.4.2 Demand-Assigned Multiple Access

DAMA is a satellite access method based on the concept of a pool of traffic channels that can be assigned on demand. When a VSAT user has traffic, the hub is petitioned for a channel. If a channel is available, the hub assigns the channel to the VSAT, which then proceeds to transmit its traffic. When the traffic transaction is completed, the channel is turned back into the pool of available channels.

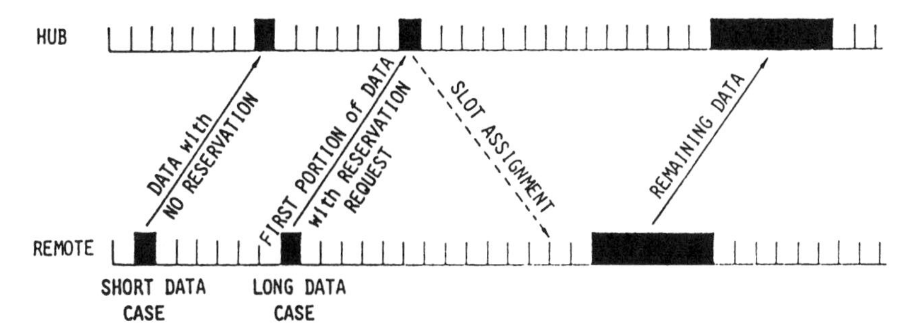

Figure 8.3. Reservation mode transmission.

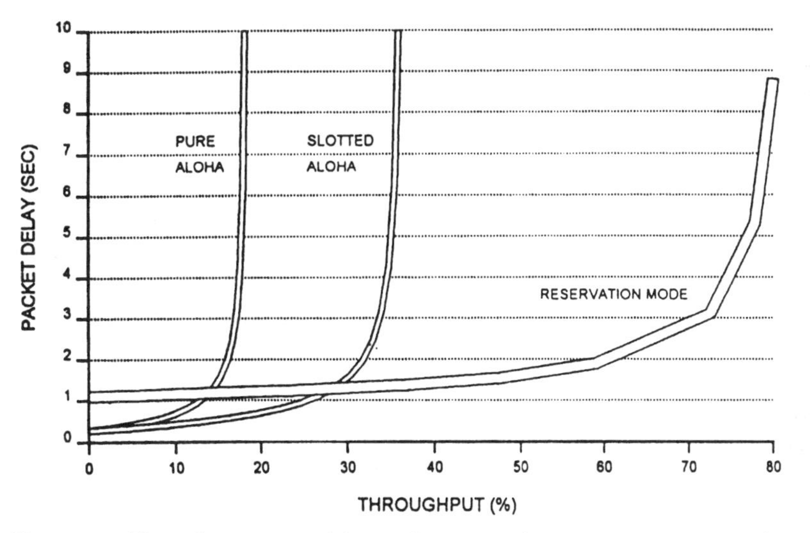

Figure 8.4. Throughput versus delay performance of some Aloha schemes. (From ITU-R Supplement 3 to *Handbook of Satellite Communications*; Ref. 3.)

These assigned channels can be on a FDMA basis or on a TDMA basis. In other words, the pool of channels may consist of a group of frequency slots or a group of timeslots. If the frequency domain is employed, it is called FDMA/SCPC, where SCPC stands for single channel per carrier. DAMA VSAT systems are attractive for voice operation in mesh networks or for voice connectivities of VSAT to hub. Of course, they are also useful when there is intense data traffic that is more or less continuous (Refs. 3 and 6).

8.4.3 Fixed-Assigned FDMA

When a nearly continuous traffic flow is expected from a VSAT to a hub, SCPC operation may be an attractive alternative. In this case, each VSAT is assigned a frequency slot on a full period basis. The bandwidth of the slot

TABLE 8.2 Comparison of Multiple-Access Techniques—Inbound

Multiple-Access Technique	Maximum Throughput	Practical Message Delay	Suitable Application
Random access			
Pure Aloha	18.4%	< 0.5 s	Interactive data
Slotted Aloha	36.8%	< 0.5 s	Interactive data
SREJ Aloha	20–30%	< 0.5 s	Interactive data
Slot reservation	60–90%	< 2 s	Batch data
Demand assigned			
FDMA, TDMA	High	0.25 s	Batch data, voice
Fixed assigned			
FDMA, TDMA	High	0.25 s	Multiplexed data, voice

Source: Table 3.1.1, ITU-R Supplement No. 3 to *Handbook on Satellite Communication* (Ref. 3).

should be sufficient to accommodate the traffic flow. Another alternative is TDMA, where a timeslot is assigned full period for the connectivity.

8.4.4 Summary

Table 8.2 presents a comparison of the several access techniques covered in Section 8.4. The choice of which technique to adopt depends on the traffic and delay requirements of the proposed VSAT system. System complexity may also be an issue: complexity not only can be costly but may also impact reliability.

8.4.5 Outbound TDM Channel

Besides a vehicle for outbound traffic, the outbound TDM channel may have other functions. Among these functions are the following:

- Provide timing to slave VSAT clocks, typically for slotted Aloha
- Channel assignment, typically for DAMA schemes, reservation Aloha
- Acknowledgment of incoming packets
- Other control functions

The TDM channel sends a continuous series of frames where data packets are inserted in the frame's information or data field. In many cases, as we pointed out earlier, the frames are formatted following the HDLC* link layer protocol (see Ref. 8 and Chapter 5). The HDLC control field often is modified to carry out VSAT control functions; the acknowledgment technique can also be patterned after HDLC. If there is no outbound message traffic to be sent, the info field can be filled with null data or supervisory

*HDLC = high-level data link control, a link layer protocol developed by the ISO. ADCCP, LAPB, and LAPD are direct derivatives of HDLC.

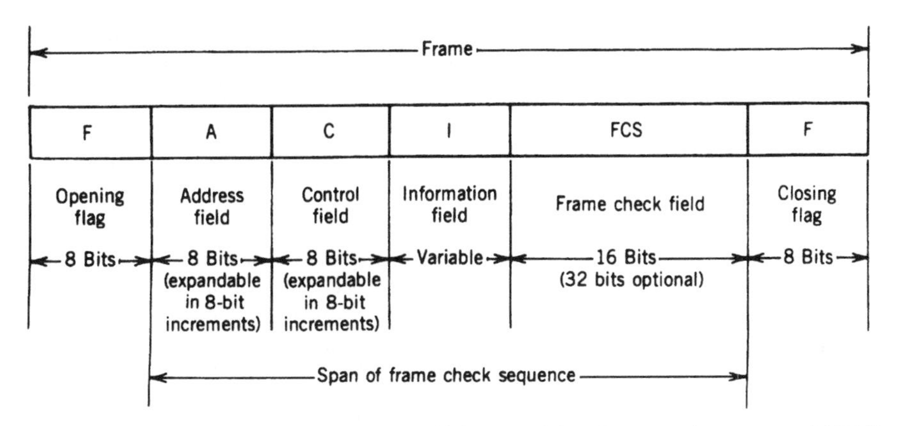

Figure 8.5. The HDLC frame format. It should be noted that there are three types of HDLC frames: information, supervisory, and unnumbered. In the latter two frame types there is no information field, and the other variant is the control field.

frames can be sent. In some implementations, a frame timing and control field, including preamble and unique word (UW), are used.

A typical HDLC frame is shown in Figure 8.5, unmodified. It should be pointed out that the address field must have sufficient length to accommodate addresses for all VSATs in the system as well as group and broadcast addresses. In some systems, an OSI layer 3 (the network layer) can be employed. Often this is based on ITU-T Rec. X.25 (see Ref. 3).

8.5 A MODEST VSAT NETWORK IN SUPPORT OF SHORT TRANSACTION COMMUNICATIONS

In this exercise, we are called upon to design a modest transaction-type VSAT network where minimal delay and cost are overriding requirements. The system will operate at Ku-band. The network is to support a chain of 50 discount department stores. The hub is located at the company headquarters. The inbound transaction messages are based on a HDLC frame 40 octets long, including overhead. The inbound transmission rate is 9600 bps with a convolutional code rate $\frac{1}{2}$, $K = 7$, 3-bit soft decision receiver with Viterbi decoding providing a coding gain of 5.3 dB (see Figure 4.19 and Table 4.7). The outbound TDM stream is 19.2 kbps with similar coding. QPSK modulation is used in both cases.

Each store can have as many as 16 checkout counters in operation at one time, and the worst case is a transaction per minute for each checkout station. Credit card verification and the actual charge transaction are carried out in one HDLC frame. Thus the worst-case frame rate for a store is 16 frames a minute at 40 × 8 or 320 bits per frame. A transaction duration, then, is 320/9600 second or 33 ms. The time on-line per store is 0.528 second per 60 seconds (i.e., 16 × 33 ms). Assuming all 50 stores have such a traffic profile, we then would have 50 × 0.528 for worst-case, peak-hour traffic intensity. This would be 26.4 seconds per minute. This value is well in excess of the 18% allowed under pure Aloha.

Three possibilities arise. (1) Use slotted Aloha with the increased cost and complexity involved. (2) Have three inbound channels, dividing the traffic up to place ourselves inside the 18% requirement of pure Aloha. Thus, on each channel, there would only be the traffic from 17 stores or 9.061 seconds per minute or around 15%. (3) Use a higher inbound bit rate, for example, 32 kbps. The decision is made by trading off recurring space segment charges based on bandwidth versus first cost of the complexity of slotted Aloha.

Slotted Aloha is selected. The outbound TDM bit stream has sufficient capacity at 19.2 kbps because the traffic predominantly is VSAT-to-hub or inbound.

Bandwidths are calculated using 1.5 Nyquist cosine rolloff. The symbol rate is twice the bit rate ($\frac{1}{2}$ rate coding). Thus the inbound channel is calculated at 10 × 2 × 1.5 or 30 kHz and the outbound channel is calculated at 20 × 2 × 1.5 or 60 kHz; both use coherent QPSK modulation.

The satellite spot beam used for the system has a +45-dBW EIRP spread uniformly over 72-MHz transponder bandwidth. The EIRP of the outbound satellite downlink carrier is +14.21 dBW. The inbound downlink carrier has +11.18-dBW EIRP. The BER, both inbound and outbound, for clear-key conditions, is 1×10^{-9}; based on QPSK modulation with coherent detection and FEC coding, the required E_b/N_0 is 8.5 dB, including 2-dB modulation implementation loss. (*Note*: There is a 5.3-dB coding gain discussed above.) As 5-dB rainfall and interference margin is required.

Elevation angle in both cases is 20°. Range to the geostationary satellite is 21,201 nm or 24,397 sm.

Frequency assignments versus functions and equivalent free-space losses are shown in Table 8.3.

For inbound uplink, the VSAT transmitter power output is assumed to be 1 watt (0 dBW). In the case of the outbound uplink (i.e., at the hub) transmitter power output is assumed also to be 1 watt. Satellite G/T in both cases is +1 dB/K.

TABLE 8.3 Frequency Assignments and Free-Space Losses

Function	Frequency	Free-Space Loss
Inbound uplink	14,400 MHz	207.50 dB
Inbound downlink	12,100 MHz	205.98 dB
Outbound uplink	14,100 MHz	207.31 dB
Outbound downlink	11,800 MHz	205.77 dB

Inbound Uplink	
Transmitter output	0 dBW
Transmission line loss	−1.0 dB
Antenna gain ($\eta = 0.65$, 6.5 ft)	+47.64 dB
EIRP	+46.64 dBW
Free-space loss	−207.50 dB
Polarization loss	−0.5 dB
Satellite pointing loss (off contour)	0.0 dB
Terminal pointing loss	−0.5 dB
Atmospheric loss	−0.3 dB
Isotropic receive level	−162.16 dBW
Satellite G/T	+1.0 dB/K
Sum	−161.16 dBW
Boltzmann's constant	−(−228.6 dBW/Hz)
C/N_0	67.44 dB
Inbound Downlink	
Satellite EIRP	+11.18 dBW
Free-space loss	−205.98 dB
Polarization loss	−0.5 dB
Satellite pointing loss (off contour)	0.0 dB
Terminal pointing loss (hub)	−0.5 dB
Atmospheric loss	−0.3 dB
Isotropic receive level	−196.1 dBW
Terminal G/T (hub, 10 ft)	+25.84 dB/K
Sum	−170.26 dBW/Hz
Boltzmann's constant	−(−228.6 dBW/Hz)
C/N_0	58.34 dB

The objective $C/N_{0(t)}$ is calculated as follows, based on an E_b/N_0 of 8.5 dB. The T_{sys} at the hub is 200 K. Thus $N_0 = -228.6 \text{ dBW} + 10\log 200 = -205.59$ dBW. E_b must be 8.5 dB above this level or −197.09 dBW. $C = \text{RSL} = -197.09 \text{ dBW} + 10\log 9600 = -157.24$ dBW. Or the objective $C/N_{0(t)}$ for the inbound links should be

$$C/N_{0(t)} = -157.24 \text{ dBW} - (-205.59 \text{ dBW}) = 48.35 \text{ dB}$$

To this C/N_0 value we add the required 5-dB rainfall and interference margin for a total of 53.35 dB.

Using equation (6.32), we next calculate the net $C/N_{0(t)}$. Calculate the numeric equivalents of each C/N_0 value:

Uplink: 67.4 dB = 5,495,408.7

Downlink: 58.34 dB = 682,338.7

$C/N_{0(t)} = 57.83$ dB

The next step is to calculate the outbound link budgets.

Outbound Uplink	
Transmit power (1 W)	0 dBW
Transmission line loss	−2.0 dB
Antenna gain (hub, 10.0 ft)	+51.19 dB
EIRP hub	+49.19 dBW
Free-space loss	−207.31 dB
Polarization loss	−0.5 dB
Satellite pointing loss (off contour)	0.0
Terminal pointing loss	−0.5 dB
Atmospheric loss	−0.5 dB
Isotropic receive level	−159.62 dBW
Satellite G/T	+1.0 dB/K
Sum	−158.62 dBW
Boltzmann's constant	−(−228.6 dBW)
C/N_0	69.98 dB
Outbound Downlink	
Satellite EIRP	+14.21 dBW
Free-space loss	−205.77 dB
Polarization loss	−0.5 dB
Satellite pointing loss (off contour)	0.0
Terminal pointing loss	−0.5 dB
Atmospheric loss	−0.3 dB
Isotropic receive level	−192.86 dBW
Terminal G/T (hub, 6.5 ft)	+21.91 dB/K
Sum	−170.95 dBW
Boltzmann's constant	−(−228.6 dBW/Hz)
C/N_0	57.65 dB

The objective C/N_0 for the outbound links is calculated as follows. The desired E_b/N_0 is 8.5 dB. T_{sys} of the VSAT receiving system is 206 K; thus N_0 for that system is $-228.6 + 10\log 206 = -205.46$ dBW. E_b must be 8.5 dB higher in level or -205.46 dBW + 8.5 dB = -196.96 dBW. C = RSL = -196.96 dBW + $10\log(19.2 \times 10^3) = -154.12$ dBW. The objective value for $C/N_{0(t)}$ is -154.12 dBW − (-205.46 dBW) = 51.33 dB. To this value we must add 5 dB of rainfall and interference margin for the final objective value of 56.33 dB.

Now we calculate the outbound $C/N_{0(t)}$ from the link budgets. First we must derive the numeric equivalents of the uplink and downlink C/N_0 decibel values:

Uplink: 69.98 dB = 9,954,054

Downlink: 57.65 dB = 582,105

$C/N_{0(t)} = 57.40$ dB

This last value is just inside the objective value of 56.33 dB. It will be noted that the hub is comparatively small; thus it probably will not require any form of automatic tracking. The inbound uplink appears overdimensioned, but the G/T of the companion downlink to the VSAT dictates an antenna diameter that the VSAT uplink must use. Likewise, the outbound uplink appears overdimensioned, but the inbound downlink requires a G/T value that needs a 10-ft antenna. It would be advisable to equip the outbound uplink with a 10-watt HPA and use the decreased output power (i.e., 1 watt). This would provide an additional 10 dB of margin on that uplink.

Any further reduction in link parameters would degrade performance below that specified at the outset.

8.6 INTERFERENCE ISSUES WITH VSATs

VSATs by definition have small antennas. As a result, they have comparatively wide beamwidths. For aperture antennas, beamwidth is related to gain. Jasik and Johnson (Ref. 9) provide the following relationship for estimating beamwidth:

$$BW_{3\text{-dB}} = 70°(\lambda/D) \tag{8.2}$$

where λ is the wavelength and D is the diameter of the antenna. Both D and λ must be expressed in the same units.

If we use the downlink Ku-band frequency of 12,000 MHz, its equivalent wavelength is 0.025 m. If we use a 1-m dish, the beamwidth will be 1.75°. This gives rise to one interference problem in installations with such small antennas. Satellites in geostationary orbit are now placed 2° apart. A beamwidth of 1.75° with a VSAT antenna pointed to one satellite will be prone to interference on the downlink from a neighboring satellite, just 2° away in the orbital equatorial plane.

For the 12-GHz frequency (i.e., $\lambda = 0.025$ m), we develop Table 8.4 for various diameters of parabolic dish antennas that may be employed in a VSAT installation. The table is based on formula (8.2).

TABLE 8.4 Antenna Beamwidths for Various Diameter Dishes

Antenna Diameter (m)	Antenna Diameter (ft)	Beamwidth (°)
0.5	1.625	3.5
0.75	2.44	2.33
1.0	3.25	1.75
1.5	4.875	1.166
2.0	6.5	0.875
2.5	8.125	0.70
5.0	16.25	0.35

TABLE 8.5 Examples of Single Entry Interference Protection Ratios for Typical Satellite Carrier Services

Fixed Satellite Service Carrier Type	Single Entry Interference Protection ratio[a]
A Frequency modulated television (FM/TV)	
Studio quality	$C/I = 28$ dB
Good quality	$C/I = 22$ dB
B Digital data channels	
Wideband, full transponder bandwidth	$E_b/I_0 = 25$ dB
Narrowband, SCPC, T1 (1.544 Mbps)	$E_b/I_0 = 20$ dB
Narrowband, SCPC (56 kbps)	$E_b/I_0 = 20$ dB
C Spread spectrum channels	$E_b/I_0 = 20$ dB
D Frequency modulated SCPC, voice interference contribution	1000 picowatts maximum in worst baseband channel
E Frequency modulated SCPC, audio program	$C/I = 24$ dB

[a] E_b/I_0 refers to the ratio of signal energy per bit per hertz of interference; C/I refers to the ratio of carrier power to interference power.

Source: Table 3.5.1, ITU-R Supplement 3 to *Handbook on Satellite Communications* (Ref. 3).

From the table we see that the larger the antenna, the narrower the beamwidth and the less the possibility of overlap from one satellite to an adjacent satellite in the geostationary orbit.

The ITU-R organization has established some interference guidelines among various telecommunication services offered by satellite relay. These guidelines are summarized in Table 8.5 based on carrier-to-interference ratios (C/I). For digital systems it is more convenient to use E_b/I_0 or energy per bit to interference spectral density ratio. The C/I and E_b/I_0 values in the table apply to the combined uplink and downlink values of the link budget process. Ku-band antenna discrimination values in decibels are given in Figure 8.6.

In a homogeneous arrangement of adjacent satellite systems providing narrowband digital services in which carrier power flux densities are approximately the same (Figure 8.6 and Table 8.4), to obtain a single entry C/I ratio (or antenna discrimination value) of about 20 dB, the VSAT antenna diameter must exceed 1.2 m for satellite separations of 2° or about 0.8 m for satellite separations of 3°. In this example it is assumed that the VSAT system employs a star network with a hub station of at least 4 m in diameter so that the hub-to-satellite link has at least 30 dB of antenna discrimination and contributes less than 0.5 dB to the overall link C/I. On the other hand, if the system is composed of VSATs only (typically with a mesh network, Figure 8.2), and neither uplink nor downlink is controlling with regard to interference, then the antenna sizes must be larger than the VSATs given in the above example. In this case, if the antennas are of the same size, the antenna discrimination value needed will more likely be on the order of 23

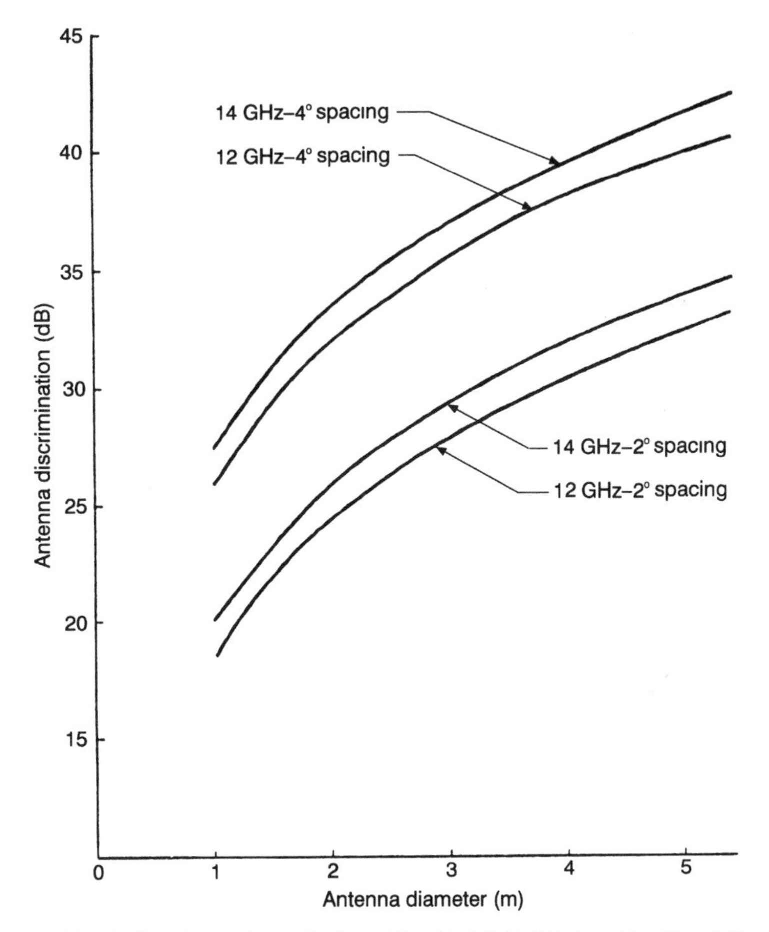

Figure 8.6. Parabolic antenna discrimination at the 14/11-12 GHz band for 2° and 4° satellite orbit spacing. (From Figure 3.5.1, ITU-R Supplement 3 to *Handbook on Satellite Communications*; Ref. 3.)

dB. From Figure 8.6, this results in an antenna diameter requirement of 1.7 m at 2° of satellite spacing or 1.1 m at 3° of satellite spacing.

Besides antenna discrimination, interference from adjacent satellite emissions can be reduced by:

- Using channelization plans in which carrier center frequencies in adjacent satellite systems are offset from each other
- Employing cross-polarization techniques
- Using FEC techniques that can reduce receiver sensitivity to interference

Unfortunately, VSAT systems often are bunched together, because many VSAT applications are found in urban areas. Thus many interference scenarios are set up such as:

- VSAT-to-VSAT
- VSAT-to-large earth station (and vice versa)
- Line-of-sight microwave-to-VSAT and vice versa

8.7 EXCESS ATTENUATION DUE TO RAINFALL

As a general rule, we have said that radio systems operating above 10 GHz must take into account excess attenuation due to rainfall and, during rain events, due to an increase in sky noise. Thus the popular Ku-band must take into account rainfall, whereas, in general, in C-band, rainfall attenuation can be neglected. These topics are dealt with in detail in Chapter 9.

PROBLEMS AND EXERCISES

1 What is the most common type of VSAT network? What other (synonymous) name is used to describe such a network?

2 For common VSAT operation, where is a common (and desirable) place to locate the hub?

3 Give an overriding reason why a hub facility is so much larger than each VSAT it serves.

4 If a VSAT network is to provide voice operation among VSATs, what type of architecture is advisable to use? Why?

5 For typical hub-and-spoke VSAT networks, there is one-way operation and two-way operation. Give at least three applications of one-way application, and give at least five applications for two-way operation VSAT networks.

6 In the case of conventional VSAT networks, the outbound link is nearly always in what type of format?

7 Name at least four possible access methods applicable to the inbound link.

8 Conventional VSATs often have a transmitter with output power in the range (watts or dBW) of what values? What is the range of values for antenna sizes?

9 What factors and parameters basically determine the size of a VSAT terminal?

10 Size a hub and its related VSATs for transmit power, receiver LNA, and antenna apertures for hub and VSAT. Use EIRP and G/T in the analysis. Outbound transmission rate is 56 kbps TDM; inbound bursty at 9600 bps. Apply reasonable BER values. Select the modulation type and FEC coding, if deemed advantageous (argue these advantages or disadvantages). Select the inbound access mode. Assign transponder bandwidth. Operate at Ku-band.

11 Why is Ku-band often more desirable than C-band operation for VSATs? What is the principal disadvantage of Ku-band when compared to C-band?

12 Why can we eliminate increased free-space loss as a factor in question 11? Think! (*Clue*: Increasing frequency affects more than free-space loss.)

13 What is the principal driver in the selection of the inbound access technique? Name two other driving factors in that selection.

14 Describe how pure Aloha operates. When does it become unwieldy and why?

15 Compare pure Aloha to slotted Aloha.

16 Describe how demand assignment TDMA works.

17 What might be a typical data link layer protocol for an outbound channel?

18 What is a principal drawback of VSAT systems considering their small antennas?

19 List three interference scenarios for VSATs operating in urban areas.

20 In large VSAT networks, why must we reduce the cost (in any way possible) of VSATs without unreasonably sacrificing performance of the VSAT network (and possibly spend more on the hub)?

REFERENCES

1. John Everett, ed., *VSATs—Very Small Aperture Earth Stations*, Peter Peregrinus/IEE, London, 1992.
2. Roger L. Freeman, *Telecommunication Transmission Handbook*, 3rd ed., Wiley, New York, 1991.
3. "VSAT Systems and Earth Stations," Supplement No. 3 to *Handbook of Satellite Communications*, ITU-Radio Communications Bureau, Geneva, 1994.
4. Dattakumar M. Chitre and John S. McCoskey, "VSAT Networks: Architectures, Protocols and Management," *IEEE Communications Magazine*, Vol. 26, No. 7, July 1988.

5. Edwin B. Parker and Joseph Rinde, "Transaction Network Applications with User Premises Earth Stations," *IEEE Communications Magazine*, Vol. 26, No. 9, Sept. 1988.
6. D. Raychaudhuri and K. Joseph, "Channel Access Protocols for Ku-Band VSAT Networks: A Comparative Evaluation," *IEEE Communications Magazine*, Vol. 26, No. 5, May 1988.
7. D. Chakraborty, "VSAT Communication Networks: An Overview," *IEEE Communications Magazine*, Vol. 26, No. 5, May 1988.
8. Roger L. Freeman, *Practical Data Communications*, Wiley, New York, 1995.
9. Henry Jasik and Richard C. Johnson, *Antenna Engineering Handbook*, 2nd ed., McGraw-Hill, New York, 1984.

9

RADIO SYSTEM DESIGN ABOVE 10 GHz

9.1 THE PROBLEM—AN INTRODUCTION

There is an ever increasing demand for radio-frequency (RF) spectrum by the nations of the world. The reason is due to the information transfer explosion in our society, resulting in a rapid increase in telecommunication connectivity, and the links satisfying that connectivity are required to have ever greater capacity. This is true even in light of the competition with fiberoptics.

The most desirable spectrum to satisfy the broadband needs is the band between 1 and 10 GHz. There are several reasons for this. For one thing, galactic noise and man-made noise are minimum. Atmospheric absorption and rainfall loss may generally be neglected. Finally, there is a mature technology with competitive pricing of equipment.

On the other side of the coin and for the reasons just cited, the band is highly congested in most countries of the world. Thus we are forced to frequencies above 10 GHz.

The 10-GHz demarcation line was arbitrarily selected. Generally, below 10 GHz, in radio system design we can neglect excess attenuation due to rainfall and atmospheric or gaseous absorption. For frequencies above 10 GHz, excess attenuation due to rainfall and atmospheric gaseous absorption can have overriding importance in the design of radio systems for telecommunications. In fact, certain frequency bands display so much gaseous absorption that they are unusable for many applications.

The principal thrust of this chapter is to describe techniques for RF band selection and link design for line-of-sight (LOS) microwave and earth–space–earth links that will operate on frequencies above 10 GHz. The chapter also has a brief discussion on certain limitations imposed on the system designer as we go higher in frequency, such as antenna gain versus beamwidth and free-space loss increase versus antenna gain increase.

9.2 THE GENERAL PROPAGATION PROBLEM ABOVE 10 GHz

Propagation of radio waves through the atmosphere above 10 GHz involves not only free-space loss but several other important factors. As expressed in Ref. 1, these are as follows:

1. The gaseous contribution of the homogeneous atmosphere due to resonant and nonresonant polarization mechanisms.
2. The contribution of inhomogeneities in the atmosphere.
3. The particulate contributions due to rain, fog, mist, and haze (includes dust, smoke, and salt particles in the air).

Under (1) we are dealing with the propagation of a wave through the atmosphere under the influence of several molecular resonances, such as water vapor (H_2O) at 22 and 183 GHz, oxygen (O_2) with lines around 60 GHz, and a single oxygen line at 119 GHz. These points and their relative attenuation are shown in Figure 9.1.

Other gases display resonant lines as well, such as N_2O, SO_2, O_3, NO_2, and NH_3, but because of their low density in the atmosphere, they have negligible effect on propagation.

The major offender is precipitation attenuation [under (2) and (3)]: it can exceed that of all other sources of excess attenuation above 18 GHz. Rainfall and its effect on propagation are the major themes of this chapter.

To better understand these loss mechanisms due to the atmosphere, including rainfall, we very briefly review absorption and scattering. It will be appreciated that when an incident electromagnetic wave passes over an object that has dielectric properties different from the surrounding medium, some energy is absorbed and some is scattered. That which is absorbed heats the absorbing material; that which is scattered is quasi-isotropic and relates to the wavelength of the incident wave. The smaller the scatterer, the more isotropic it is in direction with respect to the wavelength of the incident energy.

We can develop a formula derived from equation (1.9a) to calculate total transmission loss for a given link:

$$\text{Attenuation}_{\text{dB}} = 92.45 + 20 \log F_{\text{GHz}} + 20 \log D_{\text{km}} + a + b + c + d + e \tag{9.1}$$

where F = operating frequency in GHz
D = link length in kilometers
a = excess attenuation* (dB) due to water vapor
b = excess attenuation (dB) due to mist and fog
c = excess attenuation (dB) due to oxygen (O_2)

*"Excess" in this context means in excess of free-space loss.

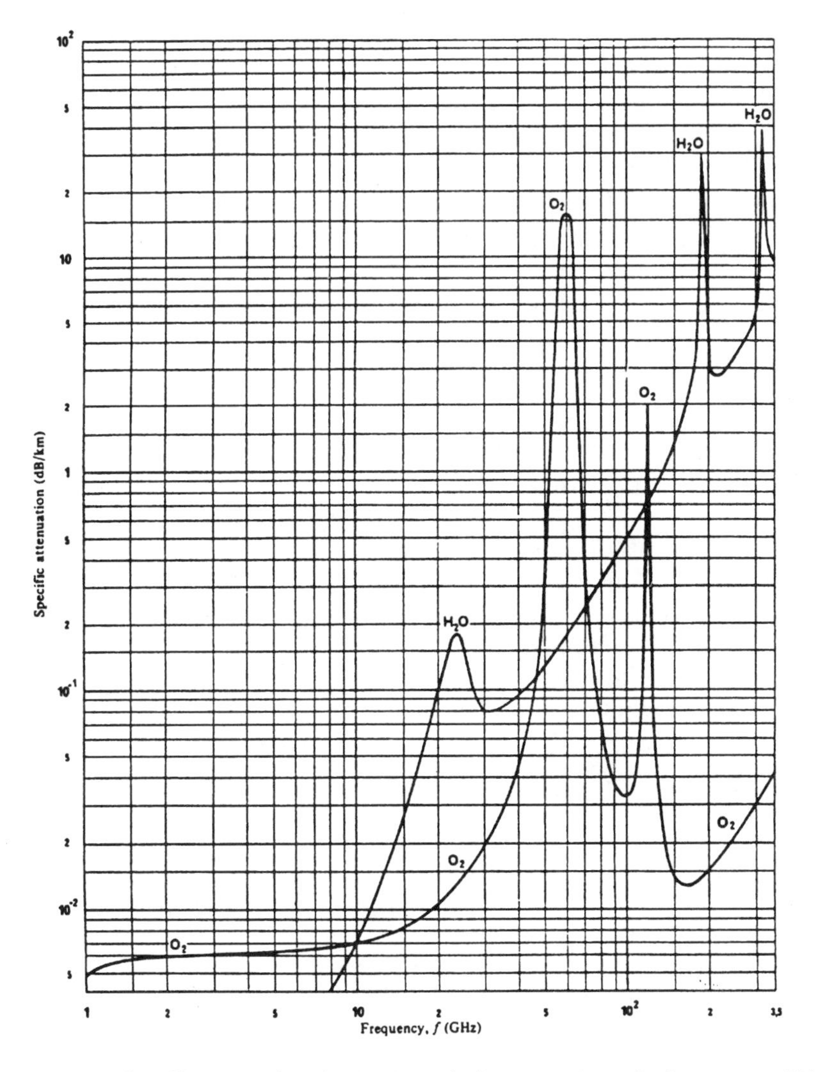

Figure 9.1. Specific attenuation due to atmospheric gases: atmospheric pressure, 1013 mb; temperature, 15°C; water vapor, 7.5 g/m^3. (From Figure 3, p. 194, CCIR Rep. 719-3, Vol. V Annex, Reports of the CCIR 1990; Ref. 2.)

d = sum of the absorption losses (dB) due to other gases
e = excess attenuation (dB) due to rainfall

Notes and Comments on Equation (9.1)

1. Parameter a varies with relative humidity, temperature, atmospheric pressure, and altitude. The transmission engineer assumes that the water vapor content is linear with these parameters and that the

atmosphere is homogeneous (actually horizontally homogeneous but vertically stratified). There is a water vapor absorption band around 22 GHz caused by molecular resonance.

2. Parameters *c* and *d* are assumed to vary linearly with atmospheric density, thus directly with atmospheric pressure, and are a function of altitude (e.g., it is assumed that the atmosphere is homogeneous).
3. Parameters *b* and *e* vary with the density of the rainfall cell or cloud and the size of the rainfall drops or water particles such as fog or mist. In this case the atmosphere is most certainly not homogeneous. (Droplets smaller than 0.01 cm in diameter are considered mist/fog, while those larger than 0.01 cm are rain.) Ordinary fog produces about 0.1 dB/km of excess attenuation at 35 GHz, rising to 0.6 dB/km at 75 GHz.

In equation (9.1) terms *b* and *d* can often be neglected; terms *a* and *c* are usually lumped together and called "excess attenuation due to atmospheric gases" or simply "atmospheric attenuation." If we were to install a 10-km LOS link at 22 GHz, in calculating transmission loss (at sea level), about 1.6 dB would have to be added for what is called atmospheric attenuation, but which is predominantly water vapor absorption, as shown in Figure 9.1.

It will be noted in Figure 9.1 that there are frequency bands with relatively high levels of attenuation per unit distance. Some of these frequency bands are rather narrow (e.g., H_2O at about 22 GHz). Others are wide, such as the O_2 absorption band, which covers from about 58 to 62 GHz with skirts down to 50 and up to 70 GHz. At its peak around 60 GHz, the sea-level attenuation is about 15 dB/km. One could ask: "Of what use are these high-absorption bands?" Actually, the 58–62-GHz band is appropriately assigned for satellite cross-links. These links operate out in space far above the limits of the earth's atmosphere, where the terms *a* through *e* may be completely neglected. It is particularly attractive on military cross-links having an inherent protection from earth-based enemy jammers by that significant atmospheric attenuation factor. It is also useful for very short-haul military links such as ship-to-ship secure communication systems. Again, it is the severe atmospheric attenuation that offers some additional security for signal intercept (LPI—low probability of intercept) and against jamming.

On the other hand, Figure 9.1 shows some radio-frequency bands that are relatively open. These openings are often called *windows*. Three such windows are suggested for point-to-point service in Table 9.1. The absorption values given in the table are nominal values at sea level.

9.3 EXCESS ATTENUATION DUE TO RAINFALL

Of the factors *a* through *e* in equation (9.1), factor *e*, excess attenuation due to rainfall, is the principal one affecting path loss. For instance, even at 22

TABLE 9.1 Windows for Point-to-Point Service

Band (GHz)	Excess Attenuation Due to Atmospheric Absorption (dB/km)
28–42	0.13
75–95	0.4
125–140	1.8

GHz for a 10-km path only 1.6 dB must be added to free-space loss to compensate for water vapor loss. This is negligible when compared to free-space loss itself, such as 119.3 dB for the first kilometer at 22 GHz, accumulating thence approximately 6 dB each time the path length is doubled (i.e., add 6 dB for 2 km, 12 dB for 4 km, etc.). Thus a 10-km path would have a free-space loss of 139.3 dB plus 1.6 dB added for excess attenuation due to water vapor (22 GHz), or a total of 140.95 dB. On the other hand, in certain regions of the world, excess attenuation due to rainfall can exceed 150 dB depending on the time availability desired for a particular path.

The calculation of excess attenuation due to rainfall is another, more complex matter. It has been common practice to express rainfall loss as a function of precipitation rate. Such a rate depends on liquid water content and the fall velocity of the drops. The velocity, in turn, depends on raindrop size. Thus our interest in rainfall boils down to drop size and drop-size distribution for point rainfall rates. All this information is designed to lead the transmission engineer to set an excess attenuation value due to rainfall on a particular path as a function of time and time distribution. This method is similar to the one used in Chapters 2, 3, and 5 for overbuilding a link to accommodate fading. Of course, the end effect on a link due to heavy rainfall is indeed fading.

As earlier approach dealt with rain on a basis of rainfall given in millimeters per hour. Often this was done with rain gauges, using collected rain averaging over a day or even periods of days. For path design above 10 GHz such statistics are not sufficient, where we may require path availability better than 99.9% and do not wish to resort to overconservative design procedures (e.g., assign excessive link margins).

As we mentioned, there is a fallacy in using annual rainfall rates as a basis for calculation of excess attenuation due to rainfall. For instance, several weeks of light drizzle will affect the overall long-term path availability much less than several good downpours that are short lived (i.e., 20-min duration). It is simply this downpour activity for which we need statistics. Such downpours are cellular in nature. How big are the cells? What is the rainfall rate in the cell? What are the size of the drops and their distribution?

Hogg (Ref. 3) suggests the use of high-speed rain gauges with outputs readily available for computer analysis. These gauges can provide minute-by-minute analysis of the rate of fall, something lacking with conventional types of gauges. Of course, it would be desirable to have several years' statistics for a specific path to provide the necessary information on fading caused by rainfall that will govern system parameters such as LOS repeater spacing, antenna size, and diversity.

A large body of information on point rainfall and rain rates has been accumulated since 1940. This information shows a great variation of short-term rainfall rates from one geographical location to another. As an example, in one period of measurement it was found that Miami, Florida, has maximum rain rates about ten times greater than those of heavy showers occurring in Oregon, the region of heaviest annual rain in the United States. In Miami, a point rainfall rate may even exceed 200 mm/h over very short time intervals. The effect of 200 mm/h on 48- and 70-GHz paths can be extrapolated from Figure 9.2. In the figure the rainfall rate in millimeters per hour extends to 100, which at 100 mm/h provides an excess attenuation of 25–30 dB/km.

When identical systems were compared (Ref. 3) at 30 GHz with repeater spacings of 1 km and equal desired signals (e.g., producing a 30-dB signal-to-noise ratio), 140 min of total time below the desired level was obtained at Miami, Florida; 13 min at Coweeta, North Carolina; 4 min at Island Beach, New Jersey; 0.5 min at Bedford, England; and less than 0.5 min at Corvallis, Oregon. Such outages, of course, can be reduced by increasing transmitter output power, improving receiver noise figure (NF), increasing antenna size, implementing a diversity scheme, and so on.

One valid approach to lengthen repeater sections (space between repeaters) and still maintain performance objectives is to use path diversity.

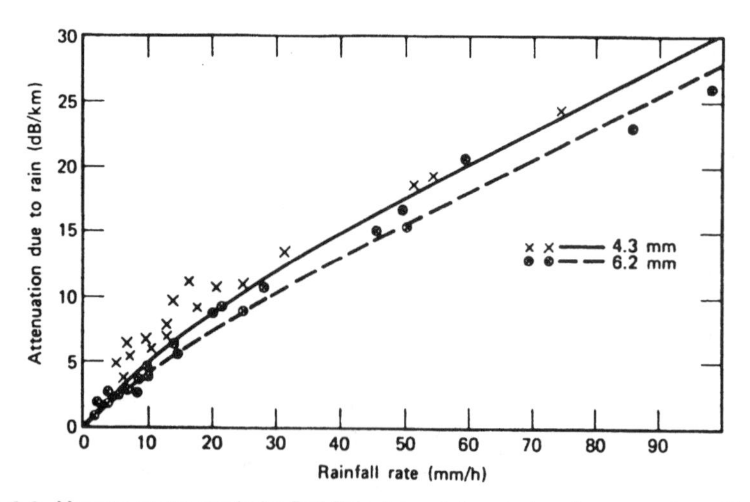

Figure 9.2. Measurements made by Bell Telephone Laboratories of excess attenuation due to rainfall at wavelengths of 6.2 and 4.2 mm (48 and 70 GHz) compared to calculated values. (From Ref. 3, copyright ©1968 AT & T.)

This is the most effective form of diversity for downpour rainfall fading. Path diversity is the simultaneous transmission of the same information on paths separated by at least 10 km, the idea being that rain cells affecting one path will have a low probability of affecting the other at the same time. A switch would select the better path of the two. Careful phase equalization between the two paths would be required, particularly for the transmission of high-bit-rate information.

9.3.1 Calculation of Excess Attenuation Due to Rainfall for LOS Microwave Paths

When designing LOS microwave links (or satellite links) where the operating frequency is above 10 GHz, a major problem in link design is to determine the excess path attenuation due to rainfall. The adjective "excess" is used to denote path attenuation in *excess* of free-space loss [i.e., the terms a, b, c, d, and e in equation (9.1) are losses in excess of free-space loss].

Before treating the methodology of calculation of excess rain attenuation, we will review some general link engineering information dealing with rain. When discussing rainfall here, all measurements are in millimeters per hour of rain and are point rainfall measurements. From our previous discussion we know that heavy downpour rain is the most seriously damaging to radio transmission above 10 GHz. Such rain is cellular in nature and has limited coverage. We must address the question of whether the entire hop is in the storm for the whole period of the storm. Light rainfall (e.g., less than 2 min/h), on the other hand, is usually widespread in character, and the path average is the same as the local value. Heavier rain occurs in convective storm cells that are typically 2–6 km across and are often embedded in larger regions measured in tens of kilometers (Ref. 4). Thus for short hops (2–6 km) the path-averaged rainfall rate will be the same as the local rate, but for longer paths it will be reduced by the ratio of the path length on which it is raining to the total path length.

This concept is further expanded on by CCIR Rep. 593-1 (Ref. 5), where rain cell size is related to rainfall rate as shown in Figure 9.3. The concept of rain cell size is very important, whether engineering a LOS link or a satellite link, particularly when the satellite link has a comparatively low elevation angle. CCIR Rep. 338-3 (Ref. 6) is quoted in part below:

> Measurements in the United Kingdom over a period of two years ... at 11, 20 and 37 GHz on links of 4–22 km in length show that the attenuation due to rain and multipath, which is exceeded for 0.01% of the time and less, increased rapidly with path length up to 10 km, but further increase up to 22 km produced a small additional effect.

The application of specific rain cell size was taken from CCIR 1978. We will call this method the "liberal method" because when reviewing CCIR Vol. V

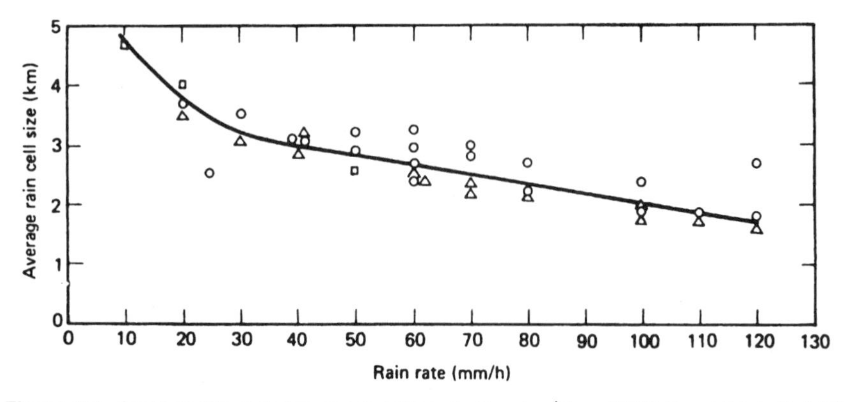

Figure 9.3. Average rain cell size as a function of rain rate. (From CCIR Rep. 593-1; Ref. 5.)

of 1990, CCIR Rep. 563-4 (Ref. 7), we can see that the CCIR has turned more cautious and conservative. Our Figure 9.3 does not appear in the report. Let us partially quote from this report:

> For attenuation predictions the situation is generally more complex (than that of interference scattering by precipitation). Volume cells are known to cluster frequently within small mesoscale areas.... Terrestrial links exceeding 10 km may therefore traverse more than one volume cell within a mesoscale cluster. In addition, since the attenuating influence of the lower intensity rainfall surrounding the cell must be taken into account, any model used to calculate attenuation must take these larger rain regions into account. The linear extent of these regions increases with decreasing rain intensity and may be as large as several tens of kilometers.

One of the most accepted methods of dealing with excess path attenuation A due to rainfall is an empirical procedure based on the approximate relation between A and the rain rate R:

$$A_{\mathrm{dB}} = aR^b \tag{9.2}$$

where a and b are functions of frequency (f) and rain temperature (T).

To calculate the value for A in decibels and then a value of excess attenuation for an entire path, we use a six-step calculation procedure based on ITU-R Rec. PN.530-5 (Ref. 8), paragraph 2.4.1.

Step 1: Derive the values of a and b from Table 9.2, interpolating where necessary for in-between frequency values in the table. Table 9.2 is based on a rain temperature of 20°C and the Laws–Parson drop-size distribution.

It should be noted that horizontally polarized waves suffer greater attenuation than vertically polarized waves because large raindrops are generally shaped as oblate spheroids and are aligned with the vertical rotation axis.

TABLE 9.2 Regression Coefficient Values for Estimating Specific Attenuation in Equation (9.2)[a]

Frequency (GHz)	a_h	a_v	b_h	b_v
1	0.0000387	0.0000352	0.912	0.880
2	0.000154	0.000138	0.963	0.923
4	0.000650	0.000591	1.121	1.075
6	0.00175	0.00155	1.308	1.265
7	0.00301	0.00265	1.332	1.312
8	0.00454	0.00395	1.327	1.310
10	0.0101	0.00887	1.276	1.264
12	0.0188	0.0168	1.217	1.200
15	0.0367	0.0335	1.154	1.128
20	0.0751	0.0691	1.099	1.065
25	0.124	0.113	1.061	1.030
30	0.187	0.167	1.021	1.000
35	0.263	0.233	0.979	0.963
40	0.350	0.310	0.939	0.929
45	0.442	0.393	0.903	0.897
50	0.536	0.479	0.873	0.868
60	0.707	0.642	0.826	0.824
70	0.851	0.784	0.793	0.793
80	0.975	0.906	0.769	0.769
90	1.06	0.999	0.753	0.754
100	1.12	1.06	0.743	0.744
120	1.18	1.13	0.731	0.732
150	1.31	1.27	0.710	0.711
200	1.45	1.42	0.689	0.690
300	1.36	1.35	0.688	0.689
400	1.32	1.31	0.683	0.684

Source: Table 1, p. 243, CCIR Rec. 838, 1994 PN Series Volume (Ref. 9).

Thus we use the subscript notations h and v for horizontal and vertical polarizations, respectively, in Table 9.2 for the values a and b.

Step 2: Obtain a value for the rain rate R (in mm/h).

Ideally, this value should be acquired from local sources such as the local weather bureau. Lacking this facility, an estimate can be obtained by identifying the region of interest from the maps appearing in Figures 9.4, 9.5, and 9.6, then selecting the appropriate rainfall intensity for the specified time percentage from Table 9.3, which gives the 14 rain regions in the maps. Select the value in the table for an exceedance of 0.01% of the time* for point rainfall rates with an integration time of 1 minute. This gives us a value for R in equation (9.2).

*"Exceedance" means a percentage of time that a value is exceeded. An exceedance of 0.01% is synonymous with time unavailability of 0.01% (rain attenuation only in this case) or a time availability of 99.99%.

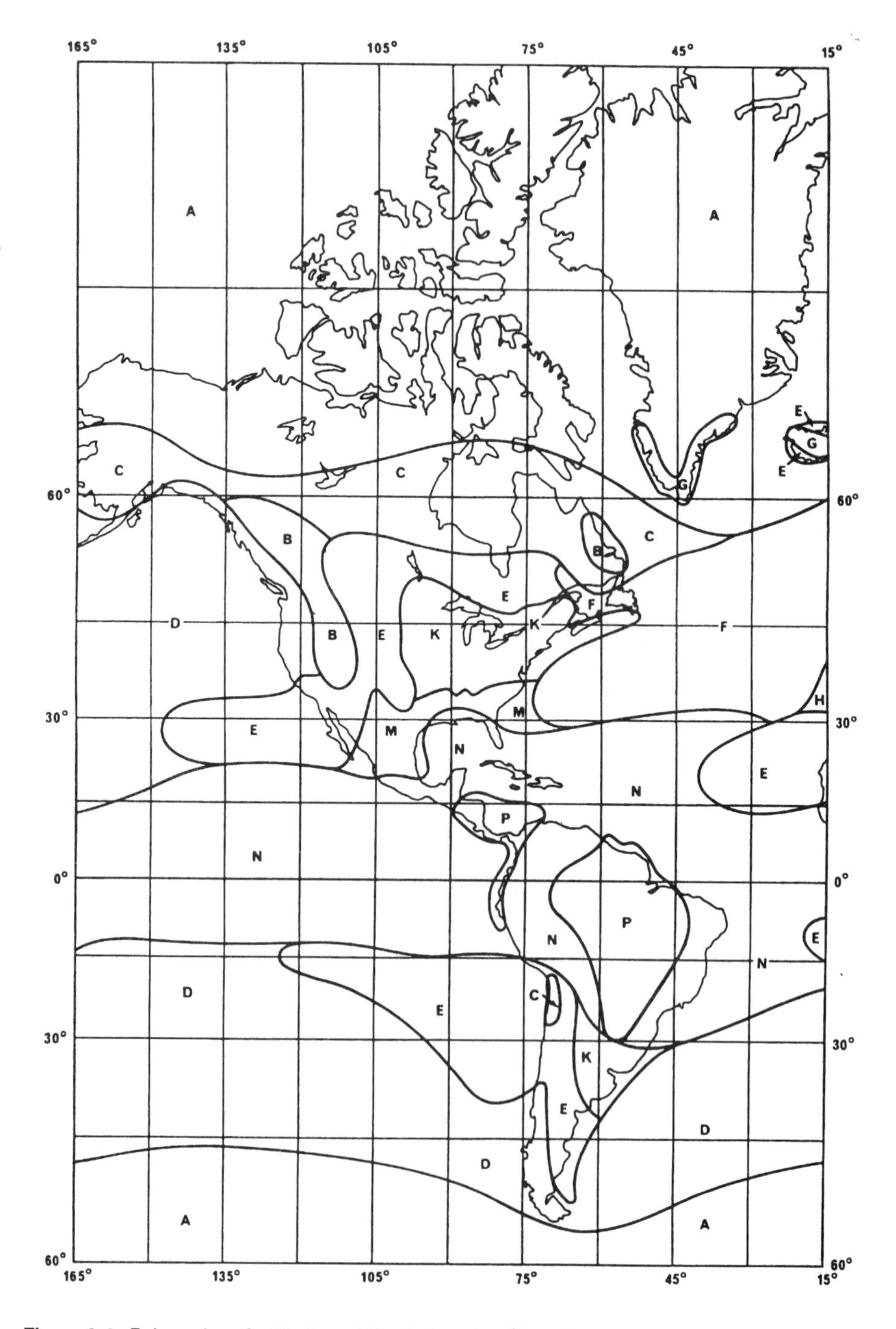

Figure 9.4. Rain regions for North and South America. (From Figure 1, ITU-R Rec. PN.837-1; Ref. 11.)

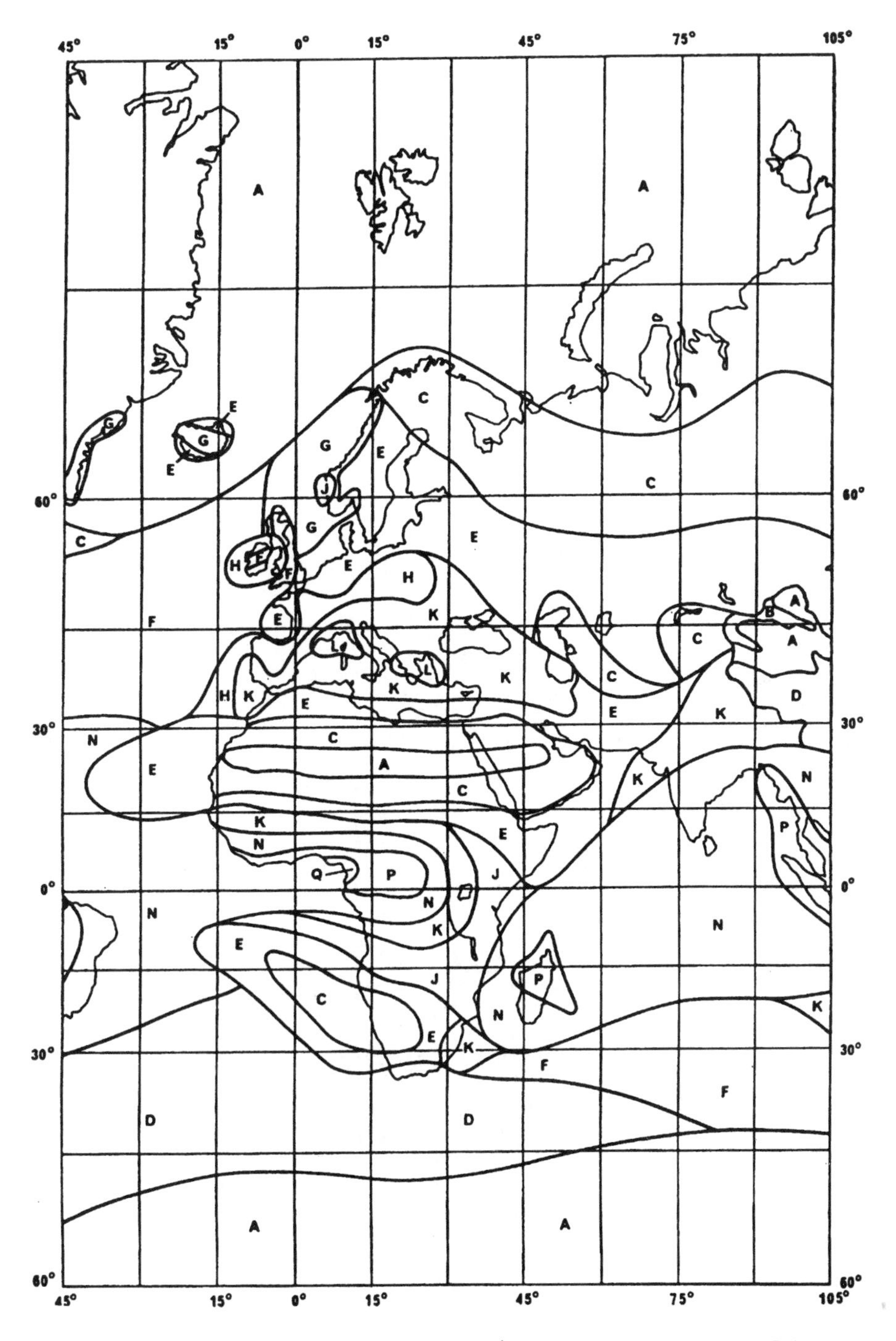

Figure 9.5. Rain regions for Europe and Africa. (From Figure 2, ITU-R Rec. PN.837-1; Ref. 11.)

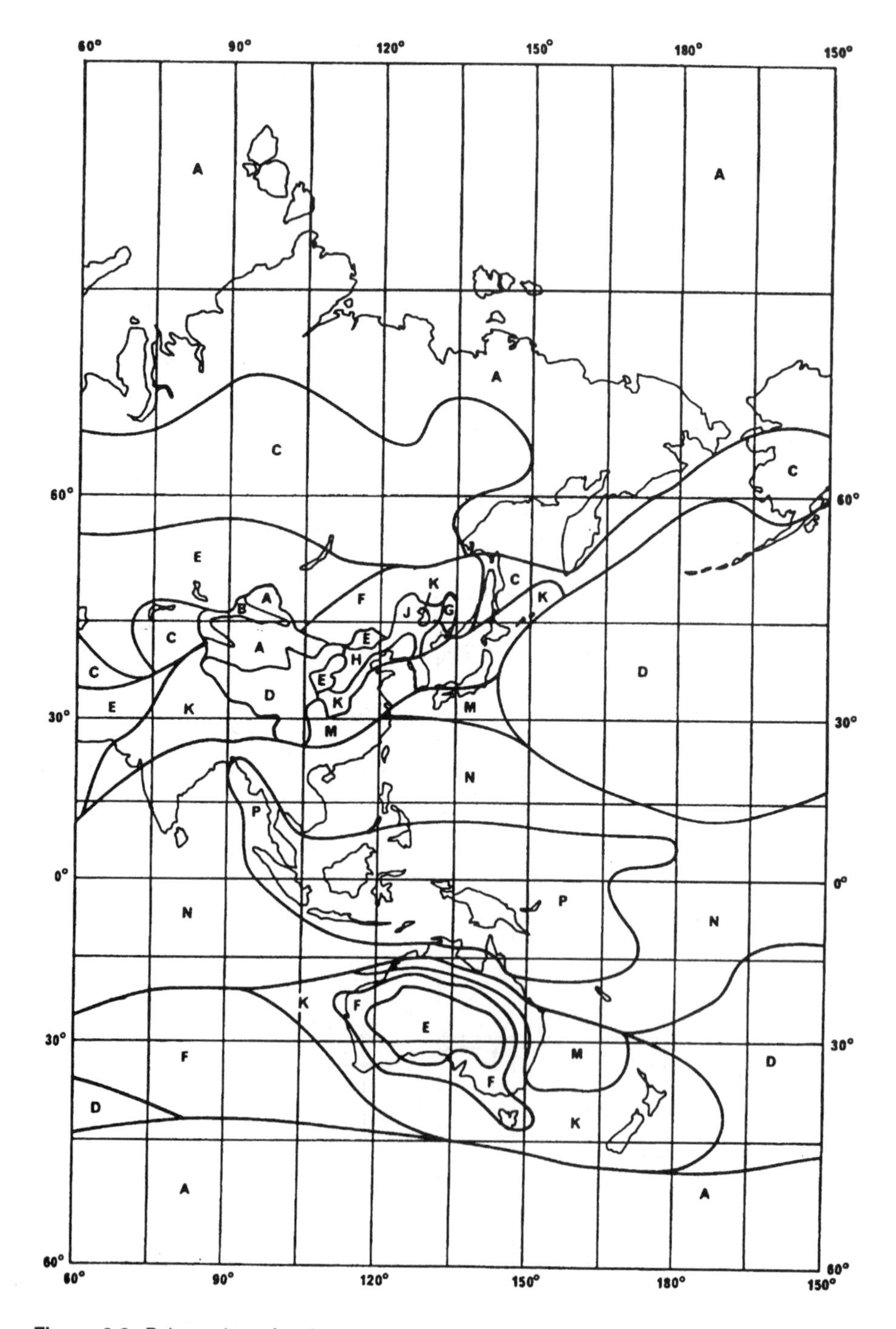

Figure 9.6. Rain regions for Asia, Oceania, and Australia. (From Figure 3, ITU-R Rec. PN.837-1; Ref. 11.)

TABLE 9.3 Rain Climatic Zones—Rainfall Intensity Exceeded (mm/h)
(See Figures 9.4, 9.5, and 9.6 for Rain Regions)

Percentage of Time (%)	A	B	C	D	E	F	G	H	J	K	L	M	N	P	Q
1.0	< 0.1	0.5	0.7	2.1	0.6	1.7	3	2	8	1.5	2	4	5	12	24
0.3	0.8	2	2.8	4.5	2.4	4.5	7	4	13	4.2	7	11	15	34	49
0.1	2	3	5	8	6	8	12	10	20	12	15	22	35	65	72
0.03	5	6	9	13	12	15	20	18	28	23	33	40	65	105	96
0.01	8	12	15	19	22	28	30	32	35	42	60	63	95	145	115
0.003	14	21	26	29	41	54	45	55	45	70	105	95	140	200	142
0.001	22	32	42	42	70	78	65	83	55	100	150	120	180	250	170

Source: Table 1, ITU-R Rec. PN.837-1 (Ref. 11).

Step 3: Calculate A using equation (9.2). This gives us a value of excess attenuation per kilometer.

Figure 9.7 and/or the nomogram in Figure 9.8 may also be used to calculate A.

Step 4: Calculate the effective path length D_{eff}.

This is obtained by multiplying the actual path length (in km) by a reduction factor r. A first estimate to calculate r is given as

$$r = \frac{1}{1 + d/d_0} \tag{9.3}$$

where, for $R_{0.01} \leq 100$ mm/h,

$$d_0 = 35e^{-0.015 R_{0.01}} \tag{9.4}$$

For $R_{0.01} > 100$ mm/h, use the value 100 mm/h in place of $R_{0.01}$.

Step 5: Calculate the total path attenuation exceeded 0.01% of the time using equation (9.5).

$$A_{0.01} = A \times d \times r \tag{9.5}$$

where d is the path length in kilometers and r is the distance factor computed from equation (9.3).

Step 6: Attenuation exceeded for other percentages of time can be calculated from the following power law:

$$A_p/A_{0.01} = 0.12p^{-(0.546+0.043 \log 10p)} \tag{9.6}$$

Reference 8 (PN.530-5) states that the formula gives factors of 0.12, 0.39, 1, and 2.14 for 1%, 0.1%, 0.01%, and 0.001%, respectively, and must be used only within this range.

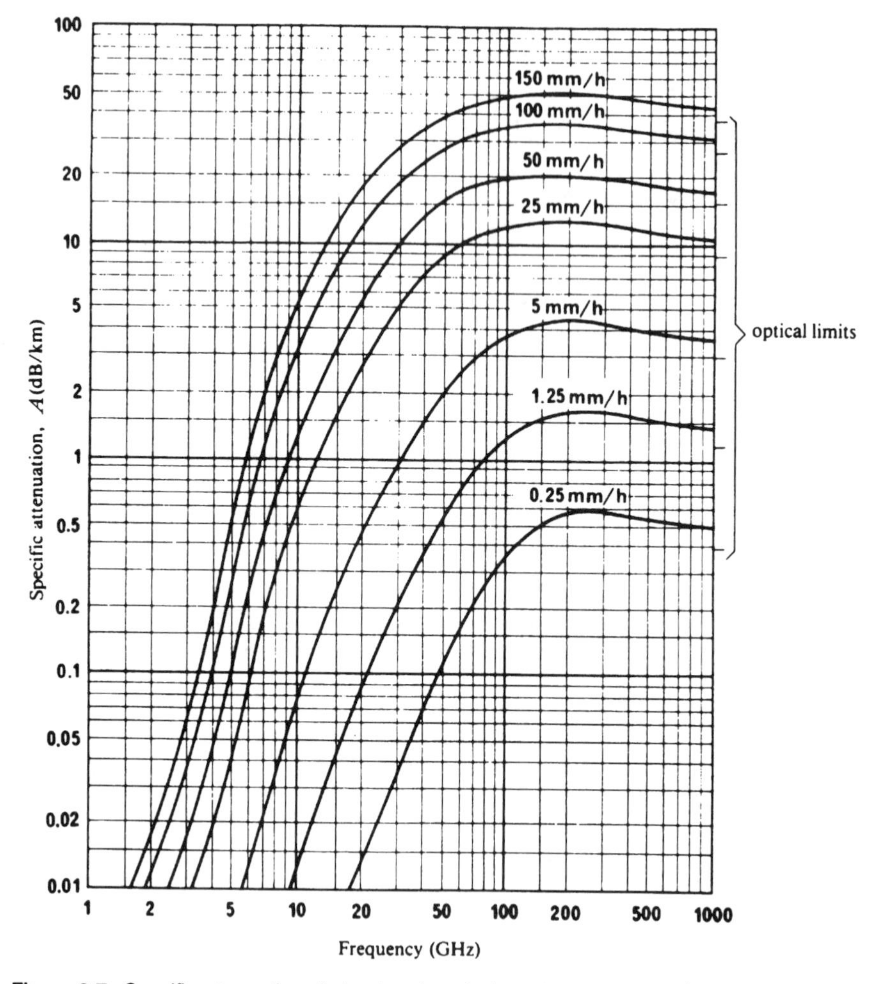

Figure 9.7. Specific attenuation *A* due to rain: raindrop-size distribution (Laws and Parsons); index of refraction of water at 20°C; spherical drops. (From CCIR Rep. 721-3, ITU-CCIR, Geneva; Ref. 12.)

The reference further advises that this procedure is valid for all parts of the world for frequencies up to 40 GHz and paths up to 60 km long.

The derived excess attenuation value is added to the free-space loss plus the atmospheric attenuation loss. We can consider the rainfall loss value as a fade margin. It is not necessary to add the multipath fade margin and the rainfall margin for a total margin value. We recommend using whichever is the higher value. Generally, during heavy rainfall events, the atmosphere becomes well mixed and multipath fading tends to be minimized when

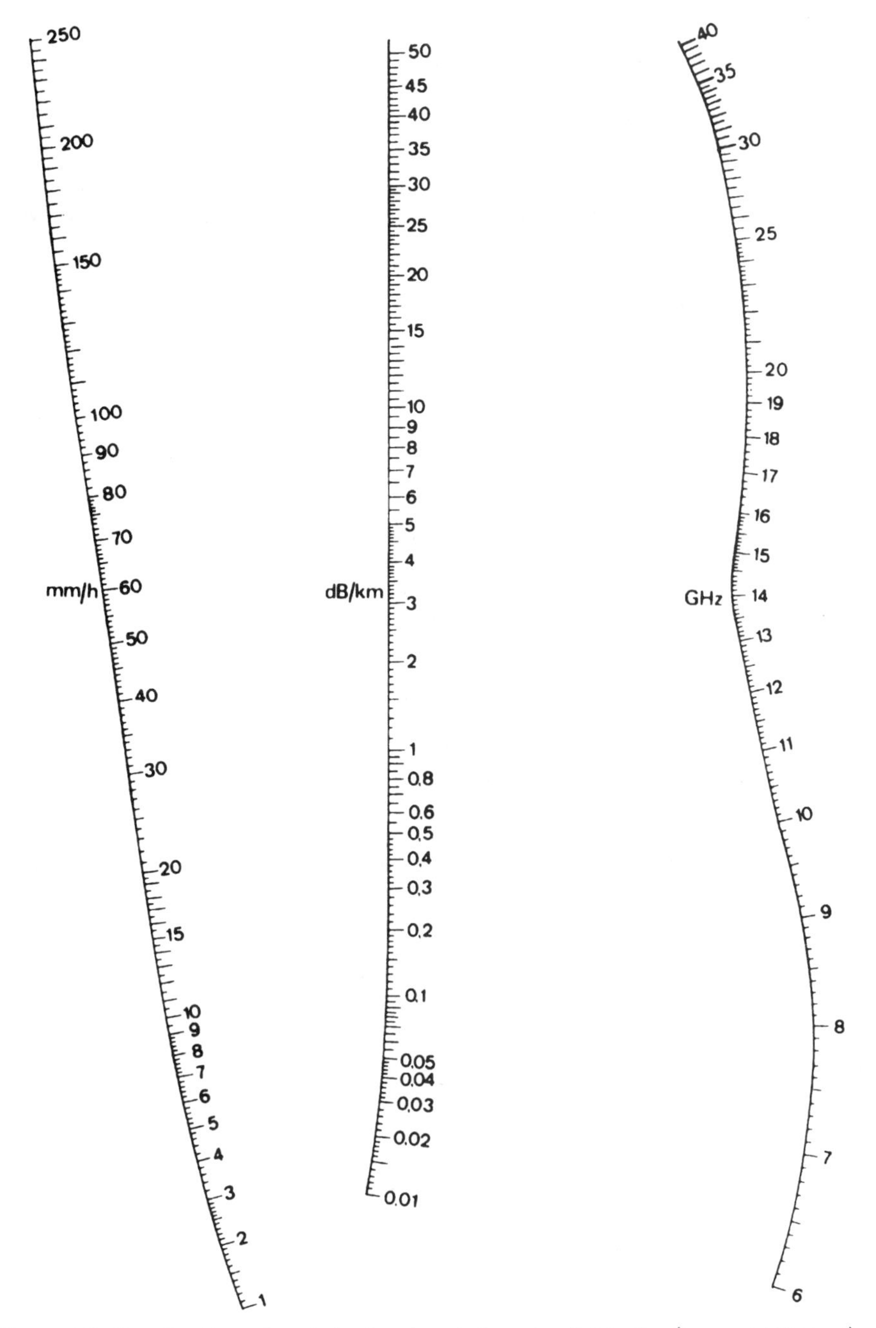

Figure 9.8. Specific attenuation *A* due to rain: raindrop-size distribution (Laws and Parsons); index of refraction of water at 18°C; spherical drops. (From CCIR Rep. 721-3; ITU-CCIR Ref. 12, Geneva.)

rainfall fading is predominant. In other words, rainfall fading and multipath fading are mutually exclusive (Ref. 10).

Example 1. A LOS microwave link 12 km long is planned for installation in Germany. The operational frequency is 38 GHz and the desired time availability is 99.999%. However, other time availabilities should also be presented: 0.1% and 0.01%. Local rain rate statistics are not available. The antenna polarization is horizontal.

Step 1: Calculate a and b based on 38-GHz, horizontal polarization. Use Table 9.2 and interpolate as follows:

35 GHz:	$a = 0.0263$	$b = 0.979$
40 GHz:	$a = 0.350$	$b = 0.939$
Difference:	0.087	0.040
38 GHz:	$a = 0.3152$	$b = 0.955$

Step 2: Obtain a value for R, the rain rate (in mm/h) for an exceedance of 0.01%. Use Table 9.3. First select the climatic region from Figure 9.5. This is determined to be region H. Thus $R = 32$ mm/h.

Step 3: Calculate A (in dB/km) using equation (9.2).

$$A = 0.3152 \times 27.38 \text{ dB/km}$$
$$= 8.63 \text{ dB/km}$$

Step 4: Calculate r, the path reduction factor. Use equations (9.4) and (9.5).

Calculate D_0 first:

$$D_0 = 35e^{-0.48}$$
$$= 21.66$$

Then

$$r = \frac{1}{1 + 12/21.66}$$
$$= 0.643$$

Step 5: Calculate the total path attenuation of an exceedance of 0.01%. Use equation (9.5).

$$A_{0.01} = 8.63 \times 12 \times 0.643$$
$$= 66.64 \text{ dB}$$

Step 6: Calculate for other time unavailabilities.

1%: $0.12 \times 66.64 = 8.0$ dB
0.1%: $0.39 \times 66.64 = 26$ dB
0.01%: 66.64 dB
0.001%: $2.14 \times 66.64 = 142.61$ dB

It is seen that the 0.01% value (99.99% time availability) may be untenable. Certainly the 0.001% (99.999% time availability) appears to be completely untenable. Here we mean that it would be difficult or impossible to achieve a fade margin of this magnitude. The use of site diversity could notably reduce such magnitudes. Site diversity is discussed in Section 9.4.3.

9.4 CALCULATION OF EXCESS ATTENUATION DUE TO RAINFALL FOR SATELLITE PATHS

A satellite path, regarding rainfall and gaseous absorption, differs from a LOS microwave path based on elevation angle. The lower that angle, the more the path approaches that of line-of-sight conditions. It simply passes through more atmosphere as the elevation angle lowers.

9.4.1 Calculation Method

Figure 9.9 is a schematic presentation of an earth–space path that we use as a model for slant-path rain attenuation. From the model we derive the distance L_G in kilometers. Thus the methodology is the same as if it were a LOS microwave path.

The following are the definitions of the parameters of interest:

$R_{0.01}$ = point rainfall rate for the location for 0.01% of an average year (mm/h)
h_s = height above mean sea level (MSL) of the earth station (km)
θ = elevation angle
φ = absolute value of the latitude of the earth station (degrees)
f = frequency (GHz)

There are eight steps in the method.

Step 1: Calculate the effective rain height, h_R, for the latitude of the station φ:

$$h_R \text{ (km)} = 3.0 + 0.028\varphi \qquad \text{for } 0 \leq \varphi < 36° \qquad (9.7a)$$

$$h_R \text{ (km)} = 4.0 - 0.075(\varphi - 36) \qquad \text{for } \varphi \geq 36° \qquad (9.7b)$$

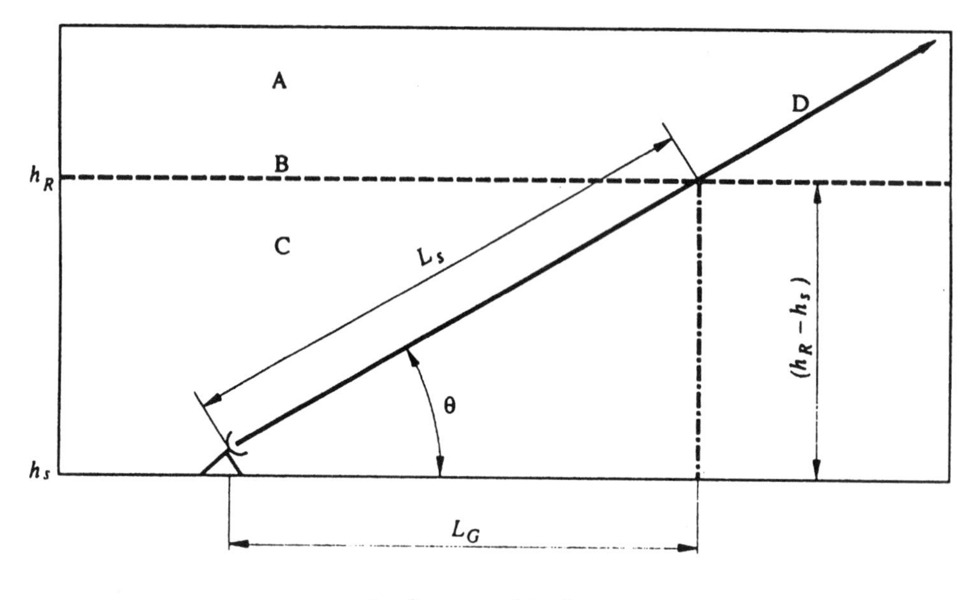

A: frozen precipitation
B: rain height
C: liquid precipitation
D: Earth-space path

Figure 9.9. Schematic presentation of an earth–space path showing the parameters to be used in the calculation method. (From Figure 1, p. 333, ITU-R Rec. PN.618-3; Ref. 13.)

Step 2: For $\theta \geq 5°$ compute the slant-path length L_S below the rain height using

$$L_s = (h_R - h_s)/\sin\theta \tag{9.8}$$

Step 3: Calculate the horizontal projection of the slant-path from (see Figure 9.9):

$$L_G = L_s \cos\theta \tag{9.9}$$

(It should be noted that Steps 4–8 are identical to those methodology steps of Section 9.3.1. Nevertheless, they are repeated below, modified with the notation used herein.)

Step 4: Obtain the rain intensity $R_{0.01\%}$ of an average year (with an integration time of 1 minute). If this information is not available from local sources, such as the local weather bureau, get an estimate from Table 9.3 using Figure 9.4, 9.5, or 9.6.

Step 5: Calculate the path reduction factor $r_{0.01}$ for 0.01% of the time for $R_{0.01} \leq 100$ mm/h:

$$r_{0.01} = \frac{1}{1 + L_G/L_0} \tag{9.10}$$

where

$$L_0 = 35e^{-0.015 R_{0.01}} \tag{9.11}$$

For $R > 100$ mm/h, use the value 100 mm/h in place of $R_{0.01}$.

Step 6: Calculate the specific attenuation A using the regression coefficients given in Table 9.2 and the rainfall rate (R) from Step 4 by the relation

$$A = aR^b \tag{9.12}$$

Step 7: The predicted attenuation exceeded 0.01% of an average year can be calculated from

$$A_{0.01} = A \times L_G \times r_{0.01} \tag{9.13}$$

Step 8: The estimated attenuation to be exceeded for other percentages of an average year, in the range of 0.001–1%, is computed from the attenuation to be exceeded by 0.01% of an average year by the relation

$$A_p/A_{0.01} = 0.12p^{-(0.546 + 0.043 \log p)} \tag{9.14}$$

This interpolation formula gives factors of 0.12, 0.38, 1, and 2.14 for 1%, 0.1%, 0.01%, and 0.001%, respectively.

Example 2. An earth station has a 13° elevation angle and operates at 20 GHz. Provide a range of margins for time availabilities of 99%, 99.9%, 99.99%, and 99.999%. The earth station latitude is 44°, located in the state of Minnesota, and its height above MSL is 500 m.

Step 1: Calculate the effective rain height. Use equation (9.7b) because the latitude is greater than 36°.

$$h_R = 4.0 - 0.075(44 - 36) \quad \text{(km)}$$
$$= 3.4 \text{ km}$$

Step 2: Compute the slant-path length L_s below the rain height. Use equation (9.8).

$$L_s = (h_R - h_s)/\sin\theta$$
$$= (3.4 - 0.5)/0.225$$
$$= 12.9 \text{ km}$$

Step 3: Calculate the horizontal projection of the slant path. Use equation (9.9).

$$L_G = 12.9 \times 0.974$$
$$= 12.57 \text{ km}$$

Step 4: Obtain the rain intensity, $R_{0.01}$. Use Figure 9.4. Applicable rain region is K. Obtain a value for R from Table 9.3. $R_{0.01} = 42$ mm/h.

Step 5: Calculate the path reduction factor, $r_{0.01}$. Use equations (9.10) and (9.11).

$$L_0 = 0.53 \times 35$$
$$= 18.64$$
$$r_{0.01} = \frac{1}{1 + 12.57/18.64}$$
$$= 0.597$$

Step 6: Calculate the specific attenuation A. Use equation (9.12). Obtain values for a and b using vertical polarization from Table 9.2, 20 GHz: $a = 0.0691, b = 1.065$; $R_{0.01} = 42$ mm/h.

$$A_{\text{dB/km}} = 3.7 \text{ dB/km}$$

Step 7: Calculate the predicted attenuation for the path for an average year. Use equation (9.13).

$$A_{0.01} = 3.7 \times 0.597 \times 12.97$$
$$= 28.65 \text{ dB}$$

Step 8: Time unavailabilities, availabilities, and excess attenuation values are as follows:

Time Unavailability	Time Availability	Excess Attenuation
1 %	99%	3.44 dB
0.1%	99.9%	10.89 dB
0.01%	99.99%	28.65 dB
0.001%	99.999%	61.31 dB

9.4.2 Rainfall Fade Rates, Depths, and Durations*

As reported in Ref. 14, fade depth varies directly with rain rate, and fade duration varies inversely with rain rate. Experimental results have shown fade durations of 4 minutes at $R = 10$ mm/h, and 1 minute at 160 mm/h and above for a COMSTAR 19-GHz downlink. Fade depths for the 19-GHz downlink are 10 dB for $R = 15$ mm/h and 20 dB for $R = 42$ mm/h; at 11

*The material provided in Section 9.4.2 is courtesy of Prof. Enric Vilar, Portsmouth Polytechnic University (UK).

GHz fade depths are less severe, on the order of 2 dB at $R = 11$ mm/h and 5 dB for $R = 29$ mm/h. Fade rate is a function of rain rate (R).

Fade duration, D (in minutes), can be calculated by a relationship given in Ref. 14 as follows:

$$D = 12.229R^{-0.469} \tag{9.15}$$

For example, if the point rain rate is 10 mm/h, then $D = 4.15$ minutes. If $R = 60$ mm/h, then $D = 1.8$ minutes.

The same reference also provides a relationship between average fade duration (D) and the average intra-exceedance (I_a).

$$I_a = 0.764DR^{0.295} \tag{9.16}$$

Intra-exceedance means the time between fades in the same rain event. For example, using D for the 10-mm/h point rain rate, $I_a = 6.25$ minutes.

Caution: Equations (9.15) and (9.16) are only valid for the 19/20-GHz region.

9.4.3 Site or Path Diversity

Excess attenuation due to rainfall often degrades satellite uplinks and downlinks operating above 10 GHz so seriously that the requirements of optimum economic design and reliable performance cannot be achieved simultaneously. Path diversity can overcome this problem at some reasonable cost compromise. Path diversity advantage is based on the hypothesis that rain cells and, in particular, the intense rain cells that cause the most severe fading are rather limited in spatial extent. Furthermore, these rain cells do not occur immediately adjacent to one another. Thus the probability of simultaneous fading on two paths to spatially separated earth stations would be less than that associated with either individual path. The hypothesis has been borne out experimentally (Ref. 15).

Let us define two commonly used terms: *diversity gain* and *diversity advantage*. Diversity gain is defined (in this context) as the difference between the rain attenuation exceeded on a single path and that exceeded jointly on separated paths for a given percentage of time. Diversity advantage is defined (in this context) as the ratio of the percentage of time exceeded on a single path to that exceeded jointly on separated paths for a given rain attenuation level.

Diversity gain may be interpreted as the reduction in the required system margin at a particular percentage of time afforded by the use of path diversity. Alternatively, diversity advantage may be interpreted as the factor by which the fade time is improved at a particular attenuation level due to the use of path diversity.

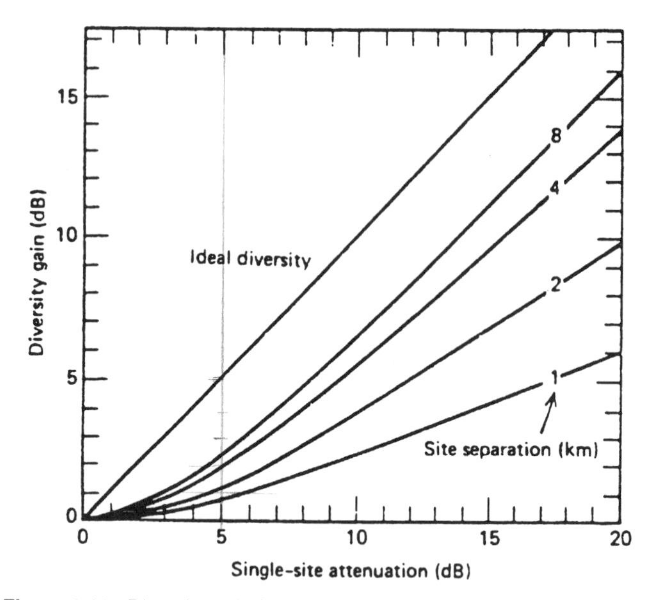

Figure 9.10. Diversity gain for various site separations. (From Ref. 17.)

The principal factor to achieve path diversity to compensate for excess attenuation due to rainfall is separation distance. The diversity gain increases rapidly as the separation distance d is increased over a small separation distance, up to about 10 km. Thereafter the gain increases more slowly until a maximum value is reached, usually between about 10 and 30 km. This is shown in Figure 9.10.

The uplink/downlink frequencies seem to have little effect on diversity gain up to about 30 GHz (Ref. 16). This same reference suggests that for link frequencies above 30 GHz attenuation on both paths simultaneously can be sufficient to create an outage. Therefore extrapolation beyond 30 GHz is not recommended, at least with the values given in Figure 9.10.

9.5 EXCESS ATTENUATION DUE TO ATMOSPHERIC GASES ON SATELLITE LINKS

The zenith one-way attenuations for a moderately humid atmosphere (e.g., 7.5 g/m^3 surface water vapor density) at various starting heights above sea level are given in Figure 9.11 and in Table 9.4. These curves were computed by Crane and Blood (Ref. 19) for temperate latitudes assuming the U.S. standard atmosphere, July, 45° N latitude. The range of values shown in Figure 9.11 refers to the peaks and valleys of the fine absorption lines. The range of values for starting heights above 16 km is even greater.

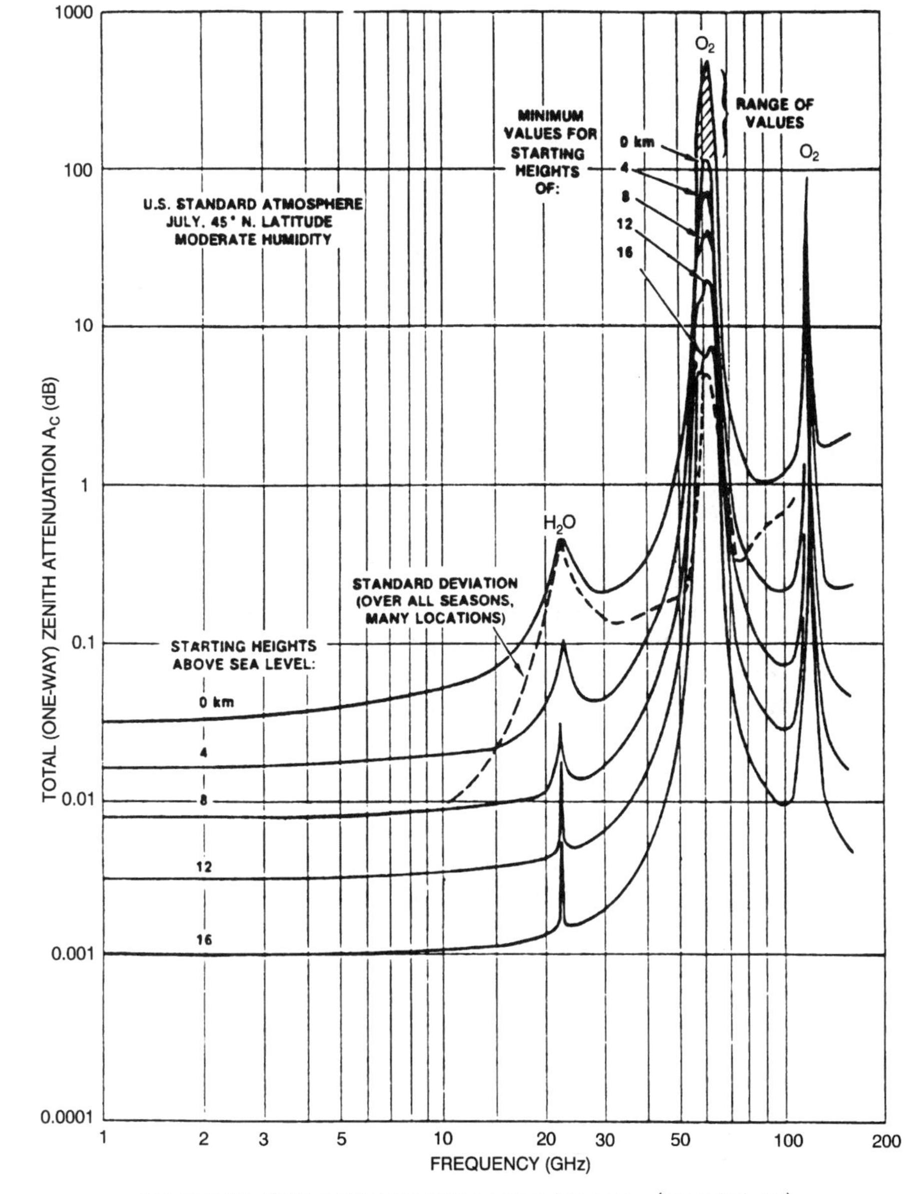

Figure 9.11. Total zenith attenuation versus frequency. (From Ref. 16.)

TABLE 9.4 Typical One-Way Clear Air Total Zenith Attenuation Values ($7.5\ g/m^3\ H_2O$, July, 45° N Latitude, 21°C)

Frequency (GHz)	Altitude				
	0 km	0.5 km	1.0 km	2.0 km	4.0 km
10	0.055	0.05	0.043	0.035	0.02
15	0.08	0.07	0.063	0.045	0.023
20	0.30	0.25	0.19	0.12	0.05
30	0.22	0.18	0.16	0.10	0.045
40	0.40	0.37	0.31	0.25	0.135
80	1.1	0.90	0.77	0.55	0.30
90	1.1	0.92	0.75	0.50	0.22
100	1.55	1.25	0.95	0.62	0.25

Source: Reference 16.

Figure 9.11 also shows the standard deviation of the clear air zenith attenuation as a function of frequency. The standard deviation was calculated from 220 measured atmosphere profiles spanning all seasons and geographical locations by Crane (Ref. 18). The zenith attenuation is a function of frequency, earth terminal altitude above sea level, and water vapor content. Compensating for earth terminal altitudes can be done by interpolating between the curves in Figure 9.11.

The water vapor content is the most variable component of the atmosphere. Corrections should be made to the values derived from Figure 9.11 and Table 9.4 in regions that notably vary from the 7.5-g/m^3 value given. Such regions would be arid or humid, jungle or desert. This correlation to the total zenith attenuation is a function of the water vapor density at the surface p_0 as follows:

$$\Delta A_{c1} = b_p(p_0 - 7.5\ \mathrm{g/m^3}) \tag{9.17}$$

where ΔA_{c1} is the additive correction to the zenith clear air attenuation that accounts for the difference between the actual surface water vapor density and 7.5 g/m^3. The coefficient b_p is frequency dependent and is given in Figure 9.12. To convert from the more familiar relative humidity or partial pressure of water vapor. refer to Section 9.5.2.

The surface temperature T_0 also affects the total attenuation because it affects the density of both the wet and dry components of the gaseous attenuation. This relation is (Ref. 19)

$$\Delta A_{c2} = c_T(21° - T_0) \tag{9.18}$$

where ΔA_{c2} is an additive correction to the zenith clear air attenuation. Figure 9.12 gives the frequency-dependent values for c_T.

The satellite–earth terminal elevation angle has a major impact on the gaseous attenuation value for a link. For elevation angles greater than about

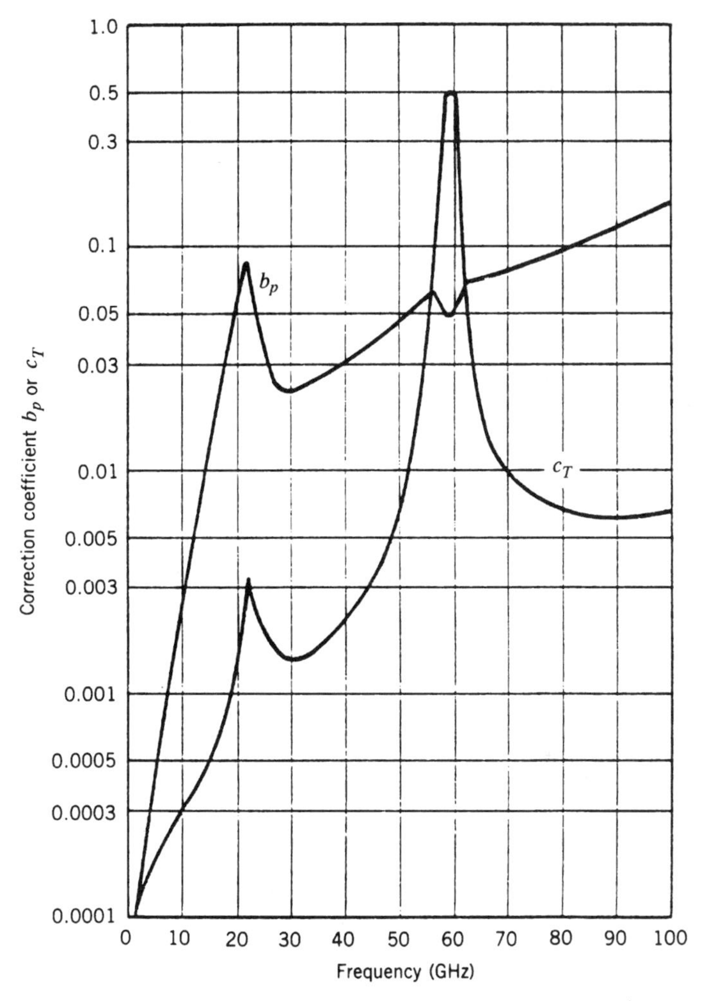

Figure 9.12. Water vapor density and temperature correction coefficients. (From Ref. 16.)

5°, the zenith clear air attenuation value A_c is multiplied by the cosecant of the elevation angle θ. The total attenuation for an elevation angle θ is given by

$$A_c = A'_c \csc \theta \tag{9.19}$$

9.5.1 Example Calculation of Clear Air Attenuation—Hypothetical Location

For a satellite downlink, we are given the following information: frequency, 20 GHz; altitude of earth station, 600 m; relative humidity (RH), 50%; temperature (surface, T_0), 70°F (21.1°C); and elevation angle, 25°. Calculate clear air attenuation.

Obtain total zenith attenuation A'_c from Table 9.4 and interpolate value for altitude: $A'_c = 0.24$ dB.

Find the water vapor density p_0. From Figure 9.13, the saturated partial pressure of water vapor at 70°F is $e_s = 2300$ N/m^2. Apply formula (9.20) (Section 9.5.2):

$$\begin{aligned} p_0 &= (0.5)2300/(0.461)(294.1) \\ &= 1150/135.6 \\ &= 8.48 \text{ g/m}^3 \end{aligned}$$

Calculate the water vapor correction factor ΔA_{c1}. From Figure 9.12 for a frequency of 20 GHz, correction coefficient $b_p = 0.05$; then [equation (9.17)]

$$\Delta A_{c1} = (0.05)(8.48 - 7.5) = 0.05 \text{ dB}$$

Compute the temperature c_T using equation (9.18). At 20 GHz, $c_T = 0.0015$. As can be seen, this value can be neglected in this case.

Calculate the clear air zenith attenuation corrected A'_c:

$$\begin{aligned} A'_c &= 0.24 \text{ dB} + 0.05 \text{ dB} + 0 \text{ dB} \\ &= 0.29 \text{ dB} \end{aligned}$$

Compute the clear air slant attenuation using equation (9.19):

$$\begin{aligned} A_c &= 0.29 \csc 25° \\ &= 0.29 \times 2.366 \\ &= 0.69 \text{ dB} \end{aligned}$$

This value would then be used in the link budget for this example link.

9.5.2 Conversion of Relative Humidity to Water Vapor Density

The surface water vapor density p_0 (g/m^3) at a given surface temperature (T_0) may be calculated from the ideal gas law:

$$p_0 = \frac{(\text{RH})e_s}{(0.461 \text{ J/g-K})(T_0 + 273)} \tag{9.20}$$

where RH is the relative humidity, and e_s (N/m^2) is the saturated partial pressure of water vapor that corresponds to the surface temperature T_0 (°C). See Figure 9.13. The relative humidity corresponding to 7.5 g/m^3 at 20°C (68°F) is RH = 0.42 or 42% (Ref. 16).

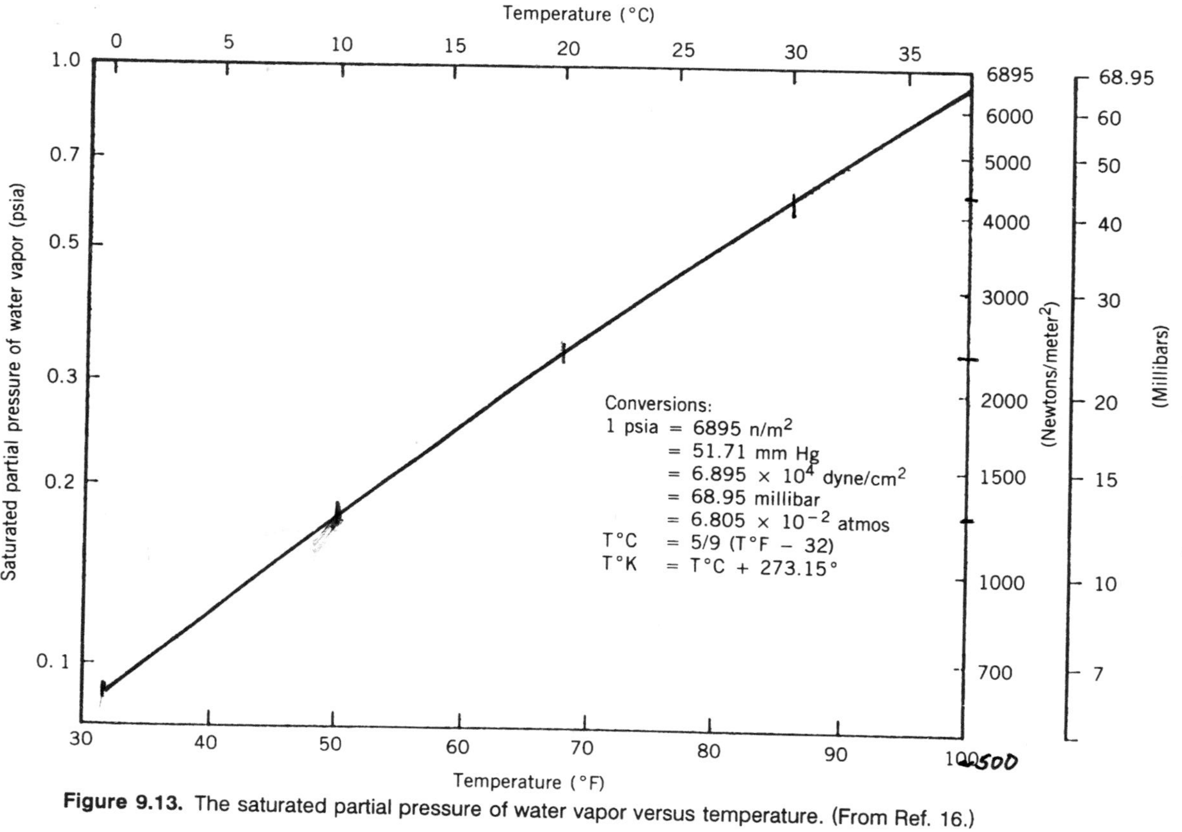

Figure 9.13. The saturated partial pressure of water vapor versus temperature. (From Ref. 16.)

9.6 ATTENUATION DUE TO CLOUDS AND FOG

Water droplets that constitute clouds and fog are generally less than 0.01 cm in diameter (Ref. 16). This allows a Rayleigh approximation to calculate the attenuation due to clouds and fog for frequencies up to 100 GHz. The specific attenuation a_c is, unlike the case of rain, independent of drop-size distribution. It is a function of liquid water content p_1 and can be expressed by

$$a_c = K_c p_1 \quad (\text{dB/km}) \tag{9.21}$$

where p_1 is normally expressed in g/m^3. K_c is the attenuation constant, which is a function of frequency and temperature, and is given in Figure 9.14. The curves in Figure 9.14 assume pure water droplets. The values for salt-water droplets, corresponding to ocean fogs and mists, are higher by approximately 25% at 20°C and 5% at 0°C (Ref. 20).

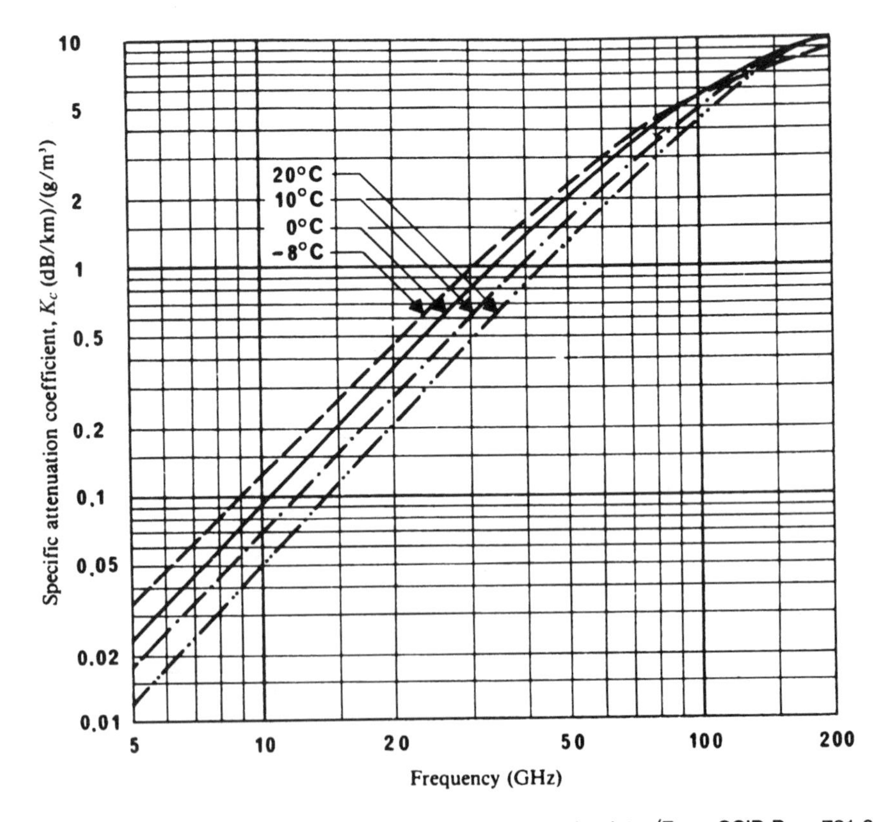

Figure 9.14. Attenuation coefficient K_c due to water vapor droplets. (From CCIR Rep. 721-3 ITU Geneva, 1990; Ref. 12.)

The liquid water content of clouds varies widely. Stratiform or layered clouds display ranges of 0.05–0.25 g/m^3 (Ref. 16). Stratocumulus, which is the most dense of this cloud type, has shown maximum values from 0.3 to 1.3 g/m^3 (Ref. 21). Cumulus clouds, especially the large cumulonimbus and cumulus congestus that accompany thunderstorms, have the highest values of liquid content. Fair weather cumulus clouds generally have a liquid water content of less than 1 g/m^3. Reference 22 reported values exceeding 5 g/m^3 in cumulus congestus and estimates an average value of 2 g/m^3 for cumulus congestus and 2.5 g/m^3 for cumulonimbus clouds.

Care must be exercised in estimating excess attenuation due to clouds when designing uplinks and downlinks. First, clouds are not homogeneous masses of air containing uniformly distributed droplets of water. Actually, the liquid water content can vary widely with location in a single cloud. Even sharp differences have been observed in localized regions on the order of about 100 m across. There is a fairly rapid variation with time as well, owing to the complex patterns of air movement taking place within cumulus clouds.

Typical path lengths through cumulus congestus clouds roughly fall between 2 and 8 km. Using equation (9.21) and the value given for water vapor density and the attenuation coefficient K_c from Figure 9.14, an added path loss at 35 GHz from 4 to 16 dB would derive. Fortunately, for the system designer, the calculation grossly overestimates the actual attenuation that has been observed through this type of cloud structure. Table 9.5 provides values that seem more dependable. In the 35- and 95-GHz bands, cloud attenuation, in most cases, is 40% or less of the gaseous attenuation values. One should not lose sight, in these calculations, of the great variability in the size and state of development of the clouds observed. Data from Table 9.5 may be roughly scaled in frequency, using the frequency dependence of the attenuation coefficient from Figure 9.14.

Fog results from the condensation of atmospheric water vapor into water droplets that remain suspended in air. The water vapor content of fog varies from less than 0.4 up to as much as 1 g/m^3.

The attenuation due to fog (in dB/km) can be estimated using the curves in Figure 9.14. The 10°C curve is recommended for summer, and the 0°C curve should be used for the other seasons. Typical liquid water content

TABLE 9.5 Zenith Cloud Attenuation Measurements

	Cloud Attenuation (dB)	
Cloud Type	95 GHz	150 GHz
Stratocumulus	0.5–1	0.1–1
Small, fine weather cumulus	0.5	0.5
Large cumulus	1.5	2
Cumulonimbus	2–7	3–8
Nimbostratus (rain cloud)	2–4	5–7

Source: Reference 12.

values for fog vary from 0.1 to 0.2 g/m^3. Assuming a temperature of 10°C, the specific attenuation would be about 0.08–0.16 dB/km at 35 GHz and 0.45–0.9 dB/km for 95 GHz. In a typical fog layer 50 m thick, a path at a 30° elevation angle would have only 100-m extension through fog, producing less than 0.1-dB excess attenuation at 95 GHz. In most cases, the result is that fog attenuation is negligible for satellite links.

9.7 CALCULATION OF SKY NOISE TEMPERATURE AS A FUNCTION OF ATTENUATION

The effective sky noise (see Section 6.3.8.1 and Figures 6.7–6.9) due to the troposphere is primarily dependent on the attenuation at the frequency of observation. Reference 23 shows the derivation of an empirical equation relating specific attenuation (A) to sky noise temperature:

$$T_s = T_m(1 - 10^{-A/10}) \tag{9.22}$$

where T_s is the sky noise and T_m is the mean absorption temperature of the attenuating medium (e.g., gaseous, clouds, rainfall) and A is the specific attenuation that has been calculated in the previous subsections. Temperatures are in kelvins. The value

$$T_m = 1.12(\text{surface temperature in K}) - 50 \text{ K} \tag{9.23}$$

has been empirically determined by Ref. 23.

Some typical values taken in Rosman, North Carolina, (Ref. 16) are given in Table 9.6 for rainfall.

TABLE 9.6 Cumulative Statistics of Sky Temperature Due to Rain for Rosman, North Carolina, at 20 GHz (T_m = 275 K)

Percentage of Year	Porint Rain Rate Values (mm/h)	Average Rain Rate (mm/h)	Total Rain Attenuation[a] (dB)	Sky Noise Temperature[b] (K)
0.001	102	89	47	275
0.002	86	77	40	275
0.005	64	60	30	275
0.01	49	47	23	274
0.02	35	35	16	269
0.05	22	24	11	252
0.1	15	17	7	224
0.2	9.5	11.3	4.6	180
0.5	5.2	6.7	2.6	123
1.0	3.0	4.2	1.5	82
2.0	1.8	2.7	0.93	53

[a]At 20 GHz the specific attenuation $A = 0.006 R_{av}^{1.12}$ dB / km and for Rosman, NC, the effective path length is 5.1 km to ATS-6.

[b]For a ground temperature of 17°C = 63°F, the T_m = 275 K.

Source: Reference 16.

Example 3. From Table 9.6 with a total rain attenuation of 11 dB, what is the sky noise at 20 GHz? Assume $T_m = 275$ K.

Use equation (9.21):

$$T_s = 275(1 - 10^{-11/10})$$

$$= 253.16 \text{ K}$$

9.8 THE SUN AS A NOISE GENERATOR

The sun is a white noise jammer of an earth terminal when the sun is aligned with the downlink terminal beam. This alignment occurs, for a geostationary satellite, twice a year near the equinoxes and, in the period of the equinox, will occur for a short period each day. The sun's radio signal is of sufficient level to nearly saturate the terminal's receiving system, wiping out service for that period. Figure 9.15 gives the power flux density of the sun as a function of frequency. Above about 20 GHz the sun's signal remains practically constant at -188 dBW/Hz-m^2 for "quiet sun" conditions.

Reception of the sun's signal or any other solar noise source can be viewed as an equivalent increase in a terminal's antenna noise temperature by an amount T_s. T_s is a function of terminal antenna beamwidth compared to the apparent diameter of the sun (e.g., 0.48°), and how close the sun approaches the antenna boresight. The following formula, taken from Ref. 16, provides an estimate of T_s, when the sun or any other extraterrestrial noise source is aligned in the antenna beam:

$$T_s = \frac{1 - \exp\left[-(D/1.2\theta)^2\right]}{f^2 D^2}\left(\log^{-1}\frac{S + 250}{10}\right) \tag{9.24}$$

where D = apparent diameter of the sun or 0.48°
f = frequency in gigahertz
S = power flux density in dBW/Hz-m^2
θ = half-power beamwidth of the terminal antenna in degrees

Example 4. An earth station operating with a 20-GHz downlink has a 2-m antenna (beamwidth of 0.5°). What is the maximum increase in antenna noise temperature that would be caused by a quiet sun transit?

Use formula (9.24) to find that:

$$T_s = 8146 \text{ K}$$

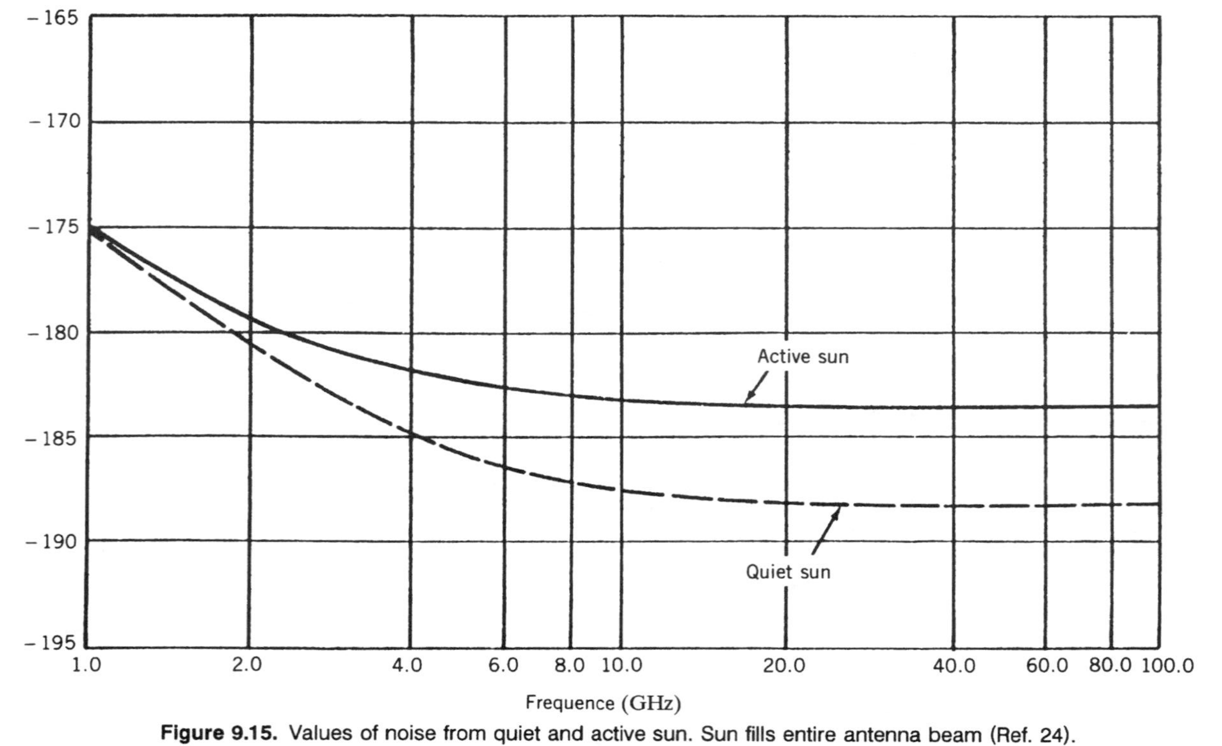

Figure 9.15. Values of noise from quiet and active sun. Sun fills entire antenna beam (Ref. 24).

9.9 PROPAGATION EFFECTS WITH A LOW ELEVATION ANGLE

As the elevation angle of an earth terminal is lowered, the ray beam penetrates an ever increasing amount of atmosphere. Below about 10°, fading on the downlink signal must be considered. Fading or signal fluctuations apply only to the ground terminal downlink because its antenna is in close proximity to a turbulent medium. The companion uplink satellite path will suffer uplink fluctuation gain degradation only due to scattering of energy out of the path (Ref. 16). Because of the large distance traversed by the uplink signal since leaving the troposphere, the signal arrives at the satellite as a plane wave and with only a small amount of angle-of-arrival effects.

Phase variations must also be expected for the low-elevation-angle condition. Phase variations arise due to the variable delay as the wave passes through a medium with variable refractivity. Phase scintillation can also occur.

9.10 DEPOLARIZATION ON SATELLITE LINKS

Depolarization is an effect wherein a satellite link's wave polarization is altered after being launched by the antenna. Some texts refer to depolarization as cross-polarization. For the case of a linearly polarized wave passing through the medium, components of the opposite polarization will be developed. For the case of circular polarization, there will be a tendency to develop into an elliptical wave. This is particularly important for the case of frequency reuse systems, where depolarization effectively reduces the polarization isolation and can tend to increase crosstalk on the signal.

Depolarization on a satellite can be caused by rain, ice, snow, multipath, effects, and refractive effects. It can also be caused by Faraday rotation.

Faraday Rotation. Analysis of the propagation of a linearly polarized high-frequency wave in the ionosphere shows that it experiences rotation of the plane of polarization such that a wave launched with vertical polarization does not remain vertical. Depending on the frequency, length of the path in the ionosphere, and orientation with respect to the earth's magnetic field, the amount of Faraday rotation may vary from negligible to 360° or many rotations.

Of most importance for satellite links is the fact that the rotation varies inversely as the square of the frequency. Typically, at 4 GHz, the rotation is only several degrees.

9.11 SCINTILLATION FADING ON SATELLITE LINKS

Scintillation, as defined by the IEEE Dictionary (Ref. 25), "is the phenomenon of fluctuation of the amplitude of a wave caused by irregular changes in the transmission path or paths with time." On satellite paths operating above 10 GHz, scintillation can become an issue as the elevation angle reaches about 15°, and its intensity grows as the angle is lowered still further. In the Canadian Arctic, fade depths have been reported up to 30 dB at 1° elevation angle and at 30 GHz (CCIR Rep. 564-4, Ref. 26).

Small-scale irregularities in the refractive index of the atmosphere can cause rapid signal level variations (i.e., scintillation fading). Tropospheric effects in the absence of precipitation are unlikely to cause serious fading in space telecommunication systems operating below 10 GHz and at elevation angles above 10°. At elevation angles below 10° and at frequencies above about 10 GHz, tropospheric scintillations can cause serious degradations in performance.

The fading periods of scintillation vary over quite a large range from less than 0.1 second to several minutes, as the fading period depends both on apparent motion of the irregularities relative to the ray path and, in the case of strong scintillation, on severity. The fading period in the gigahertz frequency range varies from 1 to 10 seconds (ITU-R Rec. PI.531-3, Ref. 27).

More in-depth information on scintillation fading may be found in ITU-R Rec. PN.618-3 (Ref. 13) and CCIR Rep. 564-4 (Ref. 26) as well as References 29 and 30.

9.12 TRADE-OFF BETWEEN FREE-SPACE LOSS AND ANTENNA GAIN

A concern that radio system engineers often face is the impact of using a higher frequency. This is a valid concern whether using LOS microwave or satellite communications. Consider the following hypothetical scenario. A certain satellite communications earth station user wishes to shift up- and downlinks from the 6/4 GHz band to the 14/12 GHz band. Leaving aside the important rainfall and gaseous absorption losses, we must pay in additional free-space loss. Use equation (1.9) and the range remains unchanged. Only the frequency term is necessary, which is $20 \log f$. For this exercise, we only consider downlinks. In the 6/4 GHz band the downlink equals 4 GHz and for the 14/12 GHz band, the downlink is 12 GHz.

$$6/4 \quad 20 \log 4000 = 72.04 \text{ dB}$$

$$14/12 \quad 20 \log 12{,}000 = 81.58 \text{ dB}$$

$$\text{Difference} = 9.54 \text{ dB}$$

It appears that an additional loss of 9.54 dB will be imposed. This is true, but we can easily compensate for this with antenna gains. Assume the antenna aperture remains unchanged, what net gain do we have? Use

TABLE 9.7 Frequency / Wavelength, Gain and Beamwidth

Frequency (GHz)	Wavelength (m)	Gain (D = 3 m)	Beamwidth (degrees)
4	0.075	39.69 dB	1.75
6	0.050	43.21 dB	1.17
12	0.025	49.23 dB	0.58
14	0.0214	50.57 dB	0.50
20	0.015	53.67 dB	0.35
28	0.0107	56.59 dB	0.25
38	0.00789	59.24 dB	0.184
55	0.00545	62.45 dB	0.127

equation (2.27). Antenna diameters and efficiencies remain unchanged, thus only the frequency term is implicated, $20 \log f$. We end up with the same results as above, in this case a net gain of 9.54 dB. However, there are *two* antennas in the link (satellite and earth station). Thus the real net gain is $2 \times 9.54 = 19.08$ dB less 9.54 dB for additional free space loss. So there is a net link gain of 9.54 dB.

The reader with insight will say that transmission line is more lossy at higher frequencies. Generally, when using these "higher" frequencies, we try to shorten waveguide runs as much as possible, and even may resort to placing the LNA right into the antenna structure. Thus that argument can be nullified.

One thing, though, we often lose sight of: beamwidth and possible penalties we have to pay with such beamwidths. Let's apply the beamwidth estimation equation (Ref. 28):

$$\text{Beamwidth } (^\circ) \approx 70^\circ(\lambda/D) \tag{9.25}$$

Let D remain constant at 3 m, and calculate beamwidths as frequency is increased (wavelength is decreased). See Table 9.7. Gain assumed at $\eta = 60\%$.

As the beamwidth gets smaller, satellite acquisition and tracking become more difficult. By narrowing the aperture, the beamwidth widens, but then the additional gain is lost. Another problem also can arise with fairly wide apertures and extremely high gains. This is the required surface tolerance of the parabolic reflector. The tolerance requirements are more rigid. With such rigid tolerances, the change in temperature from nighttime coolness to the hot daytime sun can run the surface out of tolerance. A temperature-compensated radome could mitigate the problem.

Suppose we doubled the aperture at 55 GHz to 6 m. The beamwidth would be about half of 0.127° or about 0.06°, and the gain would increase some 6 dB to about 68.45 dB.

PROBLEMS AND EXERCISES

1 Above 10 GHz, there are two additional degradations to path loss that we must take into account. What are they?

2 There are two frequency bands between 10 and 100 GHz that display high attenuation due to atmospheric gases. What bands are these? Which one displays excessively high attenuation, making it unusable for most earth-bound applications.

3 Discuss two practical applications of the high-loss band in question 2.

4 Argue why cumulative annual rainfall rates may not be used for calculation of excess attenuation due to rainfall and why, instead, we must use point rainfall rates.

5 In early attempts to build in sufficient margin on satellite and LOS microwave links operating above 10 GHz, it was found that the required margins were excessively large because we integrated excess attenuation per kilometer along the entire path (the entire path in the atmosphere for satellite links). Describe how statistics on rain cell size assisted to better estimate excess attenuation due to rainfall.

6 Why does excess attenuation due to rainfall and atmospheric gases increase as satellite elevation angle decreases?

7 When using linear polarization, one polarization displays considerably higher attenuation due to rainfall than the other polarization. Identify which polarization and explain why this is so.

8 In Chapters 2 and 3 we set a margin due to multipath fading. Is this margin additive to excess attenuation due to rainfall? Explain your answer.

9 What are the three variables we must deal with when calculating excess attenuation due to rainfall for microwave/millimeter wave LOS paths? Show how we relate these variables mathematically.

10 In rain region P for a certain location where the elevation angle was 5° and the desired time availability was 99.997%, we found excess attenuation values over 500 dB. What measures could be taken to drop this attenuation value to something more reasonable? Moving the earth station is not an option.

11 Calculate the specific attenuation per kilometer for a path operating at 30 GHz on a LOS basis with a time availability for the path of 99.9%. Neglect path length considerations, of course. The path is located in northeastern United States. Carry out the calculation for both horizontal and vertical polarizations.

12 Calculate the excess attenuation due to rainfall for a LOS path operating at 50 GHz and located in central Australia. The path length is 20 km and the desired time availability is 99.99%. Assume vertical polarization.

13 Calculate the excess attenuation due to rainfall for a LOS path 25 km long with an operating frequency of 18 GHz and for which the desired path availability (propagation reliability) is 99.99%. The path is located in the state of Massachusetts.

14 Name five ways to build a rainfall margin for the path in question 13.

15 Calculate the excess attenuation due to rainfall for a satellite path with a 20° elevation angle for a 21-GHz downlink. The earth station is located in southern Minnesota and the desired time availability for the link is 99%.

16 An earth station is to be located near Boon, Germany, and will operate at 14 GHz. The desired uplink time availability is 99.95% and the subsatellite point is 10° W. What is the excess attenuation due to rainfall?

17 An earth station is to be installed in Diego Garcia with an uplink at 44 GHz. The elevation angle is 15° and the desired path (time) availability is 99%. What value of excess attenuation due to rainfall should be used in the link budget?

18 There is an uplink at 30 GHz and the required excess attenuation due to rainfall is 15 dB. Path diversity is planned. Show how the value of excess attenuation due to rainfall for a single site can be reduced for site separations of 1, 2, 4, and 8 km.

19 For an earth station, calculate the excess attenuation due to atmospheric gases for a site near sea level. The site is planned for 30/20-GHz operation. The elevation angle is 15°. The relative humidity is 60% and the surface temperature is 70°F.

20 Calculate the sky noise contribution for the attenuation of gases calculated in question 19. Calculate the sky noise temperature due to the excess attenuation due to rainfall from question 17.

21 Why do so many satellite communication standards set minimum elevation angles for "guaranteed" performance? For example, below 10 GHz, 5°; between 10 and 20 GHz, 10°; and often above 20 GHz, 15°.

22 Name the principal cause of depolarization on satellite links. Name at least two other causes.

23 What are the principal causes and effects of scintillation on satellite links?

24 What is the main drawback in building very large aperture antennas for the higher frequencies? Consider what happens as both aperture size and frequency increase.

REFERENCES

1. H. J. Liebe, *Atmospheric Propagation Properties in the 10 to 75 GHz Region: A Survey and Recommendations*, ESSA Technical Report ERL 130-ITS 91, Boulder, CO, 1969.
2. *Attenuation by Atmospheric Gases*, CCIR Rep. 719-3, Reports of the CCIR 1990, Annex to Vol. V, XVIIth Plenary Assembly, Dusseldorf, 1990.
3. D. C. Hogg, "Millimeter-Wave Propagation Through the Atmosphere," *Science*, 1968.
4. R. K. Crane, "Prediction of the Effects of Rain on Satellite Communications Systems," *Proceedings of the IEEE*, Vol. 65, pp. 456–474, 1977.
5. *Recommendations and Reports of the CCIR*, CCIR Rep. 593-1, Vol. V, XIVth Plenary Assembly, Kyoto, 1978.
6. *Propagation Data and Prediction Methods Required for Terrestrial Line-of-Sight Systems*, CCIR Rep. 338-6, Annex to Vol. V, XVIIth Plenary Assembly, Dusseldorf, 1990.
7. *Radiometeorological Data*, CCIR Rep. 563-4, Annex to Vol. V, XVIIth Plenary Assembly, Dusseldorf, 1990.
8. *Propagation Data and Prediction Methods Required for the Design of Terrestrial Line-of-Sight Systems*, ITU-R Rec. PN.530-5, Annex 1, 1994 PN Series Volume, ITU, Geneva, 1994. (Also see Ref. 6.)
9. *Specific Attenuation Model for Rain for Use in Prediction Methods*, CCIR Rec. 838, ITU-R 1994 PN Series Volume, ITU, Geneva, 1994.
10. Private communication, Prof. Enric Vilar, Portsmouth Polytechnic University, Portsmouth, UK, Nov. 9, 1995.
11. *Characteristics of Prescipitation for Propagation Modeling*, ITU-R Rec. PN.837-1, 1994 PN Series Volume, ITU, Geneva, 1994.
12. *Attenuation by Hydrometers, in Particular Precipitation, and Other Atmospheric Particles*, CCIR Rep. 721-3, Annex to Vol. V, XVIIth Plenary Assembly, Dusseldorf, 1990.
13. *Propagation Data and Prediction Methods Required for the Design of Earth–Space Telecommunication Systems*, ITU-R Rec. PN.618-3, 1994 PN Series Volume, ITU, Geneva, 1994.
14. Enric Vilar and August Burgeño, "Analysis and Modeling of Time Intervals Between Rain Rate Exceedances in the Context of Fade Dynamics," *IEEE Transactions on Communications*, Vol. 39, No. 9, Sept. 1991.
15. D. B. Hodge, "The Characteristics of Millimeter Wavelength Satellite-to-Ground Space Diversity Links," *IEE Conference No. 98*, Apr. 1978.
16. R. Kaul, R. Wallace, and G. Kinal, *A Propagation Effects Handbook for Satellite System Design: A Summary of Propagation Impairments on 10–100 GHz Links, with Techniques for System Design*, ORI Inc., Silver Spring, MD, 1980 (NTIS N80-25520).
17. Roger L. Freeman, *Telecommunication Transmission Handbook*, 3rd ed., Wiley, New York, 1991.
18. R. K. Crane, *An Algorithm to Retrieve Water Vapor Information from Satellite Measurements*, NEPRF Tech. Report 7076, Final Report, Project No. 1423, Environmental Research and Technology, Inc., Concord, MA, 1976.

19. R. K. Crane and D. W. Blood, *Handbook for the Estimation of Microwave Propagation Effects—Link Calculations of Earth–Space Paths*, Environmental Research and Technology Report No. 1, DOC No. P-7376-TRL, U.S. Department of Defense, Washington, DC, 1979.
20. K. L. Koester and L. H. Kosowsky, "Millimeter Wave Propagation in Ocean Fogs and Mists," *Proceedings of the IEEE Antenna Propagation Symposium*, 1978.
21. B. J. Mason, *The Physics of Clouds*, Clarendon Press, Oxford, UK, 1971.
22. H. K. Weickmann and H. J. Kaumpe, "Physical Properties of Cumulus Clouds," *Journal of Meteorology*, Vol. 10, 1953.
23. K. H. Wulfsberg, *Apparent Sky Temperatures at Millimeter-Wave Frequencies*, Physical Science Research Paper No. 38, Air Force Cambridge Research Lab., No. 64-590, 1964.
24. S. Perlman et al., "Concerning Optimum Frequencies for Space Vehicle Communications," *IRE Transactions Military Electronics*, Mil-4(2-3), 1960.
25. *The New IEEE Standard Dictionary of Electrical and Electronics Terms*, 5th ed., IEEE Std 100-1992, IEEE, New York, 1993.
26. *Propagation Data and Prediction Methods Required for Earth–Space Telecommunication Systems*, CCIR Rep. 564-4, Annex to Vol. V, XVIIth Plenary Assembly, Dusseldorf, 1990. (Also see Ref. 13.)
27. *Ionospheric Effects Influencing Radio Systems Involving Spacecraft*, ITU-R Rec. PI.531-1, 1994 PI Series, ITU, Geneva, 1994.
28. Richard C. Johnson and Henry Jasik, *Antenna Engineering Handbook*, 2nd ed., McGraw-Hill, New York, 1984.
29. E. Vilar and J. R. Larsen, "Elevation Dependence of Amplitude Scintillations on Low Elevation Earth Space Paths," *6th Annual Conference on Antennas and Propagations, ICAP 89*, University of Warwick, UK, Apr. 1989.
30. O. P. Banjo and E. Vilar, "Dynamic Characteristics of Scintillation Fading on a Low Elevation Earth–Space Path," *URSI International Symposium on Millimeter Wave Propagation and Remote Sensing*, Dover, NH, July/August 1986.

10

MOBILE COMMUNICATIONS: CELLULAR RADIO AND PERSONAL COMMUNICATION SERVICES

10.1 INTRODUCTION

10.1.1 Background

The earliest application of radio was mobile communications. As we turned into the 20th century, the Marconi Company* was formed, providing ships with radio communication. In World War I aircraft were provided with radio installations, and after the war mobile radio moved into the civilian arena, particularly in the public safety sector.

Since 1980 mobile radio communications have taken on a more personal flavor. Cellular radio systems have extended the telephone network to automobiles and pedestrians. A new and widely used term in our vocabulary is *personal communications*. It is becoming the universal tether. No matter where we go, on land, on sea, and in the air, we can have near instantaneous two-way communications by voice, data, and facsimile. At some time it will encompass video.

Personal radio terminals are becoming smaller. There is the potential of their becoming wristwatch size. However, the human interface requires input/output devices that have optimum usefulness. A wristwatch size keyboard or keypad is rather difficult to operate; a hard-copy printer requires some minimum practical dimensions, and so forth.

10.1.2 Scope and Objective

This chapter presents an overview of "personal communications." Much of the discussion deals with cellular radio and extends this thinking inside

*The actual name of the company was Marconi International Marine Communications Company, which was created in London in 1900.

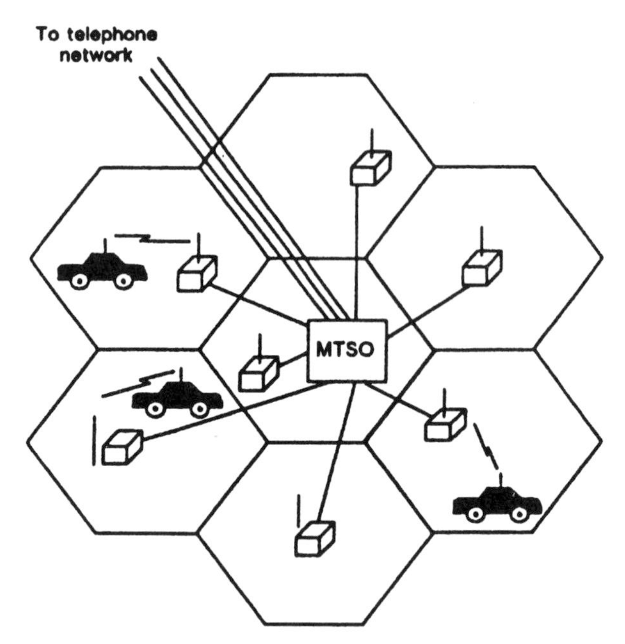

Figure 10.1. Conceptual layout of a cellular system.

buildings. The coverage most necessarily includes propagation for several environments, propagation impairments, methods to mitigate these impairments, access techniques, bandwidth limitations and ways around this problem. It will cover several mobile radio standards and compare a number of existing and planned systems. The chapter objective is to provide an appreciation of mobile/personal communications. Space limitations force us to confine our discussion to what might be loosely called "land mobile systems."

10.2 SOME BASIC CONCEPTS OF CELLULAR RADIO

Cellular radio systems connect a mobile terminal to another user, usually through the PSTN. The "other user" most commonly is a telephone subscriber of the PSTN. However, the other user may be another mobile terminal. Most of the connectivity is extending POTS* to mobile users. Data and facsimile services are in various stages of implementation. Some of the terms used in this section have a strictly North American flavor.

Figure 10.1 shows a conceptual layout of a cellular system. The heart of the system for a specific serving area is the MTSO (mobile telephone switching office). The MTSO is connected by a trunk group to a nearby

*POTS—an abbreviation for "plain old telephone service."

telephone exchange, providing an interface to, and connectivity with, the PSTN.

The area to be served by a *cellular geographic serving area* (CGSA) is divided into small geographic cells, which ideally are hexagonal. Cells are initially laid out with centers spaced about 4–8 miles (6.4–12.8 km) apart. The basic system components are the cell sites, the MTSO, and mobile units. These mobile units may be hand-held or vehicle-mounted terminals.

Each cell has a radio facility housed in a building or shelter. The facility's radio equipment can connect and control any mobile unit within the cell's responsible geographic area. Radio transmitters located at the cell site have a maximum effective radiated power (ERP*) of 100 watts. Combiners are used to connect multiple transmitters to a common antenna on a radio tower, usually between 50 and 300 ft (15 and 92 m) high. Companion receivers use a separate antenna system mounted on the same tower. The receive antennas are often arranged in a space diversity configuration.

The MTSO provides switching and control functions for a group of cell sites. A method of connectivity is required between the MTSO and the cell site facilities. The MTSO is an electronic switch and carries out a fairly complex group of processing functions to control communications to and from mobile units as they move between cells as well as to make connections with the PSTN. Besides making connectivity with the public network, the MTSO controls cell site activities and mobile actions through command and control data channels. The connectivity between cell sites and the MTSO is often via DS1 on wire pairs or on microwave facilities, the latter being the most common.

A typical cellular mobile unit consists of a control unit, a radio transceiver, and an antenna. The control unit has a telephone handset, a pushbutton keypad to enter commands into the cellular/telephone network, and audio and visual indications for customer alerting and call progress. The transceiver permits full duplex transmission and reception between mobile and cell sites. Its ERP is nominally 6 watts. Hand-held terminals combine all functions into one small package that can easily be held in one hand. The ERP of a hand-held terminal is a nominal 0.6 watts.

In North America, cellular communication is assigned a 25-MHz band between 824 and 849 MHz for mobile unit-to-base transmission and a similar band between 869 and 894 MHz for transmission from base to mobile.

The first and most widely implemented North American cellular radio system was called AMPS (advanced mobile telephone system). The original system description was contained in the entire issue of the *Bell System Technical Journal* (BSTJ) of January 1979. The present AMPS is based on 30-kHz channel spacing using frequency modulation. The peak deviation is 12 kHz. The cellular bands are each split into two to permit competition.

*Care must be taken with terminology. In this instance, ERP and EIRP are *not* the same. The reference antenna in this case is the dipole, which has a 2.16-dBi gain.

Thus only 12.5 MHz is allocated to one cellular operator for each direction of transmission. With 30-kHz spacing, this yields 416 channels. However, nominally 21 channels are used for control purposes with the remaining 395 channels available for cellular end-users.

Common practice with AMPS is to assign 10–50 channel frequencies to each cell for mobile traffic. Of course, the number of frequencies used depends on the expected traffic load and the blocking probability. Radiated power from a cell site is kept at a relatively low level with just enough antenna height to cover the cell area. This permits frequency reuse of these same channels in nonadjacent cells in the same CGSA with little or no co-channel interference. A well-coordinated frequency reuse plan enables tens of thousands of simultaneous calls over a CGSA.

Figure 10.2 illustrates a frequency reuse method. Here four channel frequency groups are assigned in a way that avoids the same frequency set used in adjacent cells. If there were uniform terrain contours, this plan could be applied directly. However, real terrain conditions dictate further geographic separation of cells that use the same frequency set. Reuse plans with 7 or 12 sets of channel frequencies provide more physical separation and are often used depending on the shape of the antenna pattern employed.

With user growth in a particular CGSA, cells may become overloaded. This means that grade of service objectives is not being met due to higher than planned traffic levels during the busy hour (BH).* In these cases, congested cells can be subdivided into smaller cells, each with its own base station, as shown in Figure 10.3. These smaller cells use lower transmitter power and antennas with less height, thus permitting greater frequency reuse. These subdivided cells can be split still further for even greater frequency reuse. However, there is a practical limit to cell splitting, often with cells with a 1-mile (1.6-km) radius. Under normal, large-cell operation, antennas are usually omnidirectional. When cell splitting is employed, 60° or 120° directional antennas are often used to mitigate interference brought about by increased frequency reuse.

Radio system design for cellular operation differs from that used for line-of-sight (LOS) microwave operation. For one thing, mobility enters the picture. Path characteristics are constantly changing. Mobile units experience multipath scattering, reflection, and/or diffraction by obstructions and buildings in the vicinity. There is shadowing, often very severe. The resulting received signal under these conditions varies randomly as the sum of many individual waves with changing amplitude, phase, and direction of arrival. The statistical autocorrelation distance is on the order of one-half wavelength (Ref. 2). Space diversity at the base station tends to mitigate these impairments.

In Figure 10.1, the MTSO is connected to each of its cell sites by a voice trunk for each of the radio channels at the site. Also, two datalinks (AMPS

*BH or busy hour is the 1-hour period during a workday with the most telephone traffic calling activity. See Ref. 1 for definitions of the busy hour.

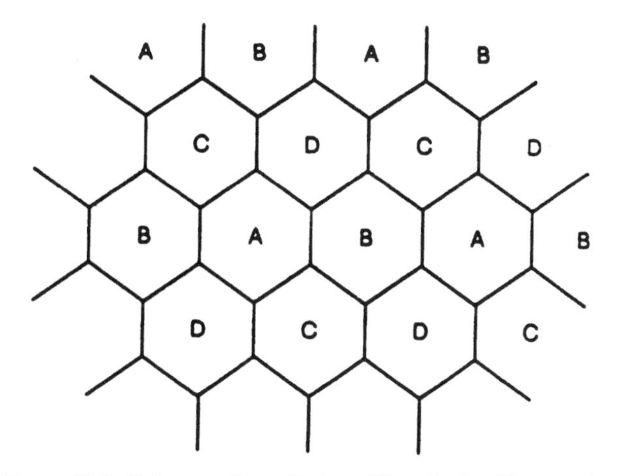

Figure 10.2. Cell separation with four different sets of frequencies.

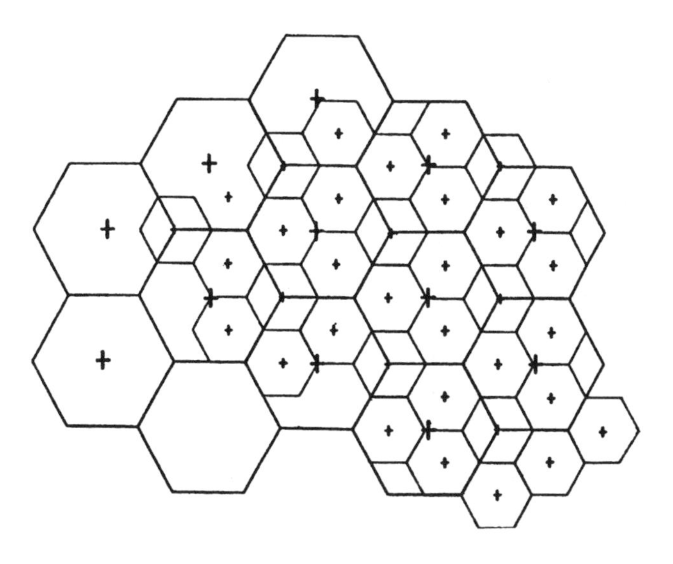

Figure 10.3. Staged growth by cell splitting (subdividing).

design) connect the MTSO to each cell site. These datalinks transmit information for processing calls and for controlling mobile units. In addition to its "traffic" radio equipment, each cell site has installed signaling monitoring equipment and a "setup" radio to establish calls.

When a mobile unit becomes operational, it automatically selects the setup channel with the highest signal level. It then monitors that setup channel for incoming calls destined for it. When an incoming call is sensed, the mobile terminal in question again samples the signal level of all appropriate setup channels so it can respond through the cell site offering the highest signal level, and then tunes to that channel for response. The responsible MTSO assigns a vacant voice channel to the cell in question, which relays this information via the setup channel to the mobile terminal. The mobile terminal subscriber is then alerted that there is an incoming call. Outgoing calls from mobile terminals are handled in a similar manner.

While a call is in progress, the serving cell site examines the mobile's signal level every few seconds. If the signal level drops below a prescribed level, the system seeks another cell to handle the call. When a more appropriate cell site is found, the MTSO sends a command, relayed by the old cell site, to change frequency for communication with the new cell site. At the same time, the landline subscriber is connected to the new cell site via the MTSO. The periodic monitoring of operating mobile units is known as *locating*, and the act of changing channels is called *handover.* Of course, the functions of locating and handover are to provide subscribers satisfactory service as a mobile unit traverses from cell to cell. When cells are made smaller, handovers are more frequent.

The management and control functions of a cellular system are quite complex. Handover and locating are managed by signaling and supervision techniques, which take place on the setup channel. The setup channel uses a 10-kbps data stream, which transmits paging, voice channel designation, and overhead messages to mobile units. In turn, the mobile unit returns page responses, origination messages, and order confirmations.

Both digital messages and continuous supervision tones are transmitted on the voice radio channel. The digital messages are sent as a discontinuous "blank-and-burst" inband data stream at 10 kbps and include order and handover messages. The mobile unit returns confirmation and messages that contain dialed digits. Continuous positive supervision is provided by an out-of-band 6-kHz tone, which is modulated onto the carrier along with the speech transmission.

Roaming is a term used for a mobile unit that travels such distances that the route covers more than one cellular organization or company. The cellular industry is moving toward technical and tariffing standardization so that a cellular unit can operate anywhere in the United States, Canada, and Mexico.

10.2.1 N-AMPS Increases Channel Capacity Threefold

N-AMPS, developed by Motorola, increases channel capacity three times that of its AMPS counterpart. Rather than 30-kHz segments of AMPS, N-AMPS assigns 10 kHz per voice channel. Its signaling and control are exactly the same as AMPS, except the signaling is one using subaudible data streams.

Of course, to accommodate the narrower FDMA channel width of only 10 kHz, the FM deviation had to be reduced. As a result, voice quality was also reduced. To counteract this, N-AMPS uses voice companding to provide a "synthetic" voice channel quieting.

10.3 RADIO PROPAGATION IN THE MOBILE ENVIRONMENT

10.3.1 The Propagation Problem

LOS microwave and satellite communications covered in Chapters 2 and 3 dealt with fixed systems. Such systems were and are optimized. They are built up and away from obstacles. Sites are selected for best propagation.

This is not so with mobile systems. Motion and a third dimension are additional variables. The end-user terminal often is in motion, or the user is temporarily fixed, but that point can be anywhere within a serving area of interest. Whereas before we dealt with point-to-point, here we deal with point-to-multipoint.

One goal in LOS microwave design was to stretch the distance as much as possible between repeaters by using high towers. In this chapter there are some overriding circumstances where we try to limit coverage extension by reducing tower heights, what we briefly introduced in Section 10.2. Even more important, coverage is area coverage where shadowing is frequently encountered. Examples are valleys, along city streets with high buildings on either side, in verdure such as trees, and inside buildings, to name just a few situations. Such an environment is rich with multipath scenarios. Paths can be highly dispersive, as much as 10 μs of delay spread (Ref. 3). Due to a user's motion, Doppler shift can be expected.

The radio-frequency bands of interest are UHF, especially around 800 and 900 MHz and 1700–2000 MHz. Some 400-MHz examples will also be covered.

10.3.2 Several Propagation Models

We concentrate on cellular operation. There is a fixed station (FS) and mobile stations (MSs) moving through the cell. A cell is the area of responsibility of the fixed station, a cell site. It usually is pictured as a hexagon in shape, although its propagation profile is more like a circle with the fixed station in its center. Cell radii vary from 1 km (0.6 mi) in heavily built-up urban areas to 30 km (19 mi) or somewhat more in rural areas.

10.3.2.1 Path Loss. We recall the free-space loss (FSL) formula discussed in Chapter 1. It simply stated that FSL was a function of the square of the distance and the square of the frequency plus a constant. It is a very useful formula if the strict rules of obstacle clearance are obeyed. Unfortunately, in the cellular situation, it is impossible to obey these rules. To what extent must this free-space loss formula be modified by atmospheric effects, the presence of the earth, and the effects of trees, buildings, and hills that exist in, or close to, the transmission path?

10.3.2.1.1 CCIR Formula. CCIR developed a simple path loss (L_{dB}) formula (CCIR Rec. 370-5, Ref. 4) for radio and television broadcasting where the frequency term has been factored out:

$$L_{\text{dB}} = 40 \log d_m - 20 \log(h_T h_R) \tag{10.1}$$

Note that the equation is an inverse fourth power law and is fundamental to terrestrial mobile radio. Distance is in meters; h_T is the height of the transmit antenna above plane earth (again in meters) and h_R is the height of the receive antenna above plane earth (in meters).

Suppose $d = 1000$ m (1 km) and the product of $h_T \times h_R$ is 10 m^2 (low antennas); then $L = 100$ dB.

Now suppose that $d = 25$ km and $h_T \times h_R = 100$ m^2, where the base station is on high ground. The loss L is 136 dB.

The CCIR formula only brings in two new, but important, variables: h_T and h_R.

10.3.2.1.2 The Amended CCIR Equation. This amended model takes the following into account:

- Surface roughness
- Line-of-sight obstacles
- Buildings and trees

The resulting path loss equation is

$$L_{\text{dB}} = 40 \log d - 20 \log(h_T h_R) + \beta \tag{10.2}$$

where β represents the additional losses listed above but lumped together.

10.3.2.1.3 British Urban Path Loss Formula. The following formula was proposed by Allesbrook and Parsons (Ref. 5):

$$L_{\text{dB}} = 40 \log d_m - 20 \log(h_T h_R) + 20 + f/40 + 0.18L - 0.34H \tag{10.3}$$

where f = frequency (in MHz)

L = land usage factor, a percentage of the test area covered by buildings of any type, 0–100%

H = terrain height difference between Tx and Rx (i.e., Tx terrain height − Rx terrain height)

Example 1. Let $d = 2000$ m, $h_T = 30$ m, and $h_R = 3.3$ m; $f = 900$ MHz, $L = 50\%$, and $H = 27$ m.

$$h_T \times h_R = 100 \text{ m}^2$$

$$L_{\text{dB}} = 132 - 40 + 20 + 22.5 + 5.4 - 9.18$$

$$= 130.76 \text{ dB}$$

10.3.2.1.4 The Okumura Model. Okumura et al. (Ref. 6) carried out a detailed analysis for path predictions around Tokyo for mobile terminals. Hata (Ref. 7) published an empirical formula based on Okumura's results to predict path loss:

$$L_{\text{dB}} = 69.55 + 26.16 \log f - 13.82 \log h_t - A(h_r) + (44.9 - 6.55 \log h_t)(\log d) \quad (10.4)$$

where f is between 150 and 1500 MHz, h_t is between 30 and 300 m, and d is between 1 and 20 km. $A(h_r)$ is the correction factor for mobile antenna height and is computed as follows. For a small or medium-size city,

$$A(h_r) = (1.1 \log f - 0.7)h_r - (1.56 \log f - 0.8) \quad \text{(dB)} \quad (10.5a)$$

where h_r is between 1 and 10 m. For a large city,

$$A(h_r) = 3.2[\log(11.75h_r)]^2 - 4.97 \quad \text{(dB)} \quad (10.5b)$$

where $f \geq 400$ MHz.

Example 2. Let $f = 900$ MHz, $h_t = 40$ m, $h_r = 5$ m, and $d = 10$ km. Calculate $A(h_r)$ and L_{dB} for a medium-size city.

$$A(h_r) = 12.75 - 3.8 = 8.95 \text{ dB}$$

$$L_{\text{dB}} = 69.55 + 72.28 - 22.14 - 8.95 + 34.4$$

$$= 145.15 \text{ dB}$$

10.3.2.2 Building Penetration. For a modern multistory office block at 864 and 1728 MHz, path loss (L_{dB}) includes a value for clutter loss $L(v)$ and is expressed as follows:

$$L_{\text{dB}} = L(v) + 20 \log d + n_f a_f + n_w a_w \quad (10.6)$$

where the attenuation (in dB) of the floors and walls was a_f and a_w, and the number of floors and walls along the line d were n_f and n_w, respectively. The values of $L(v)$ at 864 and 1728 MHz were 32 and 38 dB, with standard deviations of 3 and 4 dB, respectively (Ref. 3).

Another source (Ref. 8) provided the following information. At 1650 MHz the floor loss factor was 14 dB, while the wall losses were 3–4 dB for double plasterboard and 7–9 dB for breeze block or brick. The parameter $L(v)$ was 29 dB. When the propagation frequency was 900 MHz, the first floor factor was 12 dB and $L(v)$ was 23 dB. The higher value for $L(v)$ at 1650 MHz was attributed to a reduced antenna aperture at this frequency compared to 900 MHz. For a 100-dB path loss, the base station and mobile terminal distance exceeded 70 m on the same floor, was 30 m for the floor above, and was 20 m for the floor above that, when the propagation frequency was 1650 MHz. The corresponding distances at 900 MHz were 70 m, 55 m, and 30 m. Results will vary from building to building depending on the type of construction of the building, the furniture and equipment it houses, and the number and deployment of people who populate it.

10.3.3 Microcell Prediction Model According to Lee

For this section a microcell is defined as a cell with a radius of 1 km or less. Such cells are used in heavily urbanized areas where demand for service is high and where large cell coverage would be spotty at best. With this model, line-of-sight conditions are seldom encountered; shadowing is the general rule, as shown in Figure 10.4. The major contribution to loss in such situations is due to the dimensions of intervening buildings.

Lee's model (Ref. 9) also includes an antenna height-gain function. Reference 9 reports 9 dB/oct or 30 dB/dec for an antenna height change. This would mean that if we doubled the height of an antenna, 9-dB transmission loss improvement would be achieved.

The Lee model for a microcell breaks the prediction process down into a received signal level (dBm) for the LOS component and then the attenuation due to the building blockage component.

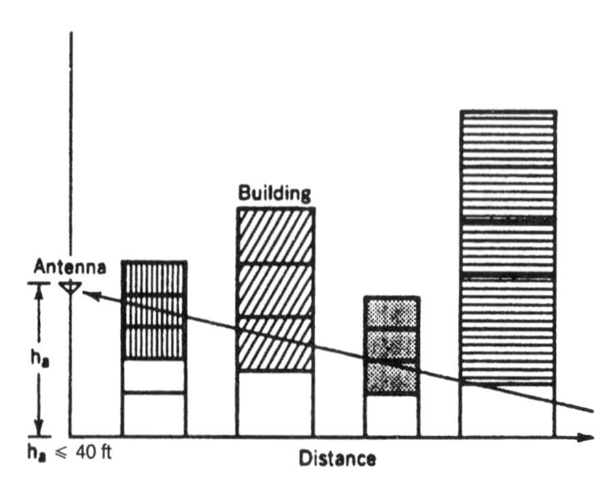

Figure 10.4. A propagation model typical of an urban microcell. (From Figure 2.24, Ref. 9.)

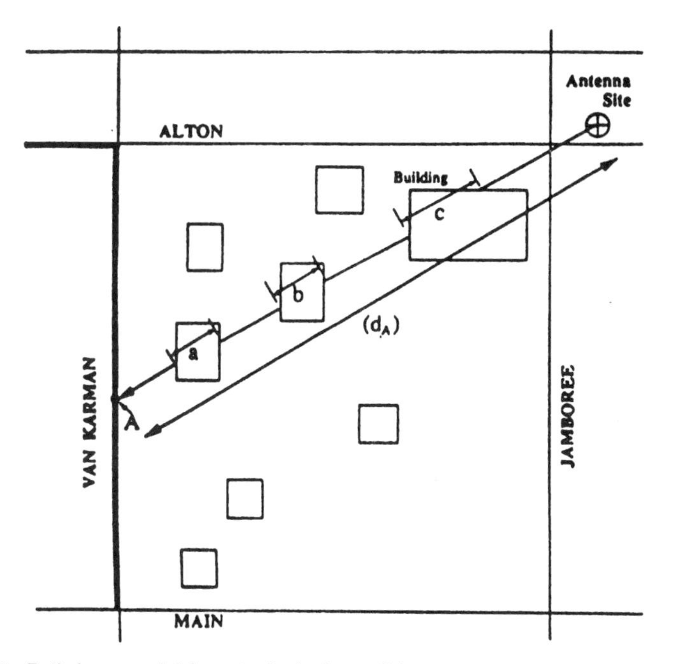

Figure 10.5. Path loss model for a typical microcell in an urban area. Mobile terminal is at location A. With antenna site so situated, there is blockage by buildings a, b, and c. Thus $B = a + b + c$. (From Figure 2.26, p. 90, Ref. 9.)

Figure 10.5 provides a microcell scenario, which we will use to understand Lee's model. The receive signal level P_r is equal to the receive signal level for LOS conditions (P_{LOS}) minus the blockage loss due to buildings, α_B. B is the blockage distance in feet. From Figure 10.5, $B = a + b + c$, the sum of the distances (in feet) through each building.

$$P_{\text{LOS}} = P_t - 77\text{ dB} - 21.5\log(d/100\text{ ft}) + 30\log(h_1/20\text{ ft}) \quad \text{for } 100\text{ ft} \leq B < 200\text{ ft} \tag{10.7a}$$

$$= P_t - 83.5\text{ dB} - 14\log(d/200\text{ ft}) + 30\log(h_1/20\text{ ft}) \quad \text{for } 200\text{ ft} \leq d < 1000\text{ ft} \tag{10.7b}$$

$$= P_t - 93.3\text{ dB} - 36.5\log(d/1000\text{ ft}) + 30\log(h_1/20\text{ ft}) \quad \text{for } 1000\text{ ft} \leq d < 5000\text{ ft} \tag{10.7c}$$

$$\alpha_B = 0 \qquad 1\text{ ft} \leq B \tag{10.8a}$$

$$\alpha_B = 1 + 0.5\log(B/10\text{ ft}) \qquad 1\text{ ft} \leq B < 25\text{ ft} \tag{10.8b}$$

$$\alpha_B = 1.2 + 12.5\log(B/25\text{ ft}) \qquad 25\text{ ft} \leq B \leq 600\text{ ft} \tag{10.8c}$$

$$\alpha_B = 17.95 + 3\log(B/600\text{ ft}) \qquad 600\text{ ft} \leq B < 3000\text{ ft} \tag{10.8d}$$

$$\alpha_B = 20\text{ dB} \qquad 3000\text{ ft} \leq B$$

where P_t is the ERP over a dipole in dBm, d is the total distance in feet, and h is the antenna height in feet.

Example 3. A mobile terminal is 500 ft from the cell site antenna, which is 30 ft high. There are three buildings in-line between the mobile terminal and the cell site antenna with cross section (in-line with the ray beam) distances of 50, 100, and 150 ft, respectively. Thus $B = 300$ ft. The ERP $= +30$ dBm (1 watt).

Use equation (10.8c):

$$\begin{aligned}\alpha_B &= 1.2 + 12.5\log(300\text{ ft}/25\text{ ft})\\ &= 1.2\text{ dB} + 13.5\text{ dB}\\ &= 14.7\text{ dB}\end{aligned}$$

The receive signal level, P_r, is

$$\begin{aligned}P_r &= -53.79\text{ dBm} - 14.7\text{ dB}\\ &= -68.47\text{ dBm}\end{aligned}$$

Of course, we must assume that the sum of the gain of the receive antenna and the transmission line loss equals 0 dB so that P_r is the same as the isotropic receive level.

Example 4. There is a 4000-ft separation between the cell site transmit antenna and the receive terminal. The transmit antenna is 40 ft high. There are four buildings causing blockage of 150 ft, 200 ft, 140 ft, and 280 ft. These distances are measured along the ray beam line. Thus $B = 770$ ft. The ERP is $+20$ dBm. What is the receive signal level (P_r) assuming no gain or loss in the receive antenna system?

Use equation (10.7c) for P_{LOS}:

$$\begin{aligned}P_{\text{LOS}} &= +20\text{ dBm} - 93.3\text{ dB} - 36.5\log(4000/1000) + 30\log(40/20)\\ &= -73.3\text{ dBm} - 21.4\text{ dB} + 9\text{ dB}\\ &= -85.67\text{ dBm}\end{aligned}$$

Now use equation (10.8d) to calculate α_B:

$$\begin{aligned}\alpha_B &= 17.95 + 3\log(770/600)\\ &= 18.28\text{ dB}\\ P_r &= -85.67\text{ dBm} - 18.28\text{ dB}\\ &= -103.94\text{ dBm}\end{aligned}$$

10.4 IMPAIRMENTS—FADING IN THE MOBILE ENVIRONMENT

10.4.1 Introduction

Fading in the mobile situation is quite different than in the static LOS microwave situation discussed in Chapter 2. Radio paths are not optimized as in the LOS environment. The mobile terminal may be fixed throughout a telephone call, but it is more apt to be in motion. Even the hand-held terminal may well have micromotion. When a terminal is in motion, the path characteristics are constantly changing.

Multipath propagation is the rule. Consider the simplified multipath pictorial model in Figure 10.6. Commonly, multiple rays reach the receive antenna, each with its own delay. The constructive and destructive fading can become quite complex. We must deal with both reflection and diffraction. Energy will arrive at the receive antenna reflected off sides of buildings, streets, lakes, and so on. Energy will also arrive diffracted from knife edges (e.g., building corners) and rounded obstacles (e.g., water tanks and hilltops).

Because the same signal arrives over several paths, each with a different electrical length, the phases of each path will be different, resulting in constructive and destructive amplitude fading. Fades of 20 dB are common, and even 30-dB fades can be expected.

On digital systems, the deleterious effects of multipath fading can be even more severe. Consider a digital bit stream to a mobile terminal with a transmission rate of 1000 bps. Assuming NRZ coding, the bit period would be 1 ms (bit period = 1/bit rate). We find the typical multipath delay spread may be on the order of 10 μs. Thus delayed energy will spill into a subsequent bit (or symbol) for the first 10 μs of the bit period and will have no negative effect on the bit decision. If the bit stream is 64,000 bps, then the bit period is 1/64,000 or 15 μs. Destructive energy from the previous bit (symbol) will spill into the first two-thirds of the bit period, well beyond the midbit sampling point. This is typical intersymbol interference (ISI), and in

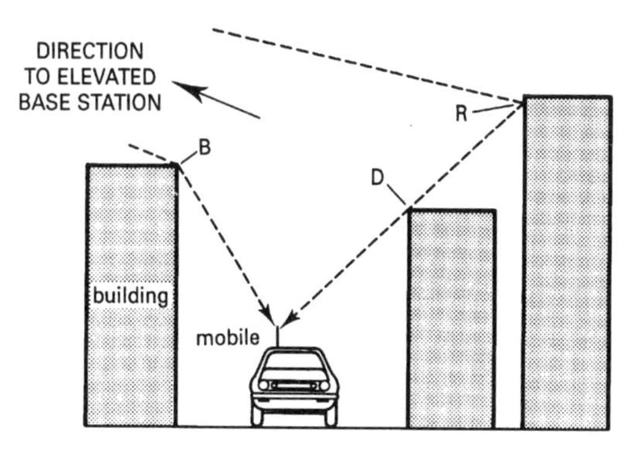

Figure 10.6. Mobile terminal in an urban setting. R = reflection, B and D = diffraction.

this case there is a high probability that there will be a bit error. The bottom line is that the destructive potential of ISI increases as the bit rate increases (i.e., as the bit period decreases).

10.4.2 Classification of Fading

We consider three types of channels to place bounds on radio system performance:

- Gaussian channel
- Rayleigh channel
- Rician channel

10.4.2.1 The Gaussian Channel. The Gaussian channel can be considered the ideal channel, and it is only impaired by additive white Gaussian noise (AWGN) developed internally by the receiver. We hope to achieve a BER typical of a Gaussian channel when we have done everything we can to mitigate fading and its results. These efforts could be diversity, equalization, and FEC coding with interleaving. The ideal Gaussian channel is very difficult to achieve in the mobile radio environment.

10.4.2.2 The Rayleigh Channel. The Rayleigh channel is at the other end of the line, often referred to as a worst-case channel. Remember, in Chapter 2, we treated fading on LOS microwave as Rayleigh fading, which gave us the very worst-case fading scenario. Figure 10.7 shows a channel approaching Rayleigh fading characteristics. Of course, we are dealing with multipath here. We showed that in the mobile radio scenario, multipath reception commonly had many components. Thus, if each multipath component is independent, the PDF (power distribution function) of its envelope is Rayleigh.

10.4.2.3 The Rician Channel. The characteristics of a Rician channel are in-between those of a Gaussian channel and those of a Rayleigh channel. The channels can be characterized by a function K (not to be confused with the K-factor in Chapter 2). K is defined as follows:

$$K = \frac{\text{power in the dominant path}}{\text{power in the scattered paths}} \tag{10.9}$$

As cells get smaller, the LOS component becomes more and more dominant. There are many cases, in fact nearly all cases where there is no full shadowing, in which there is a LOS component and scattered components. This is a typical multipath scenario. Turning now to equation (10.9), when

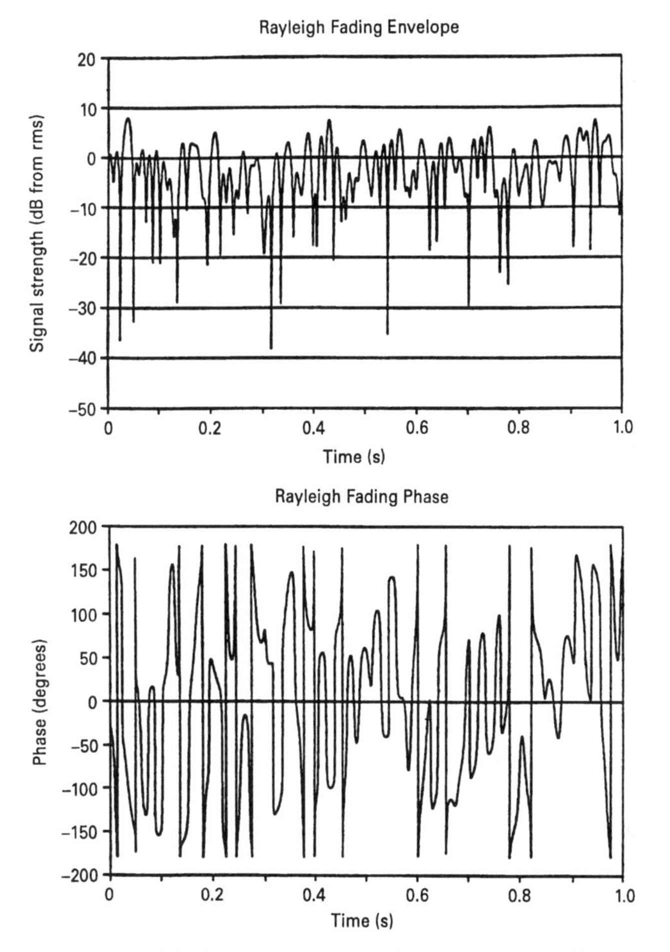

Figure 10.7. Typical Rayleigh fading envelope and phase in a mobile scenario. Vehicle speed is about 30 mph; frequency is 900 MHz. (From Figure 1.1, Ref. 3.)

$K = 0$, the channel is Rayleigh (i.e., the numerator is 0 and all the received energy derives from scattered paths). When $K = \infty$, the channel is Gaussian and the denominator is zero. Figure 10.8 gives BER values for some typical values of K. It shows that those intermediate values of K provide a superior BER than for the Rayleigh channel where $K = 0$. For a microcell mobile scenario, values of K vary from 5 to 30 (Ref. 3). Larger cells tend more toward low values of K.

There is also an advantage for Rician fading with higher values of K regarding co-channel interference performance for a desired BER. The smaller the cell, the more fading becomes Rician, approaching the higher values of K.

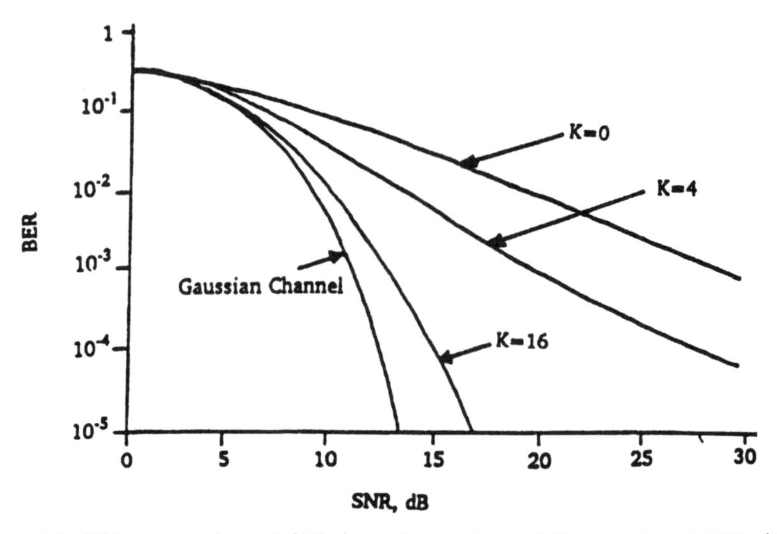

Figure 10.8. BER versus channel SNR for various values of K; noncoherent FSK. (From Figure 1.7, Ref. 3.)

10.4.3 Diversity—A Technique to Mitigate the Effects of Fading and Dispersion

10.4.3.1 Scope. We discuss diversity to reduce the effects of fading and to mitigate dispersion. Diversity was briefly covered in Chapter 2 where we dealt with LOS microwave radio system. In that chapter we discussed frequency and space diversity. There is a third diversity scheme called time diversity, which can be applied to digital cellular radio systems.

In principle, such techniques can be employed at the base station and/or at the mobile unit, although different problems have to be solved for each. The basic concept behind diversity is that when two or more radio paths carrying the same information are relatively uncorrelated, and when one path is in a fading condition, often the other path is not undergoing a fade. These separate paths can be developed by having two channels, separated in frequency. The two paths can also be separated in space, as well as in time.

When the two (or more) paths are separated in frequency, we call this frequency diversity. However, there must be at least some 2% or greater frequency separation for the paths to be comparatively uncorrelated. Because, in the cellular situation, we are so short of spectrum, using frequency diversity (i.e., using a separate frequency with redundant information) is essentially out of the question and will not be discussed further except for its implicit use in code division multiple access (CDMA).

10.4.3.2 Space Diversity. Space diversity is commonly employed at cell sites, and two separate receive antennas are required, separated in either the horizontal or vertical plane. Separation of the two antennas vertically can be impractical for cellular receiving systems. Horizontal separation, however, is quite practical. The space diversity concept is illustrated in Figure 10.9.

One of the most important factors in space diversity design is antenna separation. There are a set of rules for the cell site and another for the mobile unit.

Space diversity antenna separation, shown as distance D in Figure 10.9, varies not only as a function of the correlation coefficient but also as a function of antenna height, h. The wider the antennas are separated, the lower the correlation coefficient is and the more uncorrelated the diversity paths are. Sometimes we find that, by lowering the antennas as well as adjusting the distance between the antennas, we can achieve a very low correlation coefficient. However, we might lose some of the height-gain factor.

Lee (Ref. 9) proposes a new parameter, η, where

$$\eta = \frac{\text{antenna height}}{\text{antenna separation}} = \frac{h}{d} \tag{10.10}$$

In Figure 10.10 we relate the correlation coefficient (ρ) with η. The orientation of the antenna regarding the incoming signal from the mobile unit is α. Lee recommends a value of $\rho = 0.7$. Lower values are unnecessary because of the law of diminishing returns. There is much more fading advantage achieved from $\rho = 1.0$ to $\rho = 0.7$ than from $\rho = 0.7$ to $\rho = 0.1$.

Based on $\rho = 0.7$ and $\eta = 11$, from Figure 10.10, we can calculate antenna separation values (for 850-MHz operation). For example, if $h =$ 50 ft (15 m), we can calculate d using formula (10.10):

$$d = h/\eta = 50/11 = 4.5 \text{ ft } (1.36 \text{ m})$$

For an antenna 120 ft (36.9 m) high, we find that $d = 120/11 = 10.9$ ft or 3.35 m (from Ref. 9).

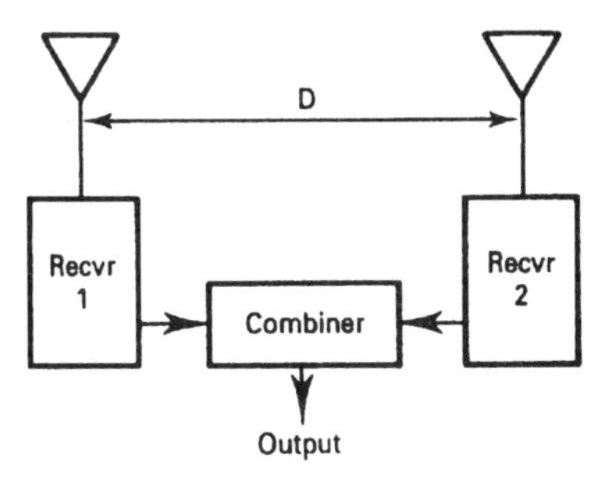

Figure 10.9. The space-diversity concept.

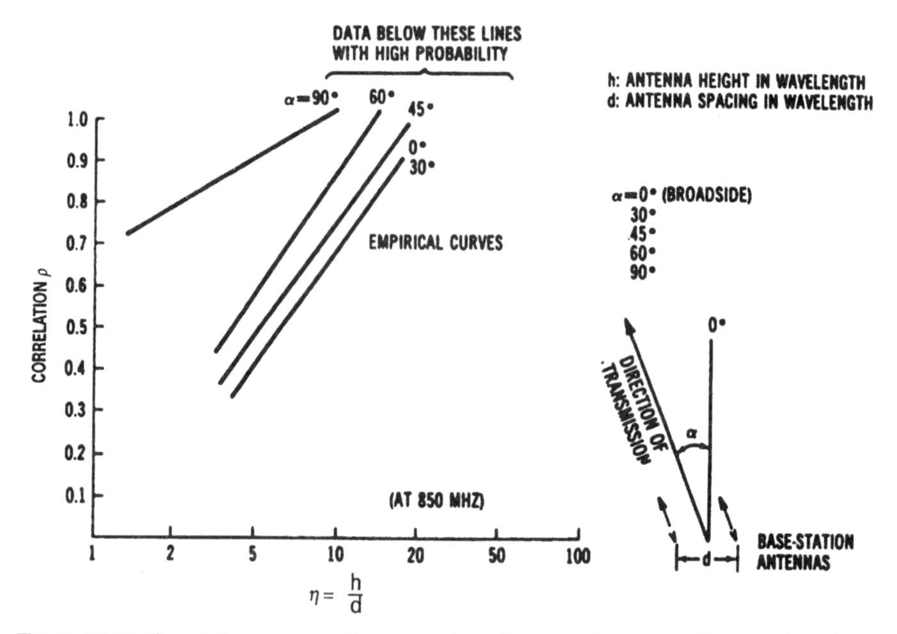

Figure 10.10. Correlation ρ versus the parameter η for two antennas in different orientations. (From Ref. 9; reprinted with permission.)

10.4.3.2.1 Space Diversity on a Mobile Platform. Lee (Ref. 9) discusses both vertically separated and horizontally separated antennas on a mobile unit. For the vertical separation case, 1.5λ is recommended, while for the horizontal separation case, 0.5λ is recommended. At 850 MHz, $\lambda =$ 35.29 cm. Then $1.5\lambda = 1.36$ ft or 52.9 cm. For 0.5λ, the value is 0.45 ft or 17.64 cm.

10.4.3.3 Frequency Diversity. We pointed out that conventional frequency diversity was not a practical alternative in cellular systems because of the shortage of available bandwidth. However, with CDMA (direct sequence spread spectrum), depending on the frequency spread, two or more frequency diversity paths are available. In most CDMA systems we have what is called *implicit diversity*, and multipaths can be resolved with the use of a RAKE filter. This is one of the many advantages of CDMA.

***10.4.3.4 Forward Error Correction*—A Form of Time Diversity.** Forward error correction (FEC) (see Chapter 4) can be used on digital cellular systems not only to improve bit error rate but to reduce fading. To reduce the effects of fading, a FEC system must incorporate an *interleaver.*

An interleaver pseudorandomly shuffles bits. It first stores a span of bits and shuffles them using a generating polynomial. The span of bits can represent a time period. The rule is that for effective operation against burst errors,* the interleaving span must be much greater than the typical fade duration. The deinterleaver used at the receive end is time synchronized to the interleaver incorporated at the transmit end of the link.

10.4.4 Cellular Radio Path Calculations

Consider the path from the fixed cell site to the mobile platform. There are several mobile receiver parameters that must be considered. The first we derive from EIA/TIA IS-19 (Ref. 10). The minimum SINAD (signal + interference + noise and distortion to interference + noise + distortion ratio) of 12 dB. This SINAD equates to a threshold of -116 dBm or 7 μV/m. This assumes a cellular transceiver with an antenna with a net gain of 1 dBd (dB over a dipole). The gross antenna gain is 2.5 dBd with a 1.5-dB transmission line loss. A 1-dBd gain is equivalent to a 3.16-dBi gain (i.e., 0 dBd = 2.16 dBi). Furthermore, this value equates to an isotropic receive level of -119.16 dBm (Ref. 10).

One design goal for a cellular system is to more or less maintain a cell boundary at the 39-dBμ contour (Ref. 11): 39 dBμ = -95 dBm (based on a 50-ohm impedance at 850 MHz). Then, at this contour, a mobile terminal would have a 24.16-dB fade margin.

If a cellular transmitter had a 10-watt output per channel and an antenna gain of 12 dBi and 2-dB line loss, the EIRP would be +20 dBW or +50 dBm. The maximum path loss to the 39-dBμ contour would be +50 dBm − (-119.16 dBm) or 169 dB.

10.5 THE CELLULAR RADIO BANDWIDTH DILEMMA

10.5.1 Background and Objectives

The present cellular radio bandwidth assignment in the 800- and 900-MHz band cannot support the demand for cellular service, especially in urban areas in the United States and Canada. AMPS, widely used in the United States, Canada, and many other nations of the Western Hemisphere, requires 30 kHz per voice channel. This system can be called a FDMA system, much like the FDMA/DAMA system described in Chapter 6. We remember that the analog voice channel is a nominal 4 kHz, and 30 kHz is seven times that value.

The trend is to convert to digital. Digital transmission is notorious for being wasteful of bandwidth, when compared to the 4-kHz analog channel. We can show that PCM has a 16-times multiplier of the 4-kHz analog channel. In other words, the standard digital voice channel occupies 64 kHz (assuming 1 bit per hertz of bandwidth).

*Burst errors are bunched errors due to fading.

One goal of system designers, therefore, is to reduce the required bandwidth of the digital voice channel without sacrificing too much voice quality. As we will show, they have been quite successful.

The real objective is to increase the ratio of users to unit bandwidth. We will describe two distinctly different methods, each claiming to be more bandwidth conservative than the other. The first method is TDMA (time division multiple access) and the second is CDMA. The former was described in Chapter 6 and the latter was briefly mentioned.

10.5.2 Bit Rate Reduction of the Digital Voice Channel

It became obvious to system designers that conversion to digital cellular required some different techniques for coding speech other than conventional PCM found in the PSTN. The following lists some of the techniques that may be considered.

- ADPCM (adaptive differential PCM). Good intelligibility and good quality 32-kbps. Data transmission over the channel may be questionable at bit rates $\geq$ 9600 bps.
- Linear predictive vocoders (voice coders). 2400 bps, adopted by U.S. Department of Defense. Good intelligibility but poor quality, especially speaker recognition.
- Subband coding (SBC). Good intelligibility even down to 4800 bps. Quality suffers below 9600 bps.
- RELP (residual excited linear predictive) type coder. Good intelligibility down to 4800 bps and fair to good quality. Quality improves as bit rate increases. Good quality at 16 kbps.
- CELP (codebook excitation linear predictive) type coder. Good intelligibility and surprisingly good quality even down to 4800 bps. At 8 kbps, near-toll quality speech.

10.6 NETWORK ACCESS TECHNIQUES

10.6.1 Introduction

The objective of a cellular radio operation is to provide a service where mobile subscribers can communicate with any subscriber in the PSTN, where any subscriber in the PSTN can communicate with any mobile subscriber, and where mobile subscribers can communicate among themselves via the cellular radio system. In all cases the service is full duplex.

A cellular service company is allotted a radio bandwidth segment to provide this service. Ideally, for full duplex service, a portion of the bandwidth is assigned for transmission from a cell site to mobile subscriber, and another portion is assigned for transmission from a mobile user to a cell site. Our goal here is to select an "access" method to provide this service given our bandwidth constraints.

We will discuss three generic methods of access: FDMA, TDMA, and CDMA. It might be useful for the reader to review our discussion of satellite access in Chapters 6 and 7 where we describe FDMA and TDMA. However, in this section, the concepts are the same, but some of our constraints and operating parameters will be different.

10.6.2 Frequency Division Multiple Access (FDMA)

With FDMA, our band of frequencies is divided into segments and each segment is available for one user access. Half of the contiguous segments are assigned to outbound cell sites (i.e., to mobile users) and the other half to inbound cell sites. A guard band is usually provided between outbound and inbound contiguous channels. This concept is shown in Figure 10.11.

Because of our concern to optimize the number of users per unit bandwidth, the key question is the actual width of one user segment. The North American AMPS system was described in Section 10.2 where each segment width was 30 kHz. The bandwidth of a user segment is greatly determined by the information bandwidth and modulation type. With AMPS, the information bandwidth was a single voice channel with a nominal bandwidth of 4 kHz. The modulation was FM and the bandwidth was then determined by Carson's rule (Chapter 2). As we pointed out, AMPS is not exactly spectrum-conservative. On the other hand, it has a lot of the redeeming features that FM provides such as the noise and interference advantage (FM capture).

Another approach to FDMA would be to convert the voice channel to its digital equivalent using CELP, for example (Section 10.5.2), with a transmission rate of 4.8 kbps. The modulation might be BPSK using a raised cosine filter where the bandwidth could be 1.25% of the bit rate or 6 kHz per voice channel. This alone would increase voice channel capacity five times over AMPS with its 30 kHz per channel. It should be noted that a radio carrier is normally required for each frequency slot.

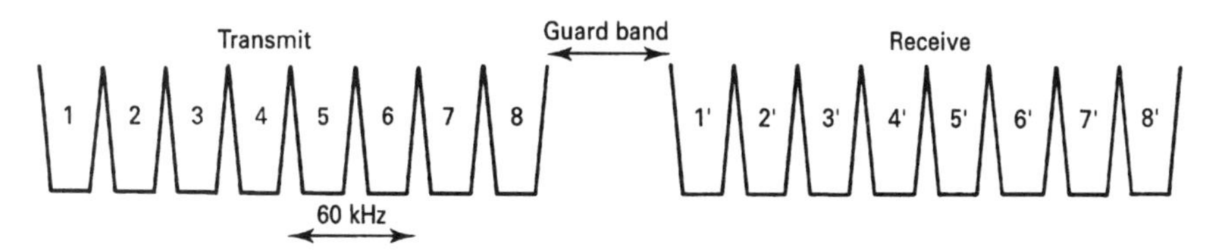

Figure 10.11. A conceptual drawing of FDMA.

10.6.3 Time Division Multiple Access (TDMA)

With TDMA we work in the time domain rather than the frequency domain of FDMA. Each user is assigned a timeslot rather than a frequency segment and during the user's turn, the full frequency bandwidth is available for the duration of the user's assigned timeslot.

Let's say that there are n users and so there are n timeslots. In the case of FDMA, we had n frequency segments and n radio carriers, one for each segment. For the TDMA case, only one carrier is required. Each user gains access to the carrier for $1/n$ of the time and there is generally an ordered sequence of timeslot turns. A TDMA frame can be defined as cycling through n users' turns just once.

A typical TDMA frame structure is shown in Figure 10.12. One must realize that TDMA is only practical with a digital system such as PCM or any of those discussed in Section 10.5.2. TDMA is a store-and-burst system. Incoming user traffic is stored in memory, and when that user's turn comes up, that accumulated traffic is transmitted in a digital burst.

Suppose there are ten users. Let each user's bit rate be R, then a user's burst must be at least $10R$. Of course, the burst will be greater than $10R$ to accommodate a certain amount of overhead bits as shown in Figure 10.12.

We define downlink as outbound, base station to mobile station(s), and uplink as mobile station to base station. Typical frame periods are:

North American IS-54	40 ms for six timeslots
European GSM	4.615 ms for eight timeslots

One problem with TDMA, often not appreciated by a novice, is delay. In particular, this is the delay on the uplink. Consider Figure 10.13, where we set up a scenario. A base station receives mobile timeslots in a circular pattern and the radius of the circle of responsibility of that base station is 10 km. Let the velocity of a radio wave be 3×10^8 m/s. The time to traverse 1 km is $1000/(3 \times 10^8)$ or 3.333 μs. Making up an uplink frame is a mobile right on top of the base station with essentially no delay and another mobile right at 10 km with 10×3.33 μs or 33.3-μs delay. A GSM timeslot is about 576 μs in duration. The terminal at the 10-km range will have its timeslot arriving 33.3 μs late compared to the terminal with no delay. A GSM bit period is about 3.69 μs so that the late arrival eats up roughly 10 bits, and unless something is done, the last bit of the burst will overlap the next burst.

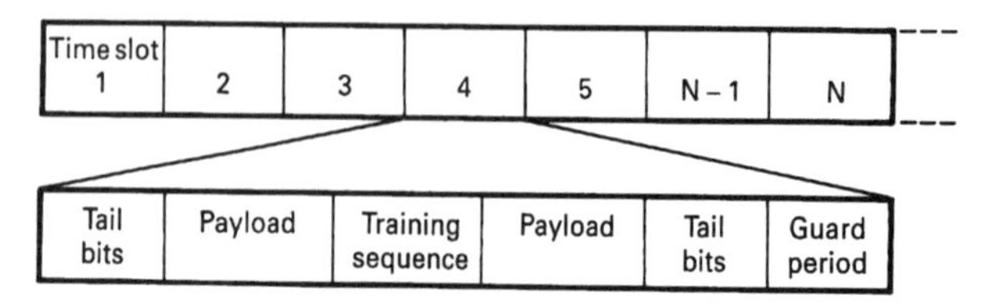

Figure 10.12. A simplified TDMA frame.

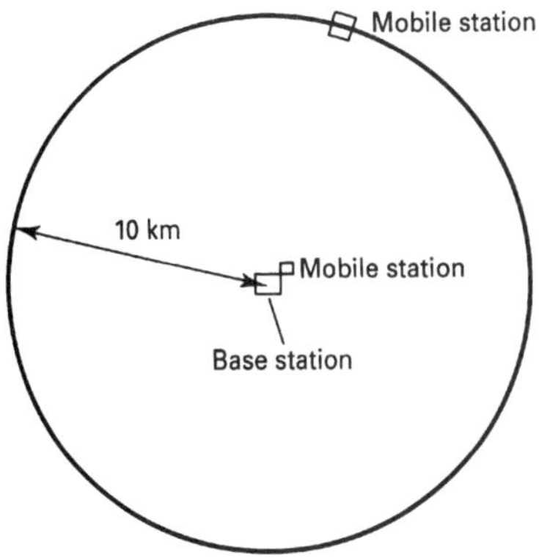
Mobile station
10 km
Mobile station
Base station

Figure 10.13. A TDMA delay scenario.

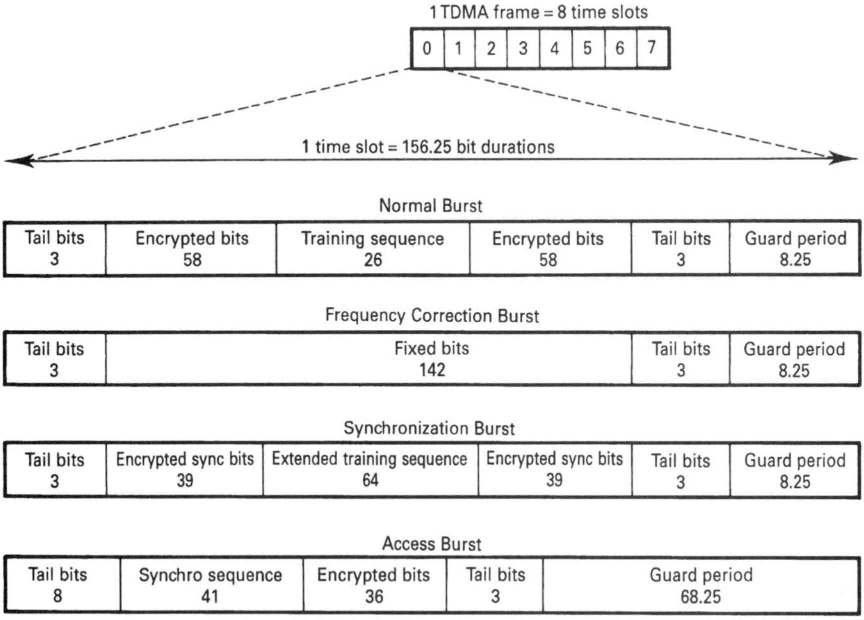
1 TDMA frame = 8 time slots
0 1 2 3 4 5 6 7
1 time slot = 156.25 bit durations
Normal Burst
Tail bits 3
Encrypted bits 58
Training sequence 26
Encrypted bits 58
Tail bits 3
Guard period 8.25
Frequency Correction Burst
Tail bits 3
Fixed bits 142
Tail bits 3
Guard period 8.25
Synchronization Burst
Tail bits 3
Encrypted sync bits 39
Extended training sequence 64
Encrypted sync bits 39
Tail bits 3
Guard period 8.25
Access Burst
Tail bits 8
Synchro sequence 41
Encrypted bits 36
Tail bits 3
Guard period 68.25

Figure 10.14. GSM frame and burst structures. (From Figure 8.7, Ref. 3; reprinted with permission.)

Refer now to Figure 10.14, which illustrates GSM burst structures. Note that the access burst has a guard period of 68.25 bit durations or a "slop" of 3.69×68.25 μs, which will well accommodate the late arrival of the 10-km mobile terminal of only 33.3 μs.

To provide the same long guard period in the other bursts is a waste of valuable "spectrum."* The GSM system overcomes this problem by using

*We are equating bit rate, or bit durations, to bandwidth. One could assume 1 bit/Hz as a first-order estimate.

adaptive frame alignment. When the base station detects a 41-bit random access synchronization sequence with a long guard period, it measures the received signal delay relative to the expected signal from a mobile station with zero range. This delay, called the timing advance, is transmitted to the mobile station using a 6-bit number. As a result, the mobile station advances its time base over the range of 0 to 63 bits (i.e., in units of 3.69 μs). By this process the TDMA bursts arrive at the base station in their correct timeslots and do not overlap with adjacent ones. As a result, the guard period in all other bursts can be reduced to 8.25×3.69 μs or approximately 30.46 μs, the equivalent of 8.25 bits only. Under normal operations, the base station continuously monitors the signal delay from the mobile station and thus instructs the mobile station to update its time advance parameter. In very large traffic cells there is an option to actively utilize every second timeslot only to cope with the larger propagation delays. This is spectrally inefficient, but in large, low-traffic rural cells, it is admissible (from Ref. 3).

10.6.3.1 Some Comments on TDMA Efficiency. Multichannel FDMA can operate with a power amplifier for every channel, or with a common wideband power amplifier for all channels. With the latter, we are setting up a typical generator of intermodulation products as these carriers mix in a comparatively nonlinear common power amplifier. To reduce the level of IM products, just like in satellite communications discussed in Chapter 6, backoff of power amplifier drive is required. This backoff can be on the order of 3–6 dB.

With TDMA (downlink), only one carrier is present on the power amplifier, thus removing most of the causes of IM noise generation. Thus with TDMA, the power amplifier can be operated to full saturation—a distinct advantage. FDMA required some guard band between frequency segments; there are no guard bands with TDMA. However, as we saw above, a guard time between uplink timeslots is required to accommodate the following situations:

- Timing inaccuracies due to clock instabilities
- Delay spread due to propagation*
- Transmission time delay due to propagation distance (see Section 10.6.3 above)
- Tails of pulsed signals due to transient response

The longer guard times are extended, the more inefficient a TDMA system is.

*Lee (Ref. 9) reports a typical urban delay spread of 3 μs.

10.6.3.2 Advantages of TDMA. The introduction of TDMA results in a much improved system signaling operation and cost. Assuming a 25-MHz bandwidth, up to 23.6 times capacity can be achieved with North American TDMA compared to FDMA (AMPS) (Table 2, Ref. 12).

A mobile station can exchange system control signals with the base station without interruption of speech (or data) transmission. This facilitates the introduction of new network and user services. The mobile station can also check the signal level from nearby cells by momentarily switching to a new timeslot and radio channel. This enables the mobile station to assist with handover operations and thereby improve the continuity of service in response to motion or signal fading conditions. The availability of signal strength information at both the base and mobile stations, together with suitable algorithms in the station controllers, allows further spectrum efficiency through the use of dynamic channel assignment and power control.

The cost of base stations using TDMA can be reduced when radio equipment is shared by several traffic channels. A reduced number of transceivers leads to a reduction of multiplexer complexity. Outside the major metropolitan areas, the required traffic capacity for a base station may, in many cases, be served by one or two transceivers. The saving in the number of transceivers results in a significantly reduced overall cost.

A further advantage of TDMA is increased system flexibility. Different voice and nonvoice services may be assigned a number of timeslots appropriate to the service. For example, as more efficient speech CODECs are perfected, increased capacity may be achieved by the assignment of a reduced number of timeslots for voice traffic. TDMA also facilitates the introduction of digital data and signaling services as well as the possible later introduction of such further capacity improvements as digital speech interpolation (DSI).

Table 10.1 compares three operational/planned digital TDMA systems.

10.6.4 Code Division Multiple Access (CDMA)

CDMA means spread spectrum multiple access. There are two types of spread spectrum: frequency hop and direct sequence (sometimes called pseudonoise). In the cellular environment, CDMA means direct sequence spread spectrum (Ref. 3). However, the GSM system uses frequency hop, but not as an access technique.

Using spread spectrum techniques accomplishes just the opposite of what we were trying to accomplish in Chapter 3. Bit packing is used to conserve bandwidth by packing as many bits as possible in 1 Hz of bandwidth. With spread spectrum we do the reverse by spreading the information signal over a very wide bandwidth.

Conventional AM requires about twice the bandwidth of the audio information signal with its two sidebands of information (i.e., approximately ± 4 kHz). On the other hand, depending on its modulation index, frequency

TABLE 10.1 Three TDMA Systems Compared

Feature	GSM	North America	Japan
Class of emission			
Traffic channels	271KF7W	40K0G7WDT	tbd[a]
Control channels	271KF7W	40K0G1D	tbd
Transmit frequency bands (MHz)			
Base stations	935–960	869–894	810–830 (1.5 GHz tbd)
Mobile stations	890–915	824–849	940–960 (1.5 GHz tbd)
Duplex separation (MHz)	45	45	130 48 (1.5 GHz)
RF carrier spacing (kHz)	200	30	25 interleaved 50
Total number of RF duplex channels	124	832	tbd
Maximum base station ERP (W)			
Peak RF carrier	300	300	tbd
Traffic channel average	37.5	100	tbd
Nominal mobile station transmit power (W): peak and average	20 and 2.5 8 and 1.0 5 and 0.625 2 and 0.25	9 and 3 4.8 and 1.6 1.8 and 0.6 tbd and tbd	tbd
Cell radius (km)			
Minimum	0.5	0.5	0.5
Maximum	35 (up to 120)	20	20
Access method	TDMA	TDMA	TDMA
Traffic channels/RF carrier			
Initial	8	3	3
Design capability	16	6	6
Channel coding	Rate $\frac{1}{2}$ convolutional code with interleaving plus error detection	Rate $\frac{1}{2}$ convolutional code	tbd
Control channel structure			
Common control channel	Yes	Shared with AMPS	Yes
Associated control channel	Fast and slow	Fast and slow	Fast and slow
Broadcast control channel	Yes	Yes	Yes
Delay spread equalization capability (μs)	20	60	tbd
Modulation	GMSK[b] (BT = 0.3)	$\pi/4$ diff.[c] encoded QPSK (rolloff = 0.25)	$\pi/4$ diff. encoded QPSK (rolloff = 0.5)
Transmission rate (kbps)	270.833	48.6	37–42
Traffic channel structure			
Full-rate speech CODEC			
Bit rate (kbps)	13.0	8	6.5–9.6
Error protection	9.8 kbps FEC + speech processing	5 kbps FEC	~ 3 kbps FEC
Coding algorithm	RPE-LTP	CELP	tbd

TABLE 10.1 (*continued*)

Feature	GSM	North America	Japan
Half-rate speech CODEC			
Initial	tbd	tbd	tbd
Future	Yes	Yes	Yes
Data			
Initial net rate (kbps)	Up to 9.6	2.4, 4.8, 9.6	1.2, 2.4, 4.8
Other rates (kbps)	Up to 12	tbd	8 and higher
Handover			
Mobile assisted	Yes	Yes	Yes
Intersystem capability with existing analog system	No	Between digital and AMPS	No
International roaming capability	Yes, > 16 countries	Yes	Yes
Design capability for multiple system operators in same area	Yes	Yes	Yes

[a]tbd = to be defined.
[b]GMSK = Gaussian minimum shift keying.
[c]diff. = differentially.
Source: Table 1, pp. 120–123, CCIR Rep. 1156 (Ref. 12).

modulation could be considered a type of spread spectrum in that it produces a much wider bandwidth than its transmitted information requires. As with all other spread spectrum systems, a signal-to-noise advantage is gained with FM, depending on its modulation index. For example, with AMPS, a typical FM system, 30 kHz is required to transmit the nominal 4-kHz voice channel.

If we are spreading a voice channel over a very wide frequency band, it would seem that we are defeating the purpose of frequency conservation. With spread spectrum, with its powerful antijam properties, multiple users can transmit on the same frequency with only minimal interference between one another. This assumes that each user is employing a different key variable (i.e., in essence, using a different time code). At the receiver, the CDMA signals are separated using a correlator that accepts only signal energy from the selected key variable binary sequence (code) used at the transmitter, and then despreads its spectrum. CDMA signals with unmatching codes are not despread and only contribute to the random noise.

CDMA reportedly provides increases in capacity 15 times that of its analog FM counterpart. It can handle any digital format at the specified input bit rate such as facsimile, data, and paging. In addition, the amount of transmitter power required to overcome interference is comparatively low when utilizing CDMA. This translates into savings on infrastructure (cell site) equipment and longer battery life for hand-held terminals. CDMA also provides so-called soft handoffs from cell site to cell site that make the transition virtually inaudible to the user (Ref. 13).

Dixon (Ref. 14) develops from Claude Shannon's classical relationship an interesting formula to calculate the spread bandwidth given the information rate, signal power, and noise power:

$$C = W \log_2(1 + \mathrm{S/N}) \tag{10.11}$$

where C = capacity of a channel in bits per second
W = bandwidth in hertz
S = signal power
N = noise power

This equation shows the relationship between the ability of a channel to transfer error-free information, compared with the signal-to-noise ratio existing in the channel, and the bandwidth used to transmit the information.

If we let C be the desired system information rate and we change the logarithm base to the natural base (e), the result is

$$C/W = 1.44 \log_e(1 + \mathrm{S/N}) \tag{10.12}$$

and, for S/N very small (e.g., ≤ 0.1), which would be used in an antijam system,* we can say

$$C/W = 1.44(\mathrm{S/N}) \tag{10.13}$$

From this equation we find that

$$\mathrm{N/S} = 1.44\, W/C \approx W/C \tag{10.14}$$

and

$$W \approx C(\mathrm{N/S}) \tag{10.15}$$

This exercise shows that for any given S/N we can have a low information-error rate by increasing the bandwidth used to transfer that information.

Suppose we had a cellular system using a data rate of 4.8 kbps and a S/N of 20 dB (numeric of 100). Then the bandwidth for this 4.8-kbps channel would be

$$\begin{aligned} W &= 100 \times 4.8 \times 10^3/1.44 \\ &= 333.333 \text{ kHz} \end{aligned}$$

There are two common ways that information can be embedded in the spread spectrum signal. One way is to add the information to the spectrum-

*If we think about it, a cellular scenario where one user is transmitting right on top of others on the same frequency, plus adjacent channel interference, is indeed an antijam situation.

spreading code before the spreading modulation stage. It is assumed that the information to be transmitted is binary because modulo-2 addition is involved in this process. The second method is to modulate the RF carrier with the desired information before spreading the carrier. The modulation is usually PSK or FSK or other angle modulation scheme (Ref. 14).

Dixon (Ref. 14) lists some advantages of the spread spectrum:

1. Selective addressing capability.
2. Code division multiplexing possible for multiple access.
3. Low-density power spectra for signal hiding.
4. Message security.
5. Interference rejection.

And of most importance for the cellular user (Ref. 14), "when codes are properly chosen for low cross correlation, minimum interference occurs between users, and receivers set to use different codes are reached only by transmitters sending the correct code. Thus more than one signal can be unambiguously transmitted at the same frequency and at the same time; selective addressing and code-division multiplexing are implemented by the coded modulation format."

Figure 10.15 shows a direct sequence (pseudonoise) spread spectrum system with waveforms.

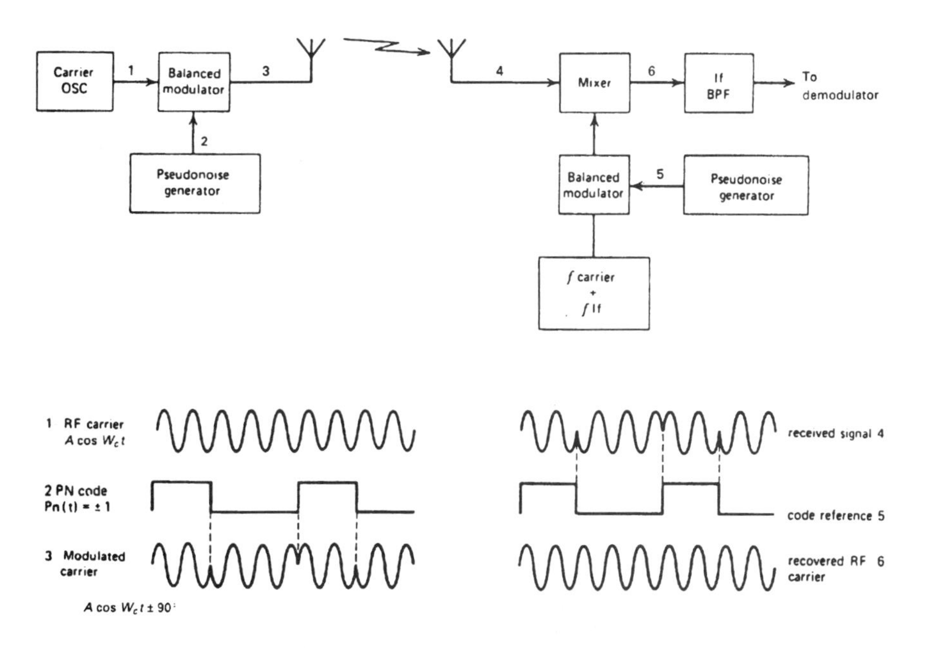

Figure 10.15. A direct sequence spread spectrum system showing waveforms. (From Figure 2.3, Ref. 14; reprinted with permission.)

Processing gain is probably the most commonly used parameter to describe the performance of a spread spectrum system. It quantifies the signal-to-noise ratio improvement when a spread signal is passed through a "processor." For instance, if a certain processor had an input S/N of 12 dB and an output S/N of 20 dB, the processing gain would then be 8 dB.

Processing gain is expressed by the following:

$$G_p = \frac{\text{spread bandwidth (in Hz)}}{\text{information bit rate}} \tag{10.16}$$

More commonly, processing gain is given in a dB value:

$$G_{p(\text{dB})} = 10\log\left(\frac{\text{spread bandwidth (in Hz)}}{\text{information bit rate}}\right) \tag{10.17}$$

Example 5. A certain cellular system voice channel information rate is 9.6 kbps and the RF spread bandwidth is 9.6 MHz. What is the processing gain?

$$\begin{aligned} G_{p(\text{dB})} &= 10\log(9.6 \times 10^6) - 10\log 9600 \\ &= 69.8 - 39.8 \quad (\text{dB}) \\ &= 30 \text{ dB} \end{aligned}$$

It has been pointed out in Ref. 1 that the power control problem held back the implementation of CDMA for cellular application. If the standard deviation of the received power from each mobile at the base station is not controlled to an accuracy of approximately ± 1 dB relative to the target receive power, the number of users supported by the system can be reduced significantly. Other problems to be overcome were synchronization and sufficient codes available for a large number of mobile users (Ref. 3).

Qualcomm, a North American company, has a CDMA design that overcomes these problems and has fielded a cellular system based on CDMA. It operates at the top of the AMPS band using 1.23 MHz for each uplink and downlink. This is the equivalent of 41 AMPS channels (i.e., 30 kHz $\times$ 41 = 1.23 MHz) deriving up to 62 CDMA channels (plus one pilot channel and one synchronization channel) or some 50% capacity increase. The Qualcomm system also operates in the 1.7–1.8-GHz band (Ref. 3). EIA/TIA IS-95 is based on the Qualcomm system. Its processing gain, when using the 9600-bps information rate, is $1.23 \times 10^6/9600$ or about 21 dB.

***10.6.4.1 Correlation*—Key Concept in Direct Sequence Spread Spectrum.** In direct sequence (DS) spread spectrum systems, the chip rate is equivalent to the code generator clock rate. Simplistically, a chip can be considered an element of RF energy with a certain recognizable binary phase

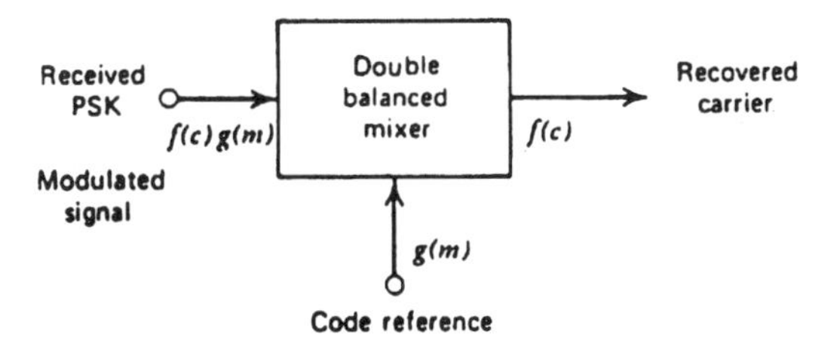

Figure 10.16. In-line correlator.

characteristic. A chip (or chips) is (are) a result of direct sequence spreading by biphase modulating a RF carrier. Being that each chip has a biphase modulated characteristic, we can identify each with a binary 1 or binary 0. These chips derive from biphase modulating a carrier where the modulation is controlled by a pseudorandom sequence. If the sequence is long enough, without repeats, it is considered pseudonoise. The sequence is controlled by a key that is unique to our transmitter and its companion remote receiver. Of course, the receiver must be time aligned and synchronized with its companion transmitter. A block diagram of this operation is shown in Figure 10.16. It is an in-line correlator.

Let's look at an information bit divided into seven chips coded by a PN sequence $+++-+--$ and shown in Figure 10.17a. Now replace the in-line correlator with a matched filter. In this case the matched filter is an electrical delay line tapped at delay intervals that correspond to the chip time duration. Each tap in the delay line feeds into an arithmetic operator matched in sign to each chip in the coded sequence. If each delay line tap has the same sign (phase shift) as the chips in the sequence, we have a match. This is shown in Figure 10.17b. As shown here, the short sequence of seven chips enhanced the desired signal seven times. This is the output of the modulo-2 adder, which has an output voltage seven times greater than the input voltage of one chip.

In Figure 10.17c we show the correlation process collapsing the spread signal spectrum to that of the original bit spectrum when the receiver reference signal, based on the same key as the transmitter, is synchronized with the arriving signal at the receiver.

Of overriding importance is that only the desired signal passes through the matched filter delay line (adder). Other users on the same frequency have a different key and do not correlate. These "other" signals are rejected. Likewise, interference from other sources is spread; there is no correlation and those signals also are rejected.

Direct sequence spread spectrum offers two other major advantages for the system designer. It is more forgiving in a multipath environment than conventional narrowband systems. No intersymbol interference (ISI) will be generated if the coherent bandwidth is greater than the information symbol bandwidth.

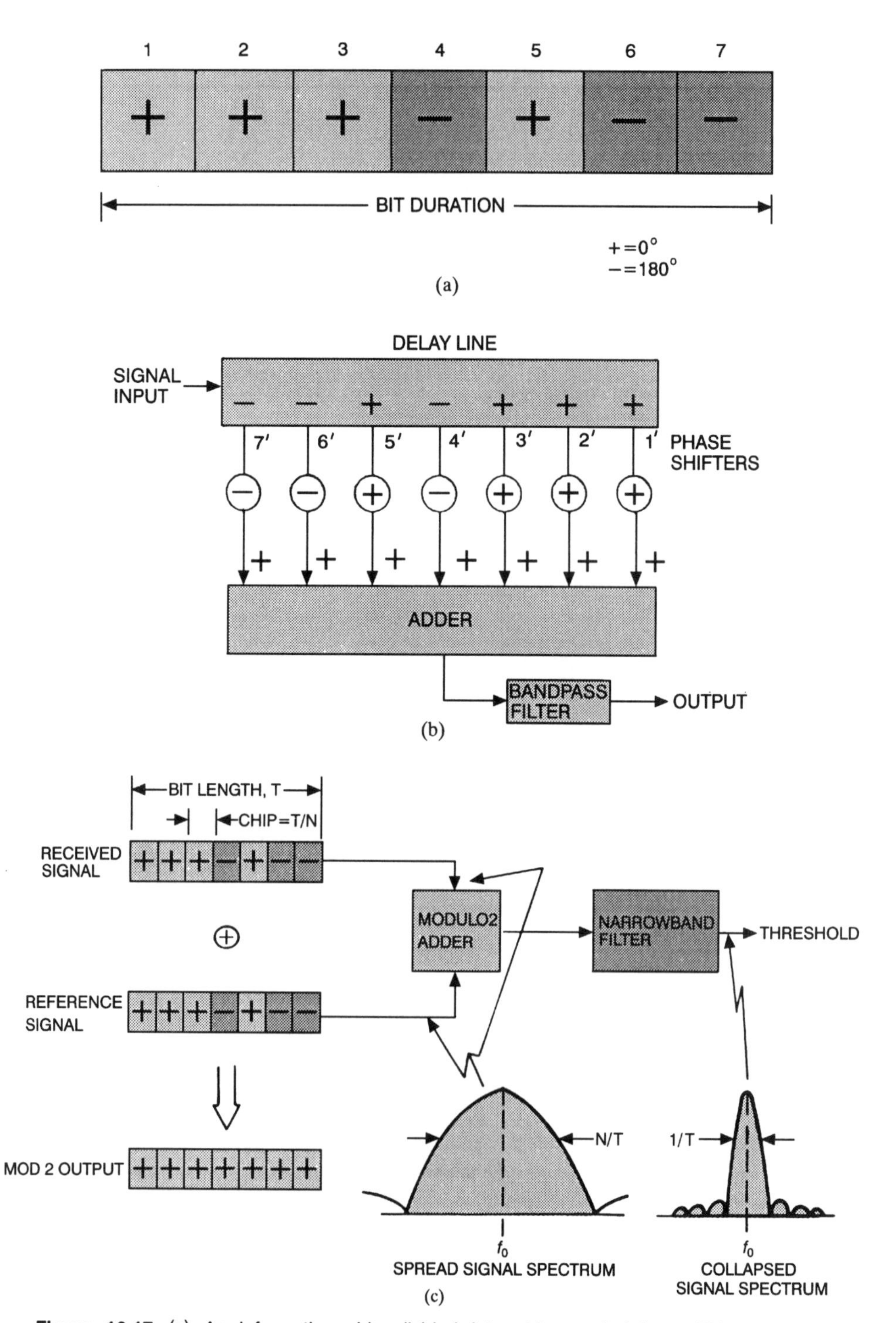

Figure 10.17. (a) An information chip divided into chips coded by a PN sequence. (b) Matched filter for seven-chip PN code. (c) The correlation process collapses the spread signal spectrum to that of the original bit spectrum.

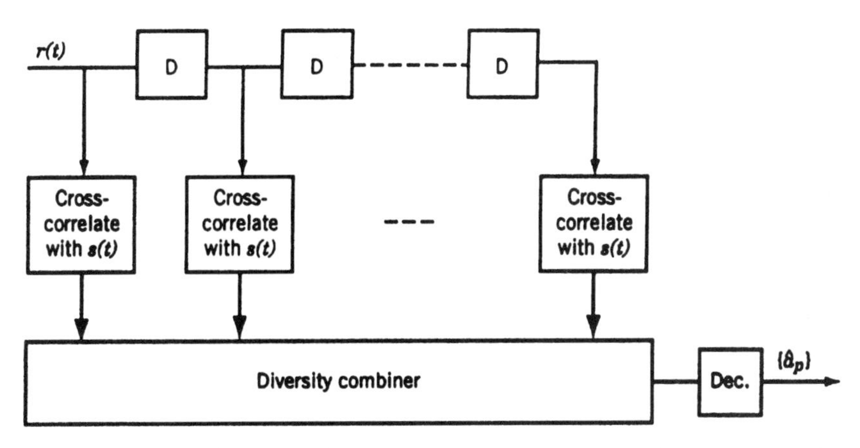

Figure 10.18. A typical RAKE receiver used for direct sequence spread spectrum reception.

If we use a RAKE receiver, which optimally combines the multipath components as part of the decision process, we do not lose the dispersed multipath energy; rather, the RAKE receiver turns it into useful energy to help in the decision process in conjunction with an appropriate combiner. Some texts call this implicit diversity or time diversity.

When a sufficiently spread bandwidth is provided (i.e., where the spread bandwidth is greater or much greater than the correlation bandwidth), we can get two or more independent frequency diversity paths as well using a RAKE receiver with an appropriate combiner such as a maximal ratio combiner. Figure 10.18 shows a RAKE receiver.

10.7 FREQUENCY REUSE

Because of the limited bandwidth allocated in the 800-MHz band for cellular radio communications, frequency reuse is crucial for its successful operation. A certain level of interference has to be tolerated. The major source of interference is co-channel interference from a "nearby" cell using the same frequency group as the cell of interest. For the 30-kHz bandwidth AMPS system, Ref. 15 suggests that C/I be at least 18 dB. The primary isolation derives from the distance between the two cells with the same frequency group. In Figure 10.2 there is only one cell diameter for protection.

Refer to Figure 10.19 for the definition of the parameters R and D. D is the distance between cell centers of repeating frequency groups and R is the "radius" of a cell. We let

$$a = D/R$$

The D/R ratio is a basic frequency reuse planning parameter. If we keep the D/R ratio large enough, co-channel interference can be kept to an acceptable level.

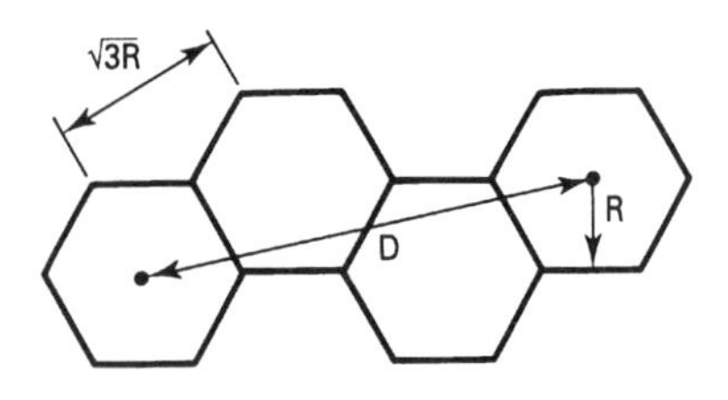

Figure 10.19. Definition of R and D.

Lee (Ref. 9) calls a the co-channel reduction factor and relates path loss from the interference source to R^{-4}.

A typical cell in question has six co-channel interferers, one on each side of the hexagon. So there are six equidistant co-channel interference sources. The goal is $C/I \geq 18$ dB or a numeric of 63.1. So

$$C/I = C/\Sigma I = C/6I = R^{-4}/6D^{-4} = a^4/6 \geq 63.1$$

Then

$$a = 4.4$$

This means that D must be 4.4 times the value of R. If R is 6 miles (9.6 km) then $D = 4.\ 4 \times 6 = 26.4$ miles (42.25 km).

Lee (Ref. 9) reports that co-channel interference can be reduced by other means such as directional antennas, tilted beam antennas, lowered antenna height, and an appropriately selected site.

If we consider a 26.4-mile path, what is the height of earth curvature at midpath? From Chapter 2, $h = 0.667(d/2)^2/1.33 = 87.3$ ft (26.9 m). Providing that the cellular base station antennas are kept under 87 ft, the 40-dB/decade rule of Lee holds. Of course, we are trying to keep below LOS conditions.

The total available (one-way) bandwidth is split up into N sets of channel groups. The channels are then allocated to cells, one channel set per cell on a regular pattern, which repeats to fill the number of cells required. As N increases, the distance between channel sets (D) increases, reducing the level of interference. As the number of channel sets (N) increases, the number of channels per cell decreases, reducing the system capacity. Selecting the optimum number of channel sets is a compromise between capacity and quality. Note that only certain values of N lead to regular repeat patterns without gaps. These are $N = 3$, 4, 7, 9, and 12, and then multiples thereof. Figure 10.20 shows a repeating 7 pattern for frequency reuse. This means that $N = 7$ or there are 7 different frequency sets for cell assignment.

Cell splitting can take place especially in urban areas in some point in time because the present cell structure cannot support the busy hour traffic load. Cell splitting, in effect, provides more frequency slots for a given area. Marcario (Ref. 15) reports that cells can be split as far down as a 1-km radius.

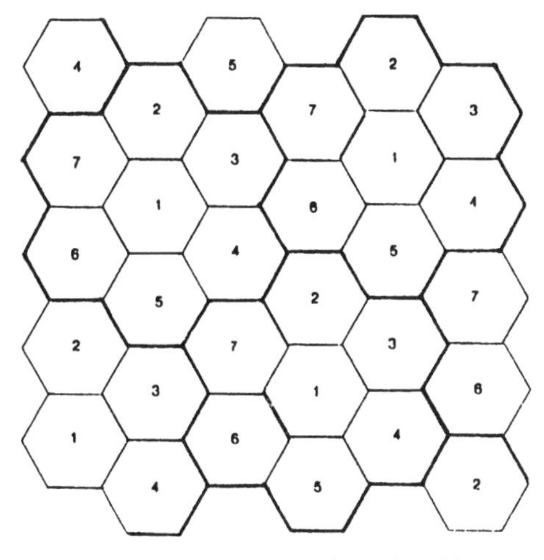

Figure 10.20. A cell layout based on $N = 7$.

Co-channel interference tends to increase with cell splitting. Cell sectorization can cut down the interference level. Figure 10.16 shows a three- and six-sector plan. Sectorization breaks a cell into three or six parts each with a directional antenna. With a standard cell, co-channel interference enters from six directions. A six-sector plan can essentially reduce the interference to just one direction. A separate channel set is allocated to each sector.

The three-sector plan is often used with a seven-cell repeating pattern, resulting in an overall requirement for 21 channel sets. The six-sector plan with its improved co-channel performance and the rejection of secondary interferers allows a four-cell repeat plan (Figure 10.21) to be employed. This results in an overall 24-channel set requirement. Sectorization requires a larger number of channel sets and fewer channels per sector. Outwardly it appears that there is less capacity with this approach; however, the ability to use much smaller cells actually results in a much higher capacity operation (Ref. 15).

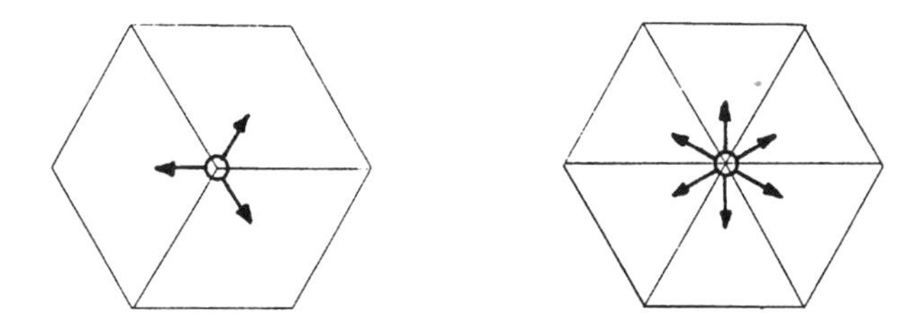

Figure 10.21. Breaking up a cell into three sectors (left) and six sectors (right).

10.8 PAGING SYSTEMS

10.8.1 What Are Paging Systems?

Paging is a one-way radio alerting system. The direction of transmission is from a fixed paging transmitter to an individual. It is a simple extension of the PSTN. Certainly paging can be classified as one of the first PCS (personal communication system) operations. The paging receiver is a small box, usually carried on a person's belt. As a minimum, a pager alerts the user that someone wishes to reach him/her by telephone. The person so alerted goes to the nearest telephone and calls a prescribed number. Some pagers have a digital readout that provides the calling number; whereas others give the number and a short message.

Most paging systems now operate in the VHF and UHF bands with a 3-kHz bandwidth. Transmitters have 1–5-watt output and paging receivers have sensitivities in the range of 10–100 μV/m (Ref. 15).

10.8.2 Radio-Frequency Bands for Pagers*

All three ITU regions have some or all of the following frequency bands allocated to mobile services:

26.1–50 MHz
68–88 MHz
146–174 MHz
450–470 MHz
806–960 MHz

10.8.3 Radio Propagation into Buildings

Measurement results submitted to the CCIR (Ref. 2) have indicated that frequencies in the range of 80–460 MHz are suitable for personal radio paging in urban areas with high building densities. It is possible that frequencies in the bands allocated around 900 MHz may also be suitable but that higher frequencies are less suitable. From measurements made in Japan, median values of propagation loss suffered by signals in the penetration of buildings (building penetration loss) have been derived. These results are summarized in Table 10.2.

10.8.4 Techniques Available for Multiple Transmitter Zones

To cover a service area effectively, it is often necessary to use a number of radio-paging transmitters. When the required coverage area is small, a single

*Sections 10.8.2–10.8.8 are based on CCIR Rep. 900-2 and Rep. 499-5 (Refs. 16 and 17).

TABLE 10.2 Propagation Loss Suffered by Signals in Penetrating Buildings

Frequency:	150 MHz	250 MHz	400 MHz	800 MHz
Building penetration loss[a]:	22 dB	18 dB	18 dB	17 dB[b]

[a]The loss is given as the ratio between the median value of the field strengths measured over the lower floors of buildings and the median value of the field strengths measured on the street outside. Similar measurements made in other countries confirm the general trend, but the values of building penetration loss vary about those shown. For instance, measurements made in the United Kingdom indicate that building penetration loss at 160 MHz is about 14 dB and about 12 dB at 460 MHz.
[b]Somewhat less accurate than the other results.

Source: Table 1, p. 59, CCIR Rep. 499-5, Vol. VIII.1 (Ref. 16).

RF channel should be used so as to avoid the need for multichannel receivers. In these circumstances, the separate transmitters may operate sequentially or simultaneously. In the latter case, the technique of offsetting carrier frequencies, by an amount appropriate to the coding system employed, is often used. It is also necessary to compensate for the differences in the delay to the modulating signals arising from the characteristics of the individual landlines to the paging transmitters. One way to do this is to carry out synchronization of the code bits via the radio-paging channel. Of course, information will be required about the bit rates, which this synchronization method would permit. It is preferable that the frequency offset of the transmitter carrier frequencies in a binary digital radio-paging system be at least twice the signal fundamental frequency. It is also preferable that delay differences between modulation of the transmitters in a binary digital paging system should be less than a quarter of the duration of a bit if direct FSK, NRZ (non-return-to-zero) modulation is employed. For subcarrier systems, the corresponding limit should be less than one-eighth of a cycle of the subcarrier frequency.

10.8.5 Paging Receivers

Built-in antennas can be designed for 150-MHz operation with reasonable efficiency. A typical radio-paging receiver antenna using a small ferrite rod exhibits a loss factor of about 16 dB relative to a half-wave dipole.

The majority of wide-area paging systems use some form of angle modulation.

Repeated transmission of calls can be used to improve the paging success rate of tone alert pagers. If p is the probability of receiving a single call, then $1 - (1 - p)^n$ is the probability of receiving a call transmitted n times, provided that the calls are uncorrelated. Correlations under Rayleigh fading conditions can largely be removed by spacing the calls more than 1 s apart. Longer delays between subsequent transmission (about 20 s) are required to improve the success rate under shadowing conditions.

Receivers with numeric or alphanumeric message displays can only take advantage of call repetitions if the supplementary messages are used to detect and correct errors.

10.8.6 System Capacity

The capacity of any paging system is affected by the following:

- Number and characteristics of the radio channels used
- Number of times each channel is reused within the system
- Actual paging location requirements of individual users
- Peak information (address and message) requirement in a location(s)
- Tolerable paging delay
- Data transmission rate
- Code efficiency
- Method of using the total code capacity throughout the system (this may also affect the system's capabilities for roaming)
- Any inefficiency introduced by battery-saving provisions
- Possible telephone system input restrictions

10.8.7 Codes and Formats for Paging Systems

The U.S. paging system broadly used across the country is popularly called a Golay Code referring to the Golay (23, 12) cyclic code with two codewords representing the address. Messages are coded using a BCH (7, 18) code. The code and format provide queueing and numeric/alphanumeric messages flexibility and the ability to operate in a mixed-mode transmission with other formats. The single address capacity of this system is up to 400,000 with noncoded battery saving and 4,000,000 with coded preamble.

Japan uses a BCH (1, 8) codeword with a Hamming distance of 7. The format gives approximately 65,000 addresses, 15 groups for battery economy, and a total cycle length of 4185 bits. Each group contains 8 address codewords headed by a 31-bit synchronizing and group-indicating signal.

The U.K. paging system employs a BCH (1, 2) code plus even-parity codeword with a Hamming distance of 6. The code format can handle over 8 million addresses and can be expanded. It can also handle any type of data message such as hexadecimal and CCITT Alphabet No. 5. It is designed to share a channel with other codes and to permit mixed simultaneous and sequential multitransmitter operation at the normal 512-bps transmission rate. This code is sometimes referred to as POCSAG and has been adopted as CCIR Radio-Paging Code No. 1 (RPC1).

10.8.8 Considerations for Selecting Codes and Formats

- Number of subscribers to be served
- Number of addresses assigned to each subscriber
- Calling rate expected, including that from any included message facility

- Zoning arrangement
- Data transmission rates possible over the linking network and radio channel(s), taking into account the propagation factors of the radio frequencies to be used
- Type of service: vehicular or personal, urban or rural

Once the data are provided from the above listing of topics, codes may be compared by their characteristics with respect to the following:

- Code address capacity
- Number of bits per address
- Codeword Hamming distance
- Code efficiency, such as number of information bits compared to the total number of bits per codeword
- Error-detecting capability; error-correcting capability
- Message capability and length
- Battery-saving capability
- Ability to share a channel with other codes
- Capability of meeting the needs of paging systems that vary with respect to size and transmission mode (e.g., simultaneous versus sequential)

10.9 PERSONAL COMMUNICATION SYSTEMS

10.9.1 Defining Personal Communications

A personal communication system (PCS) is wireless. This simply means that it is radio based. The user requires no *tether*. The conventional telephone is connected by a wire-pair through to the local serving switch. The wire pair is a tether. We can only walk as far with that telephone handset as the "tether" allows.

Both systems we have dealt with in the previous sections of this chapter can be classified as PCS. Cellular radio, particularly with the hand-held terminal, gives the user tetherless telephone communication. Paging systems provided the mobile/ambulatory user a means of being alerted that someone wishes to talk to him/her on the telephone or a means of receiving a short message.

The cordless telephone is certainly another example, which has extremely wide use around the world. By the end of 1994, it was estimated that there were 60 million cordless telephones in use in the United States (Ref. 19). We will provide a brief overview of cordless telephone developments below.

New applications are either on the horizon or going through field tests (1995). One that seems to offer great promise in the office environment is the wireless PABX. It will almost eliminate the telecommunication manager's responsibilities with office rearrangements. Another is the wireless LAN (WLAN).

Developments are expected such that PCS cannot only provide voice communications but facsimile, data, messaging, and possibly video. GSM provides all but video. Cellular Digital Packet Data (CDPD) will permit data services over the cellular system in North America.

Don Cox (Ref. 20) breaks PCS down into what he calls "high tier" and "low tier." Cellular radio systems are regarded as high-tier PCS, particularly when implemented in the new 1.9-GHz PCS frequency band. Cordless telephones are classified as low-tier.

Table 10.3 summarizes some of the more prevalent wireless technologies.

10.9.2 Narrowband Microcell Propagation at PCS Distances

The microcells discussed here have a radial range of < 1 km. One phenomenon to be considered is the Fresnel break point, which is illustrated in Figure 10.22. Signal level varies with distance R as A/R^n. For distances greater than 1 km, n is typically between 3.5 and 4. The parameter A describes the effects of environmental features in a highly averaged manner (Ref. 21).

Typical PCS radio paths can be of a LOS nature, particularly near the fixed transmitter where $n = 2$. Such paths may be down the street from the transmitter. The other types of paths are shadowed paths. One type of shadowed path is found in the highly urbanized setting, where the signal may be reflected off high-rise buildings. Another is found in more suburban areas, where buildings are often just two stories high.

When a signal at 800 MHz is plotted versus R on a logarithmic scale, as in Figure 10.22, there are distinctly different slopes before and after the Fresnel break point. We call the break distance (from the transmit antenna) R_B. This is the point for which the Fresnel ellipse about the direct ray just touches the ground. This model is illustrated in Figure 10.23. The distance R_B is approximated by

$$R_B = 4h_1h_2/\lambda \tag{10.18}$$

For $R < R_B$, n is less than 2, and for $R > R_B$, n approaches 4.

It was found that on non-LOS paths in an urban environment with low base station antennas and with users at street level, propagation takes place down streets and around corners rather than over buildings. For these non-LOS paths the signal must turn corners by multiple reflections and diffraction at vertical edges of buildings. Field tests reveal that signal level decreases by about 20 dB when turning a corner.

In the case of propagation inside buildings, where the transmitter and receiver are on the same floor, the key factor is the clearance height between the average tops of furniture and the ceiling. Bertoni et al. (Ref. 21) calls this

TABLE 10.3 Wireless PCS Technologies

	High-Power Systems				Low-Power Systems			
	Digital Cellular (High-Tier PCS)				Low-Tier PCS		Digital Cordless	
System	IS-54	IS-95 (DS)	GSM	DCS-1800	WACS/PACS	Handi-Phone	DECT	CT-2
Multiple access	TDMA/FDMA	CDMA/FDMA	TDMA/FDMA	TDMA/FDMA	TDMA/FDMA	TDMA/FDMA	TDMA/FDMA	FDMA
Frequency band (MHz) Uplink (MHz) Downlink (MHz)	869–894 824–849 (USA)	869–894 824–849 (USA)	935–960 890–915 (Europe)	1710–1785 1805–1880 (UK)	Emerging technologies[a] (USA)	1895–1907 (Japan)	1880–1900 (Europe)	864–868 (Europe and Asia)
RF channel spacing Downlink (kHz) Uplink (kHz)	30 30	1250 1250	200 200	200 200	300 300	300	1728	100
Modulation	$\pi/4$ DQPSK	BPSK/QPSK	GMSK	GMSK	$\pi/4$ QPSK	$\pi/4$ DQPSK	GFSK	GFSK
Portable transmit power, max./avg.	600 mW/ 200 mW	600 mW	1 W/ 125 mW	1W/ 125 mW	200 mW/ 25 mW	80 mW/ 10 mW	250 mW/ 10 mW	10 mW/ 5 mW
Speech coding	VSELP	QCELP	RPE-LTP	RPE-LTP	ADPCM	ADPCM	ADPCM	ADPCM
Speech rate (kbps)	7.95	8 (var.)	13	13	32/16/8	32	32	32
Speech ch./RF ch.	3	—	8	8	8/16/32	4	12	1
Channel bit rate (kbps) Uplink Downlink	48.6 48.6		270.833 270.833	270.833 270.833	384 384	384	1152	72
Channel coding	$\frac{1}{2}$ rate conv.	$\frac{1}{2}$ rate-fwd $\frac{1}{3}$ rate rev.	$\frac{1}{2}$ rate conv.	$\frac{1}{2}$ rate conv.	CRC	CRC	CRC (control)	None
Frame (ms)	40	20	4.615	4.615	2.5	5	10	2

[a]Spectrum is 1.85–2.2 GHz allocated by the FCC for emerging technologies; DS is direct sequence.

Source: IEEE Communications Magazine, Apr. 1995, Ref. 20.

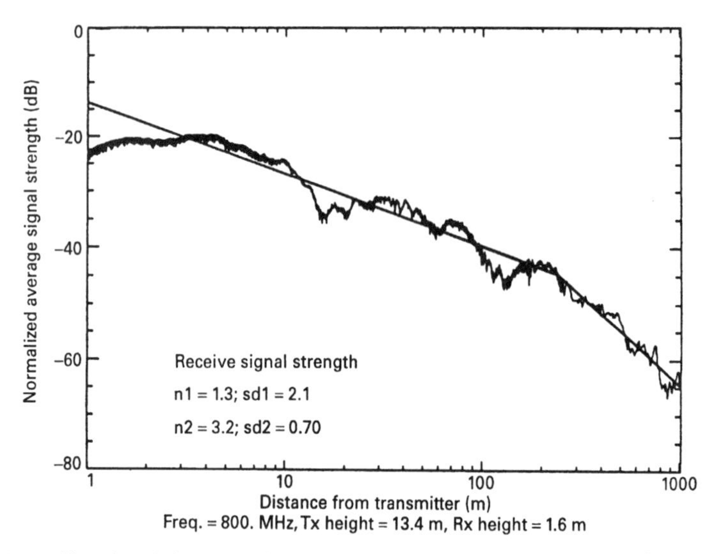

Figure 10.22. Signal variation on a line-of-sight path in a rural environment. (From Figure 3, Ref. 21.)

clearance W. Here building construction consists of drop ceilings of acoustical material supported by metal frames. That space between the drop ceiling and the floor above contains light fixtures, ventilation ducts, pipes, support beams, and so on. Because the acoustical material has a low dielectric constant, the rays incident on the ceiling penetrate the material and are strongly scattered by the irregular structure, rather than undergoing specular reflection. Floor-mounted building furnishings such as desks, cubicle partitions, filing cabinets, and workbenches scatter the rays and prevent them from reaching the floor, except in hallways. Thus it is concluded that propagation takes place in the clear space, W.

Figure 10.24 shows a model of a typical floor layout. When both the transmitter and receiver are located in the clear space, path loss can be

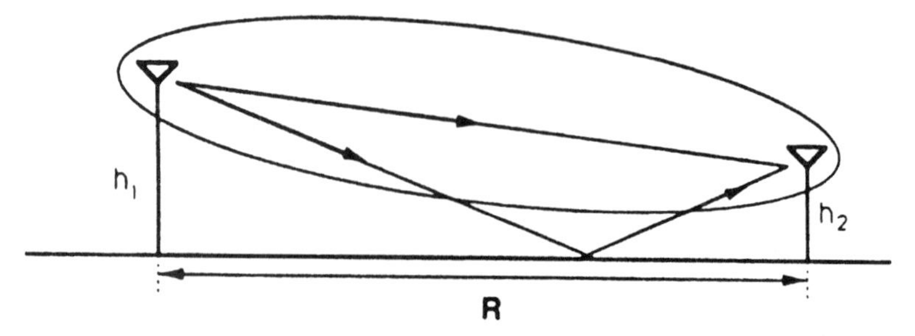

Figure 10.23. Direct and ground reflected rays, showing the Fresnel ellipse about the direct ray. (From Figure 18, Ref. 21.)

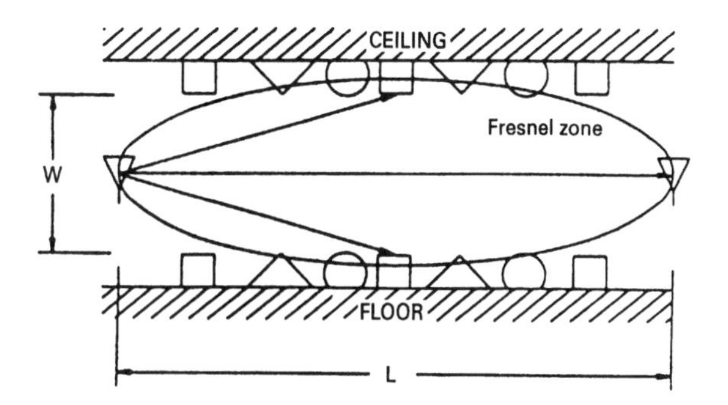

Figure 10.24. Fresnel zone for propagation between transmitter and receiver in clear space between building furnishings and ceiling fixtures. (From Figure 35, Ref. 21.)

related to the Fresnel ellipse. If the Fresnel ellipse associated with the path lies entirely in the clear space, the path loss has LOS properties $(1/L^2)$. Now as the separation between the transmitter and receiver increases, the Fresnel ellipse grows in size so that scatters lie within it. This is shown in Figure 10.25. Now the path loss is greater than free space.

Bertoni et al. (Ref. 21) report on one measurement program where the scatters have been simulated using absorbing screens. It was recognized that path loss will be highly dependent on nearby scattering objects. Figure 10.25 was developed from this program. It was found that the path loss in excess of free space calculated at 900 and 1800 MHz for $W = 1.5$ is plotted in Figure 10.24 as a function of path length L. The figure shows that the excess path loss (over LOS) is small at each frequency out to distances of about 20 and 40 m, respectively, where it increases dramatically.

Propagation between floors of a modern office building can be very complex. If the floors are constructed of reinforced concrete or prefabricated

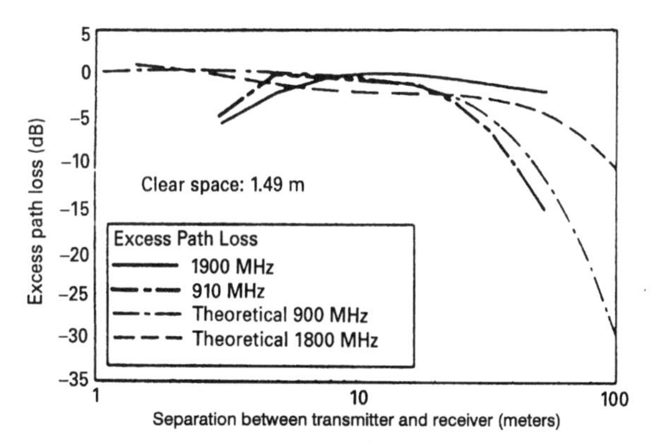

Figure 10.25. Measured and calculated excess path loss at 900 and 1800 MHz for a large office building having head-high cubical partitions, but no floor-to-ceiling partitions. (From Figure 36, Ref. 21.)

concrete, transmission loss can be 10 dB or more. Floors constructed of concrete poured over steel panels show much greater loss. In this case (Ref. 21), signals may propagate over other paths involving diffraction rather than transmission through the floors. For instance, signals can exit the building through windows and reenter on higher floors by diffraction mechanisms along the face of the building.

10.10 CORDLESS TELEPHONE TECHNOLOGY

10.10.1 Background

Cordless telephones began to become widely used in North America around 1981. By late 1994, there were over 60 million such units in use and an estimated sales rate of 5 million per year (Ref. 19).

10.10.2 North American Cordless Telephones

These telephones operate using FM with a 20-kHz bandwidth. Their ERP is on the order of 20 μW and they operate on ten frequency pairs in the bands 46.6–47.0 MHz (base transmit) and 49.6–50.0 MHz (handset transmit). Reference 19 suggests that this analog technology will continue on for some time into the future because of the telephone's low cost. It is expected that the FCC will make another 15 frequency pairs available in 1997 near 44 and 49 MHz.

10.10.3 European Cordless Telephones

The first generation European cordless telephone provided for eight channel pairs near 1.7 MHz (base unit transmit) and 47.5 MHz (handset transmit). Most of these units could only access one or two channel pairs. Some called this "standard" CT0.

This was followed by another analog cordless telephone based on a standard known as CEPT/CT1. CT1 has forty 25-kHz duplex channel pairs operating in the bands 914–915 MHz and 959–960 MHz. There is also a CT1+ in the bands 885–887 MHz and 930–932 MHz, which do not overlap the GSM allocation. CT1 is called a coexistence standard (not a compatible standard) such that cordless telephones from different manufacturers do not interoperate. The present embedded base is about 9 million units and some 3 million units are expected to be sold in 1997.

Two digital standards have evolved in Europe: the CT2 Common Air Interface and DECT (Digital European Cordless Telephone). In both standards, speech coding uses ADPCM (adaptive differential PCM). The ADPCM speech and control data are modulated onto a carrier at a rate of 72 kbps using Gaussian filtered FSK (GFSK) and are transmitted in 2-ms frames. One base-to-handset burst and one handset-to-base burst are included in each frame.

The frequency allocation for CT2 consists of 40 FDMA channels with 100-kHz spacing in the band 864–868 MHz. The maximum transmit power is 10 mW, and a two-level power control supports prevention of desensitization of base station receivers. As a by-product, it contributes to frequency reuse. CT2 has a call reestablishment procedure on another frequency after 3 seconds of unsuccessful attempts on the initial frequency. This gives a certain robustness to the system when in an interference environment.

CT2 supports up to 2400 bps of data transmission and higher rates when accessing the 32-kbps underlying bearer channels.

CT2 also is used for wireless pay telephones. When in this service it is called *Telepoint*. CT2 seems to have more penetration in Asia than in Europe.

Canada has its own version of CT2 called CT2+. It is more oriented toward the mobile environment, providing several of the missing mobility functions in CT2. For example, with CT2+, 5 of the 40 carriers are reserved for signaling, where each carrier provides 12 common channel signaling (CSC) channels using TDMA. These channels support location registration and updating, provide for paging, and enable Telepoint subscribers to receive calls. The CT2+ band is 944–948 MHz.

DECT takes on more of the cellular flavor than CT2. It uses a picocell concept and TDMA with handover, location registration, and paging. It can be used for Telepoint, radio local loop (RLL), and cordless PABX besides conventional cordless telephony. Its speech coding is similar to CT2, namely, ADPCM. For its initial implementation, ten carriers have been assigned in the band 1880–1900 MHz.

In many areas DECT will suffer interference in the assigned band, particularly from "foreign" mobiles. To help alleviate this problem, DECT uses two strategies: *interference avoidance* and *interference confinement*. The avoidance technique avoids time/frequency slots with a significant level of interference by handover to another slot at the same or another base station. This is very attractive for the uncoordinated operation of base stations because with many interference situations there is no other way around a situation but to change both in the time and frequency domains. The "confinement" concept involves the concentration of interference to a small time–frequency element even at the expense of some system robustness.

Base stations must be synchronized in the DECT system. A control channel carries information about access rights, base station capabilities, and paging messages. The DECT transmission rate is 1152 kbps. As a result of this and a relatively wide bandwidth, either equalization or antenna diversity is typically needed for using DECT in the more dispersive microcells.

Japan has developed the Personal Handyphone System (PHS). Its frequency allocation is 77 channels, 300 kHz in width, in the band 1895–1918.1 MHz. The upper half of the band, 1906.1–1918.1 MHz (40 frequencies) is used for public systems. The lower half of the band, 1895–1906.1 MHz, is reserved for home/office operations. An operational channel is autonomously selected by measuring the field strength and selecting a channel on which it meets certain level requirements. In other words, fully dynamic channel assignment is used. The modulation is $\pi/4$ DQPSK; average transmit power at the handset is 10 mW (80-mW peak power) and no greater than 500 mW (4-W peak power) for the cell site. The PHS frame duration is 5 ms. Its voice coding technique is 32-kbps ADPCM.

In the United States, digital PCS was based on the Wireless Access Communication System (WACS), which has been modified to an industry standard called PACS (Personal Access Communication System). It is intended for the licensed portion of the new 2-GHz spectrum. Its modulation is $\pi/4$ QPSK with coherent detection. Base stations are envisioned as shoebox size enclosures mounted on telephone poles, separated by some 600 m. WACS/PACS has an air interface similar to other digital cordless interfaces, except it uses frequency division duplex (FDD) rather than time division duplex (TDD) and more effort has gone into optimizing frequency reuse and the link budget. It has two-branch polarization diversity at both the handset and base station with feedback. This gives it an advantage approaching four-branch receiver diversity. The PACS version has eight timeslots and a corresponding reduction in channel bit rate and a slight increase in frame duration over its predecessor, WACS.

Table 10.4 summarizes the several types of digital cordless telephones.

TABLE 10.4 Digital Cordless Telephone Interface Summary

	CT2	CT2+	DECT	PHS	PACS
Region	Europe	Canada	Europe	Japan	United States
Duplexing	TDD		TDD	TDD	FDD
Frequency band (MHz)	864–868	944–948	1880–1900	1895–1918	1850–1910/ 1930–1990[a]
Carrier spacing (kHz)	100		1728	300	300/300
Number of carriers	40		10	77	16 pairs/ 10 MHz
Bearer channels/carrier	1		12	4	8/pair
Channel bit rate (kbps)	72		1152	384	384
Modulation	GFSK		GFSK	$\pi/4$ DQPSK	$\pi/4$ QPSK
Speech coding	32 kbs		32 kbs	32 kbs	32 kbs
Average handset transmit power (mW)	5		10	10	25
Peak handset transmit power (mW)	10		250	80	200
Frame duration (ms)	2		10	5	2.5

[a]General allocation to PCS, licensees may use PACS.

Source: Table 2, p. 35 (Ref. 19).

10.11 WIRELESS LANs

Wireless LANs, much as their wired counterparts, operate in excess of 1 Mbps. Signal coverage runs from 50 to less than 1000 ft. The transmission medium can be radiated light (around 800–900 nm) or radio frequency, unlicensed. Several of these latter systems use spread spectrum with transmitter outputs of 1 watt or less.

WLANs (wireless LANs) using radiated light do not require FCC licensing —a distinct advantage. They are immune to RF interference but are limited in range by office open spaces because their light signals cannot penetrate walls. Shadowing can also be a problem.

One type of radiated light WLAN uses a directed light beam. These units are best suited for fixed-terminal installations because the transmitter beams and receivers must be carefully aligned. The advantages for directed beam systems is improved S/N and fewer problems with multipath. One such system is fully compliant with IEEE 802.5 token ring operation offering 4- and 16-Mbps transmission rates.

Spread spectrum WLANs use the 900-MHz, 2-GHz, and 5-GHz industrial, scientific, and medical (ISM) bands. Both direct sequence and frequency hop operation can be used. Directional antennas at the higher frequencies provide considerably longer range than radiated light systems, up to several miles or more. No FCC license is required. A principal user of these higher-frequency bands is microwave ovens. CSMA and CSMA/CD (IEEE 802.3) protocols are often employed.

There is also a standard microwave WLAN (non-spread spectrum) that operates in the band 18–19 GHz. FCC licensing is required. Building wall penetration loss is high. The basic application is for office open spaces.

10.12 FUTURE PUBLIC LAND MOBILE TELECOMMUNICATION SYSTEM (FPLMTS)

10.12.1 Introduction

FPLMTS is a cellular/PCS concept proposed by the ITU-R Organization (previously CCIR). It proposes the bands 1885–2025 MHz and 2210–2200 MHz for FPLMTS service. The system includes both a terrestrial and a satellite component. Figure 10.26 shows a FPLMTS scenario for PCS terrestrial component and Figure 10.27 shows the satellite component.

10.12.2 Traffic Estimates

CCIR states that the maximum demand for PCS is in large cities where different categories of traffic will traverse the system, such as that generated

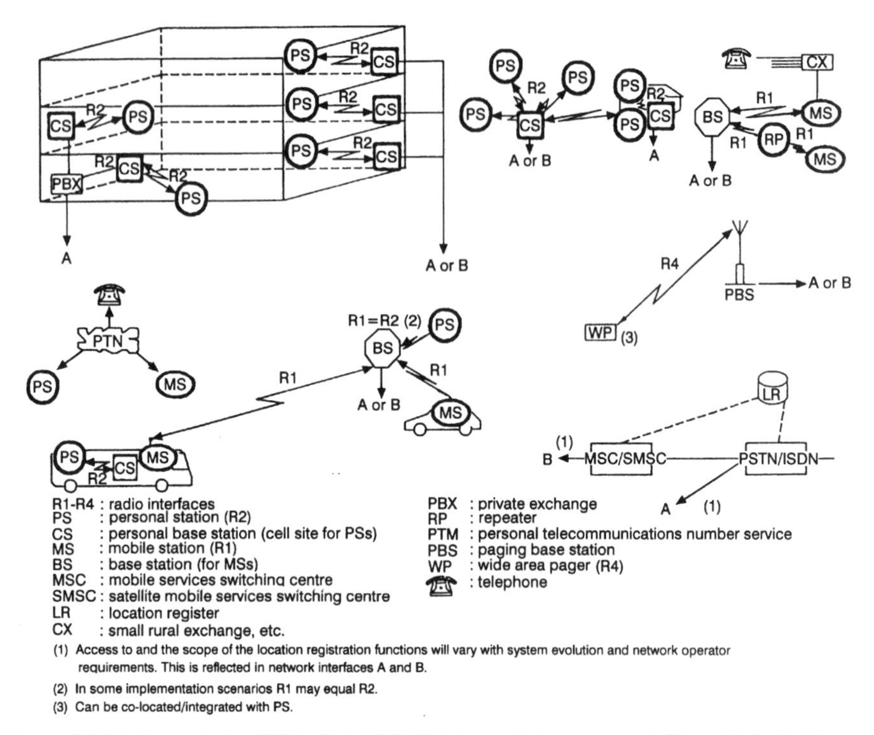

Figure 10.26. Scenario for PCS within FPLMTS, terrestrial component. (From Ref. 28; Figure 1, p. 7, CCIR Rec. 687-1, 1994 M Series Volume, Part 2.)

by mobile stations (MSs) and vehicle-mounted or portable stations (PSs), outdoors and indoors.

The worst-case scenario is during a vehicular traffic jam, where the number of vehicles per kilometer along a street is around 600 if they are stationary or 350 if they are moving slowly. Assuming a mean value of 400 vehicles per kilometer of street length, and that 50% of these vehicles are equipped with mobile stations, each generating 0.1 E, the traffic density will be 20 E/km of street length, leading to 300 E/km^2 based on typical urban street density. Adding about the same amount of traffic for portable mobile stations carried by pedestrians, the combined traffic for MSs would be around 500 E/km^2 in the more dense city areas.

The peak traffic for personal stations (PSs) is estimated at 1500 E/km^2, assuming 3000 pedestrians per kilometer of street length, 80% penetration(s) of the personal station, and 0.04 E/station. It is estimated that the peak to mean ratio of traffic for pedestrians on busy streets of large cities has a value of around 3.

For PSs indoors the traffic may increase by a factor of 10 or more in a multifloor building. CCIR estimates one station every 10^2 meters active floor

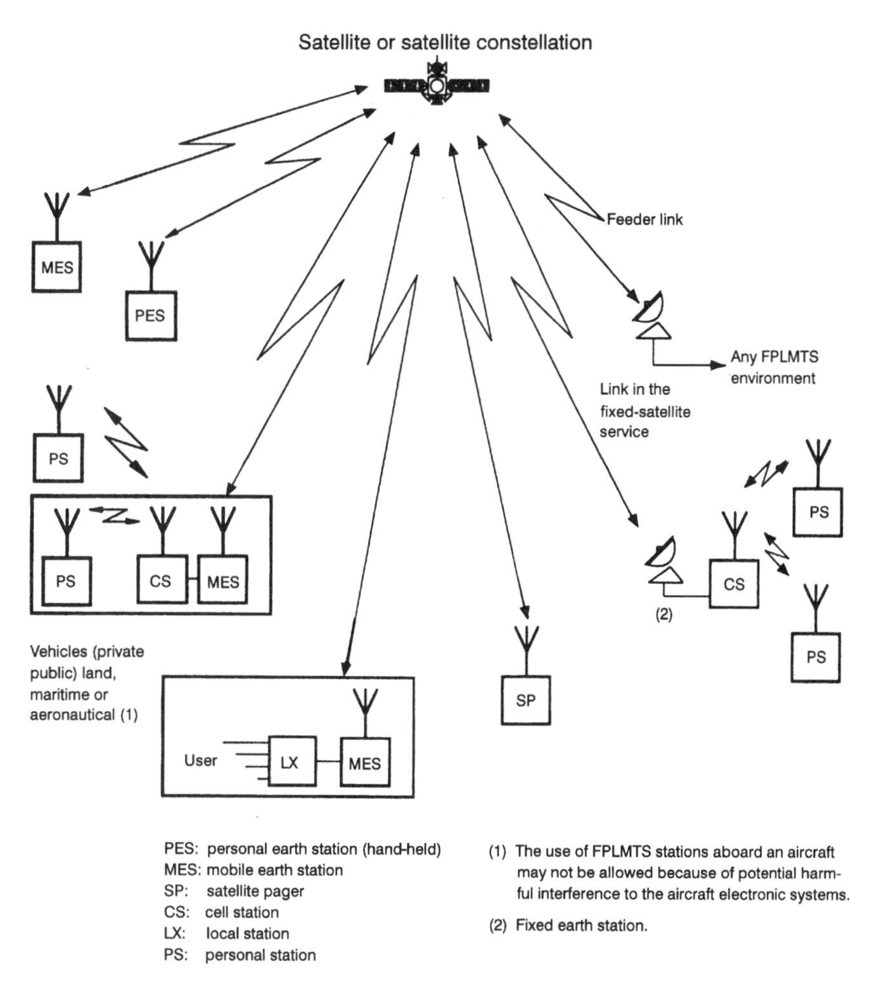

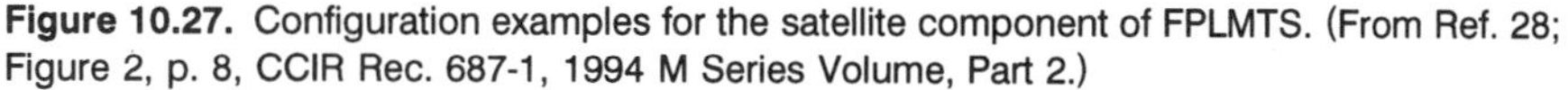

Figure 10.27. Configuration examples for the satellite component of FPLMTS. (From Ref. 28; Figure 2, p. 8, CCIR Rec. 687-1, 1994 M Series Volume, Part 2.)

area with a traffic intensity of 0.2 E/station. This corresponds to 20,000 E/km^2 per floor.

10.12.2.1 Nonvoice Traffic. Nonvoice services will constitute an increasing proportion of total traffic. Some data traffic need more bandwidth than a full-duplex voice channel, increasing spectrum estimates. If nonvoice traffic is handled by queueing procedures, less bandwidth is required, improving spectrum utilization.

One example is facsimile using circuit-switched services and is mainly relevant for vehicle-mounted MSs. If we assume 3000 terminals/km^2, of which 15% are equipped with facsimile terminals, and a call holding-time of 6 min/h per terminal, the estimated traffic amounts to 45 E/km^2.

CCIR believes that interactive data services are likely to employ packet transmission. The assumption used here is 15 s/h (15 seconds per hour) for a hand-portable (10 pages per hour with 8 kbps per page at a data rate of 4800 bps) and 30 s/h for a vehicle-mobile (4.5 and 9 mE, respectively). If we assume 5000 terminals/km^2 (i.e., 3000 vehicular and 2000 hand-portable), the estimated traffic amounts to 37 E/km^2.

10.12.2.2 PCS Outdoors. Here the assumption is a cumulative channel occupancy of 5 s/h (10 pages per hour with 8 kbps per page at a data rate of 16 kbps) that corresponds to 1.4 mE/station.

Assuming 2400 stations/km of street length (37,500 stations/km^2), as in the case for voice, the amount of traffic would be 50 E/km^2. To account for other data services, this estimation is increased by a factor of 3 to 50 E/km^2. The nonvoice traffic is then 10% of the voice traffic.

Traffic generated by circuit-switched services (e.g., fax) is considered insignificant. Thus only short, interactive data communications are considered in this scenario.

10.12.2.3 PCS Indoors. The traffic forecast for facsimile is based on 25% of stations having a fax capability and 6-minutes call-holding time per hour, per fax terminal—thus the estimated traffic is 25 mE/station (i.e., one-eighth of the voice traffic, or 2500 E/km^2).

For interactive applications, where we assume all stations are using the same application and with the assumption of 20 interactive sessions per hour, with a cumulative channel occupancy of approximately 2 seconds per station, the estimate is 0.01 E/station, which is equivalent to 1000 E/km^2.

Taking into account contention due to packet transmission, the Erlang value is doubled to 2000 E/km^2.

The grand total for nonvoice service indoors, taking into account batch data application and database retrievals with 10% overhead, a total of 5000 E/km^2 is assumed.

10.12.3 Estimates of Spectrum Requirements

Using the traffic estimates developed in Section 10.12.2, the minimum spectrum bandwidth required for voice and nonvoice services is approximately 230 MHz. The key parameters on which these estimates are based are given in Tables 10.5 and 10.6. The total requirement at the radio interface for R1 (mobile station—cellular) is 167 MHz and 60 MHz for R2 (PCS).

An important assumption used in Tables 10.5 and 10.6 was the speech coding rate: 8 kbps was assumed for the cellular environment (mobile station). Lower rates are available but the quality of transmission and delay suffer accordingly, as reported by CCIR. For the inexpensive personal stations (PCS), higher bit rates have been assumed ranging from 32 kbps down to 10 kbps.

TABLE 10.5 General Characteristics of PCS (High-Density Area) Voice Service Traffic Demands and Spectrum Requirement

Specifications	MS R1	PS R2 Outdoor	PS R2 Indoor
Radio coverage (%)	90	> 90	99
Base station antenna height (m)	50	10	< 3[a]
Base station installed: indoor/outdoor	No/yes	Yes/yes[b]	Yes/yes[b]
Traffic density (E/km^2)	500	1500	20,000[a]
Cell area (km^2)	0.94	0.016	0.0006
Blocking probability (%)	2	1	0.5
Cluster size (cell sites × sectors/site)	9	16	21 (3 floors)
Duplex bandwidth per channel (kHz)	25	50	50
Traffic per cell (E)	470	24	12
Number of channels per cell	493	34	23
Bandwidth (MHz)	111	27	24
Station[c]	Vehicle mounted or portable		
Volume (cm^3)		< 200	< 220
Weight (g)		< 200	< 200
Highest power	5 W	50 mW	10 mW

[a]Per floor.
[b]Usual case.
[c]A range of terminal types will be available to suit operational and user requirements.
Source: Ref. 28; Table 3, p. 15, CCIR Rec. 687-1, 1994 M Series, Vol. 2.

TABLE 10.6 Spectrum Estimation for Nonvoice Services

	MS Outdoor Interface R1		PS Outdoor Interface R2		PS Indoor Interface R2	
	Circuit Switched	Packet Switched	Circuit Switched	Packet Switched	Circuit Switched	Circuit Switched
Traffic density (E/km^2)	45	37	Insignificant	150	2000[a]	2500[a]
Duplex bandwidth per channel (kHz)	100	50	50	50	50	50
Bandwidth (MHz)	56		3		6	

[a]Per floor.
Source: Ref. 28; Table 4, p. 16, CCIR Rec. 687-1, 1994 M Series, Vol. 2.

The choice of network access scheme does not substantially affect the spectrum estimates given above.

The data provided in Tables 10.5 and 10.6 are based on densely populated metropolitan areas. CCIR reports that the following items were recognized but not considered in the spectrum estimates provided in the tables:

- Additional signaling traffic for system operation is expected to be significant in FPLMTS due to system complexity and quality objectives.

TABLE 10.7 Power Flux Densities for FPLMTS in an Urban Area

Stations	Base and Mobile	Personal
EIRP	10 W (base)	3 mW (indoor)
	1 W (mobile)	20 mW (outdoor)
Traffic density	582 E/km^2	25,000 E/km^2 (indoor)[a]
		1,650 E/km^2 (outdoor)
Assumed bandwidth	167 MHz	60 MHz
Estimated pfd	38 $\mu W/km^2$ Hz	1.5 $\mu W/km^2/Hz$
	-68 dB(W/m^2/4 kHz)	-82 dB(W/m^2/4 kHz)

[a]This takes into account the vertical frequency reuse of FPLMTS in buildings.

Source: Ref. 28; Table 5, p. 16, CCIR Rec. 687-1, 1994 M Series, Vol. 2.

- Road traffic management and control applications may generate additional nonvoice traffic.
- Sharing of the spectrum between several operators may result in less efficient spectrum use.

Taking these items into account, the 230-MHz estimate for FPLMTS may turn out as a lower bound.

10.12.4 Sharing Considerations

The goal is that one service does not interfere with another in excess of the CCIR limits. The basic parameters for sharing with FPLMTS are in pfd* (power/km^2/Hz) and the minimum carrier to total noise plus interference that is required. The pfd is derived from the number of terminals per square kilometer and the power for each category of station. Table 10.7 provides some estimates for an urban area.

The level of interference to FPLMTS that can be tolerated has been estimated using a link budget, which shows that PCSs are expected to be interference-limited rather than noise-limited. If we assume an allocation of 10% of the total interference budget to external interference sources, then we derive a corresponding aggregate interference power level of -117 dBm for indoor PCSs and -119 dBM for outdoor PCSs. These levels are the maximum permissible without significantly degrading service quality.

10.12.5 Sharing Between FPLMTS and Other Services

Sharing may not be feasible between FPLMTS and other services such as the fixed service, mobile-satellite services, and certain satellite TT & C† services. Operational sharing of an allocation common to FPLMTS and other services

*pfd = power flux density.
†TT & C = telemetry, tracking, and command.

requires suitable geographic separation between services or where neither service requires the total allocated band. If FPLMTS uses adaptive channel assignment, sharing will be greatly facilitated and will simplify the introduction of FPLMTS into bands currently used by other services.

Sharing is not feasible between R1 and R2 interfaces of FPLMTS and the SRS, SOS, and EESS satellite TT & C services in the 2025–2110-MHz and the 2200–2290-MHz bands.

10.13 MOBILE SATELLITE COMMUNICATIONS

10.13.1 Background and Scope

In our earlier discussions on cellular mobile radio and PCS, there seemed to be no clear demarcation where one ended and the other began. Cellular hand terminals certainly are used inside all types of buildings with some fair success, granted some of the connections are marginal. Often we speak of PCS in the bigger picture of cellular mobile radio. Even CCIR (ITU-R Organization) describes FPLMTS as an integrated system where there is no dividing line between PCS and cellular.

In this section we review satellite services that provide PCS and cellular mobile radio on a worldwide basis. We present a short overview and then discuss Motorola's IRIDIUM system in some detail.

10.13.2 Overview of Satellite Mobile Services

10.13.2.1 Existing Systems. INMARSAT [International Maritime Satellite (consortium)] has been providing worldwide full-duplex voice, data, and record traffic service with ships since the mid-1970s. It extended its service to a land-mobile market and to aircraft. At present, there are over 25,000 INMARSAT terminals. About 30% of these are land-transportable. INMARSAT satellites are in geostationary earth orbit (GEO).

INMARSAT-M systems provide service to ships and mobile land terminals. By the year 2005 some 600,000 INMARSAT-M terminals are expected to be in operation. The uplinks and downlinks for mobile terminals are in the 1.5- and 1.6-GHz bands. Services are low bit rate voice (5–8 kbps) and data operations. INMARSAT-P is a program specifically directed to the PCS market and is expected to be operational around 1998 (Ref. 22).

American Mobile Satellite Corporation (AMSC) is also providing voice, data, and facsimile service to the Americas, targeting customers in regions not served by conventional terrestrial cellular systems and terrestrial cellular subscribers who have problems "roaming." These satellites are in GEO and provide uplinks and downlinks in the L-band, much like INMARSAT.

10.13.2.2 Post-1996 Systems. Several satellite communication system companies are launching satellites for low rate services in the VHF band (148–149 MHz uplink and 137–138 MHz downlink). These systems use satellites in low earth orbit (LEO). All provide two-way message services and do not offer voice service. Some orbits are polar and some are inclined.

TRW expects to have its ODYSSEY system in operation by 1998. The satellites will be in LEOs and provide voice, data, facsimile, and paging services worldwide. Serving mobile platforms, uplinks will be in the 1610.5–1626.5-MHz band and the companion downlinks will be in the 2483.5–2500-MHz band using channelized CDMA access. The orbits will be MEOs (medium earth orbits, about 10,400 km). Twelve satellites will be in three orbital planes.

Loral and Qualcomm have joined forces to offer a LEO system called GLOBALSTAR, consisting of a network of 48 satellites in eight orbital planes. GLOBALSTAR specifically is targeting the hand-held terminal market for interconnection with the PSTN. Access is by channelized CDMA, and services offered are voice, data, facsimile, and position location. Initial operation is expected in 1997 (Ref. 22).

Constellation Communications of Herndon, Virginia (USA), plans a large LEO system called ARIES, which will provide users with voice, data, facsimile, and position location services. These LEO satellites will be in four orbital planes at an average altitude of 1020 km. CDMA access is envisioned using a 16.5-MHz segment of L-band around 1.6 GHz for uplinks and another, similar segment for downlinks around 2.5 GHz. Ten or eleven fixed earth stations are planned for connectivity to the PSTN. These facilities will use standard 30/20-GHz uplinks and downlinks. The system is expected to be operational in 1997.

ELLIPSO is another planned system employing 15 satellites in elliptical inclined orbits in three planes and up to nine satellites in equatorial orbits at maximum altitudes of about 7800 km. L-band connectivity is planned for the mobile user and C-band for the feeder uplinks and downlinks. The services offered to customers will be voice, data, facsimile, and paging. Access will be via channelized CDMA and the first satellites are now operational.

10.13.3 System Trends

The low earth orbit offers a number of advantages over the geostationary orbit, and at least one serious disadvantage.

DELAY. One-way delay to a GEO satellite is budgeted at 125 ms; one-way up and down is double this value, or 250 ms. Round-trip delay is about 0.5 second. Delay to a typical LEO is 2.67 ms and round-trip delay is 4×2.67 ms or about 10.66 ms. Calls to/from mobile users of such systems may be relayed still again by conventional satellite services. Data services do not have to be so restricted on the use of "hand-shakes" and stop-and-wait ARQ as with similar services via a GEO system.

HIGHER ELEVATION ANGLES AND "FULL EARTH COVERAGE." The GEO provides no coverage above about 80° latitude and gives low-angle coverage of many of the world's great population centers because of their comparatively high latitudes. Typically, cities in Europe and Canada face this dilemma. LEO satellites, depending on orbital plane spacing, can all provide elevation angles $> 40°$. This is particularly attractive in urban areas with tall buildings. Coverage would only be available on the south side of such buildings in the northern hemisphere with a clear shot to the horizon. Properly designed LEO systems will not have such drawbacks. Coverage will be available at any orientation.

TRACKING—A DISADVANTAGE OF LOW EARTH ORBITS AND MEDIUM EARTH ORBITS. At L-band quasi-omnidirectional antennas for the mobile user are fairly easy to design and produce. Although such antennas display only modest gain of several decibels, links to LEO satellites can easily be closed with hand-held terminals. However, large-feeder, fixed earth terminals will require a good tracking capability as LEO satellites pass overhead. Handoff is also required as a LEO satellite disappears over the horizon to another satellite just as it appears over the opposite horizon. The handoff should be seamless.

The quasi-omnidirectional user terminal antennas will not require tracking, and the handoff should not be noticeable to the mobile user.

10.13.4 IRIDIUM*

10.13.4.1 Overview. The IRIDIUM system is a LEO satellite system that will provide cellular/personal communication services worldwide. It is being developed and will be operated by the Motorola Satellite Communications Division for the system owner, Iridium, Inc. There is also participation by a number of other large North American, Asian, and European companies such as Sprint, STET, BCE, and the Raytheon Company.

Subscribers to this system will use portable and mobile terminals with low-profile antennas to reach a constellation of 66 satellites. These satellites will be interconnected by cross-links as they circle the earth in highly inclined polar orbits about 485 statute miles above the earth. The deployment of the satellites will be in six orbital planes about 31.6° apart. However, planes 1 and 6 will be only 22° apart. The delay to a satellite varies from 2.5 to 11 ms versus about 125 ms to a GEO satellite.

*IRIDIUM is a registered trademark and service mark of Iridium, Inc. Section 10.13.4 is based on Ref. 23, the original Motorola petition to the FCC for a 77-satellite configuration with seven orbital planes. It was subsequently modified by two amendments to the original petition for a 66-satellite configuration in six orbital planes (Refs. 24 and 25). The writeup was further modified by a Motorola review (Ref. 26) and a second Motorola review (Ref. 27).

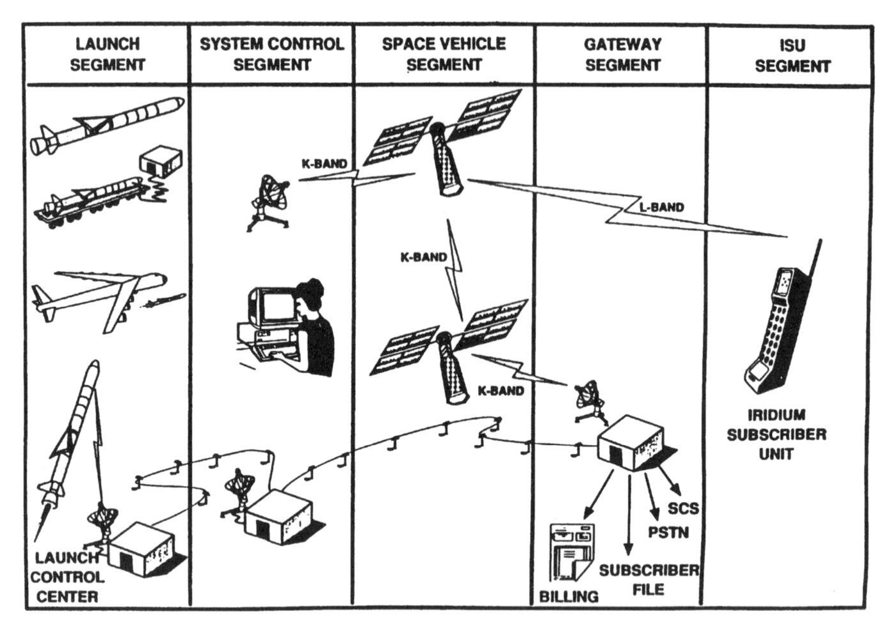

Figure 10.28. A pictorial overview of the IRIDIUM system.

The IRIDIUM system offers a wide range of mobile radio services including voice, data, and facsimile. The subscriber communication services are interconnected with the PSTN through regional gateways. The satellite–subscriber links are at L-band (1.6 GHz); gateway connectivity is via feeder links in the 30/20-GHz band, and the satellite cross-links operate in the 23-GHz band. Figure 10.28 gives a pictorial overview of the IRIDIUM system.

10.13.4.2 Space Segment. The space segment includes the 66 satellites which are in LEO, and are networked together as a switched-digital communication system utilizing cellular techniques to achieve maximum capability of frequency reuse. The subscriber uplinks and downlinks occupy the band 1616–1625.5 MHz. Each satellite will use up to 48 spot beams to form small cells on the surface of the earth. These numerous, relatively narrow beams result in high satellite antenna gains.

Taking advantage of the spatial separation of the beams allows increased spectral efficiency by means of time/frequency/spatial reuse over multiple cells, enabling many simultaneous user traffic connectivities over the same frequency channel. The constellation of satellites and its projection of cells on the earth's surface are analogous to a terrestrial cellular telephone system. With conventional cellular operations, a static set of cells services a large number of mobile/portable users. However, with the IRIDIUM system, a user's motion is relatively slow compared to that of the spacecraft.

Each of the satellites operates cross-links to support internetting. The cross-links operate in the 23-GHz band and include forward- and backward-looking links to two adjacent satellites in the same orbital plane, which are normally at a fixed angle about 2100 nautical miles away as well as cross-plane links to adjacent satellites. Up to four interplane links are maintained.

Each satellite communicates with earth-based gateways either directly or through other satellites by means of the cross-link network. Initially, there will be one gateway in the United States and about 3 to 20 gateways in other parts of the world. The system can accommodate up to 32 gateway stations. A gateway provides the interface between the IRIDIUM system and the PSTN. Table 10.8 gives a summary of major IRIDIUM satellite characteristics.

10.13.4.3 Cell Organization and Frequency Reuse

10.13.4.3.1 Satellite L-Band Antenna Pattern. Each satellite has the capability of projecting 48 L-band spot beams to form a continuous overlapping cell pattern on the earth. On a global basis, the entire constellation's beam pattern is projected on the surface of the earth. This results in approximately 2150 active beams with a frequency reuse of about 180 times. Within the contiguous United States, the system achieves up to five times frequency reuse. Each satellite is in view of a single subscriber unit (ISU) for approximately 9 minutes.

Each satellite has three multiple-beam phased-array antennas. The phased-array antennas are located on side panels of the satellite, each of which forms 16 cellular beams. Active transmit–receive (T/R) modules are used to provide power amplification for the transmit function, low noise amplification for the receive function, switch selection between transmit and receive, and digital phase control for active beam steering of both the transmit and receive beams in the phased arrays. The 16-cell pattern of each phased array is repeated for each of the three panels.

10.13.4.3.2 Cellular Pattern. IRIDIUM operates with a frequency reuse distance designed to minimize co-channel interference. A typical pattern is shown in Figure 10.29. The cells shown as A through G are covered by satellite arrays in accordance with the TDMA timing pattern and sequence shown in Figure 10.30. During each TDMA frame, satellite transmission and reception from ISUs may occur during respective transmit and receive intervals. Each satellite has multiple-beam antennas with fixed boresights to provide contiguous cell coverage. Figure 10.31 shows a typical seven-cell pattern on a satellite and how it is integrated with satellites whose antenna patterns are contiguous.

10.13.4.3.3 Frequency Plans. The frequency plan for L-band operation is shown in Figure 10.32. The total peak capacity for a satellite is 1100 channels of which 960 are full duplex. As shown in Figure 10.30, each frequency slot

TABLE 10.8 Major IRIDIUM Satellite Characteristics

Orbits (nominal)	
Number of operational satellites	66
Number of orbital planes	6
Inclination of orbital planes	86.4°
Orbital period	100 min and 28 s
Apogee	787 km
Perigee	768 km
Argument of perigee	90° ± 10°
Active service arc	360°
RAAN	0°, 31.6°, 63.2°, 94.8°, 126.3°, 157.9°
Earth coverage	5.9 million square (statute) miles per satellite
Maximum number of channels per satellite	
L-band service links	About 3840
Intersatellite links	About 6000
Gateway/TT & C links	About 3000
Frequency bands[a]	
L-band service links	1616–1626.5 MHz
Intersatellite links	23.18–23.38 GHz
Gateway/TT & C links	
Downlinks	19.4–19.6 GHz
Uplinks	29.1–29.3 GHz
Polarization	
L-band service links	Right-hand circular
Intersatellite links	Horizontal
Gateway/TT & C links	
Downlinks	Left-hand circular
Uplinks	Right-hand circular
Transmit EIRP[b]	
L-band service links	7.5 to 27.7 dBW
Intersatellite links	38.4 dBW
Gateway/TT & C links	13.5–23.2 dBW
Final amplifier output power capability[a]	
L-band service links	0.1 to 3.5 watts per carrier (burst)[c]
Intersatellite links	3.4 watts per carrier (burst)
Gateway/TT & C links	0.1 to 1.0 watts per channel
Satellite G/T	
L-band service links	−10.6 to −3.1 dBi/K
Intersatellite links	8.1 dBi/K
Gateway/TT & C	−1.0 dBi/K
Receiving system noise temperature	
L-band service links	500 K
Intersatellite	720 K (1188 K with sun.)
Gateway/TT & C	1295 K
Gain of each L-band channel	N.A. (not a transponder)

[a]Does not include circuit losses.
[b]At edge of coverage.
[c]Equipment combined power from phased-array antenna.
Source: Ref. 25.

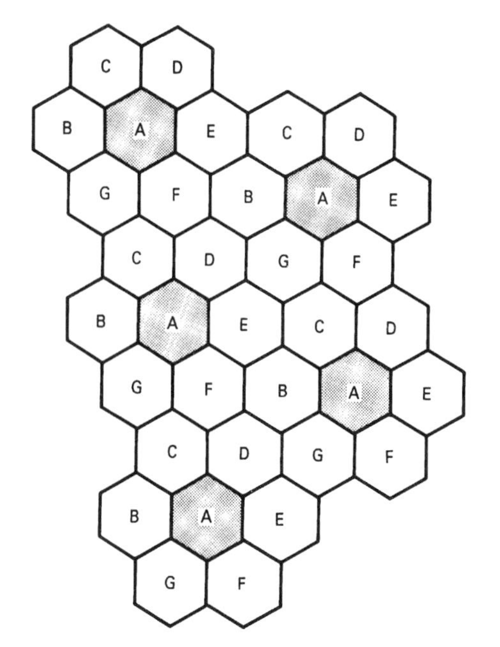

Figure 10.29. A typical seven-cell frequency reuse pattern.

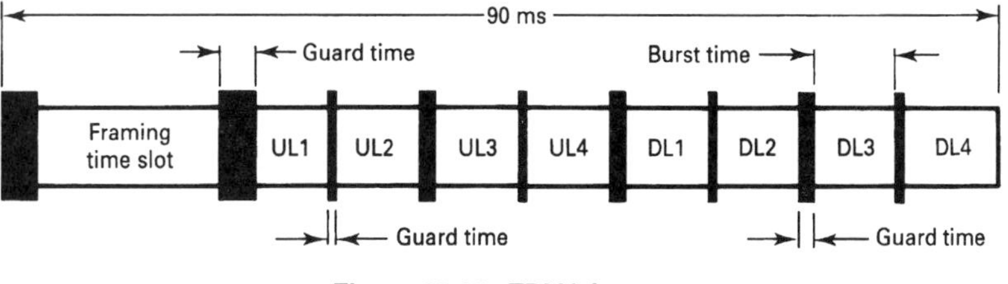

Figure 10.30. TDMA format.

has TDMA capability for four full-duplex voice channels. Without frequency reuse, the contiguous United States is covered by approximately 59 beams, which yield a maximum capacity of 3300 channels of which 2880 are full-duplex voice channels.

10.13.4.4 Communications Subsystem. The IRIDIUM communications system provides L-band communications between each satellite and individual subscriber units, K_a-band communications between each spacecraft and ground-based facilities which could be either gateway or system control facilities (TT & C) and K_a-band cross-links from satellite to satellite. The transfer of TT & C information between the system control facilities and

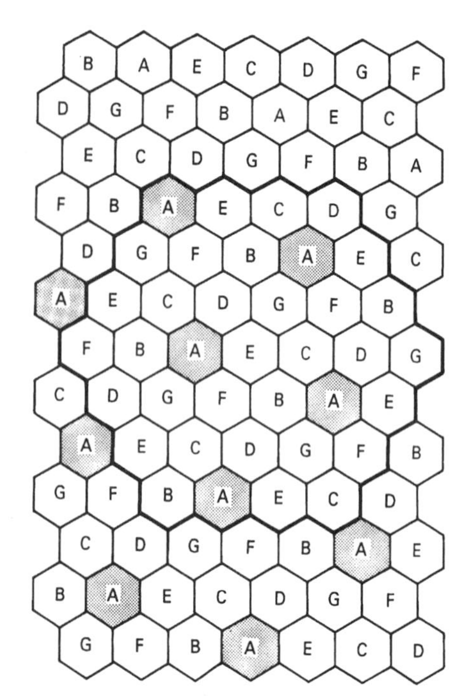

Figure 10.31. A typical reuse pattern with more than one satellite.

each satellite is generally provided via K_a-band communication links, with a dedicated link (via an omni-antenna) as backup.

10.13.4.4.1 L-Band Subscriber Terminal Links. The L-band communication subsystem is supported by an antenna complex consisting of three antenna panels that form 48 cellular beams. The 48 beams can be formed simultaneously (16 from each antenna panel). The L-band communication system is sized to provide transmit and receive capability for up to five fully loaded cells.

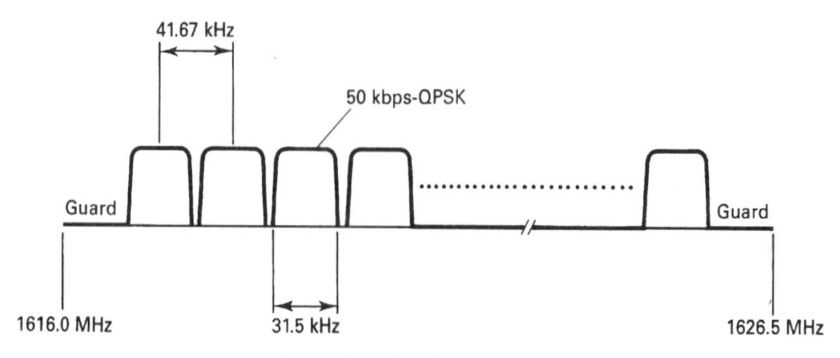

Figure 10.32. L-band uplink/downlink RF plan.

Each active beam supports an average of 236 channels. As shown in Figure 10.32, each FDMA frequency slot is 41.67 kHz wide including guard bands. Each frequency slot supports four full-duplex voice channels in a TDMA arrangement as shown in Figure 10.31. The modulation is QPSK, the coded data rate is 50 kbps per channel, and the TDMA frame period is 90 ms.

The L-band communication subsystem is designed to support a bit error rate of 2×10^{-2} end-to-end for voice operation. The improved BER required for data transmission is supported through the use of processing hardware installed in the subscriber units to apply more robust protocols and coding in order to counter deep fading experienced with L-band links in the mobile environment.

Table 10.9 gives typical L-band link budgets for uplink and downlink under shadow conditions.

10.13.4.4.2 Gateway Links. The 30/20-GHz gateway links provide full-duplex operation with two ground-based gateways or system support facilities per satellite. Satellite beam center gains for maximum range are 26.9 dBi on the downlink and 30.1 dBi on the uplink. These transmission links remain operational even with rainfall loss as high as 13 dB on the downlink and 26 dB on the uplink. Multiple earth terminal antennas spaced 34 nautical miles apart provide spatial diversity to avoid sun interference and to mitigate the effects rainfall attenuation. Time availabilities on these links are expected to be in the range of 99.8%.

Table 10.10 is a typical link budget for gateway 30/20-GHz operation. Each of the two full-duplex gateway links supports 600 simultaneous voice circuits. The frequency plan calls for the allocation of six frequencies each for uplink and downlink operation. The modulation rate in each direction is 12.5 Mbps and the channels are spaced at 15-MHz intervals. Each link supports a BER of better than 1×10^{-7}.

Figure 10.33 is a simplified functional block diagram of an IRIDIUM satellite communication payload.

10.13.4.5 Transmission Characteristics. Digital speech operation on IRIDIUM uses the AMBE* compression technique. Conventional PCM requires 64 kbps for a standard voice channel primarily to support speech operation. The AMBE compression technique used by IRIDIUM cuts this value to 2400 bps or about 27 : 1 compression.

The modulation and multiple access techniques used in the IRIDIUM system are patterned after the GSM terrestrial cellular system. A combined FDMA and TDMA access format is used along with data or vocoded voice and digital modulation techniques.

*AMBE is a registered trademark of Digital Voice Systems, Inc.

TABLE 10.9 L-Band Link Budget with Shadow Conditions

	Representative Cells			
	Cell 1	Cell 6	Cell 12	Cell 16
	Uplink			
Azimuth angle (degrees)	32.4	38.3	40.5	60.0
Ground range (km)	2215.3	1424.9	957.3	528.8
Nadir angle (degrees)	61.9	56.4	48.2	33.4
Elevation angle (degrees)	8.2	20.8	33.2	51.9
Slant range (km)	2461.7	1696.2	1278.5	960.0
IRIDIUM SUBSCRIBER UNIT				
HPA burst power (W)	3.7	3.6	1.9	3.7
(dBW)	5.7	5.6	2.9	5.7
Circuit loss (dB)	0.7	0.7	0.7	0.7
Antenna gain (dBi)	1.0	1.0	1.0	1.0
EIRP (dBW)	6.0	5.9	3.2	6.0
Uplink EIRP density (dBW/4 kHz)	−3.0	−3.1	−5.8	−3.0
PROPAGATION				
Space loss (dB)	164.5	161.3	158.8	156.3
Propagation losses (dB)	15.7	15.7	15.7	15.7
Total propagation loss (dB)	180.2	177.0	174.5	172.0
SPACE VEHICLE				
RECEIVED SIGNAL STRENGTH (DBW)	−174.2	−171.1	−171.3	−166.0
Efficient EOC antenna gain (dBi)	23.9	22.6	22.8	16.4
Signal level (dBW)	−150.3	−148.5	−148.5	−149.6
Required $E_b/(N_0 + I_0)$ (dB)	5.8	5.8	5.8	5.8
E_b/I_0 (dB)	18.0	18.0	18.0	18.0
Required E_b/N_0 (dB)	6.1	6.1	6.1	6.1
T_s (K)	500.0	500.0	500.0	500.0
Signal level required (dBW)	−148.5	−148.5	−148.5	−148.5
Link margin (dB)	−1.8	0.0	0.0	−1.1
G/T (dBi/K)	−3.1	−4.4	−4.2	−10.6
	Downlink			
Azimuth angle (degrees)	32.4	38.3	40.5	60.0
Ground range (km)	2215.3	1424.9	957.3	528.8
Nadir angle (degrees)	61.9	56.4	48.2	33.4
Elevation angle (degrees)	8.2	20.8	33.2	51.9
Slant range (km)	2461.7	1696.2	1278.5	960.0
SPACE VEHICLE				
HPA burst power (W)	3.5	2.2	1.3	3.0
(dBW)	5.5	3.5	1.2	4.8
Xtmr circuit loss (dB)	2.1	2.1	2.1	2.1
Efficient EOC antenna gain (dBi)	24.3	23.1	22.9	16.8
EIRP (dBW)	27.7	24.5	22.0	19.5
PROPAGATION				
Space loss (dB)	164.5	161.3	158.8	156.3
Propagation losses (dB)	15.7	15.7	15.7	15.7
Total propagation loss (dB)	180.2	177.0	174.5	172.0

TABLE 10.9 (*continued*)

	Representative Cells			
	Cell 1	Cell 6	Cell 12	Cell 16
IRIDIUM SUBSCRIBER UNIT				
RECEIVED SIGNAL STRENGTH (DBW)	−152.5	−152.5	−152.5	−152.5
Antenna gain (dBi)	1.0	1.0	1.0	1.0
Signal level (dBW)	−151.5	−151.5	−151.5	−151.5
Required $E_b/(N_0 + I_0)$ (dB)	5.8	5.8	5.8	5.8
E_b/I_0 (dB)	18.0	18.0	18.0	18.0
Required E_b/N_0 (dB)	6.1	6.1	6.1	6.1
T_s (K)	250.0	250.0	250.0	250.0
Signal level required (dBW)	−151.5	−151.5	−151.5	−151.5
Link margin (dB)	0.0	0.0	0.0	0.0
G/T_s (dBi/K)	−23.0	−23.0	−23.0	−23.0
SPFD at ISU (dBW/m^2 4 kHz)	−135.8	−135.8	−135.8	−135.8

Source: References 24 and 25.

TABLE 10.10 Link Budget for Gateway Operation

Item		Downlink		Uplink	
		Rain	Clear	Rain	Clear
Range	(km)	2326.0	2326.0	2326.0	2326.0
Transmitter					
Power	(dBW)	0.0	−9.7	13.0	−11.8
Antenna gain	(dB)	26.9	26.9	56.3	56.3
Circuit loss	(dB)	−3.2	−3.2	−1.0	−1.0
Pointing loss	(dB)	−0.5	−0.5	−0.3	−0.3
EIRP	(dBWi)	23.2	13.5	68.0	43.2
System					
Margin	(dB)	3.2	3.2	2.1	2.1
Space loss	(dB)	−185.8	−185.8	−189.1	−189.1
Propagation loss	(dB)	−14.2	−1.5	−30.0	−1.5
Polarization loss	(dB)	−0.2	−0.2	−0.2	−0.2
Total propagation loss	(dB)	−203.4	−190.7	−221.4	−192.9
Receiver					
Received signal strength	(dBWi)	−180.2	−177.2	−153.4	−149.7
Pointing loss	(dB)	−0.2	−0.2	−0.8	−0.8
Antenna gain	(dB)	53.2	53.2	30.1	30.1
Received signal	(dBW)	−127.2	−124.2	−124.1	−120.4
T_s	(K)	731.4	731.4	1295.4	1295.4
Noise density	(dBW/Hz)	−200.0	−200.0	−197.5	−197.5
Noise bandwidth	(dB-Hz)	64.9	64.9	64.9	64.9
Noise	(dBW)	−135.1	−135.1	−132.6	−132.6
Link E_b/N_0	(dB)	7.9	10.9	8.5	12.2
E_b/I_0	(dB)	25.0	25.0	16.0	16.0
Computed $E_b/(N_0 + I_0)$	(dB)	7.8	10.7	7.8	10.7
Required $E_b/(N_0 - I_0)$	(dB)	7.7	7.7	7.7	7.7
Excess margin	(dB)	0.1	3.0	0.1	3.0
SPFD at GW	(dBW/m^2/1 MHz)	−134.3	−131.3		

Source: References 24 and 25.

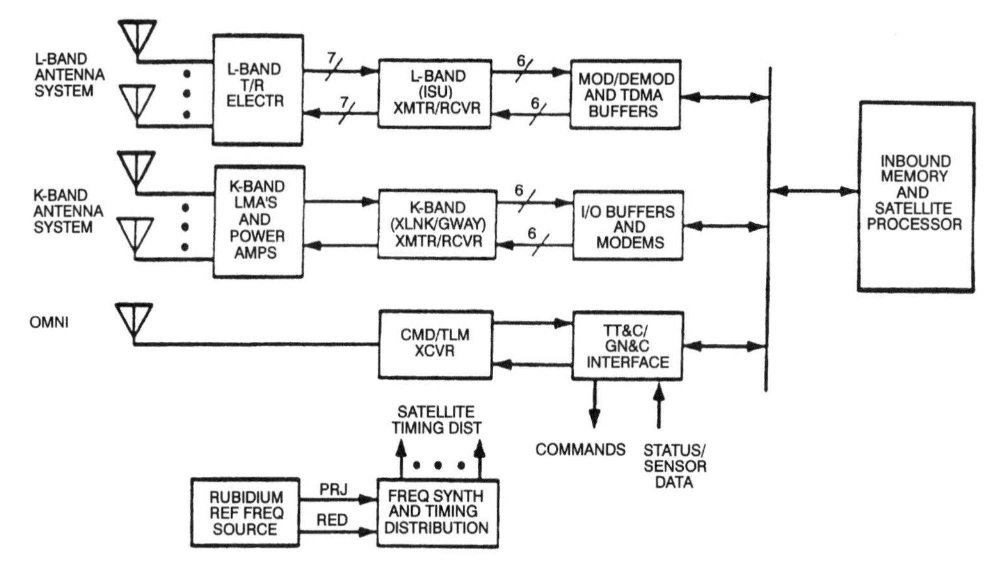

Figure 10.33. Spacecraft communications payload, simplified functional block diagram.

Each subscriber unit operates in a burst mode using a single carrier transmission. The bursts are controlled to occur at the proper time in the TDMA frame. The TDMA format is shown in Figure 10.30. The frame has four user timeslots, both uplink and downlink. Each subscriber unit will burst so that its transmission is received by the satellite in the proper timeslot, Doppler corrected. The subscriber unit is similar to a conventional hand-held unit in size and shape. It has a quadrifilar helix antenna with nearly hemispherical coverage with a gain of +1 dBi. Its peak EIRP is +8.5 dBW. The G/T is -23 dB/K. Its required E_b/N_0 is 6.3 dB.

10.13.4.5.1 Performance Objectives. The IRIDIUM system is designed to provide service to virtually 100% of the earth. However, it is recognized that it is economically and, at times, physically impossible to provide service to every single point on earth. There are practical limitations to the total number of locations that will physically be within line-of-sight of the satellites. End-to-end bit error rates will be better than 2×10^{-2} for voice transmission. The more typical bit error rates will be in the range of 1×10^{-3} to 1×10^{-4}. Coding and processing will improve these BER values to better than 1×10^{-5} for data connectivities.

PROBLEMS AND EXERCISES

1 What is the principal drawback in cellular radio considering its explosive user growth over these past several years?

2 Why is transmission loss so much greater on a cellular path compared to a LOS microwave path on the same frequency and covering the same distance?

3 What is the function of the MTSO or MSC in a cellular network?

4 What is the channel spacing in kilohertz on the North American AMPS system? On the N-AMPS system?

5 Why are antenna heights limited for cells with just sufficient height to cover the cell boundaries?

6 Why do we do cell splitting? What is approximately the minimum practical diameter of a cell (this limits splitting)?

7 When is *handover* necessary?

8 Define *roaming*.

9 Name four factors, besides distance and frequency, that cause cellular transmission loss variations.

10 Building penetration varies with what four factors?

11 Using the Okumura–Hata model, calculate the transmission loss for a 900-MHz cellular link, where the cellular mobile terminal antenna height is 3 m, the base transmitter antenna height is 35 m, and the distance is 5 km, in a medium-size city.

12 Calculate the building blockage component (dB) according to Lee for the following. The blockage distance is 250 ft and ERP = 1 watt. There are three buildings through which the transmitted wave traverses, 50 ft, 70 ft, and 80 ft. The transmit antenna height is 50 ft.

13 What are some of the fade ranges (dB) we might expect on a cellular link?

14 If the delay spread on a cellular link is about 10 μs, up to about what bit rate will there be little deleterious effects due to multipath dispersion?

15 Space diversity is commonly used at cell base sites. Like LOS microwave, antenna separation is key. This separation varies with two parameters. What are they?

16 For effective space diversity operation, there is a law of diminishing returns when we lower the correlation coefficient below what value?

17 What is the gain of a standard dipole over an isotropic radiator? What is the difference between ERP, as used in this text, and EIRP?

18 Cellular designers use a field strength contour of ___ dBμ, which is equivalent to − ___ dBm.

19 What would be the transmission loss to the 39-dBμ contour line if a cell site EIRP was +52 dBm?

20 A cell site antenna has a gain of +14 dBd. What is the equivalent gain in dBi?

21 There is not enough bandwidth assigned for cellular operation in the United States. The trend is to convert to digital operation. Why at first blush does it seem that we are defeating our purpose (i.e., have more users per unit bandwidth)?

22 What are the three access techniques that might be considered for digital cellular operation?

23 Provide the principal difference between TDMA as used in cellular service and that described in Chapter 7 (digital communications by satellite).

24 Discuss propagation delay as encountered at a TDMA cell site—the problem and its solution.

25 Analyze burst size, guard time, overhead, and training bits for a cellular TDMA system with eight users and 20-ms frame duration. Draw the frame showing user slot duration and then draw a typical user slot to show how the several fields are assigned.

26 What power amplifier advantage do we have in a TDMA system that we do not generally have with a FDMA system?

27 List four advantages of TDMA when compared to AMPS or other FDMA schemes.

28 Cellular radio, particularly in urban areas, is gated by extreme interference conditions, especially co-channel from frequency reuse. In light of this, describe how we achieve an interference advantage when using CDMA.

29 A CDMA cellular system operates at 4800 bps. It spreads this signal 10 MHz. What is the processing gain in decibels? What is the processing gain of IS-95 North American CDMA when the information rate is 9600 bps and the bandwidth is 1.23 MHz?

30 Correlation is the key to CDMA operation. Describe how it works permitting only the wanted signal to pass and rejecting other signals and interference.

31 For effective frequency reuse, the value of D/R must be kept large enough. Define D and R. What value of D/R is "large enough"?

32 In congested urban areas and where cell diameters are small, what measure do we take to reduce C/I?

33 How can paging systems use such low transmit power and achieve comparably long range?

34 PCS cells have diameters under 1 km. Give a generalized expression of path loss for such a system. Discuss the expression parameters and assign some rough values, leaving aside environmental factors.

35 For PCS (and for cellular) links, which are usually tilted in that the transmitter often is higher than the receiver, if we plot the loss curve, two slopes can be identified as we progress from transmitter to receiver. Discuss this phenomenon and its causes.

36 Name at least four different devices in a typical home that may be classified as PCS.

37 Such standards as DECT, CT1, 2, PHS, and WACS/PACS apply to what basic application of PCS?

38 Typify transmitter output for cellular compared to PCS. There should be at least two values for each.

39 Speech coding is less stringent in the several PCS scenarios (typically 32 kbps). Explain why it is much more stringent for cellular.

40 Why would CDMA be so attractive for PCS, especially in an indoor scenario?

41 WLANs use two different types of transmission media (either one or the other). What are they?

42 FPLMTS provides a skeletal guideline for PCS. In the terrestrial area, which are the three scenarios described and quantified?

43 Give two decided advantages of LEO mobile satellite systems over GEO satellites. Give one advantage of a GEO satellite.

44 LEO systems for mobile service have even more constrained allocated bandwidths than terrestrial cellular. How are they overcoming this problem to meet user demand and maintain an acceptable grade of service (i.e., blocking probability)?

45 Describe satellites and orbits of the IRIDIUM system.

46 Trace a call through the IRIDIUM system. Describe its access techniques.

47 What is the bit rate of the new voice compression technique used with IRIDIUM?

REFERENCES

1. Roger L. Freeman, *Telecommunication System Engineering*, 3rd ed., Wiley, New York, 1996.
2. *Telecommunications Transmission Engineering*, 2nd ed., Vol. 2, Bellcore, Piscataway, NJ, 1992.
3. Raymond Steele, ed., *Mobile Radio Communications*, IEEE Press, New York, and Pentech Press, London, 1992.
4. *VHF and UHF Propagation Curves for the Frequency Range from 30 MHz to 1000 MHz*, CCIR Rec. 370-5, XVIIth Plenary Assembly, Dusseldorf, 1990.
5. K. Allesbrook and J. D. Parsons, "Mobile Radio Propagation in British Cities at Frequencies in the VHF and UHF Bands," *Proceedings of the IEE*, Vol. 124, No. 2, 1977.
6. Y. Okumura et al., "Field Strength and Its Variability in VHF and UHF Land Mobile Service," *Review of Electrical Communication Laboratory (Tokyo)*, Vol. 16, 1968.
7. M. Hata, "Empirical Formula for Propagation Loss in Land-Mobile Radio Services," *IEEE Transactions on Vehicular Technology*, Vol. VT-20, 1980.
8. F. C. Owen and C. D. Pudney, "In-Building Propagation at 900 MHz and 1650 MHz for Digital Cordless Telephones," *6th International Conference on Antennas and Propagation, ICCAP '89, Part 2: Propagation*, Conf. Pub. No. 301, 1989.
9. William C. Y. Lee, *Mobile Communications Design Fundamentals*, 2nd ed., Wiley, New York, 1993.
10. *Recommended Minimum Standards for 800-MHz Cellular Subscriber Units*, EIA Interim Standard EIA/IS-19B.
11. "Cellular Radio Systems," a seminar given at the University of Wisconsin–Madison by Andrew H. Lamothe, Consultant, Leesburg, VA, 1993.
12. *Digital Cellular Public Land Mobile Telecommunication Systems (DCPLMTS)*, CCIR Rep. 1156, Vol. VIII.1, XVIIth Plenary Assembly, Dusseldorf, 1990.
13. Morris Engelson and Jim Hebert, "Effective Characterization of CDMA Signals," *Wireless Report*, Jan. 1995.
14. Robert C. Dixon, *Spread Spectrum Systems with Commercial Applications*, 3rd ed., Wiley, New York, 1994.
15. R. C. V. Macario, ed., *Personal and Mobile Radio Systems*, IEE/Peter Peregrinus, London, 1991.
16. *Radio-Paging Systems*, CCIR Rep. 900-2, Vol. VIII.1, XVIIth Plenary Assembly, Dusseldorf, 1990.
17. *Radio-Paging Systems*, CCIR Rep. 499-5, Vol. VIII.1, XVIIth Plenary Assembly, Dusseldorf, 1990.
18. A. Jagoda and M. de Villepin, *Mobile Communications*, Wiley, Chichester, UK, 1991/1992.
19. Jay C. Padgett, Cristoph G. Gunter, and Takeshi Hattori, "Overview of Wireless Personal Communications," *IEEE Communications Magazine*, Jan. 1995.
20. Donald C. Cox, "Wireless Personal Communications: What Is It?" *IEEE Personal Communications*, Vol. 2, No. 2, Apr. 1995.

21. Henry L. Bertoni et al., "UHF Propagation Prediction for Wireless Personal Communication," *Proceedings of the IEEE*, Vol. 89, No. 9, Sept. 1994.
22. William W. Wu et al., "Mobile Satellite Communications," *Proceedings of the IEEE*, Vol. 82, No. 9, Sept. 1994.
23. "Application of Motorola Satellite Communications, Inc. for IRIDIUM, A Low Earth Orbit Mobile Satellite System," before the Federal Communications Commission, Dec. 1990.
24. Minor Amendment before the Federal Communications Commission (amending Motorola's original IRIDIUM application), Aug. 1992, by Motorola Satellite Communications, Inc.
25. Minor Amendment before the Federal Communications Commission (amending Motorola's IRIDIUM system application), Nov. 1992 by Motorola Satellite Communications, Inc.
26. Private communication, Steve Clark, Motorola Chandler, Arizona, Mar. 17, 1996.
27. Private communication, Stephanie Nowack and J. Hoyt, Motorola Chandler, July 22, 1996.
28. *Future Public Land Mobile Telecommunications System (FPLMTS)*, CCIR Rec. 687-1, 1994 M Series Volume-Part 2, ITU Geneva 1992.

11

HIGH-FREQUENCY (HF) TRANSMISSION LINKS

11.1 GENERAL

Radio frequency (RF) transmission between 3 and 30 MHz by ITU convention is called high frequency (HF) or shortwave, particularly in the United Kingdom. Many in the industry extend the HF band downward to just above the standard AM broadcast band, namely, from 2.0 to 30 MHz. This text holds with the ITU convention.

HF communication is a unique method of radio transmission because of its peculiar characteristics of propagation. This propagation phenomenon is such that many radio amateurs at certain times carry out satisfactory communication better than halfway around the world with 1–2 W of radiated power. In fact, if the medium were noiseless and there were no interference, Ref. 1 states that the emitted power required could be only several milliwatts.

11.2 APPLICATIONS OF HF RADIO COMMUNICATION

HF is probably the most economic means of low data rate transmission over long distances (e.g., > 200 mi). It might be argued that meteor burst communication is yet more economic in some circumstances. Performance makes this difference. As we will show in Chapter 12, meteor burst transmission links have the disadvantage of waiting time between short data packets; HF does not.

Traditionally, since the 1930s, HF has been the mainstay of ship–shore–ship communication. Satellite communication offered by INMARSAT [International Marine Satellite (consortium)] certainly provides more reliable service, but HF continues to be the only long-distance communication means for many vessels. We will not argue rationale one way or the other.

Ship–shore HF communication service includes the following:

- CW or continuous wave, which connotes keying a RF carrier on and off, forming the dots and dashes of the international Morse code
- Selective calling teleprinter service (ITU-R Rec. 493-6, Ref. 2) using frequency shift keying (FSK) with a wide frequency shift
- Advanced automatic digital message service using ALE (automatic link establishment), with near real-time sounding, forward error correction
- Single sideband (SSB) voice telephony, often using improved Lincompex; such HF connectivities frequently access the PSTN (public switched telecommunication network)
- Simplex teleprinter service with narrow and wide shift FSK as used by many of the world's navies such as the U.S. Navy's Fox broadcast, merchant marine broadcast (MERCAST), and Hydro broadcasts (Refs. 3 and 4)

HF is widely employed for propaganda broadcast such as the U.S. Voice of America and BBC, in Cuba, Russia, and many, many other countries. It is also used for religious broadcasting such as the large HCJB facility in Quito, Ecuador. Many of these modern HF broadcast installations are very large and expensive and boast of megawatts of EIRP (effective isotropic radiated power).

The armed forces of the world make wide use of HF as a primary means of connectivity, or as an essential backup capability. In this case it can be used for voice, vocoded secure voice, record traffic, data, and facsimile. It can even be used for freeze-frame video.

HF also can provide very inexpensive point-to-point teleprinter/data service: simplex, half-duplex, full-duplex; four-channel time division multiplex (TDM); and 16-channel narrowband FDM using voice frequency carrier telegraph (VFCT) techniques.

Weather maps to ships and other entities are broadcast on HF by facsimile transmission using narrowband frequency modulation. It is also used for the transmission of time signals such as WWV on 5, 10, 15, and 20 MHz from Ft. Collins, Colorado (U.S.A.). There are numerous other HF time emitters around the world. Of course, these are eclipsed by surface and satellite systems such as Loran C (100 kHz), OMEGA (10 kHz), and GPS and GOES satellite (L-band) systems.

HF has many disadvantages. Some of these are:

- Low information rate. The maximum bandwidth by radio regulation is four 3-kHz independent sideband voice channels in a quasi-frequency division configuration with low data rates (e.g., about 2400 bps per 3-kHz voice channel).

- Degraded link time availability when compared to satellite, fiberoptic, coaxial cable, wire pair, troposcatter, and line-of-sight (LOS) microwave communication. HF link time availabilities vary from 80% up to better than 95% for some new spread spectrum wideband adaptive systems.
- Impairments include dispersion in both the time and frequency domains. Fading is endemic on skywave links. Atmospheric, galactic, and man-made interference noise are among the primary causes of low availability besides the basic propagation phenomenon itself.

11.3 TYPICAL HF LINK OPERATION, CONCEPTUAL INTRODUCTION

Figure 11.1 illustrates the operation of a HF link carrying teleprinter (or data) traffic. The upper part of the drawing shows the transmit side of the link. One-hop skywave operation is portrayed. The lower part of the drawing shows the receive side. The receiver site is at the right. Its several noise impairments are shown pictorially, such as cosmic noise, man-made noise, atmospheric noise propagated into the site from long distances, local thunderstorm noise, and interference noise from other users.

In the figure the basic elements of a HF transmitter and receiver are shown, which are described in Chapter 14. HF antennas are discussed in that same chapter.

The drawing shows a modem that converts the binary bit stream from a teleprinter keyboard to a signal compatible with the transmitter audio input. The modem develops an audio tone that is frequency shifted (FSK). A higher tone frequency represents a mark (binary 1) and the lower frequency tone represents a space (binary 0). This is further described in Chapter 14.

Another method to frequency shift the transmit frequency is to offset the transmit synthesizer frequency by one and the other tone frequency shift value (e.g., +60 and −60 Hz or +400 and −400 Hz). However, the first method is by far the most common. Several other tone formats are discussed in Section 11.12.

A companion modem is used on the receive side, which converts the audio tone output to a binary serial bit stream compatible with the receive teleprinter. The modem need not be a separate entity and may be incorporated as a printed circuit board in the transmitter exciter and in the receiver after the demodulator.

11.4 BASIC HF PROPAGATION

11.4.1 Introduction

A HF wave emitted from an antenna is characterized by a groundwave and a skywave component. The groundwave follows the surface of the earth and

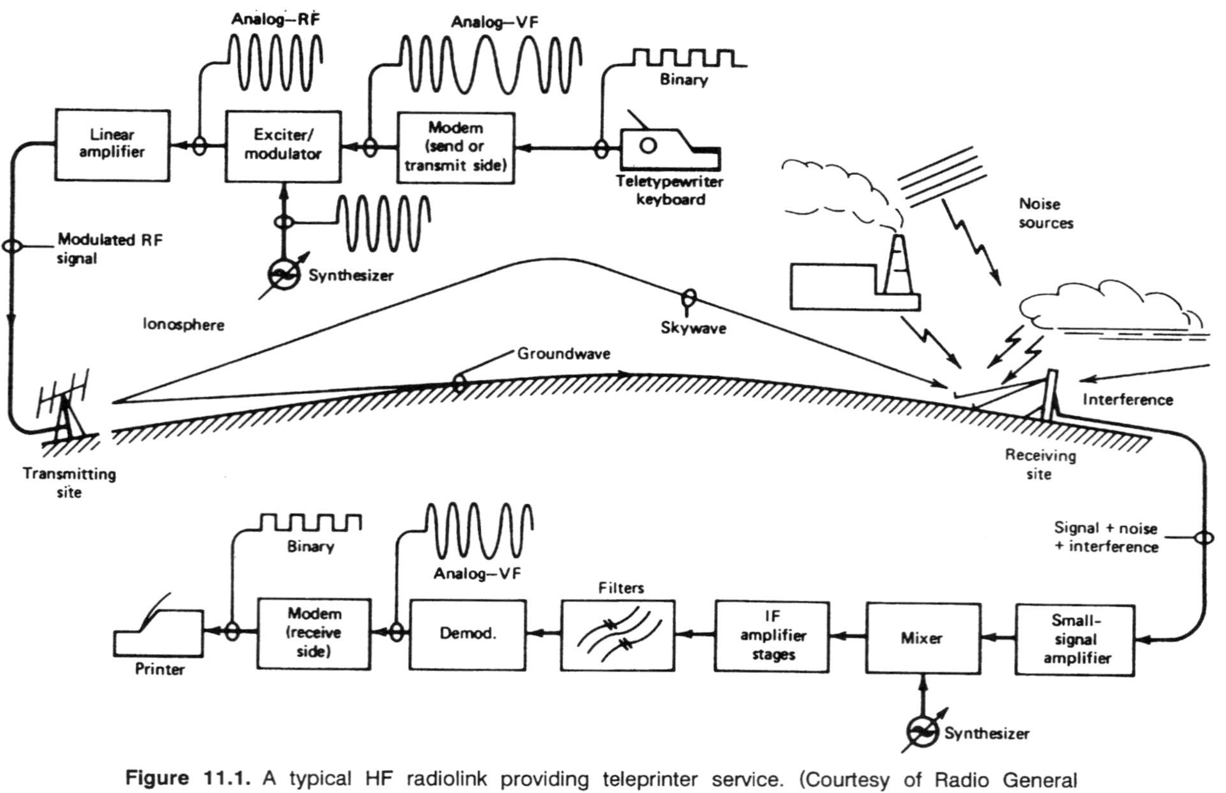

Figure 11.1. A typical HF radiolink providing teleprinter service. (Courtesy of Radio General Company.)

can provide useful communication over salt water up to about 650 mi (1000 km) and over land from some 10 mi (16 km) to 100 mi (160 km) or more, depending on RF power, the transmit frequency, antenna type and height off the ground, atmospheric noise, man-made noise, and ground conductivity. Generally, the lowest portion of the HF band is the most desirable for groundwave communication. However, as we progress lower in frequency, depending on geographical location, atmospheric noise becomes a serious impairment, forcing us to use a compromise higher frequency. Well-designed groundwave links, with their shorter range, achieve better time availabilities than skywave links.

Skywave links are used for long circuits from about 100 mi (160 km) to over 8000 mi (12,800 km). We have had successful connectivity during the IGY in 1955–1957 using CW [continuous wave (on/off keying)] from a ship in the Ross Sea (Antarctica) to WCC (Chatham, MA) in the early morning hours (local time) for over 8 consecutive weeks during the austral summer of each of the years. We also had very successful SSB voice communication from Frobisher Bay (Canada NWT) to Little America (Antarctica) in 1958 for periods of several hours a day for about $1\frac{1}{2}$ weeks. In both cases, modest antennas were used with transmitter outputs on the order of several hundred watts.

11.4.2 Skywave Transmission

The skywave transmission phenomenon of HF depends on ionospheric refraction. Transmitted radio waves hitting the ionosphere are bent or refracted. When they are bent sufficiently, the waves are returned to earth at a distant location. Often at the distant location they are reflected back to the sky again, only to be returned to earth still again, even further from the transmitter.

The ionosphere is the key to HF skywave communication. Look at the ionosphere as a layered region of ionized gas above the earth. The amount of refraction varies with the degree of ionization. The degree of ionization is primarily a function of the sun's ultraviolet (UV) radiation. Depending on the intensity of the UV radiation, more than one ionized layer may form (see Figure 11.2). The existence of more than one ionized layer in the atmosphere is explained by the existence of different UV frequencies in the sun's radiation. The lower frequencies produce the upper ionospheric layers, expending all their energy at high altitude. The higher frequency UV waves penetrate the atmosphere more deeply before producing appreciable ionization. Ionization of the atmosphere may also be caused by particle radiation from sunspots, cosmic rays, and meteor activity.

For all practical purposes four layers of the ionosphere have been identified and labeled as follows:

D Region or D-Layer. Not always present, but when it does exist, it is a daytime phenomenon. It is the lowest of the four layers. When it exists,

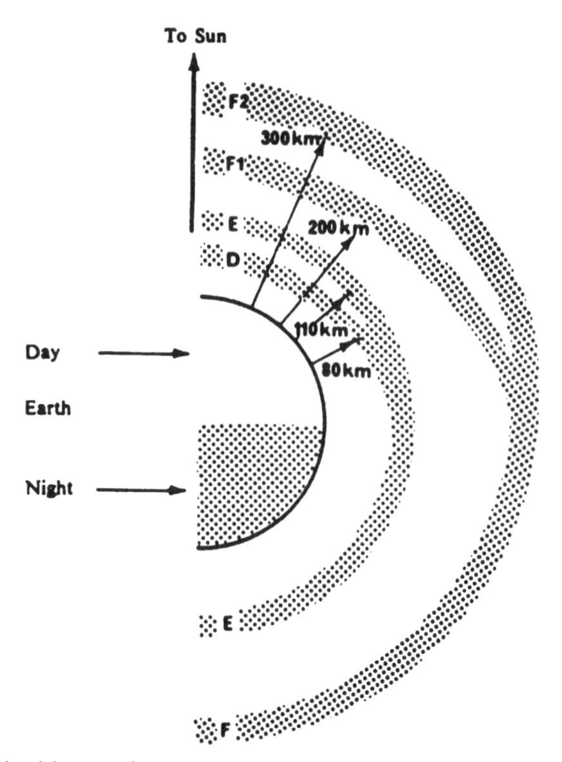

Figure 11.2. Ionized layers of the atmosphere as a function of nominal height above the earth's surface.

it occupies an area between 50 and 90 km above the earth. The D region is usually highly absorptive due to its high collision frequency.

E Region or E-Layer. A daylight phenomenon, existing between 90 and 140 km above the earth. It depends directly on the sun's UV radiation and hence it is most dense directly under the sun. The layer all but disappears shortly after sunset. Layer density varies with seasons owing to variations in the sun's zenith angle with seasons.

F1-Layer. A daylight phenomenon existing between 140 and 250 km above the earth. Its behavior is similar to that of the E-layer in that it tends to follow the sun (i.e., most dense under the sun). At sunset the F1-layer rises, merging with the next higher layer, the F2-layer.

F2-Layer. This layer exists day and night between 150 and 250 km (night) and 250 and 300 km above the earth (day). During the daytime in winter, it extends from 150 to 300 km above the earth. Variations in height are due to solar heat. It is believed that the F2-layer is also strongly influenced by the earth's magnetic field. The earth is divided into three magnetic zones representing different degrees of magnetic

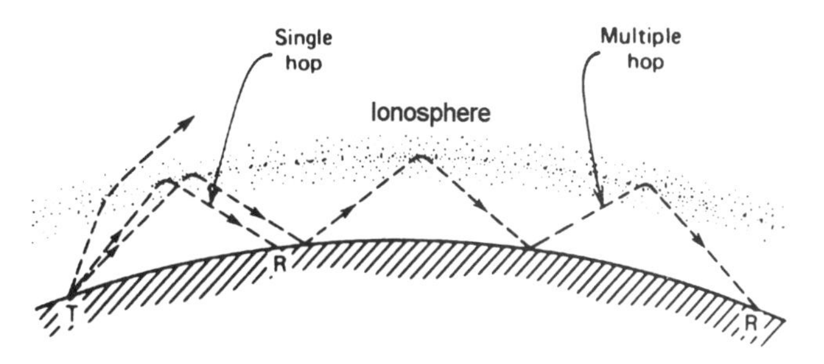

Figure 11.3. Single and multihop HF skywave transmission. T = transmitter, R = receiver.

intensity called east, west, and intermediate. Monthly F2 propagation predictions are made for each zone.* The north and south auroral zones are also important for F2 propagation, particularly during high sunspot activity.

Consider these layers as mirrors or partial mirrors, depending on the amount of ionization present. Thus transmitted waves striking an ionospheric layer, particularly the F-layer, may be refracted directly back to earth and received after their first hop, or they may be reflected from the earth back to the ionosphere again and repeat the process several times before reaching the distant receiver. The latter phenomenon is called multihop transmission. Single and multihop transmission are illustrated diagrammatically in Figure 11.3.

To obtain some idea of the estimated least possible number of F-layer hops as related to path length, the following may be used as a guide:

Number of Hops	Path Length (km)
1	< 4000
2	4000–7000
3	7000–12,000

(see Section 11.10.2.4).

HF propagation above about 8 MHz encounters what is called a *skip zone.* This is an "area of silence" or a zone of no reception extending from the outer limit of groundwave communication to the inner limit of skywave communication (first hop). The skip zone is shown graphically in Figure 11.4.

The region of coverage from a HF transmitter can be extended through the "skip zone" by a subset of skywave transmission called "near-vertical

* Monthly propagation forecasts are made by the Central Radio Propagation Laboratory (CRPL), U.S. National Bureau of Standards, now called National Institute of Standards and Technology (NIST).

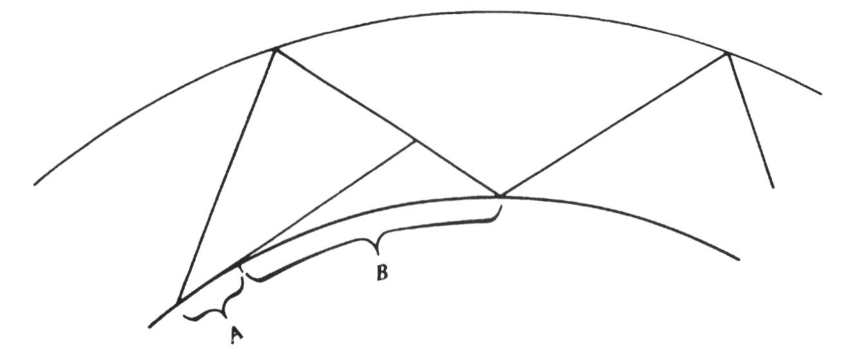

Figure 11.4. Skip zone. A = limit of groundwave communication; B = skip zone. *Note:* Good HF coverage is possible in the skip zone using near-vertical incidence skywave (NVIS) transmission.

incidence" (NVI) or "quasi-vertical incidence" (QVI). The objective is to launch a wave nearly directly upward from the antenna. Therefore special antennas that have high takeoff radiation angles are required (see Chapter 14). By radiating the HF RF energy nearly vertically, we can achieve "reflection" from the F-layer almost overhead. NVI(S) propagation provides HF connectivity from 20 mi or less to over 500 mi. The best performance for this type of operation is to use frequencies from 3 to 6 MHz. Receive signals using this mode of propagation will fade rapidly (1-Hz fade rate), producing time dispersion on the order of 100 μs and with some Doppler spread. With proper signal processing at the receiver, these impairments can be mitigated. Using some of the new wideband HF techniques that are just now emerging can fully resolve the multipath and resulting dispersion.

11.5 CHOICE OF OPTIMUM OPERATING FREQUENCY

One of the most important elements for successful operation of a HF link using skywave transmission is the selection of an operating frequency that will achieve a time availability approaching 100%, day, night, and year-round. Seldom, if ever, can this be achieved.

Optimum HF propagation between points X and Y anywhere on the earth varies with the following:

- Location, in particular, latitude
- Season
- Diurnal variations (time of day)
- Cyclical variations (relating to sunspot number)
- Abnormal (disturbed) propagation conditions

LOCATION. The intensity of ionizing radiation that strikes the atmosphere varies with latitude. The intensity is greatest in equatorial regions, where the sun is more directly overhead than in the higher latitudes.

We find that the critical frequencies for E and F1 regions vary directly with the sun's elevation, being highest in equatorial regions and decreasing as a function of the increase in latitude. The *critical frequency* is the highest frequency, using vertical incidence (transmitting a wave directly overhead, at a 90° elevation angle), at a certain time and location, at which RF energy is reflected back to earth.

The F2-layer also varies with latitude. Here the issue is more complex. It is postulated that the variance in ionization is caused by other sources, such as X-rays, cosmic rays, and the earth's magnetic field. However, F2-layer critical frequency does not have a strong variance as a function of latitude but is more associated with longitude. Critical frequencies related to the F2-layer are generally higher in the Far East than in Africa, Europe, and the Western Hemisphere.

HF transmission engineers consider regions (location) in a more general sense, again related to latitude. The most attractive region, which displays relatively low values of dispersion and Doppler spread, is the temperate region, especially overwater paths in that region. The second region covers HF paths that are transequatorial. These paths encounter *spread F* propagation, which causes relatively high time dispersion because the F-layer over the equator is more diffuse, resulting in much greater multipath effects.

The third difficult region covers those HF paths that cross the auroral oval. There are two auroral ovals, one in the Southern Hemisphere and one in the Northern Hemisphere. Each is centered on its respective magnetic pole. Transauroral paths are difficult paths in terms of meeting performance requirements and are even more difficult during magnetic storms and other solar flare disturbances.

The most difficult of all HF paths are those that cross the polar cap where Doppler spread has been measured in excess of 10 Hz and time dispersion can be over 1–2 ms (Refs. 5 and 6).

Figure 11.5 shows an ionogram* of a benign one-hop path in the evening. It shows the groundwave return appearing vertically on the far left. The one-hop F2 return(s) consist(s) of an O-ray and X-ray. O stands for "ordinary" and X for "extraordinary." How these rays are generated are discussed below. The ionogram also shows a vestige of multimode transmission appearing as a two-hop return on the far right side of the figure. Figure 11.6 shows a typical auroral oval, in this case, north of the equator. It should be kept in mind that the oval is indeed centered on the magnetic pole but that its boundary can shift; it is not fixed (Ref. 2).

*An ionogram is a graphic printout created by a type of ionospheric sounder (see Section 11.5.1.2) showing received RF energy comparative intensity and its delay.

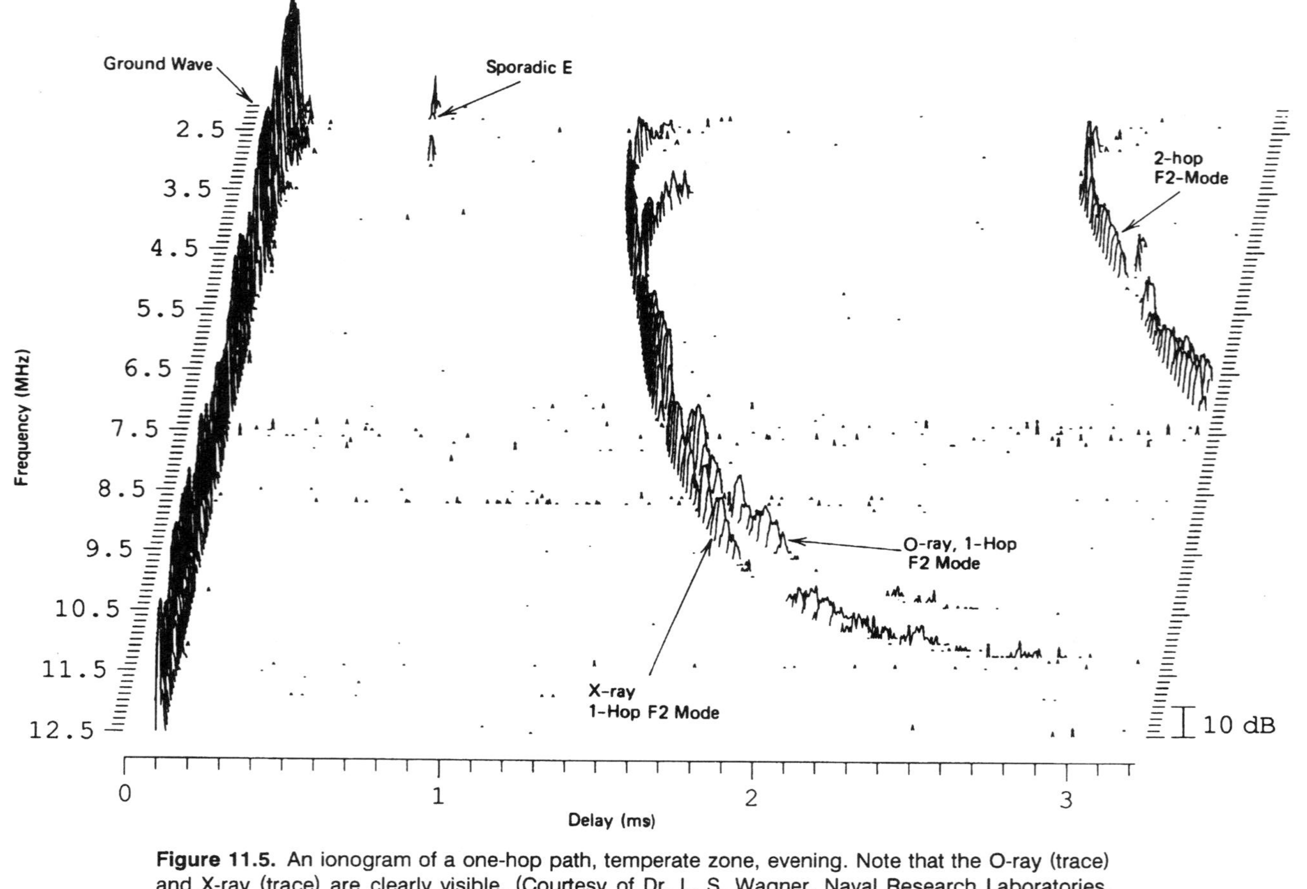

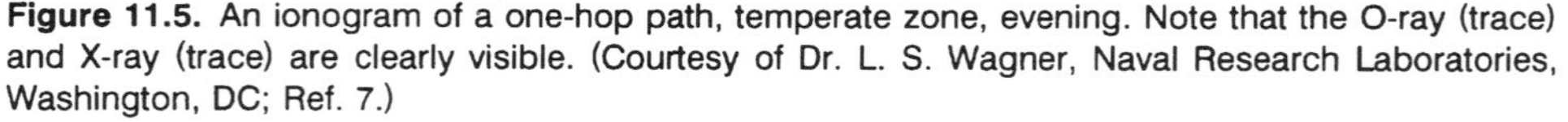

Figure 11.5. An ionogram of a one-hop path, temperate zone, evening. Note that the O-ray (trace) and X-ray (trace) are clearly visible. (Courtesy of Dr. L. S. Wagner, Naval Research Laboratories, Washington, DC; Ref. 7.)

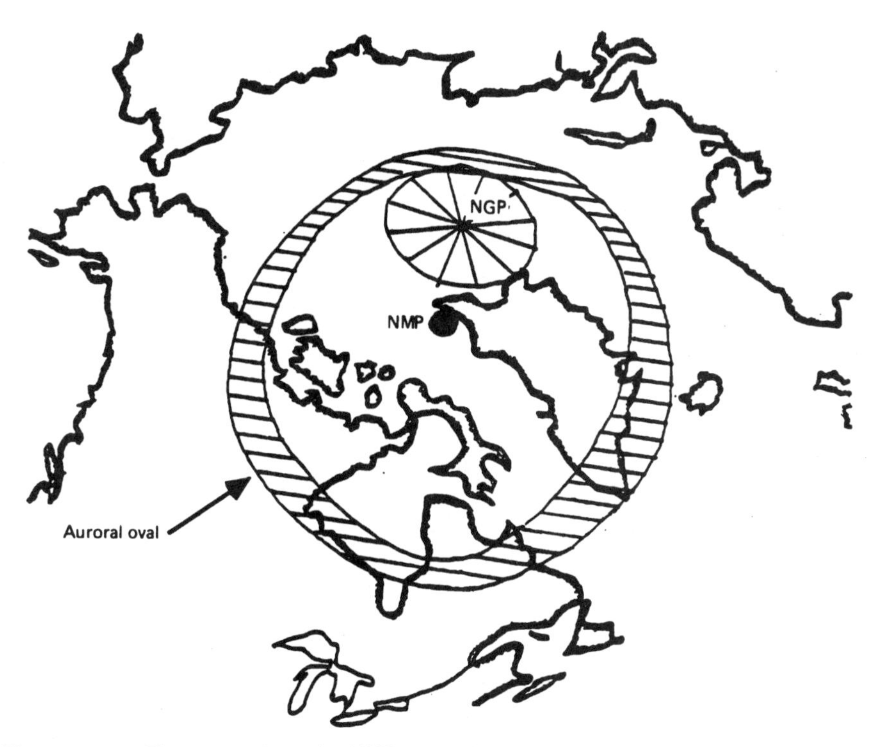

Figure 11.6. The auroral oval. NMP = northern magnetic pole; NGP = northern geographical pole.

SEASONAL FACTORS. Our earth orbits the sun with an orbital period of 1 yr. It is this orbit that brings about our seasons: spring, summer, autumn, and winter. The sun is a major controlling element on the behavior of the ionosphere. For instance, E-layer ionization of sufficient magnitude to support skywave propagation depends almost entirely on the sun's elevation in the sky; it is stronger in summer than in winter.

The F1-layer, in general, exists only during daylight hours. During the winter its critical frequency varies in a similar manner as the E-layer, being dependent on the sun's elevation. Ionization, as one might imagine, is greater in the summer months than in the winter months, when there is less sunlight. Often, in the winter months the F1-layer merges with the F2-layer and cannot be specifically identified except in equatorial regions.

The F2-layer is the reverse; daytime ionization is very intense and critical frequencies are higher than its F1 counterpart. Because of the extended periods of darkness in winter months, the ionosphere has more time to lose its electrical charge (recombination), resulting in nighttime critical frequencies dipping to very low values.

During the summer with its extended daylight hours, the F2-layer heats and expands, resulting in notably lower ionization density than in winter.

This makes summer daytime F2 critical frequencies lower than winter values. The general assumption is that the F1- and F2-layers combine at night. This certainly is true in the winter months. In summer months, because of the longer daylight hours, this recombination does not occur to the extent that it does in winter. This results in nighttime F2 critical frequencies that are significantly higher in summer than in winter. Also, the difference between day and night critical frequencies is much smaller in summer than in winter.

DIURNAL VARIATIONS (TIME OF DAY). We break the 24-h day into three generalized time periods:

1. Day
2. Transitions
3. Night

The transition periods occur twice: once around sunrise and again around sunset.

Again we are dealing with the intensity and duration of sunlight through the daytime hours. It is primarily the UV radiation from the sun that ionizes the atmosphere. During the daylight hours this radiation reaches a maximum intensity; during the hours of darkness, there is little UV radiation impinging the atmosphere and the E and F regions decrease to a relatively weak single layer through which propagate clouds with greater electron densities, which gives rise to sporadic E (E_s) propagation modes.

The F2 region critical frequency rises steeply at sunrise, reaching its maximum after the sun has reached its zenith. It starts to drop off sharply when the sun is starting to set. The transition periods, around sunrise and again around sunset, are when we encounter rapidly changing critical frequencies.

CYCLICAL VARIATIONS. The results of sunspots lead to the one phenomenon that affects atmospheric behavior more than any other. Sunspot activity is cyclical and is therefore referred to as the solar cycle. We characterize solar activity (i.e., number, intensity, and duration of sunspots) by the sunspot number. Sunspots appear on the sun's surface and are tremendous eruptions of whirling electrified gases. These gases are cooled to temperatures below those of the sun's surface, resulting in darkened areas appearing to us as spots on the sun. The whirling gases are accompanied by extremely intense magnetic fields. At times there are sudden flare-ups on the sun that some call *magnetic storms*. Magnetic storms will be discussed later.

Sunspots last from several days to several months. The sun has a rotational period of 27 days. Larger sunspots can remain visible for several rotations of the sun, giving rise to propagation anomalies with 27-day cycles.

Galileo was the first to report sunspots in modern recorded history. Some sources credit the Chinese, long before the birth of Christ, for being the first to observe sunspots.

The Zurich Observatory began reporting sunspot data on a regular basis in 1749. Here Rudolf Wolf devised a standard method of measure of relative sunspot activity. This is called the Wolf sunspot number, or the Zurich sunspot number, or just the sunspot number. The sunspot number is more of an index of solar activity than a measure of the number of sunspots observed.

The official sunspot number is derived from daily recording of sunspot numbers at a solar observatory in Locarno, Switzerland. Raw sunspot data are processed and disseminated by the Sunspot Index Data Center in Brussels. There has been a continuity of these measurements since 1749 when they were begun in Zurich. The sunspot cycle is roughly 11 years. A cycle is the time in years between contiguous solar activity minimums. A minimum sunspot number has an average value of 5 and an average maximum value of 109 (Ref. 8). Solar cycles are numbered, with cycle No. 1 starting in 1749; cycle 21 began in June 1976 and cycle 22 in September 1986, reaching a maximum value of about 190 in January 1990. Sunspot cycles are not of uniform duration. Some are as short as 9 years and others as long as 14 years (Ref. 8). Figure 11.7 shows observed and predicted sunspot numbers up through 1991 and extrapolated until 1998.

Sunspots have a direct bearing on UV radiation intensity of the sun, which, as we mentioned earlier, is the principal cause of ionization of the

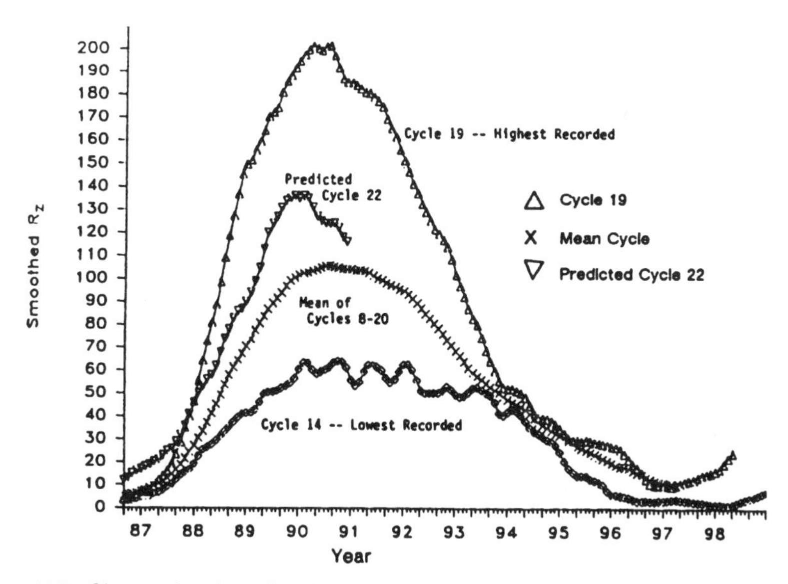

Figure 11.7. Observed and predicted sunspot numbers from McNish–Lincoln analysis. R_z is the relative sunspot number (smoothed). (From Ref. 9.)

upper atmosphere, which we call the ionosphere. It has a direct relation to the critical frequency and consequently to the *maximum usable frequency* (MUF). During periods of low sunspot numbers, the daytime MUF may not reach 17 MHz; during periods of sunspot maximum, the MUF has been known to far exceed 50 MHz. In sunspot cycle 19, state police radios operating in the 50-MHz region in Massachusetts and Rhode Island were made virtually useless by interfering signals from the West Coast, some 3000 mi away. This is one reason most dispatching systems, especially in public safety, are now using 150 MHz and some low ultrahigh frequencies (UHF). Thus they avoid the cyclical problem of those periods when sunspot numbers are high.

IONOSPHERIC DISTURBANCES. We have classified four factors dealing with skywave HF transmission. If we neglect interference, there is some predictability for the design of a HF path. The computer has helped us along the way with such excellent programs as IONCAP and PROPHET (Refs. 10–12). They take into account sunspot number, latitude/longitude, time of day, season, and system characteristics. However, the transmission engineer feels nearly helpless when faced with the unexpected, the ionospheric disturbance.

There are three types of classified ionospheric disturbances:

1. Ionospheric storm (magnetic storm)
2. Sudden ionospheric disturbance (SID)
3. Polar blackout

The author was in the Arctic during cycle 19 peak (1958), aboard ship. Whenever there was a good display of aurora borealis, the HF band would black out and would remain that way from 12 to 72 h. All one would hear is a high level of thermal noise. Curiously, it seemed that signals in the low medium-frequency (MF) and low-frequency (LF) bands were enhanced in these periods of HF blackout.

There is fair correlation with the number and intensity of SIDs, ionospheric storms, and blackouts with the sunspot cycle. The higher the sunspot number, the greater the probability of occurrence and the greater the intensity. Again, we believe the cause to be solar flare activity, a sudden change in the magnetic field around sunspots (Ref. 8). These produce high levels of X-rays, UV radiation, and cosmic noise. It takes about 8 min for the effects of this radiation to degrade the ionosphere. This is a SID. The result is heavy HF skywave signal attenuation.

Slower-traveling charged particles arrive some 18–36 h after the sudden flare-up on the sun. The result of these particles hitting the ionosphere is often blackout, especially in the higher latitudes. In other areas we notice a remarkable lowering of the MUF.

WWV transmitting from Boulder, Colorado, on 5, 10, 15, and 20 MHz is normally used for time and frequency calibration. At 18 min after each hour,

WWV of the U.S. National Institute of Standards and Technology (NIST) transmits timely solar and geomagnetic activity with updates every 6 h. WWV transmits the K index and the A index. The K index is a single digit from 0 through 9 and is a measure of current geomagnetic activity; 0 is the lowest activity factor and 9 the highest. The A index is a measure of solar flux activity.

11.5.1 Frequency Management

11.5.1.1 Definitions. In the context of this chapter, frequency management is the art/technology of selecting an optimum frequency for HF communication at a certain time of day between any two points on the earth. One important factor is omitted in the process, that is, the interference that may be present in the path, affecting one or both ends of the path.

We will be using the terms *MUF*, *LUF*, and *FOT* (*OWF*). The MUF and LUF are the upper and lower limiting frequencies for skywave communication between points X and Y. The MUF is the maximum usable frequency on an oblique incidence path; the LUF is the lowest usable frequency.

The concept of MUF is most important for the HF link design engineer for skywave links. We will find that the MUF is related to the critical frequency by the secant law. The optimum working frequency (OWF; original derivation from a French term), sometimes called FOT (from the French, *fréquence optimum de travail*), is often taken as $0.85 \times$ MUF. Little is mentioned in the literature about the LUF, since it is so system sensitive.

The MUF is a function of the sunspot number, time of day, latitude, day of the year, and so on, which are things completely out of our control. The LUF, on the other hand, is somewhat under our control. If we hold a transmitting station EIRP constant, as the operating frequency decreases, the available power at the distant receiver normally decreases owing to increased ionospheric absorption. Furthermore, the noise power increases so that the signal-to-noise ratio deteriorates and the circuit reliability decreases. The minimum frequency below which the reliability is unacceptable is called the lowest usable frequency (LUF). The LUF depends on transmitter power; antenna gain at the desired takeoff angle (TOA); factors that determine transmission loss over the path, such as frequency, season, sunspot number, and geographical location; and the external noise level. One of the primary factors is ionospheric absorption and hence, since this varies with the solar zenith angle, the LUF peaks at about noon. Consequently, in selecting a frequency, it is necessary to ascertain whether the LUF exceeds this frequency.

When applying computer prediction programs such as IONCAP, under certain situations we will find the LUF exceeds the MUF. This tells us, given the input parameters to the program, that the link is unworkable. We may be able to shift the LUF downward in frequency by relaxing link reliability (time availability), reducing signal-to-noise ratio requirements, increasing EIRP, and increasing receive antenna directional gain.

One factor in the transmission loss formula for HF is D-layer absorption. It varies as $1/f^2$. For this reason we want to operate at the highest possible frequency to minimize D-layer absorption. Suppose we operated at the MUF. It is a boundary limit and would be unstable with heavy fading. We choose a frequency as close as possible to the MUF yet stay out of boundary conditions. The operating frequency goes under two names: FOT or OWF, both defined earlier. The OWF is usually 0.85 the value of the MUF for F2 operation. Our objective is to keep our transmitter frequency as close to the OWF as possible. This is done to minimize atmospheric absorption but yet not too close to the MUF to reduce ordinary–extraordinary ray fading. As the hours of the day pass, the OWF will move upward or downward and through the cycle of our 24-h day. It could move one or even two octaves from maximum to minimum frequency. How do we know when and where to move to in frequency?

11.5.1.2 Methods. There are six general methods in use today to select the best frequency or the OWF. We might call these methods the *frequency management* function. The six methods are:

1. By experience.
2. Use of CRPL (Central Radio Propagation Laboratory) predictions.
3. Carrying out one's own predictions by one of several computer programs available.
4. Use of ionospheric sounders.
5. Use of distant broadcast facilities.
6. Self- and embedded sounding.

EXPERIENCE. Many old-time operators still rely on experience. First they listen to their receivers, then they judge if an operating frequency change is required. It is the receive side of a link that commands a distant transmitter. An operator may well feel that "yesterday I had to QSY* at 7 p.m. to 13.7 MHz, so today I will do the same." Listening to his/her receiver will confirm or deny this belief. When he/she hears his/her present operating frequency (from the distant end) start to take deep fades and/or the signal level starts to drop, it is time to start searching for a new frequency. The operator checks other assigned frequencies on a spare receiver and listens for other identifi-

*A "Q" signal is an internationally recognized three-letter operational signal used among operators. A typical Q signal is QSY, which means "change your frequency to ______." Q signals were used almost exclusively on CW (Morse) circuits and their use has extended to teleprinter and even voice circuits.

able signals near these frequencies to determine conditions to select a better frequency. If conditions are found to be better on a new frequency, the transmitter operator is ordered to change frequency (QSY at the distant end). This, in essence, is the experience method. We will address this issue further on from a somewhat different perspective.

CRPL PREDICTIONS. Here, of course, we are dealing with the use of predictions issued by the U.S. Institute of Telecommunication Sciences, located in Boulder, Colorado. The predictions are published monthly, three months in advance of their effective dates (Ref. 13).

Consider a HF circuit designed for 95% time availability* (propagation reliability) and assume a minimum signal-to-noise ratio of 12 dB Hz. The median receive signal level (RSL) must be increased on the order of 14 dB to overcome slow variations of skywave field intensity and atmospheric noise, and 11 dB to overcome fast variations of skywave field intensity. Therefore a rough order of magnitude value of signal-to-median-noise ratio with sufficient margin for 90% of the days is 37 dB for *M*-ary frequency shift keying (MFSK) data/telegraph transmission for a bit error rate (BER) of 1×10^{-4}. See CCIR Rec. 339-6 (Ref. 14).

PREDICTIONS BY PC. Quite accurate propagation predictions for a HF path can be carried out on a personal computer. A very widely accepted program is IONCAP, which stands for Ionospheric Communication Analysis and Prediction Program. It is written in Fortran (ANSI) and is divided into seven largely independent subsections (Refs. 10 and 11):

1. Input subroutines.
2. Path geometry subroutines.
3. Antenna pattern subroutines.
4. Ionospheric parameter subroutines.
5. Maximum usable frequency (MUF) subroutines.
6. System performance subroutines.
7. Output subroutines.

Table 11.1 is a listing of 30 available output methods. The IONCAP computer program performs four basic analysis tasks. These tasks are summarized below. Note that E_s indicates highest observed frequency of the ordinary component of sporadic E and HPF means highest probable frequency.

*We define "time availability" as the percentage of time a certain performance objective (e.g., BER) is met.

TABLE 11.1 IONCAP—Available Output Methods

Method	Description of Method
1	Ionospheric parameters
2	Ionograms
3	MUF–FOT lines (nomogram)
4	MUF–FOT graph
5	HPF–MUF–FOT graph
6	MUF–FOT–E_s graph
7	FOT–MUF table (full ionosphere)
8	MUF–FOT graph
9	HPF–MUF–FOT graph
10	MUF–FOT–ANG graph
11	MUF–FOT–E_s graph
12	MUF by magnetic indices, *K* (not implemented)
13	Transmitter antenna pattern
14	Receiver antenna pattern
15	Both transmitter and receiver antenna patterns
16	System performance (SP)
17	Condensed system performance, reliability
18	Condensed system performance, service probability
19	Propagation path geometry
20	Complete system performance (CSP)
21	Forced long-path model (CSP)
22	Forced short-path model (CSP)
23	User-selected output lines (set by TOPLINES and BOTLINES)
24	MUF–REL table
25	All modes table
26	MUF–LUF–FOT table (nomogram)
27	FOT–LUF graph
28	MUF–FOT–LUF graph
29	MUF–LUF graph
30	Create binary file of variables in "COMMON/MUFS/" (allows the user to save MUFs–LUFs for printing by a separate user written program)

Source: Reference 11.

1. *Ionospheric Parameters.* The ionosphere is predicted using parameters that describe four ionospheric regions: E, F1, F2, and E_s. For each sample area, the location, time of day, and all ionospheric parameters are derived. These may be used to find an electron density profile, which may be integrated to construct a predicted ionogram. These options are specified by Methods 1 and 2 in Table 11.1.

2. *Antenna Patterns.* The user may precalculate the antenna gain pattern needed for the system performance predictions. These options are specified by Methods 13–15 in Table 11.1. If the pattern is precalculated, then the antenna gain is computed for all frequencies (1–30 MHz) and elevation angles. If the pattern is not precalculated, then the gain value is determined for a particular frequency and elevation angle (takeoff angle) as needed.

3. *Maximum Usable Frequency (MUF).* The maximum frequency at which a skywave mode exists can be predicted. The 10% (FOT), 50% (MUF), and 90% (HPF, highest probable frequency) levels are calculated for each of the four ionospheric regions predicted. These numbers are a description of the state of the ionosphere between two locations on the earth and not a statement of the actual performance of any operational communication circuit. These options are specified by Methods 3–12 in Table 11.1.

4. *System Performance.* A comprehensive prediction of radio system performance parameters (up to 22) is provided. Emphasis is on the statistical performance over a month's time. A search to find the LUF is provided. These options are specified by Methods 16–29 in Table 11.1.

Table 11.2 shows a typical run for Method 16. The path is between Santa Barbara, California, and Marlborough, Massachusetts.

There are several other programs that can be run on a PC. One is Minimuf, which predicts only the MUF in midlatitudes (Ref. 15). PROPHET and Advanced PROPHET are two HF prediction codes for military application (Refs. 12 and 16). The accuracy is not as good as IONCAP, but they are a useful tool for propagation prediction and for the specific military applications for which they are designed. They provide area coverage easily, which IONCAP does not, and also allow interaction for various details of electronic warfare.

IONOSPHERIC SOUNDERS. Ionospheric oblique sounders give real-time data on the MUF, transmission modes propagating (F1, F2, one-, two-, or three-hop), and the LUF between two points that are sounder-equipped. One location has a sounder transmitter installed that sweeps the entire HF band using a FM/CW signal. Transmitter power output usually is in the range of 1–10 W. A sounder receiver and display are operated at the distant end. The receiver is time synchronized with the distant transmitter. It displays a time history of the sweep of the band. The *x*-axis of the display is the HF frequency band 2 (3) to 30 MHz and the *y*-axis is time delay measured in milliseconds. Here received power is displayed as a function of frequency and time delay. Of course, the shortest delay is that power reflected off the E-layer, the next shortest the F-layer (F1 and F2 in daytime, in that order), then multihop power from the F-layers. A multihop signal is delayed even more. Figure 11.8 shows a conceptual diagram of an oblique incidence sounder operation.

One system designed and manufactured by BR Communications is the AN/TRQ-35. This unit includes a low-power transmitter and, at the far end of the circuit, a receiver with a spectrum monitor that compiles channel occupancy statistics for 9333 channels spaced 3 kHz from 2 to 30 MHz. BR Communications calls their spectrum monitor a "signal-to-noise ratio" analyzer that determines the background noise level—usually atmospheric noise

TABLE 11.2 Samples of an IONCAP Run, Method 16

```
                              METHOD 16    IONCAP    PC.20    PAGE 2
                                   OCT 1990        SSN = 192.
SANTA BARBARA TO MARLBOROUGH                      AZIMUTHS              N. MI.      KM
               34.50 N  119.00 W - 42.50 N  71.50 W  63.19  273.90  2252.5  4171.2
MINIMUM ANGLE 1.0 DEGREE
ITS-1 ANTENNA PACKAGE
XMTR  2.0  TO  30.0  CONST. GAIN  H .00  L .00  A .0  OFF  AZ .0
RCVR  2.0  TO  30.0  CONST. GAIN  H .00  L .00  A .0  OFF  AZ .0
POWER = 1.000 KW  3 MHZ NOISE = -150.0  DBW REQ. REL = .80  REQ. SNR = 48.0
MULTIPATH POWER TOLERANCE = 6.0  DB MULTIPATH DELAY TOLERANCE = .500 MS
```

UT	MUF												
5.0	18.9	2.0	3.4	4.2	6.3	8.5	10.7	12.8	15.0	17.2	19.4	21.5	FREQ
	1F2	4 E	1F2	2ES	2F2	2F2	2F2	2F2	2F2	2F2	1F2	1F2	MODE
	3.8	8.7	3.2	1.3	11.1	11.4	12.2	14.0	19.0	19.0	3.5	3.5	ANGLE
	14.8	14.3	14.7	14.1	14.8	14.8	14.9	15.1	15.7	15.7	14.8	14.8	DELAY
	513.	103.	487.	110.	302.	309.	326.	363.	475.	475.	500.	500.	V HITE
	.50	1.00	1.00	1.00	1.00	1.00	1.00	.99	.93	.74	.41	.10	F DAYS
	151.	158.	153.	157.	143.	142.	142.	145.	177.	235.	153.	173.	LOSS
	16.	−15.	0.	2.	12.	15.	16.	14.	−16.	−73.	10.	−9.	DBU
	−116	−128	−117	−117	−110	−111	−111	−114	−147	−204	−123	−143	S DBW
	−171	−136	−143	−145	−151	−158	−163	−166	−168	−170	−172	−173	N DBW
	55.	9.	25.	28.	40.	47.	51.	51.	22.	−34.	49.	30.	SNR
	10.	48.	28.	24.	12.	5.	2.	8.	43.	99.	16.	34.	RPWRG
	.64	.00	.00	.01	.16	.45	.68	.59	.11	.00	.52	.20	REL
	.00	.00	.00	.01	.03	.00	.00	.00	.00	.00	.00	.00	MPROB
11.0	13.7	2.0	3.5	5.1	6.6	8.2	8.6	9.7	11.2	12.8	14.3	15.9	FREQ
	1F2	3 E	2ES	2F2	2F2	1F2	1F2	2F2	1F2	1F2	1F2	1F2	MODE
	4.9	4.1	1.3	16.7	15.0	4.9	3.7	17.9	1.2	2.5	4.9	4.9	ANGLE
	14.9	14.1	14.1	15.4	15.2	14.9	14.8	15.6	14.5	14.7	14.9	14.9	DELAY
	557.	89.	110.	422.	385.	559.	510.	450.	407.	457.	557.	557.	V HITE
	.50	1.00	.91	1.00	1.00	1.00	1.00	.98	.90	.68	.35	.10	F DAYS
	148.	172.	168.	147.	144.	142.	141.	150.	139.	142.	153.	169.	LOSS

	12.	−28.	−6.	7.	10.	17.	18.	8.	19.	17.	8.	−8.	DBU
	−117	−140	−123	−114	−113	−107	−107	−118	−109	−112	−122	−139	S DBW
	−167	−144	−148	−151	−154	−157	−158	−161	−164	−166	−168	−169	N DBW
	49.	4.	25.	36.	41.	49.	50.	42.	55.	54.	45.	30.	SNR
	15.	54.	31.	19.	15.	6.	5.	22.	2.	8.	19.	35.	RPWRG
	.53	.00	.01	.09	.18	.56	.61	.32	.74	.64	.43	.20	REL
	.00	.00	.00	.02	.00	.00	.00	.00	.00	.00	.00	.00	MPROB
17.0	42.3	2.0	4.6	7.1	9.7	12.2	14.8	17.4	23.0	28.7	34.3	40.0	FREQ
	1F2	3 E	2ES	2ES	2ES	2ES	4F2	2F2	2F2	2F2	2F2	1F2	MODE
	3.9	3.3	1.3	1.3	1.3	1.3	23.8	9.4	9.2	11.2	18.3	8.4	ANGLE
	14.8	14.1	14.1	14.1	14.1	14.1	15.8	14.6	14.6	14.8	15.6	15.4	DELAY
	517.	79.	110.	110.	110.	110.	261.	267.	264.	304.	459.	715.	V HITE
	.50	1.00	1.00	1.00	1.00	.92	1.00	1.00	.99	.95	.83	.61	F DAYS
	152.	626.	348.	253.	213.	195.	186.	159.	152.	151.	188.	160.	LOSS
	22.	****	****	−96.	−46.	−26.	−26.	5.	13.	16.	−21.	12.	DBU
	−117	−355	−317	−220	−172	−154	−156	−126	−120	−119	−158	−127	S DBW
	−181	−145	−155	−160	−163	−165	−167	−169	−174	−176	−179	−180	N DBW
	63.	****	****	−60.	−9.	11.	11.	42.	53.	56.	21.	54.	SNR
	1.	262.	221.	118.	67.	48.	50.	17.	8.	9.	44.	11.	RPWRG
	.78	.00	.00	.00	.00	.00	.00	.25	.62	.66	.10	.61	REL
	.00	.00	.00	.00	.00	.00	.00	.00	.00	.00	.00	.00	MPROB
24.0	38.2	2.0	4.0	6.1	10.3	14.6	18.8	23.0	27.3	31.5	35.8	40.0	FREQ
	1F2	3 E	2ES	4F2	2F2	2F2	2F2	2F2	2F2	2F2	1F2	1F2	MODE
	3.7	4.1	1.3	24.4	9.3	9.3	9.8	11.0	14.1	18.9	9.2	3.7	ANGLE
	14.8	14.1	14.1	15.9	14.6	14.6	14.7	14.8	15.1	15.7	15.5	14.8	DELAY
	510.	89.	110.	269.	266.	265.	276.	300.	366.	473.	749.	510.	V HITE

TABLE 11.2 (*continued*)

.50	1.00	.89	1.00	1.00	1.00	1.00	.99	.97	.87	.66	.31	F DAYS
157.	227.	177.	175.	151.	148.	148.	149.	155.	210.	169.	164.	LOSS
11.	−84.	−27.	−22.	8.	13.	15.	15.	11.	−43.	2.	5.	DBU
−127	−196	−146	−145	−119	−116	−117	−119	−123	−179	−136	−134	S DBW
−180	−144	−149	−153	−161	−167	−171	−174	−176	−178	−179	−180	N DBW
53.	−53.	3.	9.	42.	50.	54.	55.	52.	−2.	43.	46.	SNR
12.	111.	55.	47.	13.	6.	3.	3.	13.	67.	22.	18.	RPWRG
.60	.00	.00	.00	.22	.58	.72	.71	.57	.01	.38	.46	REL
.00	.00	.00	.00	.00	.00	.00	.00	.00	.00	.00	.00	MPROB

Notes: Header information largely supplied by user.
Line 1: Month, year, and sunspot number.
Line 2: Label as supplied by user and headings for next line.
Line 3: Transmitter location, receiver location, the azimuth of the transmitter to the receiver in degrees east of north; path distance in nautical miles and kilometers.
Line 4: Minimum radiation angle in degrees.
Line 5: Antenna subroutine used (use IONCAP antenna subroutine).
Line 6: Transmitter antenna data.
Line 6A: Receiver antenna data.
Line 7: System line that has transmitter power in kilowatts, man-made noise level at 3 MHz in dBW, required reliability, and required S/N in decibels.

System performance lines that are repeated for each hour.

Line 0, FREQ: Time and frequency line as associated with each column. The first four lines always refer to the most reliable mode (MRM). The system performance parameter usually comes from the sum of all six modes.
Line 1, MODE: E is E-layer, F1 is F1-layer, F2 is F2-layer, E_s is E_s-layer; N is a one-hop E_s with n F1 or F2 hops (MRM).
Line 2, ANGLE: The radiation angle in degrees (MRM).
Line 3, DELAY: Time delay in milliseconds (MRM).
Line 4, V HITE: Virtual height in kilometers (MRM).
Line 5, F DAYS: The probability that the operating frequency will exceed the predicted MUF.
Line 6, LOSS: Median system loss in decibels for the most reliable mode (MRM).
Line 7, DBU: Median field strength expected at the receiver location in decibels above 1 $\mu V/m$.
Line 8, S DBW: Median signal power expected at the receiver input terminals in decibels above a watt.
Line 9, N DBW: Median noise power expected at the receiver in decibels above a watt.
Line 10, SNR: Median signal-to-noise ratio in decibels.
Line 11, RPWRG: Required combination of transmitter power and antenna gains needed to achieve the required reliability.
Line 12, REL: Reliability. The probability that the SNR exceeds the required SNR. Note this applies to all days of the month and includes the effect of all mode types: E, F1, F2, E_s, and over-the-MUF modes.
Line 13, MPROB: The probability of an additional mode within the multipath tolerances (short paths only).

Source: Radio General Company, Stow, Massachusetts.

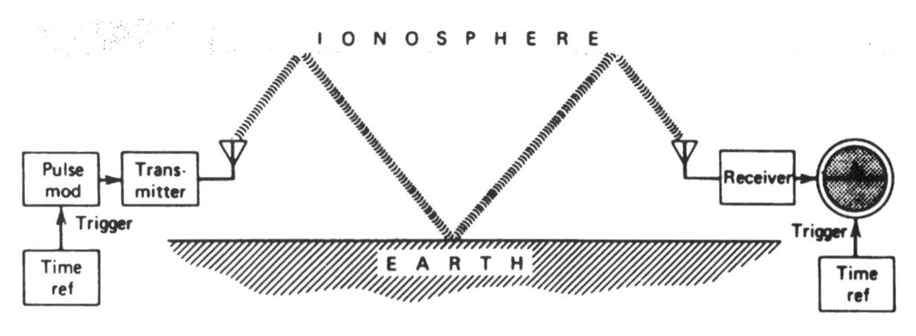

Figure 11.8. Synchronized oblique sounding.

—and then compiles statistics on the percentage of time that level is exceeded in each channel. The monitor has memory and can display any channel usage up to 30 min (Ref. 17).

A typical oblique ionospheric display (ionogram) is shown in Figure 11.9.

Another type of sounder is the backscatter or vertical incidence sounder, where the receiver and transmitter are collocated. In this case, the HF

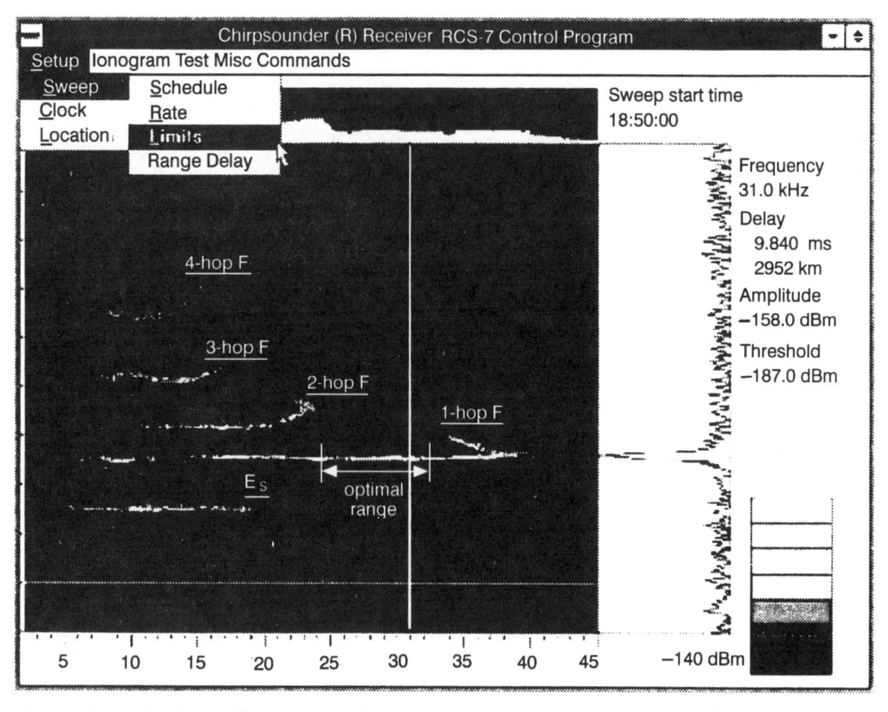

Figure 11.9. Typical HF ionogram from an oblique sounding system. (Courtesy of TCI International, Inc., Sunnyvale, CA.)

spectrum is swept as before, but the RF energy is launched vertically for refraction by the ionosphere directly overhead. Backscatter sounding provides data on the critical frequency, from which we can derive the MUF for any one-hop path. This is done with the following formula, which is based on Snell's law:

$$F_0 = F_n \sec \phi \tag{11.1}$$

where F_n = the maximum frequency (critical frequency) that will be reflected back at vertical incidence
F_0 = the maximum usable frequency (MUF) at oblique angle ϕ
ϕ = the angle of incidence (i.e., the angle between the direction of propagation and a line perpendicular to the earth)

Backscatter techniques can be used on any one-hop path where one point is the location of the backscatter sounder. On-the-other hand, oblique incidence sounders provide information on the single, equipped path, which may be one-hop or multiple hops.

Another approach that can be used on some occasions is to know the locations of other operating oblique sounder transmitters near the desired location for HF connectivity. "Near" can mean several hundred miles either side of the north–south direction and about 150 mi either side of the east–west direction of the desired distant end. Naturally, synchronization and interface data on these other sounders are necessary if we wish to make use of their signals.

USE OF DISTANT BROADCAST FACILITIES. We can obtain rough-order-of-magnitude propagation data derived from making comparative signal-level measurements on distant HF broadcast facilities "near" the desired distant point with which we wish to establish HF connectivity. Many facilities broadcast simultaneously on multiple HF bands, so a comparison can be made from one band to another on signal level, fade rate, and fade depth to determine the optimum band for operation. Many HF receivers are equipped with S-meters* from which we can derive some crude comparative level measurements.

For example, if we wish to communicate with an area in or near the U.S. state of Colorado, we can use WWV, which has excellent time and frequency dissemination service transmitting on 5, 10, 15, and 20 MHz simultaneously 24 h/day.

International broadcast such as the BBC and the Voice of America are other candidates. With present state-of-the-art synthesized HF receivers, we can tune directly to them. If one can read CW (copy international Morse code), there are hundreds of marine coastal stations that will have continuous

*An S-meter gives a comparative measurement of received signal level.

transmissions on 4, 6, 8, 12, 17, and 22 MHz (the marine bands), depending on which bands each facility considers to be "open" (useful) for the service that facility provides. These facilities will drop off automatic operation and into manual CW operation when they are "in traffic" (operating with a distant ship station). We may call this method "poor man's sounding."

EMBEDDED SOUNDING, EARLY VERSIONS. These earlier devices, still very widely used, are based on the idea that a facility or facility grouping has only a limited number of assigned operating frequencies. These devices transmit low RF energy, usually using FSK at a low bit rate, to sample each frequency to derive link quality assessment (LQA) data and, in some instances, may also use the same device for selective calling. LQA involves the exchange of link quality information on a one-on-one basis or networkwide. SELSCAN, a Rockwell–Collins trademark, is a good example of such a system. It measures signal-to-noise ratio and multipath delay distortion on up to 30 stored preset channels for use by automatic frequency selection algorithms. SELSCAN also has an ALE (automatic link establishment) feature (Ref. 18).

It is suggested that the reader consult MIL-STD-188-141A (Ref. 19) for some excellent methods of ALE and LQA. We are really dealing here with an OSI layer-2 datalink access protocol (see Ref. 20). SELSCAN, however, has been designed as an adjunct to SSB voice operation on HF for setting up a link on an optimum frequency. Harris (United States) and Tadiran (Israel), among others, have similar ALE/LQA devices.

MORE ADVANCED EMBEDDED SOUNDING. These are self-sounding systems designed for HF data/telegraph circuits that are half- or full-duplex. We will consider such circuits, for this argument, to have as near as possible to 100% duty cycle (i.e., they are active nearly 100% of the time). Such systems dedicate pure overhead, perhaps 5% or 10%, for self-sounding. Self-sounding, in this context, means that no separate sounding equipment is used and that a cooperative distant end is required.

Two cooperating facilities tied by a HF link operate on a fully synchronous basis and have their clocks, and hence time of day (TOD), slaved to a common system time or to universal coordinated time (UCT). At intervals, every 5 min, for example, both facilities take time out from exchanging traffic and search, first above their operating frequency and then below. Such systems work well with wideband HF, which operates over a full 1 MHz or 500 kHz with very low level PN spread phase-shift keying (PSK) signals. Here then each station moves up 1 MHz, for instance, then down 1 MHz in synchronism, each exchanging LQA data with the other regarding new frequencies.

LQA, for instance, can be a 3-bit group contained in the header information. A 1 1 1 exchange from each would indicate to move to the higher frequency, and a 1 1 0 to the lower frequency. Thus both stations are constantly homing in on the OWF. During transition periods, the search is carried out more frequently during calm periods, such as from, say, 10 a.m. to 3 p.m.; a longer period between searches may be used. At midpath (local time) the search may only be required every 10 min. Such systems often exchange only short message blocks and LQA data are updated in every header. As the binary number starts to drop, an algorithm kicks off a retimer to shorten the period between searches. If the first two digits hold in at binary 1 for long periods (e.g., 15 min), the timer is extended. The LQA three-digit word can be controlled by BER measurement or signal-to-noise ratio monitor (S/N). The problem with the use of S/N only is that it may not be indicative of multipath where BER or automatic repeat request (ARQ) negative acknowledgments (NACKs) (consult, Ref. 20 for ARQ) will be indicative of BER, which often can be traced to intersymbol interference, a typical effect of multipath.

11.6 PROPAGATION MODES

There are three basic modes of HF propagation:

1. Groundwave.
2. Skywave (oblique incidence).
3. Near-vertical incidence (NVI). This is a distinct subset of the skywave mode.

11.6.1 Basic Groundwave Propagation

The spacewave (not skywave) intensity decreases with the inverse of the distance, the groundwave with the inverse of the distance squared. Therefore at long distances and nonzero elevation angles, the intensity of the spacewave exceeds that of the groundwave. The groundwave is diffracted somewhat to follow the curvature of the earth. The diffraction increases as frequency decreases. Diffraction is also influenced by the imperfect conductivity of the ground. Energy is absorbed by currents induced in the earth so that energy flow takes place from the wave downward. The loss of energy dissipated in the earth leads to attenuation dependent on conductivity and dielectric constant. With horizontal polarization the wave attenuation is greater than with vertical polarization due to the different behavior of Fresnel reflection coefficients for both polarizations.

To summarize, groundwave is an excellent form of HF propagation where we can, during daylight hours, achieve 99% or better time availability. Groundwave propagation decreases with increasing frequency and with decreasing ground conductivity. As we go down in frequency, atmospheric noise starts to limit performance in daytime, and skywave interference at night is a basic limiter of groundwave performance on whichever of the lower frequencies we wish to use, providing we are not operating above the MUF.

11.6.2 Skywave Propagation

A wave that has been reflected from the ionosphere is commonly called a skywave. The reflection can take place at the E region and the F1 and/or F2 regions. In some circumstances RF energy can be reflected back from any two or all three regions at once. Figure 11.2 shows the three regions or layers of interest as well as the D-layer, which is absorptive. Figure 11.3 illustrates skywave communication.

HF skywave communications can be by one-hop, two-hops, or three-hops, depending on path length and ionospheric conditions. Figure 11.10 shows eight possible skywave modes of propagation. On somewhat longer paths ($\gtrsim$ 1000 km) we can receive RF energy from two or more modes at once, giving rise to multipath reception, which causes signal dispersion. Dispersion results in intersymbol interference (ISI) on digital circuits. Multimode reception is shown a bit later in Figure 11.12.

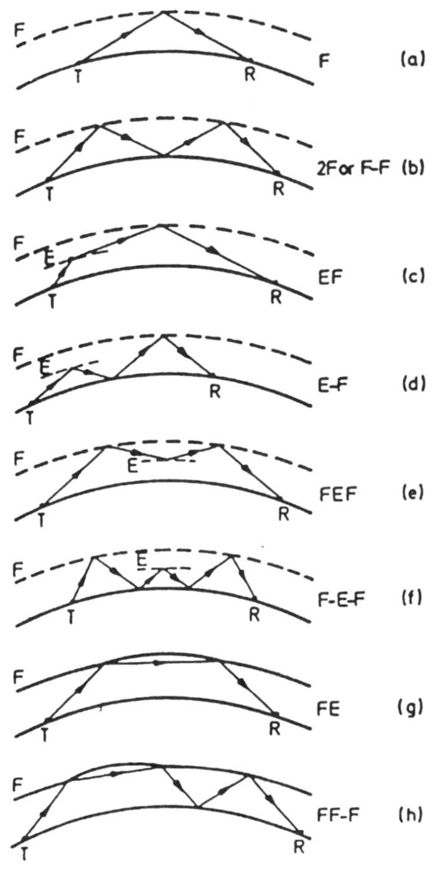

Figure 11.10. Examples of possible skywave multihop/multimode paths.

In general, we can say that multipath propagation can arise from the following:

- Multihop, especially when transmit and receive antennas have low gain and low takeoff angles
- Low and high angle paths (the low ray and the high ray)
- Multilayer propagation
- Ordinary (O) and extraordinary (X) rays from one or more paths

These effects, more often than not, exist in combination. An example of the high ray and low ray as they might be seen on an ionogram is shown in Figure 11.11. It also shows the MUF.

Typical one-hop ranges are 2000, 3400, and 4000 km for E-, F1-, and F2-layer reflections, respectively. These limits depend on the layer height of maximum electron density for rays launched at grazing incidence. The range values take into account the poor performance of HF antennas at low elevation angles.

Distances beyond the values given above can be achieved by utilizing consecutive reflections between the ionosphere and the earth's surface (see Figure 11.3 and Figure 11.10b–h). For each ground reflection the signal must pass through the absorptive D-layer twice, adding significantly to signal attenuation. The ground reflection itself is absorptive. It should also be noted that the elevation angle increases as a function of the hop number, which results in the lowering of the path MUF.

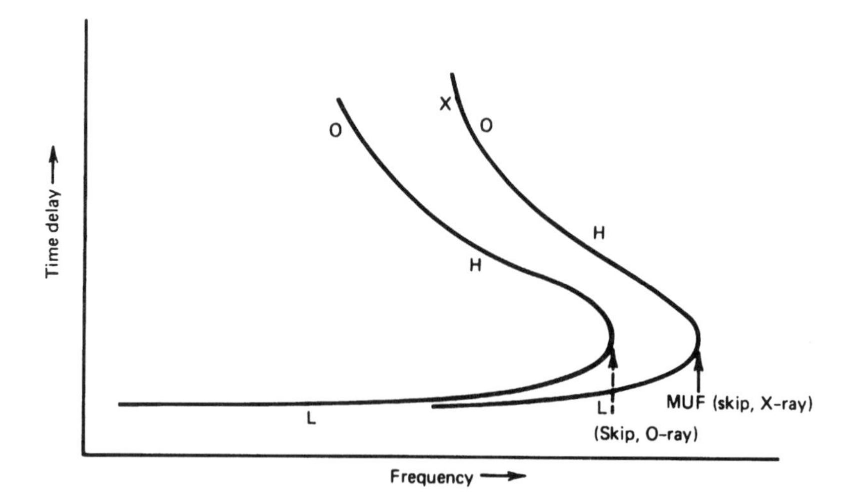

Figure 11.11. Ionogram idealized sketch. Oblique propagation along magnetic field for a fixed distance (point-to-point), F region. High ray (H) and low ray (L) and splitting into X-ray and O-ray are shown.

The skywave propagation based on F-layer modes is identified by a three-character notation such as 1F1, 1F2, 2F2, and 3F2. The first digit is the hop number and the second two characters identify the dominant mode (i.e., F1 or F2 reflection). Accordingly, 2F2 means two hops where the dominant mode is F2 reflection.

A HF receiving installation will commonly receive multiple modes simultaneously, typically 1F2 and 2F2 and at greater distances 2F2 and 3F2. The strongest mode on a long path is usually the lowest order F2 mode unless the antenna discriminates against this. It can be appreciated that higher order F2 modes suffer greater attenuation due to absorption by D-layer passage and ground reflection. The result is a lower level signal than the lower order F2 propagating modes. In other words, a 3F2 mode is considerably more attenuated at a certain location than a 2F2 mode if it can be received at the same location (assuming isotropic antennas). It has traveled through the D-layer two more times than its 2F2 counterpart and has been absorbed one more time by ground reflection.

E-layer propagation is rarely of importance beyond one hop. Reflections from the F1-layer occur only under restricted conditions, and the 1F1 mode is less common than the 1E and 1F2 modes. The 1F1 mode is more common at high latitudes at ranges of 2000–2500 km. Multiple-hop F1 modes are very rare (Ref. 21).

Long-distance HF paths ($\gtrsim$ 4000 km) involve multiple hops and a changing ionosphere as a signal traverses a path. This is especially true on east–west paths, where much of the time some portion of the path will be in transition, part in daylight, part in darkness. At the equator, a 4000-km path extends across nearly three time zones; further north or south, it extends across four and five time zones. Mixed-mode operation is a common feature of transequatorial paths. Figure 11.10 illustrates single-mode paths (a and b) and mixed-mode paths (d–f). Figure 11.10c shows a path with asymmetry. This occurs when a wave frequency exceeds the E-layer MUF only slightly, so that the wave does not penetrate the layer along a rectilinear path but will be bent downward, resulting in the asymmetry.

11.6.2.1 Ray or Wave Splitting. Because of the presence of the geomagnetic field, the ionosphere is a doubly refracting medium. Magnetoionic theory shows that a ray entering the ionosphere is split into two separate waves owing to the influence of the earth's magnetic field. One ray of the split rays is called the ordinary wave (O-wave) and the other, the extraordinary wave (X-wave). The O-wave is refracted less than its X-wave counterpart and this is reflected at a greater altitude; the O-wave will have a lower critical frequency, hence a lower MUF. Both waves experience different amounts of refraction and thus travel independently along different ray paths displaced in time at the receiver, typically from 1 to 10 μs. The O-wave suffers less absorption and is usually the stronger and therefore the more important. Figure 11.11 is a conceptual sketch of O- and X-wave propagation

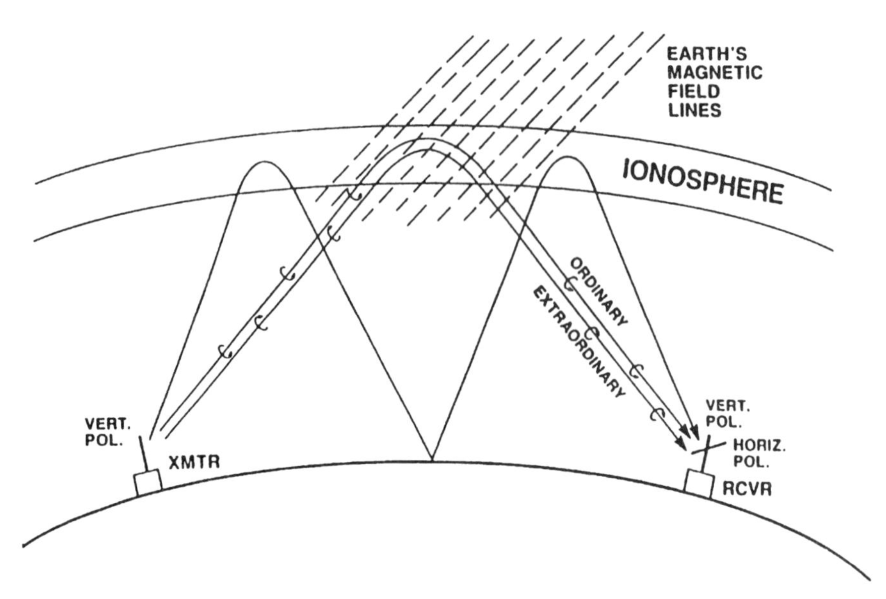

Figure 11.12. The skywave channel, showing the formation of the X- and O-waves, and simultaneous reception of one hop and two hops at the receiver.

as well as the high ray and the low ray. Figure 11.12 is a simplified diagram of the arrival at a receiver of the X-ray and O-ray (or wave).

So even if we have a one-mode path, say, somewhat under 1000 km, the received signal will still be impaired by multipath arising from the O-ray and X-ray; and if we shift down in frequency somewhat from the MUF, there is further multipath impairment of the high ray and the low ray, which are both split into O and X.

Simple multipath fading is shown in Figure 11.13, which is a snapshot sketch of 1 MHz of HF spectrum showing the effects of multipath on an emitted signal. At certain frequencies, the multipath components are in phase and the signal level constructs and builds to a higher level than predicted by calculation. At other frequencies the multipath energy components are out of phase and the signal amplitude resultant shows the out-of-phase destruction. Where multipath components are just 180° out of phase, there is full destruction and no signal is present at that snapshot moment in time at that frequency. There are five such nulls shown in the snapshot picture (Figure 11.13).

11.6.3 Near-Vertical Incidence (NVI) Propagation

NVI propagation is used for short-range HF communication. It can fill the so-called *skip zone* or *zone of silence*; here groundwave propagation is no

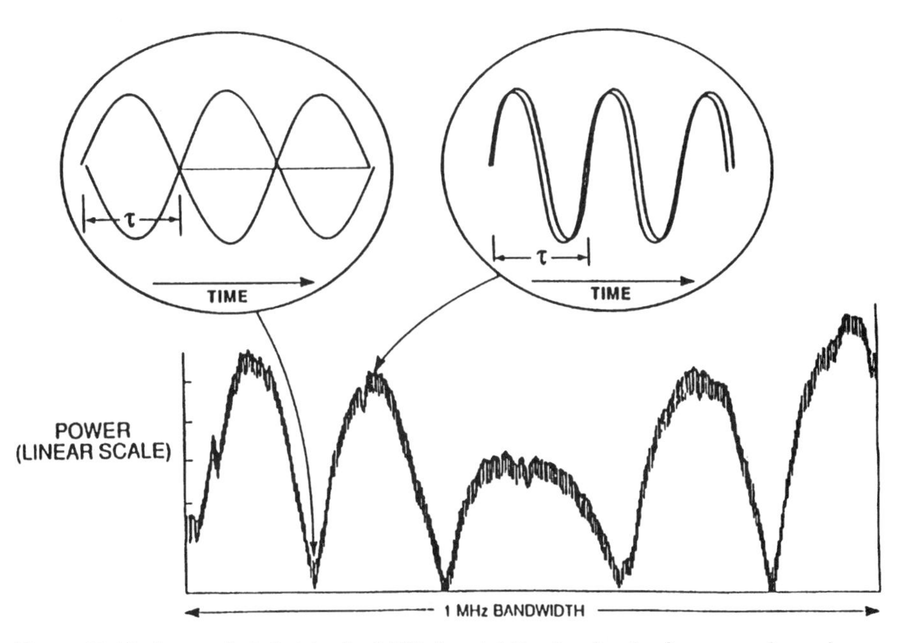

Figure 11.13. A snapshot sketch of a 1-MHz bandwidth, showing the frequency dependence of multipath effects on a transmitted wideband HF (WBHF) signal.

longer effective, to the point where one-hop skywave, using oblique incidence propagation, may be used. NVI utilizes the same skywave principles of propagation discussed earlier. The key factor in NVI operation is the antenna. For effective HF communication using the NVI mode, the antenna must radiate its main beam energy at a very high angle (TOA), near vertical, if you will.

NVI circuits suffer the same impairments as oblique skywave circuits, but in the case of NVI the fading is more severe, particularly polarization fading. One rough rule of thumb is that if an equivalent oblique path experiences, say, 5 μs of dispersion, a NVI path will experience about ten times as much, or 50 μs. NVI paths in a temperate zone have been known to suffer as much as 100 μs of dispersion. In equatorial areas such paths may experience even greater dispersion. We can also have dispersion from a second hop where the principal signal power is being derived from a 1-hop dominant mode. Figure 11.14 shows diagrammatically the operation of NVI propagation. The letter A in the figure shows the extent of useful communication by means of a groundwave component if we were to transmit with a low-elevation-angle antenna with vertical polarization, such as a whip.

We will generally favor the lower frequencies for NVI operation, from 2 to 7 MHz. Higher-elevation-angle circuits tend to lower MUFs to start with.

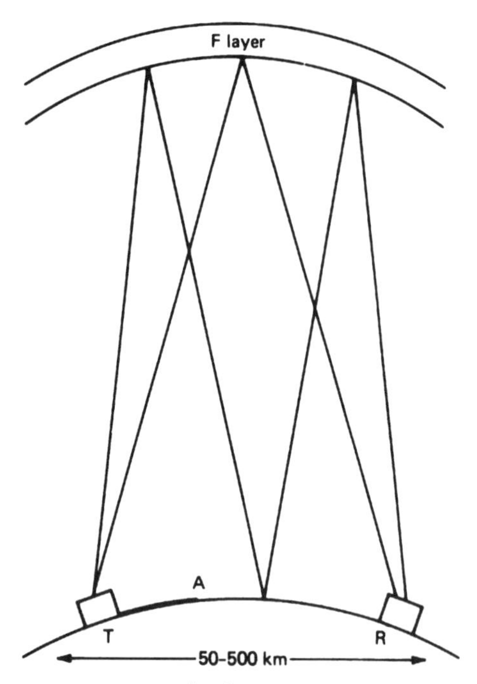

Figure 11.14. A near-vertical incidence (NVI) path showing one and two hops from the F layer. The letter A shows the maximum effective range of groundwave.

However, atmospheric noise limitations may drive us up in frequency as it does with groundwave operation. We will discuss atmospheric noise in Section 11.9.3. Atmospheric and man-made noise levels decrease with increasing operating frequencies.

11.6.4 Reciprocal Reception

Many HF system engineers assume reciprocity on a HF path. This means that if there is a good path from point X to point Y, the path is equally good from point Y to point X. This serves as the basis of one-way sounding and we do not quarrel with the assumption. There are several points that should be considered. The interference levels (including atmospheric noise) may be different at X than at Y. Also, propagation loss may differ owing to the influence of the earth's magnetic field. These two points may be particularly true on long paths. Using proper LQA procedures in both directions can alleviate such asymmetric situations by each far-end receiver optimizing its companion near-end transmitter operation in both frequency management and RF output power.

11.7 HF COMMUNICATION IMPAIRMENTS

11.7.1 Introduction

There are a number of important impairments on a HF channel that affect received signal quality. Different signals are affected in different ways and in severity. Digital data transmission can be severely affected by dispersion, both in the time domain and in the frequency domain. Analog transmission is much less affected, as is CW. FSK is much more tolerant than coherent PSK. Signal-to-noise ratio requirements are a function of signal type. Table 11.3 reviews some typical HF waveforms (signal modulation types), required bandwidths, and signal-to-noise ratios.

Much of the thrust in this section is on digital transmission. At the end of the section we will briefly discuss medium-related impairments to analog transmission.

11.7.2 Fading

HF skywave (and NVI) signals suffer from fading, a principal impairment on HF. It results from the characteristics of the ionosphere. The amplitude and phase of skywave signals fluctuate with reference to time, space, and frequency. These effects collectively are described as fading and have a decisive influence on the performance of HF radio communication systems.

We consider four types of fading here (Ref. 22):

1. *Interference Fading.* This is the most common type of fading encountered on HF circuits. It is caused by mixing of two or more signal components propagating along different paths. This is multipath fading, which may arise from multiple-mode and multiple-layer propagated rays, high- and low-angle modes, groundwaves and skywaves. This latter phenomenon is usually encountered during transition periods and at night.

2. *Polarization Fading Due to Faraday Rotation.* This is brought about by the earth's magnetic field and the split into the O-ray and the X-ray, which become two elliptically polarized components. Both components can interfere to yield an elliptically polarized resultant wave. The major axis of the resulting ellipse will have continuous changes in direction due to changes in the electron density encountered along its propagation paths. HF antennas are ordinarily linearly polarized. However, when an elliptically polarized wave with a constantly changing electric field vector shifts from perpendicular to the receiving HF antenna to parallel to the HF receiving antenna, the input signal voltage to the receiver will vary from maximum (perpendicular case) to zero (parallel case). Consequently, the receiver input voltage will vary according to the spatial rotation of the ellipse of polarization. Fading periods vary from a fraction of a second to seconds.

TABLE 11.3 Emission Types, Bandwidths, and Required Signal-to-Noise Ratios

Class of Emission	Predetection Bandwidth of Receiver (Hz)	Postdetection Bandwidth of Receiver (Hz)	Grade of Service	Audio Signal-to-Noise Ratio (1) (dB)	RF Signal-to-Noise Density Ratio (2) (3) (dB)		
					Stable Condition	Fading Condition (4) Non-diversity	Fading Condition (5) Dual Diversity
A1A telegraphy 8 bauds	3000	1500	Aural reception (6)	−4	31	38	
A1B telegraphy 50 bauds, printer	250	250	Commercial grade (7)	16	40		58
A1B telegraphy 120 bauds, undulator	600	600		10	38		49
A2A telegraphy 8 bauds	3000	1500	Aural reception (6) (19)	−4	35	38	
A2B telegraphy 24 bauds	3000	1500	Commercial grade (7) (19)	11	50	56	
F1B telegraphy 50 bauds, printer $2D$ = 200–400 Hz	1500	100	P_C = 0.01 P_C = 0.001 (8) P_C = 0.0001		45 51 (9) 56	53 63 (9) 74	45 52 (9) 59
F1B telegraphy 100 bauds, printer $2D$ = 170 Hz, ARQ	300	300	(10)		43	52	
F7B telegraphy 200 bauds, printer $2D$ = ..., ARQ			(10)				
F1B telegraphy MFSK 33-tone ITA2 10 characters / s	400	400	P_C = 0.01 P_C = 0.001 (8) P_C = 0.0001		23 24 26	37 45 (25) 52	29 34 39
F1B telegraphy MFSK 12-tone ITA5 10 characters/s	300	300	P_C = 0.01 P_C = 0.001 (8) P_C = 0.0001		26 27 29	42 49 (25) 56	32 36 42
F1B telegraphy MFSK 6-tone ITA2 10 characters/s	180	180	P_C = 0.01 P_C = 0.001 (8) P_C = 0.0001		25 26 28	41 48 (25) 55	31 35 41
F7B telegraphy							
R3C phototelegraphy 60 rpm	3000	3000			50	59	
F3C phototelegraphy 60 rpm	1100	3000	Marginally commercial (22) Good commercial (22)	15 20	50 55	58 65	
A3E telephony double-sideband	6000	3000	Just usable (11) Marginally commercial (12) Good commercial (13)	6 15 (18) 33	50 59 67 (14)	51 64 (20) 75 (14)	48 (15) 60 70 (14) (20)
H3E telephony single-sideband full carrier	3000	3000	Just usable (11) Marginally commercial (12) Good commercial (13)	6 15 (18) 33	53 62 (23) 70(14)	54 67 (20) 78 (14)	51 (15) 63 73 (14) (20)
R3E telephony single-sideband reduced carrier	3000	3000	Just usable (11) Marginally commercial (12) Good commercial (13)	6 15 (18) 33	48 57 (24) 65(14)	49 62 (20) 73 (14)	46 (15) 58 68 (14) (20)

TABLE 11.3 *(continued)*

Class of Emission	Predetection Bandwidth of Receiver (Hz)	Postdetection Bandwidth of Receiver (Hz)	Grade of Service	Audio Signal-to-Noise Ratio (1) (dB)	RF Signal-to-Noise Density Ratio (2) (3) (dB)		
						Fading Condition	
						(4)	(5)
					Stable Condition	Non-diversity	Dual Diversity
J3E telephony single-sideband suppressed carrier	3000	3000	Just usable (11)	6 (18)	47	48 (20)	45 (15) (20)
			Marginally commercial (12)	15 (18)	56	61 (20)	57 (15) (20)
			Good commercial (13)	33 (18)	64 (14)	72 (14) (20)	67 (14) (15) (20)
B8E telephony independent-sideband 2 channels	6000	3000 per channel	Just usable (11)	6 (18)	49	50 (20)	47 (15) (20)
			Marginally commercial (12)	15 (18)	58	63 (20)	59 (15) (20)
			Good commercial (13)	33 (18)	66 (14)	74 (14) (20)	69 (14) (15) (20)
B8E telephony independent-sideband 4 channels	12000	3000 per channel	Just usable (11)	6 (18)	50	51 (20)	48 (15) (20)
			Marginally commercial (12)	15 (18)	59	64 (20)	60 (15) (20)
			Good commercial (13)	33 (18)	67 (14)	75 (14) (20)	69 (14) (15) (20)
J7B multichannel VF telegraphy 75 bauds each	3000	110 per channel	$P_C = 0.01$ (8)		59 (21)	67 (21)	59 (21)
			$P_C = 0.001$ (8)		65 (21)	77 (21)	66 (21)
			$P_C = 0.0001$ (8)		69 (21)	87 (21)	72 (21)

Note:

(1) Noise bandwidth equal to postdetection bandwidth of receiver. For an independent-sideband telephony, noise bandwidth equal to the postdetection bandwidth of one channel.

(2) The figures in this column represent the ratio of signal peak envelope power to the average noise power in a 1-Hz bandwidth except for double-sideband A3E emission where the figures represent the ratio of the carrier power to the average noise power in a 1-Hz bandwidth.

(3) The values of the radio-frequency signal-to-noise density ratio for telephony listed in this column apply when conventional terminals are used. They can be reduced considerably (by amounts as yet undetermined) when terminals of the type using linked compressor–expanders (Lincompex) are used. A speech-to-noise (rms voltage) ratio of 7 dB measured at audio-frequency in a 3-kHz band has been found to correspond to just marginally commercial quality at the output of the system, taking into account the compandor improvement.

(4) The values in these columns represent the median values of the fading signal power necessary to yield an equivalent grade of service and do not include the intensity fluctuation factor (allowance for day-to-day fluctuation), which may be obtained from Report 252-2 + Supplement (published separately) in conjunction with Report 322 (published separately). In the absence of information from these reports, a value of 14 dB may be added as the intensity fluctuation factor to the values in these columns to arrive at provisional values for the total required signal-to-noise density ratios which may be used as a guide to estimate required monthly-median values of hourly-median field strength. This value of 14 dB has been obtained as follows:

The intensity fluctuation factor for the signal, against steady noise, is 10 dB, estimated to give protection for 90% of the days. The fluctuations in intensity of atmospheric noise are also taken to be 10 dB for 90% of the days. Assuming that there is no correlation between the fluctuations in intensity of the noise and those of the signal, a good estimate of the combined signal and noise intensity fluctuation factor is

$$\sqrt{10^2 + 10^2} = 14 \text{ dB}$$

(5) In calculating the radio-frequency signal-to-noise density ratios for rapid short-period fading, a log-normal amplitude distribution of the received fading signal has been used (using 7 dB for the ratio of median level to level exceeded for 10% or 90% of the time) except for high-speed automatic telegraphy services, where the protection has been calculated on the assumption of a Rayleigh distribution. The following notes refer to protection against rapid or short-period fading.

(6) For protection 90% of the time.

(7) For A1B telegraphy, 50-baud printer: for protection 99.99% of the time. For A2B telegraphy, 24 bauds: for protection 98% of the time.

(8) The symbol P_C stands for the probability of character error.

(9) Atmospheric noise (V_d = 6 dB) is assumed (see Report 322).

(10) Based on 90% traffic efficiency.

(11) For 90% sentence intelligibility.

(12) When connected to the public service network: based on 80% protection.

(13) When connected to the public service network: based on 90% protection.

(14) Assuming 10-dB improvement due to the use of noise reducers.

(15) Diversity improvement based on a wide-spaced (several kilometers) diversity.

(16) Transmitter loading of 80% of the rated peak envelope power of the transmitter by the multichannel telegraph signal is assumed.

(17) Required signal-to-noise density ratio based on performance of telegraphy channels.

(18) For telephony, the figures in this column represent the ratio of the audio-frequency signal, as measured on a standard VU-meter, to the rms noise, for a bandwidth of 3 kHz. (The corresponding peak signal power, i.e., when the transmitter is 100% tone-modulated, is assumed to be 6 dB higher.)

(19) Total sideband power, combined with keyed carrier, is assumed to give partial (two-element) diversity effect. An allowance of 4 dB is made for 90% protection (8 bauds), and 6 dB for 98% protection (24 bauds).

(20) Used if Lincompex terminals will reduce these figures by an amount yet to be determined.

(21) For fewer channels these figures will be different. The relationship between the number of channels and the required signal-to-noise ratio has yet to be determined.

(22) Quality judged in accordance with Article 23.1 of ITU publication "Use of the Standardized Test Chart for Facsimile Transmissions."

Source: Table 1, p. 22, CCIR Rep. 339-6, Vol. III, XVIth Plenary Assembly, Dubrovnik, 1986 (Ref. 14).

3. *Focusing and Defocusing Due to Atmospheric Irregularities.* These deformed layers can focus or defocus a signal wave if they encounter deformities that are concave or convex, respectively. The motion of these structures can cause fades with periods up to some minutes.

4. *Absorption Fading Is Caused by Solar Flare Activity.* This type of fading particularly affects the lower frequencies, and fades may last from minutes to more than an hour.

Of interest to the communication engineer are fading depth, duration, and frequency (fade rate). For short-term fading, the fading depth is the difference in decibels between the signal levels exceeded for 10% and 90% of the time. Measurements have confirmed that we may expect about 14 dB for a Rayleigh distribution short-term fading, which is the most common form of such fading. This value is valid for paths 1500–6000 km long and does not vary much with the time of day or season.

For long-term variations in signal level (i.e., variations of hourly median signal values of a month), the log-normal distribution provides a best fit. A good value is an 8-dB margin for HF paths below 60° geomagnetic latitude and 11 dB for paths above 60° and especially over the polar cap region (Ref. 22). This suggests that a 22-dB fade margin should provide better than a 99% time availability (for fading, not frequency management) assuming a 3-σ point and for temperate zone paths. This decibel value is referenced to the median signal level.

Some fade rate values on typical HF circuits are provided in Figure 11.15. Figure 11.16 gives a sampling of typical fade durations on a medium-to-long path.

11.7.3 Effects of Impairments at the HF Receiver

11.7.3.1 General. In Section 11.7.2 we discussed fade rate and depth. This is amplitude fading. It is accompanied by associated group path delay and phase path delay. Doppler shift arises from ionospheric movement as well as movement if one or both ends of the path are mobile (i.e., in motion).

11.7.3.2 Time Dispersion. On skywave paths the primary cause of time dispersion is multipath propagation, which derives from differences in transit time between different propagation paths, as discussed in Section 11.4. The multipath spread causes amplitude and phase variations in the signal spectrum owing to interference of the multipath wave components. When these fluctuations are correlated within the signal bandwidth (e.g., 3 kHz) and all the spectral components behave more or less in the same manner, we then call this *flat fading*. When these fluctuations have little correlation, the fading is called *frequency selective fading*. Time dispersion is characterized by a delay

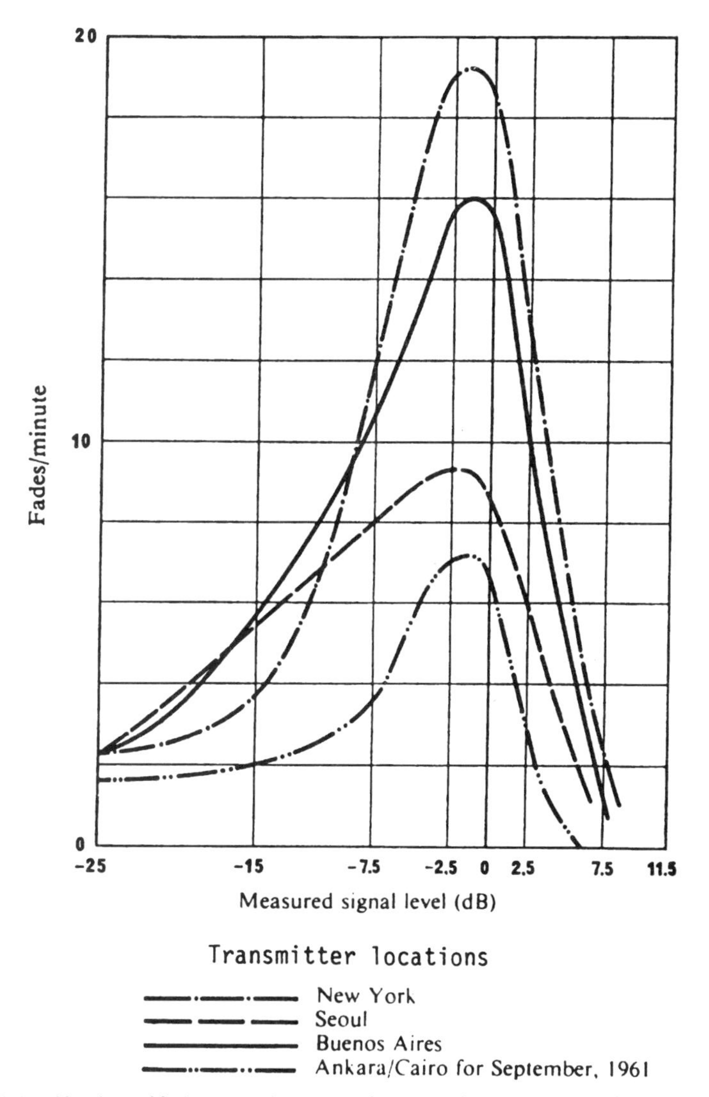

Figure 11.15. Number of fades per minute as a function of the signal level for various circuits terminating in Frankfurt, Germany. (From Figure 3, p. 261, CCIR Rep. 197-4, Vol. III, XVIth Plenary Assembly, Dubrovnik, 1986; Ref. 23.)

power spectrum and is measured as multipath delay spread in microseconds or milliseconds.

Time dispersion is an especially serious and destructive impairment to digital communication signals on HF. One rule of thumb that is useful is that if the delay spread exceeds half the time width (period) of a signal element (baud), the error rate becomes intolerable. This is one rationale for extending

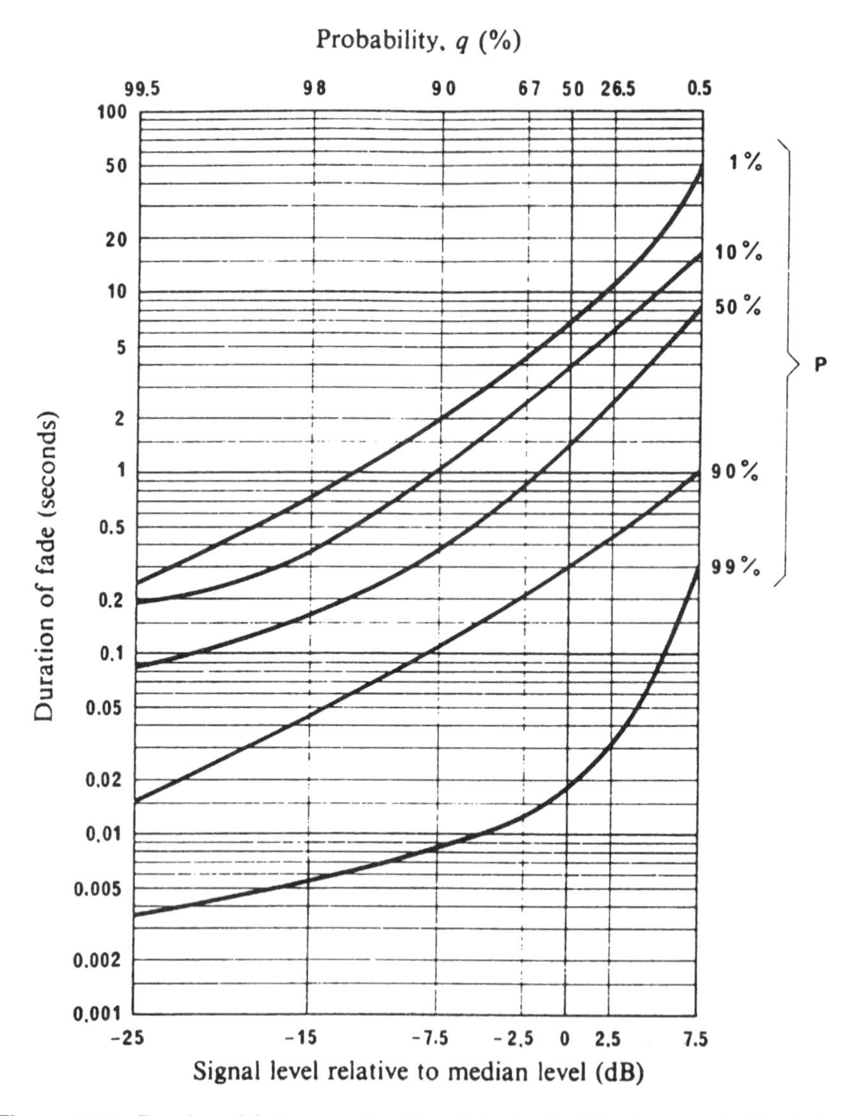

Figure 11.16. Duration of fades as a function of the level of the test signal. Circuit: New York–Frankfurt, Germany; September 14, 1961; 1100 h Central European Time; frequency, 13.79 MHz. The figures on the right-hand side of the curves represent the percentage *p* of the number of fades for which a given duration of fade is exceeded. The measured values of signal levels are shown, together with the probability *q* that these levels will be exceeded. (From Figure 4, p. 262, CCIR Rep. 197-4, Vol. III, XVIth Plenary Assembly, Dubrovnik, 1986; Ref. 23.)

the width of a signal element (e.g., by lowering the *baud* rate), to combat time dispersion. For instance, if a serial bit stream is transmitted at 100 baud, a baud period is 0.01 s (10 ms). In this case, the circuit will remain operational, although with a degraded BER, if the time dispersion remains under 5 ms (i.e., half the period of a signal element or bit). At 200 baud the half-baud period value drops to 2.5 ms and at 50 baud it is 10 ms.

Multipath has been shown to be a function of the operating frequency relative to the MUF. Multipath delay tends to approach zero as the operating frequency approaches the MUF value. Turn to Figure 11.11 and we see that as we approach the MUF moving upward in frequency, we will receive only one signal power component. (*Note*: There are really two MUFs, one for the O-ray and one for the X-ray trace. The X is usually below the O in signal strength.)

11.7.3.3 Frequency Dispersion. Experience on operating HF circuits has shown that frequency dispersion (Doppler spread) nearly always is present when there is time dispersion. But the converse does not necessarily hold true. The Doppler shift and spread are due to a drifting ionosphere. As the signal encounters elemental surfaces of the ionosphere, each with a different velocity vector, the result is a Doppler spread. Such Doppler shifts and spreads can have disastrous effects, particularly on narrowband FSK systems, which can tolerate no more than ± 2-Hz total frequency departure. If the 2-Hz value is exceeded, the BER will approach 5×10^{-1}. Typical values of Doppler spread on midlatitude paths are 0.1–0.2 Hz.

11.8 MITIGATION OF PROPAGATION-RELATED IMPAIRMENTS

The general approach microwave engineers use to overcome propagation impairments is to increase link margin (i.e., S/N). Generally, this approach has drawbacks with HF propagation because many of the impairments may be difficult to overcome with increasing power. Other measures also merit consideration.

One measure commonly used at HF installations is to employ diversity. The basis of diversity is to take advantage of signals that are not correlated. There are three basic types of diversity: space, time, and frequency. Other types of diversity are variants of the three basic types. Several types of diversity are discussed below.

Space Diversity. The same signals are received by at least two antennas separated in space. Separation should be larger than 6 wavelengths.

Polarization Diversity. Signals are received on antennas with different polarizations.

Frequency Diversity. Information is transmitted simultaneously on different frequencies. Some refer to this as in-band diversity, where subcarrier tones are separated by from 300 to 700 Hz. CCIR recommends at least 400-Hz separation between redundant FSK tones in CCIR Rec. 106-1 (Ref. 25). MIL-STD-188-342 (Ref. 24) suggests a voice frequency (VF) carrier modulation plan shown in Table 11.4. The tone pairing plan is given at the bottom of the table. In this case redundant data are transmitted on pairs of FSK tones separated by 1360 Hz.

Time Diversity. The signal is transmitted several times. Some techniques are available that can dramatically improve error performance using time diversity. These techniques not only take advantage of the decorrelation of fading with time, but also the simple redundancy.

Channel coding with interleaving is only now being exploited on HF. We discussed some typical channel coding and interleaving techniques in Chapter 4. This coupled with automatic repeat request (ARQ) schemes using short message blocks can bring error performance on HF links into manageable bounds.

TABLE 11.4 Voice Frequency Carrier Telegraph (VFCT) Modulation Plan for HF

Channel Designation	Mark Frequency (Hz)	Center Frequency (Hz)	Space Frequency (Hz)
1	382.5	425	467.5
2	552.5	595	637.5
3	722.5	765	807.5
4	892.5	935	977.5
5	1062.5	1105	1147.5
6	1232.5	1275	1317.5
7	1402.5	1445	1487.5
8	1572.5	1615	1657.5
9	1742.5	1785	1827.5
10	1912.5	1955	1997.5
11	2082.5	2125	2167.5
12	2252.5	2295	2337.5
13	2422.5	2465	2507.5
14	2592.5	2635	2677.5
15	2762.5	2805	2847.5
16[a]	2932.5	2975	3017.5
17[b]	3012.5	3145	3187.5
18[b]	3272.5	3315	3357.5

Diversity

Pair 1(9), 2(10), 3(11), 4(12), 5(13), 6(14), 7(15), and 8(16)

Note: Connect loop to lower numbered channel of each pair

[a]Marginal over HF (nominal 3-kHz) channels.

[b]Not usable over HF channels.

Source: MIL-STD-188-342 (Ref. 24).

One method of mitigation of the dispersive effects of multipath propagation by pulse width extension was discussed in Section 11.7.3.2.

11.9 HF IMPAIRMENTS—NOISE IN THE RECEIVING SYSTEM

11.9.1 Introduction

In previous chapters on radio systems, the primary source of noise was thermal noise generated in the receiver front end. Except under certain special circumstances, this is not the case for HF receiving systems. External noise is by far dominant for HF receivers. In declining importance, we categorize this noise as follows:

1. Interference from other emitters.
2. Atmospheric noise.
3. Man-made and galactic noise.
4. Receiver thermal noise.

11.9.2 Interference

The HF band has tens of thousands of users who are assigned operating frequencies by national authorities such as the FCC in the United States and by the International Frequency Registration Board (IFRB), a subsidiary organization of the ITU. The emitters operate in frequency bands in accordance with the service they perform and the region of the world they are in. The ITU has divided the world into three regions (Ref. 26). We must also take into consideration noise from ionospheric sounder transmitters, harmonics, and spurious emissions from licensed emitters. Of primary importance in establishing a HF link is a "clear" frequency. Clear means interference-free, and the next consideration is how interference-free. On several tests of HF data modems it was found that interference even 15 dB or more below desired signals corrupted BERs. If we were to try to use a 3-kHz channel, we could well find that an emitter 5000 km away can place a signal 30, 40, or more dB over our desired signal. With nominal 1-MHz wideband HF systems, it is estimated that interference from in-band emitters may contribute an integrated noise level over 30 dB above atmospheric noise (Ref. 27). A typical spectrum analyzer snapshot of a European interference spectrum is shown in Figure 11.17.

One can look at the HF spectrum between the LUF and MUF as a picket fence in which the pickets are the interferers. Unfortunately, by the very nature of HF propagation, these "pickets" randomly appear and disappear, their location changes, and their amplitudes increase and decrease not only owing to fading but also to constant changes in the ionosphere as the MUF changes with time, sporadic E propagation, transition F-layer changes, and so on.

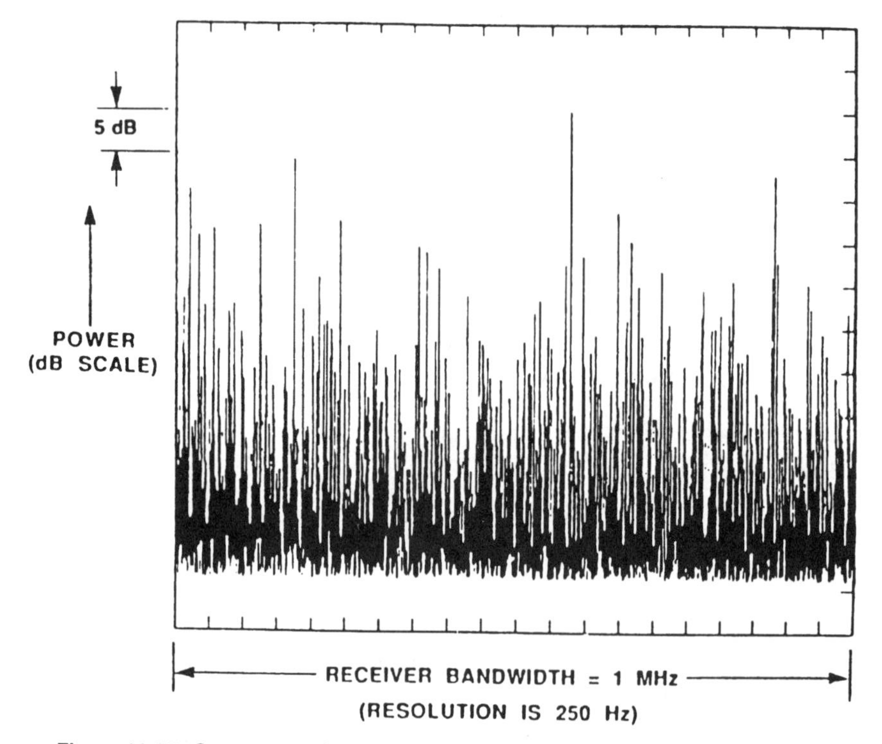

Figure 11.17. Spectrum analyzer snapshot of European HF interference spectrum.

Reference 28 gives congestion values in Manchester, United Kingdom, in July 1982 for the period between about noon and midnight local time. The HF band was scanned by stepping a receiver in 1-kHz increments and with a bandwidth of 1 kHz. It shows that the broadcast bands are the most congested and have the highest interference levels. In this study receive levels starting at −117 dBm were recorded in decades up to −77 dBm. Several examples of their measurements are given in Table 11.5, which shows the percentage of congestion values at defined thresholds.

There are several techniques to overcome or mitigate interference. The most desirable, of course, is to operate on a clear frequency with no interference at the distant-end receiver at the OWF. This is not often fully achievable.

In nearly all cases a facility is assigned a group of frequencies, usually with one of the following bandwidths, dependent on the service: 1, 3, 6, 9, or 12 kHz. Also see Table 11.3. Broadcast installations have other bandwidths. Broadcast is not a concern of this text. Distant-end receivers usually "command" near-end transmitters. The receiver site checks a new frequency for propagation and occupancy. More modern systems use advanced oblique sounders such as the AN/TRQ-35 or SELSCAN, Autolink, or MESA

TABLE 11.5 Percentage Congestion Values at Defined Threshold[a]

Frequency Band (kHz)	Service (dBm)	−117	−107	−97	−87	−77
				Day		
5950–6200	Broadcast	63	41	23	9	2
8195–8500	Maritime mobile	24	11	5	1	1
10150–10600	Fixed mobile	70	33	17	10	4
12050–12230	Fixed	82	50	24	11	5
13800–14000	Fixed/mobile	60	30	17	6	2
15000–15100	Aero mobile	57	31	10	4	1
15100–15600	Broadcast	96	80	55	33	16
17410–17550	Fixed	44	21	8	5	1
				Night		
5950–6200	Broadcast	100	100	93	75	51
8195–8500	Maritime mobile	100	92	52	29	13
10150–10600	Fixed/mobile	100	91	49	24	11
12050–12230	Fixed	100	88	60	35	16
13800–14000	Fixed/mobile	72	15	6	2	1
15000–15100	Aeronautical mobile	97	63	32	7	5
15100–15600	Broadcast	96	78	52	33	16
17410–17550	Fixed	53	32	17	7	2

[a]The value −107 dBm is equivalent here to a received field strength of 2 μV/m.

(Rockwell, Harris, and Tadiran trademarks, respectively), which also provide occupancy data. Thus modern systems have the propagation and occupancy check automated.

We can see from Table 11.5 that there is a fair probability that an optimum propagating frequency may be occupied by other users. There are several means to overcome this problem:

1. Find a clear frequency with suboptimum propagation.
2. Increase transmit power to achieve the desired signal-to-noise ratio at the distant end.
3. Use directional antennas where the interferer is in a side lobe and hence attenuated compared to the desired signal.
4. Use antenna nulling. This is a form of electronic beam steering that creates a null in the direction of the interferer.
5. Use sharp front-end preselection on receiver(s).
6. Employ wideband HF with interference excision.

Some military circuits use forward error correction (FEC) coding rich in redundancy. Rather than using code rates such as $\frac{7}{8}$, $\frac{3}{4}$, $\frac{2}{3}$, or $\frac{1}{2}$, we resort to code rates of $\frac{1}{8}$, $\frac{1}{16}$, $\frac{1}{32}$, or $\frac{1}{64}$. This highly redundant bit stream phase shifts a RF signal that is frequency hopped (spread spectrum) over a fairly wide frequency band in the vicinity of the OWF (for the desired connectivity). Typical spreads are 100 kHz, 500 kHz, up to 2 MHz. In theory, some of the hopped energy must get through the holes in the picket fence. Such coding multiplies the symbol rate of transmission. For example, if we wish to maintain 75-bps information rate, at rate $\frac{1}{16}$ we will have to transmit 16×75, or 1200 symbols per second. If we do not increase our RF power output, the energy per symbol is reduced by 16, or 6% of the energy if we had no coding. To equate energy per bit in this situation, if we use a transmitter with a 100-W output at 75 bps, we then would require a 1.6-kW transmitter output to maintain the same energy per symbol (bit) as when rate $\frac{1}{16}$ was employed.

On the other hand, the emerging wideband HF (WBHF) systems excise the high-level interferers with only a loss of several tenths of a decibel, up to 2 dB of equivalent receive power. For instance, with 30% excision, only about 1.5 dB of receive power is lost. A 1-MHz PN spread system using a chip rate of 512 kchips in effect has 512,000 pieces of redundant RF-energy-carrying information spread across 1 MHz of HF spectrum. With so much redundancy, certainly we can effectively assure some portion of the desired transmit power to successfully get through the picket fence. Such systems have very low transmitted spectral energy density and are compatible with conventional narrowband HF systems. Processing gains are on the order of 60–66 dB when information bit rates are on the order of 75–150 bps. We should not equate processing gain to power gain. In fact there is no power gain. Processing gain describes how well it will work in an interference environment.

11.9.3 Atmospheric Noise

Atmospheric noise is a result of numerous thunderstorms occurring at various points on the earth, but concentrated mainly in tropical regions. These electrical disturbances are transmitted long distances via the ionosphere in the same manner as HF skywaves. Because the resulting field intensities of the noise decrease with the distance traveled, the level of atmospheric noise encountered from this source becomes progressively smaller in the higher latitudes of the temperate zones and in the polar regions. As we are aware, skywave propagation varies with time of day and season. Hence the intensity of atmospheric noise varies with both location and time.

Since the major portion of atmospheric noise is traced to thunderstorm activity, the atmospheric noise level at a particular location is due to contributions from both local and distant sources. During a local thunderstorm, the average noise level is about 10 dB higher than the average noise of the same period in the absence of local thunderstorm activity (Ref. 29). From this it can be seen that the atmospheric noise level is related directly to weather conditions. The position of the equatorial weather front greatly affects atmospheric noise at all locations. This front varies in position from day to day and its general location seasonally moves north and south with the sun.

The degree of activity varies from time to time and from place to place, being much greater over land than over sea. The main areas of thunderstorm activity lie in equatorial regions, notably the East Indies, equatorial Africa, equatorial South America, and Central America. Thunderstorms are present about 50% of the days at locations in these equatorial belts and this activity is the principal source of long-distance atmospheric noise. It has been estimated that there are about 2000 thunderstorms in progress at each instant throughout the world (Ref. 29).

Thunderstorm activity is located over tropical land masses during local summer season and is more active over land, usually between 1200 and 1700 local time. Thunderstorm activity over the sea generally occurs at night and can last for more than a day.

Atmospheric noise from local sources shows discrete crashes similar to impulse noise, while long-distance atmospheric noise consists of rapid and irregular fluctuations with a frequency of 10 or 20 kHz per second and a damped wave train of oscillations. The amplitude of lightning disturbances varies approximately inversely with the frequency squared and is propagated in all directions both for groundwaves and skywaves.

11.9.3.1 Calculating Atmospheric Noise Using CCIR Rep. 322.

CCIR Rep. 322 (Ref. 30) is the most widely accepted data source and methodology on atmospheric noise. It has been derived from years of noise-monitoring data taken from monitoring stations around the world. The data are presented graphically as atmospheric noise contours at 1 MHz and then may be scaled graphically or by formula to the desired frequency.

CCIR divides a 24-h day into six time blocks of 4 h each starting at 12 midnight. For seasonal variations in atmospheric noise, there are charts for the four seasons. Thus there are 24 charts (6 × 4), each of which is followed by a frequency scaling chart accompanied by a noise variability chart. A sample of these is given in Figure 11.18. Figure 11.19 is for frequency scaling and Figure 11.20 is for noise variability; they accompany Figure 11.18.

It is noted in Figure 11.18 that the noise contours given represent *median* noise levels and are measured in decibels above kT_0b at 1 MHz; their notation is F_{am}. We will derive a formula for F_{am}.

We can express the antenna noise factor f_a, which expresses the noise power received from sources *external* to the antenna:

$$f_{\mathrm{a}} = \frac{P_{\mathrm{n}}}{kT_0 b} = \frac{T_{\mathrm{a}}}{T_0} \tag{11.2}$$

where
- P_{n} = the noise power available from an equivalent loss-free antenna (W)
- k = Boltzmann's constant = −228.6 dBW/Hz
- b = the effective noise bandwidth (Hz)
- T_0 = 288 K (we have used 290 K in previous chapters)

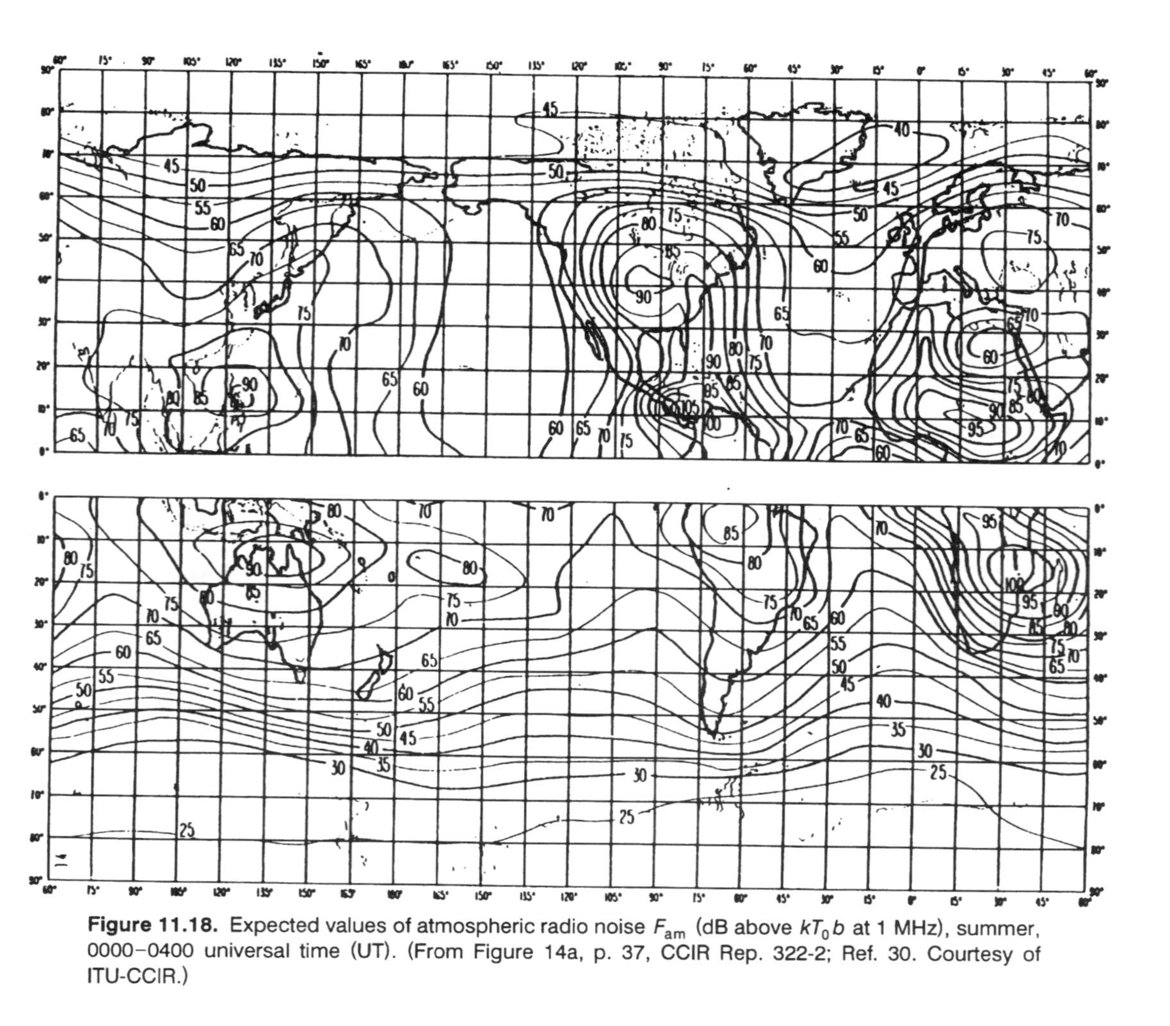

Figure 11.18. Expected values of atmospheric radio noise F_{am} (dB above kT_0b at 1 MHz), summer, 0000–0400 universal time (UT). (From Figure 14a, p. 37, CCIR Rep. 322-2; Ref. 30. Courtesy of ITU-CCIR.)

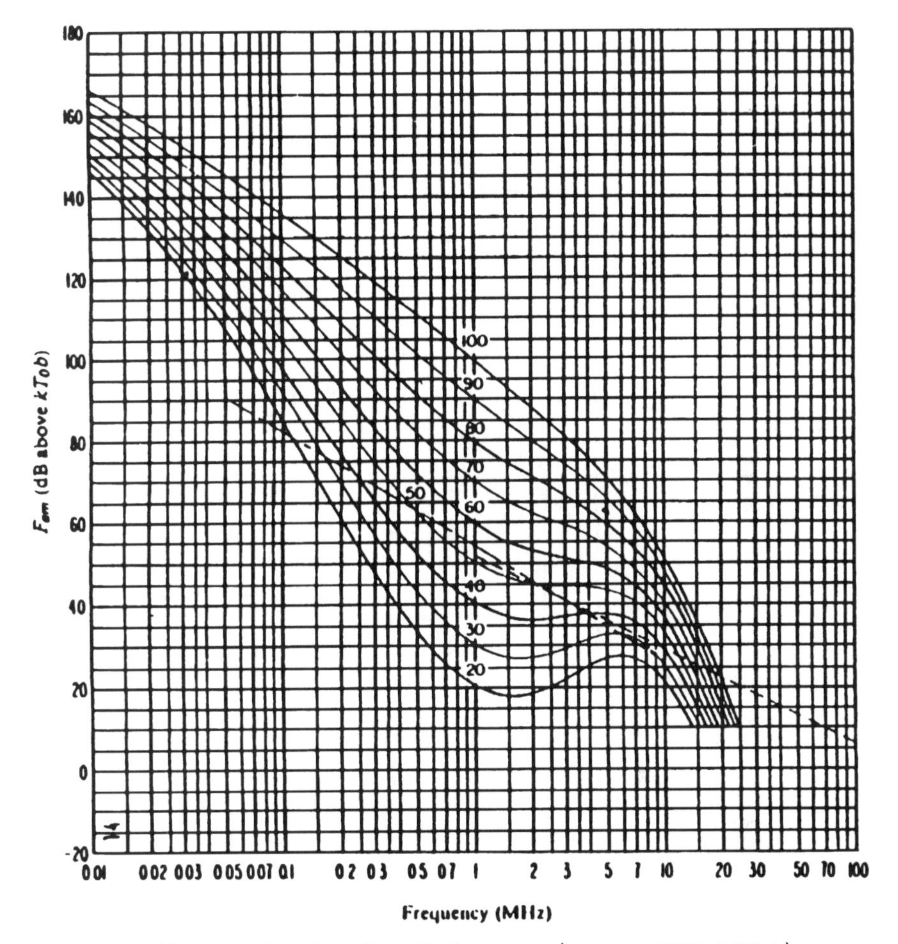

Figure 11.19. Variation of radio noise with frequency (summer, 0000–0400 h). ______, Expected values of atmospheric noise; -·-·-·, expected values of man-made noise at a quiet receiving location; and - - -, expected values of galactic noise. (From Figure 14b, p. 38, CCIR Rep. 322-2; Ref. 30. Courtesy of ITU-CCIR.)

T_a = the effective antenna temperature in the presence of external noise

$$10 \log kT_0 = -204 \text{ dBW}$$

$$F_a = 10 \log f_a$$

Both f_a and T_a are independent of bandwidth because the available noise power from all sources may be assumed to be proportional to bandwidth, as is the reference power level.

The antenna noise factor F_a, in decibels, is for a short vertical antenna over a perfectly conducting ground plane. This parameter is related rms

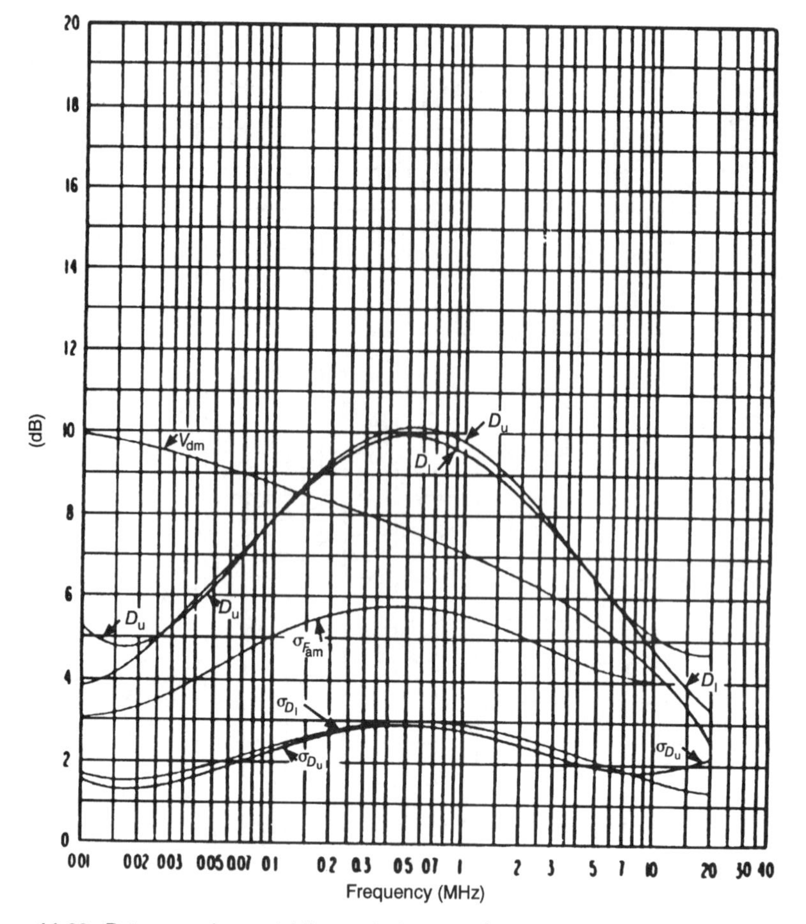

Figure 11.20. Data on noise variability and character (summer, 0000–0400 h). $\sigma_{F_{am}}$, Standard deviation of values of F_{am}; D_u, ratio of upper decile to median value, F_{am}; σ_{D_u}, standard deviation of values of D_u; D_l, ratio of median value, F_{am}, to lower decile; σ_{D_l}, standard deviation of value of D_l; V_{dm}, expected value of median deviation of average voltage. The values shown are for a bandwidth of 200 Hz. (From Figure 14c, p. 38, CCIR Rep. 322-2; Ref. 30.)

noise field strength along the antenna by

$$E_n = F_a - 65.5 - 20 \log F_{MHz} \tag{11.3}$$

where E_n is the rms field strength for a 1-kHz bandwidth (dB[μV/m]) and F_a is the noise factor for the frequency f in question (dB). F_{MHz} is the frequency.

The value of field strength for any bandwidth B_{Hz} other than 1 kHz can be derived by adding $(10 \log b - 30)$ to E_n. For instance, to derive E_n in

1 Hz of bandwidth, we subtract 30 dB. CCIR Rep. 322 (Ref. 30) cautions that E_n is the vertical component of the noise field at the antenna.

In predicting the expected noise levels, the systematic trends with time of day, season, frequency, and geographical location are taken into account explicitly. There are other variations that must be taken into account statistically. The value of F_a for a given hour of the day varies from day to day because of random changes in thunderstorm activity and propagation conditions. The median of the hourly values within a time block (the time block median) is designated F_{am}. Variations of the hourly values within the time block can be represented by values exceeded for 10% and 90% of the hours, expressed as deviations D_u and D_l from the time block median plotted on normal probability graph paper (level in dB), the amplitude distribution of the deviations, D, above the median can be represented with reasonable accuracy by a straight line through the median and upper decile values, and a corresponding line through the median and the lower decile values can be used to represent values below the median. Extrapolation beyond D_u and D_l, however, yields only very approximate values of noise.

The value of F_a is average noise power. For digital circuits, it is useful to have knowledge of the amplitude probability distribution (APD) of the noise. This will show the percentage of time for which any level is exceeded, which usually is the noise envelope described. The APD is dependent on the short-term characteristics of the noise and, as a result, cannot be deduced from the hourly values of F_a alone. The reader should consult CCIR Rep. 322 if more detailed analysis is desired in this area.

From equation (11.2) we can state the noise threshold (noise power) of a receiver (P_n):

$$P_n = F_{am} + X\sigma + 10 \log B_{Hz} \tag{11.4}$$

If $X = 1$ (one standard deviation), 68% (of the time)
$X = 2$ (two standard deviations), 95% (of the time)
$X = 3$ (three standard deviations), 99% (of the time)

Example 1. Calculate the receiver noise threshold for 95% time availability where the receiver is connected to a lossless quarter-wavelength whip antenna equipped with a good ground plane. The receiver is tuned to 10 MHz with a 1-kHz bandwidth. The receiving facility is located in Massachusetts, and we are interested in the summer time block of 0000–0400 (a.m.). From Figure 11.18 we see that the F_{am} noise contour is 85 dB at 1 MHz. We turn to Figure 11.19 and F_{am} (10 MHz) scales down to 48 dB. Figure 11.20 gives $\sigma_{F_{am}}$ as 4 dB and we wish two standard deviations (95%), which is 8 dB for the variability.

$$\begin{aligned} P_n &= 48 + 8 + 10 \log 1000 - 204 \text{ dBw} \\ &= -118 \text{ dBW in 1-kHz bandwidth} \end{aligned}$$

We have shown much of the methodology given in CCIR Rep. 322, enough for an initial system design. However, several comments are in order. F_{am} decreases rapidly with increasing frequency. Besides atmospheric noise sources, other noise sources affecting total receiver noise power must start to be considered as frequency increases. These are

f_c, the noise factor of the antenna (function of its ohmic loss)
f_t, the noise factor of the transmission line (function of its ohmic loss)
f_r, noise factor of the receiver

The operating noise factor f then equals

$$f_a - 1 + f_c f_t f_r \qquad \text{(numerics)} \tag{11.5}$$

At low frequencies (e.g., < 10 MHz) atmospheric noise predominates and will determine the value of f. Because f_a decreases with increasing frequency, transmission line noise and receiver thermal noise become more important and f_c tends to approach unity. The values of f_t and f_r can be determined from calculations involving design features of the transmission line and receiver or by direct measurement.

The effective noise factor of the antenna, insofar as it is determined by atmospheric noise, may be influenced in several ways. If the noise sources were distributed isotropically, the noise factor would be independent of directional properties. In practice, however, the azimuthal direction of the beam may coincide with the direction of an area where thunderstorms are prevalent, and the noise factor will be measured accordingly compared with an omnidirectional antenna. On the other hand, the converse may be true. The directivity in the vertical plane may be such as to differentiate in favor or against reception of noise from a strong source.

Unfortunately, CCIR Rep. 322 does not take into account the sunspot number and, as mentioned previously, horizontal polarization.

11.9.4 Man-Made Noise

Above about 10 MHz we will often find that man-made noise is predominant. Man-made noise can be generated by many sources, such as electrical machinery, automobile ignitions, all types of electronic processors/computers, high-power electric transmission lines, and certain types of lighting. As a result, man-made noise is a function of industrialization and habitation density.

The most recognized reference on man-made noise is CCIR Rep. 258 (Ref. 31). Figure 11.21 gives median values of man-made noise, expressed in

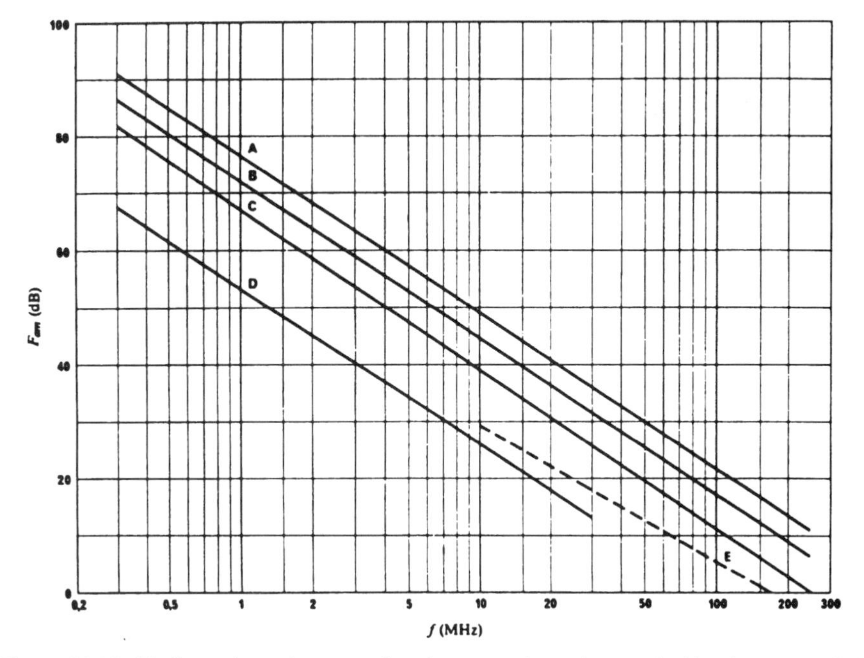

Figure 11.21. Median values of man-made noise power for a short vertical lossless grounded monopole antenna. Environmental category: A, business; B, residential; C, rural; D, quiet rural; E, galactic. (From Figure 10, p. 26, ITU-R PI 372-6; Ref. 32. Courtesy of ITU, Geneva.)

decibels, above −204 dBW/Hz. The figure shows five curves:

A. Business
B. Residential
C. Rural
D. Quiet rural
E. Galactic (extending upward in frequency from 10 MHz)

Business areas are defined as any area used predominantly for any type of business, such as stores, offices, industrial parks, large shopping centers, main streets and highways lined with various business enterprises, and so on. Residential areas are those predominantly used for single- or multiple-family dwellings of at least two single-family units per acre (five per hectare) and with no large or busy highways. Rural areas are defined as areas in which dwellings number no more than one per 5 acres (2 hectares) and where there are no intense noise sources. Minimum man-made noise may be found in "quiet" rural areas. It is in these areas where galactic noise predominates above about 10 MHz. This is the dashed line (E) shown in Figure 11.21.

TABLE 11.6 Values of Constants *c* and *d* [Equation (11.6)]

Environmental Category	c	d
Business (curve A)	76.8	27.7
Interstate highways	73.0	27.7
Residential (curve B)	72.5	27.7
Parks and university campuses	69.3	27.7
Rural (curve C)	67.2	27.7
Quiet rural (curve D)	53.6	28.6
Galactic noise (curve E)	52.0	23.0

Source: Table I, p. 26, ITU-R PI.372-6, (Ref. 32). Courtesy of ITU, Geneva.

We can also calculate the median man-made noise from the following expression:

$$F_{\mathrm{am}} = c - d \log f \tag{11.6}$$

where f is the operating frequency in megahertz. The values of the constants c and d may be taken from Table 11.6. Table 11.7 provides some selected values of F_{am} and deviations of the median value. CCIR Rep. 258 advises that equation (11.6) may give erroneous values for quiet rural (D) and galactic noise (E) environments.

Example 2. A receiver operates at 15 MHz with a 1-kHz bandwidth. The receiver is located in a rural environment and uses a lossless whip antenna one-quarter wavelength long with a ground plane. Calculate the noise power

TABLE 11.7 Representative Values of Selected Measured Noise Parameters for Business, Residential, and Rural Environmental Categories

	Environmental Category[a]											
	Business				Residential				Rural			
Frequency (MHz)	F_{am} (dB[kT_0])	D_u (dB)	D_l (dB)	σ_{NL} (dB)	F_{am} (dB[kT_0])	D_u (dB)	D_l (dB)	σ_{NL} (dB)	F_{am} (dB[kT_0])	D_u (dB)	D_l (dB)	σ_{NL} (dB)
0.25	93.5	8.1	6.1	6.1	89.2	9.3	5.0	3.5	83.9	10.6	2.8	3.9
0.50	85.1	12.6	8.0	8.2	80.8	12.3	4.9	4.3	75.5	12.5	4.0	4.4
1.00	76.8	9.8	4.0	2.3	72.5	10.0	4.4	2.5	67.2	9.2	6.6	7.1
2.50	65.8	11.9	9.5	9.1	61.5	10.1	6.2	8.1	56.2	10.1	5.1	8.0
5.00	57.4	11.0	6.2	6.1	53.1	10.0	5.7	5.5	47.8	5.9	7.5	7.7
10.00	49.1	10.9	4.2	4.2	44.8	8.4	5.0	2.9	39.5	9.0	4.0	4.0
20.00	40.8	10.5	7.6	4.9	36.5	10.6	6.5	4.7	31.2	7.8	5.5	4.5
48.00	30.2	13.1	8.1	7.1	25.9	12.3	7.1	4.0	20.6	5.3	1.8	3.2
102.00	21.2	11.9	5.7	8.8	16.9	12.5	4.8	2.7	11.6	10.5	3.1	3.8
250.00	10.4	6.7	3.2	3.8	6.1	6.9	1.8	2.9	0.8	3.5	0.8	2.3

[a] F_{am}, median value; D_u, D_l, upper, lower decile deviations from the median value within an hour at a given location; σ_{NL}, standard deviation of location variability.

Source: Table II, p. 70, CCIR Rep. 258-5, ITU Geneva, (Ref. 31).

threshold of a receiver under these circumstances. Assume a lossless transmission line.

Use equation (11.4) and replace $X\sigma_{am}$ with the expression $(D_u + \sigma_{NL})$, whose values we take from Table 11.7.

$$P_n = 34 \text{ dB} + (8.2 \text{ dB} + 4.2 \text{ dB}) + 10 \log 1000 - 204 \text{ dBW}$$

$$= -127.6 \text{ dBW}$$

11.9.5 Receiver Thermal Noise

Only under very special circumstances does receiver thermal noise become a consideration under normal HF operation. If we consider that atmospheric, man-made, and cosmic noise have a nonuniform distribution around an antenna (i.e., it tends to be directional), then the antenna gain at a certain frequency will generally favor signal and discriminate against noise.

Some HF antennas are designed with a very low gain; actually, they have a directional loss. The gain is expressed in negative dBi units. If we couple this antenna to a long or otherwise lossy transmission line, the noise at the receiver will have a significant thermal noise component. Seldom, however, is receiver-generated thermal noise a significant contributor to total HF receiving system noise. External noise is the dominant noise in determining the noise floor in the case of HF when calculating signal-to-noise ratio or E_b/N_0. In the design of HF receivers, thermal noise is not an overriding issue, and we find noise figures for these receivers are on the order of 12–16 dB.

11.10 NOTES ON HF LINK TRANSMISSION LOSS CALCULATIONS

11.10.1 Introduction

To predict the performance of a HF link or to size (dimension) link terminal equipment to meet some performance objectives, a first step is to calculate the link transmission loss. The procedure is similar to the transmission loss calculations for LOS microwave or troposcatter links. In other words, we must determine the signal attenuation in decibels between the transmitting antenna and its companion far-end receiving antenna. For a LOS link our concern was essentially free-space loss (FSL). The calculation for a HF link is considerably more involved.

11.10.2 Transmission Loss Components

The unfaded net transmission loss (L_{TL}) of a HF link can be expressed by

$$L_{TL} = L_{FSL} + L_D + L_B + L_M \quad \text{(dB)} \tag{11.7}$$

where L_{D} = D-layer absorption
L_{B} = ground reflection losses
L_{M} = miscellaneous losses
L_{FSL} = free-space loss

The free-space loss is expressed by the familiar formula

$$L_{\mathrm{FSL}} = 32.45 + 20 \log d_{\mathrm{km}} + 20 \log F_{\mathrm{MHz}} \quad (\mathrm{dB}) \tag{11.8}$$

11.10.2.1 Free-Space Loss. When we calculate L_{FSL} [equation (11.8)], d_{eff} is the total distance a signal travels from transmitter A to its far-end receiver B. For short NVI paths, d_{eff} will be notably greater than the great circle distance from A to B. As the distance from A to B increases, and we start to use "normal" skywave modes, the great circle and the effective distance (d_{eff}) between A and B start to converge. Thus for short paths (e.g., < 1000-km great circle distance), we will substitute d_{eff} in equation (11.8) in place of d. For paths greater than 1000 km, we can just use the great circle distance between transmitter A and distant receiver B (Ref. 33).

A geometric representation of a one-hop HF path is shown in Figure 11.22, where we see d_{eff} is the path APB and P is the reflection point h'_{i}

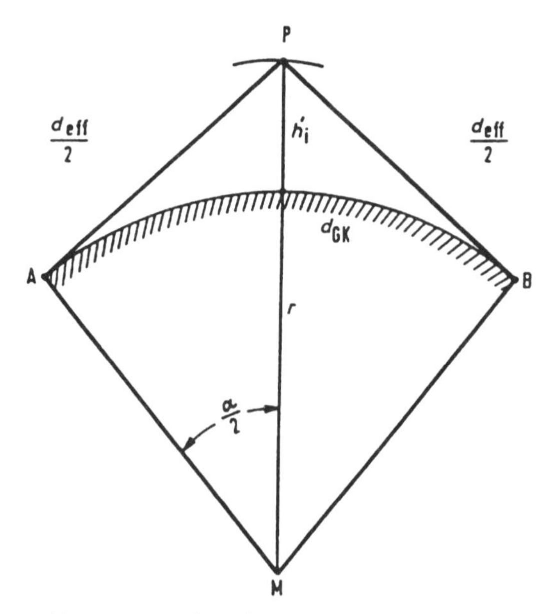

Figure 11.22. Geometric representation of a one-hop HF path where d_{eff} is the distance APB.

above the earth's surface at midpoint. The effective distance d_{eff} can be calculated:

$$d_{eff} = 2\sqrt{8.115 \times 10^7 + 12{,}740h'_i + h_i'^2 - \cos\frac{\alpha}{2} \times (8.115 \times 10^7 + 12{,}740h'_i)} \tag{11.9}$$

Equation (11.9) assumes r_i, the radius of the earth in Figure 11.22, to be 6370 km; α is the great circle arc from A to B, and h'_i is the virtual reflection height. We can calculate α from the great circle equation:

$$\cos\alpha = \sin A \sin B + \cos A \cos B \cos \Delta L \tag{11.10}$$

where α = angle of the great circle arc (see Figure 11.22)
A = latitude of station A
B = latitude of station B
ΔL = difference in longitude between stations A and B

We can also derive d_{eff} from Figure 11.23 where h' is the reflection point height above the earth's surface. The height of the reflection point h' can be derived from Figure 11.27.

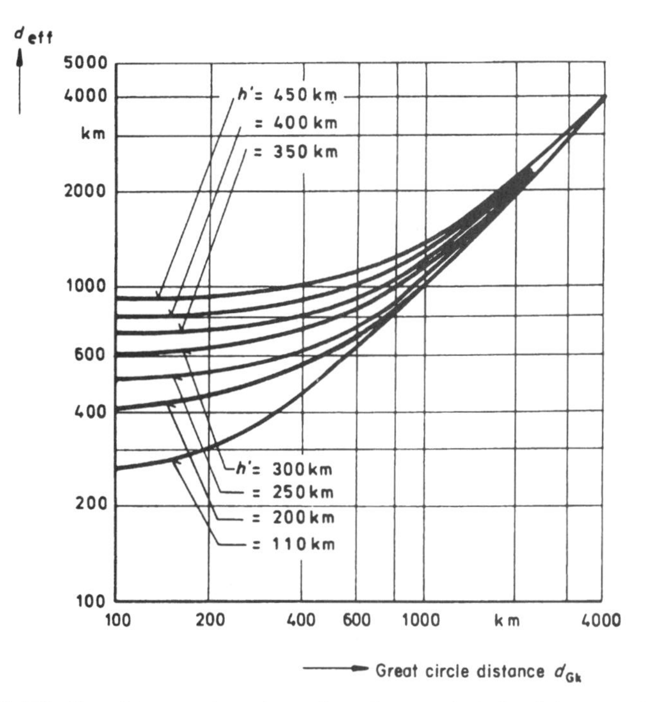

Figure 11.23. Effective distance d_{eff} when given the great circle distance and reflection height h'. (From Ref. 33.)

11.10.2.2 D-Layer Absorption Losses. D-layer absorption is a daytime phenomenon. The D-layer disappears at night. D-layer absorption varies with the zenith angle of the sun, the sunspot number, the season, and the operating frequency. In fact, it varies as the inverse of the square of the operating frequency. This latter fact is one reason we are driven to use higher frequencies for skywave links, to reduce D-layer absorption.

To calculate D-layer absorption on a particular skywave path, we first compute the absorption index I:

$$I = (1 + 0.0037R)(\cos 0.881\chi)^{1.3} \tag{11.11}$$

where R is the sunspot number and χ is the solar zenith angle of the sun. If χ is greater than 100°, it is nighttime and we can neglect D-layer absorption.

From Ref. 5, we calculate the solar zenith angle with the following formula:

$$\cos \chi = \sin \phi \sin \varepsilon + \cos \phi \cos \varepsilon \cosh \tag{11.12}$$

where ϕ = geographical latitude
ε = solar declination
h = the local hour angle of the sun measured westward from apparent noon, which is mean noon corrected for the equation of time and the standard time used at the location of interest

Tables of hourly values of $\cos \chi$ from sunrise to sunset for the 15th day of each month for most of the ionosphere vertical incidence sounding stations are given in the URSI Ionosphere Manual (Ref. 33, p. 19).

From Ref. 34, the hour angle of the sun, h, is calculated:

$$h = (\text{LST} - \text{right ascension})\left(\frac{15.0}{\pi}\right) \tag{11.13}$$

where the local sidereal time, LST, is

$$\text{LST} = \{[(\text{DN} \times 24 + \text{GMT})(1.002737909) - (\text{DN} \times 24)] - \phi\} + S \tag{11.14}$$

where ϕ = longitude of the point (in this case) (rad)
DN = the number of the day in the year
GMT = Greenwich Mean Time in decimal hours

If the local sidereal time is less than 0, then 24 h is added to it. If it is more than 24, then 24 h is subtracted from it. S is a correction factor.

Figure 11.24 can also be used to determine the solar zenith angle (Ref. 33). We now can calculate I, the absorption index, using equation (11.11). This value is then corrected for the winter anomaly, if required. These correction factors are given in Table 11.8. To use the nomogram in Figure

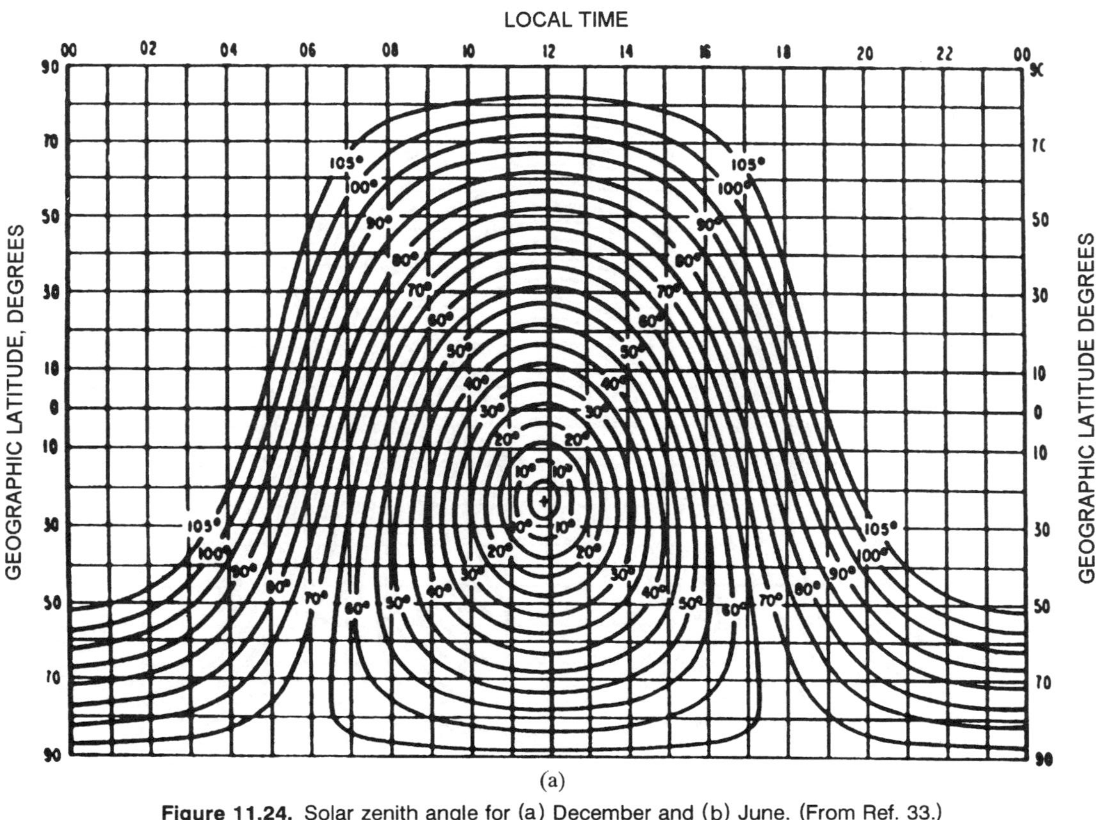

(a)

Figure 11.24. Solar zenith angle for (a) December and (b) June. (From Ref. 33.)

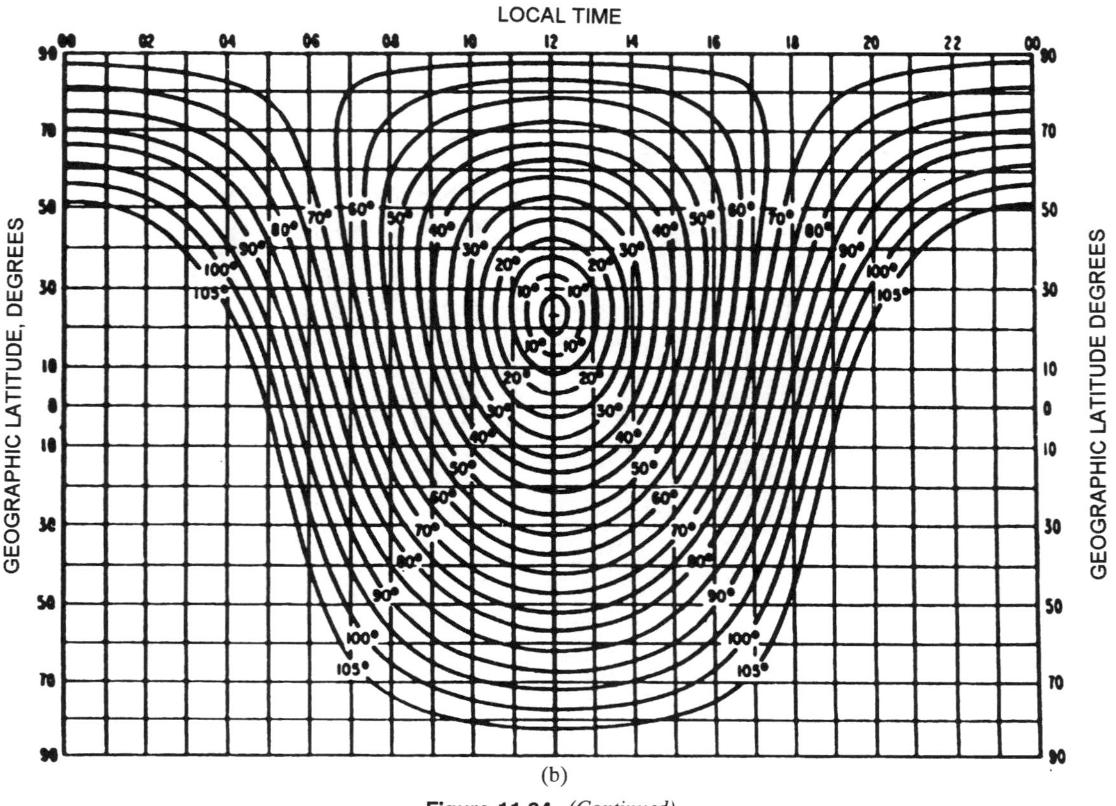

(b)

Figure 11,24. *(Continued)*

TABLE 11.8 Correction Factors of the Absorption Index to Account for the Winter Anomaly

Month		
Northern Hemisphere	Southern Hemisphere	Factor
November	May	1.2
December	June	1.5
January	July	1.5
February	August	1.2

Source: Reference 33.

11.25 effectively, we need to know the gyrofrequency for the region of interest. We can derive this value from Figure 11.26.

Example 3. A certain HF path operates during a sunspot number 100, the operating frequency is 18 MHz, and the elevation or TOA is 10°. Calculate the D-layer absorption value for 12 noon local time (at midpath) for the

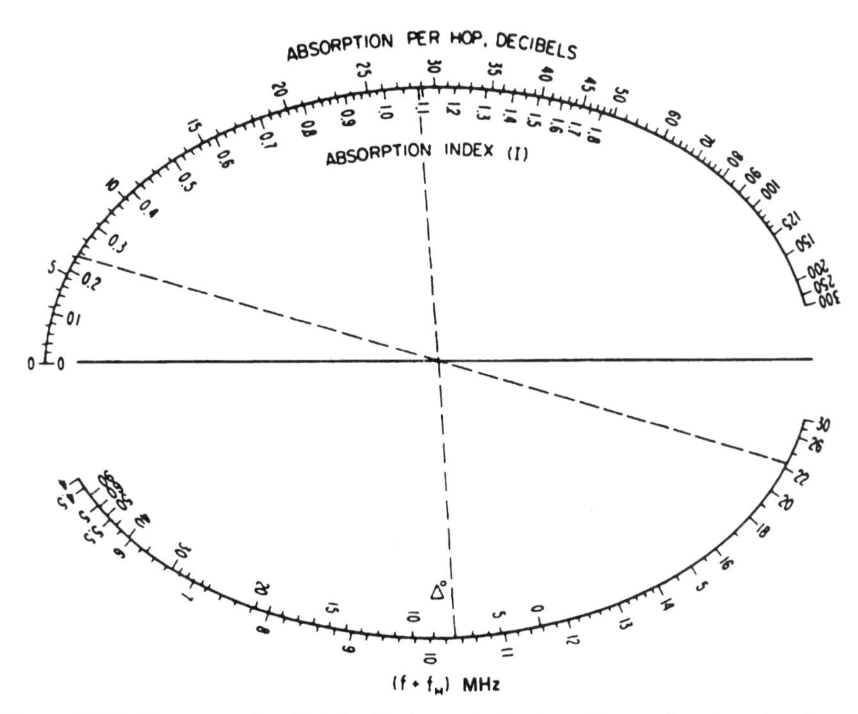

Figure 11.25. Nomogram for obtaining the ionospheric absorption per hop from the absorption index I, the effective wave frequency $f + f_H$, and the angle of elevation Δ. Example: (1) Enter with the absorption index I (= 1.09) and the elevation angle Δ (= 8°). (2) Mark the reference point of intersection with the center line. (3) With reference point and effective frequency $f + f_H$ (= 22.4), draw a straight line as far as the curve marked absorption per hop (= 6 dB). (From Ref. 33.)

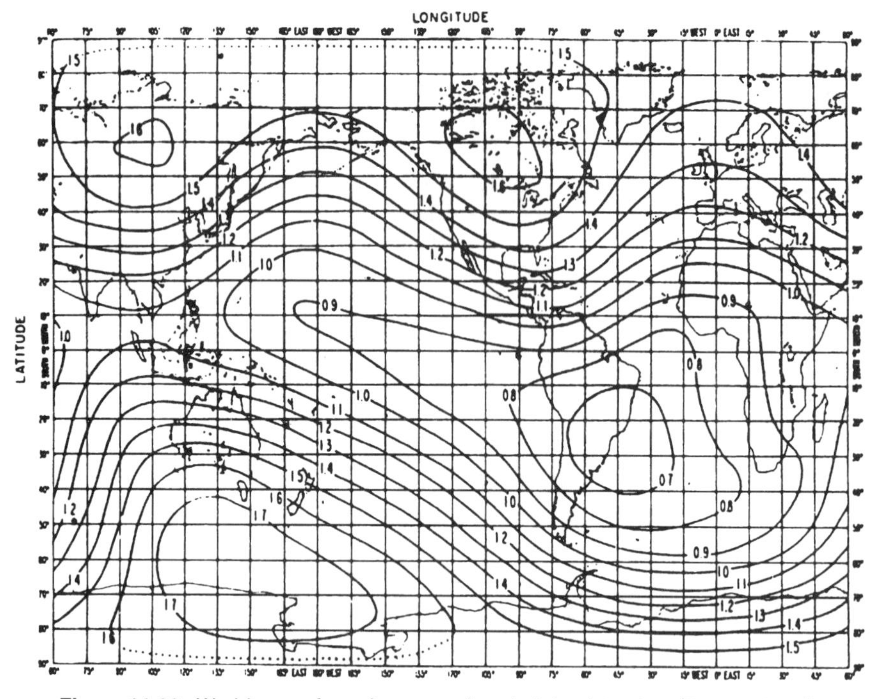

Figure 11.26. World map of gyrofrequency for a height of 100 km. (From Ref. 33.)

month of January at 40° north latitude. First we must derive the solar zenith angle from Figure 11.24a: 70°. Now we can substitute numbers in equation (11.11):

$$
\begin{aligned}
I &= (1 + 0.0037 \times 100)(\cos 0.881 \times 70°)^{1.3} \\
&= 0.52
\end{aligned}
$$

Multiply this value by the factor 1.5 taken from Table 11.8 to correct for the winter anomaly. I (corrected) is 0.78. Derive the gyrofrequency (F_H) for the latitude from Figure 11.26; it is 1.4 MHz. The value $f + f_H$ is 18 + 1.4 MHz, or 19.4 MHz. This value we use in Figure 11.25 to derive the D-layer absorption. We do this by first entering the takeoff angle (elevation angle) of the antenna, which is 10°. Now draw a line vertically to the corrected absorption index (I). We now have a reference point at the intersection of the solid horizontal lines. Through this reference point draw another line (right to left) from the frequency value of 19.4 MHz; it intersects "Absorption per Hop, Decibels" at 6 dB. We have a one-hop path, so the absorption value is 6 dB. If it were a 2-hop path with these parameters, it would be approximately 6 × 2, or 12 dB for D-layer absorption. Figure 11.27 can be used to determine the F2-layer height h'.

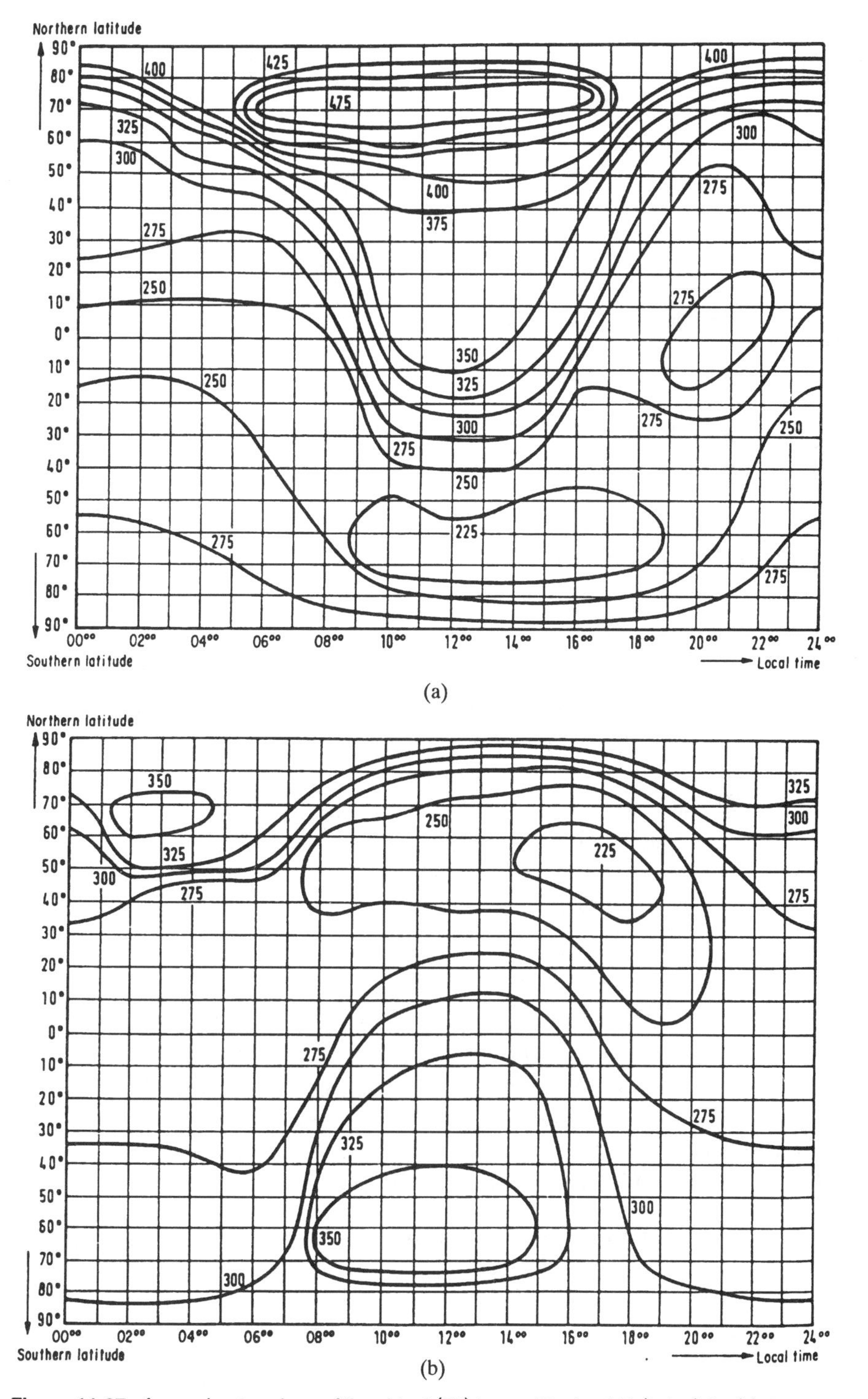

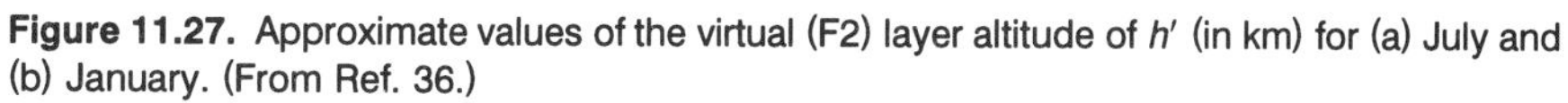
Figure 11.27. Approximate values of the virtual (F2) layer altitude of h' (in km) for (a) July and (b) January. (From Ref. 36.)

11.10.2.3 Ground Reflection Losses and Miscellaneous Skywave Propagation Losses. The following guidelines may be used to calculate ground reflection losses (Ref. 35):

One-hop = 0 dB
Two-hop = 2 dB
Three-hop = 4 dB
(etc.)

For miscellaneous skywave propagation losses, if no other information is available, use the value 7.3 dB. We will use this value for L_{M} in equation (11.7).

11.10.2.4 Guidelines to Determine Dominant Hop Mode. We only consider the lower order E- and F-layer modes (Ref. 35):

For path lengths up to 2000 km: 1E, 1F2, and 2F2
For path lengths between 2000 and 4000 km: 2E, 1F2, and 2F2
For path lengths between 4000 and 7000 km: 2F2 and 3F2
For path lengths between 7000 and 9000 km: 3F2

11.10.3 A Simplified Example of Transmission Loss Calculation

Consider a 1500-km path with 1F2 dominant mode in the temperate zone (40° N) operating over land in June with a sunspot number of 100 ($R12$) at 12 noon local time (midpath). This is an application of equation (11.7). Use Figure 11.28 to determine the radiation angle when given the great circle distance and virtual (F2) layer height. We can derive the layer height from Figures 11.27a and 11.27b. In this case, of course, we use Figure 11.27a for a value of 375 km. The elevation angle is ~ 17°. We now calculate the free-space loss using equation (11.8):

$$\mathrm{FSL}_{\mathrm{dB}} = 32.45 + 20 \log F_{\mathrm{MHz}} + 20 \log d_{\mathrm{km}}$$

The OWF is determined to be 17 MHz. We now have the distance and frequency, so we can proceed to calculate FSL, which is 120.57 dB.

We now want to calculate the D-layer absorption loss, L_{D} in equation (11.7). Use Figure 11.24b to calculate the solar zenith angle. This is 15°. Now we use formula (11.11) to calculate the absorption index (I):

$$\begin{aligned} I &= (1 + 0.0037 \times 100)(\cos 0.881 \times 15)^{1.3} \\ &= 1.37(0.97)^{1.3} \\ &= 1.45 \end{aligned}$$

Because the month of interest is June, we can neglect the winter anomaly.

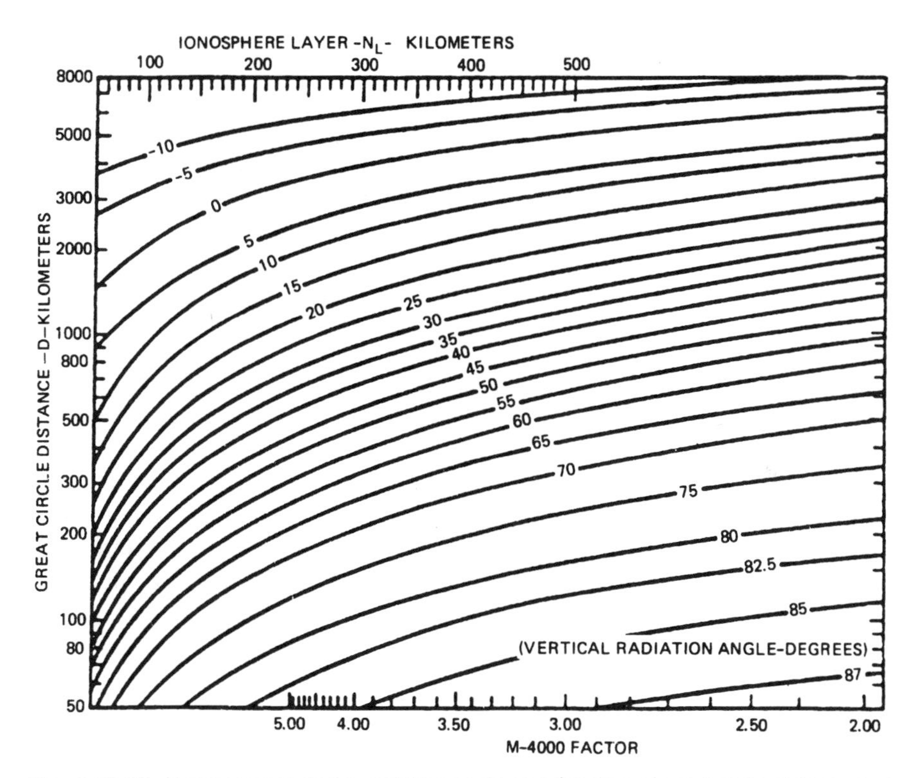

Figure 11.28. Nomogram to derive radiation angle as a function of great circle distance and ionospheric layer height. (From Ref. 33.)

The next step is to derive the gyrofrequency using Figure 11.26. The value is 1.5 MHz, so $f + f_H = 18.5$ MHz. We now derive the D-layer absorption from the nomogram, Figure 11.25. It is 8 dB. This is a one-hop path, so ground reflections are zero. L_M is 7.3 dB (see Section 11.10.2.3). Therefore the total transmission loss (unfaded) from equation (11.7) is

$$\begin{aligned} L_{TL} &= 120.57 \text{ dB} + 8 \text{ dB} + 0 \text{ dB} + 7.3 \text{ dB} \\ &= 135.87 \text{ dB} \end{aligned}$$

11.10.4 Groundwave Transmission Loss

11.10.4.1 Introduction. Groundwave or surface wave transmission by HF is particularly effective over "short" ranges. The range or distance for effective transmission decreases with increasing frequency and decreases with decreasing ground conductivity. Seawater is highly conductive and we find that ranges from 500 to 800 mi (800 to 1200 km) can be achieved during daytime. Skywave interference at night, man-made and atmospheric noise, can reduce the range. The link design engineer must find an optimum frequency trading off atmospheric noise with frequency and range. We must also note that vertical polarization is more effective than horizontal polarization for groundwave paths.

Overland groundwave path transmission range is a function of ground conductivity and roughness. The performance prediction of such paths is a rather imperfect art.

11.10.4.2 Calculation of Groundwave Transmission Loss. We use a very simple method to calculate groundwave transmission loss. We first turn to CCIR Rec. 368-7 (Ref. 37) and use the appropriate curves. Three such curves are given in Figure 11.29a–c. Figure 11.29a is for a path over seawater; Figures 11.29b and 11.29c are for overland paths with different ground conductivities. Table 11.9 will help in the selection process for the appropriate curve (ground conductivity).

We then select the operational frequency and distance, from which we will derive a field strength E, in decibels above a microvolt. To be consistent with other chapters dealing with radiolinks, we will want to calculate the transmission loss. Note 4 to CCIR Rec. 368-7 provides the following equation to derive transmission loss given the value for E:

$$L_{\mathrm{b}} = 137.2 + 20 \log F_{\mathrm{MHz}} - E \tag{11.15}$$

We illustrate this procedure by example. Suppose we have a 10-km link and have selected 3 MHz for our operational frequency over "normal" land (Figure 11.29b). We first determine the value E from the figure. It is 82.5 dB (μV/m). We apply this value of E to the formula:

$$\begin{aligned} L_{\mathrm{b}} &= 137.2 + 20 \log 3 - 82.5 \text{ dB } (\mu\text{V/m}) \\ &= 137.2 + 9.5 - 82.5 \\ &= 64.2 \text{ dB} \end{aligned}$$

Hence the transmission loss over *smooth* earth with *homogeneous* ground conductivity and permittivity is 64.24 dB. Because the intervening terrain is not smooth, we add a terrain roughness degradation factor of 6 dB, for a total transmission loss of 70.2 dB.

Some Notes of Caution. If we were to measure transmission loss on this circuit, we would probably find a much greater loss as measured at the input port of the far-end receiver. The following lists some variables we may encounter on our path and in our equipment, which may help explain such disparities:

1. The CCIR Rec. 368-7 curves are for vertical polarization. A horizontally polarized wave will have a notably higher transmission loss, on the order of 50 dB higher* (Ref. 33, p. 269).

*In theory—in practice, the gap between vertical and horizontal polarization may be much smaller.

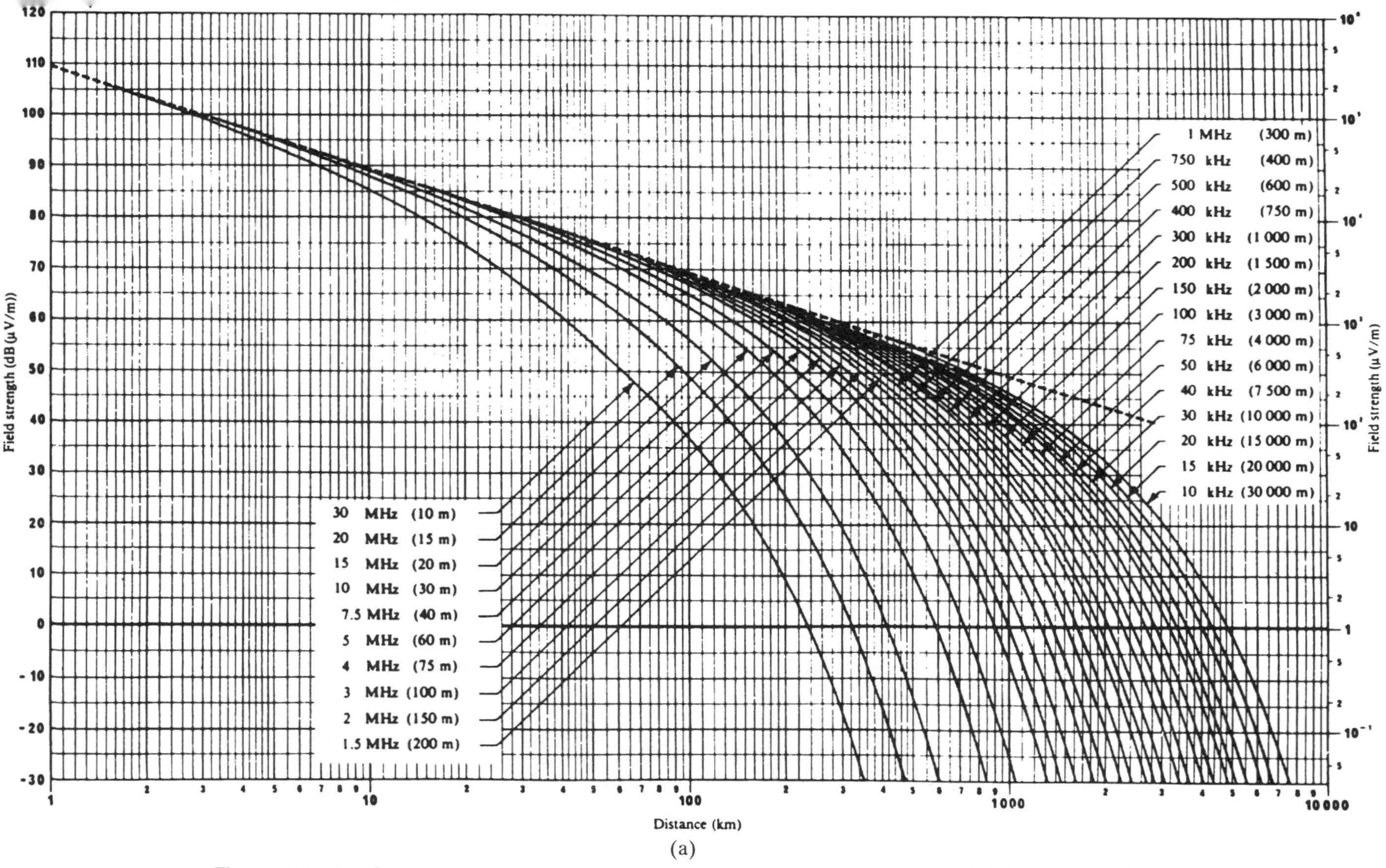

(a)

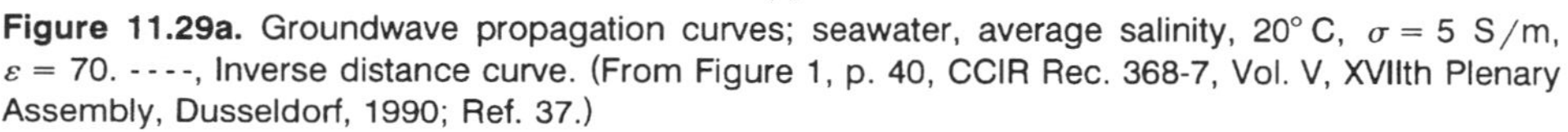

Figure 11.29a. Groundwave propagation curves; seawater, average salinity, 20° C, $\sigma = 5$ S/m, $\varepsilon = 70$. - - - -, Inverse distance curve. (From Figure 1, p. 40, CCIR Rec. 368-7, Vol. V, XVIIth Plenary Assembly, Dusseldorf, 1990; Ref. 37.)

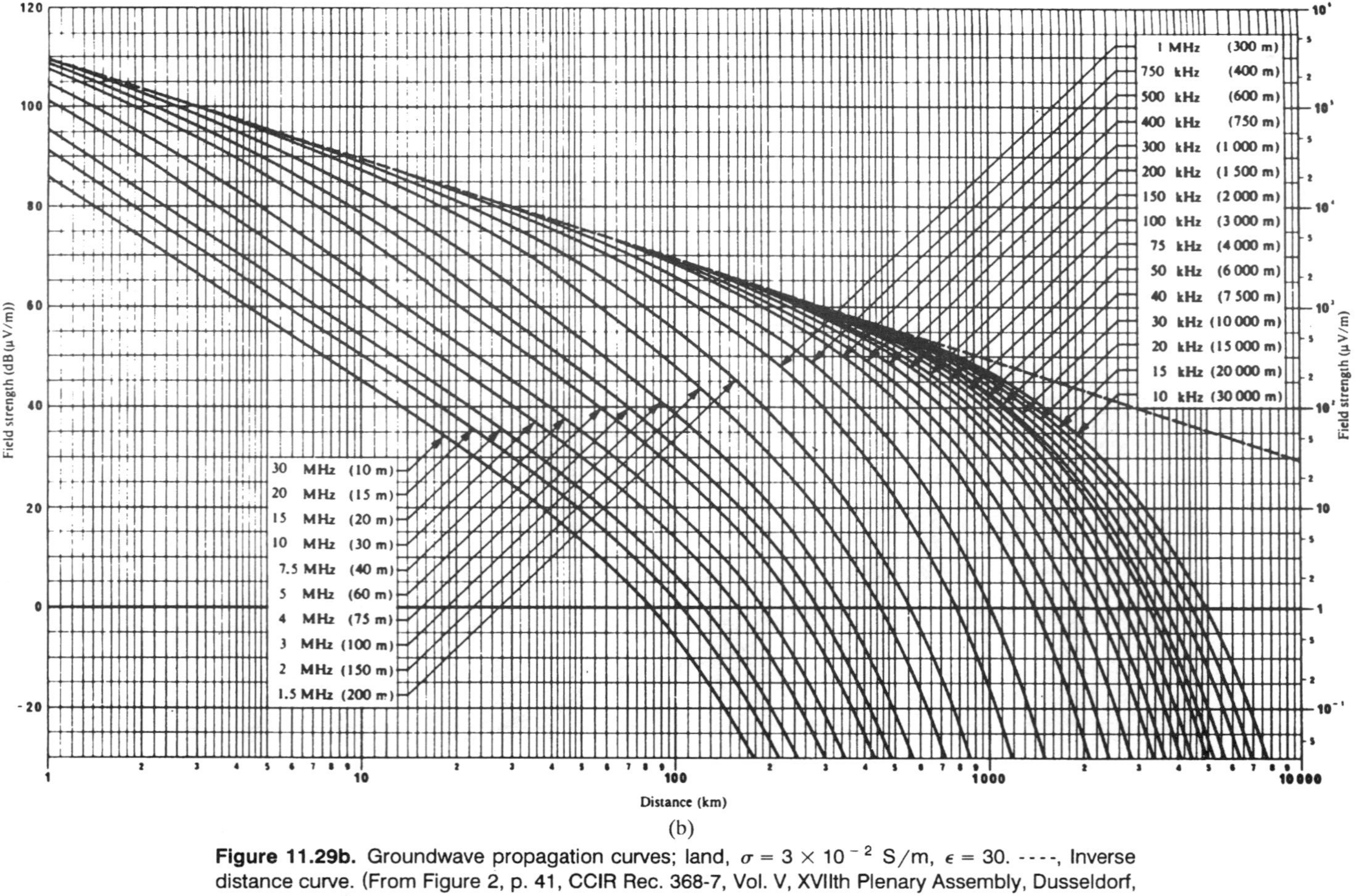

(b)

Figure 11.29b. Groundwave propagation curves; land, $\sigma = 3 \times 10^{-2}$ S/m, $\epsilon = 30$. - - - -, Inverse distance curve. (From Figure 2, p. 41, CCIR Rec. 368-7, Vol. V, XVIIth Plenary Assembly, Dusseldorf, 1990; Ref. 37.)

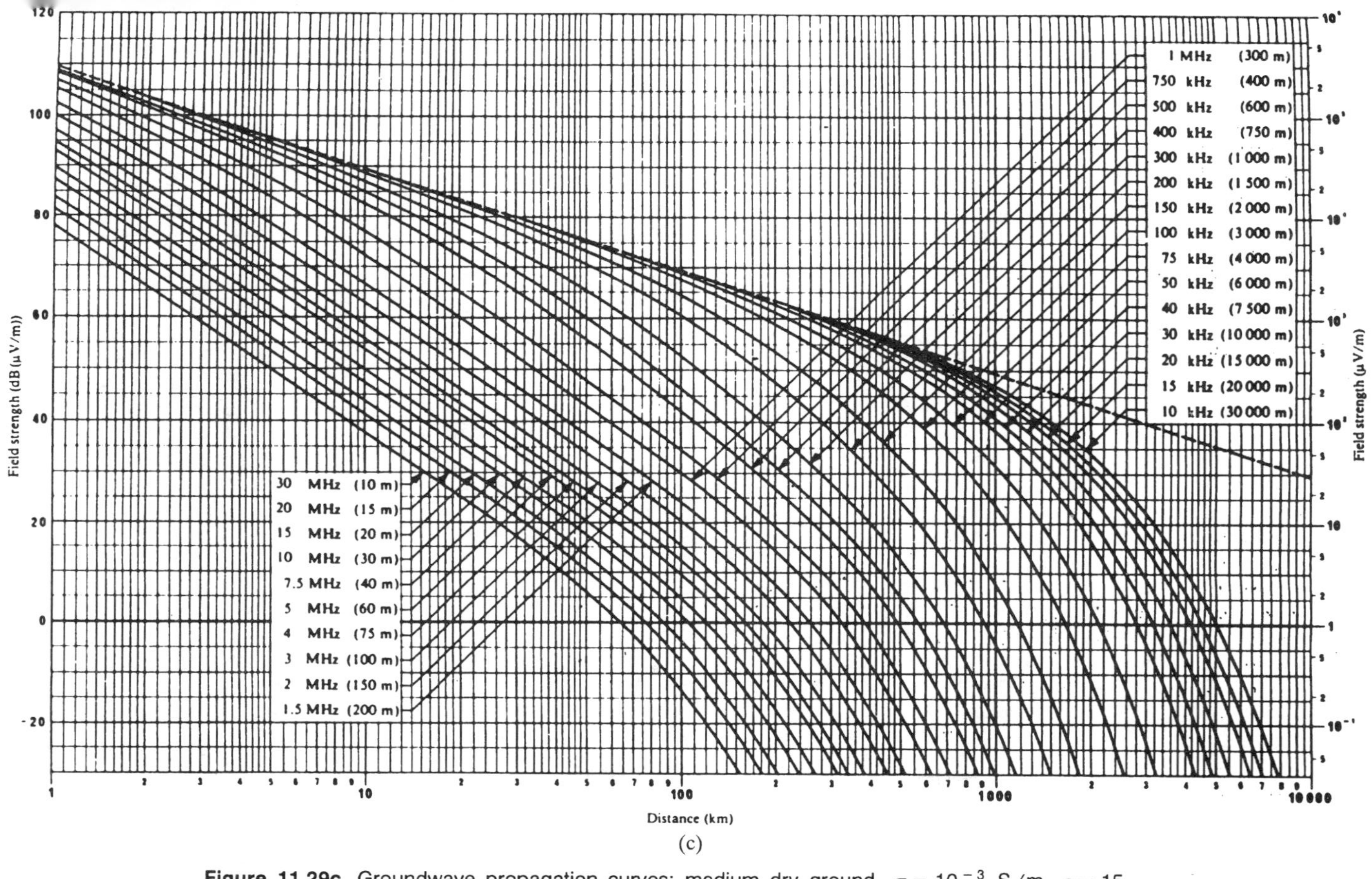

(c)

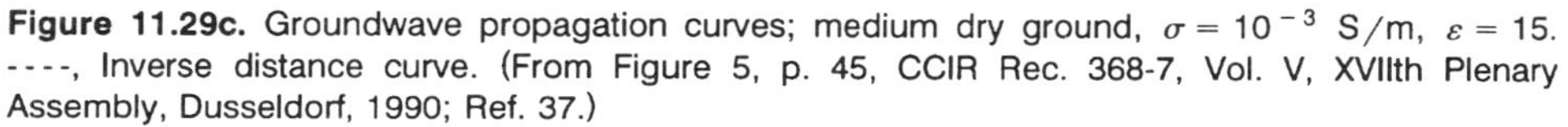

Figure 11.29c. Groundwave propagation curves; medium dry ground, $\sigma = 10^{-3}$ S/m, $\varepsilon = 15$. ----, Inverse distance curve. (From Figure 5, p. 45, CCIR Rec. 368-7, Vol. V, XVIIth Plenary Assembly, Dusseldorf, 1990; Ref. 37.)

TABLE 11.9 Guidelines in the Selection of Ground Conductivity and Dielectric Constant for Various Soil Types

Type of Terrain[a]	Conductivity σ (S/m^{-1})	Dielectric Constant ε
Seawater of average salt content	4	80
Fresh water (20° C)	3×10^{-3}	80
Moist soil	10^{-2}	30
Medium soil	10^{-3}	15
Dry soil	10^{-4}	4
Very dry soil	10^{-5}	4
	3×10^{-5}	4

[a]These types of terrain apply to the following areas:

Seawater	Practically all oceans
Fresh water	Inland waters, such as large lakes, wide rivers, and river estuaries
Moist soil	Marsh and fenland, regions with high groundwater level, floodplains, and so on
Medium soil	Agricultural areas, wooded land, typical of countries in the temperate zones
Dry soil	Dry, sandy regions such as coastal areas, steppes, and also arctic regions
Very dry soil	Deserts, industrial regions, large towns, and high mountains

Source: Reference 33.

2. With the exception of overwater paths, we will deal with ground conductivities/permittivities that can vary by more than an order of magnitude as we traverse the path.
3. The CCIR Rec. 368 curves are for a short vertical monopole at the earth's surface. The earth is assumed to be flat and *perfectly conducting* in the immediate area of the antenna but not along the path.

Thus the curves are idealized. Unless an excellent ground screen is installed under the antenna at both ends of the path, a degradation factor should be added for less than perfect ground. Degradation factors must also be added for antenna matching and coupling losses.

11.11 LINK ANALYSIS FOR EQUIPMENT DIMENSIONING

11.11.1 Introduction

In this section we will determine such key equipment/system design parameters as transmitter output power, type of modulation, impact of modulation selection on signal-to-noise ratio, bandwidth, fading and interference, use of FEC coding/interleaving, and type of diversity and diversity improvement. We assume in all cases that we are using the OWF and have calculated transmission loss for that frequency, time, season, and sunspot number. We also assume that we have calculated the appropriate atmospheric noise level for time, season, and frequency.

We will use Table 11.3 as a guide for required bandwidths and S/N requirements for some of the more common types of modulation used for HF point-to-point operation. Section 11.12 will deal with more sophisticated modulation schemes and how they can improve performance.

11.11.2 Methodology

The signal-to-noise ratio at the distant receiver can be expressed by

$$\frac{S}{N} = S_{\mathrm{dBm}} - N_{\mathrm{dBm}} \tag{11.16}$$

Substitute $\mathrm{RSL}_{\mathrm{dBm}}$ for S_{dBm}, where RSL is the receive signal level. This assumes, of course, that $S/N = C/N$ (carrier-to-noise ratio). To permit this equality, some value k is added to our transmission loss. For lack of another term, we may call this modulation implementation loss, where we sum up all the inefficiencies in the modulation–demodulation process (including coding and interleaving) to come up with a value for k. Unless we have other values, we'll use 2 dB for the value of k.

In equation (11.16) N is the noise floor of the receiver, which is probably dominated by atmospheric noise and/or man-made noise rather than thermal noise. Refer to Sections 11.9.3 and 11.9.4.

RSL is calculated in the conventional manner:

$$\mathrm{RSL}_{\mathrm{dBm}} = \mathrm{EIRP}_{\mathrm{dBm}} - L_{\mathrm{TL}} + G_{\mathrm{rec}} \tag{11.17}$$

where L_{TL} = total transmission loss in decibels, including k dB of modulation implementation loss

EIRP = effective isotropically radiated power from the transmit antenna

G_{rec} = net antenna gain at the receiver

Remember that the antenna gains used in the calculation of EIRP and the receiver antenna gain must be that gain at the proper azimuth and the elevation of the TOA (or radiation angle) (Figure 11.28).

Try the following example (refer to Section 11.10.3). There is a 1500-km path with 1F2 as the predominant mode, in the temperate zone (40° N), operating over land in June with a sunspot number of 100 ($R12$), and the time of day is noon. The receiver has a 15-dB noise figure and a 3-kHz bandwidth and operates in a rural setting. The OWF is 17 MHz and the TOA is about 17°. FSL is 120.57 dB, D-layer absorption is 9 dB, miscellaneous losses are 7.3 dB, and the modulation implementation loss is 2 dB; thus L_{TL} is 138.87 dB.

The atmospheric noise at 17 MHz is on a 10-dB contour for summer, 12 noon (Figures 11.18 and 11.19) with a variability of 8 dB for 95% of the time.

Man-made noise has a value F_{am} of 42 dB with a variability of 8 dB. The man-made noise component of the noise floor is by far the dominant noise and it is

$$\begin{aligned} p_{mm} &= 42 + 8 + 10 \log 3000 - 174 \text{ dBm} \\ &= -89.23 \text{ dBm} \end{aligned}$$

We will use this value for N_{dBm} in equation (11.16).

Let's assume that the transmitter has 1-kW RF output power and the net antenna gain at the angles of interest is 10 dB (including transmission line losses). The net receiver antenna gain is also assumed to be 10 dB. The EIRP is then

$$\begin{aligned} \text{EIRP}_{dBm} &= +60 \text{ dBm} + 10 \text{ dB} \\ &= +70 \text{ dBm} \\ L_{TL} &= 138.87 \text{ dB} \\ \text{RSL} &= +70 \text{ dBm} - 138.87 \text{ dB} + 10 \text{ dB} \\ &= -58.87 \text{ dBm} \\ \frac{S}{N} &= -58.87 - (-89.23 \text{ dB}) \\ &= 30.36 \text{ dB/3-kHz} \\ &= 65.13 \text{ dB/1-Hz} \end{aligned}$$

Suppose the link we are analyzing was going to use single-sideband (SSB) operation. Now turning to Table 11.3, we find for J3E (SSB) operation, stable condition, a S/N of 64 dB is required in 1 Hz of bandwidth. Our S/N value is for 3000-Hz bandwidth. Our margin is 65.13 − 64 dB, or 1.13 dB (unfaded condition).

Still on the J3E line in Table 11.3 we now look to the column that indicates good commercial service with fading and now we require 72 dB S/N in a 1-Hz bandwidth. We are short 72 dB − 65.13 dB ≈ 7 dB.

We can increase transmit power to 5 kW to compensate for the 7 dB shortfall.

The assumption is made, of course, that the HF facility has a frequency assignment near 17 MHz and that it is an "interference-free channel." This latter statement, in practice, is highly problematical.

11.12 SOME ADVANCED MODULATION AND CODING SCHEMES

11.12.1 Two Approaches

There are two approaches to the transmission of binary message traffic or data on a conventional HF link: serial tone transmission and parallel tone transmission. Today it is still probably more common to encounter links using binary FSK on a single channel or multiple channels in the 3-kHz passband with VFCT. (See Table 11.4.) On single-channel systems, when transmitting 150 bps or less, the center tone frequency is at 1275 Hz with a frequency shift of ± 42.5 Hz (Refs. 24 and 25). For 600-bps operation the center frequency is at 1500 Hz with a frequency shift of ± 200 Hz; for 1200 bps, the center tone frequency is at 1700 Hz shifted ± 400 Hz.

We will briefly describe two parallel tone systems and two serial tone systems. The first parallel tone system transmits 16 simultaneous tones and each tone is differentially phase shifted. The second method uses 39 tones. In either case these tones are contained in the nominal 3-kHz passband envelope. Of the two serial tone techniques, one uses 8-ary PSK and the other 8-ary FSK modulation.

For a good introductory text on data transmission, see Ref. 20.

11.12.2 Parallel Tone Operation

The first technique referenced here [MIL-STD-188-110 (Ref. 3)] operates from 75 to 2400 bps. The modulator accepts serial binary data and converts the bit stream into 16 parallel data streams. The signal element interval on each data bit stream is 13.33 ms and its modulation rate is 75 baud. The modulator provides a separate tone combination for initial synchronization and, if required, a separate tone for Doppler correction. Tone frequencies and bit locations are shown in Table 11.10. For data signaling rates of 75, 150, 300, and 600 bps at the modulator input, each data tone signal element is biphase modulated, as shown in the upper part of Figure 11.30. Each bit of the serial binary input signal is encoded, depending on the mark or space logic sense of the bit, into a phase change of the data tone signal element as listed in Table 11.11. For data signaling rates of 1200 and 2400 bps at the modulator input, each data tone signal is four-phase modulated (QPSK), as shown in the lower part of Figure 11.30. In this case, each dibit of the serial binary input signal is encoded, depending on the mark or space logic sense and the even or odd bit location of each bit, into a phase change of the data tone signal element as listed in Table 11.11. The phase change of a data tone signal element is relative to the phase of the immediately preceding signal element. This is called differential phase-shift keying. In-band diversity (see Section 11.8) is provided for all but the 2400-bps data rate.

There are two arguments in favor of the parallel tone approach. The first is that the transmitted signal element is long (13.33 ms) for all user data

TABLE 11.10 Data Tone Frequencies and Bit Locations for HF Data Modems

Tone Frequency (Hz)	Function	Even and Odd Bit Locations of Serial Binary Bit Stream, Encoded and Phase Modulated on Each Data Tone Employing:					
		Quadrature-Phase Modulation		Biphase Modulation			
		2400 bps	1200 bps	600 bps	300 bps	150 bps	75 bps
605 825[a]	Continuous Doppler Tone Synchronization Slot		←	In-Band Diversity		→	
935	Data tone 1	1st and 2nd	1st and 2nd	1st	1st	1st	1st
1045	Data tone 2	3rd and 4th	3rd and 4th	2nd	2nd	2nd	1st
1155	Data tone 3	5th and 6th	5th and 6th	3rd	3rd	1st	1st
1265	Data tone 4	7th and 8th	7th and 8th	4th	4th	2nd	1st
1375	Data tone 5	9th and 10th	9th and 10th	5th	1st	1st	1st
1485	Data tone 6	11th and 12th	11th and 12th	6th	2nd	2nd	1st
1595	Data tone 7	13th and 14th	13th and 14th	7th	3rd	1st	1st
1705	Data tone 8	15th and 16th	15th and 16th	8th	4th	2nd	1st
1815	Data tone 9	17th and 18th	1st and 2nd	1st	1st	1st	1st
1925	Data tone 10	19th and 20th	3rd and 4th	2nd	2nd	2nd	1st
2035	Data tone 11	21st and 22nd	5th and 6th	3rd	3rd	1st	1st
2145	Data tone 12	23rd and 24th	7th and 8th	4th	4th	2nd	1st
2255	Data tone 13	25th and 26th	9th and 10th	5th	1st	1st	1st
2365	Data tone 14	27th and 28th	11th and 12th	6th	2nd	2nd	1st
2475	Data tone 15	29th and 30th	13th and 14th	7th	3rd	1st	1st
2585	Data tone 16	31st and 32nd	15th and 16th	8th	4th	2nd	1st

[a]No tone is transmitted at this frequency.

Source: MIL-STD-188-110 (Ref. 3).

rates. Consequently, in theory, a system using this technique can withstand up to 6 ms of multipath dispersion (see Section 11.7.3.2). The second argument is that equalization across the 3-kHz audio passband becomes a moot point because the band is broken down into 110-Hz segments and each segment carries a very low modulation rate, so that delay and amplitude equalization are unnecessary.

The second technique uses 39 parallel tones each with quadrature differential phase-shifted modulation. A 40th unmodulated tone is added for Doppler correction. In-band diversity is available for all data rates below 1200 bps. FEC coding employing a shortened Reed–Solomon (15, 11) block code with appropriate interleaving is incorporated in the modem. A means is provided for synchronization of the signal element and interleaved block timing.

Each of the 39 tones is assigned a 52.25-Hz channel. The lowest frequency data tone is at 675.00 Hz and the highest is at 2812.50 Hz. The Doppler correction tone is at 393.75 Hz. The frequency accuracy of a data tone must be maintained within ± 0.05 Hz. For operation below 1200 bps both time and in-band frequency diversity are used.

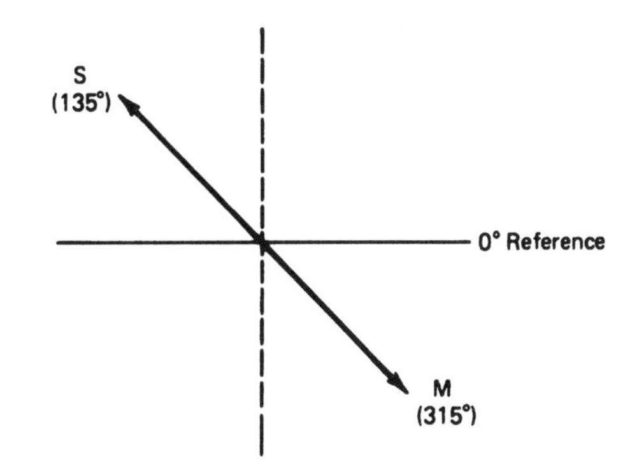

A. For data signaling rates of 75, 150, 300, or 600 bps.

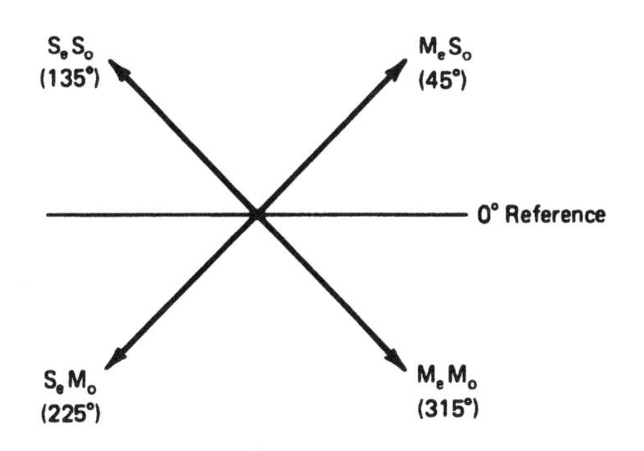

B. For data signaling rates of 1200 or 2400 bps.

Notes:
1. M = logic sense of mark; S = logic sense of space.

2. The subscripts refer to the even (e) or odd (o) bit locations of the serial binary bit stream. (see table 11.10)

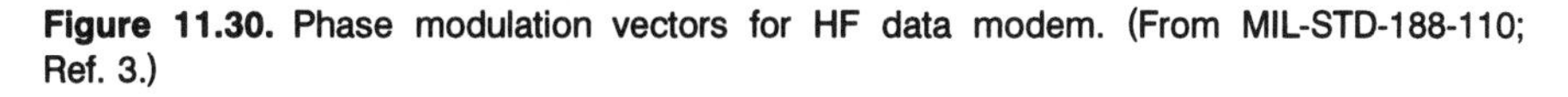

Figure 11.30. Phase modulation vectors for HF data modem. (From MIL-STD-188-110; Ref. 3.)

Table 11.12 gives bit error probabilities versus signal-to-noise ratio for the 39-tone modem based on a HF baseband simulator, assuming two independent, equal, average-power Rayleigh fading paths with a 2-Hz fading bandwidth and 2-ms multipath dispersion (Ref. 38).

11.12.3 Serial Tone Operation

The first technique we describe is covered in MIL-STD-188-110 (Ref. 3) and employs *M*-ary PSK on a single carrier frequency (or tone). The modem accepts serial binary data at 75×2^n up to 2400 bps and converts this bit stream into an 8-ary PSK modulated output signal. The serial bit stream,

TABLE 11.11 Modulation Characteristics for the 16 Parallel Tone HF Modem

Input Data Signaling Rate (bps)	Degree of In-Band Diversity Combining	Type of Modulation	Logic Sense of Dibits or Bits in Serial Binary Bit Stream Depending on:		Phase of Data Tone Signal Element Relative to Phase of Preceding Signal Element (degrees)
			Even Bit Locations	Odd Bit Locations	
2400	N/A		Mark	Space	+45
		Four-phase	Space	Space	+135
			Space	Mark	+225
1200	2		Mark	Mark	+315
600	2		Mark[a]		+315
300	4	Two-phase			
150	8				
75	16		Space[a]		+135

[a]Regardless of even or odd bit locations.

Source: MIL-STD-188-110 (Ref. 3).

TABLE 11.12 Probability of Error Versus Signal-to-Noise Ratio[a]

Signal-to-Noise Ratio (dB in 3-kHz Bandwidth)	Probability of Bit Error	
	2400 bps	1200 bps
5	8.6 E-2	6.4 E-2
10	3.5 E-2	4.4 E-3
15	1.0 E-2	3.4 E-4
20	1.0 E-3	9.0 E-6
30	1.8 E-4	2.7 E-6
Signal-to-Noise Ratio (dB in 3-kHz Bandwidth)	**Probability of Bit Error**	
	300 bps	75 bps
0	1.8 E-2	4.4 E-4
2	6.4 E-3	5.0 E-5
4	1.0 E-3	1.0 E-6
6	5.0 E-5	1.0 E-6
8	1.5 E-6	1.0 E-6

[a]Two independent, equal, average-power Rayleigh fading paths, with 2-Hz fading bandwidth and 2-ms multipath spread.

Source: MIL-STD-188-110 (Ref. 3).

TABLE 11.13 Error-Correction Coding, Fixed-Frequency Operation

Data Rate (bps)	Effective Code Rate	Method for Achieving the Code Rate
2400	$\frac{1}{2}$	Rate $\frac{1}{2}$ code
1200	$\frac{1}{2}$	Rate $\frac{1}{2}$ code
600	$\frac{1}{2}$	Rate $\frac{1}{2}$ code
300	$\frac{1}{4}$	Rate $\frac{1}{2}$ code repeated 2 times
150	$\frac{1}{8}$	Rate $\frac{1}{2}$ code repeated 4 times
75	$\frac{1}{2}$	Rate $\frac{1}{2}$ code

Source: MIL-STD-188-110 (Ref. 3).

before modulation, is coded with a convolutional code with a constraint length of 7 ($K = 7$). Table 11.13 gives the coding rates for the various input data rates. An interleaver is used with storage of 0.0-, 0.6-, and 4.8-s block storage (interleaving interval). The carrier tone frequency is 1800 Hz $\pm$ 1 Hz. Figure 11.31 shows the modulation state diagram for M-ary PSK as employed by this modem.

The waveform (signal structure) of the modem has four functionally distinct sequential transmission phases. The time phases are:

1. Synchronization preamble phase.
2. Data phase.
3. End-of-message (EOM) phase.
4. Coder and interleaver flush phase.

The length of the preamble depends on the interleaver setting. For the 0.0 setting, it is 0.0-s duration; for the 0.6 setting, 0.6-s duration; and for the 4.8-s setting, 4.8-s duration. This known preamble sequence allows the distant-end receive modem to achieve time and frequency synchronization.

During the data transmission phase, the desired data message is interspersed with known data sequences to train the far-end modem for channel equalization. We are aware, of course, that on a HF medium channel group delay and amplitude distortion are constantly changing with time. For instance, at 2400 bps 16 symbols of a known sequence (called known data) are followed by 32 symbols of user data (called unknown data). After the last unknown data bit has been transmitted, a special 32-bit sequence is sent to the coder, which performs the EOM function. It informs the distant receiving modem of end of message. The EOM sequence consists of the hexadecimal

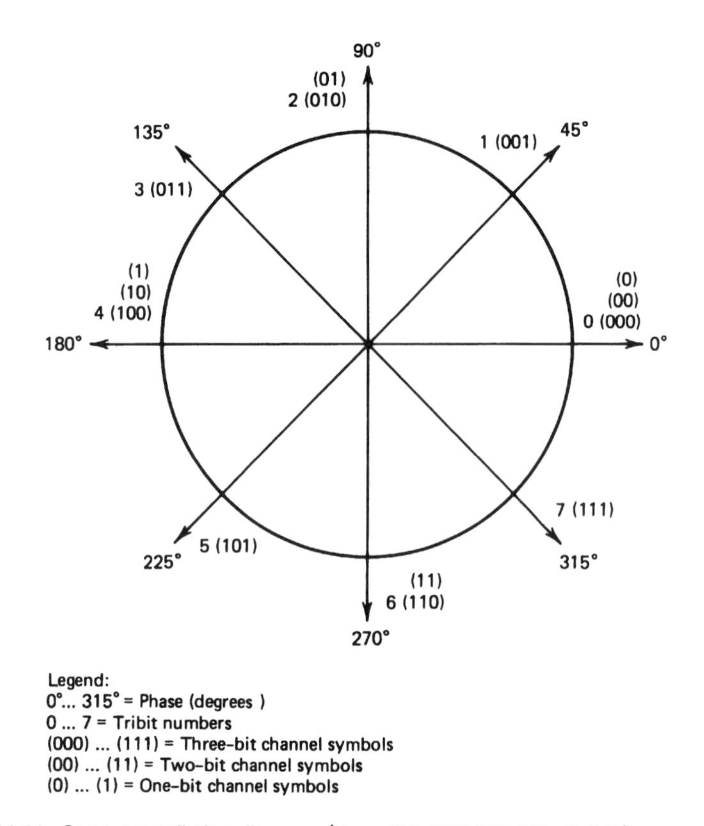

Figure 11.31. State constellation diagram. (From MIL-STD-188-110; Ref. 3.)

number 4B65A5B2. The final transmission phase is used to flush (reset) the far-end FEC decoder.

Table 11.14 gives bit error probabilities versus signal-to-noise ratio for various bit rates. This performance is based on the Waterson simulator described in CCIR Rep. 549-2 (Ref. 39). Here the modeled multipath spread values and fading (2σ) bandwidth values derive from two independent but equal average-power Rayleigh paths (Ref. 3).

The second serial tone approach is taken from Annex A of MIL-STD-188-141A (Ref. 19). The primary purpose of the standard is to automate a low-data-rate HF system, whether a single link, a star network with polling, or a large grid network. A fully detailed open system interconnection (OSI) layer 2 (datalink layer; see Ref. 20) is outlined. Two important functions are incorporated: ALE and a method of self-sounding, including a method of exchanging LQA data among network members.

The modulation is 8-ary FSK and the eight tones are as follows: 750 Hz (000), 1000 Hz (001), 1250 Hz (011), 1500 Hz (010), 1750 Hz (110), 2000 Hz

TABLE 11.14 Performance Characteristics for Serial Tone Modem Using Waterson Simulator (CCIR Rep. 549-2)

User Bit Rate	Channel Paths	Multipath (ms)	Fading[a] Bandwidth (Hz)	SNR[b] (dB)	Coded Bit Error Rate
2400	1 fixed			10	1.0 E-5
2400	2 fading	2	1	18	1.0 E-5
2400	2 fading	2	5	> 30	1.0 E-3
2400	2 fading	5	1	> 30	1.0 E-5
1200	2 fading	2	1	11	1.0 E-5
600	2 fading	2	1	7	1.0 E-5
300	2 fading	5	5	7	1.0 E-5
150	2 fading	5	5	5	1.0 E-5
75	2 fading	5	5	2	1.0 E-5

[a]Per CCIR Rep. 549-2.
[b]3-kHz bandwidth.
Source: MIL-STD-188-110 (Ref. 3).

(111), 2250 Hz (101), and 2500 Hz (100). It will be appreciated that just one tone is transmitted at a time. The tone transitions are phase continuous with a baud rate of 125 baud, which is a transmission rate of 375 coded symbols per second (1 baud = 3 symbols).

The system uses block coding FEC with the Golay (24, 12, 3) rate $\frac{1}{2}$ code (see Chapter 3). In other words, 1 data bit is represented by two coded symbols. In the data text mode (DTM), automatic message display (AMD) mode, and the basic ALE mode, an auxiliary coding is also employed: redundant $\times 32$ with $\frac{2}{3}$ majority voting (with 49 transmitted symbols).

The uncoded data rate (user data rate) is 61.22 bps in the DTM, the AMD mode, and the ALE modes. In the data block mode (DBM), the uncoded data rate is 187.5 bps (375/2). The throughput maximum data rate is 53.57 bps in the DTM, AMD, and basic ALE modes. The source coding is a subset of ASCII (see Chapter 12).

The LQA is built into the protocol header. LQA information includes BER, SINAD (signal + noise + distortion to noise + distortion ratio), and

TABLE 11.15 Probability of Linking

	Signal-to-Noise Ratio (dB)		
Probability of Linking (%)	Gaussian Noise Channel	CCIR Good Channel	CCIR Poor Channel
≥ 25	−2.5	+0.5	+1.0
≥ 50	−1.5	+2.5	+3.0
≥ 85	−0.5	+5.5	+6.0
≥ 95	0.0	+8.5	+11.0

Source: MIL-STD-188-141A (Ref. 19).

multipath value (MP). The LQA field consists of 24 bits, including 11 overhead bits, 3 multipath bits, 5 SINAD bits, and 5 BER bits. The system incorporates a sounding probe for frequency optimization/management.

Table 11.15 gives performance data regarding the probability of linking (ALE) using the standard HF simulator described in CCIR Rec. 520-2 (Ref. 38).

11.13 IMPROVED LINCOMPEX FOR HF RADIO TELEPHONE CIRCUITS

Lincompex is an acronym for *link compression and expansion.* It is a technique that provides a uniquely controlled companding (compression–expansion) function on SSB voice circuits. Performance improvement on links employing Lincompex is reported to be 22 dB minimum across time for typical speech signals under varying propagation conditions with a maximum improvement of 47 dB (Refs. 40, 41, and 42).

On systems using Lincompex, speech is compressed to a comparatively constant amplitude and the compressor control current is utilized to frequency modulate an oscillator in a separate control channel carried in a frequency slot just above the voice channel. The speech channel, which contains virtually all the frequency information of the speech signal, and the control channel, which contains the speech amplitude information, are combined for transmission on the nominal HF 3-kHz channel. As each speech syllable is individually compressed, the transmitter is more effectively loaded than with current SSB practice (without Lincompex). On reception, both the speech and control signals are amplified to constant level, the demodulated control signal being used to determine the expander gain and thus restore the original amplitude variations to the speech signal. Because the output level at the receiving end depends solely on the frequency of the control signal, which is itself directly related to the input level at the transmitting end, the overall system gain and loss can be maintained at a constant level. Operation with a slight loss (two-wire to two-wire) eliminates the need for singing suppressors, although echo suppressors will still be needed on long delay circuits.

Lincompex equipment based on ITU-R Rec. F.1111 can accommodate four different HF speech channels: 250–2500 Hz, 250–2380 Hz, 250–2000 Hz, and 250–1575 Hz; each with a control tone at 2900, 2580, 2200, or 1775 Hz, respectively. Calibrated frequency deviation on a tone is ± 60 Hz. Change in frequency for each 1-dB change in level is 2 Hz. There is a modulator frequency calibrate format that can be activated on command. It will correct errors of ± 80 Hz and bring the end-to-end frequency error down to ± 2 Hz. A Lincompex transmit unit can accommodate input levels from $+5$ dBm0 to -55 dBm0.

The maximum end-to-end frequency error of each Lincompex channel should be within ± 2 Hz if no Lincompex frequency calibrate option is available.

Performance data in this section were derived from ITU-R R. F.1111 (Ref. 43). Operational information was provided by Link Plus, Columbia, Maryland.

PROBLEMS AND EXERCISES

1 Give three advantages and three disadvantages of HF as a communication medium.

2 Identify at least five applications of HF for telecommunication that are used very widely.

3 Give at least three of the variables that affect range of HF groundwave communication links.

4 In the text, four ionospheric layers are described that affect HF propagation. What are they? One of these layers disappears at night. It affects a HF link in several ways. What are these effects and how can they be mitigated?

5 Describe two completely different ways by which we can communicate short distances by HF, say, out to 100–200 km.

6 Classify by region on the earth's surface where it is most desirable and least desirable to communicate by HF.

7 Describe how the season affects HF propagation.

8 Considering transition periods, why are north–south paths better behaved than east–west paths?

9 How does sunspot activity affect HF propagation?

10 Describe the average length of a sunspot cycle and variation in sunspot number (average) over the period.

11 List the five methods given in the text to carry out frequency management.

12 Using Table 11.2, at 1100 universal time (UT), what frequency would be recommended?

13 How can we derive the maximum usable frequency (MUF) when given the critical frequency?

14 How can one get comparative signal strength data for a particular connectivity without the use of an ionospheric sounder or by the "cut and try" method?

15 Describe how an embedded ionospheric sounding system operates. Some trademark sounding devices give other important data. Explain what some of the other data are and how they can be used to aid path performance.

16 Describe an advanced self-sounding system. What does link quality assessment (LQA) mean and how can it be used?

17 What is the weakness in the approach of basing LQA entirely on received signal strength?

18 Identify the three basic HF transmission modes. Distinguish each from the other and give their applications.

19 What detracts from groundwave operation at night?

20 What three effects give rise to multipath propagation?

21 What is the effect of D-layer absorption on multihop propagation?

22 Justify why we would be better off choosing the lowest order F2 mode on multimode reception.

23 What is the meaning of 3F2?

24 Explain the basic impairments due to propagation on transequatorial skywave paths, especially if they have only small latitude changes.

25 Explain the cause of the formation of the O-ray and X-ray. Which arrives first at the distant receiver?

26 Suppose we operate well below the MUF on a 1F2 path. How many modes could we receive? (Assume 1F2 only.)

27 What sort of antenna would we require for optimum near-vertical incidence (NVI) operation?

28 What frequency band would we favor (approximately) for NVI operation?

29 Discuss the *reciprocity* of HF operation.

30 What fade margin is recommended by CCIR on a HF skywave circuit for 99% time availability?

31 How much dispersion will a 50-baud FSK signal tolerate (i.e., binary FSK, or BFSK)?

32 Give at least three mitigation techniques that we might use to combat the effects of HF multipath.

33 Identify four types of diversity we may consider using for HF operation. Describe each in one or two sentences.

34 Why is receiver noise figure a secondary concern on HF systems?

35 There is a fair probability that the optimum operational frequency may be occupied by at least one other user. Name six methods to overcome this problem.

36 What is the cause of atmospheric noise and how does it propagate? How does it vary with frequency?

37 Calculate the receiver noise threshold at 8 MHz, summer at 7 p.m. local time in Massachusetts for a receiver with a 3-kHz bandwidth, valid for 95% of the time.

38 Above 20 MHz, what types of external noise may be predominant?

39 Man-made noise is a function of what two items?

40 What are the components (contributors) of HF link transmission loss?

41 When we calculate the free-space loss for a skywave path 100 km long, what is the most significant range component? For a 1500-km path?

42 How does D-region absorption vary with frequency? It also varies as a function of what? Give at least five items.

43 For a three-hop path, what is the estimated value of ground-reflection losses?

44 Using the text, calculate D-layer absorption for the following path: 1F2 mode, sunspot number 10, operating frequency 14.5 MHz, 12-noon midpath at 45° north latitude with an elevation angle of 9°.

45 Determine the transmission loss for a groundwave path over the ocean 100 km long at 4 MHz.

46 Calculate the optimum groundwave frequency over medium dry land in Colombia at 1400 local time assuming BFSK service, a signal-to-noise ratio (S/N) of 15 dB, a receiver noise figure of 15 dB, 75-bps operation, 1-kW EIRP, and a receiving antenna of 0 dBi, which includes line losses to receiver front end, for a time availability of 95%.

47 On long groundwave paths at night, leaving aside atmospheric noise, what must we be particularly watchful for? What are some compensating methods for this impairment?

48 Give two basic arguments in favor of parallel tone transmission (versus serial tone transmission) for 2400-bps operation. Give at least one argument against the parallel tone approach.

49 The text describes a serial tone system for synchronous transmission of digital traffic. Why is "known data" interspersed with "unknown data"?

50 What is LQA and what is its purpose? Describe how one might use a LQA sequence for a self-sounding system. Argue both sides: Using LQA, is a (PN) spread wideband system easier or more difficult than a conventional narrowband system for self-sounding?

51 Explain in six sentences or less the operation of Lincompex. Where is it applied and what approximate improvement does it provide (dB)?

52 Why is the control of out-of-band emission so important? Give two related answers.

REFERENCES

1. Kenneth Davies, *Ionospheric Radio*, Peter Peregrinus Ltd. on behalf of the Institution of Electrical Engineers, London, 1989.
2. *Digital Selective Calling System for Use in the Maritime Mobile Service*, ITU-R Rec. 493-6, ITU, Geneva, 1994.
3. *Equipment Technical Standards for Common Long-Haul/Tactical Data Modems*, MIL-STD-188-110, through Notice 2, U.S. Department of Defense, Washington, DC, 1988.
4. *Voice Frequency Telegraphy on Radio Circuits*, CCITT Rec. R.39, Fascicle VII.1, IXth Plenary Assembly, Melbourne, 1988.
5. Gerhard Braun, *Planning and Engineering of Shortwave Links*, Siemens-Heyden, London, 1982.
6. Roy F. Basler et al., *Ionospheric Distortion of HF Signals*, Final Report, Contract DNA-008-C-0155, SRI International, Menlo Park, CA, 1987.
7. Private communication, Dr. D. L. Wagner, Naval Research Laboratories, Washington, DC, Jan. 5, 1991.
8. George Jacobs and Theodore Cohen, *The Shortwave Propagation Handbook*, CQ Publishing Co., Hicksville, NY, 1979.
9. Solar Geophysical Data, Prompt Reports, National Geophysical Data Center, Boulder, CO, Mar. 1990.
10. John L. Lloyd et al., *Estimating the Performance of Telecommunication Systems Using the Ionospheric Transmission Channel—Techniques for Analyzing the Ionospheric Effects upon HF Systems*, Institute of Telecommunication Sciences, NTIA, Boulder, CO, 1983.
11. Larry Teters et al., *Estimating the Performance of Telecommunication Systems Using the Ionospheric Transmission Channel—Ionospheric Analysis and Prediction Program User's Manual*, Institute of Telecommunication Sciences, NTIA, Boulder, CO, 1983.
12. PROPHET Software Description Document, IWG Corp., San Diego, CA, 1984.
13. CRPL Predictions, NBS Circular 462, National Bureau of Standards (now NIST), Washington, DC, 1948.
14. *Bandwidths, Signal-to-Noise Ratios and Fading Allowances in Complete Systems*, CCIR Rec. 339-6, Vol. III, XVIth Plenary Assembly, Dubrovnik, 1986.

15. P. H. Levine, R. B. Rose, and J. N. Martin, "Minimuf-3—A Simplified HF MUF Prediction Algorithm," *IEE Conference on Antennas and Propagation*, Pub. No. 78, 1978.
16. *Operator's Manual for the USCG Advanced PROPHET System*, Naval Ocean Systems Command, San Diego, CA, 1987.
17. *Real-Time Frequency Management for Military HF Communication*, BR Communication Tech. Note No. 2, BR Communications, Sunnyvale, CA, June 1980.
18. Product Sheet (SELSCAN registered trademark), Collins 309L-2 & 4 and 514A-12 SELSCAN Adaptive Communications Processor, Rockwell International, Cedar Rapids, IA, 1981.
19. *Interoperability and Performance Standards for MF and HF Radio Equipment*, MIL-STD-188-141A (Fed. Std. 1045), U.S. Department of Defense, Washington, DC, Sept. 1988.
20. Roger L. Freeman, *Practical Data Communications*, Wiley, New York, 1995.
21. Klaus-Juergen Hortenbach, *HF Groundwave and Skywave Propagation*, AGARD-R-744, Cologne, Germany, Oct. 1986.
22. *Ionospheric Propagation and Noise Characteristics Pertinent to Terrestrial Radiocommunication Systems Design and Service Planning (Fading)*, CCIR Rep. 266-7, XVIIth Plenary Assembly, Dusseldorf, 1990.
23. *Factors Affecting the Quality of Performance of Complete Systems in the Fixed Service*, CCIR Rep. 197-4, Vol. III, XVIth Plenary Assembly, Dubrovnik, 1986.
24. *Equipment Technical Design Standards for Voice Frequency Carrier Telegraph (FSK)*, MIL-STD-188-342, U.S. Department of Defense, Washington, DC, Feb. 1972.
25. *Voice Frequency Telegraphy over Radio Circuits*, CCIR Rep. 106-1, Vol. III, XVIth Plenary Assembly, Dubrovnik, 1986.
26. *Radio Regulations*, ITU, Geneva, 1990, revised 1994.
27. B. D. Perry and L. G. Abraham, *Wideband HF Interference and Noise Model Based on Measured Data*, Rep. M-88-7, MITRE Corp., Bedford, MA, 1988.
28. P. J. Laycock et al., "A Model for HF Spectral Occupancy," *Fourth International Conference on HF Systems and Technology*, IEE, London, 1988.
29. *Electrical Communication System Engineering*, *Radio*, Department of the Army, Washington, DC, Aug. 1956.
30. *Characteristics and Applications for Atmospheric Radio Noise*, CCIR Rep. 322-2 (issued separately from other CCIR/ITU-R documents), ITU, Geneva, 1983. (See also Ref. 32.)
31. *Man-Made Radio Noise*, CCIR Rep. 258-5, Annex to Vol. VI. XVIIth Plenary Assembly, Dusseldorf, 1990. (See also Ref. 32.)
32. *Radio Noise*, ITU-R Rec. PI.372-6, PI Series Volume, ITU, Geneva, 1994.
33. K. Davies, *Ionospheric Radio Propagation*, NBS Monograph 80, U.S. Department of Commerce, National Bureau of Standards (now NIST), Boulder, CO, 1965.
34. *Technical Description of the Communication Assessment Program (CAP)*, Rev. 3.0, Defense Communications Agency (now DISA), Arlington Hall Station, VA, Jan. 1986.
35. *Simple HF Propagation Prediction Method for MUF and Field Strength*, CCIR Rep. 894-2, Vol. VI, XVIIth Plenary Assembly, Dusseldorf, 1990.

36. Technical Report No. 9, U.S. Army Signal Corps, Radio Propagation Agency, Ft. Monmouth, NJ, 1956.
37. *Groundwave Propagation Curves for Frequencies Between* 10 *kHz and* 30 *MHz*, CCIR Rec. 368-7, ITU, Geneva, 1995.
38. *Use of HF Ionospheric Channel Simulators*, CCIR Rec. 520-2, CCIR RF Series, Fixed Service, ITU, Geneva, 1992.
39. *HF Ionospheric Channel Simulators*, CCIR Rep. 549-2, Vol. III Annex, XVIIth Plenary Assembly, Dusseldorf, 1990.
40. *Improved Transmission System for RF Radiotelephone Circuits*, CCIR Rec. 455–22, RF Series Fixed Service, ITU, Geneva 1992.
41. *Improved Transmission Systems for Use over HF Radiotelephone Circuits*, CCIR Rep. 345–5, Vol. III, XVIth Plenary Assembly, Dubrovnik, 1986.
42. Roger L. Freeman, *Telecommunication Transmission Handbook*, 3rd ed., Wiley, New York, 1991.
43. *Improved Lincompex System for HF Radio Telephone Circuits*, ITU-R Rec. F.1111, 1994 F Series, Part 2, ITU Geneva, 1994.

12

METEOR BURST COMMUNICATION

12.1 INTRODUCTION

Meteor burst communication (MBC) can provide inexpensive,* very low data rate connectivity on links up to about 900 sm (1440 km) long. In this discussion, very low data rates are in the range of 10 to over 100 bps average throughput.

MBC utilizes the phenomenon of scattering of a radio signal from the ionization trails caused by meteors entering the atmosphere. A meteor trail must have some form of common geometry between one end of a link and the other. The usable life of a trail is short: from tens of milliseconds to several seconds. Thus a particular link can sustain useful communication with that trail for a very short period of time and then the users will have to wait for another trail entering the atmosphere with similar geometry characteristics. On a particular MBC link a transmitter bursts data when a common trail is discovered, waits for another trail, and bursts data again. The time between useful trails is called *waiting time*.

The useful radio-frequency range for meteor burst operation is between about 20 and 120 MHz. The lower frequencies are ideal and provide the best performance. As we have seen in Chapter 11, receivers in these "lower" frequencies are externally noise limited. Galactic and man-made noise are of such a magnitude in comparison with the expected signal levels of MBC systems that these lower frequencies become virtually unusable for meteor burst communications. As shown in Figure 11.21, this external noise drops off as the square of the frequency. Unfortunately, MBC performance also drops off as a function of frequency. It has been found by experience that the range 40–55 MHz is a compromise operational frequency band for MBCs.

* Inexpensive in the context of LOS microwave, troposcatter, and possibly satellite communications over equivalent distances.

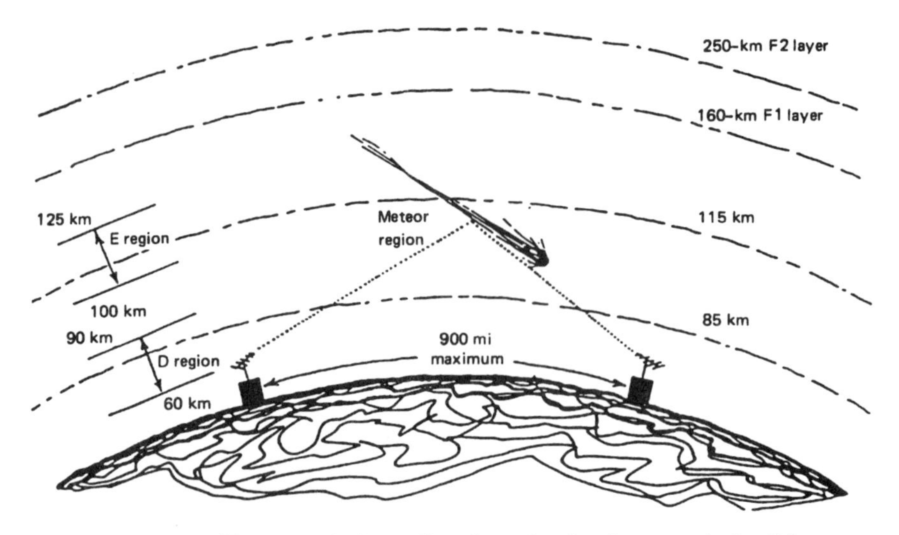

Figure 12.1. The concept of operation of a meteor burst communication link.

MBC transmitters have output powers from under 100 W to 5 kW or more. Antennas for fixed-frequency operation are usually Yagis, and horizontally polarized log periodics (LPs) if we wish to cover a bandwidth as wide as 5 MHz. Figure 12.1 illustrates the concept of a MBC system.

The implementation of MBC systems is attractive, especially from the standpoint of economy. The low data rate and the waiting times are disadvantages. One common application is remote sensing of meteorological conditions and/or reporting of seismic data. One large MBC system is installed in the Rocky Mountains in the United States to provide data on snowfall and accumulated snow. The system is aptly called SnoTel. MBC also has application in high latitudes because such systems are generally unaffected by aurora, polar cap events, or magnetic storms where HF systems often become virtually unusable.

Another application for MBC is for a "reconstitution orderwire." During natural disasters and general war, large portions of the PSTN or other networks may appear to be wiped out. Well-planned MBC systems can serve to coordinate efforts to reconstitute the PSTN or what is left of it.

12.2 METEOR TRAILS

12.2.1 General

Billions of meteors enter the earth's atmosphere every day. One source (Ref. 1) states that each day the earth sweeps up some 10^{12} objects that, upon entering the atmosphere, produce sufficient ionization to be potentially useful for reflecting/scattering radio signals.

Meteors enter the atmosphere and cause trails at altitudes of 70–140 km. The trails are long and thin, generating heat that causes the ionization. They sometimes emit visible light. The forward scatter of radio waves from these trails can support communication. The trails quickly dissipate by diffusion into the background ionization of the earth's atmosphere.

Meteor trails are classified into two categories, underdense and overdense, depending on the line density of free electrons. The dividing line is 2×10^{14} electrons per meter. Trails with a line density less than the value are *underdense*; those with a line density greater than 2×10^{14} electrons per meter are termed *overdense*. The dividing line of about 2×10^{14} electrons per meter corresponds to the ionization produced by a meteor whose weight is about 1×10^{-3} g. When averaged over 24 h, the number of meteors is almost inversely proportional to weight. As a result, we would expect that the number of underdense trails would far exceed the number of overdense trails. However, the signals reflected from underdense trails fall off roughly in proportion to the square of the weight, whereas signals from overdense trails increase only a little with weight. In practice, though, we find perhaps only 70% of received MBC signals are from underdense trails. Even so, the mainstay of a MBC system is the underdense trail.

Another interesting and useful fact about sporadic meteors is their mass distribution. This distribution is such that the total masses of each size of particle are approximately equal (i.e., there are ten times as many particles of mass 10^{-4} g as there are particles of mass 10^{-3} g). Table 12.1 lists the

TABLE 12.1 Estimate of Properties of Sporadic Meteors

Notes		Mass (g)	Radius (cm)	Number Swept Up by Earth per Day	Electron Line Density (Electrons / Meter)
Particles that survive passage through atmosphere		10^4	8	10	
	Overdense visual	10^3	4	10^2	
		10^2	2	10^3	
		10	0.8	10^4	10^{18}
		1	0.4	10^5	10^{17}
Particles totally disintegrated in upper atmosphere		10^{-1}	0.2	10^6	10^{16}
		10^{-2}	0.08	10^7	10^{15}
	Underdense nonvisual	10^{-3}	0.04	10^8	10^{14}
		10^{-4}	0.02	10^9	10^{13}
		10^{-5}	0.008	10^{10}	10^{12}
		10^{-6}	0.004	10^{11}	10^{11}
		10^{-7}	0.002	10^{12}	10^{10}
Particles that cannot be detected by radio means		10^{-8} to 10^{-13}	0.004 to 0.0002	Total about 10^{20}	Practically none

Source: Reference 2. Courtesy of Meteor Communications Corporation. Reprinted with permission.

approximate relationship between mass, size, electron density, and number (Ref. 2).

12.2.2 Distribution of Meteors

At certain times of the year meteors occur as showers and may be prolific over durations of hours. Typical of these are the Quadrantids (early January), Arietids (May, June), Perseids (July, August), and Geminids (December) (Ref. 3). However, for MBC planning purposes, we use the type of meteors discussed in Section 12.2.1, which CCIR calls "sporadic meteors."

Meteor intensity (i.e., the number of usable meteor trails per hour) varies with latitude, becoming fewer in number but more uniform diurnally for the high latitudes. With the midlatitude case, there is roughly a sinusoidal diurnal variation of incidence, with the maximum intensity about 0600 local time and the minimum some 12 hours later, or 1800 local time. The ratio of maximum to minimum is about 4 : 1. There is a seasonal variation of similar magnitude with a minimum in February and a maximum in July. Considerable day-to-day variability exists in the incidence of sporadic and shower meteors (Ref. 4).

These variabilities are explained by the earth's rotation and its orbit around the sun. The diurnal variability is a consequence of the earth's motion in its orbit around the sun. All meteor particles are in some form of orbit about the sun. At about 0600 local time a MBC site in question is on the forward side of the earth and at that time it sweeps up slower moving particles as well as a random number of particles colliding with the earth. At 1800 local time the same MBC site is on the back side of the earth, where the slower moving particles cannot catch up with the earth. Thus 1800 is the least productive time of the day. There is always the value of random meteor counts, with the earth's velocity modulating the mean, increasing the morning count and decreasing the evening count. Sometimes at midday we see a day-to-day variation of 5:1, and this is due to useful returns from sporadic E (layer) rather than just meteor intensity variation.

The season maximum in the July/August period is due to the fact that the earth's orbit takes it through a region of more dense solar orbit material.

12.2.3 Underdense Trails

MBC links basically depend on reflections from underdense trails. An underdense trail does not actually *reflect* energy; instead, radio waves excite individual electrons as they pass through the trail. These excited electrons act as small dipoles, reradiating the signal at an angle equal, but opposite, to the incident angle of the trail.

Signals received from an underdense trail rise to a peak value in a few hundred microseconds, then tend to decay exponentially (Figure 12.2). Decay times from a few milliseconds to a few seconds are typical. The decay in

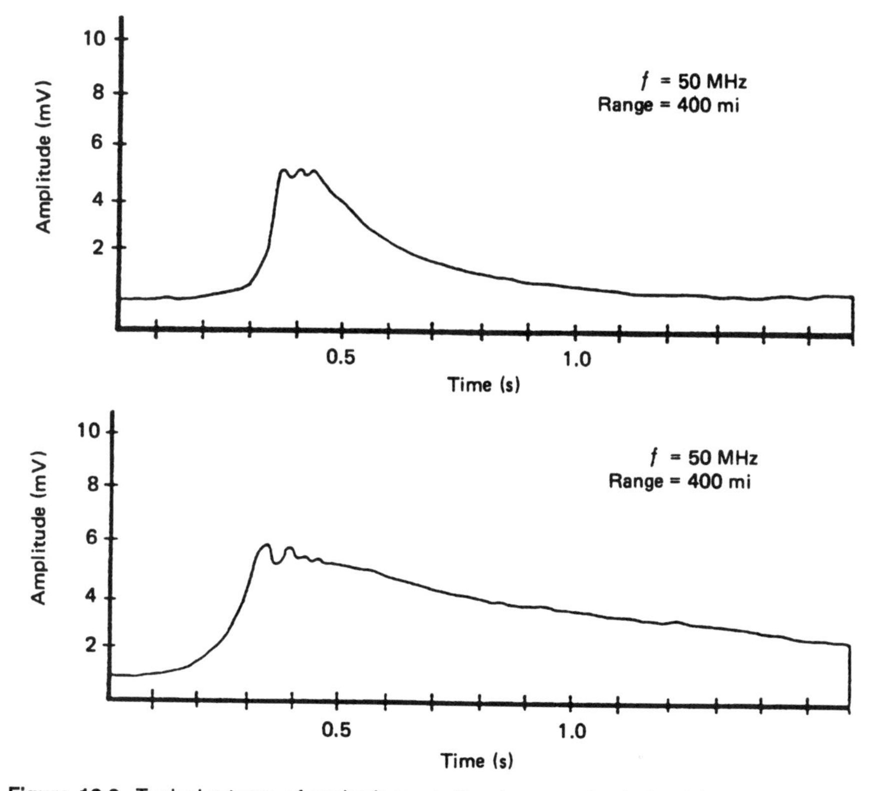

Figure 12.2. Typical returns of underdense trails where received signal intensity is plotted versus time. (From Ref. 2; courtesy of Meteor Communications Corporation. Reprinted with permission.)

signal strength results from the destructive phase interference caused by radial expansion or diffusion of the trail's electrons.

In Section 12.5.6 of this chapter we will show that the received power at a MBC terminal is proportional to $1/f^3$, which limits the maximum useful frequency to below 80 MHz. Commonly used frequencies are in the range of 40–55 MHz.

From equation (12.8) in Section 12.5.6 an amplitude–range relationship can also be found. An approximate normalized plot (Ref. 2) of range versus amplitude is shown in Figure 12.3. Reference 2 shows that range and frequency also affect usable time duration of a trail. Such a plot is shown in Figure 12.4 for four different frequencies.

12.2.4 Overdense Trails

We defined overdense meteor trails as those with electron line densities greater than 2×10^{14} electrons/meter. In this case the line density is so

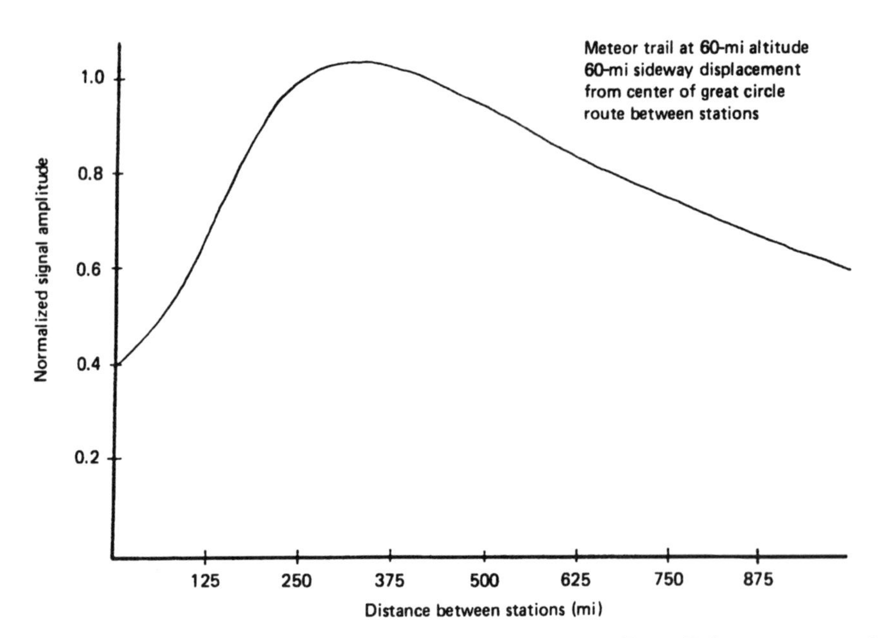

Figure 12.3. Normalized underdense reflected signal power. (From Ref. 2; courtesy of Meteor Communications Corporation. Reprinted with permission.)

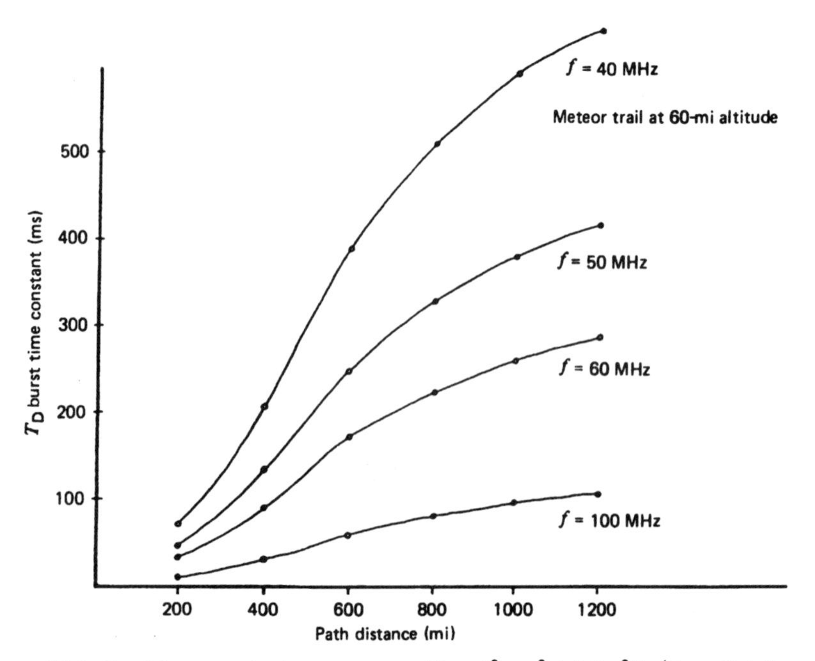

Figure 12.4. Burst time constant versus range. $T_D = \lambda^2 \sec^2 \phi / 32\pi^2 D$. (From Ref. 2; courtesy of Meteor Communications Corporation. Reprinted with permission.)

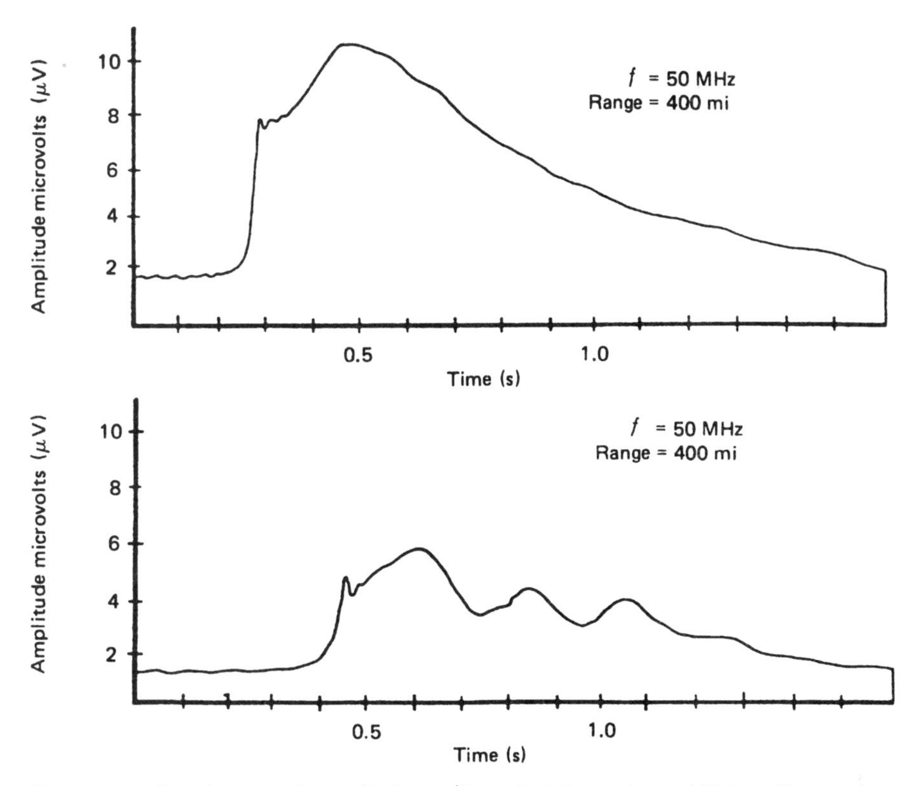

Figure 12.5. Overdense meteor reflections. (From Ref. 2; courtesy of Meteor Communications Corporation. Reprinted with permission.)

great that signal penetration is impossible and reflection occurs rather than reradiation. Donich (Ref. 2) reports that there are no distinctive patterns for overdense trails except that they may reach a higher amplitude of signal level and last longer. With long-lasting trails we can expect fading of received signals. Some of the fading may be attributed to destructive interference of reflections off different parts of the trail and the breakup of a trail during late periods because of ionospheric winds. The period of fade is over several hundred milliseconds, permitting a MBC system to operate between nulls. Figure 12.5 shows plots of received power level versus time for overdense trails.

12.3 TYPICAL METEOR BURST TERMINALS AND THEIR OPERATION

A meteor burst terminal consists of a transmitter, receiver, and modem/processor. Such a terminal is shown in Figure 12.6. For half-duplex

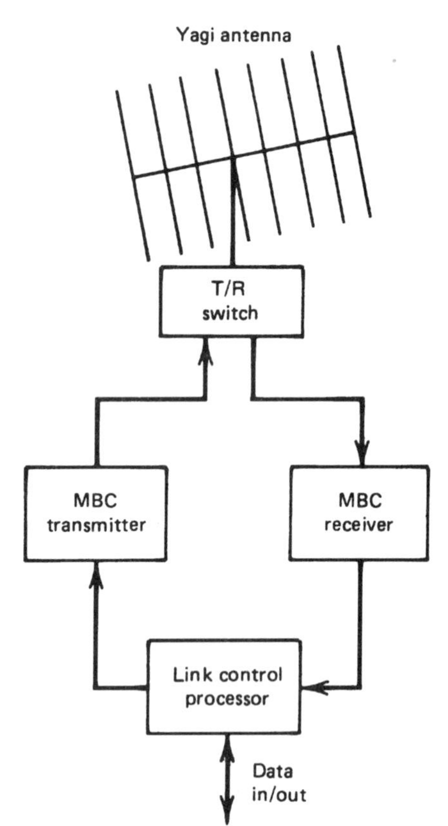

Figure 12.6. A typical meteor burst terminal (half-duplex).

operation, one antenna will suffice. However, a fast-operating T/R (transmit/receive) switch is required. In a moment, we will appreciate why the switch should be *fast*.

The most efficient modulation is phase-shift keying, either binary (BPSK) or quaternary (QPSK). Fixed-frequency (assigned-frequency) facilities commonly use Yagi antennas optimized for the frequencies of interest. If the transmit and receive frequency are separated by more than 3 or 4 MHz, a second, optimized Yagi may be desirable, because, as described in Chapter 14, a Yagi is a narrowband antenna. The transmitter and receiver in Figure 12.6 are conventional. The receiver noise figure may be in the range of 2–3 dB. In nearly all situations, external noise is dominant. In the quiet rural condition, we would expect galactic noise to predominate; in other locations, man-made noise.

The link control processor contains the software and firmware necessary to make a MBC link work. It detects the presence of the probe from the distant end, identifies that it is its companion probe, and then releases the message. Several methods of MBC link operation are described below.

There are a number of ways a meteor burst system can operate. One of the most common is the "master–remote" technique. Here a master station, usually more robust than the remote, sends a continuous probe in the direction of the remote. A probe is a transmitted signal on frequency F_1, usually with identification information. The remote is continuously monitoring frequency F_1, and when it hears the probe, properly authenticated with the identification information, it knows that a common meteor trail exists between the two and transmits a burst of data at frequency F_2.

Another method is quasi-half-duplex, where we transmit and receive on the same frequency. In this case a probe is sent in bursts, with sufficient resting time between bursts to accommodate the propagation time from the distant end. When the remote terminal hears a probe burst, it replies with its message, and the poll bursts from the master station cease.

Of course, we can expand this concept to a polling regime where a master station polls remotes, one at a time, from a large family of remotes. Such an operation is carried out on the SnoTel system mentioned earlier. A single master controls 600 remotes under its jurisdiction. One can carry this concept to full-duplex operation. In this case the transmitter and receiver each have their own antennas separated and isolated as much as possible. Co-site interference can be a major problem with this type of operation.

Another method of operation is broadcast. In this case a master station broadcasts traffic to "silent" remotes. Such broadcasts take advantage of the statistics of a MBC channel. A short message is continuously repeated over a comparatively long period of time. It is expected that the recipient will receive various pieces of the message and will have to reconstruct the message when the last piece is received.

"Message piecing" is a common technique used on MBC systems. If a message has any length at all—let's say a burst rate of 8 kbps and a burst duration of 100 ms or 0.1 s, thus a maximum message length of 800 bits (this includes all overhead)—then message piecing must occur if a message is longer than 800 bits. This is the same concept as packet transmission. Each message piece is a "packet" and these packets must be reassembled at the receive location to reconstruct the originator's complete message.

12.4 SYSTEM DESIGN PARAMETERS

12.4.1 Introduction

MBC link performance is defined by the "waiting time" required to transfer a message with a specified reliability. The principal parameters affecting performance are operating frequency, data burst rate (bps), transmitter power, antenna gain, and receiver sensitivity threshold.

12.4.2 Operating Frequency

Meteor trails will reradiate or reflect very high frequency (VHF) radio signals in the 20–200-MHz frequency range. However, since the reflected signal amplitude is proportional to $1/f^3$ and its time duration to $1/f^2$, the message waiting time increases sharply as frequency is increased. Frequencies in the 20–55-MHz range are most practical for minimum waiting time. The lower limit exists, as we mentioned previously, due to external noise conditions. External noise in this region consists of man-made and galactic. In fact, in many instances, the lower limit of 20 MHz must be increased to 30 or 45 MHz in typical urban noise environments.

12.4.3 Data Rate

Here we refer to the burst rate or burst data rate. A MBC terminal transmits data in high-rate bursts, ideally throughout the duration of the usable portion of the meteor trail event, from 0.2 to 2 s, typically. Data burst rates generally are in the range of 2–16 kbps. The upper limit may be constrained by legal bandwidth considerations dictated by national regulatory authorities such as the Federal Communications Commission (FCC). Each data burst must contain overhead information, and the amount of overhead will restrict useful data throughput. Typical modulation is coherent PSK, either BPSK or QPSK. Frequency-shift keying (FSK) may also be used.

12.4.4 Transmit Power

The higher the transmit power, the shorter the waiting time. Many MBC terminals with moderate performance features operate with radio frequency (RF) power output in the range of 100–200 W. Larger facilities operate at 0.5, 1, 5, or even 10 kW. Naturally, there is a trade-off between performance and economy.

12.4.5 Antenna Gain

As we are aware, antenna gain and beamwidth are inversely proportional. We want our MBC antennas to encompass a large portion of the sky to take advantage of as many meteor trails as possible. As we increase antenna gain, we decrease beamwidth and decrease the "slice of the sky." The trade-off between amount of sky encompassed by an antenna ray beam and antenna gain seems to be in the region of +13 dBi gain. Donich (Ref. 2) gives +16 dBi for short links (400–600 mi) and +21 to +24 dBi for long-range links (600–1000 mi).

12.4.6 Receiver Threshold

Receiver threshold can be defined as the receive signal level (RSL) required to achieve a certain bit error rate (BER). It is a function of the type of modulation used, bandwidth, and the receiver noise, which is usually externally limited. Of course, the lower the receiver threshold, the lower the waiting time. Often a receiver is noise limited by man-made noise. Methodology for calculating threshold using man-made noise as the overriding noise contributor is given in Section 11.9.4. Donich (Ref. 2) shows that a MBC system operating at 40 MHz using coherent BPSK at 2 kbps would have a noise threshold at -121 dBm for a BER of 1×10^{-3}.

12.5 PREDICTION OF MBC LINK PERFORMANCE

12.5.1 Introduction

The starting point in the methodology we present for predicting link performance is the calculation of receiver noise threshold and a threshold for the required BER.

We next provide a basic set of MBC relationships, taken from CCIR Rep. 251-5, used to calculate MBC link transmission loss. Several generalized shortcuts are also presented. Methods of calculating other prediction parameters, such as meteor rate, burst duration time, and waiting time probability, are also described.

12.5.2 Receiver Threshold

We use a similar method to calculate MBC receiver noise threshold as a HF receiver. The receiver is externally noise limited; receiver thermal noise is of secondary importance. In most cases the type of noise will be man-made. Only in a quiet rural environment will we find a receiver galactic noise limited. Therefore we turn to the methodology given in CCIR Rep. 258, which we discuss in Section 11.9.4. Use equation (11.4) modified.

$$P_{\mathrm{n}} = F_{\mathrm{am}} + (D_{\mathrm{u}} + \sigma_{\mathrm{al}}) + 10 \log B_{\mathrm{Hz}} - 204 \text{ dBW} \tag{12.1}$$

Values for D_{u} and σ_{al} are taken from Table 11.7.

Example 1. A MBC receiver operates in a residential environment at 48 MHz and its bandwidth is 8 kHz. Find the noise threshold of the receiver. Neglect transmission line losses. From Table 11.7 we find $F_{\mathrm{am}} = 25.9$ dB, $D_{\mathrm{u}} = 12.3$ dB, and $\sigma_{\mathrm{al}} = 4$ dB. Hence

$$\begin{aligned} P_{\mathrm{n}} &= 25.9 + 12.3 + 4 + 10 \log 8000 - 204 \text{ dBW} \\ &= -148.6 \text{ dBW or } -118 \text{ dBm} \end{aligned}$$

If we were to assume BPSK modulation and a 2-dB modulation implementation loss, we would require an E_b/N_0 of $9 + 2 = 11$ dB for a BER of

1×10^{-4}. (See Figure 4.12). From the example, we can calculate N_0 by subtracting $10 \log 8000$ from the P_n value or -157 dBm. Then the threshold for a BER of 1×10^{-4} must be 11 dB above the noise threshold in 1 Hz of bandwidth or -157 dBm + 11 dB = -146 dBm; or the RSL threshold value for a BER = 1×10^{-4} is -146 dBm + 39 dB or -107 dBm.

12.5.3 Positions of Regions of Optimum Scatter

The scattering of straight meteor ionization trails is strongly aspect sensitive. To be effective, it is necessary for the trails approximately to satisfy a specular reflection condition. This requires the ionized trail to be tangential to a prolate spheroid whose foci are at the transmitter and receiver terminals (Figure 12.7). The fraction of incident meteor trails that are expected to have usable orientations is about 5% in the area of the sky that is most effective (Ref. 5). Figure 12.8 shows the estimated percentages of useful trails for a terminal separation of 1000 km. This figure aptly shows that the optimum scattering regions ("hot spots") are situated about 100 km to either side of the great circle, independent of path length.

This feature, together with the fact that the trails lie mainly in the height range of 85–110 km, serve to establish the two hot-spot regions toward which both antennas should be directed. The two hot spots vary in relative importance according to time of day and path orientation (Ref. 5). Generally, antennas used in practice have beams broad enough to cover both hot spots. Thus the performance is not optimized, but on the other hand the need for beam swinging does not arise.

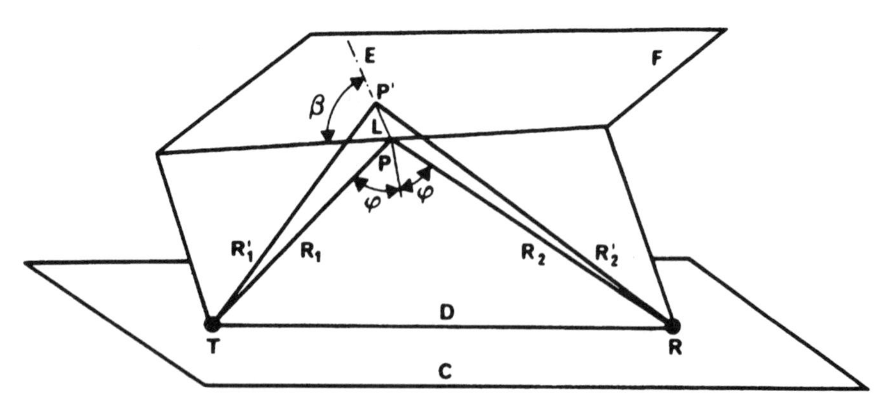

Figure 12.7. Ray geometry for a meteor-burst propagation path. C, earth's surface; D, plane of propagation; E, trail; F, tangent plane; β, angle between the trail axis and the plane of propagation; T, transmitter; R, receiver. For the terms L, P, P′, R_1, R_2, R_1', R_2', and φ, see Section 12.5.4. (From Figure 1, p. 397, CCIR Rep. 251-5, Vol. VI, XVIIth Plenary Assembly, Dusseldorf, 1990; Ref. 4.)

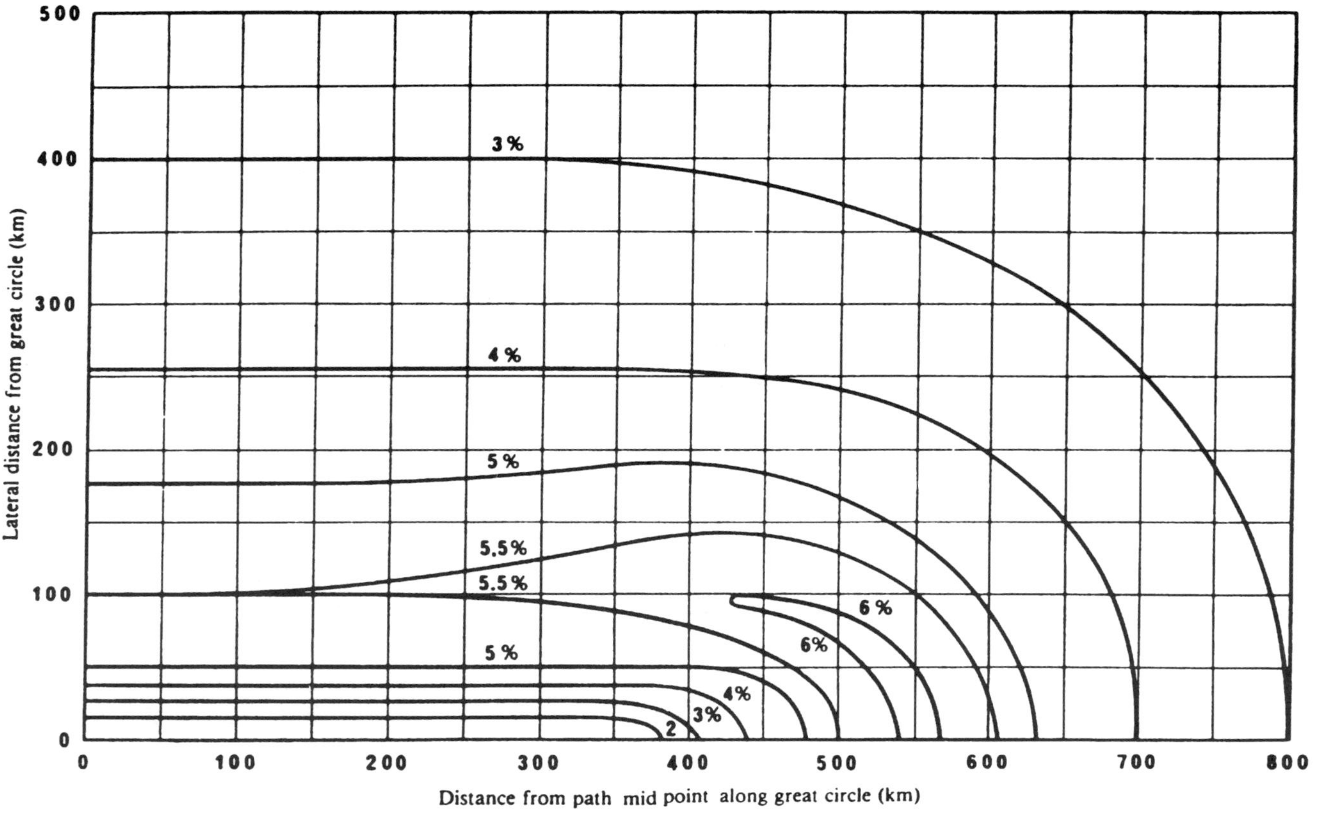

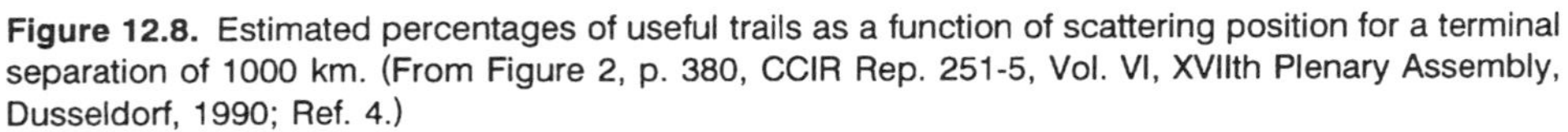
Figure 12.8. Estimated percentages of useful trails as a function of scattering position for a terminal separation of 1000 km. (From Figure 2, p. 380, CCIR Rep. 251-5, Vol. VI, XVIIth Plenary Assembly, Dusseldorf, 1990; Ref. 4.)

12.5.4 Effective Length, Average Height, and Radius of Meteor Trails*

Consider the ray geometry for a meteor burst propagation path as shown in Figure 12.7 between transmitter T and receiver R. P represents the tangent point and P′ a point further along the trail such that $(R_1' + R_2')$ exceeds $(R_1 + R_2)$ by half a wavelength. Thus PP′ (of length L) lies within the principal Fresnel zone and the total length of the trail within this zone is $2L$. Provided R_1 and R_2 are much greater than L, it follows that

$$L = \left[\frac{\lambda R_1 R_2}{(R_1 + R_2)(1 - \sin^2 \varphi \cos^2 \beta)} \right]^{1/2} \tag{12.2}$$

where φ = angle of incidence
β = angle between the trail axis and the plane of propagation
λ = wavelength

In order to evaluate the scattering cross section of the trail it is usual to assume that ambipolar diffusion causes the radial density of electrons to have a Gaussian distribution and that the volume density is reduced while the line density remains constant. These assumptions lead to an equation for the volume density N_v in electrons per cubic meter as a function of radius r and time from the instant of formation t, which is

$$N_v(r,t) = \frac{q}{\pi(4Dt + r_0^2)} \exp\left[\frac{-r^2}{(4Dt + r_0^2)} \right] \tag{12.3}$$

where q = electron line density per meter
D = ambipolar diffusion coefficient in m^2/s
r_0 = initial radius of trail in meters

Both D and r_0 are marked functions of height. From experimental results the following empirical formula for evaluating the average height of trails, which is a function of frequency, can be derived:

$$h = -17 \log f + 124 \tag{12.4}$$

where h = average trail height (km)
f = wave frequency (MHz)

The average trail height is a function of other system parameters in addition to frequency. However, equation (12.4) is a good approximation.

*This section is adapted from CCIR Rep. 251-5 (Ref. 4).

Various empirical relationships have been derived between the initial trail radius and the meteor height. An average expression is

$$\log r_0 = 0.035h - 3.45 \tag{12.5}$$

12.5.5 Ambipolar Diffusion Constant*

A good estimate of the ambipolar diffusion constant is provided by the expression

$$\log D = 0.067h - 5.6 \tag{12.6}$$

Based on the results of Greenhow and Hall the ratio of the ambipolar diffusion constant D to the velocity of the meteor V (required in the evaluation of received power) can be approximated by

$$\frac{D}{V} = \left[0.0015h + 0.035 + 0.0013(h - 90)^2\right]10^{-3} \tag{12.7}$$

where V = velocity of the meteor (m/s).

12.5.6 Received Power

12.5.6.1 Underdense Trails

$$p_R(t) = \frac{p_T g_T g_R \lambda^2 \sigma a_1 a_2(t) a_2(t_0) a_3}{64\pi^3 R_1^2 R_2^2} \tag{12.8}$$

where λ = wavelength (m)
σ = echoing area of the trail (m^2)
a_1 = loss factor due to finite initial trail radius
$a_2(t)$ = loss factor due to trail diffusion
a_3 = loss factor due to ionospheric absorption
t = time measured from the instant of complete formation of the first Fresnel zone(s)
t_0 = half the time taken for the meteor to traverse the first Fresnel zone
p_T = transmitter power (W)
$p_R(t)$ = power available from the receiving antenna (W)
g_T = transmit antenna gain relative to an isotropic antenna in free space
g_R = receive antenna gain relative to an isotropic antenna in free space
R_1, R_2 = see Figure 12.7

(Lossless transmitting and receiving antennas are assumed.)

*Note that Sections 12.5.5, 12.5.6 and subsections 12.5.6.1 and 12.5.6.2 are adapted from CCIR Rep. 251-5 (Ref. 4).

The echoing area σ is given as

$$\sigma = 4\pi r_e^2 q^2 L^2 \sin^2 \alpha \qquad (12.9)$$

where r_e = effective radius of the electron = 2.8×10^{-15} m
α = angle between the incident electric vector at the trail and the direction of the receiver from that point

Since L^2 is directly proportional to λ, the echoing area σ is also proportional to λ and hence the received power for underdense trails varies as λ^3. Horizontal polarization normally is used at both terminals. The $\sin^2 \alpha$ term in equation (12.9) is then nearly unity for trails at the two hot spots.

The loss factor a_1 is given by

$$a_1 = \exp\left(- \frac{8\pi^2 r_0^2}{\lambda^2 \sec^2 \varphi} \right) \qquad (12.10)$$

It represents losses arising from interference between the reradiation from the electrons wherever the thickness of the trail at formation is comparable with the wavelength.

The factor $a_2(t)$ allows for the increase in radius of the trail by ambipolar diffusion. It may be expressed as

$$a_2(t) = \exp\left(- \frac{32\pi^2 Dt}{\lambda^2 \sec^2 \varphi} \right) \qquad (12.11)$$

for angle φ; see Figure 12.7.

The increase in radius due to ambipolar diffusion can be appreciable even for as short a period as is required for the formation of the trail. The overall effect with regard to the reflected power is equal to that which would arise if the whole trail within the first Fresnel zone had expanded to the same extent as at its midpoint. Since this portion of trail is of length $2L$ the midpoint radius is that arising after a time lapse of L/V seconds. Calling the time lapse t_0 gives us the following for trails near the path midpoint ($R_1 \approx R_2 \approx R$):

- For trails at right angles to the plane of propagation ($\beta = 90°$),

$$t_0 \approx \left(\frac{\lambda R}{2V^2} \right)^{1/2} \qquad (12.12)$$

- For trails in the plane of propagation ($\beta = 0$),

$$t_0 \approx \left(\frac{\lambda R}{2} \right)^{1/2} \times \frac{\sec \varphi}{V} \qquad (12.13)$$

Substituting t_0 from equation (12.12) into equation (12.11) gives for the $\beta = 90°$ case

$$a_2(t_0) = \exp\left[-\frac{32\pi^2}{\lambda^{3/2}}\left(\frac{D}{V}\right)\left(\frac{R}{2}\right)^{1/2}\frac{1}{\sec^2\varphi}\right] \tag{12.14}$$

For $\beta = 0°$ the exponent in this expression is $\sec^2\varphi$ times greater.

The only time-dependent term is $a_2(t)$ and it gives the decay time of the reflected signal power. Defining a time constant T_{un} for the received power to decay by a factor e^2 (i.e., 8.7 dB) leads to

$$T_{un} = \frac{\lambda^2 \sec^2\varphi}{16\pi^2 D} \tag{12.15}$$

With reflection at grazing incidence, $\sec^2\varphi$ will be large and hence so is the echo-time constant. The echo-time constant is also increased by the use of lower frequencies.

12.5.6.2 *Overdense Trails.* The formula for the received power in the case of overdense meteor trails is usually based on the assumption of reflection from a metallic cylinder whose surface coincides with the region for which the dielectric constant is zero. The effect of refraction in the underdense portion of the trail is usually ignored. As in the underdense case, the received power varies as λ^3 and again the echo duration varies as λ^2. However, the maximum received power is now proportional to $q^{1/2}$, in contrast to q^2 for underdense trails. Thus the increase in received signal power with ionization intensity is more modest.

12.5.6.3 *Typical Values of Basic Transmission Loss.* Since any practical meteor burst communication system will rely mainly on underdense trails, the overdense formulas are of less importance. Satisfactory performance estimates can be made using formulas for the underdense case with assumed values of q in the range of 10^{13}–10^{14} electrons per meter according to the prevailing system parameters.

Basic transmission loss curves derived from equation (12.8) with $q = 10^{14}$ electrons per meter are given in Figure 12.9. As the angle β can take any value between 0° and 90° only these two extreme cases are shown. The advantage of lower propagation loss at the lower frequencies is clearly seen. Average meteor heights given from equation (12.4) have been used in deriving the curves. It should be noted that the prediction of system performance depends critically on the heights assumed.

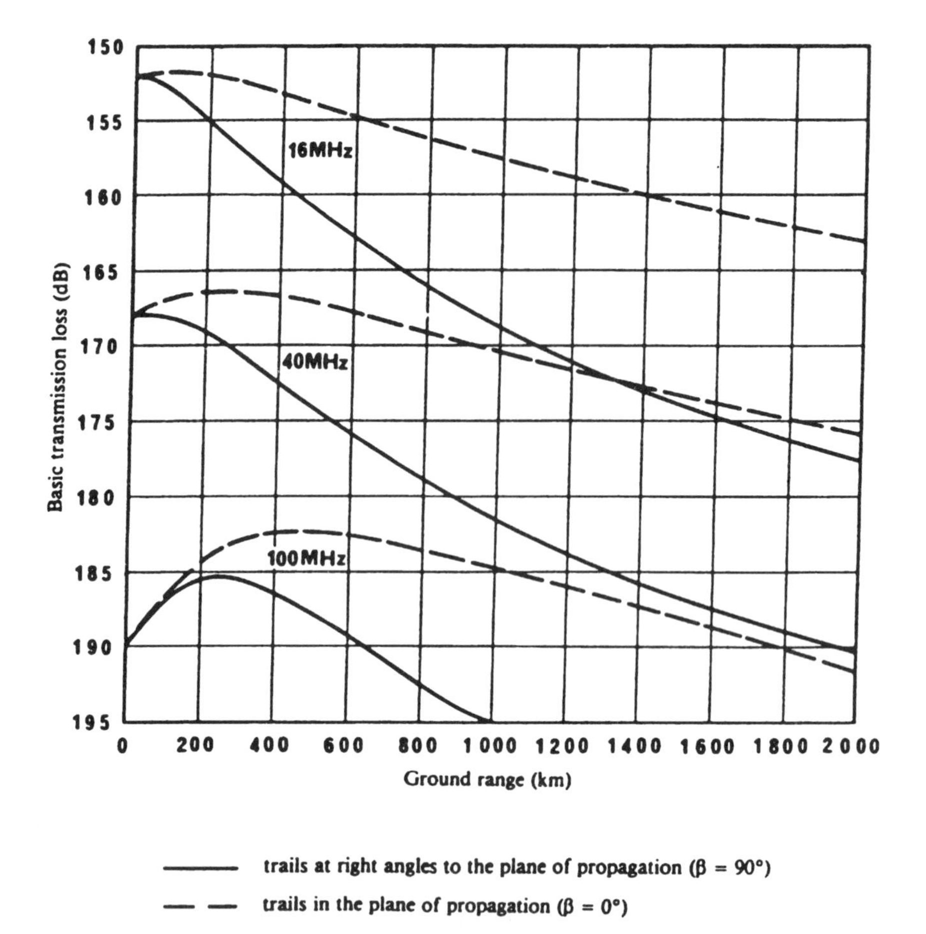

Figure 12.9. Basic transmission loss for underdense trails derived from equation (12.8) with electron density 1×10^{14} electrons / m and horizontal polarization. (From Figure 3, p. 384, CCIR Rep. 251-5, Vol. VI, XVIIth Plenary Assembly, Dusseldorf, 1990; Ref. 4.)

12.5.7 Meteor Rate

The meteor rate or meteor burst per unit time (M_c) is related to system parameters through the following expression (Ref. 6):

$$M_c = \left(P_T \times G_T \times G_R / F_c^3 \times T_R\right)^{1/2} \tag{12.16}$$

where P_T = transmitter power (W)
T_R = receiver threshold (W)
G_T = transmit antenna gain relative to an isotropic
F_c = carrier frequency (MHz)
G_R = receiver antenna gain relative to an isotropic

From this relationship and a known meteor rate M_T of a test system, operating at a known frequency and a known power level, an expression can be derived to calculate the meteor rate M_c at different values and parameters.

We now let $P = P_T G_T G_R / T_R$, where P is a power ratio. Now the meteor rate ratio of the desired system to the test system becomes

$$\frac{M_c}{M_T} = \left(\frac{P_c}{P_T}\right)^{1/2} \left(\frac{f_T}{f_c}\right)^{3/2} \tag{12.17}$$

This method advises that the higher the system power factor (PF), where PF = $10 \log P$, and/or the lower the frequency, the higher the observed meteor rate. When the power factor is expressed in terms of decibels,

$$M_c/M_T = 10^{(\mathrm{PF}_c - \mathrm{PF}_T)} + 10^{1.5 \log(f_T/f_c)} \tag{12.18}$$

where PF_c = power factor of the unknown system (dB)
PF_T = power factor of the test system (dB)
f_c = operating frequency of the unknown system
f_T = operating frequency of the test system
M_c = meteor rate of the desired system
M_T = meteor rate of the test system (see Table 12.2)

Values of M obtained from a test system operated for over 1 yr are given in Table 12.2. The test system PF was 180 dB and its operating frequency f_T was 47 MHz. The values given in Table 12.2 show the diurnal and seasonal variations. Plots of M_c/M_T versus the power factor of the desired system are given in Figure 12.10 for various operating frequencies. Thus, if the operating frequency and the power factor are known, the meteor rate M_c can be obtained. Conversely, if a desired M_c and operating frequency are given, the required power factor can be defined.

12.5.8 Burst Time Duration

Signals from underdense trails rise to an initial peak value in a few hundred microseconds, then decay exponentially in amplitude. Decay times from a few milliseconds to a few seconds are typical. This time variation must be taken into account in order to predict message waiting times; accordingly, the next step is to determine the average burst decay times (time above a threshold) for the specific MBC system in question. Eshleman's (Ref. 7) analytic expression for this varying signal strength is given as

$$V(t) = V_D e^{-(t/T_D)} \tag{12.19}$$

TABLE 12.2 Test System Data[a]

	Meteor Bursts[b] per Hour	
	Daily Average[c]	Daily Minimum
January	50	19
February	50	19
March	50	19
April	50	19
May	50	19
June	55	20
July	65	22
August	70	25
September	70	25
October	70	25
November	65	25
December	70	25
Yearly	60	23

[a]System power factor $(PF)_T$ = 180 dB; frequency f_T = 47 MHz.
[b]A burst is defined as the recognition of the coded synchronization signal from the master station.
[c]The daily average is defined by the 12 hours per day the average will be exceeded and the 12 hours per day the performance will be less than average.

Source: Reference 6. Courtesy of Meteor Communications Corporation. Reprinted with permission.

where V_D is the peak value of the signal strength and T_D is the average time constant in seconds.

T_D is related to both frequency and range between the two MBC stations, as given by

$$T_D = \lambda^2 \sec^2 \phi / 32^2 D \tag{12.20}$$

where D = diffusion coefficient, which is $8M^2/s$
$\phi = \frac{1}{2}$ the forward scattering angle as a function of range
λ = the wavelength

The value of D given above was obtained from test data given in Ref. 6. It should be noted that D exhibits a diurnal variation, with a maximum in the afternoon and a minimum in the morning. The value of $8M^2/s$ is a daily average and is the value used for all T_D calculations in this section. Plots of T_D versus range with frequency as a parameter are given in Figure 12.11. Thus once a specific operating frequency is selected and an operating range established, the average burst time can be defined (Ref. 6).

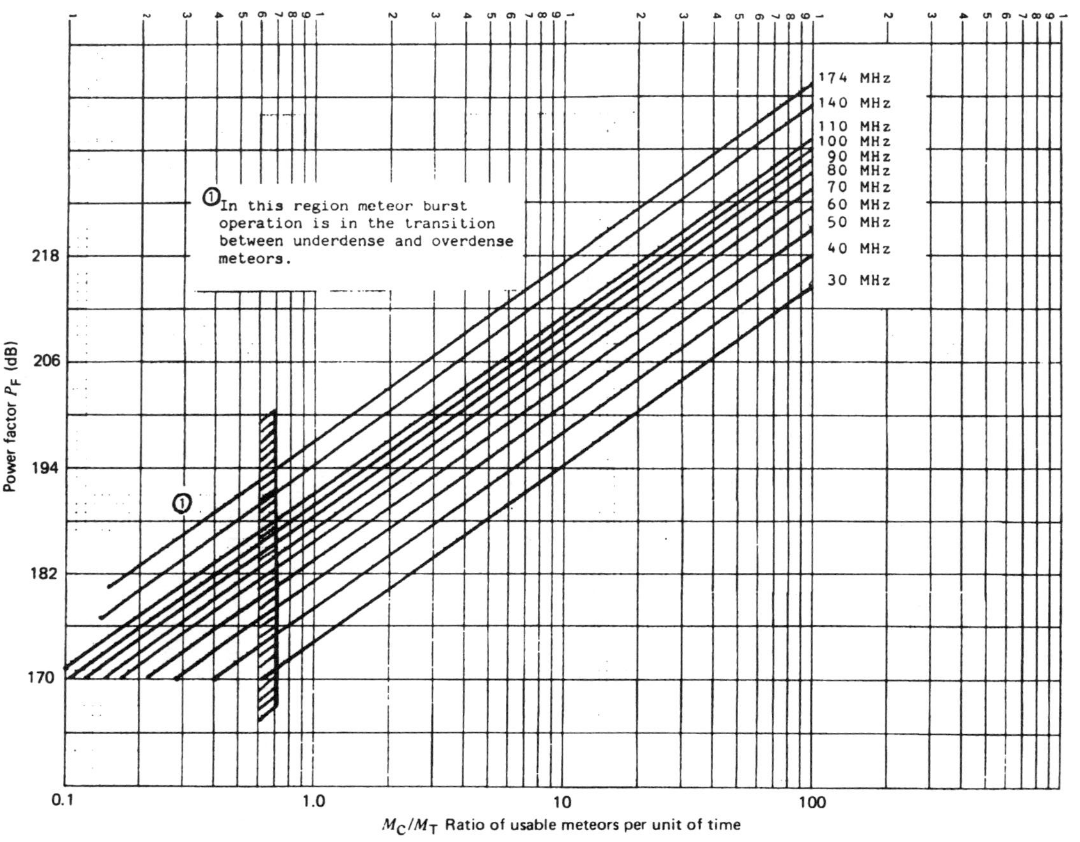

Figure 12.10. Power factor versus meteor rate. M_T is established from empirical data; P_F for test data = 180 dB; frequency for test data = 47 MHz. In the region marked meteor burst, operation is in transition between underdense and overdense meteors. (From Ref 6; Courtesy of Meteor Communications Corporation. Reprinted with permission.)

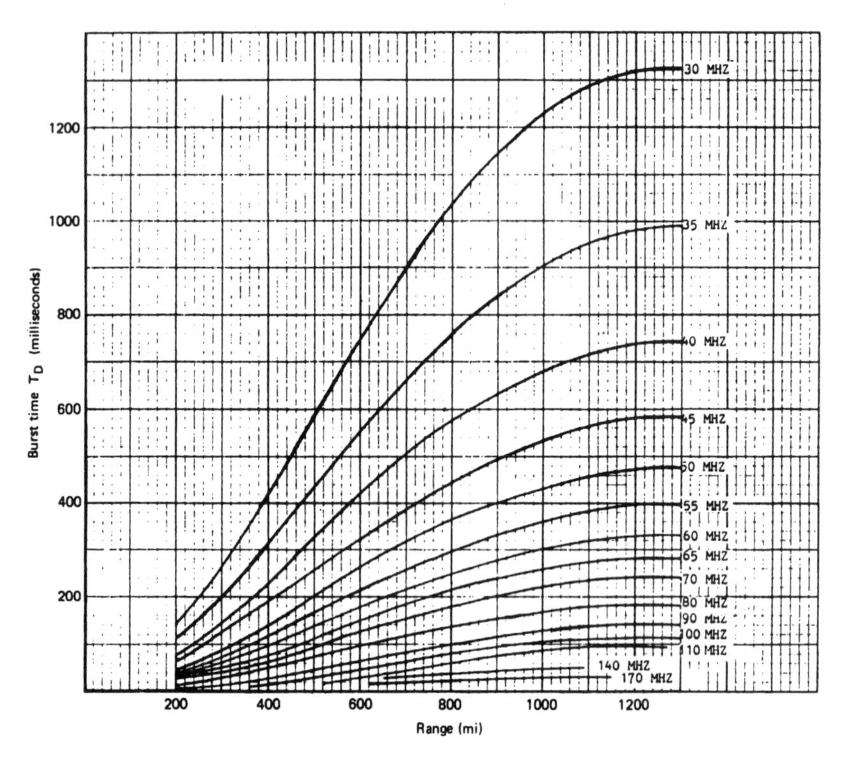

Figure 12.11. Burst time constant. (From Ref. 6; Courtesy of Meteor Communications Corporation. Reprinted with permission.)

12.5.9 Burst Rate Correction Factor

The number of bursts per hour obtained from Figure 12.10 will require a correction factor where we can calculate the number of bursts that are sufficiently long to transfer a complete message. Of course, we will have to stipulate a certain message length. The distribution of burst durations is an exponential function and can be expressed as

$$M = M_1 e^{-(t/T_D)} \tag{12.21}$$

where M = number of meteors exceeding the specified threshold for t seconds
M_1 = total number of meteors exceeding the specified threshold
T_D = burst time constant
t = time to transfer the complete message

A normalized plot of M/M_1 is shown in Figure 12.12. Therefore M/M_1 becomes a scaling factor for M_c derived from Figure 12.10 as a function of

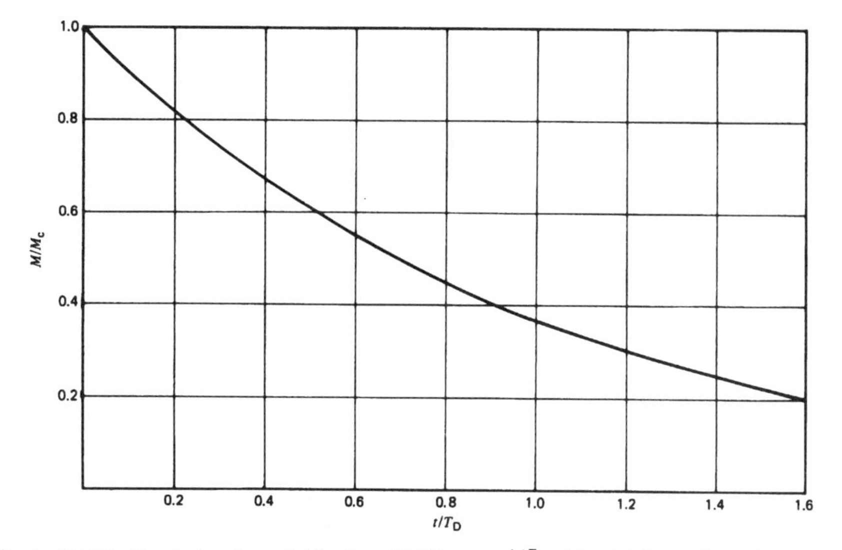

Figure 12.12. Burst duration distribution $M/M_c = e^{-t/T_D}$. M = total number of bursts/h exceeding a specified threshold for t seconds; M_c = total number of bursts/h exceeding a specified threshold (the threshold is defined as the recognition of the synchronization code from the transmitting or probing station); T_D = burst time constant; t = time (s). (From Ref. 6; Courtesy of Meteor Communications Corporation. Reprinted with permission.)

message transaction time. The value of M_c is reduced to remove bursts that have insufficient time duration to complete a message transmission by setting the value of t in equation (12.21).

12.5.10 Waiting Time Probability

Underdense meteor burst occurrences are random in nature and follow a Poisson distribution as a function of time. The fundamental Poisson equation is

$$P = 1 - e^{-Mt} \tag{12.22}$$

where P is the probability of a meteor occurrence in time t. M is the meteor density or number of bursts per hour and t is the time in hours. Equation (12.22) provides the probability relationship to derive message waiting time. If time t is given in minutes, the equation (12.22) becomes

$$P = 1 - e^{-Mt/60} \tag{12.23}$$

In the preceding paragraphs, the meteor burst communication performance prediction approach resulted in a value M. Using the Poisson distribution given above, a family of curves can be generated for a broad set of values for M. Figures 12.13–12.15 show these relationships, with primary interest, of course, where the probability is greater than 0.9 (Ref. 6).

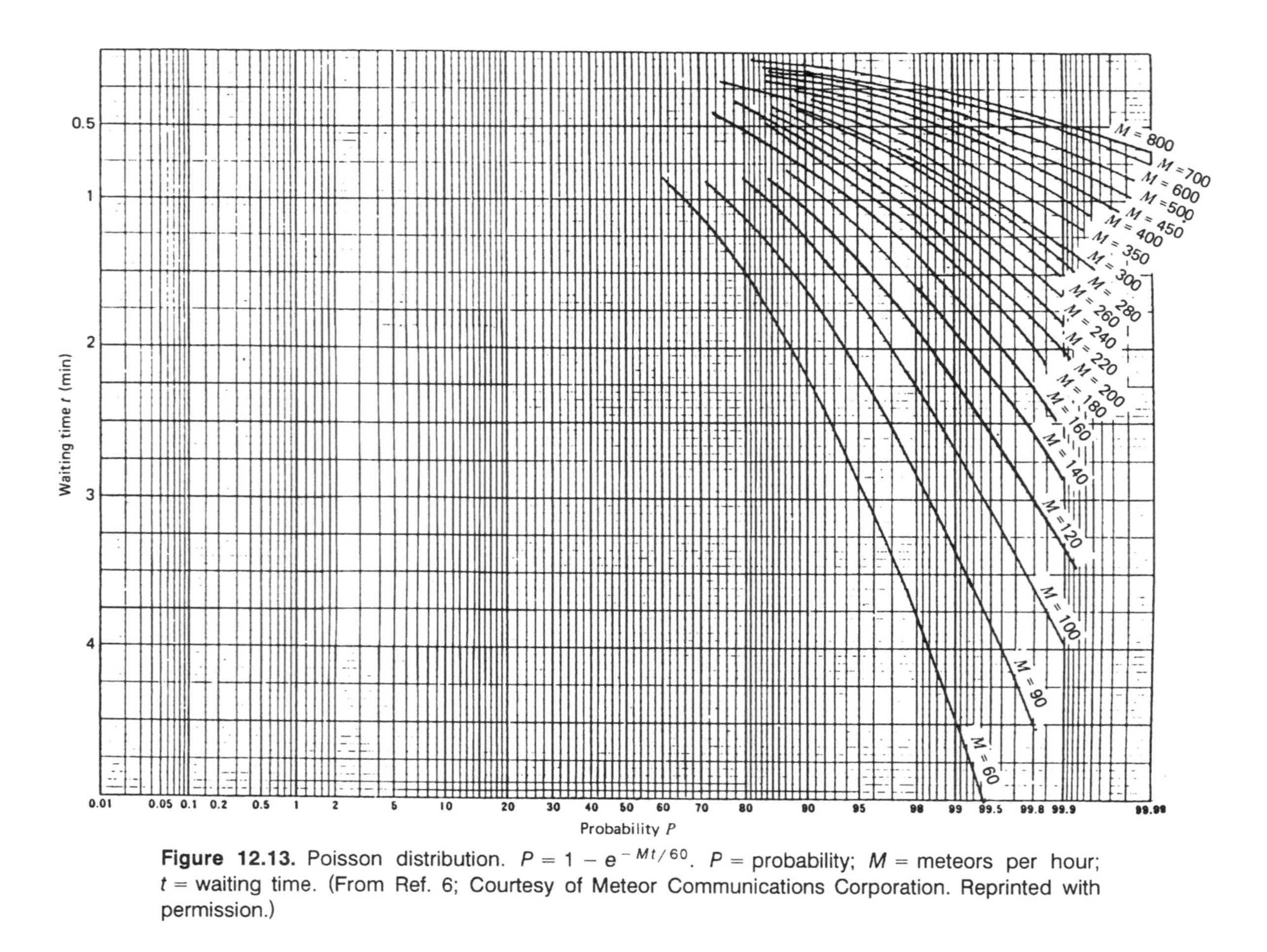

Figure 12.13. Poisson distribution. $P = 1 - e^{-Mt/60}$. P = probability; M = meteors per hour; t = waiting time. (From Ref. 6; Courtesy of Meteor Communications Corporation. Reprinted with permission.)

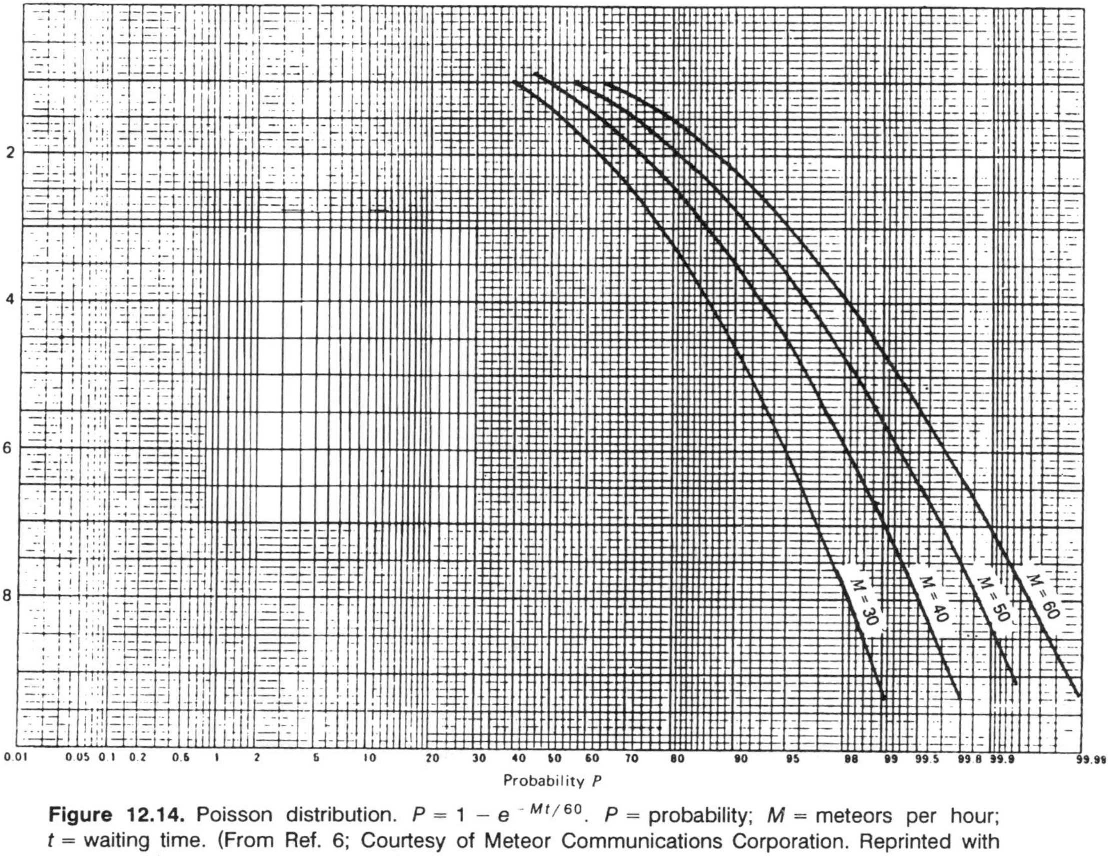

Figure 12.14. Poisson distribution. $P = 1 - e^{-Mt/60}$. P = probability; M = meteors per hour; t = waiting time. (From Ref. 6; Courtesy of Meteor Communications Corporation. Reprinted with permission.)

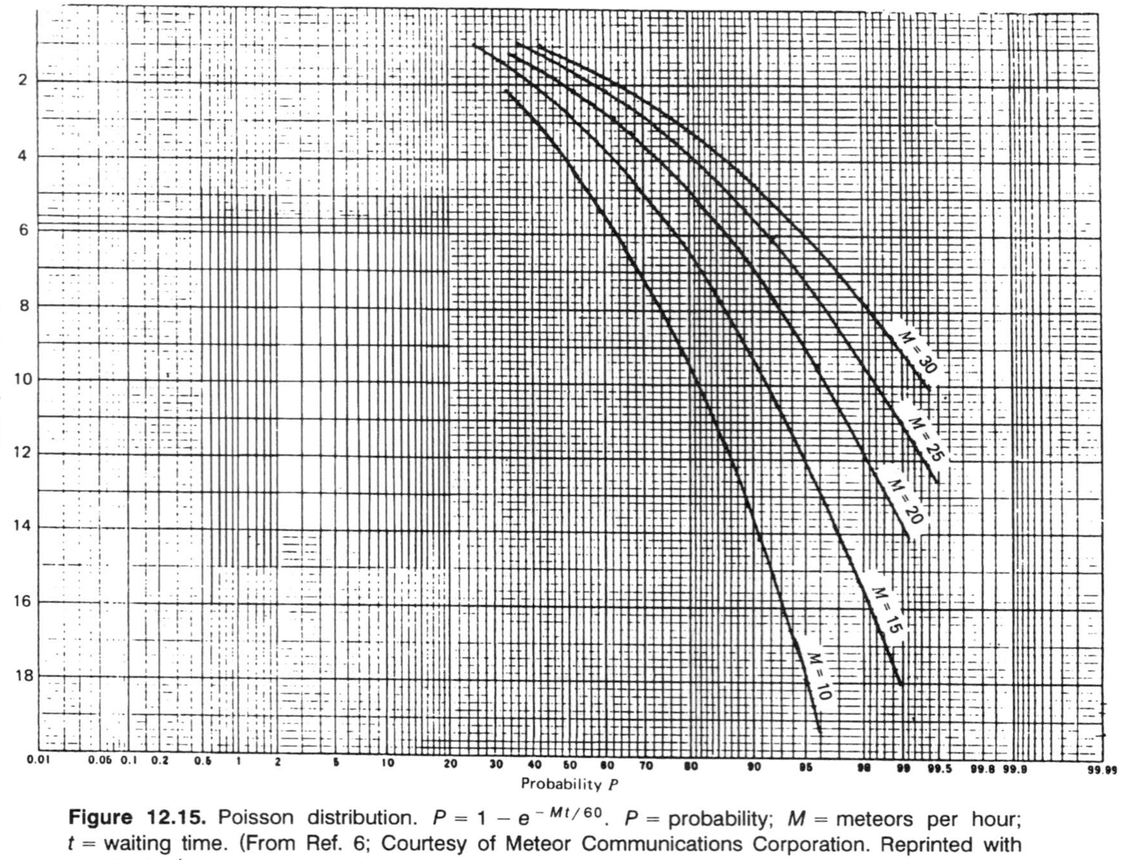

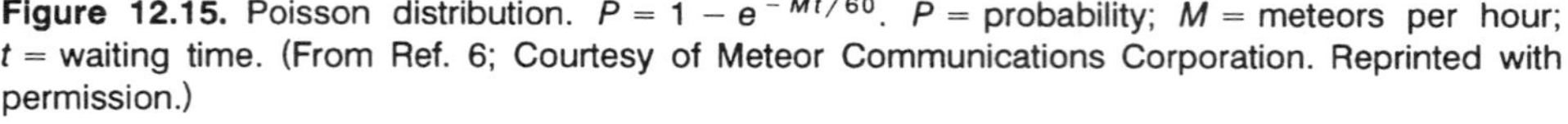

Figure 12.15. Poisson distribution. $P = 1 - e^{-Mt/60}$. P = probability; M = meteors per hour; t = waiting time. (From Ref. 6; Courtesy of Meteor Communications Corporation. Reprinted with permission.)

12.6 DESIGN / PERFORMANCE PREDICTION PROCEDURE

Donich (Ref. 6) offers the following step-by-step procedure to calculate the performance of a meteor burst link operating in a single-burst mode. One can either predict the performance given the power factor parameters (Section 12.5.7) or, in reverse, specify a desired performance level and work to the desired power factor to obtain that performance. If the power factor parameters are defined, the procedure below may be followed to predict performance within those given parameters:

1. Calculate the minimum signal levels for a BER of 1×10^{-3} by the methodology shown in Section 12.5.2 given the data rate, modulation type, receiver noise figure, and ambient external noise.
2. Determine the average M for the defined power factor parameters, including line losses, related to long-term test data. Use Section 12.5.7, Table 12.2, and Figure 12.10.
3. Calculate the average burst time constant, given the operating frequency and range. Use Section 12.5.8 and Figure 12.11.
4. Determine the message transfer time required to complete the message reception, including propagation time and message overhead requirements.
5. Remove from the value of M, obtained from step 2, the number of bursts of insufficient duration, using the message transfer time, obtained from step 4, and correct M. Use Figure 12.12.
6. Use the value of M, obtained from step 5, and determine the waiting times for required reliabilities. Use Figures 12.13–12.15. Reference 2 reports that these derived waiting times are for the average time of the year and time of day. For the worst case, multiply by 3.

12.7 NOTES ON MBC TRANSMISSION LOSS

Some insight is given in Ref. 8 regarding transmission loss at 40 MHz. Table 12.3 shows link distance; the two components of MBC transmission loss, MBC scatter loss, and free-space loss; and the total transmission loss. We see that the values range around Donich's 180-dB reference model (180 dB) (Ref. 6) and the 40-MHz values of Figure 12.9.

How will a typical link operate with such loss values when a simple link budget technique is applied? We use the receiver model described in Section 12.5.2, where the RSL threshold is -107 dBm, and there is an 8-kHz bandwidth, BPSK modulation, and a BER of 1×10^{-4}. Antenna gains at each end are 13 dB and the transmitter output power is 1 kW ($+60$ dBm). Line losses are neglected.

TABLE 12.3 MBC Transmission Loss at 40 MHz[a]

Distance (km)	MBC Scatter Loss (dB)	Free-Space Loss (dB)	Total Loss (dB)
300	52	114.03	166
500	53	118.46	171.5
1000	57	124.49	181.5
1500	58.5	128.01	186.5
2000	61.5	130.5	192

[a]The free-space loss column uses great circle distance. For links under 300 km, R_1 and R_2 should be used (the distance up and down from the reflection height). The MBC scatter loss is taken from Ref. 8.

EIRP	+73 dBm
MBC transmission loss	−180 dB
Receiver antenna gain	+13 dB
RSL	−94 dBm
Threshold	−107 dBm
Margin	13 dB

If we assume the parameters given, once the return from the trail falls below −107 dBm, the link performance is unacceptable. We also see that with this system the maximum transmission loss is 180 + 13 dB, or 193 dB.

Forward error correction (FEC) coding as well as its attendant coding gain has been offered as one means to extend a trail's useful life. FEC requires symbol redundancy and thus more bandwidth. Of course, the more we open up a receiver's bandwidth, the more thermal noise, degrading operation. Therefore coding gain derived from FEC (in decibels) must be greater than the required noise bandwidth increase (in decibels) due to redundancy. (See Ref. 9.)

Ince (Ref. 8) also aptly points out that MBC link maximum range is often more constrained by antenna gain degradation due to low takeoff angles than MBC transmission loss per se. For a 1200-km link, the takeoff angle is 5°, but for a 2000-km link the optimum takeoff angle is less than 1°. If we wish to implement such long links, we must use very elevated antennas to avoid ground reflections that cause the gain degradation. The optimum range of MBC systems is really 400–800 km.

MBC links will often experience scatter from the E-layer and such scatter is very sporadic. With E scatter, continuous connectivity can last from minutes to hours.

Another phenomenon that MBC links can experience is multipath effects (fading/dispersion) during late trail conditions. This is likely to occur on more intense trails, such as some returns from overdense trails. Such trails still give useful returns even when solar winds start to break up the trail. Multipath derives from returns from different trail segments after breakup, occurring, of course, toward the end of a trail's useful life. Coding with appropriate interleaving is one method of mitigating these multipath effects.

D-layer absorption often is neglected in link budget analyses of meteor burst links. We discuss D-layer absorption and its calculation in Chapter 11. Another source is CCIR Rep. 252-2 (Ref. 10). D-layer absorption, a daytime-only phenomenon, may exceed 3 dB at 40 MHz at midlatitudes during comparatively high sunspot number periods. It is about half this value at 60 MHz.

Yet another path loss is due to Faraday rotation. The D region and the earth's magnetic field cause a linearly polarized very high frequency (VHF) wave to be rotated both before and after meteor trail reflection. These rotations result in an overall end-to-end loss due to polarization mismatching between an incident wave and a linearly polarized receiving antenna. An excellent paper dealing with Faraday rotation effects on meteor burst communication links was published by Cannon in 1985 (Ref. 11).

As in the case of D-layer absorption, polarization rotation loss is also affected by path length (secant of the takeoff angle), the sun's zenith angle, and the sunspot number. Like D-layer absorption, Faraday rotation losses also disappear at night. Faraday rotation losses decrease rapidly as frequency increases. Cannon (Ref. 11) shows Faraday rotation losses varying from 1 dB to over 15 dB at 40 MHz (very path dependent) and dropping to a maximum of 1.6 dB at 60 MHz on the same paths.

12.8 MBC CIRCUIT OPTIMIZATION

Obviously, to help optimize MBC link operation, we want to take as much advantage of trail duration as possible. Among the details we should observe is turnaround time. Here we mean the time from receiving a valid probe to the time transmission begins. At the probe transmitter there may be a short period of receiver desensitization after transmitting. This can be minimized in the receiver design and by having a separate transmit and receive antenna. T/R (transmit/receive) switches must be fast operating. Message overhead must be as short as possible.

Adaptive trail operation has also been suggested (Refs. 12, 13, 17, and 19). Underdense meteor trails are idealized with an exponential decay. One method measures the signal-to-noise ratio (S/N) of the probe to provide a measure of trail intensity. Burst rate is adjusted for the initial intensity. More intense trails can support higher burst rates. Another system, operating in the full-duplex mode, makes periodic measures of probe intensity (which operates throughout the message transmission), periodically adjusting the burst rate accordingly. Some of the economy of MBC systems is being given up for better throughput performance. However, some argue that the performance improvement is marginal.

12.9 METEOR BURST NETWORKS

Since MBC links are limited to 1000 miles per hop, longer range communications have been achieved using several master stations "chained" together to form a network. Message piecing software has been expanded to provide routing and relay functions. Message piecing is similar to the packet communications function described in Ref. 20, Chapter 11. For example, seven MBC master stations have been chained together to provide a network stretching from Tampa, Florida (U.S.A.), to Anchorage, Alaska (USAF NORAD-SAC network—Ref. 18).

12.10 PRIVACY AND THE METEOR BURST FOOTPRINT

Meteor burst communications offer a certain element of privacy due to the footprint of a particular connectivity and the geometry of its related meteor trail. MBC footprints are comparatively small when compared to HF and

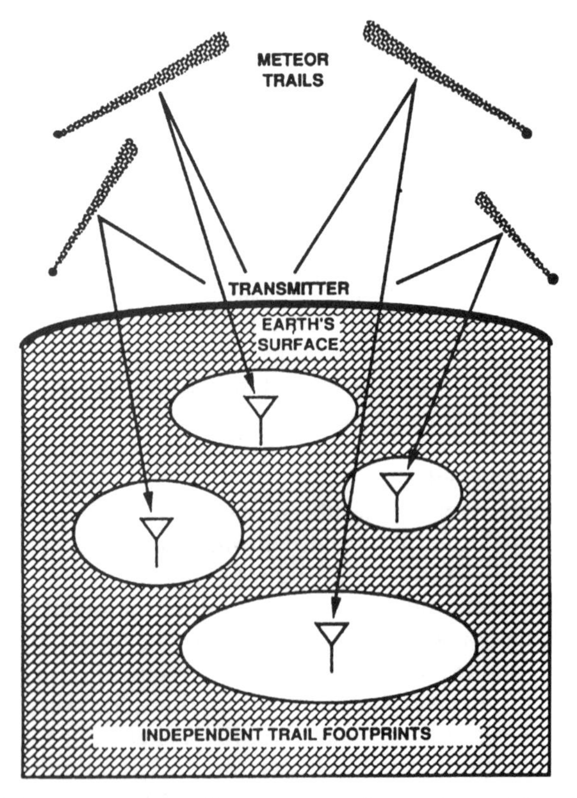

Figure 12.16. Unique footprints of MBC receivers each associated with a particular meteor trail. (From Ref. 19; reprinted with permission.)

satellite communication footprints. An artist's rendition of a footprint model is shown in Figure 12.16. Depending on many factors, particularly the geometry of a usable trail between MBC stations A and B, the size of the footprint varies. It usually is elongated with some 25–50 km in the short dimension and up to 200 km in the long direction. Stations outside the footprint cannot receive signals on that particular connectivity. Reference 19 defines a footprint as the geographic region illuminated by sufficient scattered signal energy (from a particular ionized meteor trail) to exceed the threshold RSL of any receiver in that region.

PROBLEMS AND EXERCISES

1 Describe the basis of meteor burst communication (MBC) link operation.

2 What kind of average data throughput in bits per second can we expect from a MBC link?

3 "Waiting time" describes exactly what?

4 Describe the operating radio-frequency relationship with MBC path loss. Why, then, do we go higher in frequency on most operational links when returns are more intense the lower we go in frequency?

5 For fixed-frequency operation, what is the most commonly used antenna? What is one type of antenna we might use to cover a broad band of frequencies for MBC link operation? Remember polarization.

6 What is the principal advantage of selecting MBC over other means of communication? Name at least two major disadvantages.

7 Identify the two categories of meteor trails. How are they described (i.e., in what units)?

8 Give the seasonal and daily variation in performance of a MBC link regarding time of day and months of the year. How do we explain these variations?

9 What is the range of useful time duration of an underdense meteor trail?

10 What transmission impairment can we expect from trails with comparatively long time durations?

11 What type of modulation is commonly used on MBC links?

12 Describe one of the most commonly used techniques of MBC link operation, typically from a remote sensor. (*Hint*: Consider how we know the presence of a trail and then transmit traffic.)

13 MBC links often carry very short, often "canned" messages that can be accommodated by one common trail. How are longer messages handled?

14 In urban and suburban environments, what type of external noise is predominant? In quiet, rural environments?

15 What are the two major constraints on MBC bandwidth?

16 What is the ideal gain of a MBC link antenna? Why not more gain? Include the concept of "hot spots" in the discussion.

17 Calculate the BER threshold of a MBC receiver with a 10-kHz bandwidth, BER $= 1 \times 10^{-3}$, and BPSK modulation, operating in a quiet, rural environment. Use a modulation implementation loss of 1 dB.

18 What is the range of altitude of effective meteor burst trails?

19 What is the "two hot spot" theory? How can we practically accommodate both?

20 For a MBC link operating at 40 MHz, determine the average height of useful meteor burst trails.

21 Meteor rate is a function of what parameters? It is directly proportional to three parameters and inversely proportional to two. Give all five parameters.

22 Meteor rate arrival can be defined by what (mathematical) type of distribution?

23 One simple way of viewing MBC link transmission loss is that it is made up of two loss components. What are they? Roughly what range of decibel loss values would we expect?

24 Discuss the use of forward error correction (FEC) to extend the useful life of a meteor trail. Offer some trade-offs.

25 What is a cause of multipath on a MBC link? When is multipath most likely to occur?

26 A MBC link can take advantage of other radio transmission phenomena. Name three (only one is covered in the text).

27 Name at least two additional link losses we can expect.

28 Give at least four ways to increase MBC link throughput.

29 Why would we wish to raise antennas either on towers or high ground (or both) for long MBC links?

30 What complications are seen in the design of a full-duplex MBC link besides economic ones? What advantages?

31 How does Faraday rotation loss vary with frequency? Time of day?

REFERENCES

1. D. W. Brown and W. P. Williams, "The Performance of Meteor-Burst Communications at Different Frequencies," *Aspects of Electromagnetic Wave Scattering in Radio Communications*, AGARD Conf. Proceedings 244, 24-1 to 24-26, Brussels, 1978.
2. Thomas G. Donich, *Theoretical and Design Aspects for a Meteor Burst Communications System*, Meteor Communications Corp., Kent, WA, 1986.
3. D. W. R. McKinley, *Meteor Science and Engineering*, McGraw-Hill, New York, 1961.
4. *Communication by Meteor-Burst Propagation*, CCIR Rep. 251-5, Vol. VI, XVIIth Plenary Assembly, Dusseldorf, 1990.
5. V. R. Eshleman and L. L. Manning, "Meteors in the Ionosphere," *Proceedings of the IRE*, Vol. 49, Feb. 1959.
6. Thomas G. Donich, *MCBS Design/Performance Prediction Method*, Meteor Communications Corp., Kent, WA, 1986.
7. V. R. Eshleman, *Meteors and Radio Propagation*, Stanford University Rept. No. 44, Contract N60NR-25132, Feb. 1955.
8. E. Nejat Ince, "Communications Through EM-wave Scattering," *IEEE Communications Magazine*, May 1982.
9. Scott L. Miller and Laurence B. Milstein, "Performance of a Coded Meteor Burst System," IEEE MILCOM'89, Boston, MA, Oct. 1989.
10. *CCIR Interim Report for Estimating Sky-wave Field Strength and Transmission Loss at Frequencies Between the Approximate Limits of* 2 *and* 30 *MHz*, CCIR Rep. 252-2, 1970 (out of print).
11. P. S. Cannon, *Polarization Rotation in Meteor Burst Communication Systems*, Royal Aircraft Establishment Tech. Report 85082 (TR85082), London, Sept. 1985.
12. W. B. Birkemeier et al., *Feasibility of High Speed Communications on the Meteor Scatter Channel*, University of Wisconsin, Madison, 1983.
13. *Efficient Communications Using the Meteor Burst Channel*, STS Telecom, Port Washington, NY, 1987 (NSF ISI-8660079).
14. G. R. Sugar, "Radio Propagation by Reflection from Meteor Trails," *Proceedings of the IRE*, Feb. 1964.
15. Michael R. Owen, "VHF Meteor Scatter—An Astronomical Perspective," *QST*, June 1986.
16. David W. Brown, "A Physical Meteor-Burst Propagation Model and Some Significant Results for Communication System Design," *IEEE Journal on Selected Areas in Communications*, Vol. SAC-3, No. 5, Sept. 1985.
17. Dale K. Smith and Thomas G. Dovich, *Variable Data Rate Applications in Meteor Burst Communications*, Meteor Communications Corp., Kent, WA, 1989.
18. Dale K. Smith and Richard J. Fulthorp, *Transport, Network and Link Layer Considerations in Medium and Large Meteor Burst Communications Networks*, Meteor Communications Corp., Kent, WA, 1989.
19. Donald L. Schilling, ed., *Meteor Burst Communications*, Wiley, New York, 1993.
20. Roger L. Freeman, *Telecommunication System Engineering*, 3rd ed., Wiley, NY, 1996.

13

INTERFERENCE ISSUES IN RADIO COMMUNICATIONS

13.1 RATIONALE

Interference is an onerous problem that must be dealt with. It is a fact of life and we *can* do something about it. There are interference issues with all radio systems. In essence, our concern is *electromagnetic compatibility* (EMC).

We turn to the IEEE for help from their standard dictionary (Ref. 1) to define EMC. We think the IEEE definition number 2 is appropriate:

> [EMC is] the ability of a device, equipment or system to function satisfactorily in its electromagnetic environment without introducing intolerable electromagnetic disturbances to anything in that environment.

EMC breaks down into two subsets: (1) emanation and (2) susceptibility. Nearly all electronic devices "emanate." In other words, they produce measurable RF energy, what we sometimes call electromagnetic interference (EMI) or radio-frequency interference. The path of this RF energy can be conducted or radiated. How much RF energy is produced and the harm it can do to that same device or other nearby devices are vital to good system/equipment design. The PC (personal computer) is a notorious emanator.

Electronic equipment is "susceptible" to RF energy, whether radiated or conducted. Susceptibility is defined in ANSI C63.14-1992 (Ref. 2) as "the inability of a device, piece of equipment, or system to perform without degradation in the presence of an electromagnetic disturbance." Of course, emanation at sufficient level can be called an electromagnetic disturbance.

Our concern here is much more narrow and defined. Our interests, the subjects of this text, are a wide variety of microwave communications, HF, cellular/PCS, and meteor burst communication systems.

A radio transmitter is a good example of a device where there is "wanted" radiation. There will also be some unwanted radiation. The scope of this unwanted radiation is the signal radiation from the antenna, either out the main lobe or the sidelobes. Of course, to the companion distant-end receiver, the main lobe radiation is "wanted."

We further reduce the scope because we only will consider unwanted radiation entering the receiving antenna, causing harmful interference in that facility. The system design engineer must be aware that unwanted radiation can enter practically anywhere—through power lines, as a result of a wave impinging on the equipment itself, through a poor ground, any or all signal leads, and so forth. This analysis, though, will only treat unwanted RF energy entrance through the antenna subsystem.

As a result, there are several areas where we might reduce interference to an acceptable level: (1) we can reduce the level of the unwanted energy radiated right at the source and (2) we can make a victim device less susceptible. An excellent example of both (1) and (2) is to reduce antenna sidelobes. Another is the use of filters to reduce harmonics and spurious radiation of an emitted wave at both the transmit and receive side of a radiolink. We can also reduce the EIRP to that necessary to just meet performance requirements. The trouble with this latter approach is that by reducing EIRP, we will reduce the link fade margin. However, we could use space diversity or larger antennas to compensate for the lost margin.

Our emphasis in this chapter will be interference issues of LOS microwave, satellite communication terminals, and PCS facilities. We will also deal with the cellular environment where frequency reuse and adjacent cell operations generate rich interference fields.

Before moving to our main theme, we will briefly discuss spurious response interference windows at a receiver.

13.2 SPURIOUS RESPONSE INTERFERENCE WINDOWS AT A RECEIVER

Typical receiver response windows in descending order of importance are:

- The image frequency, which is $F_0 \pm 2 \times \text{IF}$ (whether plus or minus depends on the position of the injection oscillator frequency, above or below the receive carrier)
- $F_0 \pm \frac{1}{2}$ IF (where F_0 is the receive carrier frequency)
- $F_0 \pm$ IF (where F_0 is the receive carrier frequency)

When interference enters one of these windows, we can expect an increase in baseband noise in FDM-FM systems, tonal interference in a video receiver, and threshold degradation in a digital receiver. In the case of a local (intrastation) interfering transmitter, we can expect interference of much greater magnitudes possibly making that system completely inoperable.

Paying attention to these response windows is extremely important in the frequency planning stage of a radiolink. If FCC or ITU-R frequency plans are followed, there should be no concern regarding these response windows because such plans take the response windows into account. It is when nonstandard frequency plans are used that we must concern ourselves with the windows as defined above.

13.3 TYPICAL INTERFERENCE CONTROL FOR LINE-OF-SIGHT MICROWAVE AND SATELLITE COMMUNICATION FACILITIES

13.3.1 Introduction

It was pointed out in Section 6.3.5 (Chapter 6) that the assigned satellite communication frequency bands are shared with other services, most often with wideband terrestrial line-of-sight microwave. Because of this sharing between (among) services, there is the possibility that facilities in one service will interfere with the other service. Thus we must consider two situations:

1. Terrestrial service interferes with the satellite service.
2. Satellite service interferes with terrestrial service.

There are also subsets of these, such as a satellite earth station interferes with a LOS microwave facility, and a LOS microwave facility interferes with a satellite earth station. Another set deals with the space segment, where it can interfere with LOS microwave, and even that LOS microwave or other, similar services can interfere with satellite receivers (in the space segment).

There is also the possibility that we can interfere with ourselves, or other nearby facilities can interfere with us and we with them.

We examine these situations from the point of view of two reference sources: ITU-R Organization for the generic situation and the Federal Communications Commission (FCC) for the United States. See FCC Rules and Regulations Part 25, paragraphs 25.251 through 25.256 (Ref. 3).

When considering situations of potential or actual interference, we will talk about the *offending facility* and the *offended facility*.* In the case of the offended facility, in nearly every instance we must reference the interference to the receiver of that offended facility. With the offending facility, in the great majority of cases, the problem starts with a transmitter; its signal is passed up the waveguide and then radiated by the antenna. The offending signal radiates not only through the main lobe of the antenna but also out sidelobes, and even directly out the back of the antenna. There is little or no polarization discrimination protection for radiation out antenna sidelobes.

*Often we will call the "offended facility" the *victim receiver*.

At the offended facility, in most cases, the offending signal will enter the antenna sidelobes, or possibly even through the main lobe. We are the first to admit that there are other routes of entry for offending signals. We emphasize here the importance of good facility grounding,* which will help reduce the effects of those other routes of interference. In general, good grounds provide no help at all for unwanted signal entry through the antenna system.

13.3.2 Conceptual Approach to Interference Determination

13.3.2.1 Introduction. Let's assume a new earth station is to be installed. We know its location, its antenna elevation angle and azimuth, and its antenna radiation patterns. The first step in interference determination is to carry out a survey of nearby LOS microwave facilities and other earth station facilities including VSATs that are within line-of-sight† of the new earth station and that are operating in the same frequency band.

These facilities have been licensed by a national regulatory agency and the characteristics of each emitter are known. For a candidate offending emitter, we need to know its location, height above sea level, the EIRP of its main beam, and its antenna sidelobe characteristics. If it is an earth station, we need to know its antenna azimuth and elevation angle when in operation. Of course, in each case we must also ascertain the modulation type, operating frequency bandwidth, and emission characteristics.

Let's assume, in this case, that the offender is a LOS microwave facility and the offended facility is a nearby earth station. The method we suggest here to calculate interference is quite straightforward. Once the reader grasps the concept, the more formal methods proposed by the FCC and ITU-R Organization become easier to understand. The methodology is nearly identical to the path analyses (link budgets) described in Chapters 2, 3, 4, and 6.

From the offending LOS microwave emitter we calculate an EIRP based on its appropriate antenna sidelobe gain. This is the sidelobe facing into the nearby earth station site. As the reader must imagine, we need to have the antenna radiation characteristics for 360° around that antenna. It is quite probable that the offending antenna sidelobe will hit the offended (earth station) antenna also on a sidelobe. Thus we need to know those radiation characteristics of the offended (or victim) antenna (i.e., its sidelobe pattern). The earth station sidelobe in question presents a gain (or loss) to its receiving system. We are now faced with the conventional path analysis problem. Some texts tell us that if the offending signal RSL is below the thermal noise threshold of the offended receiver, the resulting interference can be neglected. This may well not be true as we will demonstrate below.

* For information on grounding, bonding, and shielding, consult *Reference Manual for Telecommunication Engineering*, latest edition, Chapter 28 (Ref. 4).

† The FCC in rule 25.251(2) states that all facilities within 100 km of each other require coordination.

13.3.2.2 Antenna Radiation Patterns. There are three basic artifices to protect against harmful interference assuming same frequency operation:

- Free-space loss; the range (distance) separating the two facilities in question
- Beyond line-of-sight; separating the two facilities such that one is beyond line-of-sight of the other
- Antenna discrimination

*13.3.2.2.1 Antenna Discrimination—Determination of Antenna Gain Around a Parabolic Dish Antenna.** The relationship $\varphi(\alpha)$ may be used to derive a function for the horizon antenna gain, G (dB) as a function of the azimuth α, by using the actual earth station antenna pattern (or other parabolic antenna pattern), or a formula giving a good approximation. For example, in cases where the ratio between the antenna diameter and the wavelength is not less than 100, the following equation should be used:

$$G(\varphi) = G_{\max} - 2.5 \times 10^{-3}\left(\frac{D}{\lambda}\varphi\right)^2 \quad \text{for } 0 < \varphi < \varphi_m \qquad (13.1a)$$

$$G(\varphi) = G_1 \quad \text{for } \varphi_m \leq \varphi < \varphi_r \qquad (13.1b)$$

$$G(\varphi) = 32 - 25 \log \varphi \quad \text{for } \varphi_r \leq \varphi < 48^\circ \qquad (13.1c)$$

$$G(\varphi) = -10 \quad \text{for } 48^\circ \leq \varphi \leq 180^\circ \qquad (13.1d)$$

where D is the antenna diameter and λ is wavelength, expressed in the same unit.

Gain of the first sidelobe is calculated as follows:

$$G_1 = 2 + 15 \log \frac{D}{\lambda} \qquad (13.1e)$$

$$\varphi_m = \frac{20\lambda}{D}\sqrt{G_{\max} - G_1} \quad \text{(degrees)} \qquad (13.1f)$$

$$\varphi_r = 15.85\left(\frac{D}{\lambda}\right)^{-0.6} \quad \text{(degrees)} \qquad (13.1g)$$

When it is not possible, for antennas with $D/\lambda < 100$, to use the above reference antenna pattern and when neither measured data nor a relevant CCIR Recommendation accepted by the administrations concerned can be

*From *Radio Regulations*, AP-28 (Ref. 5).

used instead, administrations may use the reference diagram as described below:

$$G(\varphi) = G_{max} - 2.5 \times 10^{-3}\left(\frac{D}{\lambda}\varphi\right)^2 \qquad \text{for } 0 < \varphi < \varphi_m \qquad (13.2a)$$

$$G(\varphi) = G_1 \qquad \text{for } \varphi_m \leq \varphi < 100\frac{\lambda}{D} \qquad (13.2b)$$

$$G(\varphi) = 52 - 10\log\frac{D}{\lambda} - 25\log\varphi \qquad \text{for } 100\frac{\lambda}{D} \leq \varphi < 48° \qquad (13.2c)$$

$$G(\varphi) = 10 - 10\log\frac{D}{\lambda} \qquad \text{for } 48° \leq \varphi \leq 180° \qquad (13.2d)$$

where D is the antenna diameter and λ is wavelength, expressed in the same unit.

Gain of the first sidelobe is then calculated as follows:

$$G_1 = 2 + 15\log\frac{D}{\lambda} \qquad (13.2e)$$

$$\varphi_m = \frac{20\lambda}{D}\sqrt{G_{max} - G_1} \quad \text{(degrees)}$$

The above patterns may be modified as appropriate to achieve a better representation of the actual antenna pattern.

In cases where D/λ is not given, it may be estimated from the expression $20\log(D/\lambda) \approx G_{max} - 7.7$, where G_{max} is the main lobe antenna gain in decibels.

13.3.2.3 *Interference Evaluation.* There are two schools of thought on interference evaluation and each uses a different method to evaluate interference. One school will use the ratio I/N (interference-to-noise ratio). This may be better expressed as I_0/N_0, where the ratio is normalized to 1 Hz of bandwidth. The other school will use the ratio C/I, or carrier-to-interference ratio. This may also be expressed as C_0/I_0. We will be using the former I_0/N_0 in some cases, and C/I in other cases.

13.3.2.3.1 Examples

MODEL EARTH STATION

Transmits at 6 GHz, receives at 4 GHz (exactly)

Antenna diameter: 3 m D/λ at 6 GHz = 196.8; gain 44 dB
D/λ at 4 GHz = 131.2; gain 40 dB

Noise figure: 1 dB. $N_0 = -203/\text{Hz}$ dBW [equation (3.2)]
Transmitter output power: 100 W (+20 dBW)
Transmission line loss: 1 dB for both uplink and downlink
Modulation on downlink: QPSK occupying 7.44 MHz. Coding $R = \frac{3}{4}$.
$E_b/N_0 = 10$ dB producing a BER of 3×10^{-6} (see Figure 4.12)

MODEL LOS MICROWAVE FACILITY

Transmit and/or receive on 4 or 6 GHz (exactly)
Antenna diameter: 1 m D/λ at 6 GHz = 65.6; gain 33 dB
1 m D/λ at 4 GHz = 43.7; gain 30 dB
Noise figure: 8 dB; $N_0 = -196$ dBW
Transmitter output power: 1 W (0 dBW)
Transmission line loss: 3 dB at both frequencies
Required E_b/N_0 for 64-QAM BER = 1×10^{-7}: 20 dB (see Figure 3.12)
Bit rate = 90 Mbps
RF output is spread uniformly over 20 MHz
Fade margin: 38 dB

The facilities are separated by 4 miles (6.4 km).

We will be working in spectral signal and spectral noise densities. In other words, signal level (e.g., RSL) will be expressed in 1 Hz of bandwidth; noise level N_0 is also defined in 1 Hz of bandwidth. Interference level will be expressed in terms of I_0.

EXAMPLE ANALYSES OF SATELLITE TERMINAL RECEIVER NOISE FLOOR. Suppose that by some means an interfering signal entered the terminal at the same level as the thermal noise floor. Then $I_0 = -203$ dBW/Hz and $N_0 = -203/\text{Hz}$ dBW. They combine coherently, raising the noise floor 3 dB, or the noise floor is now -200 dBW/Hz. E_b/N_0, as a result, degrades 3 dB to 7 dB. The resulting BER = 1×10^{-3}. Figure 4.12 (Chapter 4) tells us that without interference, the BER was about 3×10^{-6} and with the interference, the BER is about 1×10^{-3}. The combining of the interference (signal) and noise (signal) is shown diagramatically as follows:

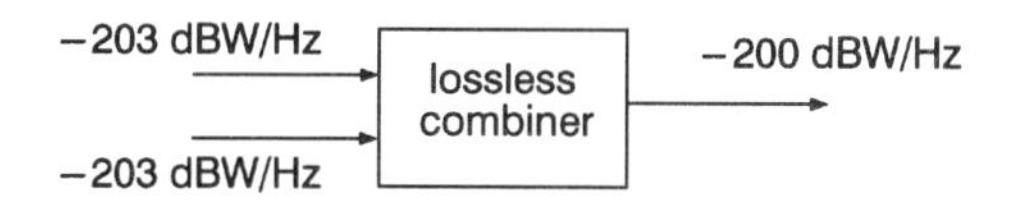

Suppose the interference level was -204 dBW/Hz (or $I_0/N_0 = -1$ dB). In a similar fashion as above, combine (coherently) the interference with the receiver thermal spectral noise level as shown below. The combined value is -200.46 dBW/Hz, degrading E_b/N_0 by 2.54 dB or the resulting E_b/N_0 is 7.46 dB. Applying this to the QPSK curve in Figure 4.12, the BER is about 4×10^{-4}.

Try now an interference level of -211 dBW/Hz (or $I_0/N_0 = -8.0$ dB), which is combined with our -203 dBW/Hz thermal spectral noise level. The combined spectral density value is -202.36 dBW/Hz. This degrades the E_b/N_0 value by 0.64 dB, or the E_b/N_0 value is 9.36 dB or a BER of 2×10^{-5}.

An interference spectral density level of -216 dBW/Hz ($I_0/N_0 = -13$ dB) produces an E_b/N_0 degradation of 0.20 dB in E_b/N_0 or an $E_b/N_0 = 9.80$ dB. This results in a BER of about 4×10^{-6}. This shows that I_0/N_0 of at least -18 dB is required for a minimal impact on BER. Yet there is still some effect due to the waterfall effect of the curves in Figure 4.12.

We now argue that an E_b/N_0 of just 10 dB is unrealistic for a LOS microwave system. But the offended system is a satellite terminal. A realistic value can be taken from INTELSAT transmission parameters for a typical satellite terminal, Table 7.6, where for the 8.448-Mbps bit stream at 4 GHz we expect a C/N_0 value of 80.3 dB. This produces an E_b/N_0 value of 80.3 dB $-$ $10 \log(8.448 \times 10^6)$ = 80.3 dB $-$ 69.26 dB = 11.04 dB. (There is a FEC rate $\frac{3}{4}$ on this signal.) The final summary: there is little margin on satellite signals and usually very large margin (e.g., 20–40 dB) on LOS microwave links to counter fading. Thus satellite receiving systems are much more vulnerable to interference than LOS microwave systems.

Example Analysis 1. A potential LOS microwave offender is 4 miles away from an earth station. The model for each is based on our model information above. The LOS microwave signal is launched at 50° off its main beam ($\delta = 50°$) and the satellite terminal receives that offending signal at 60° off-axis angle ($\delta = 60°$). For all off-axis angles greater than 48°, the antenna gain is -10 dBi [equation (13.1d)]. We calculate the EIRP in power spectral density values or that power level in 1 Hz of bandwidth. The EIRP out the backlobe is then 0 dBW $-$ 3 dB $-$ 10 dBi = -13 dBW. The power spectral density value is then -13 dBW $-$ $10 \log(20 \times 10^6)$ = -86 dBW/Hz. [The -10-dBi antenna gain value is taken from equation (13.1d).]

The EIRP out the sidelobe is then -86 dBW/Hz. The free-space loss (FSL) is

$$\begin{aligned} \text{FSL} &= 36.58 + 20 \log 4000 + 20 \log 4 \\ &= 120.66 \text{ dB} \\ \text{IRL} &= -86 \text{ dBW/Hz} - 120.66 \text{ dB} \\ &= -206.66 \text{ dBW/Hz} \end{aligned}$$

The earth station antenna gain at this sidelobe is -10 dBi [equation (13.1d)]. There is a 1.5-dB line loss; thus the $RSL_I = -218.16$ dBW/Hz.

Let's assume that the earth station had a 1-dB noise figure for its receiving system. Thus $N_0 = -204$ dBW + 1 dB, and $N_0 = -203$ dBW. Then $I_0/N_0 = -218.16 - (-203$ dBW) or $I_0/N_0 = -15.16$ dB. From the paragraphs above, this is an interference situation that will degrade the BER of the offended earth station somewhat. Separating the facilities an additional half-mile would bring us into the desired $I_0/N_0 = 18$ dB.

Example Analysis 2. In this case the earth station is the offender transmitting at 6 GHz. We use the same values of δ as before. We calculate I_0/N_0 at the microwave receiver using the models provided above.

The EIRP out the satellite antenna sidelobe is +20 dBW − 1 dB − 10 dBi = +9 dBW. The spectral density of this signal is +9 dBW − $10 \log(7.44 \times 10^6) = -59.7$ dBW/Hz.

$$\text{FSL} = 36.58 + 20 \log 6000 + 20 \log 4$$
$$= 124.18 \text{ dB}$$

The IRL at the LOS microwave antenna is −59.7 dBW/Hz − 124.18 dB = −183.88 dBW/Hz.

$$\text{RSL}_I = -183.88 \text{ dBW/Hz} - 3 \text{ dB} - 10 \text{ dBi} = -196.88 \text{ dBW}$$

The receiver $N_0 = -196$ dBW. Sum the two. The resulting noise floor is now −193.42 dBW/Hz. This results in a fade margin reduced by 2.58 dB impacting the link's time availability, not its unfaded BER requirement.

If we assume Rayleigh fading, where the original fade margin produced a 99.99% time availability, it is now reduced to about 99.975% due to the interference.

The reader is aware that we chose some worst-case circumstances allowing both facilities to be using 6 and 4 GHz (exactly). One can also see that if we allowed the offended facility to be receiving its interference from the offender's first sidelobe, and the offended facility to receive on its first sidelobe, all other parameters remaining the same, the interference level would be completely intolerable when dealing with the earth station. When the offended station is the LOS microwave facility, the interference level would also be unacceptable. Here we would use equation (13.1e).

13.3.3 Applicable FCC Rule for Minimum Antenna Radiation Suppression

Consult Table 13.1. In the table category A antennas have the most stringent requirements, and category A should be employed under all situations except in areas where there is no frequency congestion. We assume that the antenna

TABLE 13.1 FCC Minimum Radiation Suppression Values

Frequency (MHz)	Category	Maximum Beam-width to 3-dB Points (included angle in degrees)	Minimum Antenna Gain (dBi)	Minimum Radiation Suppression to Angle in Degrees from Centerline of Main Beam in Decibels						
				5°–10°	10°–15°	15°–20°	20°–30°	30°–100°	100°–140°	140°–180°
932.5–935	A	14.0	N/A	—	6	11	14	17	20	24
941.5–944	B	20.0	N/A	—	—	6	10	13	15	20
2,500–4,200	A	N/A	36	23	29	33	36	42	55	55
	B	N/A	36	20	24	28	32	32	32	32
5,925–6,425 [a]	A	N/A	38	25	29	33	36	42	55	55
	B	N/A	38	21	25	29	32	35	39	45
5,925–6,425 [b]	A	N/A	38	25	29	33	36	42	55	55
	B	N/A	38	20	24	28	32	35	36	36
6,525–6,875 [a]	A	N/A	38	25	29	33	36	42	55	55
	B	N/A	38	21	25	29	32	35	35	39
6,525–6,875 [b]	A	1.5	N/A	26	29	32	34	38	41	49
	B	2.0	N/A	21	25	29	32	35	39	45
10,550–10,680 [a,c]	A	N/A	38	25	29	33	36	42	55	55
	B	N/A	38	20	24	28	32	35	35	39
10,550–10,680 [a]	A	3.4	34	20	24	28	32	35	55	55
	B	3.4	34	20	24	28	32	35	35	39
10,565–10,615 [d]	N/A	360	N/A	N/A	N/A	N/A	N/A	N/A	N/A	N/A
10,630–10,680 [d]	N/A	N/A	34	20	24	28	32	35	36	36
10,700–11,700 [a]	A	N/A	38	25	29	33	36	42	55	55
	B	N/A	38	20	24	28	32	35	36	36
17,700–18,820	A	N/A	38	25	29	33	36	42	55	55
	B	N/A	38	20	24	28	32	35	36	36
18,920–19,700 [e]	A	N/A	38	25	29	33	36	42	55	55
	B	N/A	38	20	24	28	32	35	36	36
21,200–23,600	A	N/A	38	25	29	33	36	42	55	55
	B	N/A	38	20	24	28	32	35	36	36
31,000–31,300 [f,g]	N/A	4.0	38	N/A	N/A	N/A	N/A	N/A	N/A	N/A
Above 31,300	A	N/A	38	25	29	33	36	42	55	55
	B	N/A	38	20	24	28	32	35	36	36

[a] These antenna standards apply to all point-to-point stations authorized after June 1, 1997. Existing licensees and pending applicants on that date are grandfathered and need not comply with these standards.

[b] These antenna standards apply to all point-to-point stations authorized on or before June 1, 1997.

[c] Except for such antennas between 140° and 180° authorized or pending on January 1, 1989, in the band 10,550–10,565 MHz for which minimum radiation to suppression to angle (in degrees) from centerline of main beam is 36 dB.

[d] These antenna standards apply only to Digital Termination User Stations licensed, in operation, or applied for prior to July 15, 1993.

[e] Digital Termination User Station antennas in this band shall meet performance Standard B and have a minimum antenna gain of 34 dBi. The maximum beamwidth requirement does not apply to DTS User Stations. Digital Termination Nodal Stations need not comply with these standards.

[f] The minimum front-to-back ratio shall be 38 dB.

[g] Mobile, except aeronautical mobile, stations need not comply with these standards.

Source: FCC Rules and Regulations, part 21.108 (Ref. 3).

suppression values mean that the main axis gain is reduced by the amount shown in decibels. For example, the LOS microwave antenna shown in the model has a gain of 33 dBi at 6 GHz. Thus at 50° off the main beam we would expect the gain to be 42 dB down or 33 dBi − 42 dB = −9 dBi, where we used −10 dBi from the Radio Regulations (Ref. 5).

13.3.4 Coordination Contours

A coordination contour is a distance. One way we can mitigate the effects of interference is by free-space loss between offending and offended (victim) facilities, as we mentioned earlier. Of course, FSL is a function of distance. Inside this distance, unacceptable interference occurs; outside the distance, interference is negligible.

Coordination for terrestrial microwave systems should use a circular coordination contour with a larger radius sector extending 5° either side of the antenna main beam azimuth. These radii are referred to as the circular coordination distance (D_c) and the keyhole coordination distance (D_k). Again, D_k is ±5° about the antenna main beam. D_c is the remaining 350° circle arc about the antenna. For distribution of Prior Coordination Notices, the following circular coordination contours are to be used (from Ref. 6):

Below 15 GHz: D_c = 200 km and D_k = 400 km.
Above 15 GHz: D_c = 75 km and D_k = 140 km.

[See FCC Rules and Regulations part 21.100(d) and 94.15(b).]

The basic concept of "keyhole" coordination distance is shown in the following drawing:

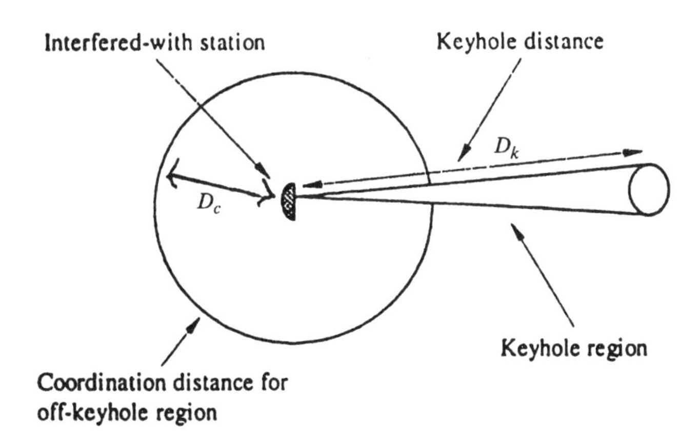

13.3.4.1 Calculating Coordination Distance.* The coordination distance can be obtained by solving the relation between the power of the interfering signal received by the victim station and the distance from the interfering station:

$$I = P_T + [G_R - D_R(\theta)] - L(d) + [G_T - D_T(\theta')] \qquad (13.3)$$

where I = power at distance d originating from the interfering station (dBm)
P_T = maximum transmitting power level (dBm) in the reference bandwidth at the input to the antenna of the interfering station
G_T = gain (dBi) of the transmitting antenna of the interfering station
G_R = gain (dBi) of the receiving antenna of the interfered-with station
D_T = antenna discrimination (dB) of the transmitting antenna (at different angles θ')
D_R = antenna discrimination (dB) of the receiving antenna (at different angles θ)
$L(d)$ = total path loss (dB) for the earth's curvature, with $K = 1.33$.

As an example for intersystem radio-relay interference analysis, a C/I ratio of greater than or equal to 65 dB can be assumed and expressed as follows:

$$C - I \geq 65 \text{ dB} \qquad (13.4)$$

where C = nominal received desired signal power (dBm)
I = maximum tolerable interfering power (dBm)

$$C \geq P_T + [G_R - D_R(\theta)] - L(d) + [G_T - D_T(\theta')] + 65 \qquad (13.5)$$

The coordination distance d, at different angles θ, can be calculated for different values of discrimination pattern $D_T(\theta')$ of the interfering station. For $D_T(\theta') = 0$, equation (13.5) represents a situation where the interfering stations are directed toward the proposed station. Coordination distance calculated under this condition will specify a region within which all interfering stations will be located. However, on some occasion, the conservative coordination distance may include a large number of stations resulting in the coordination process being quite burdensome. Under this condition, the coordination region may be reduced with a certain probability of interference.

*Section 13.3.4.1 is from ITU-R Rec. F.1095 (Ref. 7).

13.3.5 The FCC Rule—Calculation of Maximum Permissible Interference Power*

(a) The maximum permissible interference power $P_{max}(p)$ in dBW in the reference bandwidth of the potentially interfered-with station, not to be exceeded for all but a short-term percentage of the time, p, from each source of interference, is given by the general formula

$$P_{max}(p) = 10\log_{10}(kT_rB) + J + M(p) - W \quad (13.6)$$

where

$$M(p) = M(p_0/n) = M_0(p) \quad (13.7)$$

with:

k = Boltzmann's constant (1.38×10^{-23} J/K)
T_r = thermal noise temperature of the receiving system (Kelvins)
B = reference bandwidth (in Hz) (bandwidth over which the interference power can be averaged)
J = ratio (in dB) of the maximum permissible long-term interfering power to the long-term thermal noise power in the receiving system (where long-term refers to 20% of the time)†
n = number of expected entries of interference, assumed to be uncorrelated
p = percentage of the time during which the interference from one source may exceed the allowable maximum value
p_0 = percentage of the time during which the interference from all sources may exceed the allowable maximum value; since the entries of interference are not likely to occur simultaneously

$$p_0 = np \quad (13.8)$$

*Section 13.3.5 is from FCC Rules and Regulations part 25.252 (Ref. 3).
†The factor J (in dB) is defined as the ratio of total permissible long-term (20% of the time) interference power in the system, to the long-term thermal noise power in a single receiver. For example, in a 50-hop terrestrial hypothetical reference circuit, the total allowable additive interference power is 1000 pW0p (CCIR Recommendation 357-1) and the mean thermal noise power in a single hop may be assumed to be 25 pW0p. Therefore, since in a FDM/FM system the ratio of the interference noise power to the thermal noise power in a 4-kHz band is the same before and after demodulation, $J = 16$ dB. In a fixed-service satellite system, the total allowable interference power is also 1000 pW0p (CCIR Recommendation 356-2), but the thermal noise contribution of the down path is not likely to exceed 7000 pW0p, hence $J \geq 8.5$ dB. In digital systems it may be necessary to protect each communication path individually and, in that case, long-term interference power may be of the same order of magnitude as long-term thermal noise; hence $J = 0$ dB.

$M(p)$ = ratio (in dB) between the maximum permissible interference power during $p\%$ of the time for one entry of interference, and during 20% of the time for all entries of interference, respectively

$M_0(p_0)$ = ratio (in dB) between the maximum permissible interference power during $p_0\%$ and 20% of the time, respectively, for all entries of interference*

W = equivalence factor (in dB) relating the effect of interference to that of thermal noise of equal power in the reference bandwidth†

When the wanted signal uses FM modulation with rms modulation indices that are greater than unity, W is approximately 4 dB, regardless of the characteristics of the interfering signal. For low index terrestrial FDM/FM systems a very small reference bandwidth (4 kHz) should be assumed in order to avoid the necessity of dealing with a large range of characteristics of both wanted and unwanted signals upon which, for greater reference bandwidths, the value of W would depend. With this assumption, $W = 0$ dB as shown in Table [13.2]. When the wanted signal is digital, W is usually equal to or less than 0 dB, regardless of the characteristics of the interfering signal.

(b) For purposes of performing interference analyses, the maximum permissible interference power $P_{max}(20\%)$ in dBW in the reference bandwidth of the potentially interfered-with station, not to be exceeded for all but 20% of the time from each source of interference, is given by the general formula and with the remaining parameters defined in paragraph (a) of this section.

$$P_{max}(20\%) = 10\log_{10}(n_{20})(kT_r B) + J - W - 10 \qquad (13.9)$$

where n_{20} = number of assumed simultaneous interference entries of equal power level.

(c) The values of the parameters contained in the appropriate column of Table [13.2], which enter into the formulas in paragraphs (a) and (b) above, shall be used to compute the maximum permissible interference power level in all cases, unless the applicant demonstrates to the FCC that a different set

*$M_0(p_0)$ (in dB) is the "interference margin" between the long-term (20%) and the short-term ($p_0\%$) allowable interference powers. For analog radio-relay and fixed-satellite systems in bands between 1 and 15 GHz this is the ratio (in dB) between 50,000 and 1000 pW0p (17 dB). In the case of digital systems, $M_0(p_0)$ may tentatively be set equal to the fading margin for a percentage of the time, $(1 - p_0)\%$, which depends, inter alia, on the local rain climate.

†The factor W (in dB) is the ratio of RF thermal noise power to RF interference power, in the reference bandwidth, producing the same interference effect after demodulation (e.g., in FDM/FM system it would be expressed for equal voice channel performance; in a digital system it would be expressed for equal bit error probabilities). For FM signals, it is defined as follows:

$W = 10\log_{10}$ [(Interfering power in the receiving system after demodulation/Thermal noise power in receiving system after demodulation) × (Thermal noise power at the receiver input in the reference bandwidth/Interfering power at radio frequency in the reference bandwidth)

TABLE 13.2 Parameters to Be Used in the Calculation of Maximum Permissible Interference Power Level and Minimum Permissible Basic Transmission Loss

Frequency band (MHz)	3,700–4,200	5,925–6,425	6,625–7,125	10,950–11,200	11,450–12,200	12,500–12,750	14,000–14,500
Interference path	T → E	E → T	T → E	T → E	T → E	E → T	E → T
p_0 (percent)	0.03	0.01	0.03	0.03	0.03	0.01	0.01
n	3	[1]2 [2]4	3	2	2	[1]2 [2]4	[1]2 [2]4
n_{20}	3	[1]2 [2]4	3	2	2	[1]2 [2]4	[1]2 [2]5
Interference parameters and criteria							
p (percent)	0.01	[1]0.005 [2]0.0025	0.01	0.015	0.015	[1]0.005 [2]0.0025	[1]0.004 [2]0.0025
J (dB)	8.5	16.0	8.5	8.5	8.5	16.0	16.0
$M_0(p_0)$	17.0	17.0	17.0	17.0	17.0	17.0	17.0
W (dB)	4.0	0.0	4.0	4.0	4.0	0.0	0.0

Note 1: This value should be used for international systems.
Note 2: This value should be used for domestic systems.
E = earth station.
T = terrestrial station.

Frequency band	3,700–4,200	5,925–6,425	6,625–7,125	10,950–11,200	11,450–12,200	12,500–12,750	14,000–14,500
Interference path	T → E	E → T	T → E	T → E	T → E	E → T	E → T
Reference bandwidth, B (Hz)	10^6	4×10^3	10^6	10^6	10^6	4×10^3	4×10^3
System noise temperature, T_r (K)	T_r	750	T_r	T_r	T_r	1500	1500
P_t (dBW)	13	PE	9	5	5	PE	PE
G_t[3] (dBi)	42	$GE(\alpha)$	46	50	50	$GE(\alpha)$	$GE(\alpha)$
G_r[3] (dBi)	$GE(\alpha)$	45.0	$GE(\alpha)$	$GE(\alpha)$	$GE(\alpha)$	50.0	50.0
$P_{max}(\pi)$ (dBW)	$10 \log_{10} (T_r)$	−131	$10 \log_{10} (T_r)$	$10 \log_{10} (T_r)$	$10 \log_{10} (T_r) - 164$	−128	−128
L_W (dB)	0	[4]L_W	0	0	0	[4]L_W	[4]L_W
S (dBW)	—	173	—	—	—	175	175
E (dBW)	55	—	55	55	55	—	—

Note 3: $GE(\alpha)$ is the gain of the earth station antenna toward the horizon at the azimuth of interest, α, and can be derived using the methods of §25.253 (b).
Note 4: For interference analysis, actual line loss should be used, if known; if not known, assume 0 dB.

Source: Reference 3.

of values for these parameters is more appropriate for his/her particular case. Where a symbol appears in Table [13.2], the actual value of the parameter represented by the symbol is to be used.

(d) In cases where an earth station or terrestrial station may employ more than one type of emission, the parameters chosen for analysis should correspond to that pair of emissions which results in the greatest coordination distance.

If we follow FCC Rules and Regulations to the letter, we must examine all potential radio facilities within a radius of 100 km (and 200 km for the keyhole distance) of a facility in question. This could prove a monumental task if done manually. Fortunately, there are a number of special companies that can carry out the job for a fee. These companies have a running database on computer as well as a terrain database, so the whole task can be automated.

13.4 VICTIM DIGITAL SYSTEMS*

This text emphasizes digital operation. What special considerations must be taken into account for interference into such digital radio systems?

As we mentioned earlier, interference into a digital receiver causes threshold degradation. Picture this as an increase in the receiver noise level. Interference can eat into an E_b/N_0 ratio value or wipe it out completely, making the system unworkable. Looking at it another way, interference increases the N_0 value. Remember that our outage threshold on digital links is BER $= 1 \times 10^{-3}$.†

The following arguments deal with conventional digital LOS microwave.

The effect of interference on a victim digital receiver is determined from the threshold-to-interference (T/I) ratio that provides the means of specifying the sensitivity of a victim (offended) receiver to an interferer. The static (nonfaded) threshold (T) of a digital receiver is defined, for the purpose of interference (T/I) calculations, as that manually faded (with attenuators) RSL that produces a BER $= 1 \times 10^{-6}$.‡ T/I is defined as that ratio of desired (1×10^{-6} BER) to undesired (interfering) signal levels that degrades this threshold 1 dB. The 1×10^{-3} dynamic (outage) threshold is similarly degraded 1 dB by this level of interference.

The advantages of T/I are that the differences in thresholds, due to bit rate, modulation type (and bit packing), coding gain, and noise figure, are all taken into account; that the absolute level of allowable interference can easily be determined by subtracting the T/I ratio from the 1×10^{-6} static threshold of a particular digital receiver; and that this measurement can be verified in service without disrupting traffic.

The effect of interference (and the value of T/I) depends primarily on the victim's receiver bandwidth (and sharpness of the bandpass filter), the interfering signal's RF bandwidth, and the separation between their respective center frequencies.

*Section 13.4 is based on TIA TSB 10-F (Ref. 6).

†The BER $= 1 \times 10^{-3}$ value is where we expect to lose supervisory signaling, causing a circuit dropout. A user in conversation will now be cut off and will hear a dial tone.

‡Remember, this is the BER specified in ITU-T Rec. G.821 for 80% of the time.

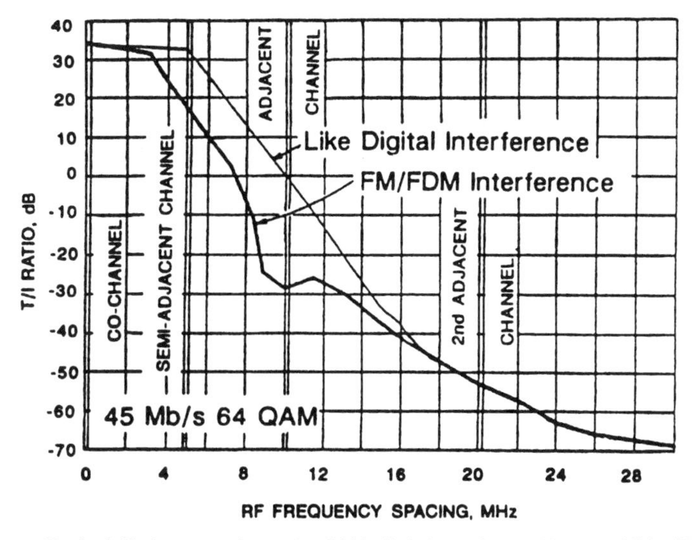

Figure 13.1. Typical T/I curves for a 64-QAM digital receiver with a 10-MHz RF channel assignment. (From Figure B-3a, p. B-7, TIA TSB 10-F; Ref. 6.)

The measurement of T/I for a specific digital radio is accomplished by fading the receiver to the point where a 1×10^{-6} BER is achieved. The RSL is then increased 1 dB and interference is injected until a BER of 1×10^{-6} is again achieved. The ratio of the initial (1×10^{-6} BER) RSL of the desired signal to the interference power, as measured, is the T/I ratio. One must, of course, take into account such variables as frequency offset, desired signal-to-interfering signal, and instances when the width of the interferer is wider than the victim's receiver bandwidth.

Reference 6 (TIA) states that, in principle, one would need to know the T/I as a function of frequency separation for all possible interferers into a digital receiver. A typical T/I curve is shown in Figure 13.1.

The acceptable value of interference (Ref. 6–TIA) in a coordination process, I_{coord} dBm, is given by

$$I_{\text{coord}} = T_s - TI_{\text{sp}}(f_d) - \text{MEA} \tag{13.10}$$

with

$$\text{MEA} = \text{the greater of } \{5\} \text{ or } 10\log(\text{BW}_V/\text{BW}_I)$$

where T_s = static threshold power of 1×10^{-6} BER (in dBm)
TI_{sp} = specific T/I curve (dB)
f_d = separation of interference and receiver center frequencies
MEA = multiple exposure allowance (dB)
BW_V = RF channel bandwidth of victim receiver (MHz)
BW_I = RF channel bandwidth of interferer (MHz)

The MEA, the multiple exposure allowance, accounts for the high usage density in the former common carrier bands and for the multiple near-adjacent channel exposures in the mixed-bandwidth frequency plans in these and other bands that were adopted in December 1993 by FCC ET Docket 92-9. There is general agreement that a 5-dB MEA is to be used (Ref. 6–TIA). There is further study underway to establish higher MEA values to account for very congested areas due to the higher probability of multiple exposures.

The preceding results can also be expressed in terms of the ratio of the nominal carrier power to the interference power, which is usually referred to as the C/I (ratio). The C/I (in dB) due to a single specific interferer separated in frequency f_d would be given by

$$\begin{aligned} C/I &= C - T_s + TI_{\text{sp}} \\ &= \text{TFM}_s + TI_{\text{sp}} \end{aligned} \tag{13.11}$$

where TFM_s is the static thermal fade margin to 1×10^{-6} BER (in dB). In reference to the U.S. FCC Rules and Regulations, prior to 1994, for common carrier bands, frequency coordination was based on meeting C/I objectives, which in the present terminology would be written

$$(C/I)_{\text{obj}} = \text{TFM}_s + TI_{\text{sp}} + \text{MEA} \tag{13.12}$$

The default static thermal fade margins (in dB) at BER $= 1 \times 10^{-6}$ upon which these C/I objectives are based were the following: 35 (2 GHz), 40 (6 and 18 GHz), and 45 (11 and 13 GHz).

13.5 DEFINITION OF C/I RATIO*

The C/I ratio, which applies to both digital and analog microwave links, is the number of decibels by which the unfaded power level of the desired signal C† at the input of the victim receiver exceeds the power level of the interference (I) at the same point.

For potential threshold degradation to all types of victim receivers, only the I or interference level must be calculated.

*Section 13.5 is based on TIA TSB 10-F (Ref. 6).
†Remember C and RSL are synonymous.

13.5.1 Example C/I Calculations Based on Ref. 6

13.5.1.1 Necessary Input Parameters. The following minimum input information is necessary for C/I (and T/I) calculations:

1. *Latitudes and Longitudes of Stations A, B, D and E (Including Datum).* With this information all of the necessary path lengths, azimuths, and discrimination angles can be calculated.
2. *Size, Types, and Polarizations of Antennas.* This information is for both systems, preferably with identification of the manufacturers and model numbers. This will allow determination of antenna gains and discrimination values.
3. *Ground Elevations and Antenna Heights at Each of the Stations.* This information is used in determining if the interference path is line-of-sight or obstructed.
4. *Nominal Receive Signal Level.* This information is used to determine C/I and fade margin.
5. *Equivalent Isotropic Radiated Power (EIRP).* The arithmetic sum of the interfering transmitter power, feeder (and pad, if any) loss, and antenna gain determines the main lobe EIRP from a microwave dish. *Note:* FCC Rules and Regulations, §21.710 and §94.79, impose EIRP limits on paths below 17 km (1.85–7.125 GHz) and 5 km (to 13.25 GHz).
6. *Transmitter Stability.* Stability of all transmitters (existing and proposed) is important in defining allowable C/I for FM-FDM into FM-FDM intercarrier beat interference.
7. *Capacity and Type of Radios.* Radio capacity (equivalent voice channels or video) type of radio, FM or digital (direct or indirect), and its electrical characteristics are important in defining the allowable C/I.

13.5.1.2 Basic C/I Calculations. Consult Figure 13.2 for a generic model for C/I calculations. First, the unfaded receiver B input carrier level (C or RSL) is calculated. Then treating D as the interfering transmitter, calculate

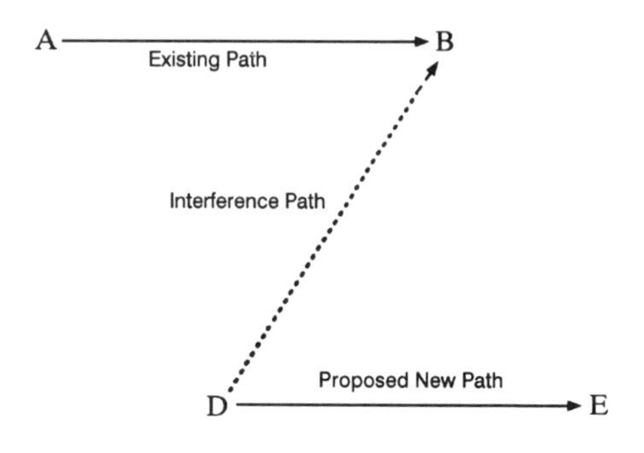

Figure 13.2. System layout for C/I example.

the unfaded interfering signal level (I) at B. With all values expressed in decibels, the C/I is the arithmetic difference between the two values, C and I.

Ignoring transmission line losses for the present,

$$\text{RSL} = C = P_A + G_A + G_B - L_{AB} \tag{13.13}$$

RSL or C and I values may be expressed either in dBm *or* dBW.

$$I = P_D + G_D + G_B - L_{DB} - M_D - M_B \tag{13.14}$$

where P = transmitter power (dBm *or* dBW)
G = antenna gains (in dBi)
M = antenna discriminations (dB)
L = path loss (dB)
I = interference level (in dBm *or* dBW)

C/I is the difference between the values calculated for C and I [equations (13.13) and (13.14)]. We now can derive C/I as that difference:

$$C/I = (P_A - P_D) + (G_A - G_P) - (L_{AB} - L_{DB}) + (M_D + M_B) \quad \text{(dB)} \tag{13.15}$$

where M_D is the discrimination of the antenna at D along the path DB that is at an angle EDB away from the main beam and M_B is the discrimination of antenna B along the path BD. These discriminations are obtained from the antenna patterns available from manufacturers. (Also, see Section 13.3.2.2 and Table 13.1.) If the antennas are known to be oppositely polarized, discrimination for one of the antennas should be taken from the main beam, and that for the other antenna from the cross-polarized pattern. This polarization discrimination choice should be the "worst case," that is, the one that gives the largest value of I for the smallest value of C/I.

When treating the interfering path as line-of-sight, the bracketed term $(L_{AB} - L_{DB})$ can be replaced by $20\log(\text{AB}/\text{DB})$ with A to B and D to B being path lengths of like distance units. Therefore equation (13.13) can be rewritten

$$C/I = (P_A - P_D) + (G_A - G_D) - 20\log(\text{AB}/\text{DB}) + (M_D + M_B) \quad \text{(dB)} \tag{13.16}$$

If greater precision is required, and if all parameters are known, an additional differential factor $(W_A - W_D)$ can be introduced into the right-hand sides of equations (13.15) and (13.16) to take into account the difference in transmission line losses between the transmitter at A and the transmitter at

D. At B there is no need for such a differential, since any transmission line losses at the receiver are identical for both paths and therefore self-canceling in C/I calculations only. For threshold degradation level calculations, the receive transmission line loss should be included.

In equation (13.13), the first two terms $(P_A + G_A)$ combined with the transmission line loss W_A [i.e., $(P_A + G_A - W_A)$] is the EIRP listed in the FCC station data. Including W_D with the first two terms of equation (13.14), $(P_D + G_D - W_D)$ is the listed EIRP for the interfering transmitter. Thus equations (13.15) and (13.16) may be rewritten:

$$C/I = (\mathrm{EIRP_A} - \mathrm{EIRP_D}) - (L_{AB} - L_{DB}) + (M_D + M_B) \quad \text{(dB)} \tag{13.17a}$$

and

$$C/I = (\mathrm{EIRP_A} - \mathrm{EIRP_D}) - 20\log(\mathrm{AB/DB}) + (M_D + M_B) \quad \text{(dB)} \tag{13.17b}$$

Consider the following specific example for calculating C/I, assuming values for a 6.7-GHz co-channel case. See Figure 13.3.

Values for this example of C/I calculations are:

$P_A = +30$ dBm Desired transmit power

$G_A = 43$ dBi Desired transmit antenna gain

$W_A = 2$ dB Desired transmit transmission line loss

$$\mathrm{EIRP_A} = (P_A + G_A - W_A)$$
$$= (30 + 43 - 2) = 71 \text{ dBm}$$

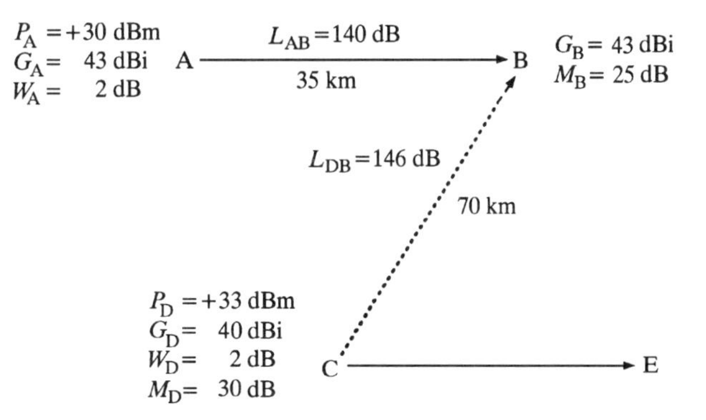

Figure 13.3. System values for C/I example.

$P_D = +33$ dBm Interfering transmit power

$G_D = 40$ dBi Interfering transmit antenna gain

$W_D = 2$ dB Interfering transmit transmission line loss

$$\text{EIRP}_D = (P_D + G_D - W_D)$$
$$= (33 + 40 - 2) = 71 \text{ dBm}$$

$G_B = 43$ dBi Victim receive antenna gain

$L_{AB} = 140$ dB AB = 35 km (22 mi) desired free-space path loss

$L_{DB} = 146$ dB DB = 70 km (44 mi) interfering free-space path loss

$M_D = 30$ dB Interfering transmit antenna discrimination

$M_B = 25$ dB Victim receive antenna discrimination

Using equation (13.11),

$$C = 30 + 43 + 43 - 140 = -24 \text{ dBM } (-54 \text{ dBW}) \tag{13.18}$$

Using equation (13.14),

$$I = 33 + 40 + 43 - 146 - 30 - 25 = -85 \text{ dBm } (-115 \text{ dBW}) \tag{13.19}$$

Thus

$$C/I = -24 - (-85) = 61 \text{ dB} \tag{13.20}$$

Or, using equation (13.17a),

$$C/I = (71 - 71) - (140 - 146) + (30 + 25) = 61 \text{ dB} \tag{13.21}$$

Then there is a third solution using equation (13.17b):

$$C/I = (71 - 71) - 20\log(35/70) + (30 + 25) = 61 \text{ dB} \tag{13.22}$$

If the path D to B is an obstructed path, further calculations should be made to determine how much the long-term path loss exceeds free-space loss (FSL). These calculations are dealt with in Chapter 5. For our example calculation that follows, an obstruction loss of 14 dB is used.

13.5.2 Example of Digital Interferer into Victim Digital System

The example model is shown in Figure 13.4. The technique employed here is explained in Section 13.4. The only calculation required is to determine the interfering signal level at the victim's receiver input.

In our example given in Section 13.5.1.2, the absolute value of the interfering level (I) was calculated as -85 dBm. In this case the victim

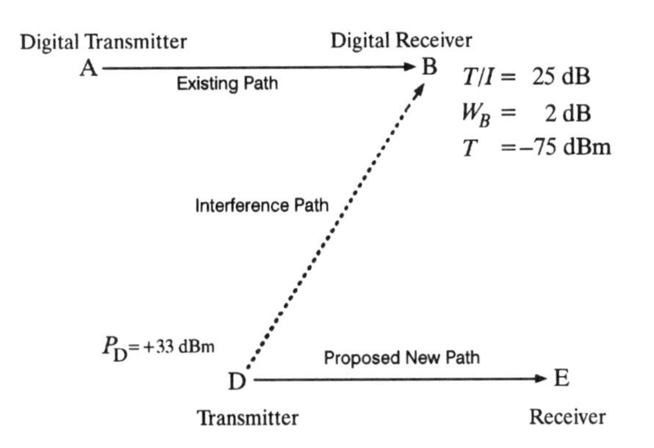

Figure 13.4. Model of digital interferer into victim digital receiver.

receiver transmission line loss (given as 2 dB in our example) is subtracted from I to determine the interfering signal level at the receiver's input port. Thus, for this example, the interfering signal level at the input of the victim receiver is -87 dBm for an unobstructed interference path.

Of course, specific data on the victim's receiver characteristics are required to accurately evaluate any potential interference case. The data required are:

Threshold (T). RSL (dBm *or* dBW) for BER $= 1 \times 10^{-6}$ (static threshold).

Threshold-to-Interference Ratio (T / I). Ratio (dB) of threshold to interfering signal that degrades the 1×10^{-6} BER static (and therefore the 1×10^{-3} dynamic or outage) threshold by 1 dB.

These values should be obtained from the equipment manufacturer and will vary depending on many factors. The co-channel T/I, which is a function only of the receiver's modulation scheme, may be taken from Table 13.3. Adjacent channel (i.e., off-frequency) interference T/I varies with interfering signal type (like digital and FDM/FM) and the victim receiver's filtering characteristics and should be taken from the receiver manufacturer's T/I chart.

In this example, let $T = -75$ dBm and assume that the T/I ratio is 25 dB. The TIA (Ref. 6) warns that *these values should not be considered typical and actual values, real values obtained from the equipment manufacturer should be used when possible.*

Using the assumed values,

$$\begin{aligned} \text{Maximum long-term } I \text{ level} &= T - T/I \\ &= -75 \text{ dBm} - 25 \text{ dB} \\ &= -100 \text{ dBm} \end{aligned}$$

TABLE 13.3 Typical Digital Microwave Noise and Spectral Characteristics

Modulation	Theoretical C/N for 10^{-4} BER[a] (dB)	Typical Co-channel T/I (dB)	Theoretical Spectral Efficiency (bps/Hz)	Typical Data Input Mbps (Bands, GHz)	b, Spectrum Bandwidth[b] (MHz) [Symbol Rate, M baud]	R, Spectrum Amplitude[c] (dB/4 kHz) Below the Average Transmit or Receive Power
4L FSK	17.6	23.6	2	13 (18)	6.5	−32.1
QPSK, OQPSK	13.5	19.5	2	45 (18)	22.5	−37.5
9 QPR	16.5	22.5	2	6, 4 (2)	3.2	−26.0
25 QPR	20.8	26.8	3.17	13 (2, 10)	4.1	−27.1
16 QAM	20.9	26.9	4	90 (6, 11)	22.5	−37.5
49 QPR	23.5	29.5	4	19 (2, 6, 10)	4.8	−27.8
				13 (2, 10)	3.3	−26.1
32 QAM	24.0	30.0	5	13 (2, 6)	2.6	−28.1
81 QPR	25.5	31.5	4.64	45 (2, 6)	9.7	−30.8
64 QAM	27.1	33.1	6	45 (2, 6)	7.5	−32.7
128 QAM	30.1	36.1	7	155 (6, 11)	22.5	−37.5
256 QAM	32.6	38.6	8	19 (2)	2.4	−27.8
512 QAM	35.5	41.5	9	155 (6)	17.2	−36.3

[a] See Table 1-A, ITU-R Recommendation F.1101 on "Characteristics of Digital Radio-Relay Systems below about 17 GHz" (from former Report 378-6, modified).

[b] b = bit rate, Mbps/efficiency, bps/Hz; QPSK and QAM: spectra 3-dB points; QPR: spectrum central lobe width.

[c] QAM/QPSK: $R = 10 \log(4/\text{kbaud})$; QPR: $R = 3 + 10 \log(4/\text{kbaud})$ (assumes 4-kHz measurement bandwidth).

Source: Table B-1, p. B-5, TIA TSB 10-F, Annex B, (Ref. 6).

This tells us that the maximum long-term I level should not exceed −100 dBm.

It should be noted that the previously calculated interfering signal level (I) of −85 dBm exceeds this maximum interfering level by 15 dB and is thus unacceptable. If we assume that the interfering path has a 16-dB obstruction loss, the interference level will be reduced by that 16 dB, thus making the path acceptable with a 1-dB margin.

13.6 OBSTRUCTED INTERFERING PATHS

Unless a microwave path strictly meets the line-of-site criterion provided in Chapter 2, then an obstruction penalty must be imposed. There are other microwave interferers that because of their geography and geometry would appear not to be interference sources. Such facts should be verified regarding "spill" by diffraction or troposcatter mechanisms into a victim receiver. Diffraction and troposcatter mechanisms are discussed in Chapter 5.

The allowable interference levels discussed so far are based on the assumption that the interfering signal at the victim receiver is essentially steady-state. When the interference path is line-of-sight, the interference level is calculated using free-space loss (FSL).

Interference paths that are normally obstructed present a more complex situation. Potential interference can exist for interference paths even beyond 200 km.* Except in mountainous terrain, which requires special treatment, interference paths that are normally obstructed have additional loss over free space, as we described in Chapter 5. Such paths normally have a much higher variability than line-of-sight paths, particularly in areas subject to significant ducting. In such cases, an obstructed interference path may have a long-term loss in excess of 30 dB above the free-space loss. However, for short periods of time this loss can approach the free-space (unobstructed) loss. Such a short-term reduction in loss can cause notably higher interference levels even though the long-term loss is satisfactory because of the large normal loss in the interference path.

It follows then that the system designer must not only be concerned about the long-term interference level but should take into account short-term (i.e., $< 0.01\%$ of the time) levels as well. TIA, in Ref. 6, points out that very short periods of higher interference level can be tolerated on most systems than for the longer periods. A differential of 10 dB is proposed by Ref. 6. This means that for very short-term interference, the C/I objective may be degraded (reduced) by 10 dB.

To reduce this problem to practical dimensions Ref. 6 suggests the following pragmatic methods:

- Paths that are line-of-sight at $k = \frac{4}{3}$, use free-space path loss and long-term criteria.
- Paths that are obstructed, use calculated obstruction loss exceeded 80% of time for long-term and exceeded 99.99% of time for short-term criteria. Both long-term and short-term criteria should be met.
- Obstructed paths longer than about 200 km need be considered only in special cases, such as stations at very high elevations or in areas of extreme ducting.

*The 200-km distances are a requirement for coordination contours discussed in Section 13.3.4.

Interference from a transmitter that is well beyond the horizon and thus obstructed from the victim receiver under normal atmospheric conditions is seldom a problem unless the antennas of the interfering transmitter and the victim receiver are pointed almost directly at each other.

Situations of this kind occur fairly frequently in backbone microwave systems following a pipeline, railroad, or the like where the points to be served follow more or less a straight line. If a four-frequency plan is followed on such a route, potential co-channel "overshoot" interference situations exist between all pairs of hops separated by only one intervening hop so cross-polarization is essential. Where zigzagging the paths to the degree necessary to break up overshoot problems is impractical, as it often is, a departure from the four-frequency plan may be necessary to avoid intolerable interference situations.

Overshoot situations are most commonly an intrasystem problem but can also occur between systems of different users, particularly where routes run parallel and close together. Flat humid areas such as those along the U.S. Gulf Coast are particularly subject to ducting, and in such areas the overshoot problem is a very important one.

13.7 CCIR APPROACH TO DIGITAL LINK PERFORMANCE UNDER INTERFERENCE CONDITIONS*

13.7.1 Gaussian Interference Environment—*M*-QAM Systems

In the case of *M*-QAM systems, which use very tight filtering (e.g., Nyquist raised cosine), the interference may be treated as Gaussian-like noise. A victim *M*-QAM receiver may be interfered with from one or several sources. The amplitude distribution of a tightly filtered interferer exhibits a high peak-to-average ratio that could be approximated by an equivalent Gaussian-like noise source. In the case of several interferers the sources of interference are considered to be independent random variables. The central-limit theorem says that, under certain general conditions, the resultant equivalent interference probability density function approaches a normal Gaussian curve as the number of sources increases. In both the single- and multiple-interference cases, the equivalent interference may be treated as Gaussian-like noise. The practical approach yields useful performance curves in which the degradation due to interference can readily be seen.

The Gaussian-like interference is combined with the assumed white Gaussian noise channel to produce a total carrier-to-noise ratio $(C/N)_T$ given by

$$(C/N)_T = (N/C + I/C)^{-1} \tag{13.23}$$

$$I/C = I_1/C + I_2/C + \cdots + I_n/C \tag{13.24}$$

*Section 13.7 is based on CCIR Rec. 766, 1994 SF Series Volume (Ref. 8).

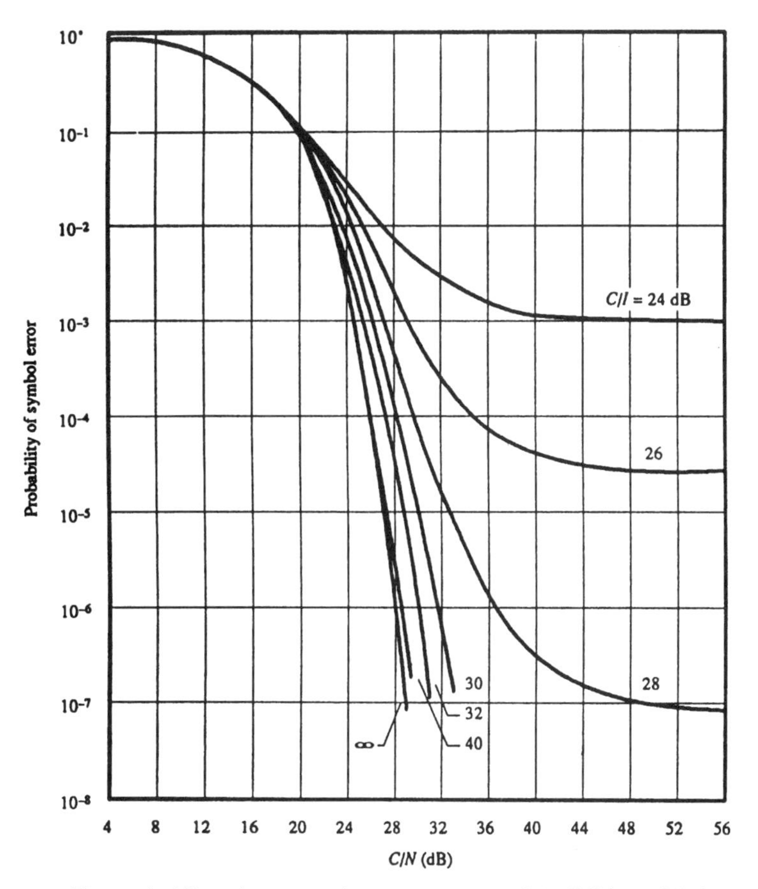

Figure 13.5. The probability of error performance curves of 64-QAM modulation system versus carrier-to-thermal-noise ratio and a carrier-to-interference ratio as a parameter (double-sided Nyquist bandwidth). (From Figure 6, p. 110, CCIR Rec. 766; Ref. 8.)

where N/C is the thermal noise-to-carrier ratio, I/C is the equivalent interference-to-carrier ratio, and I_i/C $(i = 1, \dots, n)$ is the interference-to-carrier ratio of the ith random source. There are several equations used for the calculation of error performance for coherent digital modulation schemes given C/N. These may be found in CCIR Rec. 766, pages 107 and 108 (Ref. 8). We can also calculate error performance for these systems by replacing C/N in each equation with $(C/N)_T$ and having C/I as a variable parameter. Inclusion of C/I as a variable parameter produces a series of curves shown in Figures 13.5 and 13.6.

The degradation (dB) of $(C/N)_T - C/N$ for BER $= 1 \times 10^{-6}$ versus C/I for M-QAM systems are summarized in Figure 13.7. If the carrier-to-interference ratio is at least 10 dB higher than carrier-to-thermal-noise ratio required for BER $= 1 \times 10^{-6}$, the degradation due to interference will be

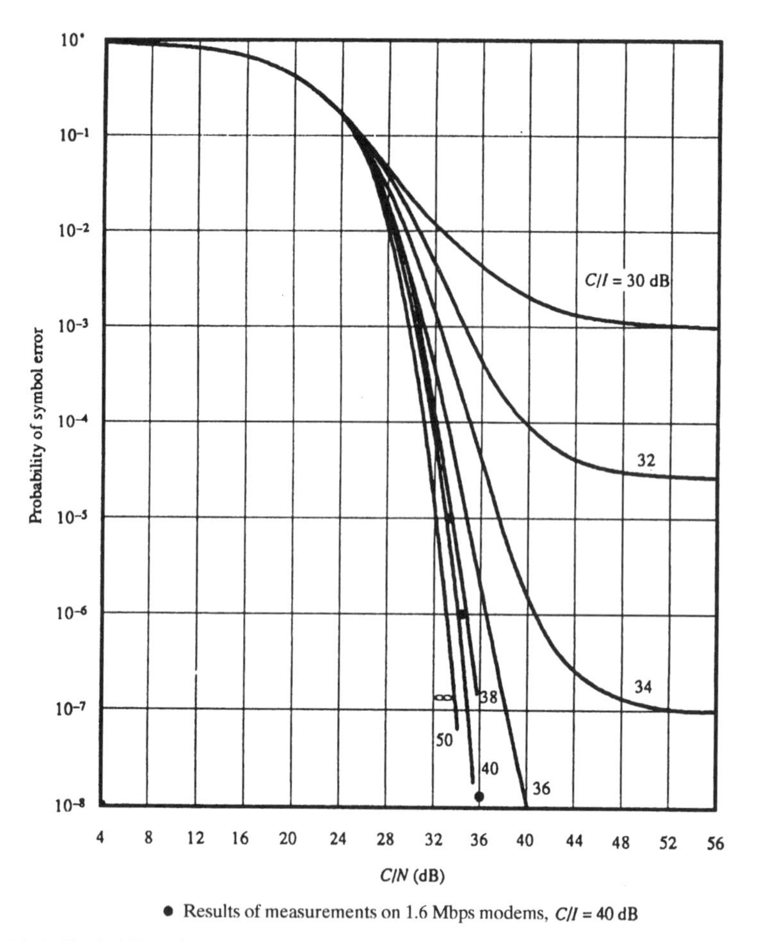

Figure 13.6. Probability of error performance curves of a 256-QAM modulation system versus carrier-to-thermal-noise ratio and a carrier-to-interference ratio as a parameter (double-sided Nyquist bandwidth). (From Figure 7, p. 111, CCIR Rec. 766; Ref. 8.)

less than 1 dB. Although not shown in Figure 13.7, it can be calculated that if the C/I is at least 6 dB higher than C/N for a BER $= 1 \times 10^{-3}$, the degradation due to interference is less than 1 dB.

CCIR Rec. 766 points out that Gaussian-like interference is not necessarily the worst case. This same reference draws the following conclusions:

- When the interfering signal power is equal to, or larger than, the thermal noise power, the effect of angle-modulation interference is considerably less than that of an equal amount of white Gaussian noise power.

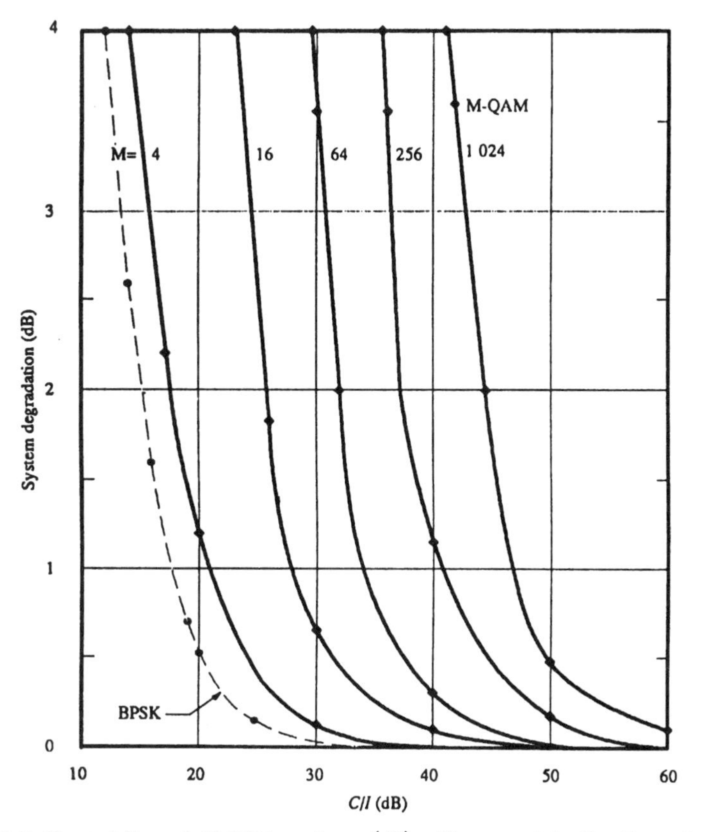

Figure 13.7. Degradation of *M*-QAM systems (dB) with respect to the theoretical value necessary to achieve a BER = 1×10^{-6} performance, versus the carrier-to-interference ratio (dB) in the double-sided Nyquist bandwidth. (From Figure 8, p. 112, CCIR Rec. 766; Ref. 8.)

- When the interfering signal power is small compared to the thermal noise power, the effect on error rate can be estimated safely by assuming that the interfering signal is equivalent to the Gaussian noise of equal power.
- At a given carrier-to-interference ratio, the vulnerability to interference increases substantially as the number of transmitted symbols, M, increases.
- For the same interfering power after filtering, interference effects tend to become larger as the frequency separation between carriers increases. These effects also tend to increase with the interfering carrier bandwidth and with the number of interfering carriers. They are approximately the same for an interfering 4-PSK or 8-PSK carrier but they increase with the number of symbols for a QAM interfering signal. All the above situations can be interpreted in terms of an increase in the

interference peak factor: for large values of frequency separation (adjacent channel interference), for large values of interfering carrier bandwidth, and for a large number of interfering carriers, the interference effect approaches that of an equal amount of white Gaussian noise.

PROBLEMS AND EXERCISES

1 Define EMC and name and discuss its two subsets.

2 Name at least three measures we can take at the outset to reduce interference and its effects on/with radio systems.

3 One area of concern is a receiver's IF. With careful frequency planning, we avoid three frequencies referenced to the IF. What are they?

4 LOS microwave and satellite communication systems share the same frequency bands. A satellite communication transponder sprays the earth with RF energy. What did we learn prior to this chapter of one measure to reduce this interference into LOS microwave and other radio facilities in these bands?

5 Before delving into interference reduction techniques developed by national/international agencies, we offered a technique or methodology to measure the level of an interfering signal in a victim receiver and its effects thereon regarding victim's performance. Describe how this is done.

6 An offending antenna has a *relative* azimuth toward a victim antenna of 38°. The offending antenna transmits at 6 GHz and its aperture is 4 ft. Calculate the "antenna gain" at that azimuth based on the ITU Radio Regulations.

7 Argue pros and cons of the two ways we can quantify the effects of an interferer: C/I and I_0/N_0.

8 A transponder in a satellite has a noise figure of 2 dB; its spot beam antenna has a 35-dB gain, line loss of 1.5 dB, and is faced in the general direction of the terrestrial LOS microwave offending facility. The satellite is 785 sm above a radio facility in the same band. The frequency is 28 GHz. The EIRP of the terrestrial transmitter in the direction of the satellite is +7 dBW. The objective I_0/N_0 is 18 dB or greater. Assume the offending emission covers the satellite receiver bandwidth completely. Is there a problem here for the victim?

9 Suppose a satellite receiver thermal noise threshold is −200 dBW and the offending signal RSL at that receiver is also −200 dBW (equivalent to white noise). These signals (i.e., signal and receiver noise) sum. What is that equivalent sum value in dBW? Suppose that satellite link had only 1 dB of margin, and the link was PSK with a BER of 1×10^{-7} for that uplink. Discuss this situation.

10 An antenna meets minimum U.S. FCC suppression values. What value would be applicable at 25°? The angle is relative to centerline main beam.

11 What is the rationale of a *coordination contour*?

12 Both ITU-R and the U.S. FCC use the term *keyhole distance* in reference to coordination contour. Define the term and discuss why this segment is selected for special treatment.

13 Define threshold-to-interference (T/I) level.

14 The effect of interference (and the value of T/I) depends largely on three factors. Name them.

15 Discuss obstructed interference paths. In this case, what does the system designer have to be particularly watchful for?

16 Which is more vulnerable to interference: satellite or LOS microwave? Give a valid reason for your selection.

17 If we wish to reduce the probability of symbol error to better than 1×10^{-8} on a link that uses 64-QAM, what C/I value must we achieve?

18 Discuss the vulnerability to interference on a digital system regarding the value of M with M-QAM modulation.

19 For purposes of coordination distance d calculate the level of interfering power I. The interfering station has +30 dBm input to its antenna, the gain of the victim receiving antenna in the direction of the offending transmitter facility is −10 dBi, the gain of the interfering transmit antenna toward the victim receive facility is +2 dBi, the distance is 35 km, and the transmit frequency is 6.7 GHz.

20 A digital receiver has a noise figure of 3 dB. An interferer has a spectral noise density in the passband of the receiver of −203 dBW/Hz and the interference signal is of a white noise nature. The victim receive facility has a 30-dB TFM and the required $E_b/N_0 = 11$ dB. Discuss this situation and, in particular, the effects on the victim facility.

REFERENCES

1. *The New IEEE Standard Dictionary of Electrical and Electronic Terms*, 5th ed., IEEE Std. 100-1992, IEEE, New York, 1993.
2. *American National Standard Dictionary for Technologies of Electromagnetic Compatibility (EMC), Electromagnetic Pulse (EMP) and Electrostatic Discharge*, ANSI C63.14-1992, IEEE, New York, 1992.

3. U.S. Federal Communications Commission Rules and Regulations, especially Parts 21, 25, and 94, Code of (U.S.) Federal Regulations 47, FCC, Washington, DC (U.S. Government Printing Office), Oct. 1994.
4. Roger L. Freeman, *Reference Manual for Telecommunications Engineering*, Wiley, New York, 1994.
5. *Radio Regulations*, Edition 1990, Revised 1994, ITU, Geneva, 1994.
6. *Interference Criteria for Microwave Systems*, Telecommunications Industry Association, TIA TSB 10-F, Washington, DC, 1994.
7. *A Procedure for Determining Coordination Area Between Radio Relay Stations in the Fixed Service*, ITU-R Rec. F.1095, 1994 F Series Volume, Part 1, ITU, Geneva, 1994.
8. *Methods for Determining the Effects of Interference on the Performance and Availability of Terrestrial Radio-Relay Systems and Systems in the Fixed Satellite Service*, CCIR Rec. 766, 1994 SF Series, ITU, Geneva, 1994.

14

RADIO TERMINAL DESIGN CONSIDERATIONS

14.1 OBJECTIVE

This chapter deals with the design factors of radio terminals that may affect the systems in which they may be incorporated. A system engineer will be concerned about equipment performance, interfaces, size, prime power requirements, and reliability and availability among other factors. Modulation has been discussed in previous chapters. The fundamental objective of this chapter is to describe the basic components of a terminal and how they relate to one another and to the system overall. We start with a description of a generic radio terminal indicating the functional blocks nearly all terminals have in common. We then progress into specific applications.

14.1.1 The Generic Terminal

For nearly every application in the point-to-point service a radio terminal consists of a transmitting and receiving subsystem, as shown in Figure 14.1. On the transmit side we can expect to find a modulator, an upconverter, and some sort of power amplifier (sometimes called a HPA or high-power amplifier). On the receive side, we find a low-noise amplifier (LNA), which is optional in some cases, a downconverter, and a demodulator. In most cases, not all, a common antenna system is used to radiate the local transmitted signal and receive the emitted signal from the distant-end transmitter.

An information baseband is the electrical representation of the intelligence we wish to transmit. It usually feeds some sort of conditioning device, which prepares the signal for modulation and transmission. For an analog system, this device could provide pre-emphasis and de-emphasis; for a digital system it could provide digital format conversion (e.g., AMI to NRZ), frame alignment, coding, and interleaving and may include provisions for a service channel. A similar conditioning device is incorporated on the companion receiving side of a link.

All systems discussed in this text employ heterodyne principles. The IEEE dictionary (Ref. 1) defines *heterodyne* as "the process occurring in a frequency converter by which the signal input frequency is changed by superimposing a local oscillation to produce an output having the same modulation information as the original signal but at a frequency which is either the sum or difference of the signal and local oscillator frequencies." Thus upconverters are used with transmitters and downconverters with receivers. The derived frequency on the receive side after downconversion is called the IF or intermediate frequency. In a transmitter, there is the same situation prior to upconversion. The most common IF is 70 MHz for all systems discussed except for HF and meteor burst, which use lower IFs. Broadband microwave, satellite, and forward-scatter systems may have higher IFs, for example, 140 MHz as well as 600, 700, and 1200 MHz. See CCIR Rec. 403 (Ref. 2).

With this background, we will proceed to discuss the following:

Topic / Discipline	Section	Comments
Analog LOS microwave	14.2	Being phased out except for TV
Digital LOS microwave	14.3	Increasing bit rates
Troposcatter, analog and digital	14.4	Reduced commercial application
Earth station, analog and digital	14.5	Digital operation replacing analog
Cellular/PCS, analog and digital	14.6	Pressure to convert to digital; TDMA or CDMA?
HF	14.7	Military application mostly
Meteor burst	14.8	Remote sensor, orderwire

14.2 ANALOG LINE-OF-SIGHT RADIOLINK TERMINALS AND REPEATERS

14.2.1 Basic Analog LOS Microwave Terminal

Analog LOS microwave systems in the context presented here provide broadband LOS communication on a point-to-point basis. Present applications in the developed world are for STL (studio-to-transmitter link) for broadcaster video transmission. Many FDM/FM systems are still operational in developing countries. We recommend that any new LOS microwave installation be digital.

The vast majority of analog microwave systems use frequency modulation (FM), although some have been fielded using single-sideband suppressed carrier (SSBSC), which are considerably more bandwidth conservative. Historically, FM systems have been favored because of their simple design, mature technology, and the advantage of trading off bandwidth for reduced thermal noise owing to the very nature of wide deviation FM.

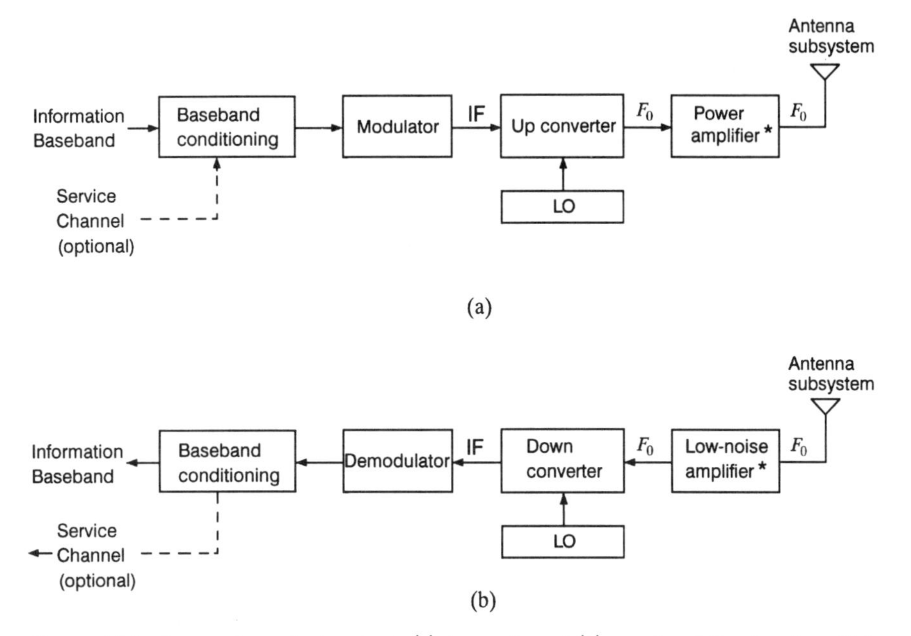

Figure 14.1. A generic radio terminal: (a) transmit and (b) receive. LO = local oscillator, IF = intermediate frequency, F_0 = operating frequency. *Optional on some systems.

Figure 14.2 is a functional block diagram of a microwave FM transmitter and receiver. The diagram is almost identical to Figure 14.1. On the transmit side (upper drawing), a terminal consists of baseband-conditioning equipment, a FM modulator, IF equipment, an upconverter,* a power amplifier (optional), and an antenna. On the receive side, there is an antenna (usually shared with the transmit side), downconverter, IF equipment, demodulator, and baseband-conditioning equipment. An optional LNA may be inserted ahead of the downconverter to improve system gain and noise performance.

FM terminals commonly transmit two different types of broadband signals: a composite FDM waveform or television. In certain implementations both waveforms may be transmitted simultaneously, on a frequency division basis. For further description of FDM consult Refs. 3 and 4.

The power output of most microwave transmitters is on the order of 1 W (0 dBW). Ten-watt HPAs are available, using either TWT (traveling-wave tube) technology or solid-state amplifiers (SSAs). Millimeter wave transmitters often have outputs in the milliwatt region. The microwave antenna subsystem will be discussed in Section 14.3 after the coverage of digital LOS microwave.

*We will use the term *mixer* interchangeably with upconverter and downconverter.

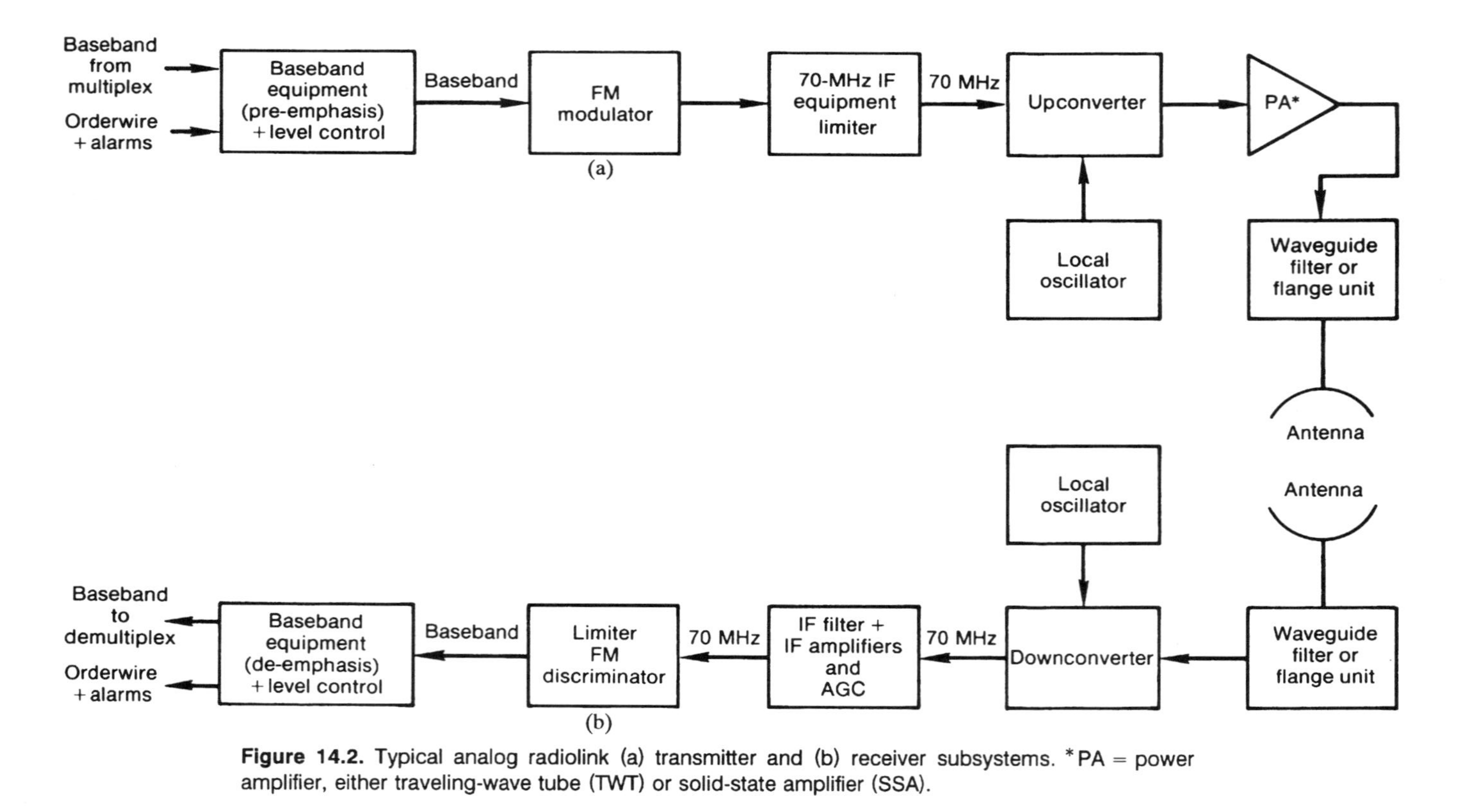

Figure 14.2. Typical analog radiolink (a) transmitter and (b) receiver subsystems. *PA = power amplifier, either traveling-wave tube (TWT) or solid-state amplifier (SSA).

14.3 DIGITAL LOS MICROWAVE TERMINALS

Figure 14.3 is a functional block diagram of a typical digital microwave radio terminal. Starting from the bottom of the figure on the transmit side (left side), there are a number of baseband processing (conditioning) functions. The line code converter takes the standard PCM line codes (e.g., DS1, DS3, E1, E2, E3, etc.), which have been implemented for good baseband transmission properties over wire systems, and converts the code, usually to a NRZ format. The resulting code is then scrambled by means of a PRBS (pseudorandom binary sequence). The scrambling tends to remove internal correlation among symbols and assures a minimum 1s-density. The resulting modulation with the PRBS tends to provide an output with a more constant power spectrum.

The signal may then be channel coded for forward error correction (FEC) (Chapter 4), although this is not often done on terrestrial LOS microwave links. Differential coding/decoding is one method to remove phase ambiguity at the receive end. A coherent phase receiving system would not require this function. However, a phase coherence reference would be required.

A serial-to-parallel converter divides the serial NRZ bit stream into two components for I and Q inputs to a phase modulator (see Figures 3.9 and 3.10). Multilevel coders (and decoders) are required for 8-ary PSK and for QAM and QPR (quadrature partial response) modulation schemes. (Refer to Section 3.3.) In the cases of M-QAM schemes where $M \geq 16$, more than two amplitudes are used per orthogonal modulation. Therefore the digital data have to be converted to multilevel logic. For instance, 64-QAM requires four amplitude levels.

The predistorter is designed to compensate for the distortion imparted to the signal by the power amplifier.

The digital modulator, of course, carries out the modulation function. Modulation to achieve spectral efficiency was discussed in Section 3.3.3. The output of the modulator is then amplified, filtered, and passed to the upconverter. The upconverter (a mixer) translates this signal from the IF to the operating frequency of the terminal. The output of the upconverter is then fed to the HPA, which amplifies the signal to the desired RF output level. Common LOS microwave transmitter outputs are 1 watt. TWT amplifiers or SSAs (solid-state amplifiers) are optional and may amplify the signal to 10 watts. The output of the HPA generally incorporates a bandpass filer to reduce spurious and harmonic out-of-band signals. The signal is then fed to the antenna subsystem for radiation to the distant end.

The received signal, starting from the antenna down in Figure 14.3, is fed from the antenna subsystem through a bandpass filter, thence through an optional LNA (low-noise amplifier) to a downconverter. Whether a LNA is used or not depends on performance versus economics. If the link in question can tolerate the additional thermal noise, some savings can be made by eliminating the LNA.

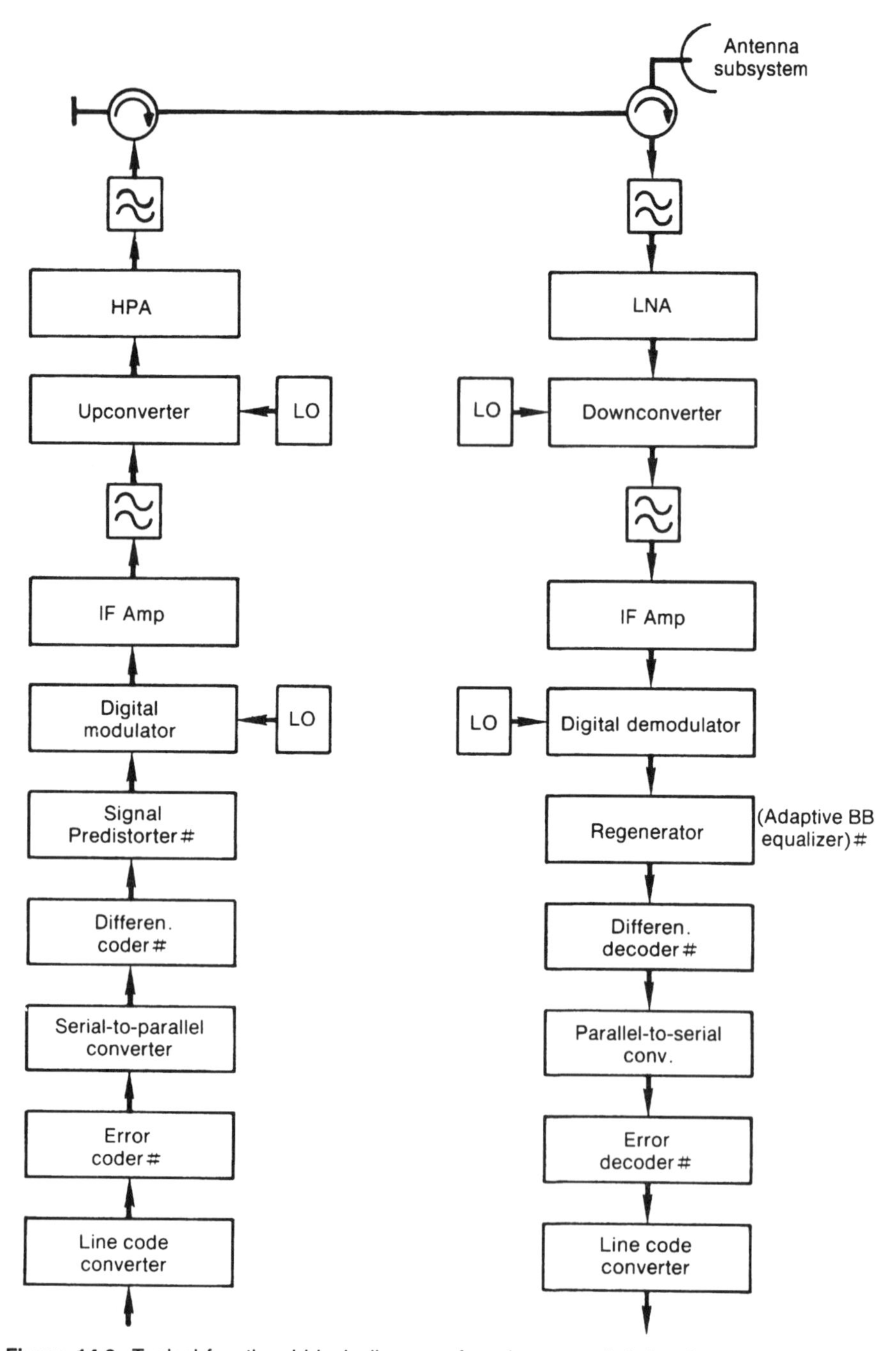

Figure 14.3. Typical functional block diagram of a microwave digital radio terminal. Typical PCM line codes: AMI, polar, BNZS, HDB3/CMI. BB = baseband, # = optional.

Common, well-designed downconverters have noise figures from 4 to 10 dB. One can estimate the noise figure of a downconverter by adding about 0.5 dB to the conversion loss. Thus a downconverter with a conversion loss of 6.5 dB will have a noise figure of about 7 dB. LNAs can display noise figures of less than 1 dB for operation below 8 GHz, and 1–3 dB for operation from 8 to 30 GHz using bipolar, GaAs FET, HEMT, and PHEMT technology. Low-noise mixers operating below 10 GHz can have noise figures from 1 to 2 dB. Remember that the system thermal noise is gated by this first active device.

The downconverter translates the incoming signal to the receive IF, which is commonly 70 MHz, or possibly 140 MHz. These are the most common IFs.

The output of the downconverter is then fed through a filter to an IF amplifier prior to being input to the digital demodulator. Excellent linearity throughout the receive chain is extremely important, particularly on high-bit-rate equipment using higher-order M-ary schemes such as 256- and 512-QAM.

Demodulation and regeneration are probably the most important elements in the digital microwave receiver. The principal purpose, of course, is to achieve a demodulator serial bit stream output that is undistorted and comparatively free of intersymbol interference (ISI).

Coherent demodulation is commonly used. The reference phase has to be maintained. The vector status of the modulated signal is compared with that of the carrier. However, since the carrier is not available in the received modulation signal, it has to be reproduced from it. One method to do this is to use a Costas loop for carrier regeneration.

After demodulation, the data clock at the incoming symbol rate is regenerated from the baseband digital bit stream. The clock is usually derived in a similar manner as the carrier. A phase-lock loop (PLL) is synchronized to the spectral component occurring at the clock rate. Here timing jitter is a major impairment that should be prevented or minimized. Specifications on jitter may be found in Update '96 to the *Reference Manual for Telecommunications Engineering* (Ref. 4), Section 7-2.11.6.

The demodulated signal is then regenerated by means of the regenerated clock and a sample-and-hold circuit is applied to regenerate the actual signal. For modulation schemes using M-PSK, the I and Q channel sequences are regenerated separately. The regenerative repeater may also incorporate a baseband equalizer to reduce signal distortion (see Section 3.5.1).

For the case of I and Q bit streams, the combining of these two bit streams into a single serial bit stream is accomplished in the parallel-to-serial converter. If FEC is implemented, the signal is then FEC decoded and the resulting signal is then conditioned for line transmission in the line converter. The signal converter (conditioner) also incorporates frame alignment circuitry to maintain such frame alignment as the signal leaves the terminal. These frame alignment circuits are, of course, peculiar to the digital format for which the equipment interfaces such as DS1, DS3, E1, E3, SONET, or SDH (synchronous digital hierarchy) waveforms.

14.3.1 Gray or Reflected Binary Codes

A certain group of codes is frequently used with higher-level M-QAM and QPR systems. These types of codes get their usefulness from the property that one and only one digit of the code changes in proceeding to or from the next higher or next lower point of the space constellation (see Figure 3.5). Thus there is no immediate instant that can be interpreted as a number that is in error by more than 1 bit in the least significant position.

A common code of this type is the reflected binary or Gray code as shown in Table 14.1. The code is useful because it is easy to convert from Gray to base-2 binary or vice versa. The two rules for conversion are the following. We define b_n and r_n to be the nth position measured from the right of the base-2 binary number and the reflected binary number, respectively. The first rule is that the most significant digit of both numbers are equal. The second rule is that the sum of $b_{n+1} + b_n + r_n$ is even. Thus a given reflected binary digit is a zero or one as the equivalent base-2 binary digit and the base-2 binary digit to its left is the same or different. In the conversion from reflected binary to base-2 binary, the leftmost digit is determined by the first rule and then the second rule is used to determine other digits in descending order of importance.

For example, in 256-QAM a noise hit causing a shift of one place in the space diagram (constellation) would cause 8 bits to be in error, whereas when Gray coded, this reduces to just 1 bit in error.

TABLE 14.1 Equivalent Numerical Representations, Binary, Reflected Binary, and Hexadecimal

Decimal	Binary	Reflected Binary	Hexa-decimal
0	0000	0000	0
1	0001	0001	1
2	0010	0011	2
3	0011	0010	3
4	0100	0110	4
5	0101	0111	5
6	0110	0101	6
7	0111	0100	7
8	1000	1100	8
9	1001	1101	9
10	1010	1111	A
11	1011	1110	B
12	1100	1010	C
13	1101	1011	D
14	1110	1001	E
15	1111	1000	F

14.3.2 The Antenna Subsystem for LOS Microwave Installations

For conventional microwave LOS radiolinks, whether digital or analog, the antenna subsystem offers more room for trade-off to meet minimum performance requirements than any other subsystem. Basically, the antenna subsystem looking outward from the transmitter must have:

- Transmission line (coaxial cable or waveguide)
- An antenna: a reflecting surface or device
- An antenna feed: a feed horn or other feeding device

In addition the antenna subsystem may have:

- Circulators or isolators
- Directional coupler(s)
- Phaser(s)
- Passive reflectors
- Radome
- Mounting device

14.3.2.1 Antennas. Below about 700 MHz, antennas used for point-to-point radiolinks are often Yagis and are fed with coaxial transmission lines. Above 700 MHz, some form of parabolic reflector-feed arrangement is used; 700 MHz is no hard-and-fast dividing line. Above 2000 MHz, the transmission line is usually a waveguide. As was previously pointed out, the same antenna is used for both transmission and reception. The essential requirements imposed on an antenna relate to the following characteristics:

- Antenna gain in the direction of the main beam. For LOS radiolinks, antenna gains of over 45 dB should be avoided because the half-power beamwidth (i.e., less than 1°) results in greatly increased requirements for tower and mounting stability and rigidity.
- Half-power beamwidth, which affects requirements for antenna and tower design.
- Sidelobe attenuation to reduce or prevent interference to/from other systems using the same frequency or adjacent frequencies.

The power radiated from or received by an antenna depends on its aperture area. The power gain G of an antenna over the area A relative to an isotropic antenna can be expressed by

$$G = \frac{4\pi\eta A}{\lambda^2} \tag{14.1}$$

where A = area of the aperture in the same units as λ
η = efficiency of the antenna aperture, usually 55% for LOS radio-links
λ = wavelength of the operating frequency (F_0)

The antenna gain in decibels is

$$G_{\text{dB}} = 20 \log F_{\text{MHz}} + 20 \log D_{\text{ft}} + 10 \log \eta - 49.92 \tag{14.2}$$

For an antenna with 55% efficiency the gain in decibels is

$$G_{\text{dB}} = 20 \log F_{\text{MHz}} + 20 \log D_{\text{ft}} - 52.5 \text{ dB} \tag{14.3}$$

where F = frequency (F_0) in megahertz
D = aperture diameter (e.g., for a parabolic dish, the diameter of the dish) in feet

Directivity is another term commonly used to describe antenna performance. Directivity is the antenna lobe pattern that actually determines the antenna gain. An antenna may radiate in any direction, but it usually suffices to know the directivity in the horizontal and vertical planes.

Beamwidth is another important parameter. Radiation patterns for antennas are often plotted in a form (simplified) as shown in Figure 14.4. The center of the graph represents the location of the antenna, and the field strength is plotted along radial lines outward from the center (on polar graph paper). The line at 0° is the direction of maximum radiation or what we have previously called the ray beam or main beam. For this simplified example at 30° either side of center, the voltage has dropped to 0.707 of its maximum

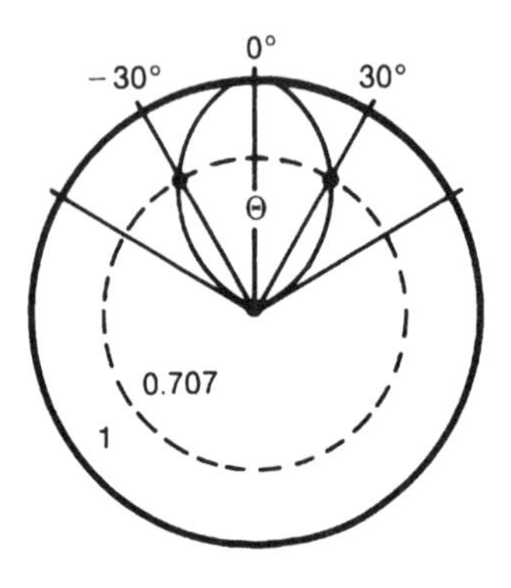

Figure 14.4. A typical (simplified) antenna pattern.

value. The decibel ratio of this voltage to the maximum is $20 \log(E_{max}/E) = 20 \log(1/0.707) = 3$ dB. These 3-dB points are considered to be a measure of the antenna directivity. In this simplified case, the antenna beamwidth $\theta_u = 2 \times 30° = 60°$. These diagrams are usually plotted directly in decibels rather than in terms of field strength.

VSWR (voltage standing wave ratio) is another important parameter used to describe antenna performance. It deals with the impedance match of the antenna feed point to the feed line or transmission line. The antenna input impedance establishes a load on the transmission line as well as on the radiolink transmitter and receiver. To have the RF energy produced by the transmitter radiated with minimum loss or the energy picked up by the antenna passed to the receiver with minimum loss, the input or base impedance of the antenna must be matched to the characteristic impedance of the transmission line or feeder.

Mismatch gives rise to reflected waves on the transmission line or standing waves. These standing waves may be characterized by voltage maxima (V_{max}) and minima (V_{min}) following each other at intervals of one-quarter wavelength on the line: VSWR = V_{max}/V_{min}. A similar parameter is the reflection coefficient (ρ), which is the ratio of the amplitude of the reflected wave to that of the incident wave. Both VSWR and ρ are representative of the quality of impedance match. They are related by

$$\rho = \frac{\text{VSWR} - 1}{\text{VSWR} + 1} \tag{14.4}$$

Return loss (RL) is another mismatch parameter. It is the decibel difference between the power incident on a mismatched discontinuity and the power reflected from the discontinuity. Return loss can be related to the reflection coefficient by

$$\text{RL}_{\text{dB}} = 20 \log(1/\rho) \tag{14.5}$$

Obviously, we would want a return loss as high as possible, in excess of 30 dB, and the reflected power as low as possible. VSWR should be less than 1.5 : 1.

Front-to-back ratio is still another measure of antenna performance. It is the ratio of the power radiated from the main ray beam to that radiated from the back lobe of the antenna. This is illustrated in Figure 14.4, where there is a small lobe extending from the back of the antenna. The ratio is expressed in decibels. For example, if an antenna radiates 20 times the power forward than back, its front-to-back ratio is 13 dB. Parabolic reflector antennas attain front-to-back ratios of 50–60 dB. The more efficient horn-reflector antennas can achieve as much as 70 dB. Figure 14.5 is a nomogram to determine gain of parabolic reflector antennas as a function of reflector diameter in feet and frequency in gigahertz.

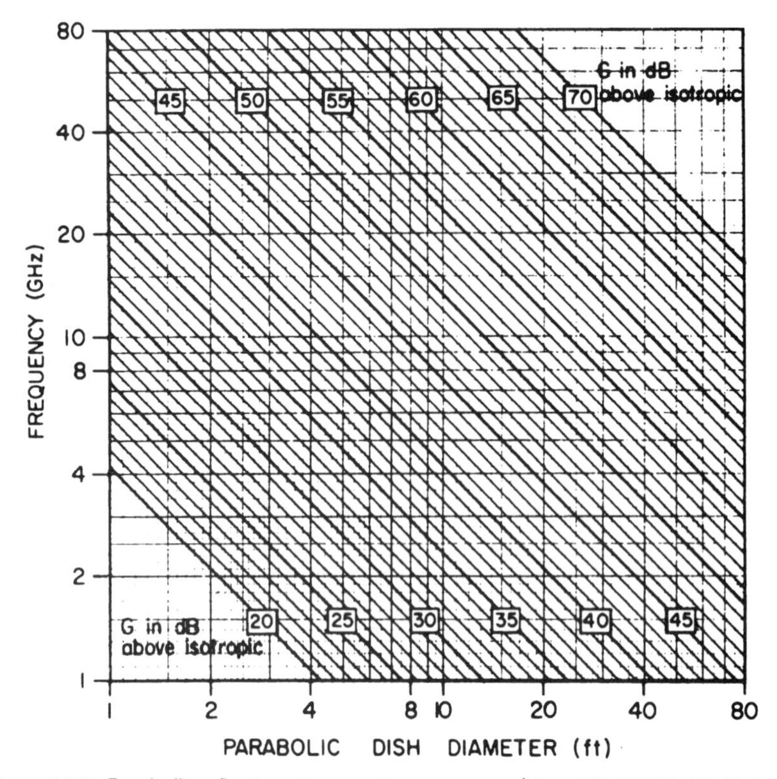

Figure 14.5. Parabolic reflector antenna gain nomogram. (From MIL-HDBK-416; Ref. 5.)

The operation of a parabolic reflector antenna is shown in Figure 14.6. The feed point is located at the focus F of the parabola. The drawing represents a cross section through a paraboloid of revolution. To calculate the beamwidth of an aperture antenna we turn to Jasik and Johnson (Ref. 6). A convenient rule-of-thumb for estimating the 3-dB beamwidth is

$$BW_{3\text{-dB}} = k\lambda/D \tag{14.6}$$

where λ is the wavelength and k is the beamwidth constant (use 70° here). D is the aperture dimension in the same unit as λ. D, in our case, is the diameter of the parabolic dish.

Example 1. A 2-meter parabolic dish operates at 6 GHz. What is its 3-dB beamwidth?

Calculate the wavelength equivalent of 6 GHz.

$$F\lambda = 3 \times 10^8 \text{ m/s} \tag{14.7}$$

$$\lambda = 0.05 \text{ m}$$

The BW = $70 \times 0.05/2 = 1.75°$.

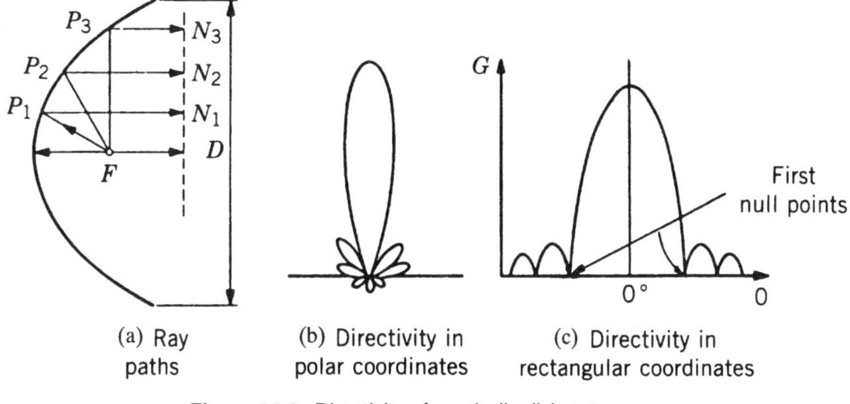

Figure 14.6. Directivity of parabolic dish antennas.

In practice, parabolic dishes are never illuminated uniformly, but the illumination tapers off toward the outer edge, reducing the overall gain somewhat. The taper acts to reduce the sidelobes, improving the front-to-back ratio, and reducing the potential for interference.

Antenna feeds are commonly waveguide horns (Figure 14.7), but dipole elements are sometimes used as radiators from about 300 MHz to approximately 3 GHz. Three types of parabolic antennas are illustrated in Figure 14.7. In Figure 14.7a, two types of feed horns are shown, the "button hook" type and the front-feed type. Such antennas permit only one polarization when a rectangular waveguide feed is used, and both horizontal and vertical polarizations with a square waveguide feed. Bandwidths are usually sufficient to cover several hundred megahertz with a fairly linear response.

The more efficient "Cassegrain" feed is shown in Figure 14.7b. This antenna is described in the earth station antenna subsection. Figure 14.7c shows a shield mounted around the edge of the dish to suppress sidelobes and back lobes. The inside surface of the shield is often lined with absorbing material to prevent reflections.

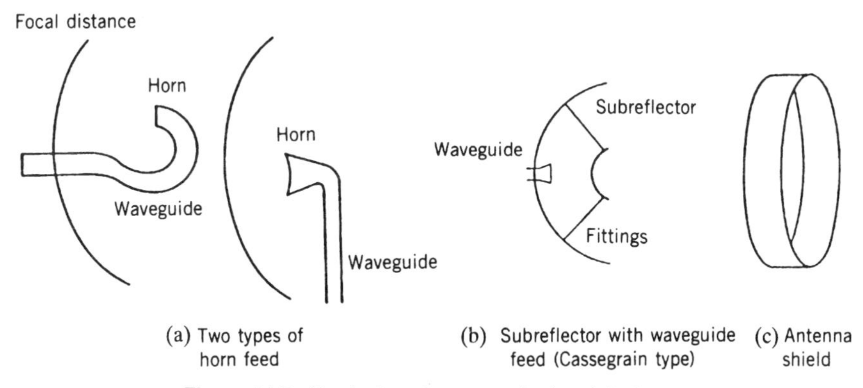

Figure 14.7. Parabolic antennas and related devices.

14.3.2.2 RF Transmission Lines. Two types of transmission line are used in radiolink terminals and repeaters: coaxial cable and waveguide. Coaxial cable, in general, is easier to install. Its loss increases exponentially with frequency, and, as a result, its upper limit of application is in the range of 2–3 GHz. Figures 14.8 and 14.9 give data on coaxial cable loss versus frequency.

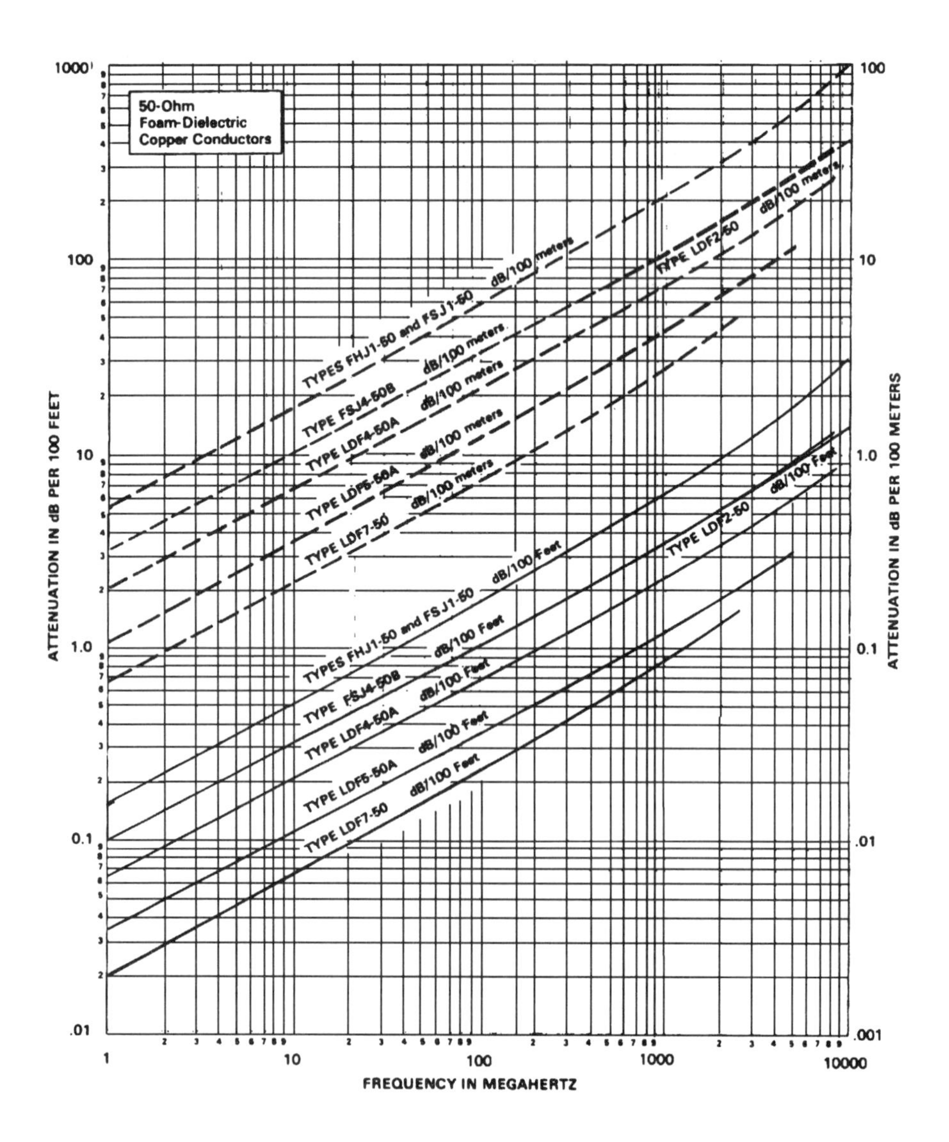

Figure 14.8. Loss versus frequency for foam dielectric coaxial cable. Attenuation curves based on: VSWR 1.0: Ambient temperature 24°C (75°F). (Courtesy of the Andrew Corporation; Ref. 7.)

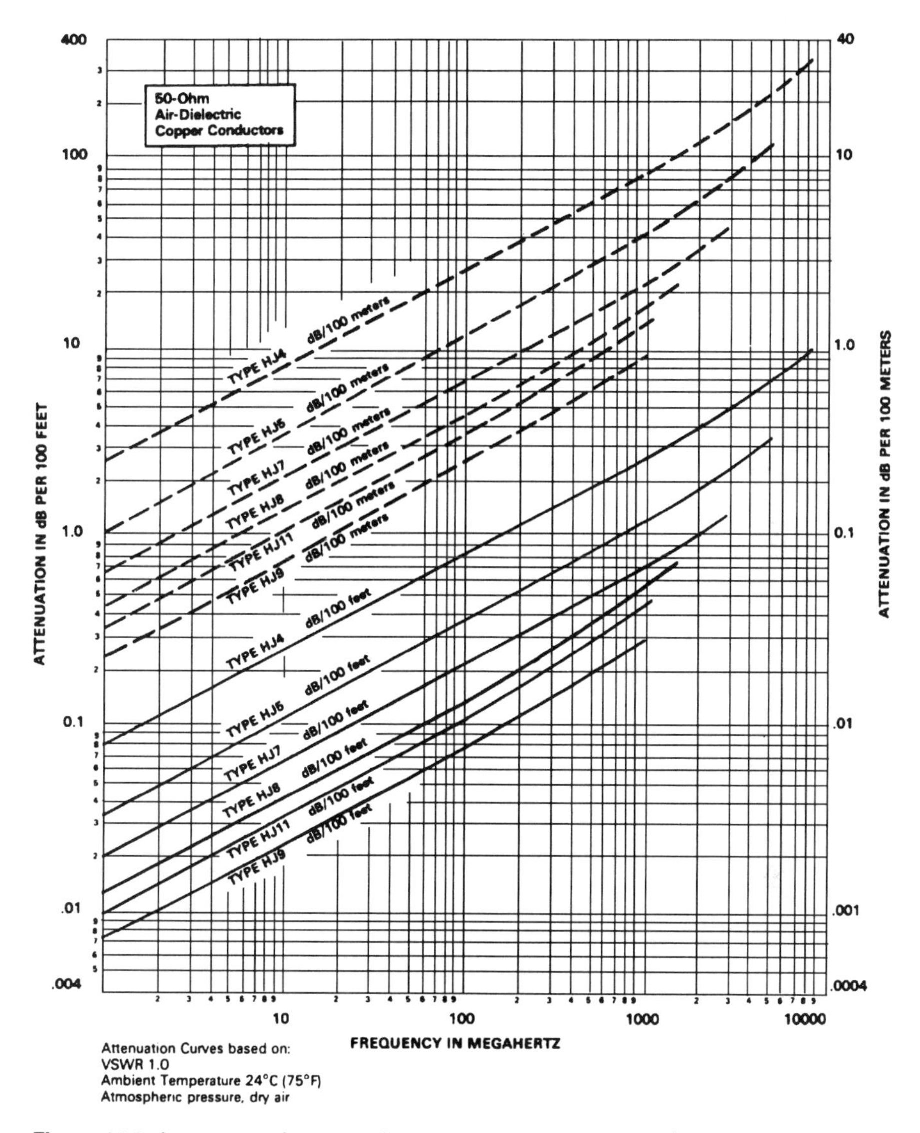

Figure 14.9. Loss versus frequency for air dielectric coaxial cable. (Courtesy of the Andrew Corporation; Ref. 7.)

There are a number of important parameters to be considered for the application of coaxial cable as a transmission line. Probably the most important for the system engineer is attenuation or loss as shown in Figures 14.8 and 14.9. Loss varies with ambient temperature. The reference value in the figures is 24° C (75° F). Figure 14.10 shows how loss varies with ambient temperature.

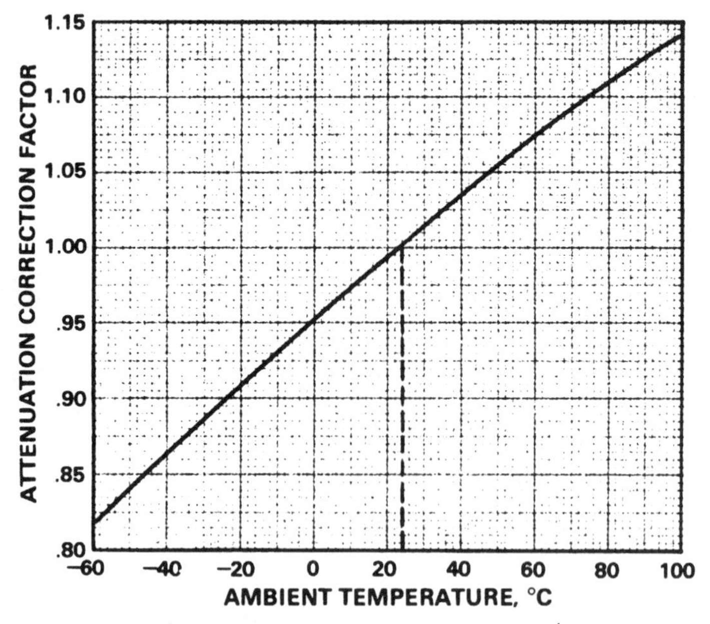

Figure 14.10. Variation of attenuation with ambient temperature. (Courtesy of the Andrew Corporation; Ref. 7.)

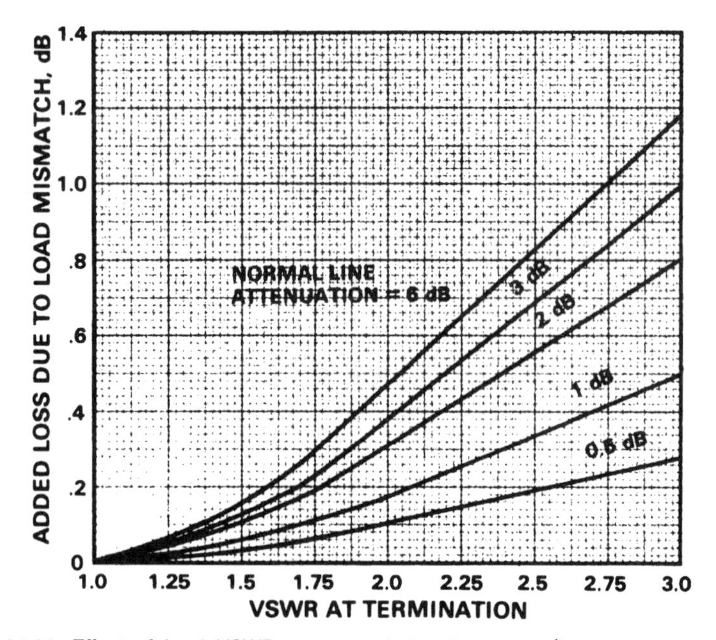

Figure 14.11. Effect of load VSWR on transmission line loss. (Courtesy of the Andrew Corporation; Ref. 7.)

VSWR is another important parameter as previously described. VSWR can effectively increase transmission line loss. Such additional loss is called mismatch loss. This is shown in Figure 14.11. However, the effect is quite small for normal operating conditions.

The power rating of the line is another important parameter. Typically, peak power ratings limit the amplitude modulation or pulsed usage, while average power ratings limit the CW usage. The peak power rating is a function of the insulation material and structure between the inner and outer conductors. Voltage breakdown is independent of frequency but varies with line pressure (see subsequent discussion) and type of pressurizing gas. Voltage breakdown can result in permanent damage to the cable.

Waveguide is superior to coaxial cable in attenuation characteristics, particularly at the higher frequencies, and will handle higher power levels. For the lower frequencies (e.g., below about 3 GHz), the choice between coaxial cable and waveguide is economic, not only for the cost of the transmission line, but its installation. There are three types of waveguide in common use: rectangular, elliptical, and circular. Rectangular waveguide is that which is most commonly associated with microwave installations. However, generally, elliptical (flex) or circular waveguides are favored because of their low-loss properties. For ease of installation, elliptical waveguide, often called "flex," is the most commonly used for installations operating below 20 GHz.

Circular waveguide displays minimum loss and is particularly suited for long vertical waveguide runs to tower-mounted antennas.

Most of the performance parameters applicable to coaxial cable are also applicable to waveguide. Figure 14.12 gives loss versus frequency for a number of the more commonly used waveguide types. Waveguide types are abbreviated: WR for rectangular, EW for elliptical, and WC for circular.

All air-dielectric waveguides, coaxial cables, and rigid lines are maintained under dry gas pressure to prevent electrical performance degradation. If a constant positive pressure is not maintained, "breathing" can occur with temperature variations. This permits moisture to enter the line causing increased loss, increased VSWR, and a path for voltage breakdown. One pressurizer/dehydrator can usually serve a number of waveguide installations in a common location.

14.3.2.3 Waveguide Devices: Separating and Combining Elements—Filters and Directional Couplers. By means of separating/combining networks, groups of several transmitters and receivers are connected to the same antenna. These include circulators, isolators, branching network, and combining networks. Figure 14.13 shows the various applications of these devices.

A waveguide circulator is used to couple two or three microwave radio equipments to a single antenna. A circulator consists essentially of three basic waveguide sections combined into a single assembly and is commonly a four-port device. The center section is a ferrite nonreciprocal phase shifter. An external permanent magnet causes the ferrite material to exhibit phase-

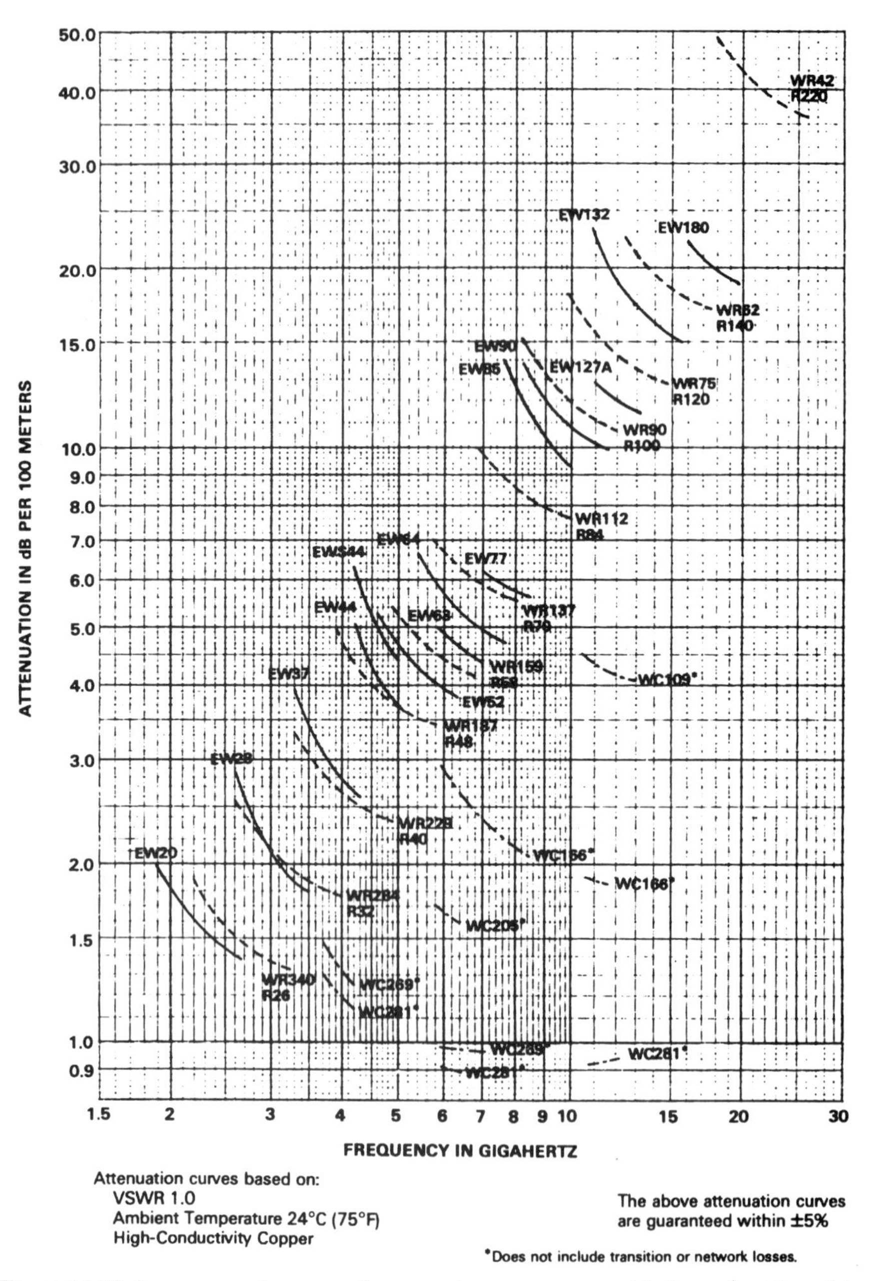

Figure 14.12. Loss versus frequency for several common waveguide types (metric units). (Courtesy of the Andrew Corporation; Ref. 7.)

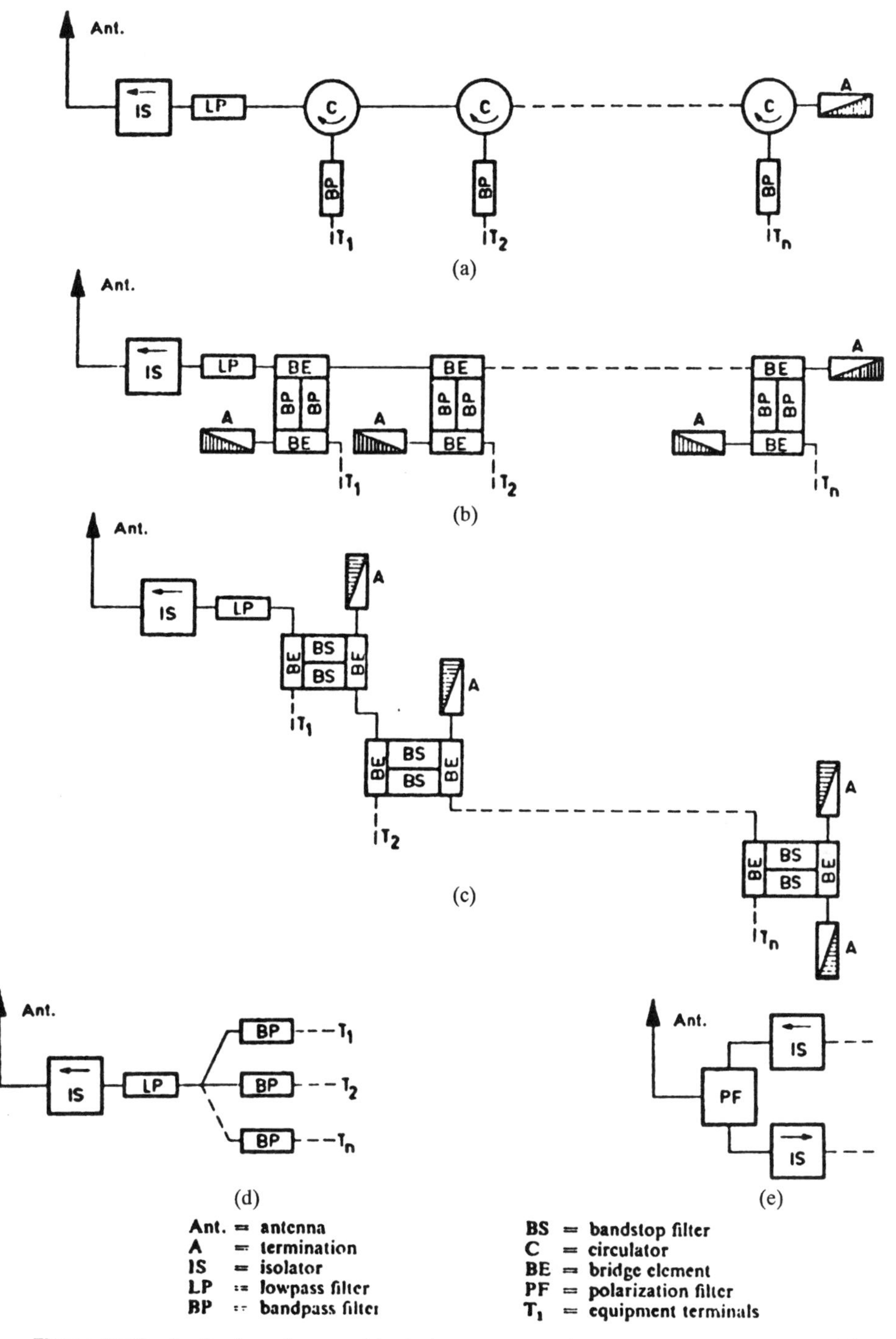

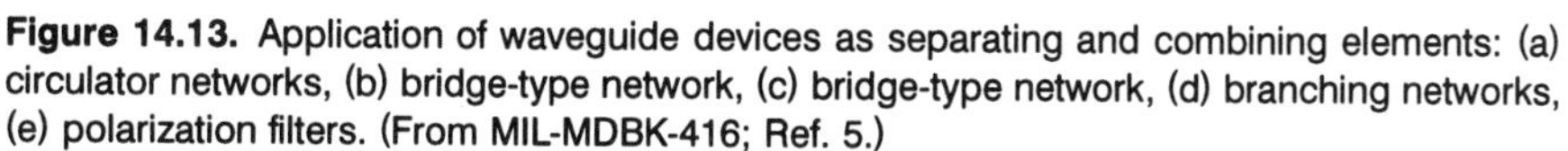

Figure 14.13. Application of waveguide devices as separating and combining elements: (a) circulator networks, (b) bridge-type network, (c) bridge-type network, (d) branching networks, (e) polarization filters. (From MIL-MDBK-416; Ref. 5.)

shifting characteristics. Normally, an antenna transmission line is connected to one arm and either three radio equipments or two equipments and a shorting plate are connected to the other three arms. Attenuation in a clockwise direction from arm to arm is low, on the order of 0.5 dB, whereas in the counterclockwise direction it is high, on the order of 20 dB. Figure 14.13a shows a typical application of circulators in a microwave radiolink antenna subsystem.

Bridge networks consist of filter networks and four-arm bridge elements such as 3-dB directional couplers or "magic tees." Two bridge elements are connected by two identical filters to produce a separating-filter element. Of the four ports of a separating-filter element, one is connected to the equipment terminal and another to a termination. The two remaining ports are connected to neighboring separating-filter elements or one of two ports is connected to the antenna transmission line or a termination. (See Figures 14.13b and 14.13c.)

Branching networks (Figure 14.13d) connect multiple equipments to a single antenna by means of bandpass filters. Polarization filters (Figure 14.13e) combine/separate polarizations from/to a common antenna.

A load isolator is a ferrite waveguide component that provides isolation between a single source and its load. A typical source is a HPA and the load is the antenna. These are commonly used in troposcatter installations. They reduce the ill effects of higher VSWRs, serve to protect the transmitter from high values of reflected power, and, in some instances, have to be cooled. Owing to the ferrite material with its associated permanent magnetic field, ferrite load isolators have a unidirectional property. Energy traveling toward the antenna is relatively unattenuated, whereas energy traveling back from the antenna undergoes fairly severe attenuation. The forward and reverse attenuations are on the order of 1 and 40 dB, respectively.

A directional coupler is a power splitter. It is a relatively simple waveguide device that divides the power on a transmission line, usually the power to/from the antenna. A 3-dB power split device divides the power in half; such a device could be used, for instance, to permit radiation of the power from a transmitter on two different antennas. A 20- or 30-dB power split has an output that serves to sample the power on a transmission line. It is most commonly used for VSWR measurements by measuring the forward and reverse power.

14.3.3 Analog Radiolink Repeaters

Radiolink repeaters amplify the signal along the radio route, providing gain on the order of 110 dB. For *analog* systems there are three types of repeaters that can carry out this function:

- Baseband repeaters
- IF repeaters
- RF repeaters

For a FDM/FM system, a block diagram of a typical baseband or demodulating repeater is shown in Figure 14.14. A baseband repeater is required, if, at the repeater relay point, there will be drops and inserts. Such repeaters are often located at or near a telephone switching center. All or part of the baseband may be dropped or inserted. A baseband repeater will insert the same amount of noise into the system as a terminal facility. To reduce some of the noise inserted by the accompanying FDM equipment, through-group and through-supergroup techniques are used for routing through traffic. Most of the gain obtained with this type of repeater is obtained at the IF and through the demodulation/remodulation process.

An IF repeater may be used when there are no drops and inserts at the repeater facility. An IF repeater inserts less noise into the system because there are less modulation/demodulation steps required to carry out the repeating process. An IF repeater eliminates two modulation steps. It simply translates the incoming RF signal to IF with the appropriate local oscillator and mixer, amplifies the derived IF, and then upconverts it to a different radio frequency. The upconverted frequency may then be amplified by a TWT or SSA (solid-state amplifier). Figure 14.15 is a simplified functional block diagram of a typical IF repeater.

With a RF repeater amplification is carried out directly at radio frequencies. The incoming RF signal is amplified, translated in frequency, and then amplified again. RF repeaters are seldom used. They are troublesome in their design with such things as sufficient selectivity, limiting and automatic gain control, and methods to correct group delay.

14.3.4 Diversity Combiners

Diversity combiners are used to combine signals from two or more diversity paths. They are also used in some hot-standby applications. Hot-standby operation improves link availability by switching in standby equipment when an on-line unit fails. The combiner removes at least one of the switching requirements. Hot standby is discussed in Section 14.2.5.

14.3.4.1 Classes of Diversity Combiners. There are two generic classes of diversity combiners: predetection and postdetection. Of course, this classification is made in accordance with where the combining function takes place, before or after detection.

A simplified functional block diagram of a predetection combiner is shown in Figure 14.16. In this case the combining is carried out at IF, and phase control circuitry is required to maintain signal coherency of the two (or more) signal paths. If selection combining is used (Section 14.2.4.2), however, this control is unnecessary, since only one signal at a time is on line.

Most systems today use postdetection or baseband combining. Figure 14.17 is a functional block diagram of a typical postdetection combiner.

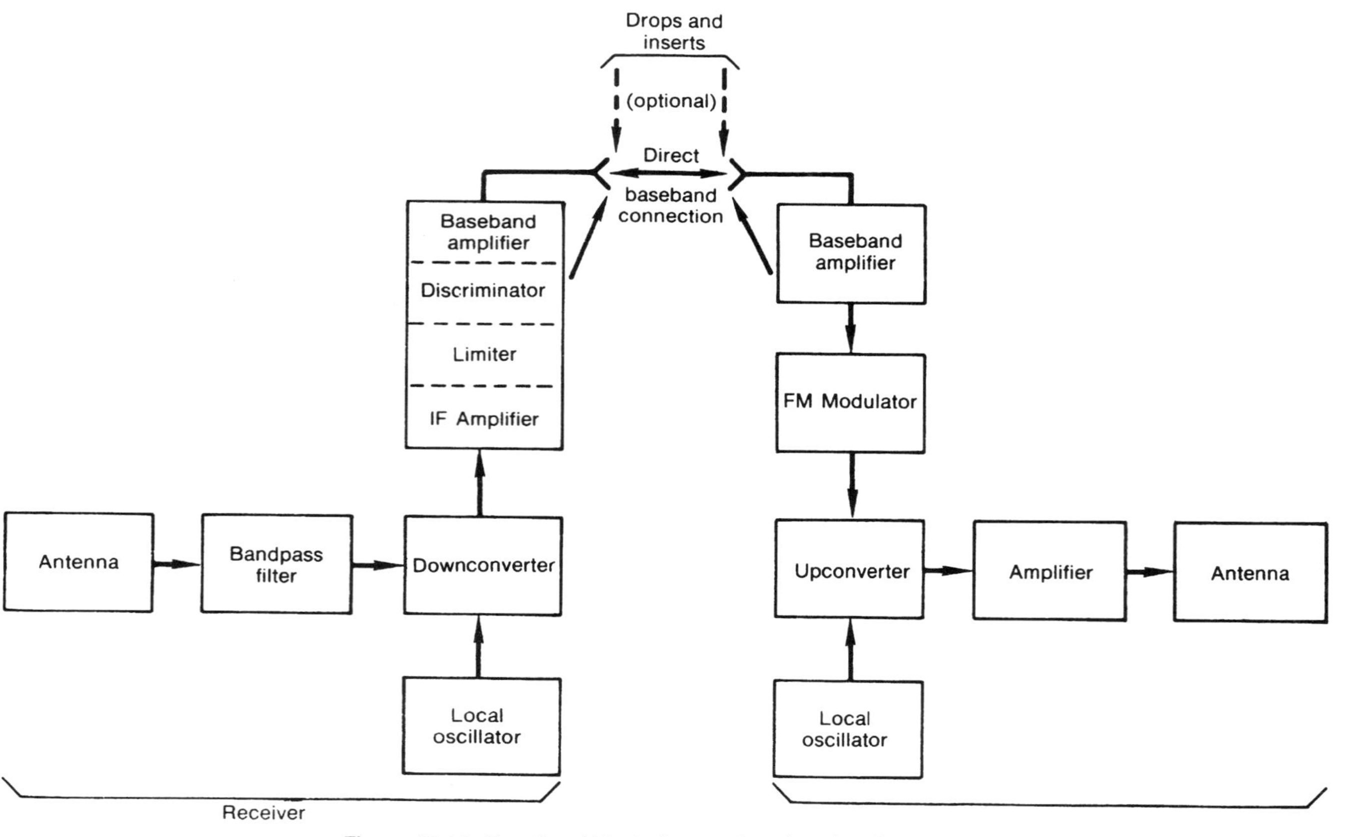

Figure 14.14. Functional block diagram for a baseband repeater.

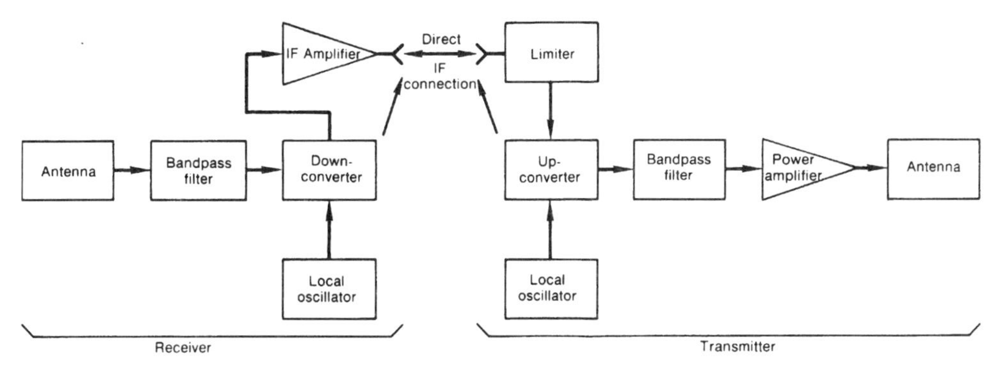

Figure 14.15. Functional block diagram of an IF repeater.

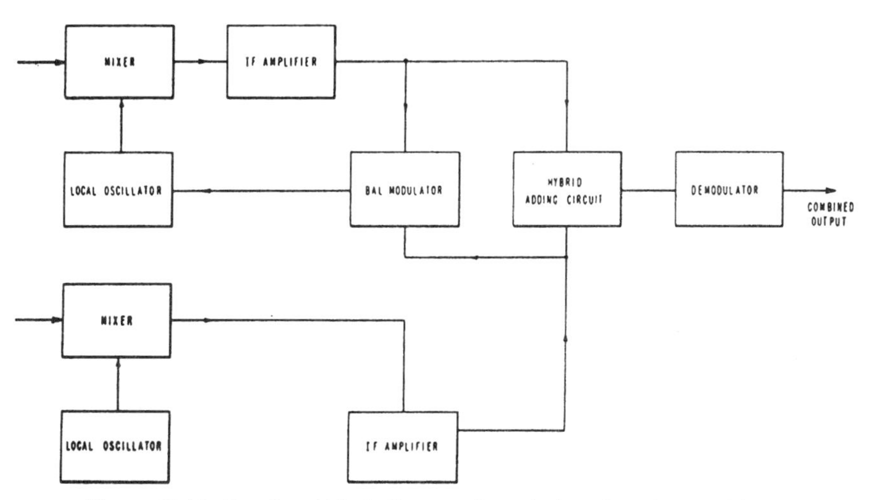

Figure 14.16. Functional block diagram of a typical predetection combiner.

14.3.4.2 Methods of Combining.* There are three general methods of combining: selection combining, equal gain combining, and maximal ratio combining. Figure 14.18 shows a simplified block diagram of each of these combining methods in a typical receiving system.

Combiner performance characteristics are comparatively illustrated in Figure 14.19. The following discussion compares the three types of combiners, and it will be assumed that in each case:

- Signals add linearly; noise adds in a rms manner.
- All receivers have equal gain.

*Section 14.3.4.2 is based on abridged material from Ref. 8.

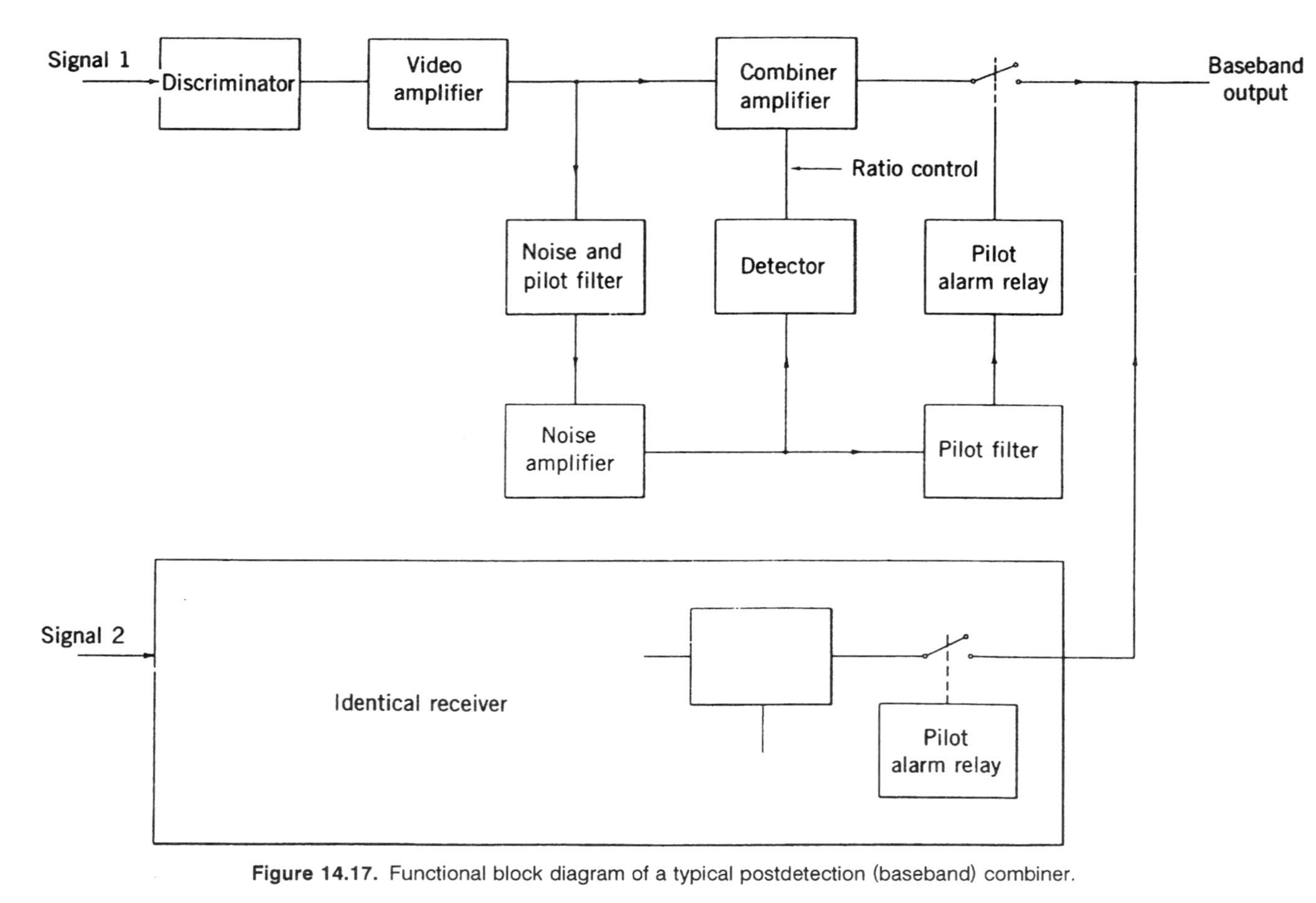

Figure 14.17. Functional block diagram of a typical postdetection (baseband) combiner.

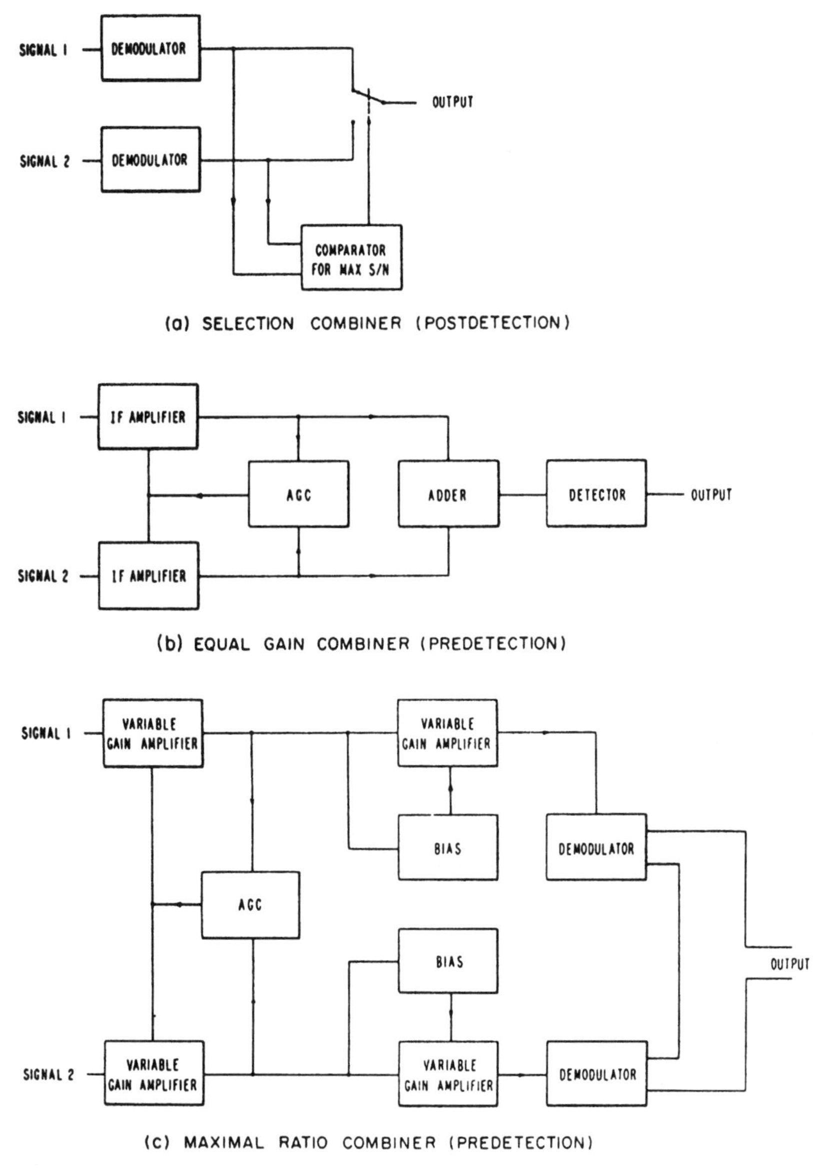

(a) SELECTION COMBINER (POSTDETECTION)

(b) EQUAL GAIN COMBINER (PREDETECTION)

(c) MAXIMAL RATIO COMBINER (PREDETECTION)

Figure 14.18.

- All receivers have equal noise outputs; the noise is random in character.
- The desired output signal-to-noise power ratio S_0/N_0 is a constant.

For the case of the selection combiner, only one receiver at a time is used. The output signal-to-noise ratio is equal to the input signal-to-noise ratio of

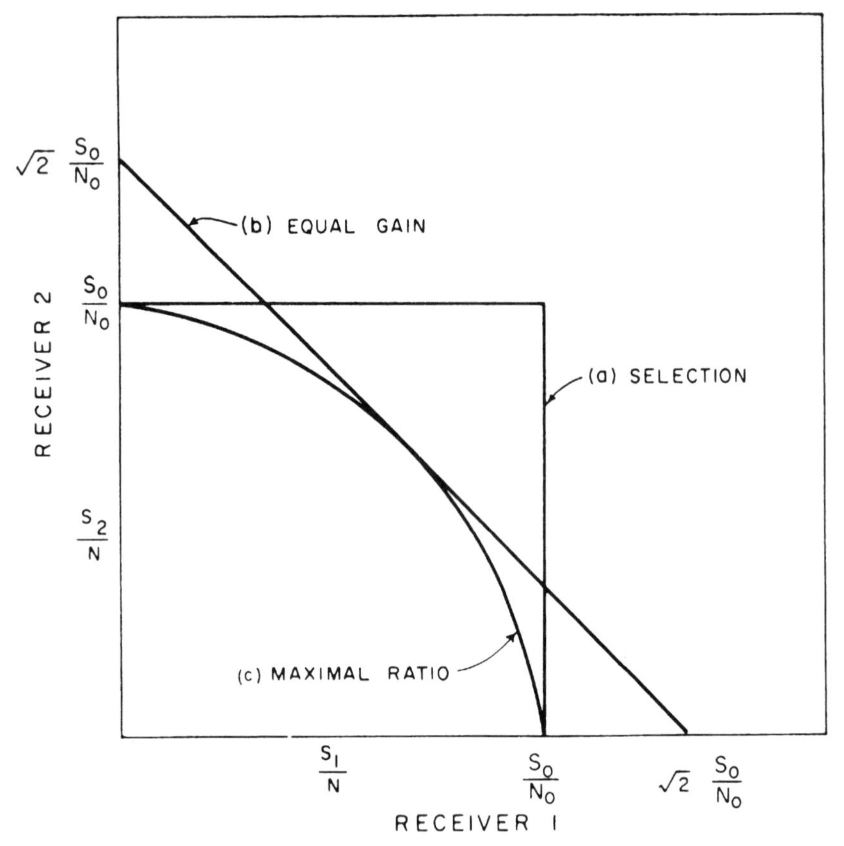

Figure 14.19. Combiner performance characteristics.

the output of the selected receiver at the time. Curve (a) refers in Figure 14.19.

With the equal gain combiner, the different inputs are only added and when two receivers are in use

$$\frac{S_0}{N_0} = \frac{S_1 + S_2}{\sqrt{N_1^2 + N_2^2}} = \frac{S_1 + S_2}{\sqrt{2}N} \tag{14.8}$$

This is illustrated in curve (b) in Figure 14.19. S and N in Figure 14.19 are the signal and noise levels, respectively, for receivers 1 and 2.

The maximum ratio or ratio-squared combiner uses a relative gain change between the output signals of the receivers in use. For example, if the stronger signal has unity output and the weaker signal has an output proportional to G, then

$$\frac{S_0}{N_0} = \frac{S_1 + GS_2}{\sqrt{N_1^2 + G^2 N_2^2}} = \frac{S_1 + GS_2}{N\sqrt{1 + G^2}} \tag{14.9}$$

Maximizing the above expression by differentiating and equating to zero yields

$$G = \frac{S_2}{S_1} \tag{14.10}$$

In other words, this signal gain is adjusted to be proportional to the ratio of the input signals. Then

$$\left(\frac{S_0}{N_0}\right)_{max} = \sqrt{\frac{S_1^2 + S_2^2}{N}} \tag{14.11}$$

$$\left(\frac{S_0}{N_0}\right)^2_{max} = \left(\frac{S_1}{N}\right)^2 + \left(\frac{S_2}{N}\right)^2 \tag{14.12}$$

which is the equation of a circle and is illustrated in Figure 14.19 by curve (c).

14.3.4.3 Comparison of Combiners. The preceding discussion indicates that the maximal ratio combiner utilizes the best features of the two other combiners. When one signal is zero, the maximal ratio combiner acts as a selector combiner. When both signals are of equal level, it acts as an equal-gain combiner. It will also be noted that the maximum ratio combiner yields the best output for any combination of S/N since the curve of operation shows that the optimum output occurs for lower values of input S/N as evidenced by its proximity to the origin.

Figure 14.20 graphically illustrates another comparison based on the difference in output to be expected from each of the three combiners as the number of independent diversity paths increases. If the signals on each of the paths are assumed to be Rayleigh distributed, then a statistical analysis of the output from each of these combiners results in the comparison illustrated in the figure. For quadruple diversity, which was discussed in Chapter 3, the *average* signal-to-noise ratio of the output using the selection combiner is about 3 dB better than the nondiversity case, the equal-gain combiner is about a 5.25-dB improvement, and the maximal ratio combiner yields about a 6-dB improvement.

If the comparison of the combiners is done on the basis of a time distribution of the output signal-to-noise ratio rather than on the basis of average value (Figure 14.20), a new significance of combiner operation is

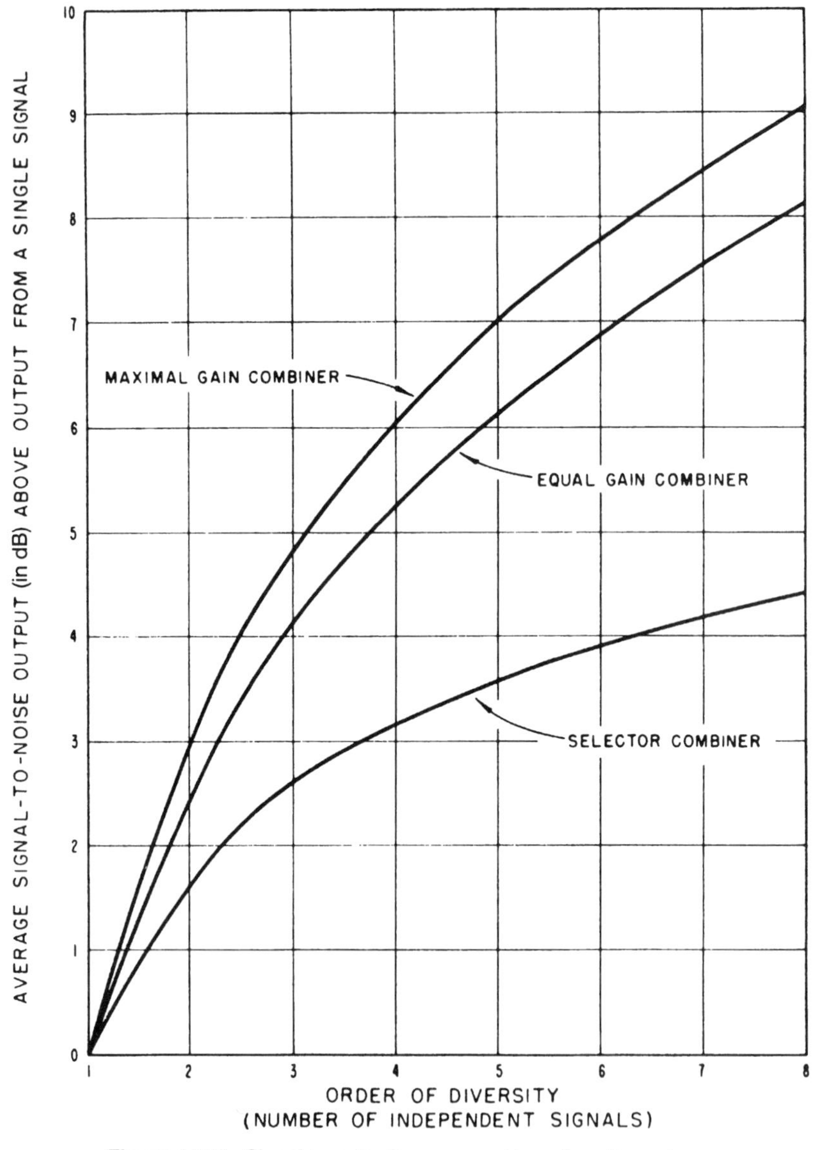

Figure 14.20. Signal-to-noise improvement in a diversity system.

revealed. Consider that a time distribution is the percentage of a time interval where the signal amplitude exceeds particular signal levels. In Figure 14.21, the Rayleigh distribution has a time distribution where the signal amplitude is approximately 5.2 dB above the median value for 10% of the time intervals of the measurement, whereas, for 90% of that interval, the signal amplitude exceeded a level that is 8.2 dB below the median. Where Rayleigh-distributed signals are combined according to the principles used in

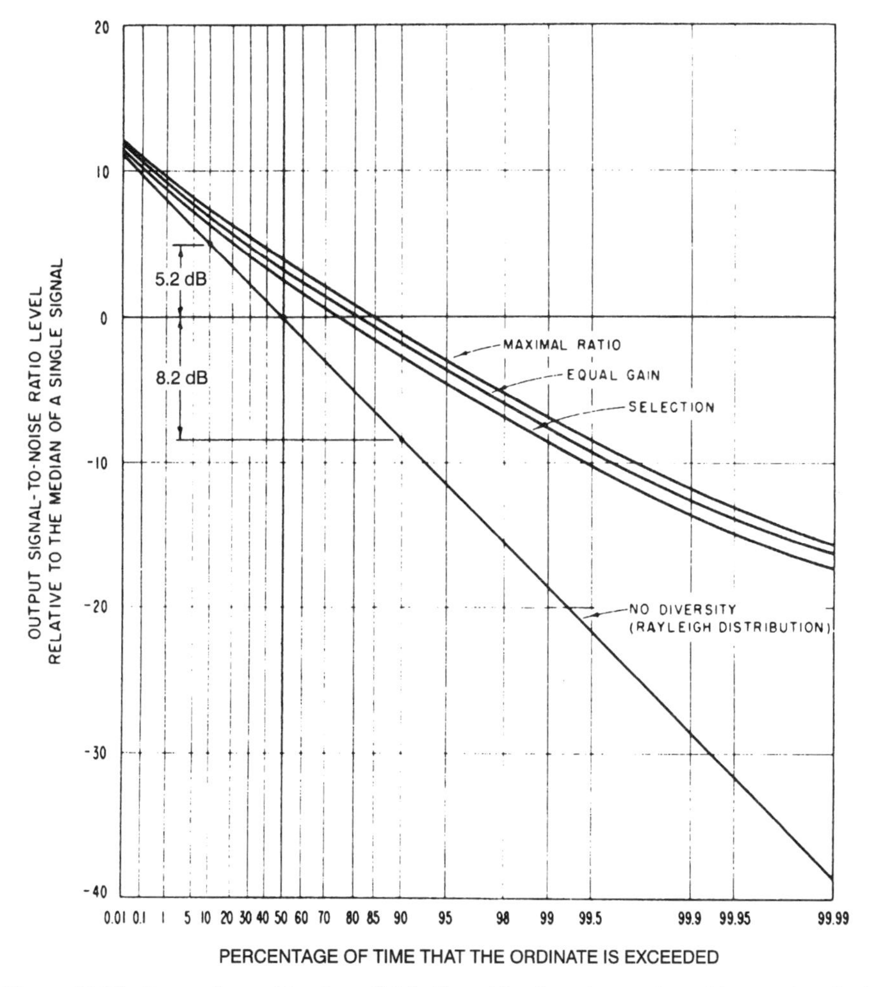

Figure 14.21. Comparison of the time distribution of the three types of combiners using dual diversity.

these three types of combiners, the output can be predicted as a function of the type of combiner and the order of diversity. Figure 14.21 shows the time distribution comparison for dual diversity with each combiner type and Figure 14.22 is for quadruple diversity. In the case of dual diversity, the signal-to-noise ratio is exceeded 99.9% of the time and is improved about 15 dB; for quadruple diversity, the SNR is improved at least 23 dB.

14.3.5 Hot-Standby Operation

Radiolinks commonly provide transport of multichannel telephone service and/or point-to-point broadcast television on high-priority backbone routes. A high order of route reliability is essential. Route reliability depends on

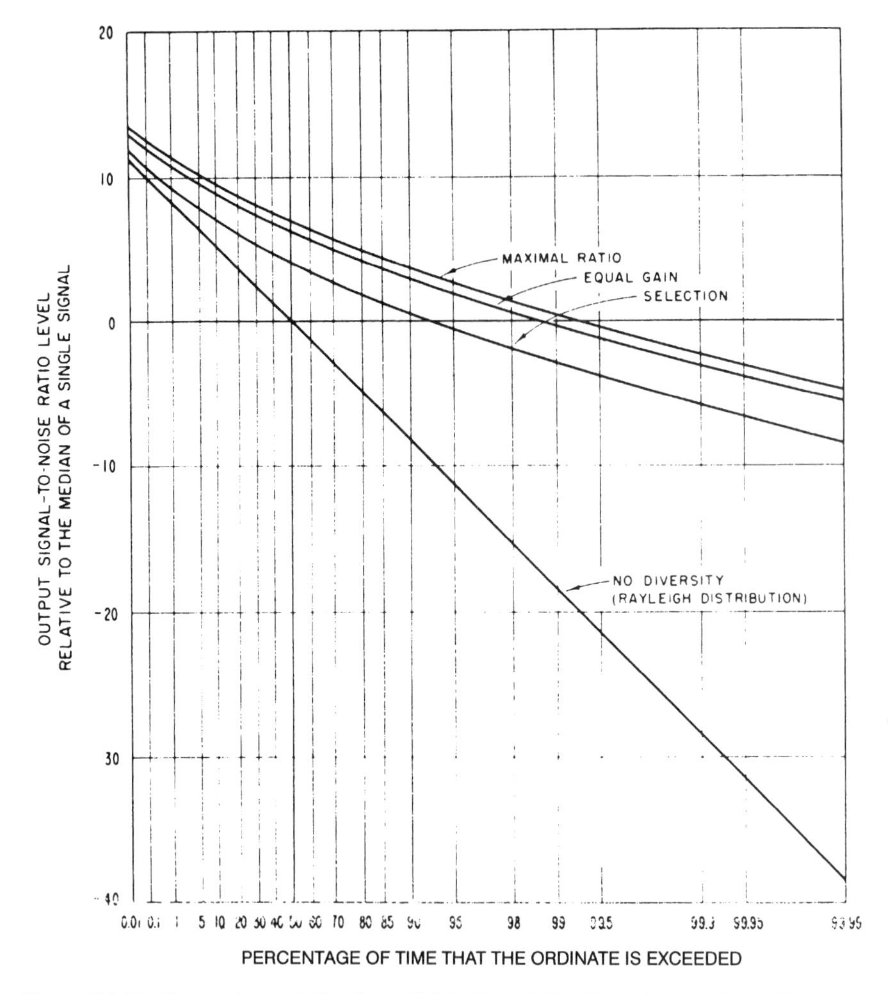

Figure 14.22. Comparison of the time distribution of the three types of combiner using quadruple diversity.

path reliability or link time availability (propagation) and equipment/system reliability. Redundancy is one way to achieve equipment reliability to minimize downtime and maximize link availability regarding equipment degradation or failure.

One straightforward way to achieve redundancy effectively is to provide a parallel terminal/repeater system. Frequency diversity effectively does just this. With this approach all equipment is active and operated in parallel with two distinct systems carrying the same traffic. This is expensive, but necessary, if a high order of link reliability is desired. Here we mean route reliability.

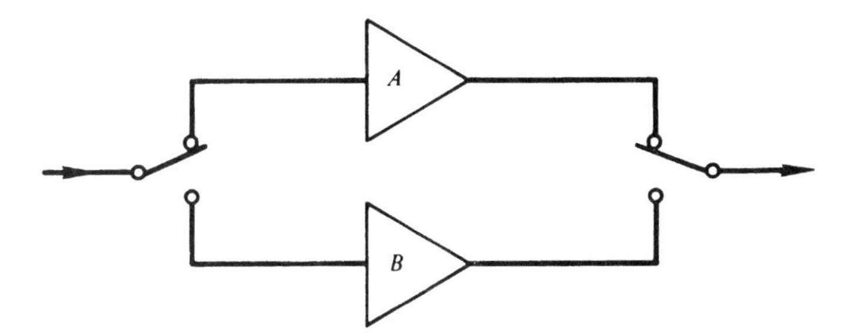

Figure 14.23. The concept of hot-standby protection.

Often the additional frequency assignments to permit operation in frequency diversity are not available. When this is the case, the equivalent equipment reliability may be achieved by the use of a hot-standby configuration. Figure 14.23 illustrates the hot-standby concept.

The equipment marked A and B in Figure 14.23 could be single modules, shelves, or groups of modules, or whole equipment racks. On a complex equipment, such as a radiolink terminal, whether digital or analog, more than one set of protection can exist.

On a radiolink terminal such as shown in Figure 14.24, the sections marked baseband and RF are both hot-standby protected. The operation of the protection system of these two sections, however, is independent. The protection system is broken down further in that the protection mechanisms for the transmit and receive paths operate independently, as shown in Figure 14.25. It should be noted that the switches ahead of each set of modules have been replaced by a signal splitter. This technique allows the signal to be fed into each set of modules simultaneously.

As the expression indicates, hot standby is the provision of parallel redundant equipment such that this equipment can be switched in to replace the operating on-line equipment nearly instantaneously when there is a failure in the operating equipment. The switchover can take place in the order of microseconds or less. The changeover of a transmitter and/or receiver line can be brought about by a change, over/under a preset amount,

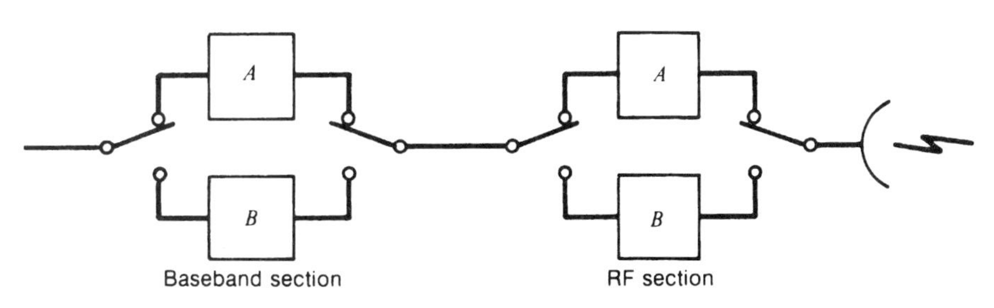

Figure 14.24. Hot-standby radio.

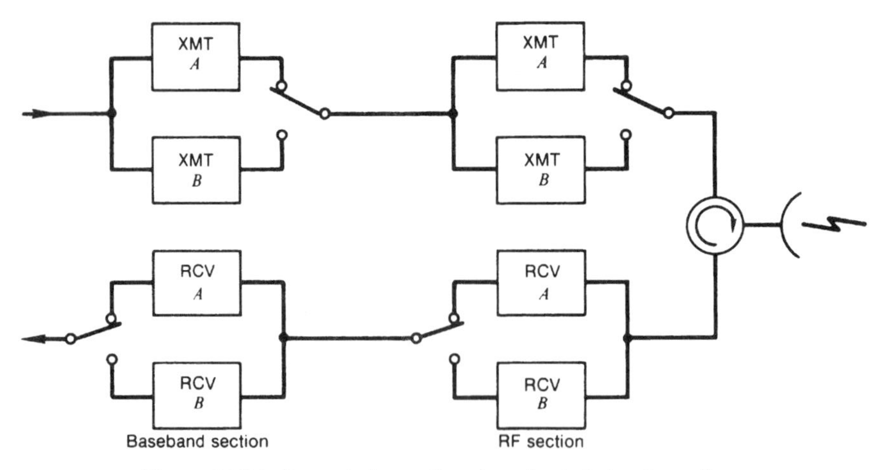

Figure 14.25. Separate transmit and receive hot-standby radio.

in one of the following values: for a transmitter,

- Frequency
- RF power
- Demodulated baseband (radio) pilot level

and for a receiver,

- AGC voltage
- Squelch
- Received pilot level
- Degraded bit error rate for a digital system
- Frame misalignment for a digital system

(*Note*: Digital systems do not use pilot tones.)

Hot-standby-protection systems provide sensing and logic circuitry for the control of waveguide switches (or coaxial switches where appropriate), in some cases IF switches as well as baseband switches on transmitters, and IF and baseband output signals on receivers. The use of a combiner on the receiver side is common with both receivers on line at once.

There are two approaches to the use of protection equipment. These are called one-for-one and one-for-*n*. One-for-one operation provides one full line of standby equipment for each operational system. See Figure 14.26. One-for-*n* provides only one full line of equipment for *n* operational lines of equipment, where *n* is greater than one. Figure 14.27 illustrates a typical one-for-four configuration.

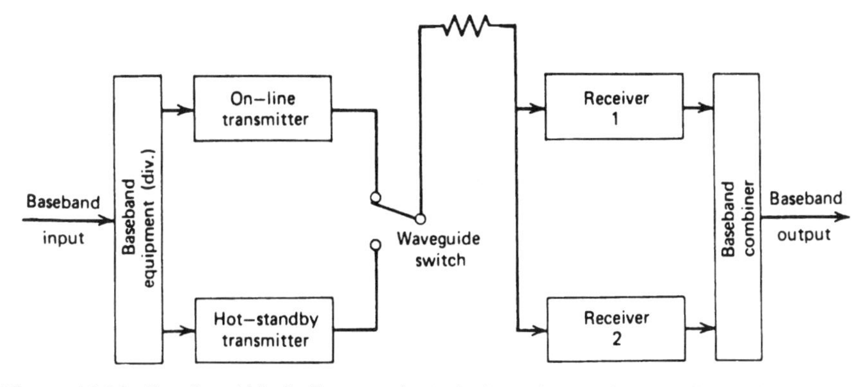

Figure 14.26. Functional block diagram of a typical one-for-one hot-standby configuration.

One-for-one is more expensive but provides a higher order of reliability. Its switching system is comparatively simple. One-for-*n* is more economic, with only one line of spare equipment for several operational lines. It is less reliable (i.e., suppose there were equipment failures in two lines of operational equipment of the *n* lines), and switching is considerably more complex.

Figure 14.28 shows a typical digital hot-standby configuration with space diversity.

14.3.6 Pilot Tones

On analog radiolinks a radio continuity pilot tone (or tones) is inserted on a link-by-link basis. These pilot tones are usually independent of multiplex pilot tones. The pilot tone is used for:

- Gain regulation
- Monitoring (fault alarms)
- Frequency comparison
- Measurement of level stability
- Control of diversity combiners

The last application involves the simple sensing of continuity by a diversity combiner. The presence of the continuity pilot tone tells the combiner that a particular diversity path is operative. The problem arises from the fact that most diversity combiners are postdetection and use noise as the means to determine the path contribution to the combined output. The path with the least noise, as in the case of the maximal ratio combiner, provides the greatest path contribution.

If, for some reason, a diversity path were to fail, it would be comparatively noiseless and would provide 100% contribution. Thus there would be no

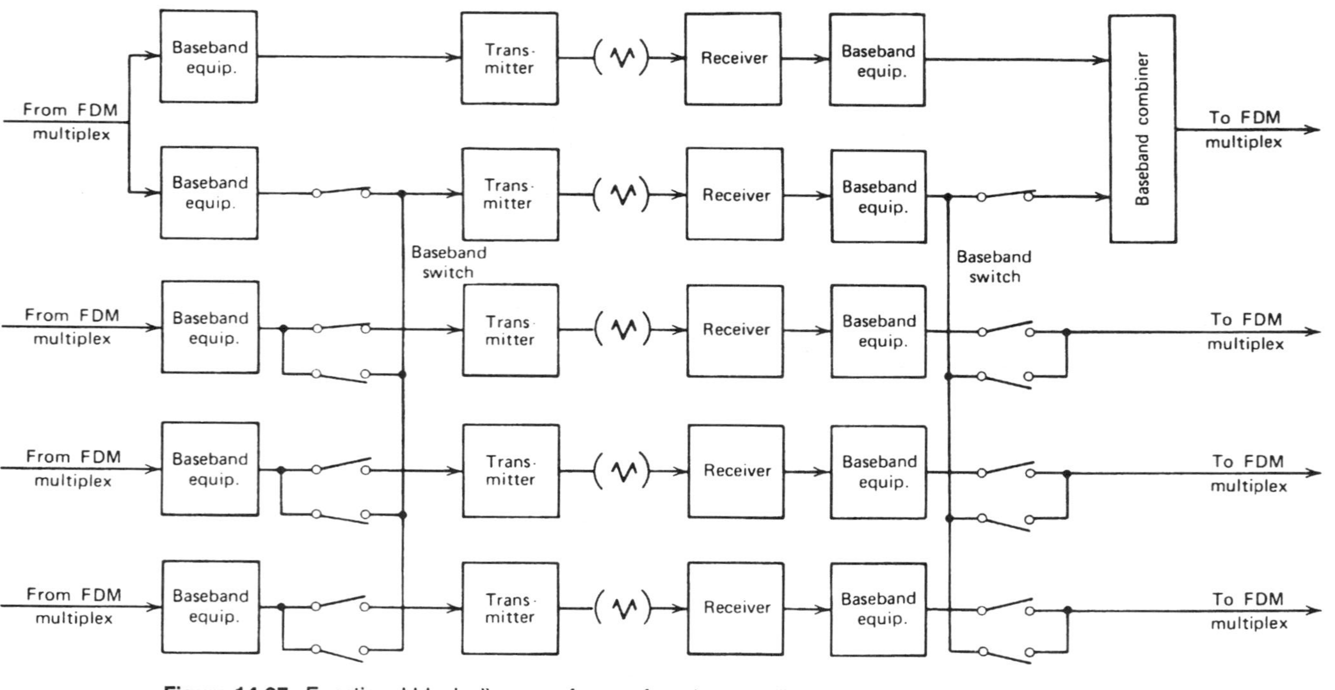

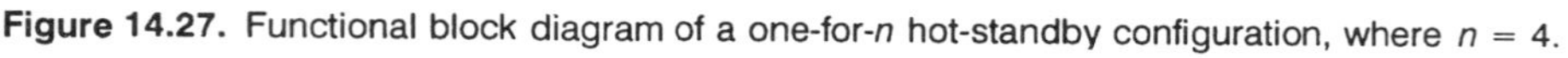
Figure 14.27. Functional block diagram of a one-for-n hot-standby configuration, where $n = 4$.

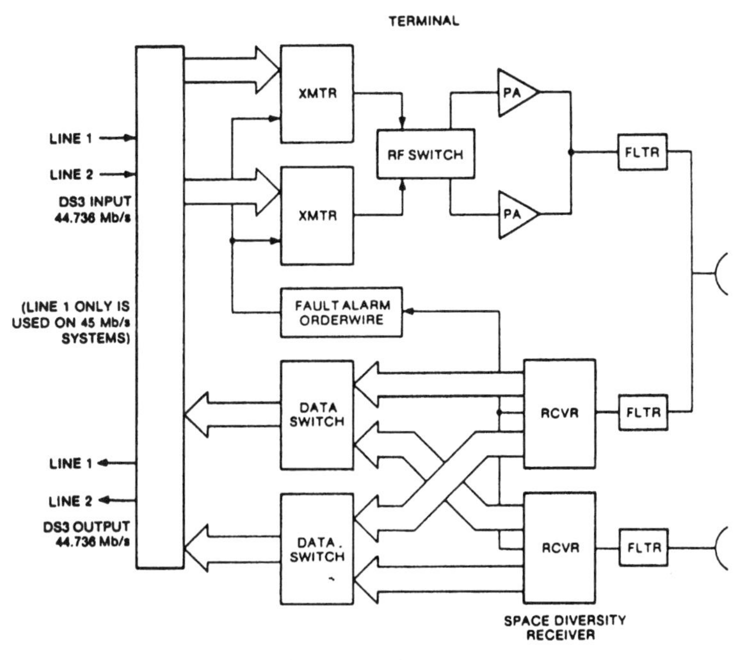

Figure 14.28. A typical hot-standby (one-for-one) digital implementation using space diversity. (Courtesy of Rockwell International, Collins Transmission Division; Ref. 9.)

signal output. To avoid such a condition, a pilot tone is inserted prior to modulation at the radiolink transmitter and deleted after the combiner. The pilot tone presence indicates to the combiner that the path is a valid path. Pilots are reinserted anew at each modulation point in a string of radiolinks in tandem. It should be noted that pilots are inserted out-of-band, generally 10% higher than the highest baseband modulating frequency. Table 14.2 provides recommended pilot tone frequencies and rms deviation produced by the pilot for 11 FDM baseband configurations and television. This table was taken from CCIR Rec. 401-2 (Ref. 10).

14.3.7 Service Channels

Service channels are separate facilities from the information baseband (on analog radiolinks) but transmitted on the same carrier. Service channels operate in frequency slots (on analog radiolinks) below the carrier for FDM telephony operation and above the video baseband for those links transmitting video (and associated but separated aural channels). Service channels are used for maintenance and link and network coordination and, in these cases, may be called "orderwire(s)." They may also be assigned to carry network status data and fault information from unmanned radio relay sites. For FDM telephony links, service channels commonly occupy the band from

TABLE 14.2 Radio Continuity Pilots Recommended by the CCIR

System Capacity (Channels)	Limits of Band Occupied by Telephone Channels (kHz)	Frequency Limits of Baseband (kHz)[a]	Continuity Pilot Frequency (kHz)	Deviation (rms) Produced by the Pilot (kHz)[b,c]
24	12–108	12–108	116 or 119	20
60	12–252	12–252	304 or 331	25, 50, 100[d]
	60–300	60–300		
120	12–552	12–552	607[e]	25, 50, 100[d]
	60–552	60–552		
300	60–1300	60–1364	1499, 3200,[f] or 8500[f]	100 or 140
600	60–2540	60–2792	3200 or 8500	140
	64–2660			
960 }	60–4028 }	60–4287	4715 or 8500	140
900 }	316–4188 }			
1260 }	60–5636 }		{ 6199	100 or 140
	60–5564 }	60–5680	{ 8500	140
1800 }	316–5564 }			
1200	312–8204	300–8248	9023	100
	316–8204			
2700	312–12,388	308–12,435	13,627	100
	316–12,388			
Television			{ 8500	140
			{ 9023[g]	100

[a]Including pilot or other frequencies that might be transmitted to line.
[b]Other values may be used by agreement between the Administrations concerned.
[c]This deviation does not depend on whether or not a pre-emphasis network is used in the baseband.
[d]Alternative values dependent on whether the deviation of the signal is 50, 100, or 200 kHz (Recommendation 404-2).
[e]Alternatively 304 kHz may be used by agreement between the Administrations concerned.
[f]For compatibility in the case of alternate use with 600-channel telephony systems and television systems.
[g]The frequency 9023 kHz is used for compatibility purposes between 1800 channel telephony systems and television systems, or when the establishment of multiple sound channels so indicates.
Source: Reference 10.

300 Hz to 12 kHz of the transmitted baseband, allowing three nominal 4-kHz channel operation.

Digital radiolinks utilize specific timeslots in the digital bit stream for service channels. At each terminal and repeater local digital service channels are dropped and inserted. An express orderwire can also be accommodated. This, of course, requires a reconstitution of the information bit stream at the terminal locations to permit the insertion of service channel information. Another method is to use a separate carrier, either analog or digitally modulated.

14.3.8 Alarm and Supervisory Subsystems

Many radiolink sites are unattended, especially repeater sites. To ensure improved system availability, it is desirable to know the status of unattended sites at a central or manned location. This is accomplished by means of a fault-reporting system. Commonly, such fault alarms are called status reports. The radiolink sites originating status reports are defined as reporting stations. A site that receives and displays such reports is defined as a supervisory location. This is the standard terminology of the industry. Normally, supervisory locations are those terminals that terminate a radiolink section. Status reports may also be required to be extended over a wire circuit to a remote location, often a maintenance center.

The following functions at a radiolink site, which is a reporting location (unmanned), are candidate functions for status reports:

Equipment Alarms

Loss of receiver signal
Loss of pilot (at receiver)
High noise level (at receiver)
Power supply failure
Loss of modulating signal
TWT overcurrent
Low transmitter output
Off-frequency operation
Hot-standby actuation

Site Alarms

Illegal entry
Commercial power failure
Low fuel supply
Standby power unit failure
Standby power unit on-line
Tower light status

Additional Fault Information for Digital Systems

Loss of BCI (bit count integrity)
Loss of sync
Excessive BER (bit error rate)

Often alarms are categorized into "major"(urgent) and "minor"(nonurgent) in accordance with their importance. For instance, a major alarm would be one where the fault would cause the system or link to go down (cease operation) or seriously deteriorate performance. A major alarm may be audible as well as visible on the status panel. A minor alarm may then show only as an indication on the status panel. On military equipment alarms are referred to as BITE (built-in-test equipment).

The design intent of alarm or BITE systems is to make all faults binary: a tower light is either on or off; the RSL (receive signal level) has dropped below a specified level, −100 dBW, for example; the transmitter power output is 3 dB below its specified output; or the noise on a derived analog channel is above a certain level in picowatts. By keeping all functions binary, using relay closure (or open) or equivalent solid-state circuitry, the job of coding alarms for transmission is made much easier. Thus all alarms are of a "go/no-go" nature.

14.3.8.1 Transmission of Fault Information. On analog systems common practice today is to transmit fault information in a voice channel associated with the service channel groupings of voice channels (Section 14.2.7). Binary information is transmitted by VF telegraph equipment using FSK or tone-on, tone-off (see Ref. 3). Depending on the system used, 16, 18, or 24 tone channels may occupy the voice channel assigned. A tone channel is assigned to each reporting location (i.e., each reporting location will have a tone transmitter operating on the specific tone frequency assigned to it). The supervisory location will have a tone receiver for each reporting (unmanned site under its supervision) location.

At each reporting location the fault or BITE points previously listed are scanned every so many seconds, and the information from each monitor or scan point is time division multiplexed in a simple serial bit stream code. The data output from each tone receiver at the supervisory location represents a series of reporting information on each remote unmanned site. The coded sequence in each case is demultiplexed and displayed on the status panel.

A simpler method is the tone-on, tone-off method. Here the presence of a tone indicates a fault in a particular time slot; in another method it is indicated by the absence of a tone. A device called a fault-interrupter panel is used to code the faults so that different faults may be reported on the same tone frequency.

On digital radiolinks fault information in a digital format is stored and then inserted into one of the service channel timeslots; a timeslot is reserved for each reporting location.

14.3.8.2 Remote Control. Through a similar system to that previously described, which operates in the opposite direction, a supervisory station can control certain functions at reporting locations via a voice frequency telegraph tone line (for analog systems) with a tone frequency assigned to each separate reporting location on the span. If only one condition is to be controlled, such as turning on tower lights, then a mark condition could represent lights on and a space for lights off. If more than one condition is to be controlled, then coded sequences are used to energize or deenergize the proper function at the remote reporting location.

There is an interesting combination of fault reporting and remote control that particularly favors implementation on long spans. Here only summary status is normally passed to the supervisory location; that is, a reporting location is either in a "go" or "no-go" status. When a "no-go" is received, that reporting station in question is polled by the supervisory location and detailed fault information is then released. Polling may also be carried out on a periodic basis to determine detailed minor alarm fault data.

On one digital radiolink equipment (Ref. 9) the following operational support system maintenance functions and performance monitoring are listed:

Maintenance Functions

Alarm Surveillance

Alarm reporting

Status reporting

Alarm conditioning

Alarm distribution

Attribute report

Control Functions

Allow–inhibit local alarms

Operate alarm cutoff

Allow–inhibit protection switching

Operate release protection switch

Remove–restore service

Restart processor

Preemptive switching—override an existing protective switch

Activate restore lockout (lockout prevents switching)

Local remote control (i.e., inhibits local operation)

Operate–release loopback

Command (control) verification (i.e., set status point)

Completion acknowledgment (i.e., completed the command received earlier)

Performance Monitoring

Report performance monitoring data such as BER, sync error, and errored second

Inhibit–allow performance monitoring data (i.e., collect but don't send unless asked)

Start–stop performance monitoring data

Initialize (reset) performance monitoring data storage registers

14.3.9 Antenna Towers—General

Two types of towers are used to mount antennas for radiolink systems: guyed and self-supporting. However, other man-made and natural structures should also be considered or at least taken advantage of. Among these are siting on the highest hill or ridge feasible, or leasing space on tall buildings or on TV towers.

One of the most desirable construction materials for towers is hot-dipped galvanized steel. Guyed towers are often preferred because of overall economy and versatility. Although guyed towers have the advantage that they can be placed closer to the equipment shelter or building than self-supporting types, the fact that they need a larger site may be a disadvantage where land values are high. The larger site is needed because additional space is required for installing guy anchors. Table 14.3 shows approximate land area needed for several tower heights.

Tower and foundation design are dependent on four main factors: (1) soil bearing capability at the specific location; (2) size and number of parabolic antennas and their location on the tower (i.e., these antenna reflectors act as wind sail devices); (3) meteorological conditions to be expected; and (4) maximum tower twist and sway that can be tolerated under worst conditions of wind loading, and ice loading, where applicable.

Tower loading is the result of all forces acting on the tower. Design of the tower must be such that with all antennas and other required items mounted on the tower, it will, when subjected to maximum specified wind- and ice-loading conditions, resist deflection or twisting beyond a specified amount.

Since the net result of all the forces acting on the tower is also, in effect, transmitted to the foundation, it in turn must be capable of distributing the force over a large enough area and depth so as not to exceed the soil bearing pressure at any point and also to resist movement in any direction. The depth of the foundation will be governed by the tower load and soil bearing characteristics, but, in colder climates, it is necessary to extend the depth of the foundation below the frost line or to firm ground.

The soil bearing capability, usually expressed as a pressure in pounds per square foot, is a determining factor in the design of a tower foundation. Table 14.4 gives the maximum soil bearing values for various types of soil conditions. It should be noted that the designation of the various soil conditions is arbitrary in nature, so the table should be used only as a rough guide for preliminary estimates. Soil borings taken at the area in question are normally required for final design.

The following information should be provided by the communication system engineer in order for the tower designer to properly design a suitable tower and foundation:

1. Size and type of antennas required and the type of radomes, if any.
2. Azimuth and elevation angles for each antenna.

TABLE 14.3 Minimum Land Area Required for Guyed Towers

Tower Height (ft)	Area Required[a] (ft)				
	80% Guyed	75% Guyed	70% Guyed	65% Guyed	60% Guyed
60	87 × 100	83 × 96	78 × 90	74 × 86	69 × 80
80	111 × 128	105 × 122	99 × 114	93 × 108	87 × 102
100	135 × 156	128 × 148	120 × 140	113 × 130	105 × 122
120	159 × 184	150 × 174	141 × 164	132 × 154	123 × 142
140	183 × 212	178 × 200	162 × 188	152 × 176	141 × 164
160	207 × 240	195 × 226	183 × 212	171 × 198	159 × 184
180	231 × 268	218 × 252	204 × 236	191 × 220	177 × 204
200	255 × 296	240 × 278	225 × 260	210 × 244	195 × 226
210	267 × 304	252 × 291	236 × 272	220 × 264	204 × 236
220	279 × 322	263 × 304	246 × 284	230 × 266	213 × 246
240	303 × 350	285 × 330	267 × 308	249 × 288	231 × 268
250	315 × 364	296 × 342	278 × 320	254 × 282	240 × 277
260	327 × 378	308 × 356	288 × 334	269 × 310	249 × 288
280	351 × 406	330 × 382	309 × 358	288 × 332	267 × 308
300	375 × 434	353 × 408	330 × 382	308 × 356	285 × 330
320	399 × 462	375 × 434	351 × 406	327 × 376	303 × 350
340	423 × 488	398 × 460	372 × 430	347 × 400	321 × 372
350	435 × 502	409 × 472	383 × 442	356 × 411	330 × 381
360	447 × 516	420 × 486	393 × 454	366 × 424	339 × 392
380	471 × 544	443 × 512	414 × 478	386 × 446	357 × 412
400	495 × 572	465 × 536	425 × 502	405 × 468	375 × 434
420	519 × 599	488 × 563	456 × 527	425 × 490	393 × 454
440	543 × 627	510 × 589	477 × 551	444 × 513	411 × 475

Tower Height (ft)	Area Required (acre)				
	80% Guyed	75% Guyed	70% Guyed	65% Guyed	60% Guyed
60	0.23	0.21	0.19	0.17	0.15
80	0.38	0.34	0.30	0.26	0.23
100	0.56	0.50	0.44	0.39	0.34
120	0.77	0.69	0.61	0.53	0.46
140	1.03	0.91	0.80	0.70	0.61
160	1.31	1.16	1.03	0.90	0.77
180	1.63	1.45	1.27	1.11	0.96
200	1.99	1.76	1.55	1.35	1.16
210	2.18	1.93	1.70	1.48	1.27
220	2.38	2.11	1.85	1.61	1.39
240	2.81	2.49	2.18	1.90	1.63
250	3.04	2.69	2.36	2.05	1.76
260	3.27	2.89	2.54	2.21	1.90
280	3.77	3.33	2.92	2.54	2.18
300	4.30	3.80	3.33	2.89	2.49
320	4.87	4.30	3.77	3.27	2.81
340	5.48	4.84	4.24	3.65	3.15
350	5.79	5.11	4.48	4.88	3.33
360	6.12	5.40	4.73	4.10	3.52
380	6.79	5.99	5.25	4.55	3.90
400	7.50	6.62	5.79	5.02	4.30
420	8.24	7.27	6.36	5.52	4.73
440	9.03	7.96	6.96	6.03	5.17

[a]Preferred area is a square using the larger of minimum area. This will permit orienting the tower in any desired position.

Source: Reference 3.

TABLE 14.4 Soil Bearing Characteristics

Material	Maximum Allowable Bearing Value (lb/ft^2)
Bedrock (sound) without laminations	200,000
Slate (sound)	70,000
Shale (sound)	20,000
Residual deposits of broken bedrock	20,000
Hardpan	20,000
Gravel (compact)	10,000
Gravel (loose)	8,000
Sand, coarse (compact)	8,000
Sand, coarse (loose)	6,000
Sand, fine (compact)	6,000
Sand, fine (loose)	2,000
Hard clay	12,000
Medium clay	8,000
Soft clay	2,000

Source: Reference 5.

3. Amount of adjustment required for antenna alignment after installation; $\pm 5^\circ$ in both azimuth and elevation is usually sufficient.
4. Height above ground level (AGL) for each antenna.
5. Location of tower with respect to the building/shelter and their corresponding orientations.
6. Beamwidths of the antennas and permissible twist and sway of the tower that can be tolerated under maximum expected wind and ice loading.
7. Type and number of waveguide runs required and power cabling for feedhorn and radome heaters.
8. Requirements for tower lighting, obstruction lights, lightning protection and grounding systems, painting.
9. Requirements for platforms to allow access to antennas for maintenance and/or alignment.
10. Required means of access up the tower including required personnel safety features.
11. Provisions for future antenna requirements.
12. Climatic conditions of the area involved.
13. Local restrictions and/or other constraints that may be involved.
14. Soil conditions.

Tower twist and sway is one of the most important requirements to be specified. As with any other structure, a radiolink tower tends to twist and sway due to wind loads and other natural forces. Considering the narrow beamwidths of radiolink antennas, with only a little imagination one can see that only a very small deflection of a tower or antennas will cause the radio ray beam to fall out of the reflection face of the distant-end corresponding antenna or move the beam out on the far-end transmit side of the link.

Twist and sway therefore must be limited. Table 14.5 sets certain limits. The table has been taken from EIA ANSI/EIA/TIA (Ref. 11). From the table we can see that angular deflection and tower movement are functions of wind velocity. It should also be noted that the larger the antenna, the smaller the beamwidth, besides the fact that the sail area is larger. Thus the larger the antenna and the higher the frequency of operation, the more we must limit the deflection.

To reduce twist and sway, tower rigidity must be improved. One generality we can make is that towers that are designed to meet required wind-load or ice-load specifications are sufficiently rigid to meet twist and sway tolerances. One way to increase rigidity is to increase the number of guys, particularly at the top of the tower. Under certain circumstances doubling the number of guys is warranted.

A number of external safety requirements are imposed on towers. A primary consideration is to prevent excessive hazards to air commerce. Therefore antenna towers must be marked in such a way as to make them conspicuous when viewed from aircraft. The type of marking to be used depends in part on the height of the structure, its location with respect to nearby objects, and its proximity to aircraft traffic routes near landing areas. In the United States the requirements and specifications for marking and lighting potential hazards to air navigation have been established through joint cooperation of the Federal Aviation Agency (FAA), Federal Communications Commission (FCC), Department of Defense (DoD), and appropriate branches of the broadcasting and aviation industries. When conducting the preliminary site survey, it is advisable to determine the prevailing ordinances concerning such structures and to discuss them with local government and building authorities. Inside the continental United States consult the latest issues of Government Rules and Regulations, FCC Form 715, FCC Rule Part 17, and FAA Standards for marking and lighting obstructions to air navigation, with all revisions.

Both the FCC and FAA lighting specifications are set forth in terms of the heights of antenna structures. The requirements for towers and obstructions are determined from FAA Specifications set forth in the latest edition of AC 70/7460-1( ). The specifications further stipulate that placement of lights on either square or rectangular towers shall be such that at least one top or side light be visible from any angle of approach. When a flashing beam is required, it shall be equipped with a flashing mechanism capable of producing not more than 40 nor less than 12 flashes per minute with a period of darkness equal to one-half the luminous period.

In the United States the FCC requires that tower lighting be exhibited during the period from sunset to sunrise unless otherwise specified. At unattended radiolink installations, a dependable automatic obstruction-lighting control device will be used to control the obstruction lights in lieu of

TABLE 14.5 Allowable Twist and Sway Values for Parabolic Antennas, Passive Reflectors, and Periscope System Reflectors

A	B	C Parabolic Antennas	D	E Passive Reflectors	F	G Periscope System Reflectors	H	I
3-dB Beamwidth 2Θ HP for Antenna Only (Note 8)	Deflection Angle at 10-dB Points (Notes 1 and 7)	Limit of Antenna Movement with Respect to Structure	Limit of Structure Movement Twist or Sway at Antenna Attachment Point	Limit of Passive Reflector Sway (Notes 4 and 5)	Limit of Passive Reflector Twist (Note 4)	Limit of Reflector Movement with Respect to Structure	Limit of Structure Twist at Reflector Attachment Point	Limit of Structure Sway at Reflector Attachment Point
Degrees	Degrees	Degrees	Degrees	Degrees	Degrees	Degrees	Degrees	Degrees
5.6	5.0	0.4	4.6	3.5	2.5	0.2	4.8	2.3
5.6	4.8	0.4	4.4	3.3	2.4	0.2	4.6	2.2
5.4	4.6	0.4	4.2	3.2	2.3	0.2	4.4	2.1
5.1	4.4	0.4	4.0	3.0	2.2	0.2	4.2	2.0
4.9	4.2	0.4	3.8	2.9	2.1	0.2	4.0	1.9
4.7	4.0	0.3	3.7	2.8	2.0	0.2	3.8	1.8
4.4	3.8	0.3	3.5	2.6	1.9	0.2	3.6	1.7
4.2	3.6	0.3	3.3	2.5	1.8	0.2	3.4	1.6
4.0	3.4	0.3	3.1	2.3	1.7	0.2	3.2	1.5
3.7	3.2	0.3	2.9	2.2	1.6	0.2	3.0	1.4
3.5	3.0	0.3	2.7	2.1	1.5	0.2	2.8	1.4
3.4	2.9	0.2	2.7	2.0	1.45	0.1	2.8	1.3
3.3	2.8	0.2	2.6	1.9	1.4	0.1	2.7	1.3
3.1	2.7	0.2	2.5	1.8	1.35	0.1	2.6	1.25
3.0	2.6	0.2	2.4	1.8	1.3	0.1	2.5	1.2
2.9	2.5	0.2	2.3	1.7	1.25	0.1	2.4	1.15
2.8	2.4	0.2	2.2	1.6	1.2	0.1	2.3	1.1
2.7	2.3	0.2	2.1	1.6	1.15	0.1	2.2	1.05
2.6	2.2	0.2	2.0	1.5	1.1	0.1	2.1	0.1
2.5	2.1	0.2	1.9	1.4	1.05	0.1	2.0	0.95
2.3	2.0	0.2	1.8	1.4	1.0	0.1	1.9	0.9
2.2	1.9	0.2	1.7	1.3	0.95	0.1	1.8	0.85
2.1	1.8	0.2	1.6	1.2	0.9	0.1	1.7	0.8
2.0	1.7	0.2	1.5	1.1	0.85	0.1	1.6	0.75
1.9	1.6	0.2	1.4	1.1	0.8	0.1	1.5	0.7
1.7	1.5	0.2	1.3	1.0	0.75	0.1	1.4	0.65
1.6	1.4	0.2	1.2	0.9	0.7	0.1	1.3	0.6
1.5	1.3	0.1	1.2	0.9	0.65	0.1	1.2	0.55
1.4	1.2	0.1	1.1	0.8	0.6	0.1	1.1	0.5
1.3	1.1	0.1	1.0	0.7	0.55	0.1	1.0	0.45
1.2	1.0	0.1	0.9	0.7	0.5	0.1	0.9	0.4
1.1	0.9	0.1	0.8	0.6	0.45	0.1	0.8	0.35
0.9	0.8	0.1	0.7	0.5	0.4	0.1	0.7	0.3
0.8	0.7	0.1	0.6	0.4	0.35	0.1	0.6	0.25
0.7	0.6	0.1	0.5	0.4	0.3	0.1	0.5	0.2
0.6	0.5	0.1	0.4	0.3	0.25	0.1	0.4	0.15
0.5	0.4	0.1	0.3	0.2	0.2	0.07	0.3	0.13
0.3	0.3	0.05	0.25	0.2	0.15	0.05	0.25	0.10
0.2	0.2			0.14	0.1			
0.1	0.1			0.7	0.05			
See Notes below						Only for configuration where antenna is directly under the reflector		

TABLE 14.5 (*continued*)

Notes:

1. If values for columns *A* and *B* are not available from the manufacturer(s) of the antenna system or from the user of the antenna system, then values shall be obtained from Figure C1, C2, or C3 [in the reference publication].
2. Limits of beam movement for twist or sway (treated separately in most analyses) will be the sum of the appropriate figures in columns C and D, G and H, and G and I. Columns G, H, and I apply to a vertical periscope configuration.
3. It is not intended that the values in this table imply an accuracy of beamwidth determination or structural rigidity calculation beyond known practicable values and computational procedures. For most microwave structures it is not practicable to require a calculated structural rigidity of less than 0.25° twist or sway with a 50-mi/h (22.4-m/s) basic wind speed.
4. For passive reflectors the allowable twist and sway values are assumed to include the effects of all members contributing to the rotation of the face under wind load. For passives not elevated far above ground [approximately 5–20 feet (1.5–6 m) clearance above ground], the structure and reflecting face supporting elements are considered an integral unit. Therefore separating the structure portion of the deflection is only meaningful when passives are mounted on conventional microwave structures.
5. The allowable sway for passive reflectors is considered to be 1.4 times the allowable twist to account for the amount of rotation of the face about a horizontal axis through the face center and parallel to the face compared to the amount of beam rotation along the direction of the path as it deviates from the plane of the incident and reflected beam axes.
6. Linear horizontal movement of antennas and reflectors in the amount experienced for properly designed microwave antenna system support structures is not considered a problem (no significant signal degradation attributed to this movement).
7. For systems using a frequency of 450 MHz, the half-power beamwidths may be nearly 2Θ degrees for some antennas. However, structures designed for microwave relay systems will usually have an inherent rigidity less than the maximum 5° deflection angle shown on the chart.
8. The 3-dB beamwidths, 2Θ HP in column A are shown for convenient reference to manufacturers' published antenna information. The minimum deflection reference for this standard is the allowable total deflection angle Θ at the 10-dB points.

Source: Ref. 11, Appendix C. Reprinted with permission of EIA/TIA

manual control. This requirement is met in microwave installations by employing a tower-lighting kit. These kits apply power to the lights when the north skylight intensity is less than approximately 35 foot-candles and disconnect the power when the north skylight intensity is greater than approximately 58 foot-candles.

14.3.10 Waveguide Pressurization

All external waveguide runs should be pressurized with dry air or an inert gas such as nitrogen. While standard tanks of nitrogen may be used for this purpose, the use of automatic dehydrator-pressurization equipment is generally preferred. The pressure should not exceed the manufacturer's recommended values for the particular waveguide involved. This is generally on the order of 2–10 lb/in.^{2} Pressure windows are available to isolate pressurized from unpressurized portions of the waveguide.

14.4 TROPOSPHERIC SCATTER AND DIFFRACTION INSTALLATIONS: ANALOG AND DIGITAL

Tropospheric scatter/diffraction installations are configured in such a way as to (1) meet path performance requirements and (2) be economically viable. These installations can become very costly, between $500,000 and many millions of dollars (1997 value) per terminal.

There is little difference between an analog and a digital installation. Of course, as we know from Chapter 5, the overriding impairment with digital tropo is dispersion, some ten times greater than we would expect on a LOS microwave path. Thus the equipment/system must compensate for dispersion, either through adaptive equalizers or by means of special waveforms, as described in Chapter 5 in the description of the AN/TRC-170(V) with its DAR (digital adaptive receiver), or a combination of both. Thus very specialized receivers or transmitter–receiver combinations are required.

All tropospheric scatter installations use some form of diversity, more often quadruple diversity. Antennas usually have considerably larger apertures than their LOS counterparts. Antenna apertures range from 6 to 120 ft. Towers are seldom used. Transmitter power ranges from 1 to over 50 kW, and unlike LOS installations, low-noise receiving systems are the rule.

A typical quadruple-diversity tropospheric scatter terminal layout is shown in Figure 14.29. It is made up of the following identifiable subsystems:

- Antennas, duplexer, and transmission lines
- Modulators/exciters and power amplifiers
- Preselectors, receivers, and threshold extension devices (optional)
- Diversity combiners

Through the proper selection and sizing of these devices, viable tropospheric/diffraction systems can be implemented for links up to 250 statute miles (400 km).

Tropospheric scatter links commonly operate in the following frequency bands:

350–450 MHz
755–985 MHz
1700–2400 MHz
4400–5000 MHz

The reader should also consult ITU-R Rec. F.698-2 (Ref. 12) and CCIR Rep. 285-7 (Ref. 13).

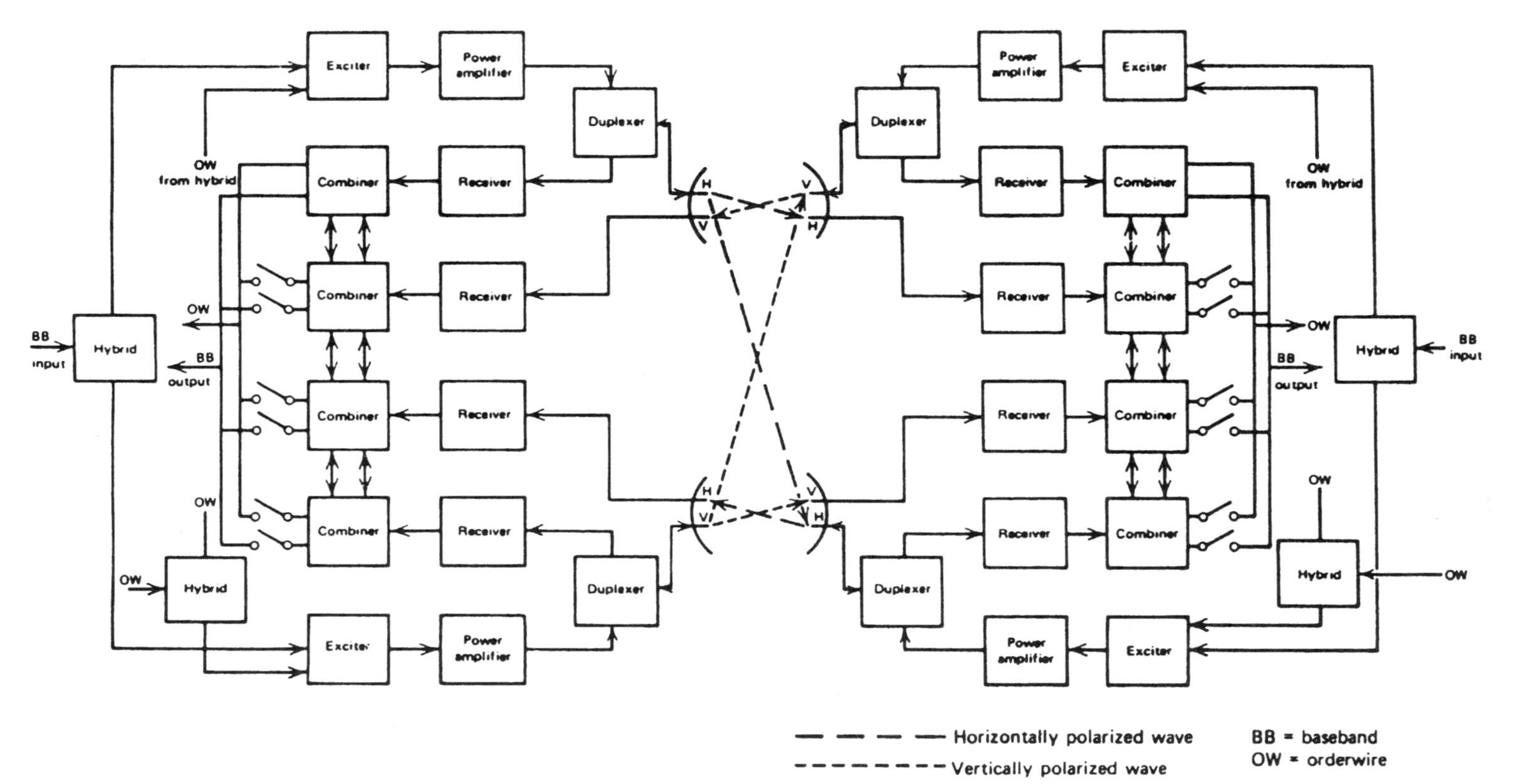

Figure 14.29. Simplified functional block diagram of a quadruple-diversity tropospheric scatter configuration for analog operation.

14.4.1 Antennas, Transmission Lines, Duplexer, and Related Transmission Line Devices

The antennas most commonly used in tropospheric scatter/diffraction installations are broadband, high-gain parabolic reflector devices. The antennas described here are similar in many respects to those discussed in Section 14.2.2 but have higher gain and, in some cases, better efficiency and therefore are larger and considerably more expensive. In some installations they are the major cost driver.

As we have discussed previously in this text, the gain of this type of antenna is a function of the reflector diameter or aperture; 1- or 2-dB additional gain can be obtained by improved efficiency, particularly through the selection of the feed type. Improved antenna feeds illuminate the reflector more uniformly and reduce spillover.

It is desirable, but not always practical, to have the two antennas (space diversity) of a terminal spaced not less than 100 wavelengths apart to ensure proper space diversity operation. Antenna spillover (i.e., radiated energy in the sidelobes and back lobes) must be reduced not only to improve radiation efficiency but also to minimize interference with your own installation, with full duplex operation, and with other services. The first sidelobe should be down (attenuated) as referenced to the main beam at least 23 dB, and the remaining sidelobes should be down at least 40 dB from the main lobe. Antenna alignment is extremely important because of the narrow antenna beamwidths. These beamwidths are usually less than 2° and often less than 1° at the half-power (3-dB) points.

A good VSWR is also important, not only from the standpoint of improving system efficiency, but also because the resulting reflected power with a poor VSWR may damage components further back in the transmission chain. Often, load isolators are required to minimize damaging effects of reflected waves (e.g., reflected power). In high-power tropospheric scatter systems these devices may even require a cooling system to remove the heat generated by the reflected energy.

A load isolator is a ferrite device with approximately 0.5-dB insertion loss. The forward wave (e.g., the energy radiated toward the antenna) is attenuated 0.5 dB; the reflected wave (e.g., the energy reflected back from the antenna) is attenuated more than 20 dB.

Another important consideration in planning a tropospheric scatter/diffraction antenna system is polarization (see Figure 14.29). For a common antenna the transmit wave should be orthogonal to the receive wave. This means that if the transmitted signal is horizontally polarized, the receive signal should be vertically polarized. The polarization is established by the feed device, usually a feed horn. The primary reason for using opposite polarizations is to improve isolation, although the correlation of fading on diversity paths may also be reduced. A figure commonly encountered for isolation between polarizations on a common antenna is 35 dB.

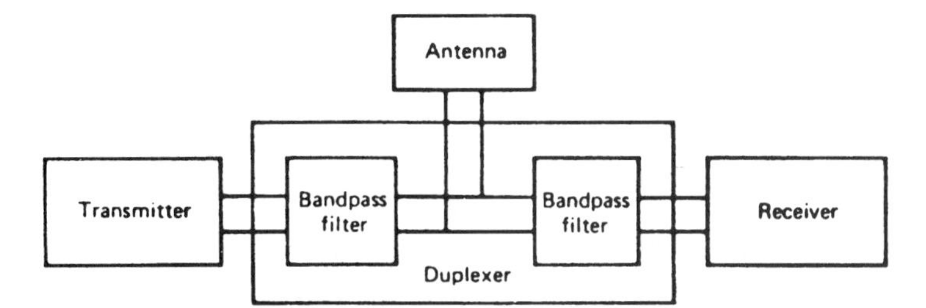

Figure 14.30. Simplified functional block diagram of a duplexer.

In selecting and laying out transmission lines for tropospheric scatter installations, it should be kept in mind that losses must be kept to a minimum. That additional fraction of a decibel is much more costly in tropospheric scatter/diffraction installations than in radiolink installations. The tendency therefore is to use waveguide on most tropospheric scatter installations because it displays lower loss than coaxial cable. Waveguide is universally used above 1.7 GHz.

Transmission line runs should be less than 200 ft (60 m). The attenuation of the line should be kept under 1 dB, whenever possible, from the transmitter to the antenna and from the antenna to the receiver, respectively. To minimize reflective losses, the VSWR of the line should be 1.05 : 1 or better when the line is terminated in its characteristic impedance. Figure 14.12 shows some waveguide types commercially available.

The duplexer is a transmission line device that permits the use of a single antenna for simultaneous transmission and reception. For tropospheric scatter/diffraction application, a duplexer is a three-port device (see Figure 14.30) so tuned that the receiver leg appears to have an admittance approaching (ideally) zero at the receiving frequency. To establish this, sufficient separation in frequency is required between transmit and receive frequencies. Figure 14.30 is a simplified block diagram of a duplexer. The insertion loss of the duplexer in each direction should be less than 0.5 dB. Isolation between the transmitter port and the receiver port should be better than 30 dB. High-power duplexers are usually factory tuned. It should be noted that some textbooks call the duplexer a diplexer.

14.4.2 Modulator–Exciter and Power Amplifier

The type of modulation used on tropospheric scatter/diffraction links is commonly FM, but digital tropo is now taking on a particularly important role in military systems. We should also keep in mind as this discussion develops that tropospheric scatter/diffraction systems are high-gain, low-noise extensions of radiolink (LOS microwave) systems.

The tropospheric scatter/diffraction transmitter is made up of a modulator–exciter and a power amplifier (see Figure 14.29). Power outputs have been fairly well standardized at 1, 2, 10, 20, and 50 kW. For most commercial applications the 50-kW installation is often not feasible from an economic viewpoint. Installations that are 2 kW or below are usually air-cooled. Those above 2 kW are liquid-cooled, usually with a glycol–water solution using a heat exchanger. When klystrons are used in high-power amplifiers, they are about 33% efficient. Thus a 10-kW klystron will require at least 20 kW of heat exchange capacity.

14.4.3 FM Receiver Group

The receiver group in a FM tropospheric scatter installation usually consists of two or four identical receivers in dual- or quadruple-diversity configurations, respectively. Receiver outputs are combined in maximal ratio-squared combiners or in other types of combiners such as selection combiners. (See Section 14.2.4.) A simplified functional block diagram of a typical quadruple-diversity receiving system is shown in Figure 14.31.

Unlike their microwave LOS counterpart, tropo receiving systems universally use low-noise front ends or LNAs.

14.4.4 Diversity Operation

Some form of diversity is mandatory in a tropospheric scatter receiving system, primarily to mitigate Rayleigh (short-term) fading. Most systems employ quadruple diversity. There are several ways to obtain some form of

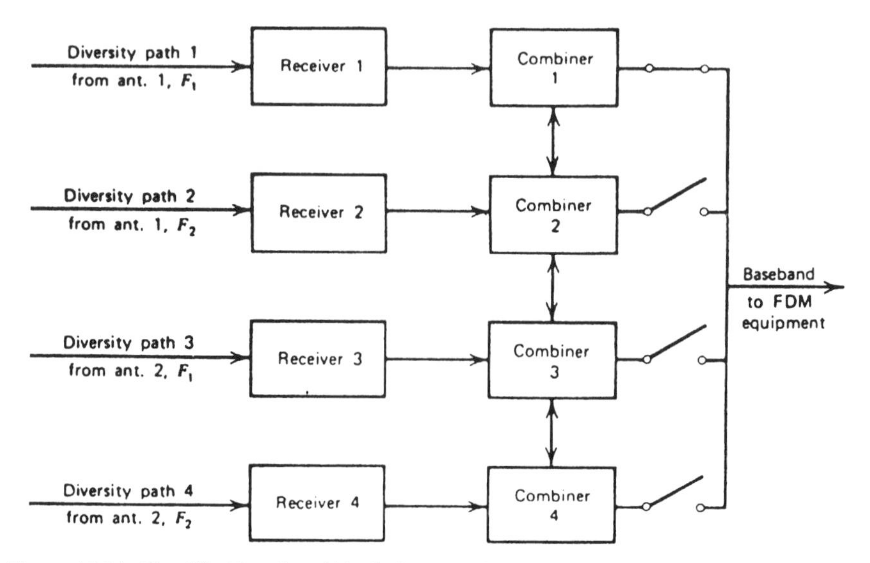

Figure 14.31. Simplified functional block diagram of a quadruple-diversity troposcatter receiving system. F_1 = frequency 1 and F_2 = frequency 2.

quadruple diversity. One of the most desirable ways is shown in Figure 14.29, where both frequency and space diversity are utilized. For the frequency-diversity section, the design engineer must consider frequency separation, sufficient to reduce signal correlation between paths enough for the desired performance enhancement. As a minimum, at least 2% separation in frequency is required. That means that at 4 GHz we would calculate the frequency separation as 0.02 × 4000 MHz or 80 MHz. The ideal separation is 10%, but such a separation is difficult to obtain from regulating authorities.

Space diversity is easier to implement and is used universally on troposcatter systems. The physical separation of the antennas is the critical parameter. Separation is normally in the horizontal plane with a separation distance from 100 to 150 wavelengths.

Frequency diversity, although very desirable, often may not be permitted due to a shortage of frequency assignments in the desired band or due to RFI considerations. Another form of diversity to obtain the four orders of diversity making up quadruple diversity is based on polarization or what some call "polarization diversity." However, this is actually another form of space diversity and has been found not to provide a complete additional order of diversity. Nevertheless, it will often make do when the additional frequencies are not available to implement frequency diversity.

Polarization diversity is often used in conjunction with conventional space diversity. The four space paths are achieved by transmitting signals in the horizontal plane from one antenna and in the vertical plane from the second antenna. On the receiving end two antennas are used, each antenna having dual-polarized feed horns for receiving signals in both planes of polarization. The net effect is to produce four signal paths that are relatively independent.

A discussion of diversity combiners is given in Section 14.3.4.

14.4.5 Isolation

An important factor in tropospheric scatter/diffraction installation design is the isolation between the emitted transmit signal and the receiver input. Normally we refer to the receiver sharing a common antenna feed with the transmitter.

For this discussion we can say that a nominal receiver input level is −80 dBm. If a tropo transmitter has an output power of 10 kW or +70 dBm, and transmission line losses are negligible, then overall isolation should be greater than 150 dB (e.g., 80 + 70).

To achieve the isolation necessary such that the transmitted signal interferes in no way with receiver operation during full duplex operation, the following items contribute to the required isolation when there is sufficient frequency separation between transmitter and receiver:

- Polarization
- Duplexer

- Receiver preselector
- Transmit filters
- Normal isolation from receiver conversion to IF

14.5 SATELLITE COMMUNICATIONS, TERMINAL SEGMENT

14.5.1 Functional Operation of a "Standard" Earth Station

14.5.1.1 The Communication Subsystem. Figure 14.32 is a simplified functional block diagram of an earth station showing the communication subsystem only. We shall use this figure to trace a signal through the equipment chain from antenna to baseband. Figure 14.33 is a more detailed functional block diagram of a typical earth station. By "standard" we can assume typically INTELSAT A, B, or C service or regional/national domestic satellite service.

The operation of an earth station communication subsystem in the FDMA/FM mode really varies little from that of a LOS radiolink system as

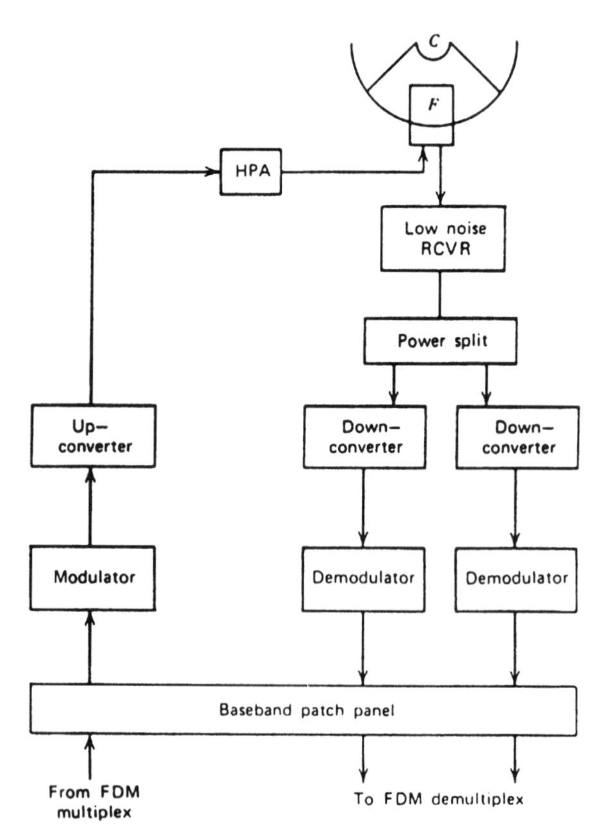

Figure 14.32. Simplified functional block diagram of an earth station communication subsystem. *F* = feed, HPA = high-power amplifier, C = Cassegrain subreflector.

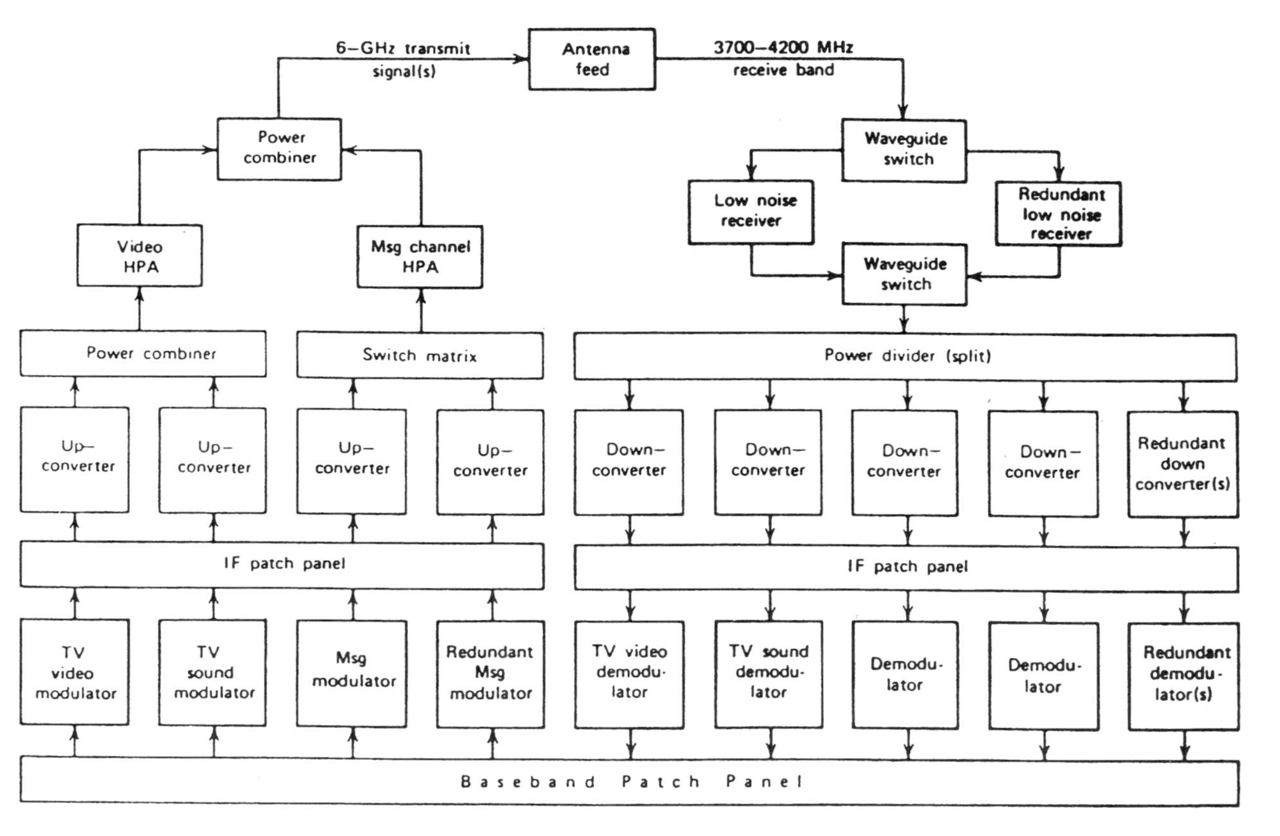

Figure 14.33. More detailed functional block diagram of a communication subsystem, typical of an INTELSAT Standard A earth station.

shown in Figure 14.1. The variances are essentially these:

- Use of low-noise front ends on receiving systems, GaAs FETs, HEMT, and PHEMT technology
- A HPA with a capability of from 200- to 8000-watt output
- Larger high-efficiency antennas, feeds
- Careful design to achieve as low a noise as possible
- Use of a signal-processing technique that allows nearly constant transmiter loading (e.g., spreading waveform), FM systems
- Use of threshold extension demodulators in some cases (FM systems only)
- Use of forward error correction on many digital systems, and above 10 GHz with interleaving to mitigate rainfall fading

Now let us trace a signal through the communication subsystem typical of Figure 14.32. On the transmit side the FDM baseband is fed from the multiplex equipment through the baseband patch facility to the modulator. A spreading waveform is added to the very low end of the baseband to achieve constant loading. The baseband signal is then shaped with a pre-emphasis network (see Section 2.6.5). The baseband so shaped frequency modulates a carrier, and the resultant is then upconverted to a 70-MHz IF. Patching facilities usually are available at the IF to loop back through the receiver subsystem or through a test receiver for local testing or troubleshooting. The 70-MHz IF is then fed to an upconverter, which translates the IF to the output frequency (6 or 14 GHz). The signal is then amplified by the HPA, filtered by a low-pass filter, directed to the feed, and radiated by the antenna.

For reception, the signal derives from the feed and is fed to a low-noise receiver. In the case of an INTELSAT Standard A earth station, the low-noise receiver looks at the entire 500-MHz band (i.e., in the case of 4-GHz operation, the band from 3700 to 4200 MHz), amplifying this broadband signal 20–40 dB. When there are long waveguide runs from the antenna to the equipment building, the signal is amplified still further by a low-level TWT or SSA called a driver. The low-noise receiver is placed as close as possible to the feed to reduce ohmic noise contributions to the system. GaAs and PHEMT LNAs have noise temperatures between 30 and 50 K and 25–35 dB gain.

The comparatively high-level broadband receive signal is then fed to a power split. There is one output from the power split for every downconverter–demodulator chain. In addition, there is often a test receiver available as well as one or several redundant receivers in case of failure of an operational receiver chain. It should be kept in mind that every time the broadband incoming signal is split into two equal-level paths, there is a 3-dB loss due to the split, plus an insertion loss of the splitter. A splitter with eight outputs will incur a loss of something on the order of 10 dB.

A downconverter is required for each receive carrier, and there will be at least one receive carrier from each distant end. Each downconverter is tuned to its appropriate carrier and converts the signal to the 70-MHz IF. In some instances dual conversion is used.

The 70-MHz IF is then fed to the demodulator on each receive chain. The resulting demodulated signal, the baseband, is reshaped in the de-emphasis network (see Section 2.6.5) and spreading waveform signal is removed. The resulting baseband output is then fed to the baseband patch facilities and thence to the demultiplex equipment.

14.5.1.1.1 Digital Operation. Present-day configurations of digital earth stations are very much like their analog counterparts. FM modulators/demodulators are replaced by digital modulators/demodulators. The modulation is BPSK or QPSK. Heavy bit packing radios such as encountered in line-of-sight microwave are as yet to be applied to satellite links. Forward error correction is commonly applied, often with convolutional coding rate $= \frac{3}{4}$.

With digital operation, equalization for group delay distortion is particularly important. For example, INTELSAT sets the following requirements given in Table 14.6.

The information transmitted over a digital satellite link enters the earth station in either the DS1 or E1 family of digital (PCM) formats basically covered in CCITT Rec. G.703. With some fairly simple bit packing schemes, we see the lower rates of SONET and SDH being transported over satellite links as well. This would be STS-3 or STM-1 both at the same 155-Mbps data rate.

These digital bit streams enter an earth station in a BNZS or HDB3 format, which is then converted to a NRZ format, then FEC coded, and then scrambled prior to modulation. Scrambling of a digital waveform for satellite

TABLE 14.6 Earth Station Equalization Required for Satellite Group Delay

Equalized Bandwidth (MHz)	Linear Equalization (n/MHz)	Parabolic Equalization[a] (ns/MHz^2)
0.0 < BW < 10.0	Not required	Not required
10.0 ≤ BW < 15.75	0 to ±5	0 to 0.5
15.75 ≤ BW < 22.5	0 to ±3	0 to 0.5
22.5 ≤ BW < 30.0	0 to ±2	0 to 0.5
30.0 ≤ BW ≤ 45.0	0 to ±1	0 to 0.25
54.0 (full transponder)[b]	0 to ±0.2	0 to 0.1
72.0 (full transponder)[b]	0 to ±0.2	0 to 0.05

[a]By convention, the sign of the parabolic component of the satellite group delay is positive, and therefore the earth stations should insert a negative value to achieve equalization.

[b]These parameters apply if group delay compensation is provided over the full transponder, rather than on a per-carrier basis. Typical transponder group delay characteristics can be supplied upon request for transponder bandwidth units greater than 72 MHz.

Source: Table 4, INTELSAT IESS-308 (Rev. 7A) (Ref. 14).

communications performs a function similar to energy dispersal of analog transmission. On the receive side, the reverse of these functions is carried out. However, the receive side also frame aligns and regenerates the digital bit stream. Care must be taken with the frame alignment function. Frame alignment is different for the European E1 series of formats when compared to North American DS1 (T1) formats. For a description of these formats, consult Refs. 3 and 4.

Digital operation can be FDMA and/or TDMA. Digital operation with FDMA is nearly identical to that of its analog counterpart. Typical is the INTELSAT IDR (intermediate data rate) operation described in Chapter 7, where all of the functions mentioned in the previous paragraph are carried out.

TDMA, as we know, is quite different. First, TDMA is only feasible with digital operation. Comparatively larger digital buffer storage is required. The operation is basically store and burst (see Section 7.2.3).

Suppose a TDMA frame is of 2-ms duration and there are five burst slots, each of 0.4-ms duration. Bursting is at 120 Mbps or, for 0.4 ms, 48 kb per burst. In this perfect TDMA system, there is no guard time or overhead; all bits are traffic bits. Before its next burst, the station has to wait four slot periods, so it must store $4 \times 48{,}000$ or 192 kbps. This allows no operational slack at all, and in practical systems more bits would be stored, probably at frame boundaries. Keep in mind that there is a downside to buffer storage. Each time we store, we add the delay equivalent of milliseconds of storage—here it was 1.6 ms. Delay is particularly critical on voice circuits where total round-trip delay is limited to 400 ms.

Timing and synchronization are critical for TDMA operation. As described in Section 7.2.3, a TDMA reference station is used. Such a reference station will have an alternate reference station for added reliability. Of course, a traffic station must start its burst exactly where its timeslot begins. We assume that it monitors the reference station and tags its *reception* of start-of-burst. This reception leading edge is delayed from the actual start-of-burst of the reference station by the propagation time to the satellite and back down to the traffic station in question. These time values will be different for each user. Then there are the variables caused by the motion of the satellite in the small figure-8 suborbits. The satellite is also being pushed and pulled by the earth's gravity and solar winds.

A TDMA-capable earth station must also be able to reset its burst boundaries in very short time periods to carry out the exercise. It needs this capability for several reasons. First, it is for the addition or deletion of other TDMA users on the frame. The second reason is to adjust burst duration to meet traffic requirements. As we are aware (see Ref. 15), these requirements vary throughout a typical 24-hour period. The greatest traffic demand is during the busy hour; the least from about 8 p.m. to 5 a.m. local time. The variation could be as great as 100 : 1. These changes in slot duration are coordinated via an orderwire/service channel. Thus an automated orderwire capability is a requirement for a TDMA-equipped station.

14.5.2 The Antenna Subsystem

The antenna subsystem is one of the most important component parts of an earth station, since it provides the means of radiating signals to the satellite and/or collecting signals from the satellite. The antenna not only must provide the gain necessary to allow proper transmission and reception, but it also must have the radiation characteristics that discriminate against unwanted signals and minimize interference into other satellite or terrestrial systems.

Earth stations most commonly use parabolic dish reflector antennas or derivatives thereof. Dish diameters range from 1 to 30 m.

The sizing of the antenna and its design are driven more by the earth station required G/T than the EIRP. The gain is basically determined by the aperture (e.g., diameter) of the dish; but improved efficiency can also add to the gain on the order of 0.5–2 dB. For this reason the Cassegrain feed technique is almost always used on larger earth terminal installations. Smaller military terminals also resort to the use of Cassegrain. In some cases efficiencies as high as 70% and more have been reported. The Cassegrain feed working into a parabolic dish configuration (Figure 14.34) permits the feed to look into cool space as far as the spillover from the subreflector is concerned.

Antennas of generally less than 30-ft diameter often use a front-mounted feed-horn assembly in the interest of economy. The most common type is the prime focus feed antenna (Figure 14.35). This is a more lossy arrangement, but, since the overall requirements are more modest, it is an acceptable one.

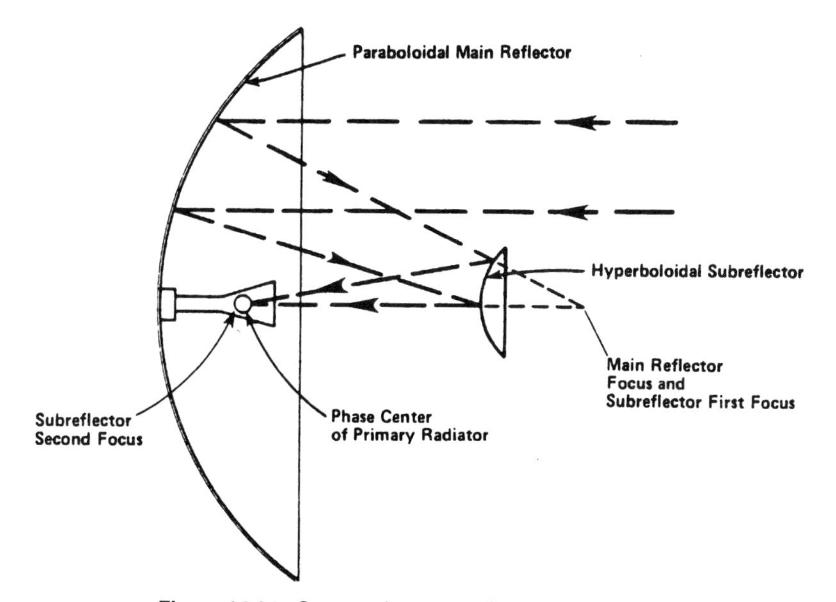

Figure 14.34. Cassegrain antenna functional operation.

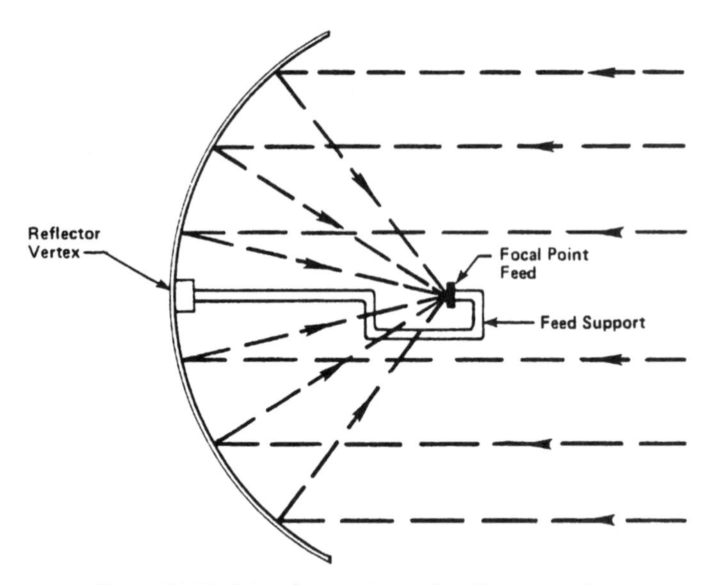

Figure 14.35. Prime focus antenna, functional operation.

In order to keep interference levels on both the uplink and downlink to acceptable levels, antenna sidelobe envelope limits (in dB) of $-32 - 25 \log \theta$, relative to the main beam maximum level, have been internationally adopted; θ is the angular distance in degrees from the main beam lobe maximum.

Figure 14.34 shows the functional operation of a Cassegrain fed antenna. Such an antenna consists of a parabolic main (prime) reflector and a hyperbolic subreflector. Here, of course, we refer to truncated parabolic and hyperbolic surfaces. The subreflector is positioned between the focal point and the vertex of the prime reflector. The feed system is situated at the focus of the subreflector, which also determines the focal length of the system. Spherical waves emanating from the feed are reflected by the subreflector. The wave then appears to be emanating from the virtual focus. These waves are then, in turn, reflected by the primary reflector into a plane wave traveling in the direction of the axis of symmetry. The size of the aperture (diameter) of the prime reflector determines the gain. The gain for a parabolic dish reflector antenna can be calculated from equation (14.2) (Section 14.3.2.1).

It can be shown that a uniform field distribution over the reflector gives the highest gain of all constant-phase distributions. The aperture efficiency can be shown to be a function of the following:

- Phase error loss
- Illumination loss due to a nonuniform amplitude distribution over the aperture

- Spillover loss
- Cross-polarization loss
- Blockage loss due to the feed, struts, and subreflector
- Random errors over the surface of the reflector (e.g., surface tolerance)

As we mentioned, nearly all very-high-gain reflector antennas are of the Cassegrain type. Within the main beam the antenna behaves essentially like a long focal-length front-fed parabolic reflector. Slight shaping of the two reflector surfaces can lead to substantial gain enhancement. Such a design also leads to more uniform illumination of the main reflector and less spillover. Typically, efficiencies of Cassegrain type antennas are from 65% to 70%, which is at least 10% above most front-fed designs. It also permits the LNA to be placed close to the feed, if desired. The ability of the antenna to achieve these characteristics rest largely with the feed-horn design. The feed-horn radiation pattern has to provide uniform aperture illumination and proper tapering at the edges of the aperture.

The simplest type of feed for the antenna is a waveguide, which can be either open ended or terminated with a horn. Both rectangular and circular waveguides have been used with the circular considered superior, since it produces a more uniform illumination pattern over the aperture and provides better cross-polarization loss characteristics. The illumination pattern should taper to a value of -10 dB at an angle corresponding to the edge of the reflector (subreflector). This results in an asymmetric feed radiation pattern causing a loss in efficiency, increased cross-polarization losses, and moderate reflector spillover. Cross-polarization loss can be reduced by careful selection of waveguide radius, while improvement in efficiency can be achieved by using a corrugated horn. The positive aspect to 10-dB taper is to reduce sidelobe levels.

Smaller earth stations with less stringent G/T requirements resort to using the less expensive prime-focus-fed parabolic reflector antenna. The functional operation of this antenna is shown in Figure 14.35. For intermediate aperture sizes, this type of antenna has excellent sidelobe performance in all angular regions except the spillover region around the edge of the reflector. Even in this area a sidelobe suppression can be achieved that will satisfy FCC/CCIR pattern requirements. The aperture efficiency for apertures greater than about 100 wavelengths is around 60%. Therefore it represents a good compromise choice between sidelobes and gain. For aperture sizes less than approximately 40 wavelengths, the blockage of the feed and feed support structure raises sidelobes with respect to the peak of the main beam such that it becomes exceedingly difficult to meet the FCC/CCIR sidelobe specification. However, the CCIR specification can be met since it contains a modifier that is dependent on the aperture size.

14.5.2.1 Polarization. By use of suitable geometry in the design of an antenna feed-horn assembly, it is possible to transmit a plane electric wavefront in which the E and H fields have a well-defined orientation. For linear polarization of the wavefront, the convention of vertical or horizontal electric (E) field is adopted. The generation of linearly polarized signals is based on the ability of a length of square-section waveguide to propagate a field in the $TE_{1,0}$ and $TE_{0,1}$ modes, which are orthogonally oriented. By exciting a short length of square waveguide in one mode by the transmitted signal and extracting signals in the orthogonal mode for the receiver, a means is provided for cross-polarizing the transmitted and received signals. The square waveguide is flared to form the antenna feed horn. Similarly, by exciting orthogonal modes in a section of circular waveguide, a left- and right-hand rotation is imparted to the wavefront, providing two orthogonal circular polarizations.

The waveguide assembly used to obtain this dual polarization is known as a diplexer or an "orthomode junction." In addition to generating the polarization effect, the diplexer has the advantage of providing isolation between the transmitter and the receiver ports that could be in the range of 50 dB if the antenna presented a true broadband match. In practice, some 25–30 dB or more of isolation can be expected.

For linear polarization, discrimination between received horizontal and vertical fields can be as high as 50 dB, but diurnal effects coupled with precipitation can reduce this value to 30 dB. To maintain optimum discrimination, large antennas are equipped with a feed-rotating device driven by a polarization-sensing servo loop. Circularly polarized fields do not provide much more than 30-dB discrimination, but they tend to be more stable and do not require polarity tracking. This makes circular polarization more suitable for systems that include mobile terminals. One of the important parameters of antenna performance is how well cross-polarization is preserved across the operating spectrum.

Polarization discrimination can be used to obtain frequency reuse. Alternate transponder channel spectra are allowed to symmetrically overlap and are provided with alternate polarization. The channelization schemes of INTELSAT V, VI, and VII are typical. The number of transponders in the satellite is effectively doubled by this process. Polarization discrimination can also provide a certain degree of interference isolation between satellite networks if the nearest satellite (in terms of angular orbit separation) uses orthogonal uplink and downlink polarizations (from Ref. 16).

14.5.2.2 Antenna Pointing and Tracking. Satellites orbiting the earth are in motion. They can be in geostationary orbits and inclined orbits. Those in geostationary orbits appear to be stationary with respect to a point on earth. Those that are in inclined orbits are in motion with respect to a point on earth. All satellite terminals working with this latter class of satellite require a tracking capability.

Even though we have said that geostationary satellites appear stationary relative to a point on earth, they do tend to drift in small suborbits (figure-8).

However, even with improved satellite stationkeeping, the narrow beamwidths encountered with large earth station antennas such as the INTELSAT Standard A (~ 50 ft or 15 m) require precise pointing and subsequent tracking by the earth station antenna to maximize the signal on the satellite and from the satellite. The basic modes of operation to provide these capabilities are:

- Manual pointing
- Programmed tracking (open-loop tracking)
- Automatic tracking (closed-loop tracking)

Pointing deals with "aiming" the antenna initially on the satellite. Tracking keeps it that way. Programmed tracking (open-loop tracking) may assume both duties. With programmed tracking, the antenna is continuously pointed by interpolation between values of a precomputed time-indexed ephemeris. With adequate information as to the actual satellite position and true satellite terminal position, pointing resolutions are in the order of 0.03–0.05°.

Manual pointing may be effective for initial satellite acquisition or "capture" for later active tracking (closed-loop tracking). It is also effective for wider beamwidth antennas, where the beamwidth is sufficiently wide to accommodate the entire geostationary satellite suborbit. Midsized installations may require a periodic trim up, and some smaller installations need never be trimmed up, assuming, of course, that the satellite in question is keeping good stationkeeping.

We will now discuss three types of active or closed-loop tracking: monopulse, step-track, and conscan.

MONOPULSE TRACKING. Monopulse is the earliest form of satellite tracking, and today it is probably still the most accurate. In monopulse tracking, multiple antenna feed elements are used to obtain multiple received signals. The relative signal levels received by the various feed elements are compared to provide azimuth and elevation-angle-pointing error signals. The error signals are used to control the servo system, which operates the antenna drive motors.

Monopulse has taken its name from radar technology, and it derives from the fact that all directional information is obtained from a single radar pulse. Beam switching or mechanical scanning is not necessary for its operation. In a three-channel monopulse system, RF signals are received by four antenna elements, usually horn feeds, located symmetrically around the boresight axis as an integral part of the antenna feed system. The multiple RF receive signals derived from these horns are combined in a beam-forming network (hybrid comparators) to produce sum and difference signals simultaneously in orthogonal planes. Figure 14.36 is a functional block diagram of the front end of a typical monopulse tracking system. The sum of the four-element radiation pattern is characterized by a single beam whose maximum lies on the antenna axis. The difference patterns are characterized by a null on the antenna axis with the lobes on opposite sides in antiphase. Because the difference radiation pattern is in phase with respect to the sum pattern on one side of the antenna axis and out of phase on the other side, bearing angle sense information can be derived. By applying the sum signal to one input of a detector and the difference signal to the other input, an error voltage is produced that is proportional to the angle off-center and whose polarity is determined by the direction off-center.

In large earth stations such as INTELSAT Standard A, the sum and the two difference RF signals are kept separate. They are downconverted and applied to a three-channel tracking receiver to generate azimuth and eleva-

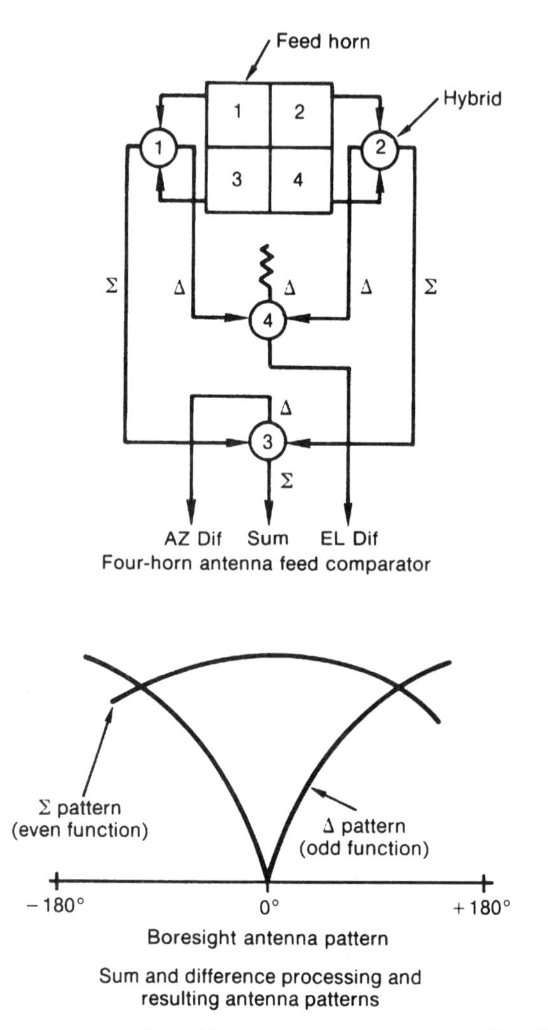

Figure 14.36. Functional operation of three-channel monopulse tracking. (From Ref. 16.)

tion error signals for the antenna servo drive system. In other satellite terminals employing the monopulse technique, the azimuth and elevation difference signals are commutated onto a single channel, often by ferrite switches controlled by a digital scan generator. The commutated output of the ferrite switches is added to the communications (sum) channel by a directional coupler. Since the difference and sum signals are phase coherent, this has the effect of amplitude modulating the sum signal. This latter implementation of monopulse reduces the needed equipment from a three-channel tracking receiver to a single-channel receiver.

The satellite signal used for tracking is usually the satellite beacon channel. Thus, in this single-channel monopulse design, the modulated sum channel is amplified by a LNA, downconverted, and demodulated in the beacon receiver. The signals are next decommutated and phase-detected to obtain error voltages that are used to drive the antenna so that the azimuth and elevation difference signals are minimized. Figure 14.37 is a functional block diagram of a typical satellite terminal monopulse tracking receiver subsystem.

The complexity and cost of monopulse feeds arise from the fact that they must be packaged into a small volume and, at the same time, provide low mutual coupling among the units without obstructing the illumination characteristics of the communications feed horn.

With monopulse tracking the beam scanning can be performed at almost any arbitrarily high rate, thus providing the potential for high tracking rates. On the other hand, with step-track the tracking rate is limited by the dynamics of offsetting the antenna.

STEP-TRACK. Despite the limitation of step-tracking, it is cost effective where there are low dynamic tracking requirements such as with geostationary antennas of medium aperture size. It does not require the complex feed

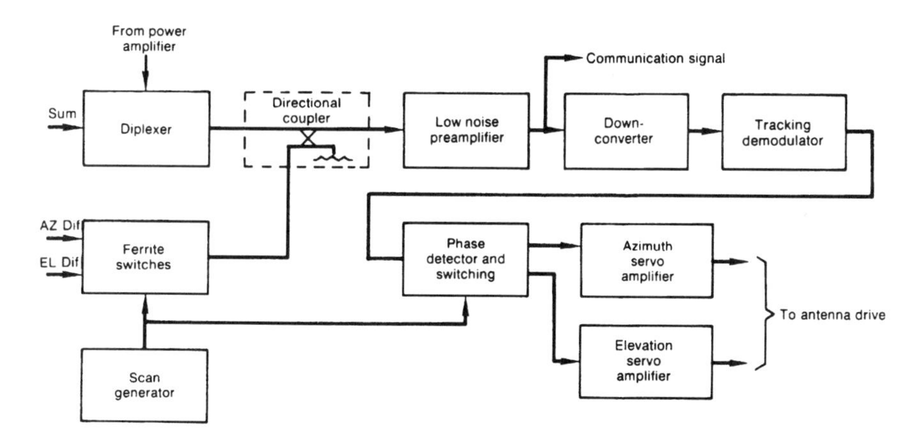

Figure 14.37. Functional block diagram of a satellite terminal monopulse tracking receiver subsystem.

arrangement of the monopulse system and requires only simple, low-cost electronics. The only input signal required from the satellite terminal is the AGC voltage or other DC signal proportional to the received RF signal level such as a signal level indication from a communications demodulator or from a beacon receiver. The output of the step-track processor algorithm can be as simple as a periodic step function for each antenna axis or as complex as a pseudorandom sinusoid. This type of output applied to the antenna servo drive subsystem will result in smooth, continuous antenna motion. Figure 14.38 is a simplified functional block diagram of a step-track system.

Step-tracking is a considerably lower-cost tracking system when compared to its monopulse counterpart. It is also less accurate. It lends itself to midsize satellite terminal installations operating with geostationary satellites and to tracking some inclined satellites with relatively slow orbital motion.

In the step-track technique the antenna is periodically moved a small amount along each axis, and the level of the received signal is compared to its previous level. A microprocessor, or part of the terminal control processor, provides processing to convert these level comparisons into input signals for the servo system, which will drive the antenna in directions that maximize the received signal level. In contrast to the monopulse technique, which seeks the null of the antenna difference pattern, step-tracking seeks the signal peak. Locating a beam maximum can never be as accurate as finding a sharp null as with the monopulse tracking technique.

CONICAL SCAN TRACKING (CONSCAN). Conical scan tracking is a refinement of the old antenna lobing technique used in World War II radars. Some of these original tracking radars used an array of radiating elements that could be switched in phase to provide two beam positions for the lobing operation. The radar operator observed on his/her display the same target side by side, which were the returns of the two beam positions. When the target was on-axis, the two pulses were of even amplitude, and when moved off-axis, the two pulses became unequal. To track a remote target, all the operator did

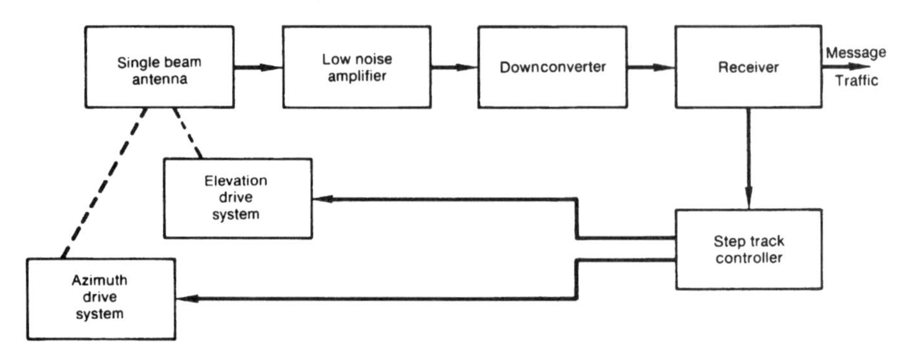

Figure 14.38. Functional block diagram of a step-track system.

was to maintain a balance between the two pulses by steering the antenna correctly.

This lobing technique was refined to a continuous rotation of the beam around the target, which is the basis of conscan tracking. Angle-error-detection circuitry is provided to generate error voltage outputs proportional to the tracking error and with a phase or polarity to indicate the direction of the error. The error signal actuates a servosystem to drive the antenna in the proper direction to null the error to zero.

One method to accomplish this continuous beam scanning is by mechanically moving the antenna feed, since the antenna beam will move off-axis as the feed is moved off the focal point. The feed is typically moved in a circular path around the focal point, causing a corresponding movement of the antenna beam in a circular path around the satellite to be tracked.

The feed scan motion may be by either a rotation or a nutation. A rotating feed turns as it moves with a circular motion, causing the polarization to rotate. A nutating feed does not rotate with the plane of polarization during the scan; it has an oscillatory movement of the axis of a rotating body or wobble. This is sometimes accomplished in the subreflector. The rotation or wobble modulates the received signal. The percentage of modulation is proportional to the angle tracking error, and the phase of the envelope function relative to the beam-scanning position contains direction information. In other words, this modulation is compared in phase with quadrature reference signals generated by the nutating mechanism to obtain error direction, and the amplitude of the modulation is proportional to the magnitude of the error. Figure 14.39 shows the conscan tracking technique and Figure 14.40 is a simplified functional block diagram of a conscan subsystem.

Conscan tracking may be used to track geosynchronous or polar orbit satellites with high or low target dynamics. Its principal advantages are low cost, only one RF channel requirement, and a single-beam feed. Its principal

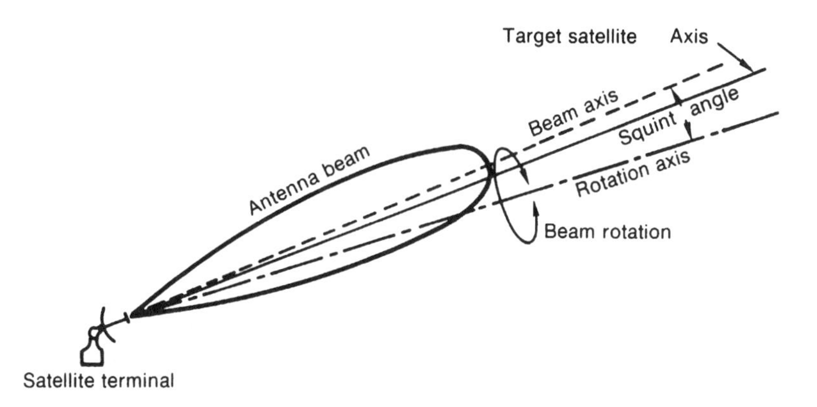

Figure 14.39. Conscan tracking operation.

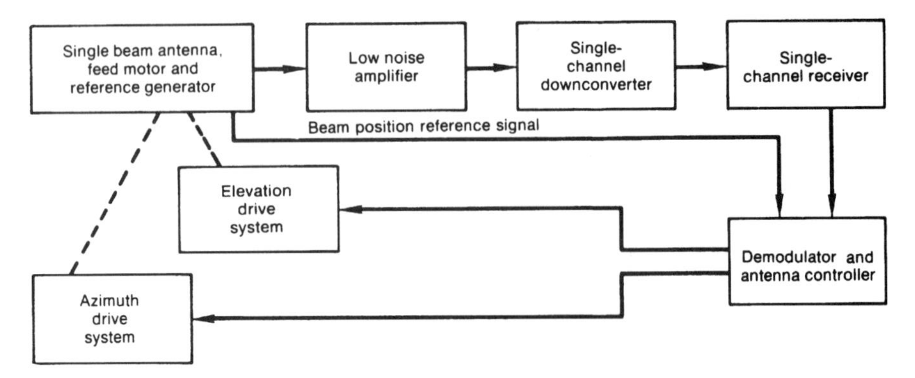

Figure 14.40. Simplified functional block diagram of a conical scanning subsystem.

disadvantages are that it requires four pulses or a short-duration continuous signal to obtain tracking information, is subject to mechanical reliability problems due to continuous feed rotation, and is subject to tracking loss due to propeller and other low-rate modulation sources.

PROGRAM TRACKING (OPEN-LOOP TRACKING). Program tracking is a processor-based tracking system where the processor calculates the required azimuth and elevation angles as a function of time. Such a processor is called an ephemeris processor. Ephemeris (plural: ephemerides) refers to a tabulation of satellite locations referenced to a time scale. The ephemeris processor uses an algorithm that calculates the relative direction of the satellite with respect to the terminal on a continuous and real-time basis. It contains in memory the forecast satellite location with respect to time, which requires periodic updating every 30 or 60 days. For fixed terminal sites, site location latitude and longitude are programmed into the processor at installation; for mobile terminals the processor requires continuous positional updates, often provided by an inertial navigation system.

Since program tracking is an open-loop process, it is subject to several sources of error that are automatically corrected in autotrack (closed-loop) systems. Some of these error sources are atmospheric refractions; structural deformation due to wind, ice, and gravity loads on the antenna; misalignment of mechanical axes; errors in axis angle measuring devices; and errors in input data relating to terminal position, satellite ephemeris, and absolute time.

In practice, a method of closed-loop tracking is usually included as well, since the accuracy required in the program tracking system is greatly relaxed if the last few tenths of a decibel in tracking accuracy are not required. In such cases program tracking is primarily used as an aide to initial satellite acquisition. However, for a geostationary satellite, acquisition causes little difficulty. In most cases a "look-up table," such as shown in Figure 6.5, suffices. Program tracking therefore usually has application for terminals that have to rapidly acquire nongeostationary satellites or rapidly slew from one satellite to another, as in the case of many military applications. Military ephemeris processors may have in memory ephemeris data for up to 20 satellites.

14.5.3 Very Small Aperture Terminals (VSATs)

VSAT networks are specialized satellite networks principally used for data traffic. In Chapter 8, Figure 8.1, we saw that a VSAT network most commonly consists of a hub and a series of small outstation satellite terminals, which we call VSATs.

Small VSATs have parabolic dish antennas with diameters ranging from 0.5 m for a very modest installation to up to 2.4 m for a much higher capacity installation. Hubs are much larger facilities to compensate somewhat for the small disadvantaged VSATs. One hub may serve more than 2000 VSAT outstations. Hub antennas range from 3 to 10 m in diameter. Their HPAs range from 20 to 200 watts of output power from 6/4-GHz operation and from 3 to 100 watts for 14/12-GHz operation.

VSAT outstations are optimized for cost because there are so many, in one system over 2000. Thus, in that system, the cost multiplier is 2000. For example, a variance in cost of $1000 multiplied by 2000 would become an increment in cost of $2 million.

Both a VSAT and a hub each consist of an indoor unit and an outdoor unit. A block diagram of a typical hub is shown in Figure 14.41 and of a typical VSAT in Figure 14.42. For the VSAT, the outdoor unit is miniaturized and jammed up into the parabolic reflector shell to make waveguide runs as short as possible, 1 or 2 inches. This reduces loss and noise due to ohmic noise; it also can save money.

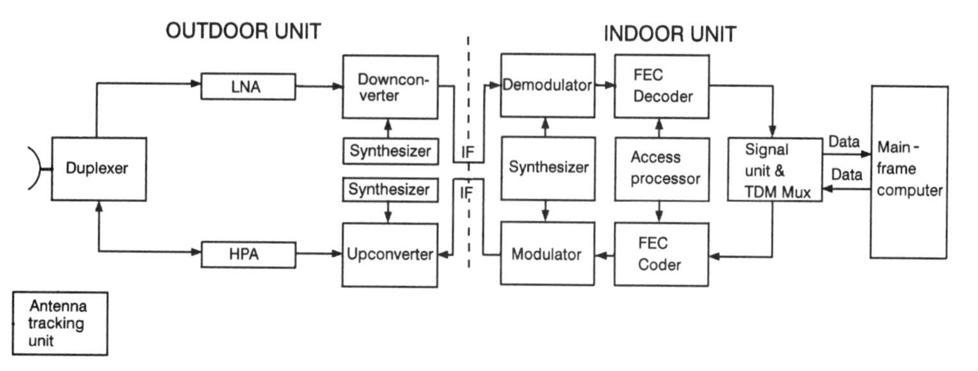

Figure 14.41. Hub unit, part of VSAT network.

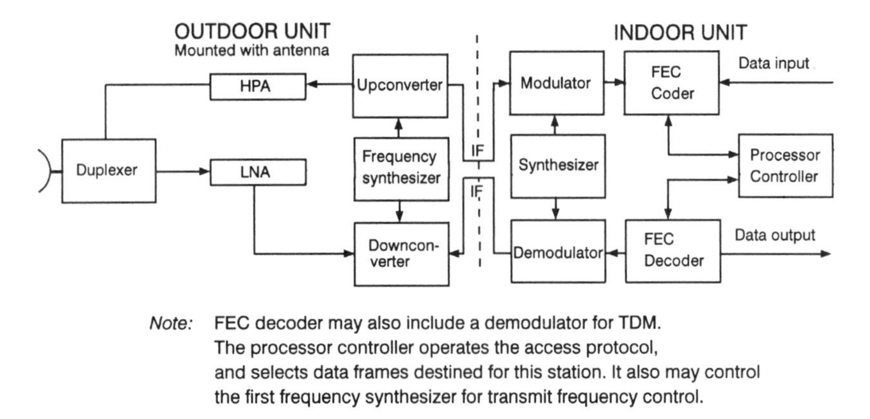

Note: FEC decoder may also include a demodulator for TDM. The processor controller operates the access protocol, and selects data frames destined for this station. It also may control the first frequency synthesizer for transmit frequency control.

Figure 14.42. A typical VSAT block diagram.

Hubs with larger diameter antennas will require tracking. Step-tracking is a cost-effective alternative. The tracking requirement will depend largely on the stationkeeping capabilities of the satellite accessed.

Figure 14.41 shows a hub configured for a very widely used application, that is, to interexchange data messages with VSAT outstations. The outdoor unit will probably be located in the antenna pedestal, with the indoor unit in a standard rack. Both Figure 14.41 and 14.42 show a configuration for the outbound link as a single thread high rate TDM data stream. The inbound link may use any one of several access protocols, including DAMA, as described in Chapter 8 (Ref. 17).

14.6 CELLULAR AND PCS INSTALLATIONS: ANALOG AND DIGITAL

14.6.1 Introduction

The goal of this section is to describe cellular installations and PCS facilities, which we will consider as an extension of cellular. In Chapter 10, three basic, generic modulation/access schemes were described: FDMA typified by AMPS and N-AMPS, TDMA typified by the North American IS-54 standard and the European GSM standard, and, finally, CDMA typified by the North American IS-95 standard. In large cellular schemes all three access/modulation types require cells or base stations with their radio towers and equipment, in most cases operating in the 800–960-MHz band and some kind of central control facility variously called a MTSO (mobile telephone switching office) or MSC (mobile switching center). Base stations or cells are connected to the MSC or MTSO usually via microwave carrying DS1 or E1 formats. However, wire pair and optical fiber are alternatives, if they are cost effective. The MTSO/MSC provides the interface with the PSTN. This concept is shown in Figure 14.43. Figure 14.44 illustrates the basic functional

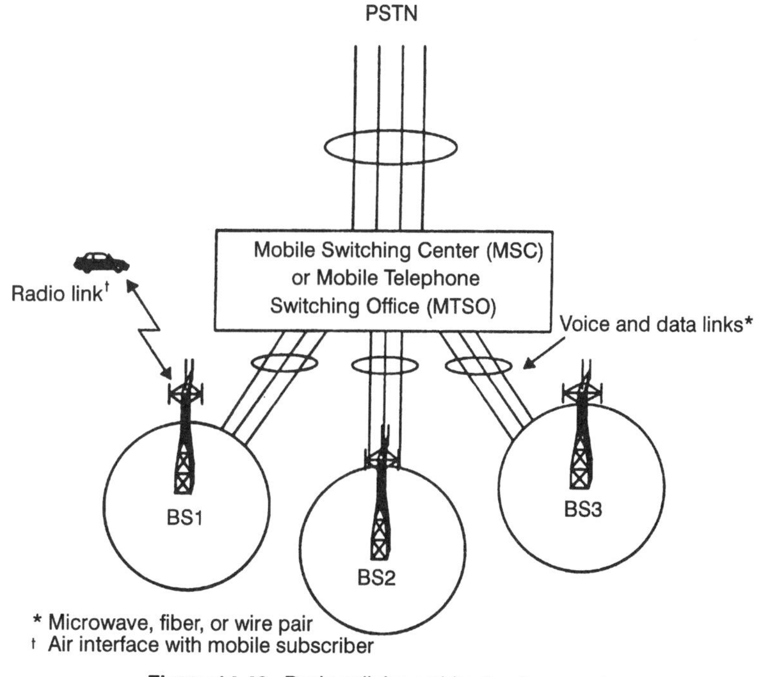

Figure 14.43. Basic cellular architectural concept.

blocks of a generic cellular system as envisioned by CCIR. This diagram is interchangeable with the European GSM digital TDMA cellular system.

14.6.2 Base Station or Cell Design Concepts

A base station consists of a shelter, a tower with antennas at or near its tops, transmission lines, and microwave/fiber/wire line terminations. There are also transmitters, receivers, and control and signaling shelves. Figure 14.45 shows a typical 800-MHz U.S. AMPS cell facility. Receivers commonly work in a space-diversity configuration. A receive multicoupler connects multiple receivers to a single antenna. And a second, similar, multicoupler is connected to the companion diversity antenna. There are then two receivers for each frequency and their outputs are combined by means of a selective combiner. The multicoupler/diversity concept is shown in Figure 14.46.

Each cell site (base station) has a controller for local control of operations. A conceptual overview of a controller is shown in Figure 14.47.

FCC Parts 22.904/22.905 (Ref. 18) limit cellular transmit power to 500 watts ERP.* The ERP value is scaled to HAAT (height above average

*ERP = effective radiated power. This is *not* EIRP, which references the power to an isotropic antenna. Here the reference is a dipole. This ERP value is 2.15 dB greater than its equivalent EIRP. Thus, to calculate EIRP given ERP, add 2.15 dB to the ERP value in dBm or dBW.

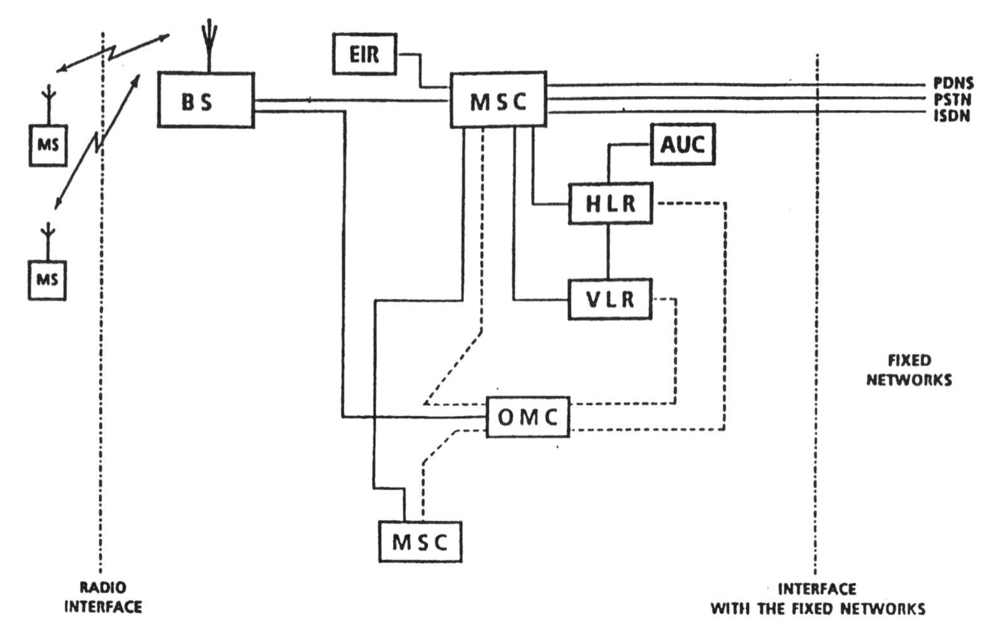

Figure 14.44. Functional blocks for a generic digital cellular system. (From Figure 1, p. 118, CCIR Rep. 1156, Reports of the CCIR 1990, Annex 1 to Vol. VIII; Ref. 19.) ——— Physical connections, ----- logical relationships, MS = Mobile station, BS = base station, MSC = mobile services switching center, HLR = home location register, VLR = visitor location register, OMC = operation and maintenance center, EIR = equipment identity register, AUC = authentication center.

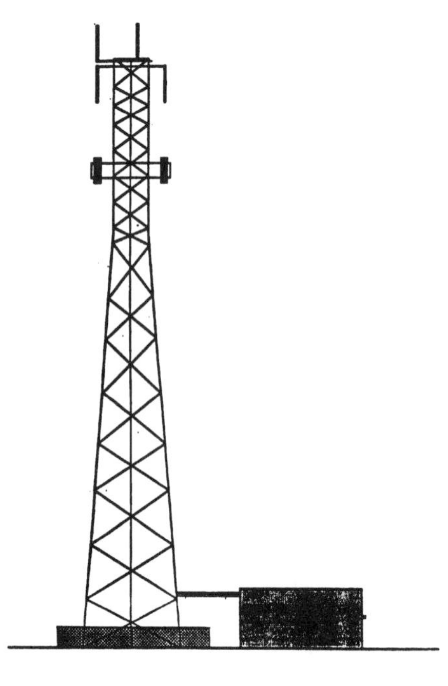

Figure 14.45. A typical U.S. AMPS cell facility: shelter, tower, and antennas.

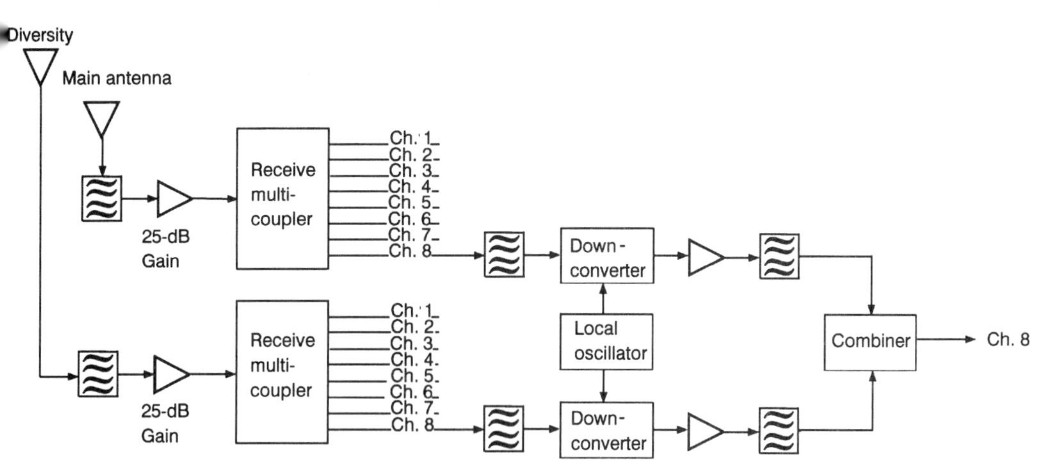

Figure 14.46. Cellular eight-channel receiver showing multicoupler operation in a space-diversity configuration. Only channel 8 diversity combining is illustrated.

terrain). The 500-watt value is for towers 500 ft and below. For towers in excess of 500 ft, less ERP is permitted. It should be noted that as cells get smaller, RF power is reduced to help mitigate frequency-reuse interference.

Figure 14.48 illustrates a typical cellular transmitter with 100-watt power amplifiers. There are two approaches to power amplifier design. The first is shown in Figure 14.48. A second approach would be to combine at a lower level and have just one comparatively broadband amplifier.

14.6.3 The MTSO or MSC

The mobile telephone switching office (MTSO), which is also called mobile switching center (MSC), provides the brains for cellular operations in a given

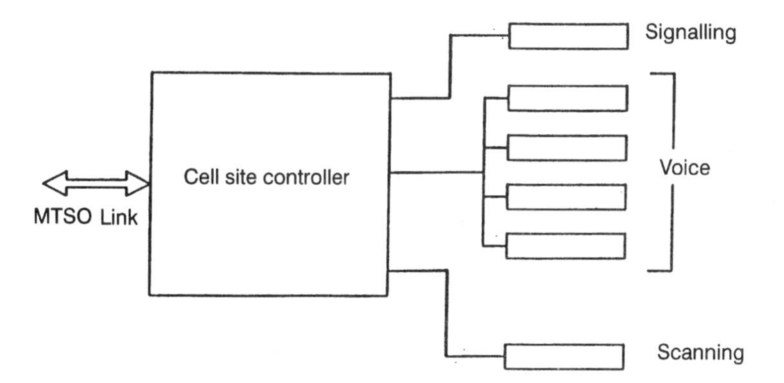

Figure 14.47. A cell site controller.

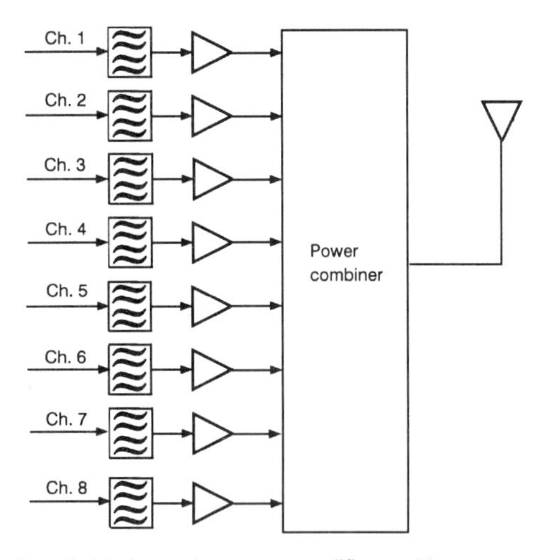

Figure 14.48. Cell site, eight-channel power amplifiers with power combiner. Transmitter output at combiner input port is typically +50 dBm.

area. It is responsible for several or many cell sites, cell/channel assignment for a particular mobile subscriber, handover to another cell when appropriate, and handover to another MTSO or another cellular system included under the blanket term "roaming." Control information exchanged between/among MTSOs is carried out by data circuits based on ITU-T Rec. X.25.*

Unless a MTSO shares a cell site, it does not have any direct radio functions. In other words, there are no cellular radios installed in a MTSO. One would expect to find microwave equipment used for links connecting the MTSO to cell sites and from the MTSO interconnecting with the PSTN. It is common practice that links connecting the MTSO with cell sites use the DS1 format (in the United States and Canada). In Europe, South America, and Africa, we would expect these links would use the E1 format and we would use the term MSC rather than MTSO. As one might expect, a MTSO has signaling, monitoring, and supervisory equipment, related to the setup, active supervision (including handover), and takedown of cellular telephone calls. If data service is available, the MTSO would also be responsible for these connectivities.

The MTSO is also responsible for billing of mobile subscribers and the maintenance of a "home location register (HLR)," which maintains records of those portable units permanently residing in the responsible area, and a visitor location register (VLR). Ideally, it should have communication capability with other HLRs to service "roaming" customers operating outside their "home" area to update their VLR.

* For a good description of ITU-T Rec. X.25, consult *Practical Data Communications* (Ref. 20).

14.6.4 Personal Communication Services (PCS)

PCS is difficult to strictly define. It is a radio system; it is small; it can provide multiple services; and it often can be considered an extension of cellular radio. Consider that paging, various types of remote control devices, cordless telephones, wireless PBXs, and wire loop replacement and wireless LANs may be considered PCSs. In our brief discussion below, we will consider only wireless telephones, which are operated in a cellular-like manner. Of course, in this case, cells are much smaller. We might call them microcells or picocells.

There are two standards that have developed in the recent past: Digital European Cordless Telephone (DECT) and Personal Access Communication System (PACS), a North American initiative (see Chapter 10). We briefly describe PACS hardware implementation in this subsection.

Some key features of PACS are:

- Small, inexpensive radio ports (RPs) and small coverage area per port
- Low complexity per-circuit signal processing
- Low transmit power and small batteries for subscriber units (SUs)
- Capability to provide PSTN access comparable to wire line
- Optimized to provide service to and in-building, and for pedestrian and city traffic operating environments
- Cost effective to service high traffic capacities

Figure 14.49 illustrates the PACS functional architecture. We make comparisons to cellular functions where appropriate.

The radio port (RP in Figure 14.49) can be likened to a cell site. This is a radio unit consisting of a digital transceiver. Its transmitter has 200-mW RF peak power output; its average power is 25 mW. RPs are powered by wire pair from a central power source. The radio interface uses $\pi/4$ quadrature phase-shift keying (QPSK) modulation in a TDMA/TDM type format where the downlink is TDM. The radio frame is 2.5 ms in duration with 8 bursts/frame. A RP range is about 650 ft, so for full coverage, the maximum RP spacing is about 1300 ft.

The RPCU (radio port control unit) may roughly be likened to a combination of cell site control and MTSO control and carries out functions of both. An access manager (AM) (Figure 14.49) can support multiple RPCUs with network-related tasks such as querying remote databases for visiting users, assisting in network call setup and delivery, coordinating link transfer between RPCUs, and multiple RP management. Between the PACS AM and RPCU, nearly all of the MTSO responsibilities are covered.

The RPCU connects to the serving PSTN switch or PABX by ISDN BRI.* Data are transmitted using the X.25 protocol with link access by LAPB.

*BRI = basic rate interface. It consists of two 64-kbps clear channels and a 16-kbps signaling channel. For a good overall description of ISDN and BRI, consult *Practical Data Communications* (Ref. 20).

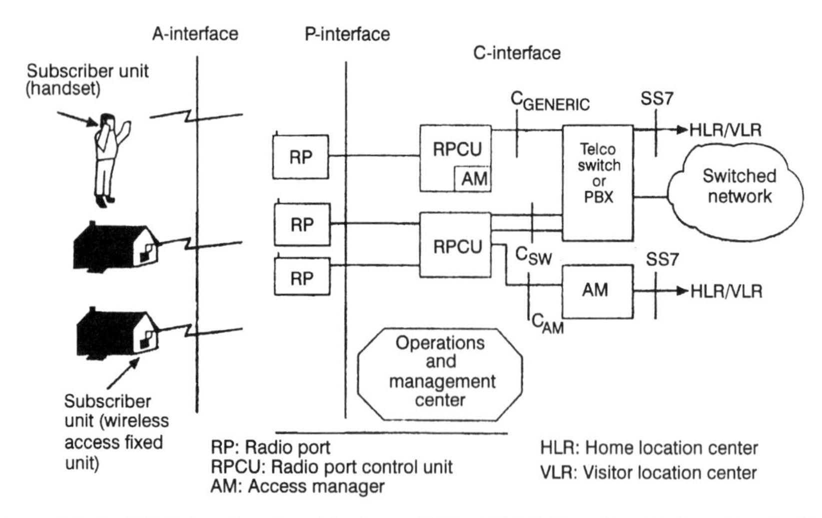

Figure 14.49. PACS functional architecture. SS7 = CCITT Signaling System No. 7. (From Figure 1, "PACS: Personal Access Communications System—A Tutorial," IEEE Personal Communications, June 1996; Ref. 21.)

PACS can operate as a FCC-licensed operation or in the PACS-UA or UB mode, in unlicensed operation. When operating as unlicensed, the frequency band authorized by the FCC is the 1920–1930-MHz band. Otherwise, as licensed, it may operate in the 1850–1910-MHz and 1930–1990-MHz bands.

14.7 HF TERMINALS AND ANTENNAS

14.7.1 Introduction

HF terminals can vary greatly in complexity, size, and cost. Such terminals can be small, transportable, or fixed in a 19-inch rack, taking up no more than 6 inches of rack space. Its transmitter may have no more than 10 or 100 watts of output. A small terminal such as this may cost no more than $1000 (1997 dollars). They can also be multiple transmitter/receiver installations occupying several sites, physically separated by some miles or kilometers. In this case transmitter RF outputs may be 1, 10, 25, 50, or even 100 kW and the cost could be multiple millions of 1997 dollars.

Most HF terminals use single-sideband suppressed carrier (SSBSC) modulation and can provide long-range voice, data, facsimile, and freeze-frame video service.

HF is an extremely congested medium. For this reason, there are very strict bandwidth, spurious, and harmonic radiation requirements.

14.7.2 Composition of Basic HF Equipment

A HF installation for two-way communication consists of one or more transmitters and one or more receivers. The most common type of modulation/waveform is single-sideband suppressed carrier (SSBSC). The operation of this equipment may be half- or full-duplex. We define half-duplex as the operation of a link in one direction at a time. In this case, the near-end transmitter transmits and the far-end receiver receives; then the far-end transmitter transmits and the near-end receiver receives. There are several advantages with this type of operation. A common antenna may be shared by a transmitter and receiver. Under many circumstances, both ends of the link use the same frequency. A simplified diagram of half-duplex operation with a shared antenna is shown below.

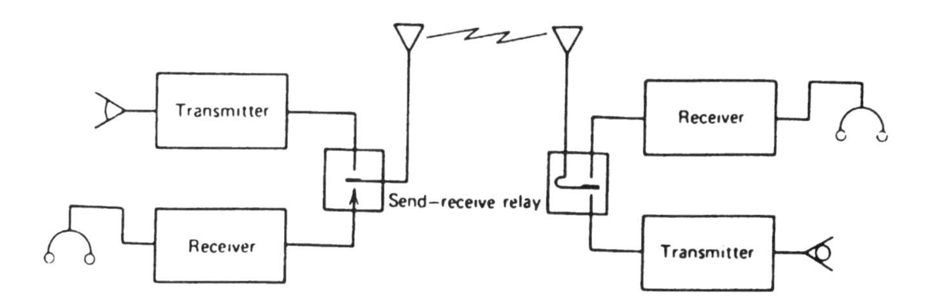

Full-duplex is when there is simultaneous two-way operation. At the same time the near end is transmitting to the far end, it is receiving from the far end. Usually, a different antenna is used for transmission than for reception. Likewise, the transmit and receive frequencies must be different. This is to prevent the near-end transmitter from interfering with its own receive frequency. Typical full-duplex operation is shown below.

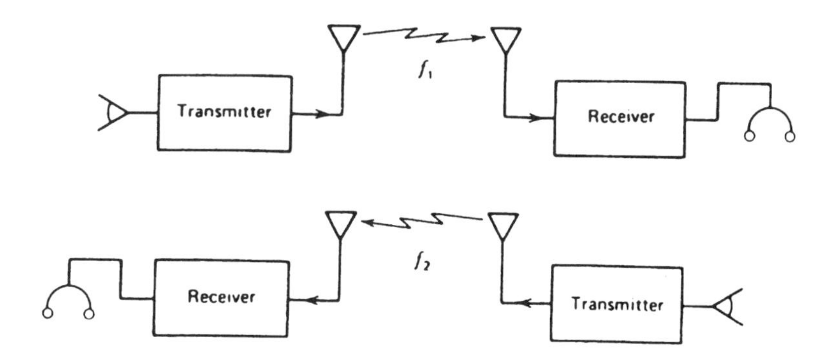

Being interfered with by one's own transmitter is called *co-site interference.* Care must be taken to assure sufficient isolation. There are many measures to be taken to mitigate co-site interference. Among those that should be considered are frequency separation, receiver selectivity, use of separate antennas and their sufficient separation, shielding, filtering transmit output, power amplifier linearity and grounding, bonding, and shielding. The worst environments for co-site interference and other forms of electromagnetic interference (EMI) include airborne and shipboard situations, particularly on military platforms.

When there is multiple transmitter and receiver operation required from a common geographical point, we may have to resort to two- or three-site operation. With this approach, isolation is achieved by physical separation of transmitters from receivers, generally by 2 km or more. However, such separations may even reach 20–30 km. At more complex installations, a third site may serve as an operational center, well isolated from the other two sites. In addition, we may purposely search for a "quiet" site to locate the receiver installation; "quiet" in this context means quiet regarding man-made noise. Both the transmitter site and receiver site may require many acres of cleared land for antennas.

The sites in question are interconnected by microwave LOS links and/or by coaxial cable or fiberoptic cable. Where survivability and reliability are very important, such as at key military installations, at least two distinct transmission media may be required. Generally, one is LOS microwave and the other some type of cable.

14.7.3 Basic Single-Sideband (SSB) Operation

Figure 14.50 shows the simplified functional block diagrams of a typical HF SSB transmitter and receiver.

TRANSMITTER OPERATION. A nominal 3-kHz voice channel amplitude modulates a 100-kHz* stable intermediate-frequency (IF) carrier; one sideband is filtered and the carrier is suppressed. The resultant sideband couples the RF spectrum of 100 + 3 kHz or 100 − 3 kHz, depending on whether the upper or lower sideband is to be transmitted. This signal is then upconverted in a mixer to the desired output frequency, F_0. Thus the local oscillator output to the mixer is F_0 + 100 kHz or F_0 − 100 kHz. Note that frequency inversion takes place when lower sidebands are selected. The output of the mixer is fed to a linear power amplifier and then radiated by an antenna. Transmitter power outputs can vary from 10 W to 100 kW or more.

RECEIVER OPERATION. The incoming RF signal from the distant end, consisting of a suppressed carrier plus a sideband, is received by the antenna

*Other common IF frequencies are 455 and 1750 kHz.

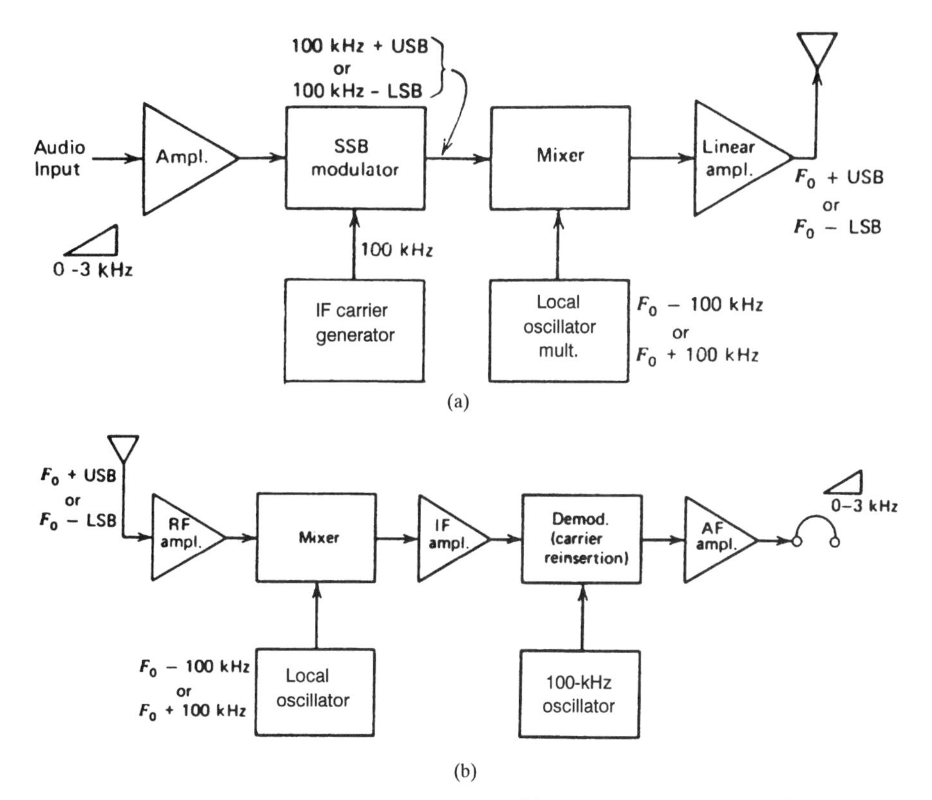

Figure 14.50. Simplified functional block diagram of (a) a typical SSB transmitter (100-kHz IF) and (b) a typical SSB receiver (100-kHz IF). USB = upper sideband; LSB = lower sideband; F_0 = operating frequency; IF = intermediate frequency.

and filtered by a bandpass filter, often called a preselector. The signal is then amplified and mixed with a stable oscillator to produce an IF (assume again that the IF is 100 kHz*). Several IF amplifiers increase the signal level. Demodulation takes place by reinserting the carrier at IF. In this case, it would be from a stable 100-kHz source, and detection is usually via a product detector. The output of the receiver is the nominal 3-kHz voice channel.

14.7.4 SSB System Considerations

One of the most important considerations in the development of a SSB signal at the near-end transmitter and its demodulation at the far-end receiver is accurate and stable frequency generation for use in frequency sources.

Generally, SSB circuits can maintain tolerable intelligibility when the transmitter and companion far-end receiver have an operating frequency

*Other common IF frequencies are 455 and 1750 kHz.

difference no greater than 50 Hz. Narrow shift FSK operation such as voice frequency carrier telegraph (VFTG) will not tolerate frequency differences greater than 2 Hz.

Prior to 1960, end-to-end frequency synchronization for high-quality SSB circuits was by means of a pilot carrier. At the far end, the receiver would lock on the semisuppressed transmitted carrier and slave its local oscillators to this circuit.

Today, HF transmitters and receivers use synthesizers as frequency sources. A synthesizer is a tunable, highly stable oscillator. It gives one or several simultaneous sinusoidal RF outputs on *discrete* frequencies in the HF range. In most cases it provides the frequency supply for all RF carrier needs in SSB applications. For example, it will supply the following:

- Transmitter IF carrier
- IF carrier reinsertion supply
- Transmitter local oscillator supply
- Receiver local oscillator supplies

The following are some of the demands we must place on a HF synthesizer:

- Frequency stability
- Number of frequency increments
- Spectral purity of RF outputs
- Frequency accuracy
- Supplementary outputs
- Capability of being slaved to a frequency standard

Synthesizer RF output stability should be better than $1.5-10^{-8}$. Frequency increments may be in 1-kHz steps for less expensive equipment and ranges down to every 10 Hz for higher quality operational equipment.

14.7.5 Linear Power Amplifiers

The power amplifier in a SSB transmitter raises the power of a low-level signal with minimum possible added distortion. That is, the envelope of the signal output must be as nearly as possible an exact replica of the signal input. Therefore, by definition, a linear power amplifier is required. Such power amplifiers used for HF communication display output powers in the range of 10 W to 100 kW or more. More commonly, we would expect to find this range narrowed to 100–10,000 W. Power output of the transmitter is one input used in link calculations and link prediction computer programs described later in the chapter.

The trend in HF power amplifiers today, even for high-power applications, is the use of solid-state amplifier modules. A single module may be used for the lower power applications and groups of modules with combiners for higher power use. One big advantage of using solid state is that it eliminates the high voltage required for vacuum tube operation. Another benefit is improved reliability and graceful degradation. Here we mean that solid-state devices, especially when used as modular building blocks, tend to degrade rather than suffer complete failure. This latter is more the case for vacuum tubes.

14.7.5.1 Intermodulation Distortion. Nonlinearity in a HF transmitter results in intermodulation (IM) distortion when two or more signals appear in the waveform to be transmitted. Intermodulation noise or intermodulation products are discussed in Chapter 1.

IM distortion may be measured in two different ways:

1. Two-tone test.
2. Tests using white noise loading.

The two-tone test is carried out by applying two tones simultaneously at the audio input of the SSB transmitter. A 3 : 5 frequency ratio between the two tones is desirable so that the IM products can be identified easily. For a 3-kHz input audio channel, a 3 : 5 frequency ratio could be tones of 1500 and 2500 Hz.

The test tones are applied at equal amplitude, and their gains are increased to drive the transmitter to full power output. Exciter or transmitter output is sampled and observed on a spectrum analyzer. The amplitudes of the undesired products and the carrier products are measured in terms of decibels below either of the equal-amplitude test tones as they appear in the exciter or transmitter output. The decibel difference is the signal-to-distortion ratio (S/D). This should be 40 dB or better. As one might expect, the highest level product is the third-order product. This product is two times the frequency of one tone minus the frequency of the second tone. For example, if the two test tones are 1500 and 2500 Hz, then

$$2 \times 1500 - 2500 = 500 \text{ Hz} \quad \text{or} \quad 2 \times 2500 - 1500 = 3555 \text{ Hz}$$

and, consequently, the third-order products are 500 and 3500 Hz. The presence of IM products numerically lower than 40 dB indicates maladjustment or deterioration of one or several transmitter stages, or overdrive.

The white noise test for IM distortion more nearly simulates operating conditions of a complex signal such as voice. The approach here is similar to that used to determine noise power ratio (described in Chapter 2). The 3-kHz audio channel is loaded with uniform amplitude white noise and a slot is cleared, usually the width of a VFTG channel (e.g., 170 Hz). The signal-to-distortion ratio is the ratio of the level of the white noise signal outside the slot to the level of the distortion products in the slot.

14.7.6 HF Configuration Notes

In large HF communication facilities, space-diversity reception is the rule. A rule of thumb for space diversity is that the antennas must be separated by a distance greater than 6λ. Each antenna terminates in its own receiver. The outputs of the receivers are combined either at the receiver site or at the operational center.

One receive antenna can serve many receivers by means of a multicoupler. A multicoupler, in this case, is a broadband amplifier with many output ports to which we can couple receivers.

A major impairment at HF receiver sites is man-made noise. For this reason, we select receiver sites that are comparatively quiet or quiet rural, as described in ITU-R Rec. PI.372.6, paragraph 5 (Ref. 22).

14.7.7 HF Antennas*

14.7.7.1 Introduction. The HF antenna installation can impact link performance more than any other system element of a HF system. It is also often the least understood and appreciated.

The antenna subsystem can be a "force-multiplier" if you will. A 10-dB net gain antenna system can make a 1-kW RF power output behave like 10 kW. It can attenuate interfering signals entering sidelobes. With an arrayed antenna system using advanced interference nulling techniques, interference rejection can be even more effective.

The selection of a particular type of HF antenna is application driven. We consider three generalized applications: (1) point-to-point, (2) near-vertical incidence (NVI), and (3) multipoint and subsets of skywave and groundwave. Table 14.7 reviews these applications and some antenna types appropriate to the application.

Important parameters in the selection of antennas are the following:

- Radiation patterns: azimuthal and vertical (elevation)
- Directive gain (receiver)
- Power gain (tuner efficiency, transmission line loss—transmitter)
- Polarization
- Impedance
- Ground effects
- Instantaneous bandwidth

Other important factors are size and cost.

* The material in Section 14.7.7 is primarily based on Ref. 23.

TABLE 14.7 Antenna Applications

Application	Antenna Type	Notes
Point-to-point skywave, one-hop	Horizontal LP[a]	
	Terminated long wire	
	Dipole and doublet	
Point-to-point skywave, multihop	Vertical LP[a]	
	Rhombic	
	Sloping vee	
Point-to-point groundwave	Whip	Vertical polarization
	Tower	generally preferred
NVI	Doublet	Reasonable gain
	Frame	at high TOAs[b]
Multipoint skywave, one-hop	Conical monopole	
	Discone	
	Cage antenna	Broadband
Multipoint skywave, multihop	LP[a] rosette	omnidirectional
	Array of rhombics	(azimuth)
Multipoint groundwave	Whip with counterpoise	
	Tower	

[a]Log periodic.
[b]Takeoff angles.

In previous chapters we have used the isotropic radiator as the reference antenna, where gain is expressed in dBi (dB related to an isotropic). The isotropic antenna has a gain of 1 or 0 dB and radiates uniformly in all directions in free space.

HF engineers often use other antennas as reference antennas. Table 14.8 compares some of these with the isotropic.

14.7.7.2 Antenna Parameters

14.7.7.2.1 Radiation Patterns. Radiation patterns are usually provided by the antenna manufacturer for a specific model antenna. These are graphical plots, one for the horizontal plane or azimuthal and one for the vertical plane

TABLE 14.8 Reference Antennas

Antenna Type	Gain (dBi)	Notes
Isotropic	0	In free space
Half-wave dipole	2.15	In free space
Full-wave dipole	3.8	In free space
Short vertical	4.8	In free space
Vertical	5.2	Quarter wave on perfectly conducting flat ground; lossless tuning device

Source: DCAC 330-175-1 (Ref. 23).

(elevation). Figures 14.51a and 14.51b show typical plots. The vertical plane should have some or total correspondence with the calculated elevation angle(s) or TOAs (takeoff angles).

One quantity determined uniquely by the radiation pattern is the antenna *directivity* that an antenna can attain. It is defined as the ratio of the maximum radiated power density to the average radiated power density. Whether a gain is in fact attained depends on losses with the antenna system.

14.7.7.2.2 Polarization. Polarization of the radiation from an antenna is produced by virtue of the fact that the current flow direction in an antenna is a vector quantity to which spatial orientation of the electric and magnetic fields are related. Single linear antennas in free space produce linear polarized waves in the far field, with the electric vector in a plane parallel to and passing through the axis of the radiating element. Antennas containing radiating elements with different spatial orientations and time phases produce elliptically polarized waves, with the ellipticity being a function of direction.

A linear antenna in free space used for receiving responds most strongly to another antenna with the same polarization and not at all to one polarized at 90° to the first antenna. Right-hand circularly (RHC) polarized antennas produce maximum receiving response when receiving transmission from a right-hand polarized antenna, and none from a left-hand circularly (LHC) polarized antenna.

Several other effects are also important. A circularly polarized antenna receiving a signal from a linear polarized antenna delivers 3 dB less power for a given transmitting power than would be received from a circular transmitted wave of the correct rotation sense. Conversely, of course, the same received signal loss applies to a linear receiving antenna and a circularly polarized transmitting antenna.

Vertically polarized waves radiated by a vertical antenna over imperfect ground are elliptically polarized because of the presence of a small electric field along the ground in space quadrature and a different time phase from the major electric component.

For HF skywave propagation, the received signal is usually elliptically polarized even when the transmitting antenna produces linear polarization. This condition arises from the effect of the earth's magnetic field in the ionized layers, which splits the incident linearly polarized signal into ordinary and extraordinary waves. The two waves travel at different velocities and experience different polarization rotations. The resulting received signal is elliptically polarized.

14.7.7.2.3 Impedance. The impedance of an antenna depends on the radiation resistance, the reactive storage field, antenna conductor losses, and coupled impedance effects from nearby conductors. For simple antennas such as dipoles, integration of the power pattern over a spherical surface will

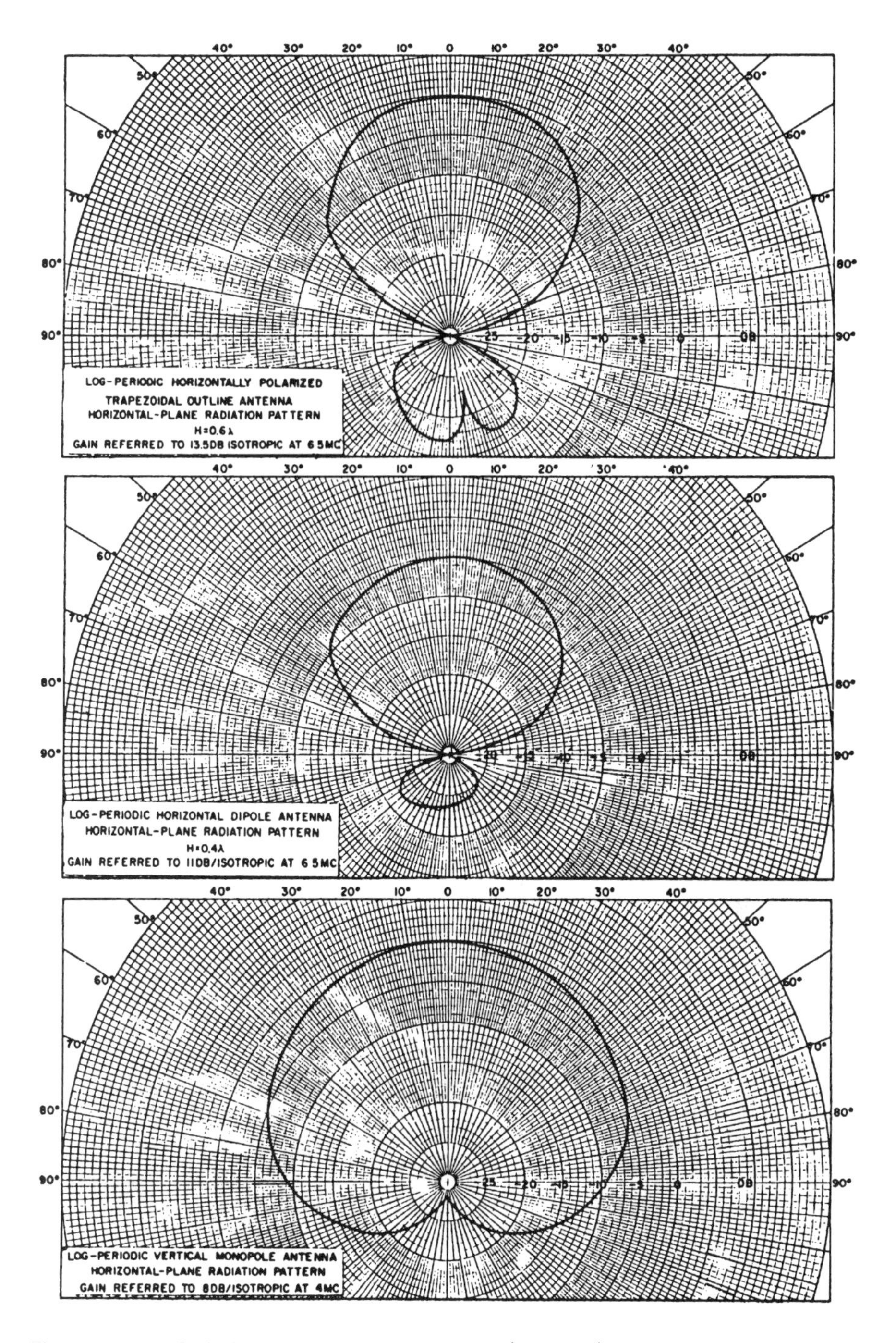

Figure 14.51a. Radiation patterns, horizontal plane (azimuthal), log-periodic antennas. For directive values (dB): upper curve, add 13.5 dB; middle curve, add 11 dB; and lower curve, add 8 dB to readings. (From DCAC 330-175-1; Ref. 23.)

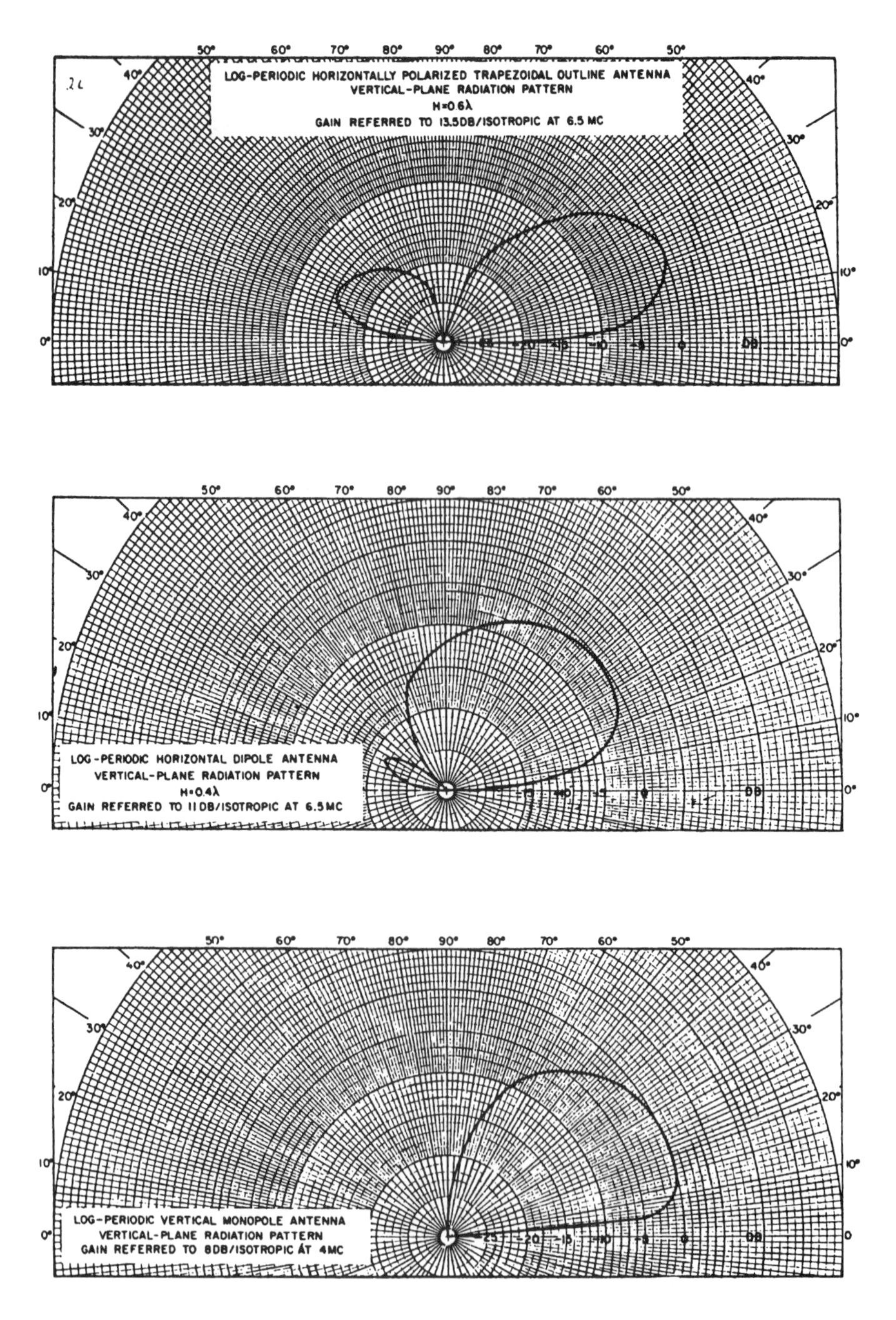

Figure 14.51b. Radiation patterns, vertical plane (elevation or TOA), log-periodic antennas. For directive gain values (dB): upper curve, add 13.5 dB; middle curve, add 11 dB; and lower curve, add 8 dB to the readings. (From DCAC 330-175-1; Ref. 23.)

yield a quantity called the radiation resistance, which appears as a multiplying factor for the antenna current in the integrated power pattern. The radiation resistance obtained from the antenna impedance can be measured by a RF bridge placed at the current reference terminals of the antenna.

The antenna impedance as measured with a bridge will contain the radiation resistance, a resistance proportional to antenna conductor losses and, if the antenna is near other conducting objects, a component representing an interaction between the antenna and the nearby conductors. This latter component may be either positive or negative resistance and will usually have a reactive component.

The reactive storage field of an antenna depends to a considerable extent on the antenna diameter and physical shape. The storage field represents energy that flows into the antenna and then is returned to the generator during each cycle.

Antennas that have small diameters in terms of wavelength have large storage fields, whereas thicker antennas have lesser storage energy. Since the storage field represents circulating energy, which is analogous to that flowing in a physical reactance, the storage field contributes a reactive component to the antenna impedance.

Dipole antennas and other balanced standing wave antennas have reactive behavior similar to that of an open-circuited transmission line with a length equal to half the antenna length. For example, the reactive component of the impedance of a dipole is capacitive for lengths shorter than half-wavelength, is zero at about half-wavelength, and is inductive for lengths between one-half and one wavelength. The antenna reactance continues to alternate cyclically in a half-wavelength period for longer lengths. For convenience in matching the standing wave antenna, it is usual at high frequencies to use antennas at a frequency where resonance occurs, that is, where the reactance is zero.

Traveling-wave antennas, such as rhombics and terminated vees, have wide impedance bandwidths. In fact, for these antennas, impedance bandwidths usually exceed radiation pattern bandwidths. Log-periodic (LP) antennas achieve wide impedance bandwidth by selectively choosing only elements that are nearly resonant at a specific frequency.

Nominal impedance values of some common HF antennas are as follows (Ref. 23):

Horizontal rhombic and terminated vee	600 Ω
Horizontal LP	50/300 Ω
Vertical (dipole and monopole)	50 Ω
Yagi and half-wave dipole	50 Ω
Conical monopole, discone, and inverted discone	50 Ω
Vertical tower	50 Ω

14.7.7.2.4 Gain and Bandwidth. The gain of an antenna is defined as the ratio of the maximum power density radiated by the antenna to the maximum

TABLE 14.9 Antenna Comparison

Type	Power Gain[a] (dB)	Usable Radiation Angle[b] (deg)	Bandwidth[c] Ratio	Horizontal Beamwidth	Sidelobe Suppression
Horizontal rhombic	8–23	4–35	≥ 2 : 1	6–26°	< 6 dB
Terminated vee	4–10	4–35	≥ 2 : 1	8–36°	> 6 dB
Horizontal LP	10–17	5–45	≥ 8 : 1	55–75°	> 12 dB
Vertical LP (dipole)	6–10	3–25	≥ 8 : 1	90–140°	> 12 dB
Horizontal half-wave dipole	5–7	5–80	≥ 5%	80–180° /lobe	NA
Discone	2–5	4–40	≥ 4 : 1	NA	NA
Conical monopole	−2 to +2	3–45	≥ 4 : 1	NA	NA
Inverted monopole	1–5	5–45	≥ 4 : 1	NA	NA
Vertical tower	−5 to +2	3–30	≥ 4 : 1	NA	NA

[a] Typical power gains are gains over good earth for vertical polarization and poor earth for horizontal polarization.
[b] Usable radiation angles are typical over good earth for vertical polarization and poor earth for horizontal polarization.
[c] Normal bandwidth is the ratio of the two frequencies within which the specified voltage standing wave ratio (VSWR) will not be exceeded or within which the desired pattern will not suffer more than 3-dB degradation.

Source: DCAC 330-175-1 (Ref. 23).

power density radiated by a reference antenna (see Table 14.8). The directivity of an antenna, which is sometimes confused with antenna gain, is the ratio of the maximum power density radiated by the antenna to the *average power* radiated by the antenna. The distinction between the two terms arises from the fact that directivity will exceed antenna gain. Since all antennas have some losses, the directivity will exceed antenna gain. The directivity of an antenna can be obtained from the antenna radiation patterns alone, without consideration of antenna circuit losses. Because directivity is a ratio, absolute power values are not required in determination of directivity, and convenient normalizing factors can be applied to radiation power values.

Rhombic and vee antennas, which dissipate portions of the antenna input power in terminators, will have lower gain values than directivity values by about the termination losses. Co-phased dipole arrays have low conductor losses, and directivity, as a result, only slightly exceeds gain values for such antennas. LPs and Yagis can have 1-dB or more difference between directivity and gain. Table 14.9 compares power gain, usable radiation angles, nominal bandwidth, horizontal beamwidth, and sidelobe suppression.

14.7.7.2.5 Ground Effects. The free-space radiation pattern efficiency and impedance of an antenna are modified when the antenna is placed near ground. The impedance change is small for antennas placed at least one wavelength above ground, but the change becomes increasingly greater as that height is reduced. Since the ground appears as a lossy dielectric medium for HF, location of the antenna near ground may increase the losses a considerable extent unless a ground plane (a wire mesh screen or ground radials) is used to reduce ground resistance. Vertical monopole antennas, which are often fed at the ground surface, require a system of ground radials extending from the antenna to a sufficient distance to provide a low-resistance ground return path for the ground currents produced by the induction fields. For short vertical antennas, the radial length should be approximately $\lambda/2\pi$ long; for longer antennas the length should be approximately the antenna height. In addition, near the antenna base a wire mesh ground screen is recommended to reduce I^2R losses. For medium- or long-distance HF links, horizontal antennas should be mounted higher than one-half wavelength above ground and usually do not require a ground screen or ground plane.

In addition to losses, the presence of ground causes a change in antenna impedance. The change is brought about by the interaction between the antenna fields and fields of the ground currents.

Ground-reflected energy combines with the direct radiation of an antenna to modify the significantly vertical radiation pattern of the antenna. The magnitude of the image antenna current is equal to the magnitude of the real antenna current multiplied by the ground-reflection amplitude coefficient. An additional phase difference between the direct and the ground-reflected field at a distant point results from the difference of the two path lengths.

14.7.7.2.6 Bandwidth. Antenna bandwidth is specified as the frequency band over which the voltage standing wave ratio (VSWR) criteria are met and the radiation pattern provides the required performance. The frequency band is, to some extent, determined by the application, since greater deterioration of antenna characteristics can be tolerated for receiving use than for transmitting use. Antenna bandwidth is usually limited by the change of either or both the radiation shape and impedance with change of frequency.

Simple antennas, such as dipole, when in free space, have a figure-8 pattern broadside to the antenna. For dipoles, the pattern maxima are oriented normal to the dipole with lengths up to approximately $1\frac{1}{4}$ wavelengths. With greater lengths, the pattern maxima may not be oriented in this direction.

Linear conductors with traveling-wave distributions have a continuous shift of the distribution in the direction of radiation pattern maximum, and the maximum approaches the axis of the conductor with increasing frequency. For this type of antenna and all other linear antennas, the pattern maximum can never fall exactly along the conductor axis because of the inherent null of the conductor in the direction of current flow, except when located near the earth.

When simple antennas are arrayed with other like antennas, the pattern changes increase with frequency because the changing electrical spacing between the arrayed elements also becomes a factor in the determination of the radiation pattern. In an array with discrete elements, an additional complication is produced by grating lobes. These lobes can occur whenever the interelement spacing exceeds a half-wavelength. If the element spacing equals one wavelength, the grating lobes can have intensities equal to the main radiation lobe. These lobes arise because the wide interelement spacing allows all element radiations to combine in phase in more than one direction.

Another factor in pattern change with frequency is produced by antenna energy, which is reflected from the ground and combined with the direct radiated energy. Since the electrical path difference between the direct and the ground-reflected energy is proportional to the frequency, the elevation pattern of an antenna above ground is a function of frequency.

In a communication circuit, two degrees of severity in pattern change with frequency can occur. In one the pattern shape changes, but the maximum is in the required radiation direction. In the second the maximum is deflected from the desired direction with frequency. The first can be tolerated at the expense of increased circuit interference and noise. The second can cause complete circuit outages with frequency change.

Considerations of pattern change with frequency usually limit the use of broadband monopoles or dipoles to no more than two octaves. This usable frequency range is reduced to no more than one octave when these elements are arrayed. Rhombic antennas have a tolerable pattern shift, if used within a frequency range of less than one octave. LP antennas escape the directivity limitations of antennas using conductors of fixed physical length by selectively changing the active conductor length with frequency. However, unless the LP antenna is specially designed for use over ground, the antenna will have a pattern shift caused by ground reflection.

Input impedance change of an antenna may limit the coverage of an antenna to a smaller frequency range than would be expected from the antenna radiation pattern. This is particularly true of transmitting antennas when the antenna supplies the load for the transmission line and transmitter, and any inability to load the transmitter. Input impedance requirements for receiving antennas are not so severe because the receiver is the termination for the line and the antenna VSWR causes only a reduction in received power transfer.

For a description of the various types of HF antennas covered here, consult the latest edition of *Telecommunication Transmission Handbook*, (Ref. 3).

14.8 METEOR BURST INSTALLATIONS

Meteor burst communication (MBC) terminals operate in the 40–60-MHz region of the spectrum. They consist of a transmitter, receiver, control processor, and antenna(s). MBC systems are transmit data/telemetry and are commonly used for remote readout of field sensors, such as new snowfall and total accumulation in the U.S. Rocky Mountains. Payload traffic in this case is unidirectional. There are a master station and literally hundreds of slave

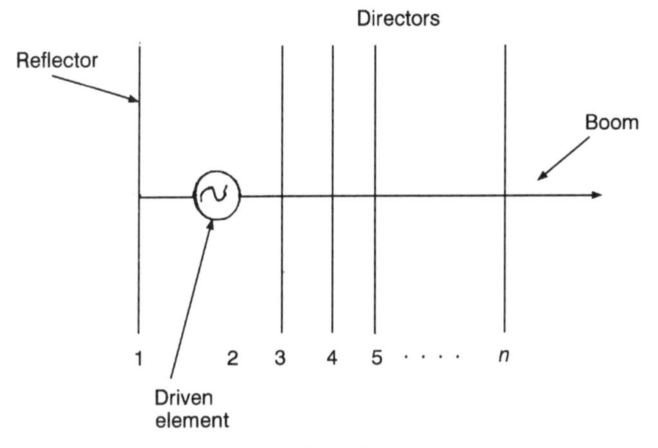

Figure 14.52. Outline of a Yagi antenna.

field stations. A master station polls outstations, which reply with traffic. Master stations are often comparatively large, transmitting kilowatts of RF power. The field (slave) stations are small and modest, transmitting with RF power between 10 and 200 watts, and derive their prime power from batteries. Antennas used on MBC links are almost universally Yagis. Gains are in the range of 10–16 dB.

There is disagreement in the meteor burst community as to the ideal gain for a MBC antenna. As we are aware, beamwidth is a function of gain. The more the gain, the narrower the beamwidth. As beamwidths are decreased there is less and less of the sky encompassed. As a result, less and less usable trails are available. Of course, systems work better with higher gain antennas. But this does us little good if we begin missing usable meteor trails. The following subsection describes the important antenna subsystem, the Yagi antenna.

14.8.1 Yagi Antennas*

The typical Yagi antenna† consists of many dipole antennas with different lengths and spacings. These antennas are mounted on a supporting boom as shown in Figure 14.52. Only one of the dipoles is driven, while the remainder are parasitic elements. The element behind the driven element is the reflector, and there is usually only one. The elements in front of the driven element are called directors. The reflector is the longest element, which is $\lambda/2$ in length. The director elements are always shorter than the driven

*Section 14.8.1 is based on material from Ref. 6.
†The Yagi antenna if often called the Yagi–Uda antenna. The inventor was S. Uda, and the first description in English of the antenna was provided by H. Yagi.

element in length. The reflector is usually spaced $\lambda/4$ to the rear of the driven element. A gain of +10 dBi can be realized with only a moderate number of dipoles. An 18-element Yagi array can display about +16.5 dBi power gain.

PROBLEMS AND EXERCISES

1 Draw a simplified functional block diagram of a basic generic radio communications terminal. (It should have, as a minimum, nine blocks.)

2 Why are almost all LOS microwave facilities being installed today digital rather than analog (aside from STL)?

3 List similarities and differences between analog LOS microwave terminals and digital terminals.

4 Calculate the gain of a radiolink parabolic dish antenna with a 4-ft aperture at 12 GHz. Assume a 55% efficiency. Increase the efficiency to 65%. What is the gain now?

5 Why is sidelobe attenuation (antenna discrimination) so important, whether LOS microwave or satellite communications? If the return loss of an antenna is 30 dB, what is the equivalent reflection coefficient?

6 What is the equivalent beamwidth of a parabolic antenna with a 6-ft diameter at 6 GHz?

7 A radiolink transmitter operates at 7 GHz. Why would I rather use waveguide for the transmission line than coaxial cable? What type of waveguide should I consider, taking into account all factors, including the practical? What is the loss of a waveguide length of 120 ft that might be selected?

8 Where is a circulator used in a radiolink terminal and how does it work?

9 What is the function of a load isolator? Why is its use so important when transmit power is high (e.g., > 100 watts) as might be found on certain satellite terminals and on all troposcatter terminals?

10 Why would an IF repeater not be appropriate for digital radiolinks?

11 Generally, U.S. FCC regulations regarding bandwidth–bit rate can be accommodated with 64-QAM. Why, then, are we trending to higher and higher M-values for M-QAM, such as $M = 512$ or even 1024?

12 Why is space diversity deemed even more important on digital LOS radiolinks than on analog radiolinks?

13 Compare one-for-one and one-for-n hot-standby operation on LOS radiolinks.

14 Name at least three parameters/values that would cause hot-standby switchover for a transmitter chain and for a receiver chain.

15 What is the function of a line-code converter on a digital LOS microwave terminal?

16 Why would one want to specify twist and sway limits for a LOS microwave tower?

17 For what reason do we convert from binary codes to Gray codes at the digital radiolink transmitter?

18 What is the difference between *gain* and *directivity* in an antenna?

19 Name at least two of the purposes of service channels on radiolinks.

20 Make a comparison chart of troposcatter terminals and LOS microwave termninals. The chart should compare at least five items.

21 What is the function of a duplexer in a tropo terminal?

22 How is sufficient isolation achieved on a troposcatter terminal when the same antenna transmits and receives simultaneously?

23 Explain why LNAs are used universally on satellite earth stations and not necessarily universally on LOS microwave facilities.

24 Why would a TDMA-equipped earth station have to be able to set and reset burst time boundaries rapidly?

25 At the higher frequencies (i.e., at Ku-band and above) why is it so important to place the HPA and LNA as close as possible to the antenna or even right in the antenna structure?

26 Many widely used antennas have no autotracking capability at all. Explain in your own words why antenna tracking becomes less and less stringent as antenna sizes get smaller (i.e., less gain).

27 For linear polarization of a satellite communications antenna, what would be a good practical value for polarization isolation?

28 Compare a prime focus antenna with a Cassegrain antenna.

29 What is the purpose of an orthomode junction?

30 Name and compare the two basic methods of tracking a satellite.

31 Why is step-tracking attractive, where can it be used, and what are some of its limitations?

32 Consider even very small parabolic dish antennas for LEO satellites. Do we have an option of no tracking with these antennas (see question 26)?

33 As a system designer of VSAT systems, the major objective in the design of the outstation VSATs is to reduce costs just as much as possible. What is the overriding reason for this?

34 What range of power outputs would we expect from a VSAT outstation transmitter?

35 What are the two basic approaches to the design of cell site transmitters?

36 What is a common value for cell site transmitter power output? Under what circumstances should we consider lowering this value?

37 What are the functions of a cell site controller?

38 Make a comparison chart of the characteristics and attributes of (a) cellular radio and (b) PCS as we described it. PACS might be a good model.

39 A top-of-the-line, complex, multichannel HF facility consists of three sites. Identify the function of each site.

40 What is the most common type of modulation used on HF?

41 Modern HF circuits can tolerate up to how many hertz of frequency difference between transmitter and receiver for voice operation? How many hertz for data operation?

42 Name at least four frequency supply functions of a synthesizer.

43 What is a major advantage of using solid-state amplifiers in high-power HF transmitters?

44 Either with vacuum tube or solid-state amplifier operation, the level of IM products is a major concern. How do we test for these IM products?

45 For space-diversity operation on HF, what is a rule-of-thumb separation distance for antennas?

46 Name two popular types of broadband HF antennas that might be used for long-range skywave transmission.

47 What is the gain of an isotropic antenna? What is the gain of a standard half-wave dipole antenna?

48 What are the two parameters that are required to specify bandwidth?

49 Name and describe the two different element types used with a Yagi antenna. What range of gain can we expect from a single-array Yagi antenna?

REFERENCES

1. *The New IEEE Standard Dictionary of Electrical and Electronic Terms*, 5th ed., IEEE Std. 100-1992, IEEE, New York, 1993.
2. *Intermediate-Frequency Characteristics for the Interconnection of Analog Radio-Relay Systems*, CCIR Rec. 403-3, Vol. IX, Part 1, XVIIth Plenary Assembly, Dusseldorf, 1990.
3. Roger L. Freeman, *Telecommunication Transmission Handbook*, 3rd ed., Wiley, New York, 1991.
4. Roger L. Freeman, *Reference Manual for Telecommunications Engineering*, 2nd ed., Wiley, New York, 1994.
5. *Design Handbook for Line-of-Sight Microwave Communication Systems*, MIL-HDBK-416, U.S. Department of Defense, Washington, DC, Nov. 1977.
6. Henry Jasik and Richard C. Johnson, *Antenna Engineering Handbook*, 2nd ed., McGraw-Hill, New York, 1984.
7. *System Planning, Product Specifications, Services*, Andrew Corporation, Catalog No. 36, Orland Park, IL, 1994.
8. A. P. Barkhausen et al., *Equipment Characteristics and Their Relationship to Performance for Tropospheric Scatter Communication Circuits*, Technical Note 103, U.S. National Bureau of Standards (now NIST), Boulder, CO, Jan. 1962.
9. *Transmission Systems Engineering Symposium*, Rockwell International, Collins Transmission Systems Division, Dallas, TX, Sept. 1985.
10. *Frequencies and Deviations of Continuity Pilots for Frequency Modulation Radio-Relay Systems for Television and Telephony*, CCIR Rec. 401-2, XVIIth Plenary Assembly, Dusseldorf, 1990.
11. *Structural Standards for Steel Antenna Towers and Antenna Supporting Structures*, EIA/TIA-222-E, Electronics Industries Association, Washington, DC, Mar. 1991.
12. *Preferred Frequency Bands for Trans-horizon Radio-Relay Systems*, ITU-R Rec. 698-2, 1994 F Series Volume, Part 1, ITU, Geneva, 1994.
13. *Propagation Effects on the Design and Operation of Trans-horizon Radio-Relay Systems*, CCIR Rep. 285-7, Reports of the CCIR 1990, Annex to Vol. IX, Part 1, XVIIth Plenary Assembly, Dusseldorf, 1990.
14. *Performance Characteristics for Intermediate Data Rate (IDR) Digital Carriers*, INTELSAT Doc. IESS-308, Rev. 7A, INTELSAT, Washington, DC, 1994.
15. Roger L. Freeman, *Telecommunication System Engineering*, 3rd ed., Wiley, New York, 1996.
16. *Satellite Communications Reference Data Handbook*, Computer Sciences Corp., Falls Church, VA, Mar. 1983, DCA contract DCA100-81-C-0044.
17. G. Maral, *VSAT Networks*, Wiley, Chichester, U.K., 1995.
18. Code of Federal Regulations, Telecommunication, 47, Parts 20 to 39, Revised as of October 1, 1994. More commonly known as *FCC Rules and Regulations*, U.S. Government Printing Office, Washington, DC, 1994.
19. *Digital Cellular Land Mobile Telecommunication Systems (DCLMTS)*, CCIR Rep. 1156, Reports of the CCIR 1990, Annex 1 to Vol. VIII, XVIIth Plenary Assembly, Dusseldorf, 1990.

20. Roger L. Freeman, *Practical Data Communications*, Wiley, New York, 1995.
21. "PACS: Personal Access Communications System—A Tutorial," *IEEE Personal Communications*, June 1996.
22. *Radio Noise*, ITU-R Rec. PI.372-6, 1994 PI Series Volume, ITU, Geneva, 1994.
23. "MF/HF Communications Antennas," DCAC 330-175-1, Addendum No. 1 to *DCS Engineering—Installation Standards Manual*, Defense Communications Agency (now DISA), Washington, DC, May 1996.

Appendix 1

AVAILABILITY OF A LINE-OF-SIGHT MICROWAVE LINK

A1.1 INTRODUCTION

The IEEE (Ref. 1) defines availability as "the long-term average fraction of time that a system is in service satisfactorily performing its intended function." Availability, in our context, is usually expressed as a percentage or decimal. It defines the time a system, link, or terminal is meeting its operational requirements.

Equipment availability is expressed by the familiar equation

$$A = \text{MTBF}/(\text{MTBF} + \text{MTTR}) \qquad \text{(A1.1)}$$

where MTBF is mean time between failures and MTTR is mean time to repair. Both are measured in hours.

As one can see, equation (A1.1) only treats equipment failure and its repair time; it does not reflect outage due to fading. By restating the equation, we can cover the general case:

$$A = \text{uptime}/(\text{uptime} + \text{downtime}) \qquad \text{(A1.2)}$$

Example. If a system has an uptime of 10,000 hours and a downtime of 10 hours, then

$$A = 10{,}000/(10{,}000 + 10) = 0.999 \text{ or } 99.9\%$$

In our discussion we will be working with *unavailability*, which we can call "a state of nonservice that occurs due to failure, other outage or degraded BER performance of at least 10 seconds duration." Our notation of unavailability is U, and

$$U = 1 - A \tag{A1.3}$$

or

$$U = 1 - [\text{uptime}/(\text{uptime} + \text{downtime})]$$

Applying this to the above example, we find

$$\begin{aligned} U &= 1 - 0.999 \\ &= 0.001 \text{ or } 0.1\% \end{aligned}$$

With radiolinks, we will deal with one-way (e.g., west-to-east) and two-way availabilities. If the unavailability objective of a two-way channel is 0.02%, and the outage probabilities in the two directions are independent, the objective for a one-way channel is 0.01% or about 105 minutes/year for a two-way system, and 53 minutes/year for an equivalent one-way system.

The importance of *availability* cannot be understated. It tells us how well we can depend on the system, link, or terminal to do its job. In this appendix we discuss availability from several aspects and point up some weaknesses in traditional arguments on the subject.

A1.2 CONTRIBUTORS TO UNAVAILABILITY

CCIR Rep. 445-3 (Ref. 2) lists five major contributors to outage on radio-relay systems:

1. Equipment
   a. Failure/degradation of radio equipment such as modulators and demodulators.
   b. Failure of auxiliary equipment, such as switchover equipment.
   c. Failure of primary power (equipment).
   d. Failure of antenna or feeder.
2. Propagation
   a. Deep fading causing noise to exceed a certain limit. This may be due to ducting and usually lasts for a fairly long time.
   b. Excessive precipitation attenuation that is caused mainly by heavy rainfall and, in some cases, heavy snowfall. Generally, the effect lasts a fairly long time.
   c. Fading causing short interruptions (dispersive fading) (includes ISI degradation and outage.
3. Interference: noise in excess of a certain limit caused by interference sources that may exist within or outside the system.

4. Support facilities: collapse of towers or buildings in disastrous circumstances.
5. Human error: this includes maintenance downtime/outages.

The items listed above are in order of greatest to least contributors to unavailability. Equipment failure/degradation is certainly the greatest contributor.

A1.3 AVAILABILITY REQUIREMENTS

The AT & T unavailability objective is 0.01% for a one-way channel over a 4000-mi route (Ref. 3). The equivalent availability is 99.99%.

In Canada a tentative objective of 99.97% is used for a 1000-mi one-way radio system. This corresponds to 99.95% availability on a 2500-km base.

The current United Kingdom availability objective for bidirectional transmission is 99.994% per 100 km, which corresponds to 99.84% for a 2500-km circuit (Ref. CCIR Rep. 445-2) (Ref. 2).

A1.4 CALCULATION OF AVAILABILITY OF LOS RADIOLINKS IN TANDEM

In this section we derive per-hop availability given a system availability consisting of n hops in tandem with independent outages on different links. Often such a system availability will derive from the CCIR hypothetical reference circuit, which is 2500 km long. Panter (Ref. 4) assumes that such a 2500-km circuit consists of 54 hops each 30 mi (48 km) long.

We describe the procedure by an example. If a one-way circuit requires an availability of 99.95% over a 2500-km reference circuit, what is the required per-hop availability? First calculate the system unavailability, which, in this case, is $1 - 0.9995 = 0.0005$. Divide this unavailability by 54 or $0.0005/54 = 0.000009259$. This then is the unavailability for one hop. Its availability is $1 - 0.000009259 = 99.999074\%$.

ATT uses a 4000-mi reference circuit with a required availability of 99.99% or an unavailability of 0.0001 or 1×10^{-4}. If we assume as above that each hop is 30 mi long, then there are 133 hops in the reference circuit. We now divide 1×10^{-4} by 133 and the resulting unavailability per hop is 0.000075%. The equivalent availability per hop is then 99.999925%.

A1.4.1 Discussion of Partition of Unavailability

At first glance we could apportion half the outage (unavailability) to equipment failure and half to propagation outage. Thus, in the case of ATT, the unavailability per hop would be 0.000075%/2 or 0.0000375% for equipment and 0.0000375% for propagation outages.

White of GTE-Lenkurt argues against this approach (Ref. 5). Let us consider a 1-year or 8760-h interval. A year has 525,600 min or 31,536,000 s. What is the annual expected outage when the unavailability is 0.000075%?

$$8760 \text{ h} \times 0.00000075 = 0.00657 \text{ h}$$

$$525{,}600 \text{ min} \times 0.00000075 = 0.3942 \text{ min}$$

$$31{,}536{,}000 \text{ s} \times 0.00000075 = 23.652 \text{ s}$$

If we assign half of this number to equipment outage and half to propagation outage, we then have 11.8 s of outage per year for each.

The next step is to apply the conventional formula for availability [equation (A1.1)] and calculate MTBF in hours:

$$0.000075\% = \frac{\text{MTBF}}{\text{MTBF} + \text{MTTR}}$$

However, we first must assign a reasonable value for MTTR or repair time.

Consider that most LOS radiolink sites are unattended, and when a failure occurs, a technician must be sent to that site. He/she must be alerted, gather up tool kit and required parts, and travel to that site, possibly 60–100 mi away, and, of course, time must be allowed to carry out the repair. Values for MTTR in the literature for this application are from 2 to 10 hours. We use the worst case, then

$$99.999925\% = \frac{\text{MTBF}}{\text{MTBF} + 10}$$

$$\text{MTBF} = 1.32 \times 10^6 \text{ h or } 150.68 \text{ yr}$$

If we allotted half the outage to equipment failure and half to propagation, we must double the MTBF, requiring a MTBF of about 301 years!

The argument then follows that propagation outage, say 32 s/yr, might consist of many events (short fades) in 1 year, whereas with equipment reliability we are dealing with one event every 301 years.

It would follow, then, that we treat these two types of outages separately and independently for they are truly not summable. It is like summing 6 apples and 4 oranges resulting in 10 lemons. What is driving us to these large values of MTBF is the large values for MTTR. On sophisticated military radio terminals (non-LOS radiolink), MTTR runs at about 0.3–0.5 h. It is assumed that the technician is on site and on duty.

A1.4.2 Propagation Availability

We then treat propagation availability separately, but it too requires some special considerations of reasonableness. Again let us turn to an example. We use the Canadian values (Section A1.3) of 99.95% for a 2500-km reference circuit and the equivalent unavailability is 0.0005. Assume, again, 54 hops in tandem for the 2500-km reference circuit. The unavailability per hop is then $0.0005/54 = 0.000009259$ or an availability of 99.9990741%. This, of course, assumes a very worst-case fading where all hops fade simultaneously but independently. This is unrealistic.

Panter (Ref. 4) reports a more reasonable worst case where we would allow only one-third of the hops to fade at once, $54/3 = 18$; thus we can say that the unavailability due to propagation (multipath fading) is $0.005/18 = 0.0000278$ or an availability of 99.972 would be required per hop. This would be nearly in keeping with CCIR Rec. 395 paragraph 1.2 or $(50/2500) \times 0.1$ of any month (not year) where the availability required is 99.998%/month to a reference level of 47,500 pWp. We essentially derive the same value for a 2500-km reference circuit where $L = 2500$; then the unavailability is $0.001/54$ or 99.998% availability per hop, following CCIR Rec. 395 to the letter.

Again the assumption is made that all hops in tandem are simultaneously subject to fading. Following Panter's reasoning, the divisor would be 18 rather than 54 or a required availability per hop of 99.994% to the reference level of 47,500 pWp.

A1.5 IMPROVING AVAILABILITY

Examining equation (A1.1), we can improve availability by increasing MTBF (improving reliability) and decreasing MTTR (mean time to repair).

MTBF can be increased by using Hi-rel (high-reliability) components, particularly for those components that have a history of numerous failures. Another approach is to use redundancy, such as hot-standby operation. If the necessary automatic changeover circuits are employed, redundancy (i.e., have two identical equipments with automatic changeover on failure) squares the MTBF of the combination. For example, if a transmitter has a MTBF of 10,000 hours and we equip with a second, identical, transmitter with changeover, the resulting MTBF is $10{,}000 \times 10{,}000$ hours or 10^8 hours.

The difficult and ambiguous part of the availability equation is MTTR. Unattended repeater (or drop and insert) sites require that a technician travel to that site when there is a failure or degradation. How long does it take to get there? Did he/she bring the right part, card, or subassembly with him/her? If not, is the card, part, or subassembly available in the storeroom? If not, it must be ordered from the equipment manufacturer, and suppose the manufacturer does not have the part.

It is not feasible from an economic sense to have every single card, part, and subassembly available as a spare part in the storeroom. Most operations only stock high-failure items. Some do not stock any and depend on fast turnaround from the manufacturer. For this latter approach MTTR can extend over 24 hours.

How long does it take the technician to find the failed part? This can be improved by using BITE (built-in test equipment), and if this information can be remoted to the servicing terminal via a service channel, all the better. BITE identifies on a go/no-go basis (binary) if a card, circuit, or subassembly is working. It may be working, but how well? These BITE circuits can be set for a certain threshold before kicking in an alarm. Thus they can alarm at a certain degradation point.

CCIR Rep. 445-3 (Ref. 2) uses 10 hours for a MTTR value. The U.S. military often uses 20 minutes. Of course, the U.S. military assumes a technician on duty and that the spare part is available. Nearly all military electronic equipment have excellent BITE, so troubleshooting can be easy and sure. Note how availability drastically improves with a 20-minute (0.333-h) MTTR.

Every item mentioned above costs money—the price goes up. It boils down to how much we are willing to pay for good availability values.

A1.6 APPLICATION TO OTHER RADIO MEDIA

The same approach to availability and its calculation and improvements can be applied directly to over-the-horizon links such as troposcatter and diffraction. It can also be applied to satellite communications links. For cellular/PCS, HF, and meteor burst systems, modifications would have to be made for propagation effects. Certainly for HF and meteor burst systems, propagation takes on a much more important role, probably a dominant role.

REFERENCES

1. *The New IEEE Dictionary of Electrical and Electronic Terms*, 5th ed., IEEE Std 100-1992, IEEE, New York, 1993.
2. *Availability and Reliability of Radio-Relay Systems*, CCIR Rep. 445-3, Annex to Vol. IX, XVIIth Plenary Assembly, Dusseldorf, 1990.
3. *Transmission Systems for Communications*, 5th ed., Bell Telephone Laboratories, Holmdel, NJ, 1982.
4. P. F. Panter, *Communication Systems Design for Line-of-Sight Microwave and Troposcatter Systems*, McGraw-Hill, New York, 1972.
5. R. F. White, *Reliability in Microwave Systems—Prediction and Practice*, GTE-Lenkurt, San Carlos, CA, 1970.

Appendix 2

REFERENCE FIELDS AND THEORETICAL REFERENCES; CONVERTING RF FIELD STRENGTH TO POWER

A2.1 REFERENCE FIELDS—THEORETICAL REFERENCES*

In many propagation problems, the methods of solution are based on a transmitting antenna that will produce some standard value of field intensity at a standard distance (e.g., MF, HF, some VHF/UHF, broadcast, and cellular). This standard field may be one of several produced by one of several types of reference antenna. Three of the most commonly used reference antennas (with this application in mind) are the omnidirectional radiator (isotropic source) in free space, the omnidirectional radiator over perfectly conducting earth, and the short lossless vertical antenna over perfect earth. Each field is expressed in reference to 1-kW power input at a standard distance of 1 mi of 1 km.

The most often used reference antenna is the omnidirectional radiator, a theoretical antenna that radiates equally well in all directions. The field intensity produced by this antenna may be found by equating the radiated power to the surface integral of a uniform field over a spherical surface surrounding the radiator:

$$E^2 = \frac{P_t \eta}{4\pi d^2} \quad \text{mV/m} \tag{A2.1}$$

where P_t = the transmit power (W)
η = the impedance of free space
d = the distance (km)

*Section A2.1 is based on material from "MF/HF Communication Antennas," DCAC 330-175-1, Addendum No. 1 to *DCS Engineering—Installation Manual*, Defense Communication Agency, Washington, DC, 1966.

Thus if the transmit power P_t is 1 kW and the distance is 1 km, the field intensity is found by

$$E = \left(\frac{1000\eta}{4\pi}\right)^{1/2} \quad \text{mV/m} \tag{A2.2}$$

and since $\eta = 120\pi$ (impedance of free space),

$$\begin{aligned} E &= [1000(30)]^{1/2} \\ &= 173.2 \text{ mV/m at 1 km} \end{aligned}$$

If this omnidirectional antenna is placed on the surface of perfectly conducting earth, we obtain

$$\begin{aligned} E &= 173.2\sqrt{2} \\ &= 245.0 \text{ mV/m at 1 km} \end{aligned}$$

as the field intensity, since all the power is radiated in a hemisphere and, therefore, the power per unit area is doubled.

The field of a short vertical antenna over perfect earth may be found by

$$E_\Delta = \frac{60\pi I}{d}\left(\frac{l}{\lambda}\right)\cos\Delta\left(1 + e^{-j4\pi h/\lambda \sin\Delta}\right) \tag{A2.3}$$

where l/λ = length in wavelengths
Δ = vertical radiation angle
I = input current
h/λ = effective height in wavelengths
d = the distance in kilometers

But since the height of the differential element for a ground-based antenna is zero, the field intensity in the ground plane over perfect earth is

$$E_\Delta = \frac{120\pi I}{d}\left(\frac{l}{\lambda}\right) \tag{A2.4}$$

The radiation resistance R_r is

$$R_r = 160\pi^2\left(\frac{l}{\lambda}\right)^2 \tag{A2.5}$$

and since $I = \sqrt{P_t/R_r}$ then

$$I = \frac{\sqrt{P_t}}{\sqrt{160\pi^2(l/\lambda)^2}} - \frac{\sqrt{P_t}}{4\sqrt{10}\pi l/\lambda} \tag{A2.6}$$

and the field intensity becomes

$$E_\Delta = \frac{120\pi}{d}\left(\frac{l}{\lambda}\right)\frac{\sqrt{P_t}\cdot\lambda}{4\sqrt{10}\,\pi l} \tag{A2.7}$$

and

$$E_\Delta = \frac{30}{d}\sqrt{\frac{P_t}{10}} \tag{A2.8}$$

where d is in kilometers
P is in watts
E_Δ is in mV/m

Solving for a power input of 1 kW at a distance of 1 km in the ground plane ($\Delta = 0°$), we find

$$E_\Delta = 30\sqrt{\frac{1000}{10}} = 300 \text{ mV/m}$$

A2.2 CONVERSION OF RADIO-FREQUENCY (RF) FIELD STRENGTH TO POWER*

Many radio engineers are accustomed to working in the power domain (e.g., dBm, dBW). For example, we may wish to know the receive signal level (RSL) at the input to the first active stage of a HF receiver. In the power domain, the characteristic impedance is not a consideration, by definition.

HF engineers traditionally work with field strength usually expressed in microvolts per meter (μV/m). When we convert μV/m to dBm, characteristic impedance becomes important. We remember the familiar formula

$$\text{Power (W)} = \frac{E^2}{R} = I^2R \tag{A2.9}$$

where E is expressed in volts and I in amperes, and we can consider R to be the characteristic impedance.

*Section A2.2 is based on *Technical Issues* #89-1 (*Dave Adamy*), Association of Old Crows, Alexandria, VA, 1989.

Carrying this one step further,

$$\text{Power (W)} = \frac{[E(\text{V/m})]^2[\text{effective antenna area (m}^2)]}{(\text{impedance of free space})} \qquad \text{(A2.10)}$$

The impedance of free space is 120π or 377 Ω.

The effective antenna area is

$$A\ (\text{m}^2) = \frac{G\lambda^2}{4\pi} \qquad \text{(A2.11)}$$

where A = effective antenna area (m^2)
G = antenna gain (numeric, *not* dB)
λ = wavelength (m)

If $G = 1$ (0 dBi), then

$$P\ (\text{W}) = \frac{E^2\lambda^2}{377 \times 4\pi} \qquad \text{(A2.12)}$$

We rewrite equation (A2.12) expressed in frequency rather than wavelength:

$$P\ (\text{W}) = \frac{\dot{E}^2(c^2)}{4737.5(f^2)} \qquad \text{(A2.13)}$$

where c = velocity of propagation in free space, or 3×10^8 m/s. Convert to more useful units: express P in milliwatts, E in microvolts per meter, and f in megahertz. Then

$$P\ (\text{mW}) = \frac{1.89972 \times 10^{-8}(E)^2}{f^2\ (\text{MHz})} \qquad \text{(A2.14)}$$

Here E is expressed in microvolts per meter. We now derive

$$P\ (\text{dBm}) = -77 + 20\log(E) - 20\log(f) \qquad \text{(A2.15)}$$

where E is the field strength in microvolts per meter and f is expressed in megahertz. To convert back to field strength

$$E = 10^{[P+77+20\ \log(f)]/20} \qquad \text{(A2.16)}$$

Appendix 3

GLOSSARY OF ACRONYMS AND ABBREVIATIONS

A

ACK	Acknowledgment
ACTS	Advanced Communications Technology Satellite (NASA)
ADPCM	Adaptive differential pulse code modulation
AGC	Automatic gain control
AIFM	Adjacent channel interference fade margin
ALE	Automatic link establishment
ALOHA	Random access technique developed by the University of Hawaii
AM	Amplitude modulation; access management
AMD	Automatic message display
AMI	Alternate mark inversion
AMPS	Advanced mobile telephone system
AMSSB	Amplitude-modulation single sideband
ANSI	American National Standards Institute
ARQ	Automatic repeat request (from old-time telegraphy, used in error correction)
ASK	Amplitude-shift keying
ATB	All trunks busy
ATM	Asynchronous transfer mode
AUC	Authentication center
AWGN	Additive white Gaussian noise
AZD	Ambiguity zone detection

B

BBE, BBER	Background block error, background block error ratio

BCH	Bose–Chaudhuri–Hocquenghen (a family of cyclic error-correcting codes, named for the inventors)
BCI	Bit count integrity
BCM	Block-coded modulation
BER	Bit error rate or bit error ratio
BF	Bandwidth (correction) factor
BFSK	Binary frequency-shift keying
BH	Busy hour
BINR	Baseband intrinsic noise ratio
BITE	Built-in test equipment
BLOS	Beyond line-of-sight
BNZS	Binary *N*-zeros substitution
BP	Bandpass
bps	Bits per second
BPSK	Binary phase-shift keying
BRI	Basic rate interface
BSS	Broadcast satellite service
BSTJ	Bell System Technical Journal
BTR	Bit timing recovery
BW	Bandwidth
BWR	Bandwidth ratio

C

CATV	Community antenna television (same as "cable" television)
CCIR	International Consultive Committee (on) Radio, now called ITU-R, more formally, ITU Radiocommunications Sector
CDC	Control (and) delay channel
CDMA	Code division multiple access (spread spectrum)
CDPD	Cellular digital packet data (a standard)
CELP	Codebook excitation linear predictive (coder)
CFDM	Compandered frequency division multiplex
CFM	Composite fade margin, compandered FM
CGSA	Cellular geographic serving area
CMI	Coded mark inversion
C/I	Carrier-to-interference ratio
C/N, C/N_0	Carrier-to-noise ratio, carrier-to-spectral noise ratio (i.e., in 1 Hz of bandwidth)
CODEC	Coder–decoder (in PCM and similar)
COMSAT	Communication Satellite (Corp.)
CONSCAN	Conical scanning (tracking system)
CR	Carrier recovery
CRC	Cyclic redundancy check

CRPL	Central Radio Propagation Laboratory
CSC	Common channel signaling
CSMA, CSMA/CD	Carrier sense multiple access, carrier sense multiple access with collision detection
CONUS	Continguous United States
CT	Cordless telephone
C/T	Carrier-to-thermal noise ratio
CVSD	Continuous variable slope delta modulation

D

DAR	Distortion adaptive receiver
DAMA	Demand assignment multiple access
DA/TDMA	Demand assignment TDMA
dB	Decibel
dBi	Decibels referenced to an isotropic (antenna)
dBm, dBm0p	Decibels referenced to a milliwatt; dBm psophometrically weighted, referenced to the 0 test level point
DBPSK	Differential binary phase-shift keying
dBrnC	Decibels reference noise, C-message weighted, a North American noise measurement unit
dBW	Decibels referenced to a watt
DC, dc	Direct current
DCME	Digital circuit multiplication equipment
DDF	Digital distribution frame
DECT	Digital European Cordless Telephone (a standard)
DFM	Dispersive fade margin
DFMR	Reference dispersive fade margin
D/L	Downlink
DM	Degraded minute, delta modulation
DNI	Digital noninterpolated
DoD	Department of Defense (U.S.)
DQPSK	Differential quadrature phase-shift keying
DS	Direct sequence (spread spectrum)
DS1, DS1C, DS2 . . .	North American plesiochronous digital hierarchy
DSCS	Defense Satellite Communication System
DSI	Digital speech interpolation
DTM	Data text mode

E

EB	Errored block
E_b/N_0	Energy per bit-to-noise spectral density ratio
EC	Earth coverage
EI	Errored interval

EIA	Electronics Industries Association (U.S.)
EIFM	External interference fade margin
EIR	Equipment identity register
EIRP	Effective (also equivalent) isotropic radiated power
EMC	Electromagnetic compatibility
EMI	Electromagnetic interference
EOC	End of coverage
EOM	End of message
ERP	Effective radiated power (2.15-dB difference with EIRP)
ES, ESR	Errored second, errored second ratio
ESC	Engineering service circuit
EUTELSAT	European telecommunications satellite
EW	Elliptical waveguide; electronic warfare

F

FAA	Federal Aviation Administration (U.S.)
FCC	Federal Communications Commission
FCS	Frame check sequence
FDD	Frequency division duplex
FDM	Frequency division multiplex
FDMA	Frequency division multiple access
FEC	Forward error correction
FET	Field effect transistor
FLTSAT	Fleet Satellite (a family of U.S. Navy UHF satellites)
FM	Frequency modulation
FOT	Frequencè optimum de travail (French; same as OWF, optimum working frequency)
FPLMTS	Future Public Land Mobile Telecommunication System
FSK	Frequency-shift keying
FSL	Free-space loss
FSS	Fixed satellite service

G

GEO	Geostationary earth orbit
GFSK	Gaussian frequency-shift keying
GHz	Gigahertz
GMSK	Gaussian minimum shift keying
GPS	Geographical positioning system
GSM	Ground Systeme Mobile (European digital cellular standard)

G/T	Figure of merit of earth station or satellite receiving systems. $G/T = G_{dB} - 10 \log T_{sys}$. G is net gain of receiving antenna; T_{sys} is equivalent receiving system noise temperature.
GTE	General Telephone & Electronics

H

HAAT	Height above average terrain
HC	Hemispherical coverage
HDB3	High-density binary 3 (European type of B3ZS for digital waveforms)
HDLC	High-level datalink control
HEMT	High electron mobility (transistor)
HF	High frequency; the radio-frequency band from 3 to 30 MHz
HLR	Home location register
HPA	High-power amplifier
HPF	Highest probable frequency
HRDP	Hypothetical reference digital path
HRP	Hypothetical reference path

I

IBPD	In-band power difference
IBS	International business service (offered by INTELSAT)
I/C	Interference-to-carrier ratio
IDR	Intermediate data rate (offered by INTELSAT)
IEE	Institution of Electrical Engineers (U.K.)
IEEE	Institute of Electrical and Electronics Engineers (U.S.)
IF	Intermediate frequency
IFRB	International Frequency Registration Board (ITU)
IGY	International geophysical year
INMARSAT	International Marine Satellite (consortium)
INTELSAT	International Telecommunication Satellite (consortium)
IONCAP	Inonospheric communication analysis prediction program
IM, IM noise	Intermodulation; intermodulation noise
I & Q	In-phase & quadrature
IRL	Isotropic receive level
ISC	International switching center
ISDN	Integrated services digital network

ISI	Intersymbol interference
ISL	Intersatellite link
ISO	International Standards Organization
ISU	Iridium subscriber unit
ITU	International Telecommunications Union
ITU-T, ITU-R	ITU Telecommunications Standardization Sector (previously CCITT); ITU Radiocommunications Sector (previously CCIR)

K

Ka-Band	28–30-GHz satellite uplink band, microwave LOS band
kbps	kilobits per second
km	kilometer
kHz	kilohertz
Ku-band	14/12/11-GHz satellite communication band; microwave LOS band

L

LAD	Linear amplitude dispersion
LAN	Local area network
LAPB	Link Access Protocol B-channel
LEO	Low earth oribt
LHC	Left-hand circular (polarization)
Lincompex	Link compressor–expander
LNA	Low-noise amplifier
LO	Local oscillator
LORAN	Long-range (radio) navigation
LOS	Line-of-sight
LP	Log-periodic (antenna)
LPI	Low probability of intercept
LQA	Link quality assessment
LRE	Low-rate encoding
LSB	Lower sideband
LST	Local sidereal time
LUF	Lowest usable frequency

M

M-ary	Refers to multilevel signaling where $M > 2$; M = number of levels such as 64-QAM where $M = 64$
MBA	Multiple beam antenna
MBC	Meteor burst communication(s)

Mbps	Megabits per second
MEA	Multiple exposure allowance
MEO	Medium earth orbit
MERCAST	Merchant marine broadcast (under auspices of U.S. Navy)
MES	Mobile earth station
MF	Medium frequency (300–3000 kHz)
MHz	Megahertz
MLCM	Multilevel coded modulation
MLSE	Maximum-likelihood sequence estimation
MPE	Multipulse excitation
M-PSK	Multilevel PSK or *M*-ary PSK
M-QAM	Multilevel quadrature amplitude modulation (e.g., 256-QAM)
MSC	Mobile switching center
MSL	Mean sea level
MTBF	Mean time between failures
MTSO	Mobile telephone switching office
MTTR	Mean time to repair
MUF	Maximum usable frequency
mW	Milliwatt
M/W	Microwave

N

NACK	Negative acknowledgment
N-AMPS	Narrowband advanced mobile telephone system
NASA	National Aeronautics and Space Administration (U.S.)
NATO	North Atlantic Treaty Organization
NBS	National Bureau of Standards (now called NIST)
N/C	Thermal noise-to-carrier ratio
NF	Noise figure
NFD	Net filter discrimination
NGP	Northern geographical pole
NIST	National Institutes of Standards and Technology (U.S.)
NLR	Noise load ratio
NMP	Northern magnetic pole
NRZ	Non-return-to-zero (a family of baseband wave-forms)
ns or nsec	Nanosecond
NTSC	National Television Systems Committee (U.S.)
NVI, NVIS	Near-vertical incidence, near-vertical incidence skywave

O

Octal-PSK	The same as 8-ary PSK or 8-level PSK
OH	Overhead
OMC	Operations and maintenance center
OOS	Out-of-service
OPSK, OQPSK	Offset PSK, offset QPSK
OWF	Optimum working frequency (see FOT)

P

PABX	Private automatic branch exchange
PACS	Personal Access Communication System
PAL	Phase alternation line, one of two European color TV standards, the other being SECAM
PBS	Paging base station
PCM	Pulse code moculation
PCS	Personal communication service or personal communication system
PES	Personal earth station
PDH	Plesiochronous digital hierarchy (vis-à-vis SDH)
PF	Peaking or peak factor
PHEMT	Pseudomorphic high electron mobility (LNA)
PHS	Personal handy phone system (Japan)
PLL	Phase-lock loop
PM	Phase modulation
POCSAG	Post Office Code Standardization Advisory Group (a paging standard)
POTS	"Plain old telephone service"
PRB	Primary reference burst
PRBS	Pseudorandom binary sequence
PRV	Partial response violation
PSK	Phase-shift keying
PSTN	Public switched telecommunication network
pW, pWp	Picowatts, picowatts psophometrically weighted

Q

QAM	Quadrature amplitude modulation
QPR	Quadrature partial response
QPSK	Quadrature phase-shift keying
QSY	A "Q" signal ordering a frequency change
QCELP	Qualcomm code excited linear predictive (coder)
QVI	Quasi-vertical incidence (also see NVIS)

R

RBER	Residual BER
RB	Reference burst
recvr, recver	Receiver
RELP	Residual excited linear predictive (coder)
RF	Radio frequency
RFP	Request for proposal
RH	Relative humidity
RHC	Right-hand circular (polarization)
RL	Return loss
RLL	Radio local loop
rms	Root mean square
RP	Radio port
RPC-1	Radio Paging Code-1
RPCU	Radio port control unit
RPE	Regular pulse excitation
RSL	Receive signal level
Rx	Receive

S

SAR	Search and rescue
SATCOM	Satellite communications
SAW	Sample assignment word
SBC	Subband coding
SCADA	Supervisory control, data acquisition, and automatic control
SCPC	Single channel per carrier
S/D	Signal-to-distortion ratio
SDH	Synchronous digital hierarchy
SDP	Severely disturbed period
SECAM	Sequential color with memory, one of two European color television standards, the other is PAL
SELSCAN	Selective scanning (Rockwell patent for HF)
SER	Symbol error rate
SES	Severely errored second(s)
SESR	Severely errored second ratio
SFF	Single-frequency fade
SIC	Station identification code
SID	Sudden ionospheric disturbance
SINAD	Signal plus interference plus noise and distortion to interference plus noise plus distortion ratio
SMSC	Satellite mobile services switching center

S/N, SNR	Signal-to-noise ratio
SOF	Start of frame
SONET	Synchronous Optical Network (U.S.)
SRB	Secondary reference burst
SREJ	Selective reject (from HDLC protocol)
SSA	Solid-state amplifier
SSB, SSBSC	Single-sideband, single-sideband suppressed carrier
SS/TDMA	Switched-satellite TDMA
STL	Studio-to-transmitter link
STM (STM-1, STM-2 . . .)	Synchronous transport module (SDH)
STS	Synchronous transport signal (SONET)

T

TASI	Time-assigned speech interpolation
tbd	To be determined
TCM	Trellis-coded modulation
TDD	Time division duplex
TDMA	Time division multiple access
TelCo	Telephone company
TFM	Thermal fade margin
T/I	Threshold-to-interference ratio
TOA	Takeoff angle
TOD	Time of day
Topo	Topographic (maps)
T/R	Transmit/receive
Tropo	Tropospheric (scatter)
Trspdr	Transponder
TRT	Timing reference transponder
TSM	TDMA system monitor
TT & C	Telemetry, tracking & command (control)
TWT	Traveling-wave tube
Tx	Transmitter

U

UHF	Ultra-high frequency (the band from 300 to 3000 MHz)
U/L	Uplink
USAF	United States Air Force
UTC	Coordinated universal time
UV	Ultraviolet
USB	Upper sideband
μs or μsec	microsecond(s)
UW	Unique word

V

VF	Voice frequency
VFCT, VFTG	Voice frequency carrier telegraph
VHF	Very high frequency (30–300 MHz)
VLR	Visitor location register
VOW	Voice orderwire
VSAT	Very small aperture (satellite) terminal
VSELP	Vector sum excited linear predictive (coder)
VSWR	Voltage standing wave ratio

W

WACS	Wireless access communication system
WARC	World Administrative Radio Congress
WBHF	Wideband HF
WC	Circular waveguide
WCC	Call letters of a marine radio coastal station at Cape Cod (CC), Massachusetts
W/G	Waveguide
WLAN	Wireless local area network
WR	Rectangular waveguide
WWV	Call letters of U.S. National Institutes of Standards and Technology at Ft. Collins, transmission of time signals at 5, 10, 15, and 20 MHz

X

X.25	CCITT/ITU-T recommendation (standard) for packet data
XCVR, xcvr	Transceiver
XIF	XPD improvement factor
xmtr	Transmitter
XPD	Cross-polarization discrimination

Z

ZC	Zone coverage

INDEX

An *italic* term denotes a definition. **Boldface** denotes extensive coverage of a subject.